Encyclopedia of
ENVIRONMENTAL
BIOLOGY

VOLUME 3
O – Z, Index

Encyclopedia of
ENVIRONMENTAL BIOLOGY

VOLUME 3
O – Z, Index

Editor-in-Chief

William A. Nierenberg

Scripps Institution of Oceanography
University of California, San Diego

ACADEMIC PRESS

San Diego   New York   Boston   London   Sydney   Tokyo   Toronto

This book is printed on acid-free paper. ∞

Academic Press, Inc.
A Division of Harcourt Brace & Company
525 B Street, Suite 1900, San Diego, California 92101-4495

United Kingdom Edition published by
Academic Press Limited
24-28 Oval Road, London NW1 7DX

Library of Congress Cataloging-in-Publication Data

Encyclopedia of environmental biology / edited by William A.
   Nierenberg.
        p.   cm.
     Includes bibliographical references and index.
     ISBN 0-12-226730-3 (set). -- ISBN 0-12-226731-1 (v. 1). -- ISBN
0-12-226732-X (v. 2) -- ISBN 0-12-226733-8 (v. 3)
     1. Ecology--Encyclopedias.   2. Environmental sciences-
-Encyclopedias.   I. Nierenberg, William Aaron, date.
QH540.4.552   1995
574.5'03--dc20   94-24917
                                                   CIP

PRINTED IN THE UNITED STATES OF AMERICA
95 96 97 98 99 00 EB 9 8 7 6 5 4 3 2 1

Contents

C

D

E

VOLUME 2

F

How to Use the Encyclopedia

The *Encyclopedia of Environmental Biology* is intended for use by both students and research professionals. Articles have been chosen to reflect important areas in the study of the environment, topics of public and research interest, and coverage of environmental issues vital to lawyers. Each article provides a comprehensive overview of the selected topic to satisfy readers from students to professionals.

The *Encyclopedia* is designed with the following features to allow maximum accessibility. Articles are arranged in alphabetical order by subject. A complete table of contents appears in the front matter of each volume. This list of titles represents topics carefully selected by our esteemed editorial board. Here, such general topics as "Acid Rain" and "Speciation" are listed.

A 10,000 entry Subject Index is located at the end of Volume 3. The index is the most direct way to access the *Encyclopedia* to find specific information. Although the article "Evolution and Extinction" is thorough and complete, additional text on evolution can be found in articles such as "Galápagos Islands" and "Bird Communities." Subjects in the index are listed alphabetically and indicate the volume and page number where corresponding information can be found.

The Index of Related Titles appears in Volume 3 following the Subject Index. This index presents an alphabetical list of the articles as they appear in the *Encyclopedia*. Following each article title is a group of related articles that appear in the encyclopedia.

Articles contain an outline, a glossary, cross-references, and a bibliography. The outline allows a quick scan of the major areas discussed within the article. The glossary contains terms that may be unfamiliar to the reader, with each term defined in the *context of its use in that particular article*. Thus, a term may appear in the glossary for another article defined in a slightly different manner or with a subtle nuance specific to that article. For clarity, we have allowed these differences in definition to remain so that the terms are defined relative to the context of each article.

Each article contains cross-references to other *Encyclopedia* articles. Cross-references are found at the end of the paragraph containing the first mention of a subject covered more fully elsewhere in the *Encyclopedia*. By using the cross-references, the reader gains the opportunity to find additional information on a given topic.

The bibliography lists recent secondary sources to aid the reader in locating more detailed or technical information. Review articles and research articles that are considered of primary importance to the understanding of a given subject area are also listed. Bibliographies are not intended to provide a full reference listing of all material covered in the context of a given article, but are provided as guides to further reading.

Ocean Ecology

Daniel Pauly

International Center for Living
Aquatic Resources Management, Manila, Philippines
and Fisheries Centre, University of British Columbia,
Vancouver, Canada

I. INTRODUCTION

Ocean ecology is the subdiscipline of oceanography dealing with the environmental adaptations and mutual interactions of the organisms populating the world's oceans, from the recently discovered, heat-loving bacteria of deep sea vents to the majestic blue whales.

Ecology, the systematic study of the interrelationships of organisms, is only one century old, as is oceanography, which began with the circumglobal *Challenger* Expedition (1872–1876); ocean ecology is therefore a new field, with many scattered observations still awaiting synthesis, perhaps similar to geology before the advent of plate tectonics.

This brief account presents some of the elements that will have to be incorporated in such synthesis, with emphasis on the processes that generate biomass that can be exploited to meet human needs. It concludes with the requirements—and a plea—for a mode of exploitation that can ensure the sustainability of the living resources of the ocean.

II. MAJOR ECOLOGICAL ZONES OF THE OCEANS

The ecological zones of the oceans can be defined according to different sets of criteria, depending on the specific discipline of the scientists performing the classification. One type of classification, by depth zone and distance from the coast, is illustrated in Fig. 1, which also defines a number of terms used in this article.

An alternative classification is presented in Fig. 2, based on surface features of the ocean and work performed by fisheries oceanographers mindful of biogeography. The marine ecosystems thus identified are large, but perhaps still homogeneous enough for management by appropriate international management bodies. (Unfortunately, these areas do not overlap with the 15 "FAO Areas" used by the Food and Agriculture Organization of the United Nations to present global fisheries catches.)

Another, more basic classification scheme would consist of differentiating the entire world ocean into an enormous body of cold water ($-2°$ to

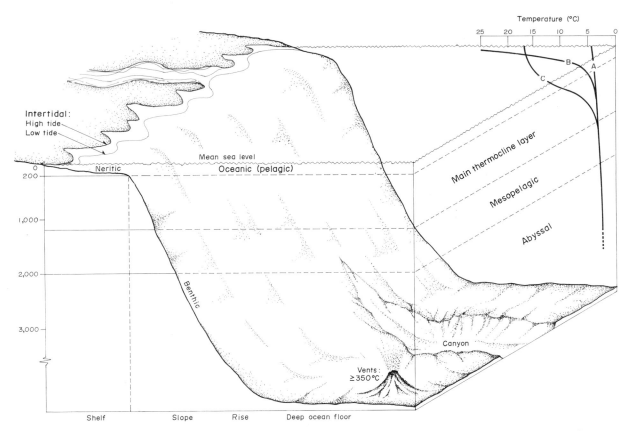

FIGURE 1 Definition of major zones of the ocean, in terms of depth and distance from the coast—the major factors affecting aquatic plant and animal distribution. The temperature profiles to the right, pertaining to the winter season, identify a high-latitude area (A), a low-latitude area (B), and a temperate area (C). Note in B and C the existence of well-marked thermoclines, that is, transition layers separating the warm surface from the cold deep water.

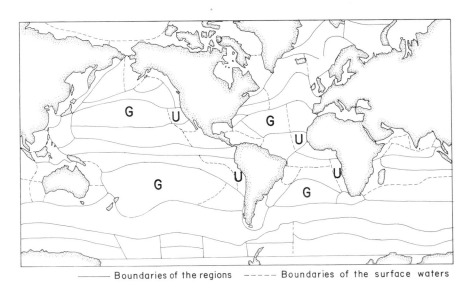

— Boundaries of the regions   ----- Boundaries of the surface waters

FIGURE 2 Major regions of the world oceans, defined from physical features of the surface layer, largely corresponding to floral and faunal regions, and defining Large Marine Ecosystems. U refers to areas of strong upwelling, G to central gyres. [Modified from I. Hela and T. Laevastu (1982). "Fisheries Oceanography." Fishing News Books, London.]

20°C), on top of which floats a thin but wide lens of warmer water, deepest in the tropics, and whose borders seasonally oscillate along subtropical shores.

The cold waters are generally rich in nutrients—mainly nitrates, silicates and phosphates—but do not receive enough sunlight for them to support all the photosynthesis, or primary production, they otherwise could: the nutrients are too deep for light to reach them, or in high latitudes they receive light only during a short but hectic summer season. Conversely, the warm waters receive lots of light, but generally lack nutrients—hence the desertlike nature of the central gyres of the oceans, for example, the Sargasso Sea.

Massive primary production—mainly by phytoplankton, the drifting organisms of the sea, but also by treelike macroalgae reaching 30 m in length—occurs wherever nutrients and light are brought together by mixing processes. Typically this is in the uppermost 10 to 100 m of the oceans. In high-latitude, cold-water areas, much of the primary productivity occurs during the well-lit summer period, when storms destratify the upper layers of the sea, thus regenerating the nutrients built up during the winter storms but depleted by the first massive populations of phytoplankton, the "spring bloom."

Stratification and superficial nutrient depletion are far more of a problem in low-latitude areas, where the warm surface waters are much lighter than the waters in deeper layers and are thus harder to destratify. There, massive production occurs only when a mechanism stronger than the occasional storm breaks the thermocline (see Fig. 1) and pumps a regular supply of nutrient-rich, deep water to the surface, a phenomenon known as "upwelling." This occurs mainly on the eastern boundaries of the oceans, where regular, equatorward winds exert stress on the coastal waters, which are deflected offshore by the Coriolis force and replaced by upwelled water (Fig. 3). Other, smaller upwellings occur where local conditions maintain at least a seasonal supply of upwelled waters, for example, off Somalia, or in the Bali Strait, Indonesia.

All upwelling systems of the world support massive but highly variable populations of anchovies, sardines, and related fishes, which feed on the large populations of phyto- and zooplankton that occur in these areas, and themselves are preyed upon by mackerels (*Scomber japonicus*), the horse mackerels (*Trachurus* spp.), fish-eating birds (cormorants, boobies, pelicans), and various pinnipeds.

The central area of the warm-water lens alluded to earlier harbors another type of system in which light and nutrients are combined to yield extremely high rates of organic production: coral reefs, which are specialized in trapping nutrients from the surrounding waters. Successful trapping of nutrients leads to more coral growth, and the deposition of calcareous skeletons that ensues builds structures capable of trapping even more nutrients. This feedback loop enables some coral reefs to stand out, cathedral-like, in otherwise barren expanses of highly stratified tropical seas. [*See* CORAL REEF ECOSYSTEMS.]

However, coral reefs can only grow in shallow waters. Where warm, nutrient-poor water seals off a deep basin, as in the central gyres of the oceans (Fig. 2), primary and secondary production remain low, in spite of intricate adaptations by numerous species of phyto- and zooplankton to these impoverished habitats.

Foremost among these adaptations are those that enable nutrients to be recycled quickly and economically within the euphotic layer. This reduces the leakage of detritus from the surface to the deeper layers, and ultimately to the sea bottom, resulting in lower biomasses of benthic, or bottom-living, organisms below the central gyres (Fig. 4).

Another adaptation of central gyre organisms, but one suitable only to fast-swimming animals, is to range over a large area and to feed opportunistically wherever food patches occur. This defines the niche of tropical tuna, dolphinfish (*Coryphaena*), and other large pelagic (i.e., free swimming) fish whose biomass per unit of ocean surface is always small but that can have a large absolute biomass because of their capacity to tap into the production of entire ocean basins.

Combining the classification scheme in Fig. 1 with that of Fig. 2 would yield yet another, rather detailed classification scheme that defines numerous subecosystems scattered across latitudinal and

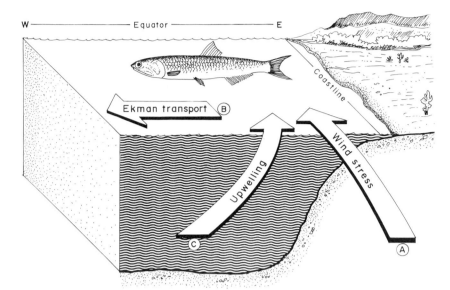

FIGURE 3 Schematic representation of the mechanism driving Southern Hemisphere upwelling systems off Peru/Chile and Southwest Africa: a steady equatorward wind (A) causes, via Ekman transport and the Coriolis force, an offshore transport of surface water (B), which is replaced by nutrient-rich deep water (C). Only the direction of wind stress arrow needs to be inverted for this graph to depict a Northern Hemisphere system off California or Northwest Africa. The fish drawing represents the Peruvian anchoveta, *Engraulis ringens*.

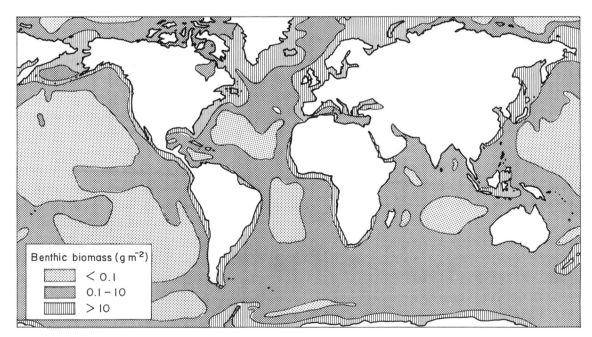

FIGURE 4 Biomass of benthic organisms in the world ocean, indicating areas of low (central gyres) and high productivity (upwellings, equatorial divergence, well-mixed temperate areas). Adapted from several sources.

longitudinal gradients. Each of these subecosystems would define a set of ecological conditions appropriate for certain types of organisms, drawn from a pool of species "placed there" by their highly contingent evolutionary history. Thus, not all potential "niches" would be filled in all subsystems. For instance, there are no autotrophic giant clams (Tridacnidae) on Caribbean reefs.

III. TROPHIC ENERGY FLOWS IN OCEAN COMMUNITIES

A. The Fate of Primary Production

Living organisms consist of molecular structures, and energy is required both to create and to maintain these structures. In the autotrophs, or "self-feeding" organisms of the sea, this energy is derived either from sunlight, through the photosynthesis typical of algae, or through the reduction of sulfuric compounds, or chemotrophy, typical of the bacteria inhabiting deoxygenated sediments under strong upwelling plumes, deep-sea vents, or other extreme habitats. The organic substances thus synthesized are either:

1. used by the autotrophs themselves to build up and maintain their biomass;
2. leaked out of the autotrophic cells to feed a host (e.g., coral polyps or *Tridacna,* the giant clams of Indo-Pacific coral reefs, or *Calyptogenia,* the clams of deep-sea vents), or a population of free-living bacterioplankton;
3. consumed as live biomass by some herbivore or omnivore;
4. broken down and consumed as dead biomass by a detritivore; or
5. allowed to sediment and fossilize, leading after millions of years and complex chemical and physical processes to either recoverable organic substances (e.g., oil) or sedimentary rocks.

Items (1) to (4), and vast numbers of predators feeding on the consumers, themselves eventually being eaten by second-order predators, and by de-

tritivores feeding on their excreta and carcasses, imply the existence in the sea of vast and intricate cycles that release nutrients back to the primary producers; this is not a single food "chain" with a well-defined beginning and ending, but a food web (Fig. 5).

B. Major Strands of the Ocean's Food Web

It would be wrong to think of the food web ultimately linking all organisms of the ocean as being particularly efficient—even though it has had millions of years to evolve. Indeed, many key cycles and links in the ocean's food web are leaky, a fact previously explained by, among other things, the difficulties that floating or weakly swimming organisms have in maintaining themselves at certain depth and/or locations. The classic textbook "The Oceans" by H. U. Sverdrup *et al.* suggests that for phytoplankton "the means of adaptation are mostly along the lines of increased length of appendages, of spine or bristles, or of dorsoventral flattening of the body."

Recent research led by V. Smetacek and collaborators has demonstrated, however, that the spiky shape of many phytoplanktonic algae, previously thought to work against their sinking and ultimate sedimentation, may in fact enable them to hook up with each other, which, along with their tendency to become "sticky," allows them to form aggregates that *increase* their sinking speed. This leads to a phenomenon known as "marine snow," whose largest "flakes" sink toward the seafloor. The marine snow phenomenon was first discovered by Japanese researchers in the 1950s, but was largely ignored as the snow easily disintegrates with sampling and is best observed by divers or through underwater cameras. The snow may consist of many sorts of materials, and only recently have studies shown that marine snow can consist of apparently intact planktonic algae.

The clumping mechanism leading to marine snow is triggered off when nutrient depletion in the euphotic zone prevents phytoplankton populations from growing fast enough to match the grazing pressure of zooplankton, and seems to be particu-

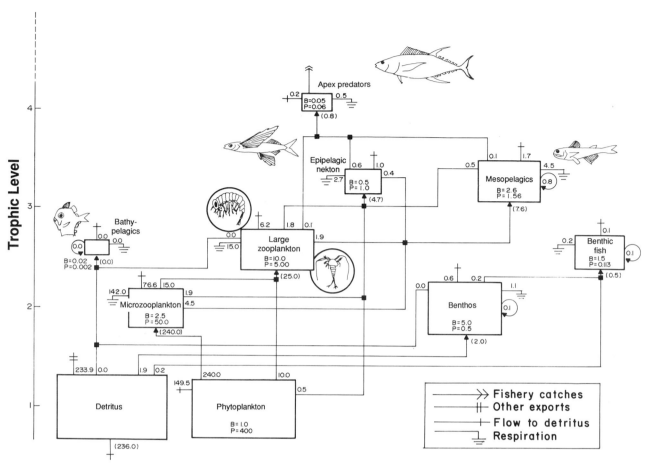

FIGURE 5 Food web (ECOPATH II) model of a low-productivity area, the central gyre of the South China Sea. Only the flows up the food web are shown (in tons km^{-2} year^{-1}), that is, the backflows to the detritus are omitted for clarity. The areas of the boxes correspond to the logarithms of the biomass (B, in tons km^{-2}); P refers to production (tons km^{-2} year^{-1}). [From D. Pauly and V. Christensen (1993). *In* "Large Marine Ecosystems: Stress, Mitigation and Sustainability" (K. Sherman *et al.*, eds.), pp. 148–174, AAAS Press, Washington, D.C.]

larly strong in diatoms, which are a major component of single-cell algal communities around the world.

Explaining this sophisticated "behavior" is straightforward, as diatoms are now known to be the descendants of spore-bearing species that survived the great Cretaceous extinction, most likely caused by the impact of a large meteor 65 million years ago at Chicxulub, Yucatan, in Mexico. This event blotted out sunlight long enough for most phytoplankton without resting spores to become extinct, leaving the field wide open for diatoms and other forms whose dormant stages had "preadapted" them for such an event. Resting spores are also a fine adaptation permitting some planktonic

algae species to increase their biomass rapidly when nutrient conditions are optimal.

Hence, the classic accounts of a phytoplankton spring bloom being, in temperate areas, grazed down by gradually increasing zooplankton populations may not apply, the interactions between phyto- and zooplankton being rather a game of "hide and seek" in which the algae stay in the well-lit surface waters only as long as nutrient levels are high, and then escape through a single "spring cleaning" of the water column to the bottom, where they form resting spores. Working against this, the zooplankton organisms try to maintain biomass near the surface by ingesting cells as fast as they can, assimilating only a small fraction of what they consume and releasing feces that may still be consumed and that will slowly release nutrients in the surface layer—"creative inefficiency" as it were.

C. The Higher Trophic Levels

Although some of the biomass synthesized by primary producers is consumed via routes other than through zooplankton grazing on phytoplankton (e.g., coral polyps and clams feeding on the exudates of captive zooxanthellae), this route is the one through which most secondary production occurs. Thus, zooplankton serves as the key link to the fish; most fish feed on zooplankton, either when their larvae graduate from the less challenging capture of single-cell algae or as adults specialized in filtering or particulate feeding on zooplankton.

The ubiquitous sardines and their relatives are zooplanktivores, along with less well-known groups such as the fusiliers or banana fish (Caesionidae) of Indo-Pacific coral reefs. The most spectacular zooplankton feeders are the basking and whale sharks and the baleen whales. All of these also ingest small and imprudent zooplanktivores—anchovies and sardines—when they swoop on a concentration of large zooplankton.

Whether a patch of phytoplankton will be grazed or not may thus depend not only on the fortuitous development of an overlapping zooplankton population, but on the even more fortuitous occurrence of a large zooplankton feeder, that is, quite literally on the fluke of a whale.

Needless to say, zooplankton organisms cannot evolve individual defense mechanisms against such large consumers. Rather, populations of weak and small organisms can maintain themselves only because of the very patterns of occurrence/nonoccurrence of their consumers, and because large consumers and predators, when feeding on patches, cannot afford the energetic costs of "finishing them off" down to the last individuals, or to return to freshly grazed areas.

Indeed, life can be precarious for large predators and consumers, which generally will not find enough food in any small area to maintain a viable population (e.g., of Loch Ness monsters), especially because our present fisheries require, to sustain their catches, a large fraction of the primary production of marine food webs (see Table I). Large predators must consequently undertake feeding migrations of various length. The most spectacular among such migrations are those of blue whales, which in the Austral summer move deep into Antarctic waters to feed on krill (*Euphausia superba*), or of gray whales swimming deep into the North Pacific, then returning, hulled in fat, to Mexican lagoons to reproduce.

The best swimmers of all are the warm-blooded tropical tunas (e.g., *Thunnus albacares, Katsuwonus pelamis*) whose entire life, once they have left the food patch in which they hatched, consists of nothing but high-speed, indeed frenetic, hopping from one food patch to the next, and suffering quick death by starvation if they fail to reach, in time, a new patch of life-sustaining, high-density food.

Here again, sedimentation—or at least a rain of dead carcasses reaching the ocean floor—will be the result of the essential inefficiency of the pelagic food webs, further proven by the existence of organisms (large amphipods, marcrourid fishes) specialized in feeding on such carcasses.

One can only wonder at the many ways the ocean's organisms have evolved to make a living. Yet comparative studies across a wide range of taxonomic groups and ecosystems have confirmed that the ratio of food consumed to flesh produced is only about 10% for *all* consumers, at all trophic levels.

This implies another form of inefficiency: the energetic loss of 90% of the biomass consumed at each trophic level within oceanic food webs. This evidently is why animals feeding high on the food web, three or four trophic levels above the primary producers, have low densities—or to paraphrase P. Colinveaux, "why big fierce [aquatic] animals are rare."

Thus, the production of reef sharks feeding on groupers that feed on fusiliers feeding on zooplankton can only approximate one-thousandth of the production of the zooplankton, and one-ten-thousandth of the phytoplankton upon which that specific food chain rests. With such low production, reef shark biomass must remain low, even if they live long and incorporate many years worth of their prey's production. So, with big fierce animals such as sharks being relatively rare, trophic levels higher than five or six are not well represented in the oceans, and usually include only the parasites of these predators, often strongly modified zooplankton organisms.

Another reason for the absence of very high trophic levels is the (necessary!) breadth of the food spectrum of large predators, which may include animals themselves relatively high in the food chain (e.g., groupers), but also herbivores, thus reducing the (fractional) trophic level that may be computed for such predators (Fig. 5 and Table I).

Given the pyramidlike nature of marine ecosystems, one obvious temptation with marine resource exploitation is to suggest that the fisheries can increase their yield—now about 0.025% of primary production (see Table I)—by moving down the food webs, that is, that humans substitute themselves for the upper elements of the trophic pyramids leading to the presently observed yields. Like all good ideas that are simple, this idea has a downside, namely, variability, which increases as one goes down the food web and increasingly precludes large sustained harvests of any one species. This is the topic of the next section.

IV. PROCESSES INDUCING VARIABILITY

A. Causes of Variability

The basic cause for the high variability of biological production in the oceans relative to that of terrestrial systems is that the bulk of the primary production depends on single-celled algae, which have little or no impact on their habitat, in contrast with terrestrial plants, which usually create conditions capable of sustaining increasing biomasses, eventually modifying their own (micro-) climates, for example, tropical rain forests. Only isolated belts of macroalgae (e.g., the giant kelp *Macrocystis*) along the coast of upwelling systems may have an analogous effect on their habitat (e.g., by reducing wave action and producing enough detritus to support a distinct food web, shaped by the dynamics of sea urchins, abalone, and sea otters).

Phytoplankton also affects the physical properties of its habitat: blooms increase light absorption in the upper water layers, and hence should increase surface temperatures, a process shown to impact strongly on the food webs of some lakes. Such impacts have not been shown unequivocally in marine environments, whose dynamics are usually described as if they were free of life.

TABLE I

Primary Production of the World Oceans, by Major Environments, Related to Mean Catch for the Late 1980s and Early 1990s[a]

Features	(Units)	Open oceans	Tropical shelves	Nontropical shelves	Upwelling areas	Estuaries, bays, etc.	Total or mean
Area	(%)[b]	91.7	2.4	5.1	0.2	0.5	100
Primary production	(g C m^{-2} year^{-1})	100	300	300	1,000	1,000	122
Total production	(tons $\times$ 10^6 year^{-1})	332,000	25,800	55,200	8,000	20,000	441,000
Fish catches[c]	(tons $\times$ 10^6 year^{-1})	3 (0.52)	18 (5.8)	30 (13)	17 (2.7)	16 (5.0)	84 (27)
System efficiency[c]	(%)	0.001	0.092	0.078	0.246	0.105	0.025
Trophic levels		4	3.3	3.5	2.8	2.5	3.3
Tribute to fisheries	%[e]	1.1	27.6	39.0	19.7	5.3	7.9

[a] Adapted from FAO Fisheries Statistics and data from the works cited in the Bibliography.

[b] Percentage of 362 $\times$ 10^6 km^2.

[c] Figures in brackets are discards, mainly from shrimp fisheries, which must be added to catches to assess tribute to fisheries. (Adapted from D. L. Alverson, M. H. Freeberg, J. G. Pope, and S. A. Murowski (1994). "A Global Assessment of Fisheries Bycatch and Discards," FAO Fish Tech. Rep. No. 339. FAO, Rome.

[d] Catch $\times$ 100/total production.

[e] Percentage of primary production required to sustain the fisheries catches.

Thus, we can assume that much of the high-frequency physical variability of the ocean, including its chaotic elements, will be mirrored by single-species phytoplankton populations rather than dampened, and that only the gross features of primary productivity, integrated over numerous phytoplankton species and over larger areas and periods, will generate predictable patterns.

B. Within-Year Variability

The smallest physical features inducing ocean variability are minute turbulent vortices, lasting perhaps a few seconds, used by single-celled algae to break the "skin" of nutrient-free water surrounding them and by zooplankton organisms to transport food particles within reach of their grasping appendages. These vortices, occurring at different scales, and only recently discovered to play a crucial role in interlinking molecules, food aggregates, and organisms, represent a new area of research driven by the realization of the fractal nature of the marine realm, in which these turbulences are but small copies of large eddies and even of the global current system spanning the world's ocean.

The next most important high-frequency scale is the 24-hr rhythm of day and night. During daytime, given the availability of nutrients and sunlight, primary producers engage in bouts of primary production, accumulation of synthesis products, and the production of O_2. At night, these processes are inverted: synthesized matter is respired, O_2 is absorbed, and CO_2 is excreted, with the balance over a 24-hr cycle being usually positive, so that overall about 440 billion tons (wet weight) of plant biomass are synthesized every year (Table I).

For many visual feeders, daytime is the time when guts are filled, whereas the night provides protection from predators, and even time for sleep (e.g., in herrings). However, some taxa depart from this (anthropomorphically) intuitive scheme and have turned the succession of days and night into a resource that they actively use.

For example, tropical and subtropical lantern fish (Myctophidae) spend most of the day in cold, deep water, typically 5° to 10°C at 300 to 1000 m, migrating at dusk to feed in the mixed layer surface (30–100 m), where the water is warmer (25–30°C) and rich in zooplankton that they can see despite limited moonlight or starlight. Then at dawn, they migrate back to their mesopelagic night-time habitat, where the low temperature reduces their metabolic requirements—an adaptation that, along with their exploitation of the surface layers, has made these fishes the most abundant in the world in terms of total biomass. However, because of their low density (few mg per m^3), myctophids have yet to become the target of a major fishery.

Daily migrations such as this become less marked as one moves from the tropics to the poles, both because the annual cycles of changing daylight hours would make such adaptation precarious for short-lived animals and because of the decline of the temperature gradient that makes such migration energetically profitable. Indeed, at higher latitudes, daily cycles generally become confounded with the predominant annual cycle, both ultimately culminating near the poles in a single, long night called winter—which introduces low-frequency variability.

The adaptations displayed by plants and animals in dealing with long winter periods are simple: the short-lived organisms usually die, leaving spores, eggs, or other dormant stages that should bloom or hatch in spring. In contrast, the longer-lived organisms either leave (as do most whales), or adapt by shedding their filtering apparatus (as do basking sharks) and living off their accumulated liver oil or body fat, or change from an active to an energetically less demanding life-style. An example of life-style change is when southern krill (E. superba) switch in winter from its pelagic mode to a more sedentary life of feeding on algae under the Antarctic shelf ice, protected—at least in part—from some of its open-water predators—whales, "crabeater" seals, penguins, and notothenid fish.

The strong summer–winter differences of temperate latitudes may be tracked by long-lived organisms, such as most fish, in the phenology of their life processes, from the spawning and fertilization of eggs, to their hatching and drifting of the larvae to a nursery, and the subsequent growth of the juveniles and adults. Indeed, the precise timing of such a sequence of events may be crucial for the reproductive success of a fish population, hence the scientific interest in D. H. Cushing's "match–mismatch hypothesis": the notion that a chronological mismatch of perhaps a few days between the hatching of larvae and the bloom of their planktonic food offers a major explanation for the interannual variability of temperate fish stocks.

Mismatched as they might be, these seasonal events are sufficiently regular to induce annual marks on the hard parts (vertebrae, scales, and other bones) of fish, and whose straightforward interpretation has allowed fisheries scientists studying temperate stocks to structure their discipline around annual rhythms and perhaps, in the process, neglect intraannual variations. The perceptional bias this has caused, combined with the dominance in marine sciences of scientists from temperate countries, delayed for decades the coherent interpretation of observations on intraannual rhythms in tropical fishes, shrimps, and other marine organisms.

Important rhythms of this sort, sometimes overriding any single annual cycle, are those linked with the tides, for example, the monthly (lunar) pulses of recruitment and settlement of invertebrates and fish onto reefs and tropical shelves, as well as those linked with the monsoons, which cut the year into two uneven parts (usually 7 and 5 months) each with distinct peaks of recruitment and settlement (Fig. 6).

Even so, annual cycles remain crucial, especially as they entrain and "reset" most intraannual rhythms, thereby providing stability and long-term self-similarity to ecosystems. This point must be made lest variability is overemphasized and one forgets that most marine ecosystems retain their integrity from one year to the next.

C. Between-Year Variability

Among the major interannual events affecting marine ecosystems, El Niño is the most prominent. El Niño causes a massive intrusion of warm oceanic water into coastal upwelling systems and, by interrupting its supply of cold, deep, nutrient-rich water, starves these systems of the high throughput of nutrients that they require.

The El Niño events that have struck the Peruvian upwelling system at intervals of 3–7 years have been increasingly well described since the 1920s, and although many accounts have emphasized their distinctive effects, recent studies have shown them to be a natural phenomenon of this system, far more benign than chronic overfishing, against which biotic responses cannot evolve fast enough.

El Niño events are part of even larger oceanic and atmospheric processes interlinking entire ocean basins, notably in the Pacific, and causing similar or complementary effects on far-away continents. Thus, the rains associated with El Niño events off Peru and California generally imply droughts in Australia and the Philippines. The opposite happens during cold La Niña periods.

The changes typically induced by the warm waters of El Niño events on the Peruvian upwelling ecosystems are:

- reduction of *new* primary production due to reduced upwelling (whereas gross production, based on nutrients regenerated within the warm surface layer, does not diminish appreciably);
- submergence of anchovies and other pelagic fishes that crowd themselves into the thin, cool, oxygenated water layer between the warm covering layer and the deep deoxygenated waters typical of upwelling systems; and
- inability of air-breathing piscivores (mainly cormorants, boobies, pelicans, sea lions, and fur seals) to prey on the sinking anchovies, leading to their young being abandoned and the adults migrating southward (especially the fish-eating birds) and/or starving.

Recovery from El Niño events, which usually last 2–3 months at most, involves swift re-establishment of primary and zooplankton production from spores and eggs in upwelled waters, followed

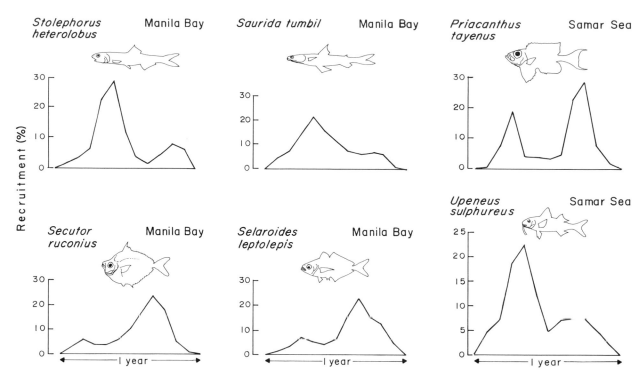

FIGURE 6 Bimodal patterns of recruitment in six species of Philippine fishes, corresponding to the seasonal changes of monsoon winds. [From J. Ingles and D. Pauly (1984). "ICLARM Tech. Rep. No. 13." ICLARM, Manila; used with permission.]

by reconstitution of the anchoveta population, principally from residual pockets of cold waters or from farther south. Overall, recovery is rather rapid, as anchoveta predation by birds is low due to the absence of young birds, the first to die during El Niño events.

Interestingly, the Peruvian upwelling ecosystem appears to occur in two states, "high" and "low." Following an El Niño event, the system stabilizes at high or low biomass, until the next El Niño event again resets everything. What sets the system "high or low" is not clear, and neither is it known why certain El Niño events appear to have helped the system switch from one state to the other (e.g., that of 1971–1972 from high to low), whereas others did not lead to a switch; for instance, the 1965 event occurred while the system was set high, and the very strong event of 1982–1983 occurred when the system was low.

Such considerations have led J. McGlade to suggest that upwelling systems such as that off Peru incorporate a "strange attractor," and oscillate seasonally within either of two states, while occasion-ally (and unpredictably) switching from one state to the other, with the whole robust ensemble of states and cycles maintaining its self-similarity through evolutionary time.

Other temporal scales of variability have been investigated, notably a scale approximately double that of El Niño events and corresponding to the 11-year sunspot cycle. This explanation, however, lacks causal links and even a time series unequivo-cally depicting the required periodicity.

Longer-term cycles spanning hundreds (from commercial fisheries) or even thousands of years (from sedimented plankton) have been docu-mented and used for hind- and forecasting climatic changes, but this issue will not be pursued here, because it requires an extensive account of paleobi-ology and global climate modeling. Rather, we shall now combine our previous consideration of trophic mechanisms with our discussion of vari-ability to address the biology and dynamics of fish populations, thus providing a basis for presenting some of the requirements for sustained fishing from the oceans.

V. MARINE FISH AND FISHERIES

A. Morphological Constraints to the Ecology of Fish

Ever since the days when natural historians sought God's design in biological adaptations, the ease with which fish flow through their watery medium has been paradigmatic, with more beautiful examples of structure matching function emerging from each study. Paradoxically, these marvels of adaptation have obscured the constraints imposed on fish by their anatomy, and the truly wondrous processes by which they still manage to overcome their "imperfection" and to achieve their huge biomasses. [See FISH ECOLOGY.]

The main problem of fish is that they live in water—a highly viscous medium (compared to air) requiring a high degree of streamlining, but containing (again as compared to air) very small amounts of dissolved oxygen, which diffuses through water 300,000 times more slowly than in air. Fish therefore devote a large fraction of the energy they consume (often about 10%, and up to 30% in some species) to extracting from the surrounding water the oxygen needed for their metabolism. This process takes place through gills that cannot grow indefinitely because little space is available in the streamlined head and because gills can easily clog up or be damaged by parasites, silt, or osmotic stress.

However, the main reason why fish gills cannot grow enough to completely avoid respiratory constraints is that any surface, even one that is highly convoluted, cannot keep up with a growing volume. Hence, in any fish, gill area per unit weight decreases as weight increases (Fig. 7). This simple fact has numerous implications for the ecology of fishes, and here we shall consider those implications that can be used to explain some aspects of the differential distribution of fish species and/or of various life stages of the same species, and some observed patterns of growth and natural mortality in fishes.

The first distributional pattern to be examined in the light of respiratory constraints is the well-documented observation that generally the small

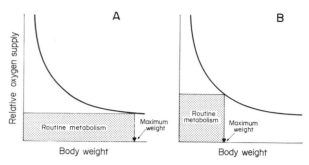

FIGURE 7 Relationship between relative oxygen supply (i.e., gill surface/body weight) and the body weight of growing fish. Note that (in A) growth in weight must cease when, at maximum size, O_2 supply is equal to the requirements for routine metabolism, and that (in B) any factor increasing these requirements (elevated temperature, reduced food availability, stress, etc.) will tend to reduce the weight at which O_2 supply becomes limiting to further growth and hence reduces the maximum size that can be reached.

young individuals of a fish population occur in shallow waters, whereas the large, old individuals occur in deeper water. This is interpreted here as a response to the fact that large fish, given their lower relative gill area, move into cooler waters where their oxygen requirements will be lower (Fig. 8). The resulting depth/size gradients, first discussed by F. Heincke for North Sea plaice, are particularly evident in tropical tunas, whose high oxygen requirements restrict the adults to a narrow range of relatively low temperatures. Thus, they are able to come near the warm surface only during short feeding or spawning forays, or when large oceanographic structures push the inhabitable cool water layers close to the surface, for example, around the Costa Rica Dome, a major fishing ground for Eastern Central Pacific tuna.

Respiratory constraints also limit shark sizes, which is why attacks on humans by large tiger and white sharks occur mainly in subtropical and rarely in tropical waters. These constraints are implicit in the daily vertical migrations of myctophids outlined earlier and in the response of Peruvian anchoveta to El Niño events. Respiratory constraints can also be evoked to explain the seasonal migrations of pelagic fish populations off Northwest Africa, whose extensive movements neatly track seasonal shifts of temperature along that same coast, from Morocco to Guinea. A particularly interesting feature is that the extent of the migratory shift is

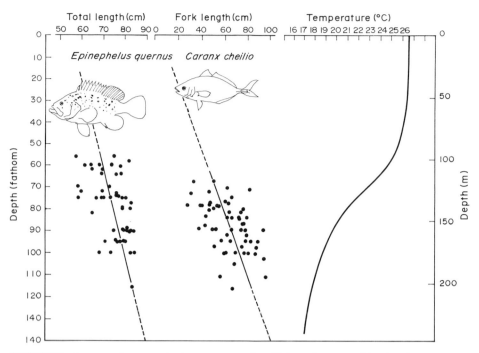

FIGURE 8 Increase of average size in two species of reef fish as a function of depth and temperature at French Frigate Shoals, Hawaiian Islands. [From S. Ralston and J. J. Polovina, (1987). *In* "Ecology of Tropical Oceans" (A. R. Longhurst and D. Pauly, eds.). Academic Press, San Diego, Calif.]

directly related to the size of the fish, with the larger individual, for whom the respiratory constraint is more severe, undertaking the more extensive migrations.

The implications of respiratory constraints for fish growth and mortality are numerous, and only a few of these can be presented. Generally, fish from (sub-) tropical environments remain smaller than their temperate counterparts: halibut (*Hippoglossus hippoglossus*) reaches lengths of 4 m off Greenland, whereas the Queensland halibut (*Psettodes erumei*) reaches 40 cm; North Sea mackerel (*Scomber scombrus*) reaches 50 cm and Indian mackerel (*Rastrelliger kanagurta*) reaches 30 cm; (North) Atlantic menhaden (*Brevoortia tyrannus*) reaches 50 cm and Gulf (of Mexico) menhaden (*Brevoortia patronus*) reaches 25 cm.

However, the different populations of the same species and their size difference along latitudinal gradients provide the best evidence of temperature (and hence respiratory) effects on fish. Thus, Atlantic menhaden commonly reach 8–10 years and 40 cm off New York (with the single largest specimen of 50 cm, reported from farther north), 4–6

years and 35 cm off North Carolina, and 2–3 years and 30 cm off eastern Florida, and western Florida and the rest of the Gulf Coast are home to the aforementioned *B. patronus,* whose growth characteristics simply extend the tend for *B. tyrannus.*

Obviously, a fish requiring several years to reach maturity (i.e., its age at first spawning, t_m) cannot sustain the same (natural) mortality (M) as a fish reaching t_m within a year, and hence strong inverse relationships exist between t_m and M, documented for hundreds of species ranging from sharks to myctophids. Thus, predatory fishes whose food requires them to be large and that therefore will also tend to have few predators, at least in the adult phase, will tend to have a reproductive output suited to compensate only for low mortalities. The implications for fisheries are obvious.

B. Gross Trends in Fish Distribution

Through evolutionary time, respiratory constraints have led to strong links between, on the one hand, the swimming and growth performance of fish, and, on the other hand, their natural mortality and reproductive output. These links are essential to their ability to maintain populations in spite of a variable food supply and hordes of predators. However, these respiratory constraints, while shaping the life-history parameters of constituent species, have not necessarily shaped the species *composition* of ocean fish communities. This is determined, as mentioned before, mainly by a contingent history, that is, by the succession of random events extinguishing certain lineages in certain places or allowing others to diversify.

Still, some gross taxonomic patterns do emerge when one examines maps of the distribution of fish in the oceans. The first is obviously the high species diversity in low-latitude areas. Another is the importance of highly derived perciforms (snappers, wrasses, and related fishes) and balistiforms (trigger fish) in tropical seas relative to the importance of the more primitive clupeiforms (sardines, herrings, anchovies) and gadiforms (cod, pollack) in temperate and/or upwelling ecosystems.

It is tempting to attribute these patterns to the long-time persistence of the ancient and benign Thetys Sea and its tropical descendants, which contrasts with the frequent catastrophes (e.g., the ice ages) that so profoundly affected temperate systems. However, evolutionary biologists would not fail to identify exceptions to these patterns (e.g., the high species numbers of tropical anchovies or of temperate wrasses). Thus, we leave this issue unresolved, but not without pointing out that there are many superb books on the fish fauna of various countries for amateur divers, interested laypersons, and scientists alike that may be viewed as complementary to this account, which can only hint at the diversity and beauty of fish, especially in the tropics.

VI. SUSTAINABLE USE OF THE OCEAN'S LIVING RESOURCES

A. Small-Scale Fisheries

Humans have exploited the ocean fish resources, especially coastal resources, from time immemorial, and here we should not speak of fisher*men*, as

recent anthropological research has made abundantly clear that fishing by women (and children) occurs or occurred in virtually all cultures. Indeed, this activity, concentrating on shallow-water invertebrates and small fishes, and using digging sticks, small traps, and other inconspicuous gears, may have contributed to most of the food consumed in fisher communities, with the men hunting big fish and often coming back empty-handed—a situation analogous to what may have prevailed among hunters/gatherers. This would make traditional or "local" knowledge of the habits of resource species a highly gender-specific affair, a fact often overlooked by those who attempt to "manage" small-scale traditional fisheries.

These fisheries may in fact best be left to themselves, as they almost always rely on community-enforced rules for limiting access to the resources and thus precluding the rapid buildup of fishing effort that leads to stock collapses—not to mention their high ecological efficiency (Fig. 9).

B. Large-Scale Fisheries

Ecological or even commercial efficiency is not a characteristic of industrial fisheries, whose growth in the last 100 years paralleled the destruction of fish populations previously thought to be too large to be ever affected by human depredations (e.g., the Atlanto-Scandian herring, the California Sardine, the Peruvian anchoveta, and countless other fishes). [*See* FISH CONSERVATION.]

Not much was learned from these collapses: presently, the overwhelming majority of fish stocks monitored worldwide by the Food and Agriculture Organization of the United Nations (FAO) are considered to be in a state of economic or biological overfishing or depletion resulting from population collapse. This trend is aggravated by destructive developments such as trawl fishing for shrimps, which usually involves discarding about 90% of the catch (the nonshrimp component, usually consisting of perfectly edible fish), and "fishing" with drift nets tens or even hundreds of kilometers long, which kill mammals, seabirds, and nontarget fish (Fig. 9 and Table I).

Not only is the resource devastation caused by industrial fishing thoroughly known, but the economic and social mechanisms driving overfishing are known as well. Foremost among them are government subsidies, direct and indirect, and the pernicious linkage between *common property,* which is what fish resources are throughout most of the world, and *open access,* the presumed right to pillage these resources, given enough capital to purchase suitable gear and to hire crew. [*See* COMMUNITY-BASED MANAGEMENT OF COMMON PROPERTY RESOURCES.]

The link between these two concepts implying that common resources must *ipso facto* be open is a recent and questionable development. There are still numerous cultures, notably in the South Pacific, where access to reefs—a common property—is not open.

C. Prerequisites for Sustained Harvests

The Law of the Sea (LOS) treaty that emerged in the 1980s, following 20 years of negotiations involving practically all countries in the world, established that the continental shelves, from which most of the ocean's catch originates (see Table I), now belong to coastal countries and form the core of their Exclusive Economic Zones (EEZ). Even so, access continues to be largely open, especially along the coastlines of developing countries that are eager to earn whatever little cash fishing licenses can provide and/or that are unable to patrol their water and to apprehend illegally operating vessels. Moreover, even under the LOS treaty, with its provisions regulating access to a nation's EEZ by foreign fleets, access is still generally open from *within* countries. These realities, combined with the worldwide economic crisis that came along with the end of the Cold War, the global tendency toward deregulation of most industries and capital flows, and the runaway demographic growth of entire continents, will continue to maintain fishing effort on most marine fisheries at excessive levels—globally two to three times that required to extract present catches.

However, if humankind can resolve on land the tremendous problems of living in peace, and sustainably producing enough food for all, we can hope that responsible stewardship of marine fisheries will also emerge—especially because marine ecosystems are—except for coral reefs—generally more robust than their terrestrial counterparts. The emergence of sustainable fisheries systems on both coastal and oceanic scales will, however, require some or all of the following elements:

- explicit property rights, enabling real (collective or private) owners (but not governments) to take effective control of the resources, and hence to become materially interested in their sustainability;
- increased use of sanctuary areas where no fishing is allowed—to replenish adjacent, exploited areas and to maintain among- and within-species biodiversity;
- bans on fishing and marketing various species, ranging from endangered forms (various large whales) to protected forms (e.g., selected reef species); and
- a cultural shift away from game fishing (e.g., billfish, shark) as happened for lion and tiger hunting, now widely considered irresponsible, not "manly."

Further, coastal fisheries resources will have to be allocated principally to small-scale, artisanal fishers: their operation is ecologically far more sound than that of the large-scale, industrial fleets and the social benefits are far greater (Fig. 9).

Moreover, and especially for tropical areas, where demographic pressure generates both a huge demand for fish products and large numbers of excess fishers, development that does not rely on short-term extractive activities and population growth are issues that will have to be tackled, not only talked about. If we do all this, the world oceans will reward us by continuing to provide us with *some* of the food we need. But we cannot continue increasing the pressure; if we do, many of the intricate cycles described here will unravel, and we will lose all they can produce for us.

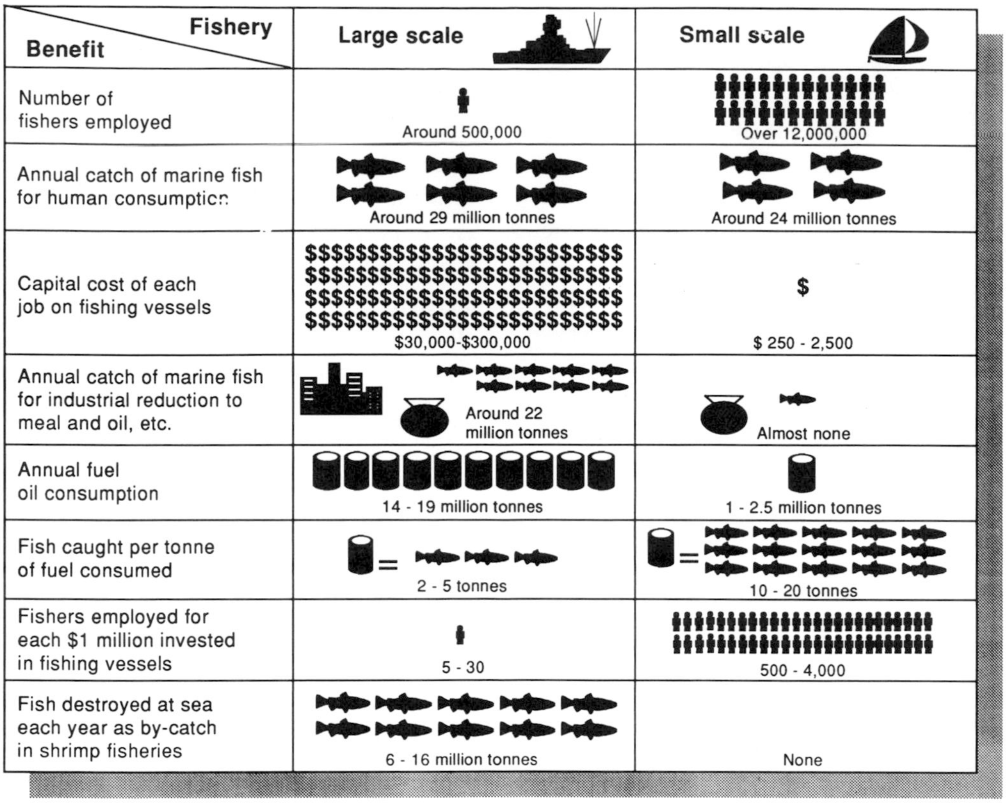

Benefit / Fishery	Large scale	Small scale
Number of fishers employed	Around 500,000	Over 12,000,000
Annual catch of marine fish for human consumption	Around 29 million tonnes	Around 24 million tonnes
Capital cost of each job on fishing vessels	$30,000-$300,000	$ 250 - 2,500
Annual catch of marine fish for industrial reduction to meal and oil, etc.	Around 22 million tonnes	Almost none
Annual fuel oil consumption	14 - 19 million tonnes	1 - 2.5 million tonnes
Fish caught per tonne of fuel consumed	2 - 5 tonnes	10 - 20 tonnes
Fishers employed for each $1 million invested in fishing vessels	5 - 30	500 - 4,000
Fish destroyed at sea each year as by-catch in shrimp fisheries	6 - 16 million tonnes	None

FIGURE 9 How large-scale (industrial) and small–scale (artisanal) fisheries compare in terms of catches, ecological impact, and social benefits. [From D. Thomson and FAO (1988). *Naga, The ICLARM Quarterly* **11**(3), 17; used with permission.]

Acknowledgment

I thank Villy Christensen for his help with various parts of this contribution, notably the section on marine snow and Table I. This is ICLARM Contribution No. 979.

Glossary

Benthos Community of organisms living in, on, or near the sea bottom.

Biomass Combined weight (in dry or wet weight units) of organisms belonging to one or several species and/or populations. Generally expressed on a per area basis, as average over a conventional period, for example, one year.

Detritus Ensemble of dead particulate or dissolved organism matter, consisting of carcasses and excreta, that is suspended, in solution, or lying at the bottom of the ocean, and that may be consumed by detritivores, especially bacteria.

Ecosystem Ensemble of plants and animals of a given area, interacting among themselves and with their physical environment such that persistent biomass flows and cycles emerge. A well-defined ecosystem, for example, an almost closed coastal lagoon, will tend to have stronger flows within itself than between itself and adjacent ecosystems.

Efficiency (of food conversion) Dimensionless number expressing the ratio of food ingested to biomass produced, always less than one, and often near 0.10.

Euphotic zone That part of the surface layer of the ocean whose depth depends on plankton biomass and detritus concentration, and in which there is enough light to sustain photosynthesis; generally defined as reaching to the depth where the light is 1% of that at the water surface.

Metabolism In the widest sense, all process by which living cells maintain themselves and grow; in the narrow sense, the oxygen consumption required for these cellular processes.

Photosynthesis Process by which plants use energy derived from sunlight to synthesize organic compounds from CO_2, water, and nutrients, thereby releasing O_2. Most primary production, that is, production from nonliving matter, occurs as photosynthesis, whereas chemosynthesis, relying on the chemical energy in certain organic compounds, is important in some sediments and in deep-sea vent ecosystems.

Plankton Community of living plants (phytoplankton) and animals (zooplankton) whose lack of powerful propulsive organs forces them to drift with the water body in which the vagaries of turbulences placed them.

Production Sum of all growth increments of the animals or plants of a population over a defined period, including the growth of individuals that may not have survived to the end of that period.

Shelf That part of the sea bottom that is not deeper than 200 m and surrounding continents and islands, and from which most ocean fish catches originate.

Trophic level Number expressing how many intermediate steps separate a consumer organism, for example, a predator or parasite, from the herbivores at the base of the food web (conventionally assigned a trophic level of 1). This number may be an integer, implying a straight (hypothetical) food chain leading to a given predator, or fractional (e.g., 2.7), indicating that the prey of that predator occur at different trophic levels, as in real food webs.

Bibliography

Berger, W. H., Smetacek, V. S., and Wefer, G., eds. (1989). "Productivity of the Ocean: Present and Past." Chichester, England: John Wiley & Sons.

Christensen, V., and Pauly, D., eds. (1993). "Trophic Models of Aquatic Ecosystems," ICLARM Conference Proceedings No. 26. Manila: ICLARM.

Longhurst, A. R., ed. (1981). "Analysis of Marine Ecosystems." London: Academic Press.

Longhurst, A. R., and Pauly, D. (1987). "Ecology of Tropical Oceans." San Diego, Calif.: Academic Press.

Sherman, K., Alexander, L. M., and Gold, B. D., eds. (1993). "Large Marine Ecosystems: Stress, Mitigation and Sustainability." Washington, D.C.: AAAS Press.

Valiela, I. (1984). "Marine Ecological Processes." New York: Springer-Verlag.

Packrat Middens, Archives of Desert Biotic History

Scott A. Elias
University of Colorado

Packrat middens are caches of objects, including edible plants, cactus spines, insect and vertebrate remains, small pebbles, and feces, brought to the den site for a variety of reasons, including food, curiosity, and den protection; then they are cemented into black tarry masses by packrat urine. Once dried, the exterior of the midden hardens into a protective coating that can preserve a paleoecological record for thousands of years in dry rock shelters of the arid West. The study of fossil records derived from middens is expanding our knowledge of the paleoecology of this region by providing factual data on the nature and timing of biotic responses to environmental changes. Midden data are being used to test biogeographic, ecological, paleoclimatic, archaeological, and evolutionary hypotheses.

I. PACKRATS AND THEIR MIDDENS

Packrats, or woodrats, are the native North American rats of the genus *Neotoma*. There are 21 species of packrats; they are broadly distributed, ranging from Arctic Canada to Nicaragua. They are a very successful group and inhabit a variety of habitats. Although packrats are not strict *herbivores,* plants are by far the most important part of their diets. Packrats are den builders. Regardless of size, each den is occupied by one individual, or the female and her young. These dens range from piles of sticks under shrubs or trees to less elaborate shelters wedged in caves and rock shelters. In the arid Southwest, these dens often include tightly packed bundles of cactus stems (for instance, cholla), which serve as an effective deterrent to predators. The walls of many dens consist of finely shredded plant material that serves as thermal insulation. All packrats require *succulent plants* as food, and they must have shelter, as they do not have the metabolic mechanisms for water retention found in some other rodents. The dens, especially those in rock shelters, crevices, and caves, create a microclimate that is significantly buffered from ambient climatic conditions. The seasonal variation of winter temperatures in packrat dens from middle to high latitudes has been shown to be as little as 15°C.

Besides bedding material and food plants, packrats also accumulate other objects from the surrounding landscape, including bones, scat, feathers, or anything else that captures their attention (dens have been found to contain false teeth, silverware, cigarette lighters, aluminum foil, corn cobs, etc.). These objects, plus packrat feces, accumulate in middens that are cemented together by layer upon layer of crystallized packrat urine. In caves or rock shelters, this matrix of urine-soaked material dries to form a hardened mass, called amberat. When protected from moisture amberat is extremely resistant to weathering and decomposition. It locks biological specimens (plant parts, bones, insects, etc.) into a kind of dessicated time capsule. Middens of late Holocene age (i.e., the last 4000 yr) have been found throughout the western United States and southwestern Canada. In the southwestern United States and northern Mexico, packrat middens are known to remain intact for more than 45,000 yr (i.e., beyond the limit of the radiocarbon dating method).

Packrat middens accumulate as long as a den is occupied. In rock shelters, caves, or crevices, these middens build up for centuries and millennia, as favorable denning sites are inhabited by generation after generation of packrats. Packrat middens are an unusual source of fossils. Lakes, ponds, and streams accumulate the remains of plants and animals that are washed or blown into them, but the packrat is the primary agent selecting most of the materials that end up in middens (Fig. 1). They eat a wide variety of plants, and unconsumed pieces of these plants accumulate in middens, forming the bulk of the material in midden records (No. 1 in Fig. 1). The urination perch (Fig. 1, No. 2) is the site where the rat urinates and defecates. They urinate frequently, keeping the surface of the perch moist and sticky. This surface acts as a trap for airborne pollen. Dung beetles and other dung-feeding arthropods are attracted to it, and other arthropods also get trapped in the sticky matrix of feces and urine. Vertebrates living outside of caves and rock shelters sometimes fall in through cracks or other openings to the surface, or seek shelter there (Fig. 1, No. 3). The bones of mammals and reptiles are often found in packrat middens, as well as the bony plates and scales of reptiles.

Owls and other birds of prey sometimes nest in caves and rock shelters. They leave behind the undigested fur and bones of their prey (mostly small rodents), and these may also end up in middens. Sometimes packrats pick these up from the cave floor and add them to the accumulation of midden debris. Most insects that are preserved in middens represent species that live outside of caves and rock shelters (Fig. 1, No. 5). These are mostly predators and scavengers that come into the packrat's den to make use of food resources (dung beetles, nest parasites, scavenging insects that remain in and around the nest, etc.). In cold months of the year, "outside" insects are apparently attracted to packrat dens because of their warmth. The remains of other cave and rock shelter inhabitants, such as mice, may end up in packrat middens (Fig. 1, No. 6). This sometimes happens by accident, or the packrat may add these remains to its den.

A. How Middens Are Analyzed

Packrat middens are common in caves and rock shelters throughout arid and semiarid regions of North America. Most ancient middens are *indurated* and resemble blocks of asphalt. The amberat (dried urine) cements the midden together and forms a tough, outer rind that resists erosion and decomposition. Because packrat urine contains toxic compounds from food plants, it also repels living insects and other scavengers that might burrow into middens. The stickiness of amberat often acts to cement middens to the walls of caves or rock shelters. Midden samples are obtained by the removal of slabs representing discrete horizontal sections of middens. This usually requires hammers and chisels. Each sample must be radiocarbon-dated to determine its age, because successive layers in a midden may have been deposited at widely spaced time intervals, or occasionally in random order. In the laboratory, midden samples are soaked in water to dissolve the amberat. Once this is done, the samples are wet-screened to remove fine sediments, then dried and picked through under a low-power microscope. At this stage, plant *macrofossils* (stems, seeds, buds, spines, etc.) are picked out and sorted from the matrix, as well as the *exoskeletons* of insects and other arthropods, and any vertebrate remains, including bones, teeth,

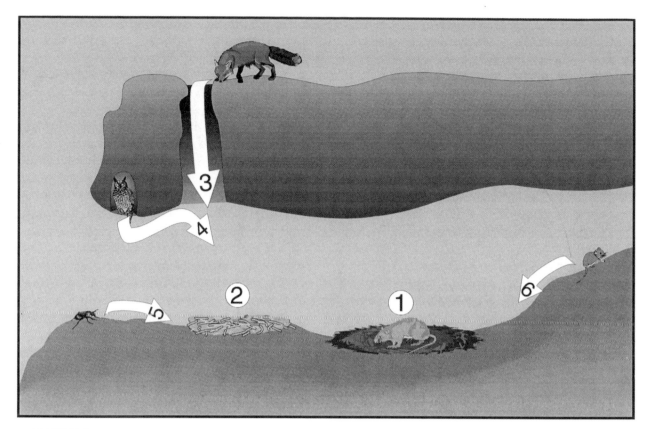

FIGURE I Diagrammatic drawing of a rock shelter, showing the main sources of fossil materials preserved in packrat middens (see discussion of numbered sources in text).

hair, scales and bony plates of reptiles, and bird bones and feathers. Radiocarbon dates are often obtained from packrat fecal pellets in midden samples, or from plant macrofossils. Macrofossils from individual species of plants are sometimes submitted for *accelerator mass spectrometer (AMS) radiocarbon dating*.

Plant macrofossils and vertebrate bones are frequently identified to genus or species level; about half of the arthropod remains can be identified to species, the rest to genus or family. Because of packrat collecting bias in midden materials, analysis of macrofossils (plant, arthropod, or vertebrate remains) from midden samples is generally done on a presence/absence basis for each identified taxon. Occasionally, a subjective quantification of the remains is conducted, either through application of a relative abundance scale (rare, common, etc.) or through specimen counts and dry weights of materials. Because plant remains are the most abundant

macrofossils in midden samples, the results of their analyses are often tabulated in terms of relative species abundance (e.g., separated into categories such as *rare, common, abundant,* and *very abundant*). Ancient pollen extracted from midden samples is tabulated either by relative percentage (percent composition of pollen types per sample) or by pollen concentration (number of pollen grains of a given type per unit of sample weight).

B. Distribution of Fossil Sites in the Arid Southwest

Figures 2 through 5 locate sources of published packrat midden records from the Chihuahuan, Sonoran, Mojave, and Great Basin deserts, and the semiarid Colorado Plateau. Dozens of sites have been studied in each of these regions, and more than a thousand radiocarbon dates have been obtained from midden samples. Many midden samples date either to the late Holocene (the last few thousand years) or to the end of the last glaciation [12,000–9000 years before present (yr B.P.)]. Some are older than radiocarbon dating can measure (i.e., >45,000 yr B.P.). Most midden samples have been taken from broken terrain in cliffs, canyons, and rocky hillsides, just above the desert floor. No Pleistocene middens have been reported at elevations greater than 2500 m. The elevational distribution of midden samples reflects regional physiographic differences. Sites south of 36° latitude generally occur at elevations below 1500, whereas more northerly sites reflect a more even range of elevations.

II. REGIONAL STUDIES

A. Chihuahuan Desert

The Chihuahuan Desert is an interior continental desert, stretching from Arizona in the northwest to northern Zacatecas, Mexico, in the southeast (Fig. 2). The desert lies in the rain shadow of the Sierra Madre Occidental to the west and the Sierra Madre Oriental to the east. Subfreezing temperatures are not uncommon in winter, and Arctic air mass incursions, "blue northers," may bring very cold weather to the southernmost parts of the desert. The southern regions support subtropical vegetation, rich in succulent scrub species. These taxa diminish to the north, in favor of more cold-tolerant desert grassland. The fossil record developed from packrat middens shows that the Chihuahuan Desert experienced a series of environmental changes through the Pleistocene and the Holocene. These changes brought about wide-scale shifts in the distributions of Chihuahuan plants and arthropods, both within the desert (i.e., elevational shifts) and outside its boundaries (i.e., regional extirpations and establishment of species new to the region), whereas the small vertebrate fauna (e.g., rodents, rabbits, lizards, and snakes) was essentially the same as the modern fauna. Occassionally bones of large mammals are found in middens. The midden samples studied from this desert range in age from >43,300 yr B.P. to recent. [*See* ECOPHYSIOLOGY.]

In the northern part of the desert, pinyon–juniper–oak woodland dominated limestone slopes during the middle to late Wisconsin interval (42,000 to 10,800 yr B.P.). An early Holocene oak–juniper woodland gave way to desert–grassland after about 8250 yr B.P. At about 4200 yr B.P., the common elements of the Chihuahuan Desert flora migrated into the northern Chihuahuan Desert region, forming a relatively modern desert–scrub vegetation on rocky slopes. This vegetation remains regionally dominant today, and in historic times shrubs such as honey mesquite and creosote bush have invaded much of the grassland regions.

In the central part of the Chihuahuan Desert, in and around Big Bend National Park, the vegetation during the middle of the last glaciation (40,000–22,000 yr B.P.) was a pinyon–juniper–oak woodland on the uplands and juniper grassland on the lowlands. By late Wisconsin times (prior to 21,000 yr B.P.), papershell pinyon became important in the upland flora, and pinyon–juniper–oak woodland spread into lowland regions, indicating relatively mild winters, cool summers, and greater precipitation than today. The boundary between late Wisconsin and early Holocene environments (between 11,200 and 10,300 yr B.P.) is well marked in the vegetation, with the demise of papershell pinyon and increase of scrub oak before 8200 yr B.P., followed by a succulent desert–scrub vegetation, including lechuguilla, catclaw acacia, honey mesquite, and prickly pear cactus. Vegetation diversity increased in the middle Holocene, indicating subtropical climatic conditions. More mesic vegetation communities have developed here in the last 1000 yr. Middle Holocene vegetation in the lowlands of Big Bend was essentially modern Chihuahuan desert–scrub. Lowland midden samples from 1700 to 1600 yr B.P. contained a very xeric flora, and the modern vegetation is as xeric as any in the Holocene.

In the southern Chihuahuan Desert of Mexico, the late Wisconsin vegetation on limestone slopes was a woodland dominated by juniper and papershell pinyon in association with succulents such as sotol and beaked yucca. The abundance of pinyon pine declined after 13,000 yr B.P. as succulents

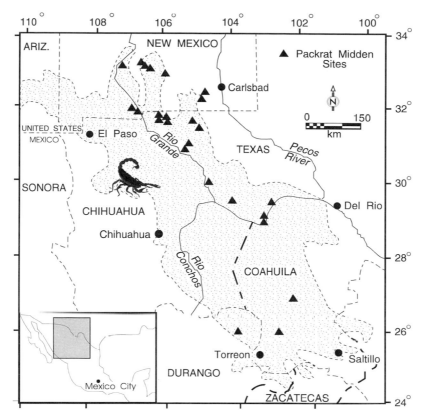

FIGURE 2 Map of the Chihuahuan Desert region, showing location of packrat middens from which fossil analyses have been published (triangles).

increased. Some time between 11,730 and 9360 yr B.P. the regional vegetation shifted to a Chihuahuan desert–scrub without a transitional woodland. The paleobotanical record from this region shows unique combinations of floristic elements. Late glacial floras include both conifers and desert succulents, but temperate plants drop out of regional records during the middle and late Holocene. A mosaic of temperate and xeric habitats has been available to the regional biota, even during the last 1000 yr, when other regions of the Chihuahuan Desert have experienced extremes of aridity.

B. Sonoran Desert

The Sonoran Desert is the arid, subtropical region centered around the head of the Gulf of California (Fig. 3), encompassing parts of Sonora and Baja California in Mexico and Arizona and California in the United States. All parts of the Sonoran Desert are occasionally subject to winter freezes, although rarely for more than one night at a time. Most of the large, columnar cacti (e.g., saguaro and organ pipe cacti) that typify the Sonoran Desert are not able to tolerate prolonged intervals below freezing. Rainfall ranges from a biseasonal pattern with summer monsoon moisture in Sonora and Arizona to a winter rainfall pattern along the west coast of Baja California. The Sonoran Desert, like the Chihuahuan, was remote from glacial ice advances during the Wisconsin Glaciation, but changes in large-scale atmospheric circulation patterns during the last glaciation brought marked changes in regional climates, especially in precipitation patterns and frequency of winter freezes.

The plant macrofossil record from Sonoran Desert packrat middens indicates that regional vegetation succession from the late Wisconsin through the Holocene made a single, continuous progression toward modern plant communities. The

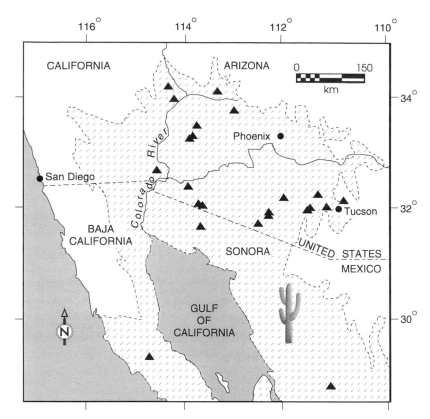

FIGURE 3　Map of the Sonoran Desert region, showing location of packrat middens from which fossil analyses have been published (triangles).

chronological sequence is similar to the modern vegetational gradient from mountain tops to valley floors. These events are summarized in Table I. Central and southern Arizona were covered with a variety of woodlands in the late Wisconsin. In upland regions, juniper or juniper–oak woodlands persisted through the early Holocene. Juniper was extirpated from the Sonoran Desert after 8900 yr B.P. Lowland regions have had desert–scrub vegetation since the early Holocene. Desert–scrub communities, including species found today in the Mojave Desert, persisted in the harsh lowlands of the lower Colorado River valley throughout the late Pleistocene; this region may have served as a core for North American desert plants for much of the Quaternary, although typical Mojave Desert species moved north along the Colorado River and crossed over it only during the warmest (interglacial) periods. The boundary or ecotone between the Sonoran and Mojave deserts has essentially remained in its current position since 8900 yr B.P.

After 4000 yr B.P., the modern vegetation became established throughout the Sonoran Desert.

Based on the paleobotanical evidence, the late Wisconsin climate in the Sonoran Desert was characterized by cooler-than-modern summers and a shift to winter precipitation. Though winters may also have been generally cooler, it appears that freezing conditions were rarer than today. Early Holocene climates were transitional between late Wisconsin and modern climates. Summer temperatures probably remained cooler, based on the persistence of conifers in the uplands. Estimates suggest that parts of south-central Arizona received up to 30% more precipitation in the early Holocene compared with today. A relatively modern climate was established after 9000 yr B.P. characterized by hot summers. Middle Holocene plant assemblages reflect greater species richness than today, probably because of increased rainfall. By 4000 yr B.P., a completely modern climatic regime was established.

TABLE I

Summary of Changes of Sonoran Desert Vegetation Inferred from Packrat Midden Data

Region	Time interval	Vegetation
Central Arizona		
Uplands (550–1550 m)	Late Wisconsin	Pinyon–juniper–oak woodland
	Early Holocene	Juniper–oak woodland with Sonoran Desert–scrub
	Late Holocene	Sonoran Desert–scrub with increasing subtropical elements
Lowlands (<600 m)	Late Wisconsin	Juniper woodlands and/or Mohave Desert–scrub
	Early Holocene onward	Sonoran Desert–scrub
Southern Arizona		
Uplands (550–1550 m)	Late Wisconsin	Pinyon–oak–juniper woodland with Joshua tree
	After 11,000 yr B.P.	Xeric juniper woodland with shrub oak or chaparral
	After 8900 yr B.P.	Juniper disappears from entire region
Lowlands (300–600 m)	Late Wisconsin	Xeric juniper woodland with Joshua tree, yucca, beargrass
	Early Holocene	Desert–scrub common
	After 4000 yr B.P.	Modern Sonoran Desert–scrub
Baja California		
Lowlands (0–400 m)	Early Holocene	Abundant Boojam tree; buckwheats and Mormon tea in desert–scrub community
	Late Holocene	Boojam tree less abundant; buckwheats and Mormon tea disappear; increasing succulents

C. Mojave Desert

The Mojave is a small but topographically complex desert, situated in southern Nevada and adjacent California and Arizona (Fig. 4). The southern Mojave (south of 36°N latitude) consists of large outwash slopes (bajadas) and low valleys, dissected by many small mountain ranges. The vegetation of this region is typified by creosote bush, white bursage, and pygmy cedar. The northern Mojave region is generally higher in elevation, and the thermophilous desert–scrub vegetation of the south is replaced here by temperate desert–scrub, including blackbrush, Joshua tree, ground-thorn, boxthorn, and goldenbushes. The higher mountains in southern Nevada (the Spring and Sheep ranges) support pinyon–juniper woodlands above 1800 m elevation, giving way to fir–pine forest above 2200 m. Bristlecone pine dominates subalpine landscapes above 2700 m.

The late Wisconsin and Holocene vegetation changes within the Mojave Desert affected all elevations (Table II). More studies have been done in the northern part of the desert than in the south. During the late Wisconsin period, woodlands extended across the lowlands of the Mojave Desert region. These woodlands are described by paleobotanists as "pygmy" woodlands; they were composed primarily of juniper (mixed with pinyon pine at elevations below 1500 m), shrub oak, and Joshua tree in some regions. At elevations above 1600 m in the northern part of the Mojave Desert, limber pine, replaced at times by white fir, characterized the last glacial maximum. The lower (downstream) end of the Grand Canyon lies in the southeastern corner of the Mojave Desert. This region supported more succulent species in the late Wisconsin, including barrel cacti and yuccas.

By 10,000 yr B.P., desert–scrub vegetation spread across regions below 1000 m elevation, at the expense of semiarid woodland. The transition from glacial to postglacial vegetation was not smooth; it was a time of multiple incursions of new taxa and extirpations of others. Not all of the newly invading species have remained throughout the Holocene. The immigration of desert thermophiles from their refugia took place in stages, with some species arriving much later than others. On the other hand, widespread desert vegetation appears to have developed earlier in the Mojave than in the Sonoran or Chihuahuan deserts. This vegetation type occurred sporadically as early as 15,000 yr B.P. By 7800 yr B.P., woodlands were replaced by desert vegetation in all low-desert regions. Creosote bush

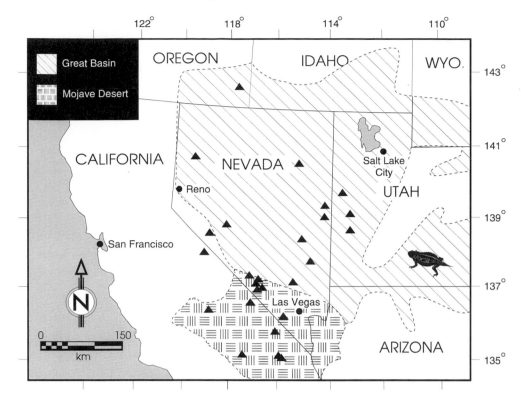

FIGURE 4　Map of the Mojave Desert and Great Basin regions, showing location of packrat middens from which fossil analyses have been published (triangles).

TABLE II
Summary of Changes of Mojave Desert Vegetation Inferred from Packrat Midden Data

Region	Time interval	Vegetation
Southern and Central Mojave		
Lowlands (<1000 m)	Late Wisconsin	Juniper woodland
	Early Holocene	Desert–scrub widespread by 10,000 yr B.P.; succulents and grasses increasing
Uplands (1000–1800 m)	Late Wisconsin	Juniper–pinyon woodland
	Early Holocene	Woodland persists at higher elevations and moist slopes
	Middle Holocene	Desert–scrub rich in species
	Late Holocene	Downward shift of blackbrush desert–scrub
Northern Mojave		
Uplands	Late Wisconsin	Desert–scrub in dry locations, with Mormon tea, shadscale, rubber rabbitbrush, and snowberry; juniper woodland in most regions; at higher elevations, limber pine in drier sites and white fir in mesic sites
	Early Holocene	Woodland persists on some slopes; after 7800 yr B.P., replacement by creosote bush and Joshua tree scrub
	Middle Holocene	Desert–scrub lacking in diversity
	Late Holocene	Rich desert–scrub communities at lower elevations and pinyon–juniper woodlands at higher elevations

spread across the Mojave from the south, reaching its northernmost limit by about 5500 yr B.P. Middle Holocene aridity is marked by vegetation changes in most regions of the Mojave. Regional packrat midden evidenced from the desert, pollen records and discharge histories from desert springs, and tree ring evidence from the White Mountains all suggest that effective moisture increased between 3800 and 1500 yr B.P. This increase was probably due to a combination of increased precipitation and decreased temperatures. Conifers retreated up mountain slopes in the Mojave region during the last 1500–500 yr, signaling a return to hotter, drier conditions.

D. The Colorado Plateau

The Colorado Plateau is a well-marked physiographic province covering nearly 400,000 km² of the Four Corners region (Fig. 5). Elevations on the plateau range from 1000 to 4600 m, with much of the plateau below 1800 m. Regional climates are strongly affected by topography, both now and in the late Quarternary. The interior regions fall in the rain shadow of highlands to the south, east, and west. Modern vegetation zones are shown in Fig. 6. These zones do not fall in elevational bands across the entire region, because local bedrock types strongly affect vegetation types. In general, both Pleistocene and modern vegetation of the plateau suggest wetter conditions than the vegetation of similar elevations farther west.

More than 180 packrat middens have been analyzed from sites on the Colorado Plateau. The middens occur only in regions with sandstone and limestone bedrock; paleoenvironmental reconstructions are largely limited accordingly. Late Wisconsin vegetation reconstructions are summarized in Fig. 6. Upper treeline in the platcau region was significantly depressed, and alpine tundra regions may

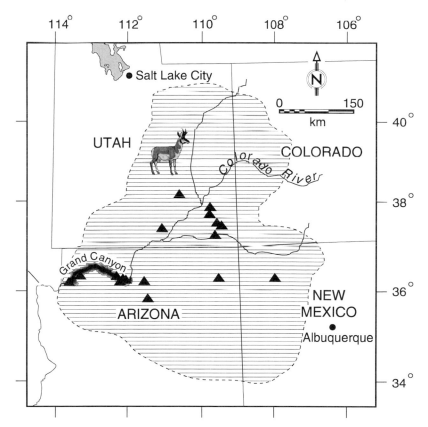

FIGURE 5 Map of the Colorado Plateau region, showing location of packrat middens from which fossil analyses have been published (triangles).

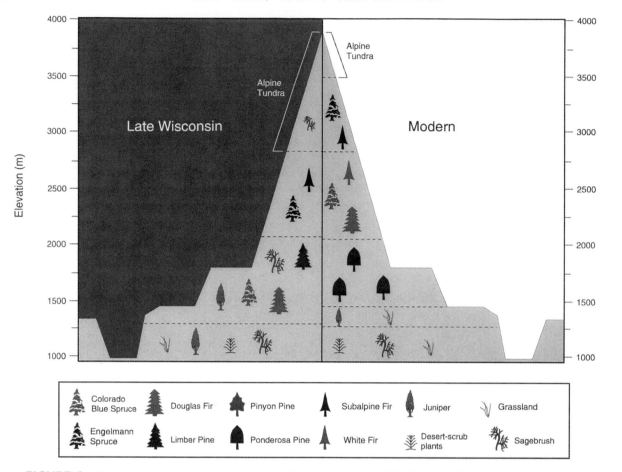

FIGURE 6 Diagrammatic drawing showing principal vegetation zones of the Colorado Plateau during Late Wisconsin and modern times.

have supported more sagebrush cover than is found there today. Alpine tundra is found today in only three high mountain regions of the plateau, all at elevations above 3600 m. Late glacial coniferous forests extended down to 2000 m elevation in mesic canyon habitats. Coniferous woodlands with limber pine, Douglas fir, junipers, and Colorado blue spruce extended down to 1300 m elevation on rocky sites near the San Juan Mountains and in Glen Canyon. Canyons and other lowlands below 1300 m elevation were covered by desert–scrub with junipers and sagebrush.

Evidence from the Grand Canyon suggests that plant migration lagged significantly behind climate change at the end of the Wisconsin Glaciation. For instance, the upslope immigration of species to their modern zones was not completed in some cases until the middle Holocene; some plants, such

as ocotillo, may still be adjusting to past climate fluctuations. The unexpected associations of plants in early Holocene midden assemblages may be due in part to unique climatic conditions, involving the dynamics of monsoonal moisture and shifts in the timing of summer warmth.

E. Great Basin

The Great Basin is a large, arid region between the Rocky Mountains and the Sierra Nevada (Fig. 4). It is made up of broad valleys of sagebrush steppe, dissected by more than one hundred narrow mountain chains that trend north to northeast. During the last glaciation, immense lakes covered the valleys of the Great Basin, and glaciers flowed down from high mountain valleys. In the modern vegetation, sagebrush and shadscale dominate steppe communities at low elevations. Pinyon–juniper woodland covers most low mountain slopes in the southern two-thirds of the region, whereas juniper woodland characterizes the northern third. Montane forests of ponderosa pine, white fir, and Douglas fir grow in the mountains of the eastern and southern sectors. The montane zone in the central and northern regions of the Great Basin is covered by sagebrush grassland. Some isolated mountain ranges are covered by mountain mahogany woodland. Subalpine forests of bristlecone pine, limber pine, and Englemann spruce occur in the higher mountain ranges, and small pockets of alpine tundra grow on isolated peaks. Botanical studies suggest that the Great Basin flora shows more affinities to the Rocky Mountain flora than to that of the Sierra Nevada.

The history of regional vegetation during the late Pleistocene and Holocene is summarized in Table III. Many plant species that lived in the Great Basin during the Pleistocene have their northern limits in the modern Mojave Desert; other plants grew as much as 1000 m downslope from their modern ranges. Unlike some other desert regions, there is no evidence that *extra-limital* plant species immigrated into the Great Basin during the late Pleistocene. Rather, regional floras were impoverished of species, as many of the modern taxa were absent during that time.

Subalpine conifers, especially limber pine (and occasionally bristlecone pine in the central Great Basin and whitebark pine in the west), expanded their ranges downslope by as much as 1000 m during the late Wisconsin interval. Curiously, the conifer species now in the montane zone (spruce and fir) are not recorded in the region for much of the late Pleistocene. Many persisted in more southern localities, such as southern Nevada and southeastern California. Likewide, woodland plants grew in more southerly locations and at lower elevations than they occur in today. In the western Great Basin, juniper

TABLE III
Summary of Changes in Great Basin Vegetation Inferred from Packrat Midden Data

Time interval	Vegetation
Late Wisconsin (22,000–14,000 yr B.P.)	Uplands: Subalpine conifers grow as much as 1000 m downslope from modern limits; dominant trees: bristlecone pine, limber pine, prostrate juniper, and Englemann spruce in some locations; many shrubs and trees common today are absent, including Douglas fir, white fir, and Rocky Mountain juniper. These shifted south, into the Mojave Desert and elsewhere. Steppe regions: Sagebrush and other steppe species present across entire range of elevations in Great Basin, up to subalpine (as ground cover under coniferous forests). Steppe species include sagebrush, rabbitbrush, horsebrush, snake-weed, and snowberry. Steppe species spread south into Mojave and Sonoran deserts, and other regions.
Late Glacial (14,000–10,000 yr B.P.)	Limber pine, prostrate juniper, Englemann spruce, and some shrubs decline from 13,000 to 11,000 yr B.P., replaced by snake-weed and rock spiraca. Bristlecone and limber pine persist at late Wisconsin levels until after 10,000 yr B.P. Mesic-adapted plants at many elevations replaced by thermophiles starting at 13,000 yr B.P. In southern Great Basin, a major shift to desert vegetation before 11,000 yr B.P.
Early Holocene (10,000–7000 yr B.P.)	Bristlecone and limber pines absent from western Utah and eastern Nevada by 8600 yr B.P. Limber pine and Douglas fir replaced by Utah juniper and Gambel oak in southeastern Nevada by 8900 yr B.P.
Middle Holocene (7000–4000 yr B.P.)	Rapid transition to modern vegetation zones between 7000 and 6000 yr B.P. Pinyon–juniper woodland established in several mountain ranges, characterized by Utah juniper and single-needle pinyon pine. Midden assemblages include a number of species that are rare or absent from the Great Basin today.
Late Holocene (4000 yr B.P.–present)	Few changes in regional vegetation. Ranges of sagebrush and shadscale steppe contract; upper treeline retreats to middle Holocene levels. Lower treeline drops about 170 m between 4000 and 2000 yr B.P., then retreats to modern levels during last 2000 yr.

dominated lower-elevation woodlands in areas adjacent to pluvial lakes. Although Utah juniper appears to have been the dominant species at these sites, possible hybrids of western juniper, Rocky Mountain juniper, and California juniper have been found in some middens. Sagebrush and other steppe species expanded their ranges in the late Wisconsin, both latitudinally and altitudinally. They spread south into the Mojave and Sonoran deserts, and into the adjacent Sierras and probably the western Rockies. More xeric areas in the northern Great Basin were dominated by shadscale.

The downslope extension of bristlecone and limber pines might have produced a contiguous forest zone stretching from the eastern shore of pluvial Lake Lahontan to the Rockies, if not for edaphic factors that probably precluded conifer establishment in some regions. At any rate, what today are small habitat islands of subalpine forest surrounded by a "sea" of sagebrush steppe once supported many more pines (limber and bristlecone pine). This floristic domain became fragmented again as the lower treeline crept upslope in response to climatic warming at the end of the Pleistocene. The shift from Pleistocene to Holocene vegetation regimes took place in different ways at different sites. In the boundary region between the Mojave Desert and Great Basin, the major shift took place by 11,000 yr B.P. To the north, the changes were completed between 9000 and 7000 yr B.P.

The Great Basin has a large number of endemic herbaceous plants. The packrat midden record suggests that this is due to a combination of factors. One of these is range fragmentation that permitted genetic isolation and rapid evolution, at least at the subspecific level. The other is the opening up of new habitats as Pleistocene lakes receded and climates changed. During the Holocene, semiarid lowlands have acted as biotic barriers, much as pluvial lakes did during the Pleistocene.

III. MAJOR RESULTS AND IMPLICATIONS OF STUDIES

A. Paleobotanical Studies

The most diverse and abundant fossil record from packrat middens has come from plant macrofossils.

The plant macrofossil record is providing valuable data on past and present ecosystems of the arid Southwest. Modern ecologists generally study the desert and ecosystems on a short-term basis. In a sense, they are trying to understand nature's story by reading only the last page of the book. The fossil record has shown that it is impossible to discern long-term patterns from the modern data alone. The fossil record shows that the story is very long, complicated, and surprising. For instance, modern ecological theory holds that most biological communities converge to stable equilibria following massive perturbation, and that this state is maintained by competition and predation. This view tends to exclude history, chance, and long-term environmental change as major factors. The fossil record shows that environmental change during the Pleistocene was substantial, even in regions such as the arid Southwest that were far removed from continental ice sheets. Equilibrium theory also postulates that the species of a given community are able to respond to environmental changes and shift their distributions more or less in synchroneity with them. The fossil record suggests otherwise, in many cases. The midden record offers good evidence for individualistic responses to major climate changes that brought about uneven displacement of species, and anomalous associations in which some species assemble in ways unknown historically.

The age and origin of southwestern deserts are important questions in the field of biogeography. About 1200 of the 5500 species of southwestern plants are considered endemic to that region, a far higher proportion than elsewhere in North America. Packrat midden data suggest that many of these plants evolved locally, but that desert plant communities are very recent and very dynamic. One of the reasons the midden data are so important is that they offer tangible evidence of the history of species in this region. For instance, based on modern data, creosote bush was thought by some to have immigrated from South America during the Holocene. However, the fossil evidence shows that it resided along the Arizona–Sonora border country during the middle of the last glaciation. We are not currently able to verify the exact whereabouts of southern refugia for the desert biota in Mexico, because many regions hypothesized to have been refugia are now in the tropical or subtropical zones, where preservation of middens is extremely limited (these areas are sufficiently moist that middens decompose).

Many intriguing questions remain to be addressed by the plant macrofossil record. New research tools are being developed that will be helpful. For instance, during the late Wisconsin, pygmy conifer communities with pinyon pine, junipers, and oaks covered much of the lower elevations across the American Southwest. These woodlands remained intact until after 11,000 yr. Was this long-term vegetational stability a product of climatic stability across a wide region? What brought about the demise of the lowland pygmy forests? One new way of approaching these questions is the analysis of stable isotopes trapped in the cellulose of woody plant macrofossils. In particular, analysis of changes in the concentration of the hydrogen isotope deuterium may help to clarify the climate regimes under which these plants grew, and to define the climatic changes that forced their retreat from the deserts' floor. Another difficult question concerns a major switch in pinyon pines at the end of the Pleistocene. Colorado pinyon and single-needle pinyon (*Pinus monophylla*) became the dominant pinyon pine species in the Holocene, at the expense of papershell pinyon and another single-needle pinyon, *Pinus californiarum* var. *fallax*. The latter two were dominant in the Pleistocene. There may be climatic significance to this reversal of roles, but modern physiological studies must be done before this question can be answered. Another approach to problems concerning shifting distributions and population dynamics of pinyon pines is the study of ancient DNA. Gene amplification techniques are being applied to pinyon wood, cones, and needles preserved in middens to unravel the genetic relationships between ancient populations and their modern descendants. Pinyon pine distributions remain far from static. Midden data indicate that both single-needle pinyon and Colorado pinyon are still migrating. The northernmost population of Colorado pinyon became established in north-central Colorado only 400 years ago. The same is true of single-needle pinyon, which arrived at its current northwestern limit (near Reno, Nevada) within the last 700 years.

B. Vertebrate Studies

On the basis of packrat midden data, small mammals in the arid Southwest were generally less affected by glacial climates than were plants. The Chihuahuan Desert middens have yielded numerous remains of small reptiles and vertebrates, especially small mammals. These include the remains of shrews, bats, rabbits, ground squirrels, prairie dogs, pocket gophers, kangaroo rats, voles, mice, rats, porcupine, and deer. Reptile remains from middens include tortoises, lizards, and snakes. In general, the Wisconsin glacial-age vertebrate faunas of the northern Chihuahuan Desert were less like modern regional faunas than were the faunas living farther south in the desert. Some of the species identified were living in plant communities they do not inhabit today. This is perhaps to be expected, because the plant communities of the late Pleistocene occupied very different regions than they do now. Shifts in temperate animal species' distributions progressed from south to north. In the southern Chihuahuan Desert, this process was essentially complete by 10,000 yr B.P. In the Holocene, desert-dwelling animals expanded their ranges to higher elevations and northward.

During the late Wisconsin, the small vertebrate fauna of the Chihuahuan Desert was essentially the same as the modern fauna, with most fossil records occurring close to where the species are found today. Some temperate-zone animals, such as the large mountain cottontail, were more widespread during the late Pleistocene than today. Its range shifted progressively northward in the Holocene. The botanical evidence for cooler, moister climates in the Chihuahuan Desert during the late Pleistocene is also substantiated by regional fossil records of voles and prairie dogs, animals adapted to temperate-zone grasslands and meadows. The small mammals found in the Chihuahuan middens fit into a general latitudinal pattern of decreasing differences between Wisconsin and modern faunas at lower latitudes of the desert bordering the Rio

Grande. This is the hottest part of the desert today, and this trend extends back into the Pleistocene. Sonoran Desert packrat middens have yielded mostly reptiles; these species are still found in the Sonoran Desert today.

In the Great Basin, pika remains have been found in a 16,000-yr-old midden in the Las Vegas Valley. Pikas are found today in the subalpine and alpine zones, among rocks in fellfields where grasses are available. The closest modern populations live in the Sierra Nevada, well above the desert floor. This is another example of dramatic shifts in regional distribution of both plants and animals since the late Pleistocene.

C. Arthropod Studies

In the last ten years, insect fossils from packrat middens have received intensive study. Although fossil insect research is now under way for sites in the Sonoran Desert and the Great Basin, the most intensively studied region (191 midden insect fossil samples from 27 sites) is the Chihuahuan Desert. Late Pleistocene faunas were mixtures of temperate and desert species not seen in any one region today. Since the end of the Pleistocene, some of these species have become established in different regions of the Chihuahuan Desert. Others now live outside this desert. The fossil insect record indicates that even sedentary, flightless beetles (such as the heavy-bodied weevils) have undergone marked distributional shifts in the American Southwest within the space of a few centuries. Moreover, even highly specialized cave-dwellers have somehow managed to move from one cave system to another in response to changes in late Wisconsin and Holocene environments.

The fossil insect assemblages from the southern Chihuahuan Desert contain mixtures of desert- and temperate-zone species in almost every interval from the late glacial through the late Holocene. Midden assemblages from locations farther north in the Chihuahuan Desert are generally separated into glacial-age faunas with temperate-zone affinities and postglacial faunas with desert-zone affinities. The "no modern analog" faunal assemblages indicate that the late Quarternary environments in

this part of the desert were unlike any that exist today. This conclusion is also borne out by the plant macrofossil record.

The insect faunas from the Big Bend region (Fig. 1) suggest greater effective moisture from 30,000 to 12,000 yr B.P. In the late Wisconsin, many temperate grassland species lived in the Big Bend region. After 12,000 yr B.P., most of these species were replaced by either desert species or by more cosmopolitan taxa. The faunal change suggests a climatic shift from cool, moist conditions of Wisconsin glacial times to hotter, drier conditions of latest Wisconsin and early Holocene.

In the northern Chihuahuan Desert, full-glacial (22,000–18,000 yr B.P.) arthropod records suggest widespread coniferous woodland at elevations as low as 1200–1400 m above sea level. These woodland environments persisted until 11,000 yr B.P., but the insect data suggest considerable open ground, with grasses at least locally important at the midden sites. The grassland nature of the arthropod fauna was also suggested in the regional vertebrate record. The transition from the temperate Wisconsin fauna to the more xeric postglacial fauna started by 12,500 yr B.P., the timing of this faunal change being essentially synchronous throughout the northern Chihuahuan Desert. A major difference between the Big Bend and northern Chihuahuan Desert scenarios is the nature of this faunal change. In the Big Bend region, the transition was characterized by the disappearance from the record of all but one of the temperate insect species at about 12,000 yr B.P. However, the xeric-adapted fauna did not appear in the Big Bend records until about 7500 yr B.P. In the northern Chihuahuan Desert assemblages, the xeric species first appeared at 12,500 yr B.P., and several of the temperate grassland species from the Wisconsin interval persisted well into the Holocene. This mixture of xeric and temperate elements makes sense from an ecological standpoint, because these northern faunas were living close to the edge of the Chihuahuan Desert. The gradual shifting of northern desert boundaries in the Holocene probably created many marginal habitats for temperate species in ecotones between grassland and desert–scrub. By about 7500 yr B.P., the appearance of more xeric species indicates establishment of desert environments, including desert grasslands, throughout the Chihuahuan Desert region. After 2500 yr B.P., the last of the temperate species was replaced by species associated with desert–scrub communities.

Recent work on late Quarternary insect faunas from the Colorado Plateau region suggests that late Wisconsin climatic conditions were cooler and moister than present, and that the plateau supported a mosaic of grassland and shrub communities without modern analog.

Insect faunas have also been studied from several sites in the Sonoran Desert. Unlike the Chihuahuan Desert insect fauna, the Sonoran Desert fauna indicates little change from the late Pleistocene through modern times. All taxa found in the Sonoran middens probably live within a few kilometers of the midden sites today. The Sonoran arthropod fauna showed a marked increase in diversity during the late Holocene, as more subtropical plants and warmer climates became established about 4000 yr B.P. Many warm-stenothermic insects probably dispersed into southern Arizona from Sonora, Mexico, during the last 4000 yr. The increase in diversity in fossil arthropod assemblages has been correlated with the frequency of winter freezes, and secondarily with the amount of summer precipitation. This late Holocene peak in species richness is in sharp contrast to the Chihuahuan Desert insect record, which showed the least number of species in late Holocene samples. If the Sonoran Desert insect fauna truly was stable through the late Quarternary, then this represents a significant difference between the Sonoran and Chihuahuan desert insect faunal histories.

IV. SUMMARY AND CONCLUSIONS

The fossil record developed from packrat middens in the arid Southwest and apparently in dry regions on other continents is a unique resource for investigating the pace and direction of biotic changes. Indeed, without the packrat midden data, we would know little about the biotic history of this region, where lake sediments are scarce and peat

bogs almost nonexistent. The midden data have shown that late Pleistocene landscapes of the Southwest were radically different than today, with coniferous woodlands covering much of the lowland regions that now support only desert–scrub. This change was not just a downslope shift of intact communities; it brought together new associations of plants and animals in an ever-changing dynamic that is still at work today. Arid-land biotic communities viewed as "stable" from our short-term perspective are actually in a constant state of flux. Each new environmental change brings a different cast of characters to the biological stage. In some regions, like the Great Basin, the same players appear in succeeding intervals. In other regions, such as the Sonoran and Chihuahuan deserts, new players appear from time to time, often at the expense of others. The fossil record has much to say to both the paleontological and modern ecological communities; we are just starting to understand its message.

A. Paleoclimatic Implications

The fossils found in the packrat middens record serve as *proxy data* for the reconstruction of ancient climates. Much of this work is based on comparisons between the ancient and modern distributions of the species found in the midden records. Evidence from plants, vertebrates, and insects alike suggests that during the last glaciation, the climates of the arid Southwest were substantially cooler than modern climates. Paleoclimate modelers hypothesize that the jet stream split in two for much of the Wisconsin glaciation, with one airstream flowing north of the continental ice sheets and the other flowing south of them. The southern flow probably brought greater moisture to much of the Southwest, as attested to in the fossil record and in the growth of large *pluvial* lakes in the lowlands of the Great Basin. The fossil data from packrat middens indicate that late Wisconsin climates in the arid Southwest were unlike any seen today and that, as recently as the middle Holocene, the seasonality of moisture and warmth may have been substantially different from today. The middle fossils show that changes in climate over the last 40,000 years were far more complex and more regionally diverse than a simple pattern of "cold, wet Pleistocene" and "warm, dry Holocene."

B. Biogeographic Implications

A whole host of biogeographic issues have arisen from the study of packrat middens. At the same time, biogeographic theories, such as the theory of island biogeography (as it applies to habitat islands), have been tested with fossil midden data. At issue here is the fragmentation of montane, subalpine, and alpine habitats across the isolated mountain ranges of the arid Southwest. Some mountain chains, such as the Sierras and the Rockies, have perpetuated nearly contiguous bands of these habitats through a series of shifting elevations during much of the Pleistocene and the Holocene. Other, smaller ranges are separated from their nearest neighbors by broad lowland regions. As we have seen from the Great Basin midden record, pygmy conifer woodland in the south and subalpine woodland in the north may have spanned the gulf between many if not most of its mountain ranges during the late Pleistocene. Apparently, not all conifer species benefited by such avenues of dispersal, however. A case in point is ponderosa pine, absent from most of the western interior during the last glaciation. There are now five distinct races of this tree, two of which meet in central Montana. Even this limited contact probably did not occur during the Pleistocene. Prolonged separation of the various gene pools, punctuated by east–west migrations, may explain the differentiation of these races. In contrast to this, bristlecone and limber pine were widespread across the central Great Basin region during the late Pleistocene; these two species do not have separate races on either side of the Great Basin today.

The question of endemism is a recurring theme in southwestern biology. However, the case of the cave-dwelling ground beetles from Chihuahuan Desert middens is a cautionary tale for modern biologists. It would be very tempting, for instance, to label as endemic a cave-dwelling beetle species found today only in Carlsbad Caverns, if it were not for the fossil data showing that in the Pleisto-

cene it also lived several hundred kilometers distant from Carlsbad, in the Big Bend region of Texas.

C. Ecological and Evolutionary Implications

In the past few decades the discovery of a well-preserved fossil record in ancient packrat middens has greatly enriched paleoecological research in the arid Southwest. The new findings bear on the subjects of biogeography, biotic response to climate change, and genetic isolation of populations across the western region through the late Pleistocene. All of these topics are intertwined; it is impossible to discuss one without involving the others. The development of an awareness of these interconnections is one of the most important contributions that packrat middens have made to southwestern biology. The desert ecosystems are dynamic, responding to environmental changes on century and millennial time scales. These are the important time scales for ecosystem establishment and dissolution, for the development of biological communities, and the reshuffling of species in those communities. The ancient history of the deserts cannot and should not be separated from the modern ecosystems; they form a seamless continuum. Understanding derived from stable isotope analyses may provide clues to the ways in which organisms adapt to changing climate, and how organisms interact in communities. Such studies may help us to understand ancient communities for which there is no modern analog.

Theories of speciation, adaptation, and extinction are also being tested using packrat midden data. This process has only just begun, but it holds great potential. One new tool is the study of ancient DNA preserved in midden fossils. With the wealth of fossil materials from middens, each identified, genetically fingerprinted, and dated to a resolution of plus or minus 100 years in time and plus or minus 50 m in space, we may examine migrations across the deserts in great detail, and find out where give populations came from, how they got there, and when. Midden analyses may begin to test the punctuated equilibrium theory of evolution, in which evolution takes place in bursts, separated by long periods of stasis. This could be examined by looking at populations of plants, insects, or other biota preserved in middens over time spans of 500, 2000, or 10,000 years. Middens are also being used to test environmental impacts of ancient human populations, such as whether the Anasazi inhabitants of Chaco Canyon and Mesa Verde overharvested firewood. Packrat middens are the most likely source of the type of detailed, consistent records needed to do these kinds of analyses. This is the kind of exciting, interdisciplinary research that allows science to advance in many different directions.

Glossary

AMS radiocarbon dating Method of radiocarbon dating using an accelerator mass spectrometer to measure the concentrations of the carbon ions (^{14}C, ^{13}C, and ^{12}C) in a sample. The ions are accelerated in a cyclotron or tandem accelerator chamber to extremely high velocities. Then they are passed through a magnetic field that separates the different ions, allowing them to be distinguished from each other and counted individually.

Exoskeleton External skeleton of insects, other arthropods, and crustaceans.

Extra-limital species Species occurring in a fossil assemblage that have a modern distribution that does not include the region of the fossil locality.

Herbivore Organism that eats plants.

Indurated midden Midden that is hardened to near the consistency of rock.

Macrofossils Fossils that are visible to the naked eye.

Pluvial lake Lake formed during a period of exceptionally heavy rainfall, especially during times of glacial advance during the Pleistocene. These lakes have now disappeared, or only smaller remnants remain.

Proxy data In Quarternary studies, data from fossil organisms, sediments, ice cores, etc., used to reconstruct past environments; proxy data serve as substitutes for direct measurements of such phenomena as past temperatures, precipitation, and sea level.

Succulent plants Plants that contain much tissue rich in cell sap and therefore are fleshy or juicy, for example, cactus.

Bibliography

Betancourt, J. L., Mitton, J. R., and Anderson, R. S. (1991). Fossil and genetic history of a pinyon pine (*Pinus edulis*) isolate. *Ecology* **72,** 1685–1697.

Betancourt, J. L., Van Devender, T. R., and Martin, P. S., eds. (1990). "Packrat Middens, the Last 40,000 Years of Biotic Change." Tucson: University of Arizona Press.

Elias, S. A. (1991). Late Quarternary zoogeography of the Chihuahuan Desert insect fauna, based on fossil records from packrat middens. *J. Biogeogr.* **19,** 285–297.

Grayson, D. K. (1993). "The Desert's Past: A Natural Prehistory of the Great Basin." Washington, D.C.: Smithsonian Institution Press.

Tausch, R. J., Wigand, P. E., and Burkhardt, J. W. (1993). Viewpoint: Plant community thresholds, multiple steady states, and multiple successional pathways: Legacy of the Quarternary? *J. Range Management* **46,** 439–447.

Van Devender, T. R., Rea, A. M., and Hall, W. E. (1991). Faunal analysis of late Quarternary vertebrates from Organ Pipe Cactus National Monument, southwestern Arizona. *Southwestern Naturalist* **36,** 94–106.

Wright, K. (1993). Revelations of rat scat. *Discover,* Sept., 64–71.

Paleolimnology

P. E. O'Sullivan
University of Plymouth, United Kingdom

I. INTRODUCTION

Paleolimnology is the study of the *ontogeny* (development through time) of lakes and their watersheds via analysis of lacustrine sediments. These are composed of material that originates both within lakes and in their watersheds. Therefore, the information obtained records not only lake but also catchment history. Thus paleolimnology may be used to study not only lacustrine but also terrestrial ontogenies, and the important interactions between them. [*See* LIMNOLOGY, INLAND AQUATIC ECOSYSTEMS.]

Paleolimnology *sensu stricto* began at Yale University during the 1930s among G. E. Hutchinson and his students, notably Edward S. Deevey, Jr., who systematized the subject as *biostratonomy*. This (characteristically) is a more elegant term, but probably does not adequately reflect the broader scope of modern paleolimnology.

This account summarizes the main principles and applications of paleolimnology, with a focus on its biological aspects. However, it must be emphasized that paleolimnology is interdisciplinary, involving elements of physics, chemistry, and biology, as well as anthropology, archeology, and history.

Paleolimnology is nowadays used mainly to study the impact of human beings on nature. It provides a time perspective on "environmental problems" and can frequently be used to identify their origin(s) and, perhaps, their solution(s).

II. SOURCES OF LAKE SEDIMENTS

The sources of material incorporated in lake sediments (Tables I and II) are very diverse. They contain much more information than can presently be extracted from them, using even the most modern analytical techniques.

Allochthonous matter (Table I) originates outside the lake basin, mainly in the catchment, but also in the atmosphere above it. A very wide range of inorganic and organic substances, in dissolved, colloidal, and suspended forms, is involved.

Diagenesis begins as this material leaves its place of origin, so that substances that begin their journey to the lake in one particular form may arrive there

TABLE I

General Composition of Sediments Generated
by Catchments[a]

	Inorganic	Organic
Dissolved	Major ions (Na^+, K^+, Ca^{2+}, Mg^{2+}, Cl^-, SO_4^{2+}, $[HCO_3]^-$), minor ionic constituents such as NO_3^-, NH_4^+, and PO_4^-, with sources in, for example, farmland, sewage works, industry, etc.; other metal ions, for example, Fe^{3+} and heavy metals; gases	Organic products of the terrestrial environment (e.g., sugars, alcohols, other lipids such as soluble animal and human wastes, e.g., urea, uric acid)
Colloidal	Clay minerals from soils; industrial wastes of clay mineral size (e.g., china clay waste)	Humic and fulvic acids; organic wastes from farms and sewage works; petrochemicals
Suspended	Sand, silt, and clay from streambanks; soil particles; solid waste from, for example, mines, spoil heaps, and quarries	Particulate organic wastes from eroding soils, farm drains, sewage outfalls, etc.; leaf litter
Bedload	Pebbles, cobbles, boulders; discarded consumer durables	Branches, logs, tree trunks, animal carcasses

[a] Note that during transport, material may pass from one fraction to another via processes such as decomposition, mineralization, denitrification, and so on.

in another, having been used during transit by the biota of the fluvial ecosystem.

Autochthonous matter (Table II) is contributed to the sediment by the lake itself. Thus it consists

TABLE II

General Composition of Sediments Generated by Lakes

Biogenic silica

Biogenic carbonates

Organic copropel

Vegetable matter—dead and decaying algal and macrophytic tissues

Flocculated organic (i.e., humic and fulvic) and inorganic matter (e.g., metal ions and phosphorus adsorbed onto humic and clay particles)

partly of material generated by organisms and partly of substances precipitated chemically within the lake waters.

III. RELEVANCE OF LAKE SEDIMENTATION PROCESSES

Material may pass through several processes before being sedimented (Fig. 1). Inert allochthonous matter (e.g., quartz) takes no part in lake biology or chemistry. It therefore proceeds directly to the sediments. Dissolved allochthonous matter is taken up by primary production and passes along the food chains, ultimately to be returned to the water as dead algal or other plant matter, or as animal excretion products. It then falls as *detritus* to the lake bottom, and reaches the *sediment–water interface* (SWI). Here most of the decomposition and mineralization of detrital material in the lake takes place, and nutrients are recovered for reuse by the ecosystem. In this sense, the SWI fulfills the same role in lakes as the soil surface in terrestrial environments. Detritus may also be resuspended and shifted elsewhere within the lake, only to be remineralized, passed again around the ecosystem, and returned once more to the SWI. Some material may follow these pathways several times before being sedimented.

In many lakes, fine material is slowly shifted from the shallower waters into the *profundal* zone in a process known as *sediment focusing*. Sediment accumulation rates are seldom uniform across the bottom of lakes, except in exceptional circumstances.

Allochthonous particles often arrive in forms that can be utilized by detritus feeders. They also therefore become incorporated in the lake ecosystem and take part in the processes just described. Alternatively, on reaching the lake, allochthonous detritus is decomposed and then becomes ready for biological uptake. However, it may also be rendered chemically inert and pass directly to the SWI.

IV. OPERATIONAL FRACTIONS OF LAKE SEDIMENTS

Although the allochthonous/autochthonous subdivision is sufficient for purposes of discussion of

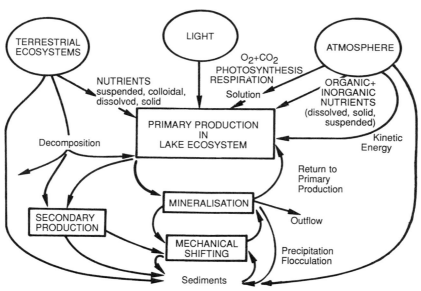

FIGURE I Pathways of sedimentation in freshwater lakes. [From a diagram by Dr. Maureen Longmore.]

sediment *origins,* in analytical terms, further fractions are defined that allow more sophisticated interpretation of results.

The basic model of lake sediment formation still used by paleolimnologists was developed during the 1960s by F. J. H. Mackereth, working in the English Lake District. He regarded lake sediments as a series of soils washed into the lake, made up of stable, oxidized material. He confined his analyses to determination of total element *concentration* ($mg\ g^{-1}$ dry sediment, or $mg\ g^{-1}$ mineral matter).

During the 1980s, Mackereth's model was refined by Daniel R. Engstrom. It is therefore now customary for some paleolimnologists to distinguish three sediment fractions: the *authigenic, biogenic,* and *allogenic* components (Fig. 2). The authigenic fraction is precipitated and flocculated chemically within the water column, whereas biogenic material passes through the food chains and is therefore fixed by organisms. Allogenic material is mainly minerogenic and largely crystalline. These divisions are somewhat arbitrary in that authigenic and biogenic matter is ultimately allogenic. As Fig. 2 implies, movement between the biogenic and authigenic fractions is frequent, and in the case of some elements (e.g., phosphorus) rapid. Diagenesis also continues after burial, creating geochemically important changes in sediment compo-

sition. It may therefore be necessary eventually to define a further *endogenic* fraction.

Engstrom's model represents a significant advance in paleolimnological method in that it allows much more sophisticated identification of sediment provenance, and thus of changing sediment *source.* It enables investigators to distinguish between sedimentary matter formed in the catchment and that produced (in its present form) exclusively within that lake.

The model therefore enables us to study either lake or catchment history, as preferred, but also the all-important interaction between processes on the catchment surface and their effects "downstream" within the lake. Allied with modern dating techniques (see Section VI,D), it enables expression of results as sediment *influx* (mg matter deposited cm^{-2} of sediment surface $year^{-1}$), which unlike concentration ($mg\ cm^{-3}$, see above) is not perturbed by changes in sediment accumulation rate.

V. CORING METHODS

In order to study lake sediments, it is first necessary to obtain cores that preserve intact the stratigraphy of the deposit. Devices fall into four main categories.

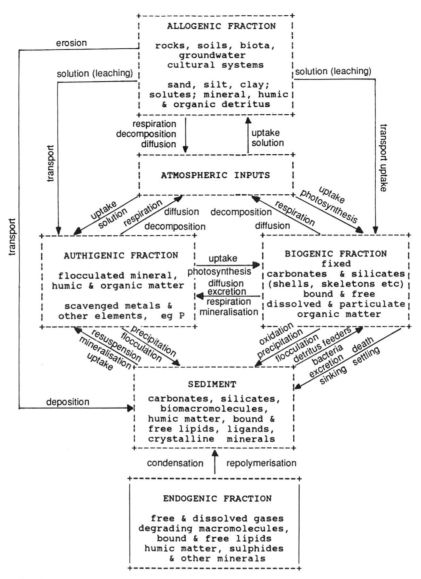

FIGURE 2 The operational fractions of lake sediments. [After Daniel R. Engstrom.]

A. Hand-Operated Corers

Hand-operated corers are developed on the same principle as peat samplers. They are therefore employed using extension rods, which limits the depth of water in which they can be operated. A lightweight raft anchored over the sampling site is also usually required.

The most popular device in this category is the "piston" corer designed by Daniel A. Livingstone (Fig. 3a). This successfully recovers sediments up to 10–15 m deep, in 1- or 1.5-m increments, according to size of core chamber. Often used in Europe is the "Russian" peat sampler (Fig. 3b), with which the author has recovered 11 m plus of compact, clay-rich lake sediment, in 0.5-m increments. The "square rod" corer developed at the University of Minnesota can recover up to 12 m of sediment in a single drive.

B. Free-Fall Corers

In larger (or deeper) lakes, free-fall samplers are used. Most investigators employ a version of the Kullenberg corer developed for oceanographic

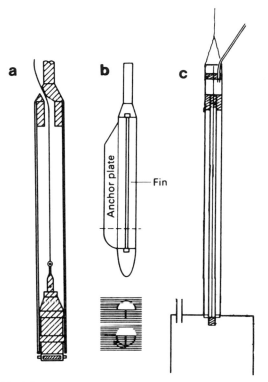

FIGURE 3 Some commonly used sampling/coring devices for lake sediments: (a) the Livingstone corer, (b), the "Russian" peat sampler, and (c) the Mackereth corer.

work. Such devices do not recover the SWI intact, but do provide long, continuous cores up to about 20 m in length. They are therefore particularly appropriate for work in African, North American, European, and other great lakes. In small lakes, the 0.2 to 0.5 m of mobile material at or just below the SWI may be recovered using the Kayak sampler, which operates by means of a trigger and messenger arrangement.

C. Pneumatic Corers

The pneumatic corers, designed by Mackereth (see Section IV), are used mainly in Britain. These are normally built in two sizes, 1-m versions that recover the SWI intact (Fig. 3c) and 6-m corers designed to sample the entire Holocene (Fig. 4). Other versions and sizes have also been built. Mackereth corers operate via a piston that pushes the core tube into the sediment by forcing it downward with compressed air. The 1-m version is then simply pulled or winched to the surface. Longer

Mackereth corers also use the compressed air to return to the surface. Inevitably this involves a certain amount of "overshoot," which causes the corer to "leap" out of the water, especially in shallower lakes. Plenty of rope (and air-line) are needed to retire to a safe distance (cf. Fig. 4).

D. Freezer Samplers

In the case of annually laminated (*varved*) sediments, samplers that retrieve a crust of sediment frozen *in situ* on the outside of the device (Fig. 5) are employed. These come in one of three designs, all of which employ a principle developed by Joseph Shapiro. Solid carbon dioxide ("dry ice") is used as a cooling agent and liquids with low freezing point (e.g., acetone) as dispersants.

The "icy-finger" corer developed in the United States by Albert M. Swain (and in Finland by Matti Saarnisto) recovers a hollow cylindrical crust of frozen sediment (Fig. 5a). The Finnish "box" sampler devised by Pertti Huttunen and Jouko Meriläinen (Fig. 5b) recovers a "slab" rather than a core. The wedge-shaped sampler developed by Ingemar Renberg (Fig. 5c) collects a hollow V-shaped crust.

E. Sediment Core Storage

Cores returned to the laboratory are stored at low temperature (4–5°C). Some samplers (e.g., the Mackereth) recover sediment in perspex or plastic tubes in which they are retained until required. Other cores need to be kept wrapped during storage to avoid desiccation and oxidation. Freezer samples must be kept frozen during transit and in the laboratory. A chest freezer and a box in which to transport dry ice and frozen cores are basic requirements. A 0°C room in which to work is desirable.

VI. METHODS OF ANALYZING LAKE SEDIMENTS

Numerous methods of analysis, too many to describe in detail here, have been developed. Tables III–VIII set out the main techniques currently em-

FIGURE 4 Six-meter Mackereth corer in use at Loch Morlich, Scotland. [Photograph by P. Thompson.]

ployed. These will be considered under the headings of physical, chemical, biological, and chronological methods.

A. Physical Methods

Physical methods are among the basic techniques applied to most lake sediment cores (Table III). Determination of water or dry matter content is a means of identifying layers of greater density, which usually relate to inwashes caused by disturbance of catchment soils or other allochthonous sources. Ignition of dried sediment in air at 550°C removes combustible matter, leaving "ash" or mineral content. This again may highlight peaks in mineral matter influx.

Other basic physical methods include recording of sediment stratigraphy, which may reveal epi-

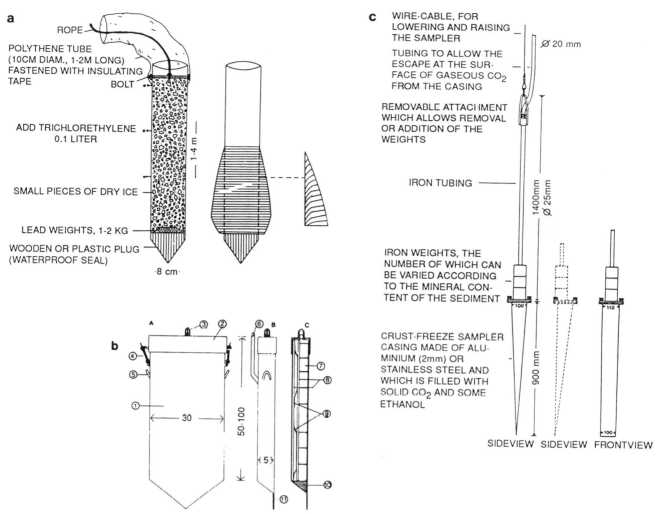

FIGURE 5 Types of freezer samplers: (a) the Shapiro/Swain/Saarnisto "icy finger" corer, (b) the Huttunen/Meriläinen "box-freezer" sampler, and (c) Renberg's wedge-shaped sampler.

sodes of variation in sediment quality. Some investigators use the "objective" description system of J. Troels-Smith or Munsell soil color charts. A few have studied detailed mineralogy to determine whether the matrix has been generated by allochthonous or autochthonous sources.

More rapid physical methods of analysis, applicable to other environmental materials, involve determination of mineral magnetic properties, a field developed by Frank Oldfield, John Dearing, and their colleagues. Sophisticated analyses of changing allochthonous mineral matter sources to lakes, in turn related to landscape and other environmental changes, have been developed. This approach is particularly suited to studying the impact on terrestrial environments of powerful agencies such as

human action and climatic change, and their consequences for soil profile development and for mineral cycling in catchments. Therefore it lies mainly outside the scope of this article.

B. Chemical Methods

I. Inorganic Chemistry

As stated, inorganic chemical analysis of lake sediments was begun by Mackereth, who applied his theories of lake sediment formation to studies of the bottom deposits of lakes in the English Lake District. His ideas regarding the information obtained from each determinand are set out in Table IV. Fractionating sediments and calculating

TABLE III
Physical Methods of Lake Sediment Analysis

Method	Determinand	Information
Wet/dry mass	Water/dry matter content	Compaction/fluidity
Ignition loss	Ash content	Mineral matter
Sediment matrix	Sand, silt, clay content	Erosion episodes
Mineralogy	Identification of mineral species	Allochthonous/autochthonous ratios
Mineral magnetism	Susceptibility (K or X)	Intercore correlation, sediment source
	Remanent magnetization (IRMs)	Ratio hematite/magnetite, sediment source

changes in both concentration *and* influx (see Section IV) improve our ability to interpret such results and allow differentiation between those forms of each element that enter the lake in detritus and those that arrive in solution. These ideas are set out in Table V.

2. Organic Chemistry

With a few notable exceptions, organic geochemical analyses of lake sediments are not as well developed as those of marine deposits (Table VI). Nevertheless, certain groups of organic compounds may be used to study aspects of paleolimnology, notably the lipids. These make up only a small fraction (1–5%?) of total sedimentary organic matter (SOM) but have attracted the most attention. An

TABLE IV
Information Gained by Inorganic Chemical Analysis of Lake Sediments (According to Mackereth)

Determinand	Information
Na, K, Mg, Ca	Erosion indicators
Fe, Mn, Fe/Mn	
Unproductive lake	Catchment soil redox
Productive lake	Lake redox
C, N, C/N, P	Lake productivity, eutrophication
Halogens (I)	Precipitation
Heavy metals	Erosion

important class of nonlipid compounds also extractable from lake sediments are plant pigments and their degradation products. Terrestrial pigments do not normally survive transport, so that determination of sedimentary concentrations refers to autochthonous inputs and to lake productivity.

C. Biological Methods

1. Allochthonous Materials

The main allochthonous biological materials present in lake sediments and used in paleoenvironmental reconstructions are the pollen and spores of higher plants. These are washed into lakes by surface waters or blown in by wind. Owing to their strong resistance to chemical and bacterial attack, they are readily sedimented and preserved in most lake sediments. The literature on this subject is vast. Fortunately for this author, it is more properly the province of paleoecology than paleolimnology.

Palynology, also called "pollen analysis," is the main paleoenvironmental technique available for reconstruction of catchment vegetation history (Table VII). Its results are therefore of interest to paleolimnologists and are employed by them to investigate terrestrial events forcing changes in the ecology of lakes. Similarly, macroscopic remains (seeds, fruits) of terrestrial plants (and some animals, e.g., Coleoptera) are present in some lake sediments. These are also used, mainly by paleoecologists, to investigate catchment history. Most plant macrofossils in lake sediments are, however, autochthonous, and so refer to lake ontogeny and hydroseral succession.

2. Autochthonous Materials

a. Diatoms

i. Introduction The main autochthonous microfossils extracted from lake sediments by paleolimnologists are the siliceous skeletons ("frustules") of the yellow-brown algae known as diatoms (the Bacillariophyceae or Bacillariophyta). Thousands of members of this order are recognized by taxonomists, many of which, owing to their distinctive morphology, are in theory identifiable in fossil form. A leading exponent of diatom analy-

TABLE V
Information Obtained by Fractional Inorganic Chemical Analysis of
Lake Sediments[a]

| Determinand | Fraction | | |
	Authigenic	Biogenic	Allogenic
Na, K, Mg			Erosion
Ca	Lake chemistry		
Mg/Ca	Paleosalinity		
Fe, Mn, Fe/Mn (unproductive lake	Soil redox		Erosion
Fe, Mn, Fe/Mn (productive lake	Lake redox		Erosion
C, N, P	Eutrophication	Lake productivity	Agricultural soil input (P)
S	Lake redox		
Al	Acidification		Erosion
Si		Lake productivity	Erosion
Heavy metals	Industrial input		Erosion, aerial input

[a] After Daniel R. Engstrom & Herbert E. Wright, Jr. (1984).

sis is on record, however, as stating that "Diatom taxonomy is not easy!" The present author is not a diatomist, and so only brief attention will be paid here to taxonomic matters.

Basically, diatoms may be divided into disc-shaped genera (the Centrales, or centric diatoms) and linear, "cigar-shaped" forms (the Pennales, or pennate diatoms; Fig. 6). Further taxonomic differentiation is made on the basis of shape (e.g., length/breadth ratio), the presence and form (in the case of the Pennales) of the longitudinal "raphe," and the nature, distribution, and density of surface ornamentation. Diatoms in lakes may be classified according to habitat (pelagic, periphytic, littoral, benthic), habit (planktonic, epiphytic, epilithic, epipsammic, epipelic), and response to salinity, pH, and nutrient status. Composition of subfossil diatom assemblages may therefore reflect lake nutrient status, water quality, and productivity.

Using diatoms, paleolimnologists have studied topics such as lake ontogeny, climatic and water level changes, human impact on lakes (eutrophication, acidification), and paleosalinity. The last may be related to climatic change, evaporation, and lake level, as well as to marine incursion and shoreline displacement.

ii. The Method and Its Limitations Diatoms are extracted from the sediment matrix by removal of calcareous, organic, and coarse mineral matter using dilute hydrochloric acid, hydrogen peroxide, and flotation or sieving, respectively. Strong alkalis, in which the opal silica of the frustule is soluble, must be avoided.

Extracts are mounted on slides in a matrix of some substance with a low refractive index. Counting may be conducted in terms of percentages, concentration (number of frustules per unit volume of sediment), or diatom biovolume accumulation rate (DAR), where concentrations are divided by the sediment accumulation rate and adjusted for variation in frustule size between taxa. Information on total amorphous/biogenic silica influx (see Section VI,B) is of help in interpreting results.

Identifications are made with the help of taxonomic information from floras compiled over many years by workers such as Cleve-Euler, Foged, Germain, Huber-Pestalozzi,

TABLE VI

Information Derived by Organic Geochemical Analysis of Lake Sediments

Determinand	Relevant property	Information
n-Alkanes	Chain length C_{16}–C_{18} = aquatic C_{24}–C_{30} = terrestrial	Sediment source (allochthonous/autochthonous ratio)
CPI	Low = bacterial High = terrestrial	Sediment source
C/N ratio	Aquatic, 4–10 Terrestrial, 20–80	Sediment source
δ^{13}C (‰)	Terrestrial (-20) Aquatic (-25 to 30) Bacterial (-30)	Sediment source
Fatty acids	Chain length Terrestrial (24 : 0–28 : 0) Endogenic (15 : 0–18 : 0)	Sediment source
Monocarboxylic acids	Concentration	Lake productivity
Sterol/stanol ratio	Carbon number C_{27} = aquatic C_{29} = terrestrial	Sediment source/ diagnostic history
α/β stanols	Ratio	Hypolimnetic oxygen
Hopanoids	Concentration	Bacterial input
PAHs	Concentration	Industrial input
Plant pigments	Concentration	Lake productivity

Hustedt, Molder and Tynni, and Schmidt, and for North America, Patrick and Reimer. Reference collections, drawings, and photographs of each particular flora are essential. Access to scanning electron microscopy (SEM) is an advantage.

TABLE VII

Biological Methods of Lake Sediment Analysis

Allochthonous matter	Pollen and spores	Vegetation history
	Plant and animal macrofossils	Catchment history
Autochthonous material	Diatoms	Lake history, trophic status, pH, salinity
	Chrysophyte cysts	Lake history
	Sponge spikules	Lake history
	Cladocera(ns)	Lake history, trophic status, pH
	Chironomids	Trophic history
	Ostracods	Trophic history

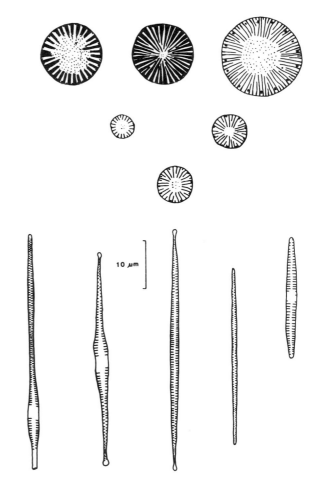

10 μm

FIGURE 6 Some common European diatom species. [After H. Barber and E. Y. Haworth.]

Most diatomists hold considerable reservations about their technique, in that preservation of critical taxonomic features varies from site to site, identification is difficult, and much regarding the synecology of diatoms, and the autecology of individual species, is poorly understood. Similarly, the representativity of subfossil diatom assemblages may be questioned on the grounds of the nonlinear response of diatoms to ecological changes. At high nutrient concentrations they are replaced by cyanobacteria and in acid lakes by cryptophyta.

Subfossil diatom assemblages are also subject to error from losses caused by dissolution, especially in alkaline, calcareous, acid, and chemically rich meromictic lakes, and by recycling of silica where supply becomes limiting. Further losses may be the result of grazing by zooplankton and washing of diatoms through the outflow of the lake. Finally, littoral diatoms are often underrepresented in profundal cores compared to planktonic forms. Although percentages may be stable across the lake bottom, concentrations are not.

iii. Applications Lake Ontogeny

Broadly, centric diatoms consist mostly of planktonic taxa, whereas pennate diatoms, with certain exceptions, are found in the periphyton, or as benthos in shallow waters. The ratio of planktonic to periphytic forms is therefore used to investigate lake ontogeny, especially the effects of shallowing. As lakes fill with sediment, greater numbers of periphytic forms are washed into the profundal zone and sedimented there. Similarly, in shallow lakes, epiphytic forms increase as macrophyte growth expands. However, eutrophication (not to be confused with lake ontogeny!) often produces the opposite effect (see later section on lake productivity).

pH F. Hustedt, studying distribution of modern taxa, classified diatoms according to the pH of surface waters (Table VIII). This system was used by G. Nygaard, and improved by Meriläinen, to construct an index of lake pH, which is calculated from the relative contribution of each category to the diatom flora.

Studies of recent lake acidification have led to further modifications by Renberg. His Index B is calculated by

TABLE VIII
Classification of Diatoms, According to pH[a]

Category	Range
Alkalibiontic (alkb)	Living at pH > 7
Alkaliphilous (alkp)	Living at about pH 7, but distributed more widely at >7
Indifferent (ind)	Living at about pH 7
Acidophilous (acp)	Living at about pH 7, but distributed more widely at <7
Acidobiontic (acb)	Living at pH < 7, with optimum conditions at pH 5.5 and below

[a] After F. Hustedt.

$$\text{Index B} = \frac{\% \text{ ind} + (5 \times \% \text{ acp}) + (40 \times \% \text{ acb})}{\% \text{ ind} + (3.5 \times \% \text{ alkb}) + (108 \times \% \text{ alkb})}.$$

This is used to reconstruct lake pH (Fig. 7) and to show how, in many cases, lakes in vulnerable areas have experienced greater decline in pH over the past 50–150 years than during several previous millennia, sometimes by one or more orders of magnitude (see Section VII, C).

Lake Productivity Indices of lake productivity, again based on composition of subfossil diatom assemblages, have also been developed. However, they are not so easy to apply as indices of pH. For example, many investigators use the ratio of planktonic to periphytic species in the opposite way to that discussed earlier. Expansion of planktonic forms relative to periphytic is taken to indicate increased nutrient status, and hence rising productivity. Thus, in shallow lakes, changing abundance of Centrales relative to Pennales (the C/P ratio) is sometimes related to variations in water clarity, a switch from benthic and littoral species to planktonic, and the shading out of macrophytes by phytoplankton, all of which may denote eutrophication. [*See* EUTROPHICATION.]

In deep lakes, however, as observed by Nygaard, changes in productivity resulting from increased or decreased nutrient loading are recorded by variations *within* the planktonic dia-

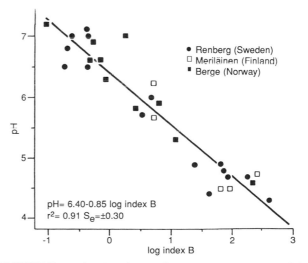

FIGURE 7 Renberg's index B of pH versus composition of the diatom assemblage.

tom community. Thus the C/P ratio rises as productivity increases, so long as only planktonic forms are considered. John Stockner, however, suggested that eutrophication could be detected by the increasing ratio of Araphidinae (one form of pennate diatoms) to Centrales (the A/R ratio). This is almost the opposite of Nygaard's C/P ratio, and appeared to successfully describe the increased productivity of Lake Washington (see Section VIII, B).

It now appears that the A/R ratio is only useful in identifying the early stages of eutrophication in large, originally unproductive lakes of low alkalinity. It cannot be applied to already naturally productive systems, where advanced eutrophication is associated with increase in abundance of small centric genera. Some of these, for example, *Stephanodiscus,* may be used as indicator taxa.

As stated by Richard Battarbee, using diatom biovolume accumulation rate is therefore a more conservative approach to the study of productivity, in that it is independent of floristic changes. During changes in lake ecology, the relative contribution to the assemblage of some taxa may *decline,* while their DAR *increases.* Even then, DAR must also be used cautiously, because it is affected by factors such as silica limitation, dissolution, and recycling, and is sensitive to changes in sediment accumulation rate (see Section VI, D) and to sediment redistribution within the lake.

Salinity Again according to Battarbee, salinity is the strongest factor influencing diatom distribution. Two categories, thalassic (coastal) and anthalassic (inland salt lake) environments, whose characteristic ionic concentrations are quite different, must be distinguished.

In thalassic systems, Cl^- is the main anion present. Kolbe classified the diatoms of such environments in terms of salt concentration as in Table IX. This system was then modified by Hustedt as shown in Table X. R. Simonsen pointed out that each category contains taxa of widely varying tolerance limits. He maintained that all brackish water (mesohalobous) diatoms are in fact euryhaline (tolerant of a wide range of salinity conditions), but that marine and freshwater taxa could be separated

TABLE IX
Classification of Diatoms in Terms of Salinity (According to Kolbe)

Category	Range
Euhalobous (saline)	30–40‰[a]
Mesohalobous (brackish)	5–20‰
Oligohalobous[b] (fresh)	<5‰

[a] ‰ = per mill.
[b] The last category is also subdivided into halophilous, indifferent, and halophobous.

into much more critical tolerance categories. His rather complex classification scheme is shown in Fig. 8.

Many studies using the Kolbe–Hustedt system have been conducted in the Baltic region and on the west coast of Scandinavia regarding land uplift, shoreline displacement, and isolation from the sea recorded in lake sediments. In the broader, paleoecological sense, land and sea level changes related to glacioeustatic and other variations in ocean volume, and hence to climatic change, including global warming, may be studied using this methodology.

In anthalassic systems, the key anions may be Cl^-, CO_3^-, or SO_4^{2-}. They are therefore quite different from thalassic environments, and the Kolbe–Hustedt classification is not applicable. Fresh and brackish taxa may be present, but there are no marine diatoms. These environments have been used to study climatic change, principally of

TABLE X
Classification in Terms of Salinity of Habitat (According to Hustedt)

Category	Range
Polyhalobous (widely distributed)	>30‰
Mesohalobous (brackish)	
Euryhaline	0.2–3‰
α-Mesohalobous	>10‰
β-Mesohalobous	0.2–10‰
Oligohalobous (freshwater)	
Halophilous	
Indifferent	<5‰
Halophobous (ultrafresh)	

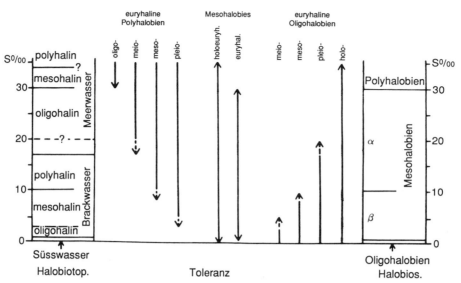

FIGURE 8 Simonsen's classification of diatoms with respect to salinity of the water. [From Battarbee.]

course in arid and semiarid regions such as East Africa and the North American Great Plains. Variations in lake level are denoted by changes from fresh to brackish taxa and back again.

b. Cladoceran Analysis

i. Principles   According to David G. Frey, the most important microscopic animal remains recoverable from lake sediments are those of the Cladocera. Along with Rotifera and Copepoda, these constitute part of the offshore (limnetic) zooplankton and are also abundant in the littoral zone.

The limnetic taxa, which include the Bosminidae and the Daphnidae, are mostly filter feeders that graze the phytoplankton. The littoral taxa, which are mainly members of the Chydoridae, either live at the SWI or are associated with macrophytes. As they grow, cladocera shed their chitinous exoskeletons, which then disintegrate into their component parts (exuviae). Cladoceran remains in sediments consist of headshields, shells, abdomens, postabdomens, postabdominal claws, antennae, mandibles, trunks, eppiphia, and copulating hooks. The number of exuviae found depends on the number of molts that each individual undergoes, the amount of postmolt damage, and the preservation of chitin. Some families are well represented and others only poorly. Most informa-

tion is obtained from chydorids and bosminids. The equally important Daphnidae are difficult to recognize except with SEM.

Each milliliter of sediment contains several thousand exuvial fragments, from which they are removed by a combination of dilute alkali, mineral acids (including, if needed, hydrofluoric acid, with which extreme precautions must be used), and sieving. Formulae are available for calculating numbers of individuals from fragments recorded. These assume that numbers of molts per species are standard. A sum of chydorids is used for expressing quantitative relationships. No specific fauna for cladoceran analysis exists. According to Frey, species identification varies from Europe to North America, so that the field is in need of taxonomic revision.

ii. Interpretation   The remains of chydorids are carried from the littoral zone and sedimented in deeper waters. Their relative abundance may therefore be a good descriptor of lake condition, but concentration decreases markedly offshore. In contrast, sedimentary abundance of pelagic species peaks in the deepest waters and declines inshore. Distribution of pelagic taxa in the sediments therefore runs opposite to sediment focusing (see Section III). Thus variations in exuvial concentration are difficult to relate to changes in lake productivity (see following discussion). Annual accumulation rates can be calculated, but they are not easy to interpret.

Results of analysis of sediment cores usually include the ratio of planktonic to littoral species, and the number of chydorids. Perturbation of the zooplankton is usually recorded by a lowering of diversity, with the most common species becoming more abundant. Consequently, changes are usually discussed in terms of values of diversity such as the Shannon–Weiner index or the Spearman rank correlation. Heavy grazing pressure by fish or other predators usually leads to a decrease in the size of individual animals. Diversity of chydorids is greatest at high water transparency and low phytoplankton productivity. It is also related to structural diversity of the habitat, in that as macrophytes decrease or increase, so do the chydorids.

iii. Applications

The greatest amount of information is usually obtained from the Bosminidae, which exhibit a high diversity of morphotypes. The replacement of *B. longispina* by *B. longirostris* has traditionally been associated with eutrophication, although it can also be related to predation. *Chydorus sphaericus* is one of only a few species that are clear indicators of eutrophication.

As lake productivity increases, biomass and hence sediment concentration rise. Diversity of the cladoceran community, however, declines as rising turbidity and shading of macrophytes by phytoplankton increase.

The planktonic/littoral ratio among Cladocera may be used to study eutrophication only with caution. It is more likely, especially over longer time scales, to be related to lake ontogeny. As lakes become shallow, the ratio is likely to decline, but such changes may also be related to variations in habitat diversity. Cladoceran remains have been used to study the results of climatic variation, vegetation change and soil erosion, lake ontogeny, water level changes, and human impact on lakes. They have also been applied to the topic of lake acidification.

c. Chironomid Analysis

i. Principles

The remains of chitinous head capsules of the larvae of midges called the Chiro-

nomidae are also preserved in lake sediments. The larvae are members of the bottom fauna of lakes. As such, the profundal species are indicative mainly of hypolimnetic oxygen conditions, which in turn *may* be related to lake productivity, trophic status, and ontogeny. Littoral species have been studied to a much lesser degree.

Capsules are separated from the matrix of the sediment by the action of hot alkali (KOH) and sieving. Several keys for identification exist. Again, taxonomy is complex, especially as the larval stage of the midge's life cycle involves several instars. Similar measurements of diversity as performed on subfossil cladoceran remains are used to process the data (see Section VI, C, 2, b, ii). Influx data are, however, reliable in this case.

ii. Interpretation and Application

In a classic early limnological study, August Thienemann used living chironomid species to characterize unproductive, oxygen-rich, clear-water lakes as *Tanytarsus* lakes, and productive, oxygen-deficient systems as *Chironomus* lakes. Unfortunately this was something of a misnomer, as most members of the genus *Tanytarsus* are littoral species (although one, *T. lugens,* is found in the profundal of unproductive lakes and may be used to identify phases of oxygen richness in the history of *deep* lakes). Results of chironomid analysis of lake sediments may be applied to the study of topics such as lake eutrophication, lake ontogeny, climatic change, lake acidification, and paleosalinity.

d. Other Autochthonous Microfossils

Numerous other lacustrine microfossils are available to the paleolimnologist. Among these are remains of algae themselves, the cysts (resting stages), scales, and bristles of the class of algae called the Chrysophyta, the spores of cyanobacteria, the spikules of freshwater sponges, and the shells of the microscopic crustaceans known as ostracods. Among the algal remains most commonly studied are those of green algae, especially *Botryococcus, Pediastrum,* and *Scenedesmus.* Some evidence suggests that the first two are more sensitive to eutrophication than the third.

Chrysophyte cysts are difficult to identify under light microscopy only, but their scales and bristles can be referred to species and even subspecies, and so if present may provide valuable information. They too seem sensitive to eutrophication by increased nutrient influx. Ostracod remains refer to conditions in the profundal of lakes. They are also sensitive to changes in lake salinity, trophic status, and oxygen circulation.

D. Chronology

Paleolimnologists require chronological information on a number of time scales. For lake ontogeny, which operates over millennia, standard techniques of Pleistocene geology will often suffice. For more recent events, techniques peculiar to paleolimnology are required (Table XI). Such information is essential, because without it, investigators cannot identify the time span over which events occur. Equally, without sufficient chronological information, we cannot assess the *rate* at which processess operate or have operated.

I. Radioisotopic Analyses

a. Cesium-137 (Half-life, or T/2 = 33 years)

One technique for dating sediments from the past 50 years involves identifying the peak of environmental ^{137}Cs associated with large-scale testing of nuclear weapons in the atmosphere during the period 1954–1963. The technique is based not on the decay rate of the isotope, but on the history of its addition to nature (Fig. 9).

In some profiles, not only 1963, the peak year of environmental ^{137}Cs concentration, but also

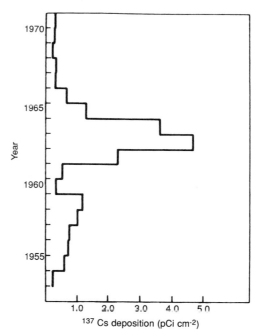

FIGURE 9 Deposition history of 137 Cs at a site in Europe during the middle twentieth century. [After Pennington, Cambray, and Fisher (1973). *Nature.*]

1954, the beginning of its increase, and even 1959, a minor peak associated with U.S. weapons testing, can be identified. However, in many sediments, detail is lacking owing to mixing at the SWI, vertical diffusion of the isotope, or dilution by other material. The main horizon identified is therefore normally 1963 only.

In parts of Northwest Europe, ^{137}Cs from the Chernobyl accident of 1986 (separable from "weapons" ^{137}Cs by its ^{134}Cs/^{137}Cs ratio) has been observed in lake sediments. In these regions, it will form a second marker peak of ^{137}Cs in future decades.

Other isotopes added to nature during atmospheric weapons testing include ^{241}Am (T/2 = 432 years). According to Peter Appleby, sedimentary in-growth of this isotope has now reached a point where it may be used instead of ^{137}Cs, especially in soft-water lakes. This is an important advance, because in poorly buffered waters ^{137}Cs appears to diffuse readily through the sediment profile, thus invalidating the basis of the technique.

b. Lead-210 (T/2 = 22.26 years)

A naturally occurring isotope comprising part of the end of the uranium/lead series (and used for

TABLE XI
Methods of Establishing Recent Chronology in Lake Sediments

Isotopic analyses	^{137}Cs	1963 peak, last 50 years
	^{210}Pb	Last 100–150 years
	^{14}C	500–40,000 B.P.
Paleomagnetism (NRM)	Declination/ inclination	0–10,000 + B.P.
Varves	Ferrogenic Biogenic Calcareous Clastic	Time period of formation of deposit
Historical records		Duration of record

dating by, among others, Oldfield, Appleby, and their colleagues) is ^{210}Pb. This is formed, via a series of short-lived progeny, by disintegration of atmospheric ^{222}Rn (T/2 = 4 days), the daughter of ^{226}Ra (T/2 = 1620 years). Thus ^{210}Pb is deposited on the soil, whence it is washed into lakes to join that directly falling on their surface (Fig. 10). In the sediments, ^{210}Pb disintegration continues. The time-depth relationship is based on knowledge of the decay rate of the isotope and measurement of variation in activity/concentration with depth.

Unfortunately, the technique is not as simple as implied, as ^{226}Ra present in the sediments also decays (ultimately) to ^{210}Pb. Thus, while inwashed ^{210}Pb declines in concentration, ^{210}Pb formed *in situ* remains a constant. However, this amount, the *supported* ^{210}Pb, can be estimated and subtracted from the total to leave the *unsupported* portion of the isotope. Dates are calculated from variation in concentration of this component with depth.

Lead-210 profiles are subject to sources of error, mainly inwash of radiometrically inert material. Several models of ^{210}Pb accumulation in lake sediments have been developed, principally those involving CIC (constant initial concentration) or CS/CRS (constant supply/constant rate of supply) assumptions. Discussion of the applicability of these models to specific ^{210}Pb profiles is outside the scope of this article.

c. Radiocarbon Dating

For dating longer-term processes, for example, lake ontogeny, isotopes such as ^{14}C (T/2 = 5730 years) may be used. As in the case of palynology, this technique is more the province of paleoecology than paleolimnology, so that only brief attention will be devoted to it here.

Owing to combustion in the atmosphere of radiometrically inert carbon from fossil fuels since around AD 1700, radiocarbon dating cannot be used to date modern materials. In lake sediments, however, a more serious source of error exists that extends in some regions, further back into the past.

Once human cultures practicing agriculture inhabit a catchment, inwash of "old" carbon, principally from eroding soils, results in dilution of contemporary lake carbon and the production of artificially "old" dates. Radiocarbon dates from lake sediments deposited after the advent of the first farmers (in Europe some 6000 years ago) may therefore be subject to such errors and must be interpreted accordingly.

2. Paleomagnetism/Natural Remanent Magnetization (NRM)

Records of variations in the earth's magnetic field (NRM) are preserved in the sediments of some lakes, especially large ones. The technique was originally invented during the early 1970s by

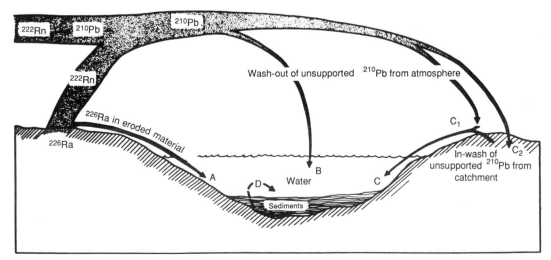

FIGURE 10 Pathways by which ^{210}Pb reaches lake sediments. [After Oldfield and Appleby (1980). *Catena*.]

Mackereth. His results indicated that a record of cyclical variations in both the horizontal (*declination*) and the vertical (*inclination*) components of the earth's magnetic field could be identified in lake sediments. Since then, master curves for seven geomagnetic provinces have been developed by Roy Thompson. Variations in declination are labeled a, b, c, d, etc., and inclination α, β, γ, δ, etc. (Fig. 11). These chronologies are, however, calibrated by dating with ^{14}C. The dating of profiles from a new site involves comparing them with the appropriate master curve. In recent paleolimnology, the declination peak labeled a, dating from AD 1820, is of particular use.

3. Varved (Annually Laminated) Sediments

Varved sediments are discussed in greater detail in Section VII. Their value lies in their great precision, which allows the investigator to identify the specific year in which a particular event is recorded in the sediment profile, even several millennia in the past.

4. Historical Records

In many sediments, horizons related to specific events in lake or watershed history may be identified. Such events may include, for example, raising or lowering of lake level, canalization of inflowing

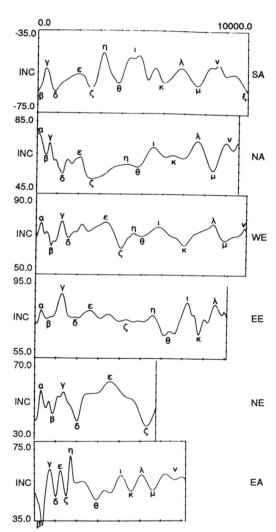

FIGURE 11 Master curves of declination/inclination of natural remanent magnetization for various geomagnetic provinces. [After Thompson (1983). *Hydrobiologia* **103**.] SA, South America; NA, North America; WE, Western Europe; EE, Eastern Europe; NE, Northern Europe; EA, East Africa.

streams, and influx of mineral matter associated with floods, deforestation, land use change, mining, or construction.

Dating is achieved by matching known events to specific horizons, and preferably by cross-matching with other techniques such as ^{137}Cs, ^{210}Pb, or NRM. In some countries, detailed records cover several centuries and are used for more detailed paleoenvironmental reconstruction. In the American Midwest, expansion of ragweed (*Ambrosia*) pollen in the uppermost sediments denotes the arrival of European agriculture. Consultation of land use records for the relevant county usually provides a precise date for this horizon.

5. Soot Particle Chronologies

Recently, Renberg and co-workers have developed chronologies for the uppermost parts of lake sediment profiles based on abundance of particles of soot emitted by fossil fuel combustion, especially by power stations. These are used to produce chronologies in the same way as ^{137}Cs, in that the history of emission in particular regions must be known. Soot chronologies are especially appropriate for studying lake acidification, as the same power stations that emit the soot are often responsible for the acid emissions!

VII. THE SPECIAL CASE OF ANNUALLY LAMINATED (VARVED) SEDIMENTS

A. Formation

During the formation of "normal," unlaminated sediments (Fig. 12), in-lake processes such as water circulation and the activities of benthos mix sedimentary materials and generalize the sedimentary record. Layers representing individual events, or seasons, are rare (see Section VI,D,4). In physically "deep" lakes (i.e., in relation to surface area) or in deep basins of great lakes, circulation processes are weak or ineffective. Bottom waters rarely come into contact with oxygenated surface layers, or only incompletely. In such lakes, the seasonal cycle of production of biological and other particulate material (*seston*) in the water column is reflected in rhythmic changes in sediment composition. This produces a sequence of laminae (Fig. 13), each of which represents a season.

If the signal(s) generating this sequence can be identified, and the number of laminae representing a cycle (or varve) determined, they may be counted and used to provide a very precise chronology of sedimentation. Although this may extend for millennia, it may still be resolved in terms of individual years.

In lakes of the temperate zone, varves partly composed of spring layers made up of diatoms are called biogenic laminations (Fig. 12). If spring is marked by calcite layers produced during high rates of photosynthesis, they are called calcareous. Finally, if they are composed of alternating black and brown layers, rich in ferrous sulfides and ferric hydroxide, respectively, and formed under conditions of seasonally variable redox in the bottom waters of the lake, they are referred to as ferrogenic laminations.

In some lakes, clastic varves, composed at least partly of detrital allochthonous material, may also be formed. Here, laminations are produced by seasonal variability of sediment quality, especially particle size and sediment *supply*. Such deposits are characteristic of near-glacial environments, where stream discharge varies strongly on a seasonal basis according to freezing and melting of surface water. They are also found in semiarid regions, where discharge varies seasonally and where much loose soil material is available for transport as stream sediment during rainy seasons.

The sediments of tropical mountain lakes may show laminations that represent the effects of the phenomenon known as El Niño–Southern Oscillation. Other tropical lakes contain sediments that are true varves, that is, they are annual rather than supraannual in their nature.

B. Distribution

At present, varved lake sediment sequences have been found in four main areas: the Great Lakes region of North America; northern Sweden, south and central Finland, and Estonia; the Alpine region of central Europe; and the African Great Lakes.

FIGURE 12 Pathways by which annually laminated (varved) sediments are formed in freshwater lakes.

However, this is not a random distribution and is best explained by the observation that most sequences occur in areas where paleolimnologists have looked for them. Many more therefore remain to be discovered.

C. Applications

The main use of varved lake sediments is to provide chronology for paleolimnological and paleoecological investigations. They have also been used to calibrate other dating techniques, especially, so far, ^{210}Pb and Holocene paleomagnetism. Several laboratories (e.g., Krakow, Zurich, Umeå, Plymouth, and Trier) are currently working on the varve calibration of ^{14}C dating.

Varved sediments have been put to a wide range of other paleoecological uses, including determining (1) the duration of interglacials, (2) immigration rates of tree species during interglacials and duration of periods of forest ecosystem stability during such periods, (3) the rate of shoreline displacement

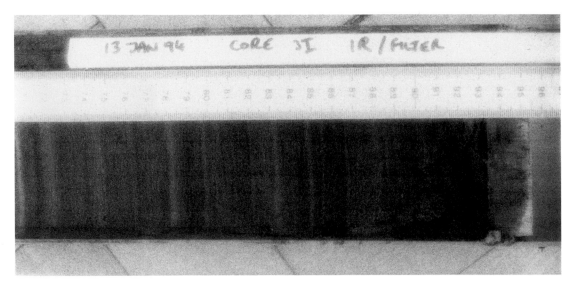

FIGURE 13 Laminated sediments from Loch Ness, Scotland. [Photograph by B. Fox.]

in formerly glaciated regions, and (4) the frequency of fires and their effects on forest composition in boreal environments.

Their main paleolimnological application has been in providing chronologies for studies of human impact on natural systems, including the effects of swidden cultivation in boreal forests and their relation to population pressure, eutrophication of lakes by sewage inputs, agricultural land use changes, and peatland drainage.

VIII. CASE STUDIES

A. The Ontogeny of Pickerel Lake, South Dakota

The Late Pleistocene and Holocene ontogeny of Pickerel Lake in northeastern South Dakota was investigated by Elizabeth Haworth. Using pollen analysis, five phases of vegetation development around the site were identified (Table XII).

Diatom evidence (Fig. 14) suggests that during phase 1, lake pH was acid to neutral, under the influence of soil formation under the spruce/tamarack forests. More alkaline conditions developed with the immigration of mixed deciduous woodland (phase 2). With the development of prairie

TABLE XII

Phases in the Ontogeny of Pickerel Lake, South Dakota[a]

Phase	Years before present	Catchment vegetation
5		European agriculture
	200	
4		Prairie with some woodland
	4,200	
3		Bluestem grass prairie
	9,400	
2		Mixed deciduous woodland
	10,500	
1		Spruce/tamarack forest

[a] After Haworth.

(phase 3), the lake became shallower, slightly more saline, somewhat more turbid, and much more productive. About 4200 years ago (phase 4), development of woodland led to stabilization of the soil and a return to less alkaline, less productive lake conditions. Finally, with the arrival of European agriculture (phase 5), further eutrophication occurred, but not to the extent experienced during phase 3.

The study of Pickerel Lake therefore illustrates the difference between lake ontogeny, which is a response to development under a stable ecological watershed regime, and eutrophication (both natural

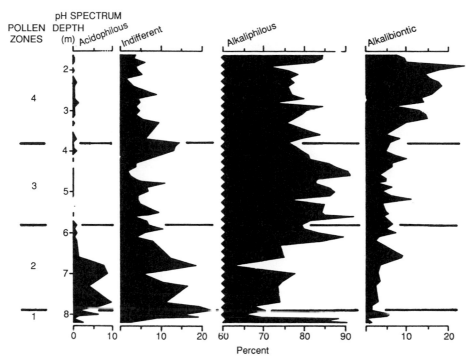

FIGURE 14 pH spectrum of diatoms in a core from Pickerel Lake South Dakota. [After Haworth.]

and artificial), which takes place (in this case) as a result of catchment soil instability.

B. The Eutrophication of Lake Washington

Lake Washington is a large, deep lake now mostly surrounded by the city of Seattle. During the late 1950s, blooms of blue-green algae (*Oscillatoria rubescens*) appeared in the lake, leading to considerable loss of amenity and hazard to the local population. Eutrophication of the lake and its recovery have been well documented by W. T. Edmondson. It was also studied using paleolimnological methods during the 1960s by Stockner.

A sand layer at some 14–18 cm depth in the sediments was correlated with the year 1916, when the lake was connected to the sea by a canal and level was lowered by some 3 m. Using this as a marker, the investigators were able show that changes in the ratio of araphic to centric diatoms (see Section VI,C,2,b) correlated very precisely

with the history of sewage disposal from Seattle to the lake (Fig. 15).

During the period 1913–1930 (peaking in 1920), the proportion of araphic diatoms sedimented increased. This coincided with dumping of raw sewage in the lake, which began around 1910 and reached a peak in 1924. Araphic species then declined in sediments dating from the 1930s, but rose again from about 1940 onward as the city of Seattle increased in size and greater amounts of now treated sewage were introduced into the lake.

Paleolimnological evidence clearly showed that increased nutrient loadings responsible for eutrophication were derived from sewage. Agitation for restoration, in the form of a pressure group called the METRO, then began. Restorative diversion of sewage effluent commenced in 1963 and was completed in 1968. Since then, Lake Washington has been experiencing sustained recovery.

Comparison of the results from Lake Washington and from Pickerel Lake illustrate the difference between lake ontogeny, which occurs under condi-

PER CENT DIATOM ABUNDANCE

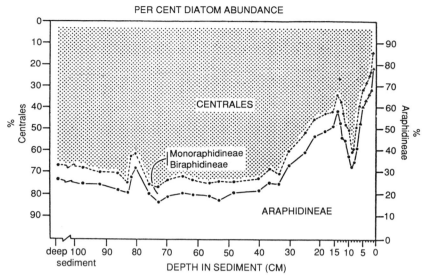

FIGURE 15 Changes in the ratio of Araphidiniae to Centrales among diatom species in the recent sediments of Lake Washington. [After Stockner and Benson.]

tions of stable nutrient loadings, and eutrophication, which represents perturbation of a steady state.

C. The Acidification of Lakes in the Adirondack Mountains, New York

Paleolimnology has been used to study lake acidification in the Adirondack Mountains of northern New York State by Donald F. Charles and colleagues. Here, diatom analysis of near-surface sediments, and calculations of indices of lake pH from studies of composition of surface diatom assemblages, showed that clear-water lakes with a present pH of <5.5 became acidified in the period 1900–1925, following the beginning of large-scale combustion of fossil fuels in the late nineteenth and early twentieth centuries.

Similar sequences, identified using the same techniques, may be defined for Sweden and the rest of northern Europe, and also for Britain (the results of the SWAP, or Surface Water Acidification Project). The recent changes are much greater (sometimes by one of more orders of magnitude) and much more rapid than the long-term downward drift of pH that has occurred in European soft-water lakes throughout the Holocene. Recovery of pH in some

lakes, produced by reduction in emissions, can also be detected.

D. Paleolimnology and the Recent Eutrophication of Lake Victoria, East Africa

Analysis of cores (by e.g., Robert Hecky) of very recent sediments from this African Great Lake of international importance shows that it began to change to its present highly productive, oxygen-deficient state during the 1920s, with decline in the abundance of green algae and increase in planktonic diatoms. During the 1960s, these were replaced by blue-green algae, coinciding with an observed decline in the nitrogen/phosphorus ratio of lake waters. In the uppermost sediments, biogenic silica influx (see Sections VI, B and VI, C, 2, b) increases, indicating a further rise in productivity.

Correlation of these observations with events in the watershed is complicated by the influence of climatic factors on the lake ecology. Biomass rises during periods of calm weather and lake stratification and falls during phases of increased storminess and mixing. Furthermore, lake ecology was severely affected during the 1950s and early 1960s by introduction of Nile perch (*Lates niloticus*) and other exotic species.

Clearance of native forests in the watershed for plantation forestry during the period 1930–1960 led to transport of large quantities of nutrients and sediments to the lake. At present, burning of native vegetation for agriculture is increasing aerial input of nitrogen, sulfur, and phosphorus.

The ecological problems of Lake Victoria, and other lakes in Africa and other developing countries, are unlikely to be solved unless the economic circumstances of riparian states (in this case Kenya, Rwanda, Tanzania, and Uganda) improve. Cancellation of their so-called "debts" to Western banks, accumulated over several decades of postcolonial economic exploitation, would be a good way to begin!

Acknowledgments

I would like to acknowledge the help of Dr. Elizabeth Haworth of the U.K. Institute of Freshwater Ecology, and of Mr. Mick Cooper of the University of Plymouth, in the preparation of this article. I would also like to record my debt to Dr. Maureen Longmore of the University of Adelaide, who provided me with Fig. 2 when she was an undergraduate student at the University of Ulster some 24 years ago. I hope she is not too embarrassed by my publication of it at this rather late stage.

This article is sincerely dedicated to my partner, Mhairi Margaret Mackie, who put up with several even more stressful weeks than usual during its production.

Glossary

Allochthonous Originating outside (the lake basin) (cf. autochthonous, allogenic).

Allogenic Fraction of lake sediment contributed in mineral form by erosion from watershed (cf. authigenic, biogenic, allochthonous).

Authigenic Fraction of lake sediment formed and/or precipitated chemically within the lake water column (cf. biogenic, allogenic, autochthonous).

Autochthonous Formed within the lake basin (cf. biogenic, authigenic).

Benthic Bottom dwelling.

Biogenic Fraction of lake sediment precipitated within the lake water column by the action of lake biota (cf. authigenic, allogenic, autochthonous).

Catchment See watershed.

Clastic Material contributed to lakes by erosion of catchment rocks and soils.

Copropel Lake sediment formed in an oxygen-rich lake, of which the feces of lake organisms, especially zooplankton (see planktonic), forms an important component.

Detritus Particulate matter contributed to lakes and lake sediments by catchment and lake ecosystems, consisting of both mineral and dead organic materials.

Diagenesis Alteration (mainly chemical and bacterial) of substances contributing to sediments that begins on leaving the parent organism and continues after sedimentation, until stable molecules are formed.

Drainage basin See watershed.

Endogenic Fraction of lake sediment formed *in situ,* mainly by diagenesis.

Epilithic Living on the surface of stones.

Epipelic Living on the sediment surface.

Epiphytic As in the terrestrial environment, the habit of smaller plants of dwelling on the outside of other, larger plants.

Epipsammic Living on the surface of sand grains.

Euryhaline Tolerant of a wide range of salinity.

Eutrophication Increased nutrient loading in an ecosystem. As a consequence, there is an increase in productivity and a change in community structure. Not to be confused with ontogeny, succession, or infilling.

Half-life Time taken for the radioactivity of a given isotope to decay by 50%; symbolized by $T/2$. It is used as a means of expressing longevity or otherwise of isotopes, in that "full life" may never be reached and is practically impossible to detect.

Hydrosere Successional infilling of small lakes by accumulation of autochthonous organic material; part of lake ontogeny. Not to be confused with eutrophication.

Influx Accumulation of sediment per unit area (cm^2) per unit time (year).

Lipid Relatively simple organic compound (nC normally < 40) soluble in common organic solvents (methanol, dichloromethane); formerly called "fats."

Littoral Pertaining to the marginal, shallow zone of lakes.

Macrophytes Higher and larger plants living in fresh waters ("pondweeds").

Meromictic Lake in which circulation of waters is incomplete and confined to the surface layers (the mixolimnion), leaving a bottom layer that is rarely if ever mixed with the upper parts of the water column (the monimolimnion).

Mineral magnetism Response of a sample to being deliberately magnetized in the laboratory; not to be confused with paleomagnetism.

Ontogeny Development of systems, in this case lakes, through time; not to be confused with eutrophication.

Paleomagnetism Analysis of the changing signal of the earth's magnetic field (natural remanent magnetization, NRM); not to be confused with mineral magnetism.

Pelagic Pertaining to the surface waters of the deep zones of lakes (cf. littoral); also called the limnetic zone.

Periphytic Pertaining to the periphyton, the plants (both microscopic and macroscopic) that live at the margins of lakes.

Planktonic Living in the surface waters of lakes, either (in the case of microscopic photosynthetic organisms, the phytoplankton) by means of maintaining buoyancy or (in the case of animals, the zooplankton) by weakly swimming.

Profundal Pertaining to the bottom waters of the deep parts of lakes (cf. littoral, pelagic, limnetic).

Redox Changing availability of oxygen in lake waters, leading to alternating oxidation or reduction of chemical species.

Sediment–water–interface (SWI) Contact between the waters and the sediments of a lake.

Swidden Cultivation adapted to forest environments, consisting of temporary clearance and recovery.

Varve Sequence of two or more sedimentary microlaminae (thickness ca. 0.5–5 mm) making up the deposition of a single year.

Watershed Area delivering inflowing water to a lake or other water body (cf. catchment, drainage basin).

Bibliography

Berglund, B. E., ed. (1986). "Handbook of Holocene Palaeoecology and Palaeohydrology." London/New York: John Wiley & Sons.

Davis, R. B. (1990) "Palaeolimnology and the Reconstruction of Ancient Environments." Boston: Kluwer Academic Press.

Haworth, Y., and Lund, J. W. G., eds. (1984). "Lake Sediments and Environmental History." Leicester, England: Leicester University Press.

Lerman, A., ed. (1978). "Lakes: Geology, Chemistry, Physics." London/New York: Springer-Verlag.

Löffler, H., ed. (1986). "Palaeolimnology IV." Boston: Kluwer Academic Press.

Meriläinen, J., Huttunen, P., and Battarbee, R. W., eds. (1983). "Palaeolimnology." The Hague: Dr. W. Junk.

Smith, J. P., Appleby, P. G., Battarbee, R. W., Dearing, J. A., Flower, R., Haworth, E. Y., Oldfield, F., and O'Sullivan, P. E., eds. (1991). "Environmental History and Palaeolimnology." Boston: Kluwer Academic Press.

Parasitism, Ecology

John Jaenike
University of Rochester

Parasites live and feed on or within other, larger organisms. Most types of parasites rarely kill their hosts directly, but often reduce the fitness of their hosts, through depletion of host nutrients, damage to host tissues, causing the host to change its behavior, or production of toxins. Because such reductions in host fitness can lower the basic reproductive rate of a host population, parasites can play a role in regulating populations of their hosts. Parasites can also affect the outcome of interactions between other species and thereby have significant effects on the structure of ecological communities.

I. INTRODUCTION

Ecological interactions between parasites and their hosts are extraordinarily diverse. Parasitic lineages have evolved in many groups of organisms (see Appendix), including some that are exceptionally speciose. Many of the species in the most diverse groups of insects (Leipidoptera, Diptera, Hymenoptera, Coleoptera, and Homoptera) feed in a parasitic manner on plants. Because the shift to a parasitic life-style entails retention of some characteristics of the free-living ancestors from which parasitic groups ultimately evolved, the ecology of parasites is extremely diverse, reflecting the numerous times it has arisen in many major taxonomic groups. The diversity of parasites is equaled or exceeded by the diversity of hosts parasitized; no group of organisms is free from parasitism.

Finally, parasites exhibit a greater variety of life cycles than do free-living organisms. Some, like cold viruses, are simply transmitted directly from one host to another of the same species. Others require passage through several hosts, completing different stages of their life cycles in different host species. The plasmodia that cause malaria in humans and other vertebrates alternate between vertebrate and insect hosts. Some tapeworms pass successively through crustaceans, fish, and vertebrate hosts at different stages in their life cycles. Many rust fungi alternate between unrelated species of plant to complete their life cycles. Interactions between the different hosts of a parasite's life cycle can thus affect the interactions between these hosts and the parasites. Some parasites are highly host specific, whereas others infect a broad range of host species.

Despite the diversity and ubiquity of parasites, ecologists have accorded them far less attention than they have given to competitors or predators. There are several reasons for this. First, parasites are unapparent to field workers. Detection of most parasites requires either that the host be dissected or studied in the hand, or that the blood or feces be examined microscopically. Thus, study of parasites generally requires laboratory study, and this is not always feasible for field ecologists. Second, the effects or parasites on their hosts may be rather subtle. Modest reductions in host mating success, fertility, or survival caused by parasites are much less noticeable than a lion or a spider capturing prey, or antagonistic interactions between competitors contesting a limiting resource.

Finally, a general and tractable conceptual framework for modeling ecological interactions between hosts and parasites was not available until the late 1970s. The theoretical studies of competition and predation by A. J. Lotka and V. Volterra in the 1920s and the empirical work of G. F. Gause in the 1930s laid the foundation for much of contemporary thought in population ecology. Thus, the role of competitors and predators in natural communities has long been recognized by ecologists. Several developments in the 1970s paved the way for the current interest in the ecology of parasitism. P. R. Price argued in his 1980 monograph "The Evolutionary Biology of Parasites" that a major fraction of species on earth are parasites, encompassing many taxa and ecologically diverse lifestyles, and that these species differ from free-living species in their population biology and modes of speciation. The potential importance of parasites as selective agents in the maintenance of genetic

variation in their host species was becoming widely recognized at this time. Finally, R. M. Anderson and R. M. May published a series of mathematical models of host–parasite interactions that were general and tractable. Rather than focusing on a specific type of parasite and its host, they considered generic parasites sharing certain life-history characteristics. They categorized parasites very broadly into macroparasites and microparasites and developed general models for each category. Thus, they emphasized functional rather than phylogenetic differences among parasite types. [*See* EQUILIBRIUM AND NONEQUILIBRIUM CONCEPTS IN ECOLOGICAL MODELS.]

The major distinctions between micro- and macroparasites are outlined in Table I. Because microparasites can multiply rapidly within a host, infected hosts typically harbor millions or more individual parasites. Thus, hosts can be classed as either infected or not infected. The noninfected individuals may either be susceptible to infection or recovered and immune to future infection. Vertebrates are capable of mounting a defensive immunological reaction to many microparasites. An effective immunological response allows the host to recover from infection quickly and to develop long-lasting immunity to subsequent infections.

Macroparasites are larger than microparasites and do not generally complete more than one round of multiplication within a host. For a variety of reasons, including immunological host mimicry by the parasites and location of the parasites in the gut

TABLE I
Functional Differences between Microparasites and Macroparasites

Characteristic	Microparasites	Macroparasites
Reproduction	Direct within host	Does not complete life cycle within host
Size	Small	Large
Length of infection	Transient	Persistent
Immunity	Long-lasting	Weak immune response
Number of parasites per host	Many (often millions)	Few
Host categories	Discrete variation: susceptible, infected, and recovered	Continuous variation: dependent on parasite numbers per host
Generation time	Short	Long
Representative groups	Viruses, bacteria, protozoa	Helminths, fungal pathogens of plants, phytophagous insects and mites, ectoparasites

or the epidermis, hosts typically mount weaker immunological responses to macroparasites than to microparasites. Thus, hosts do not develop lasting immunity to these parasites and are therefore subject to continual reinfection. The pathology associated with infection by macroparasites is generally dependent on the number of parasites harbored by an individual host. Heavily infected individuals suffer greater reductions in fitness than do lightly infected hosts.

This article focuses on the ecology of interactions between hosts and parasites. However, I will not specifically consider parasitoids or phytophagous arthropods, as each of these topics could be the subject of an entire article. Most of the examples will be drawn from parasites of animals, with the empasis on wild as opposed to domestic species. I will consider: (1) how parasites affect the ecology of individuals, specifically their behavior; (2) theoretical and empirical studies of the effects of parasites on the population dynamics of hosts and parasites; and (3) how parasites may affect the structure of ecological communities.

II. EFFECTS ON THE BEHAVIOR OF HOST INDIVIDUALS

Any ecological effects of parasites must ultimately be mediated through their effects on individual parasitized hosts. Parasite-induced reductions in host fitness will be treated in the next section. Here I consider how parasites affect the day-to-day activities of their hosts. The effects of parasites on host behavior can be viewed from several perspectives.

1. What specific aspect of an animal's behavior is affected by parasitism? Important features include activity level, mate acquisition, habitat selection, foraging behavior, tendency to disperse, and time of activity.

A number of studies show that parasitism can affect an animal's *temperature preference*. In some species, parasitized animals prefer warmer temperatures than unparasitized ones, a phenomenon referred to as *behavioral fever*. Such behavioral fevers have been found in variety of poikilothermic organisms, including various species of insects, crusta-

ceans, fishes, amphibians, and reptiles. The opposite pattern is seen in the bumblebee *Bombus terrestris,* in which workers parasitized by a conopid fly prefer cooler temperatures than do unparasitized bees. The parasitized bees also spend the night in the field, rather than returning to the nest.

Hosts may also alter their *foraging behavior* in response to parasitism. For nest material, starlings choose plants with volatile compounds that adversely affect the ectoparasitic mites and lice residing within the nest. In lakes where baterial pathogens are more prevalent, tiger salamanders are less likely to cannibalize conspecifics; these pathogens can spread from one salamander to another via cannibalism. Infections of arthropod vectors can result in altered patterns of host feeding. For instance, parasitism of sand flies and tsetse flies with *Leishmania* and *Trypanosoma,* respectively, is associated with elevated levels of probing when these flies bite vertebrates.

The effects of parasites on *mate acquisition* have been investigated in a wide variety of species. It is often found that parasitized males have reduced mating success relative to unparasitized males. Such effects have been recorded in insects, fish, frogs, lizards, a variety of birds, and deer.

Changes in *habitat selection* can also be correlated with parasitism. These are generally caused by alterations in the hosts' responses to environmental stimuli, such as light. Several examples are listed in Table II. Acanthocephalan worms are particularly noteworthy in altering the behavior of their arthropod hosts.

2. Why do parasitized individuals behave differently? Parasitized individuals may behave differently from unparasitized ones for at least four different reasons.

2.1. Alterations in host behavior may increase transmission rates of parasites to the next host, thus being of direct benefit to the parasite. J. C. Holmes and W. M. Bethel described four types of altered host behavior that increase parasite transmission rates. All of these mechanisms serve to increase predation upon parasitized hosts, with the predators being the next hosts utilized by parasites with complex life cycles. These mechanisms include: (1) reduction in prey stamina, so that parasitized

TABLE II
Examples of Changes in Habitat Preference of Hosts Caused by Parasitism

Host species	Parasite species	Habitat preference of parasitized hosts relative to unparasitized hosts
Notropis cornutus (shiners)	*Ligula intestinalis* (plerocercoid)	Warm, shallow water near shore
Ilyanassa obsoleta (estuarine snail)	*Gynaecotyla adunca* (trematode)	Higher levels in the intertidal zone
Camponotus spp. (carpenter ants)	*Brachylecithum mosquensis* (trematode)	Open, rocky areas
Gammarus lacustris (crustaceans)	*Polymorphus paradoxus* (acanthocephalan)	Surface waters (positively phototactic)
Armadillidium vulgare (pill bug)	*Plagiorhynchus cylindricus* (acanthocephalan)	Less likely to hide under shelter

hosts are easier to capture by their predators; (2) increase in host conspicuousness, making them easier to find by their predators; (3) host disorientation, another mechanism making prey easier to capture; and (4) altered host responses to environmental stimuli, placing the parasitized hosts in habitats where they are more likely to be preyed upon.

For example, individuals of the estuarine snail *Ilyanassa obsoleta* that are parasitized by the trematode *Gynaecotyla adunca* seek higher levels in the intertidal zone, often becoming stranded on the beach after high tides. The stranded snails are then preyed upon by semiterrestrial amphipods and fiddler crabs, which serve as the next host in the life cycle of these parasites. All acanthocephalans that have been studied change the habitat preferences of their invertebrate hosts in a way that increases their chance of being preyed upon by the parasites' definitive hosts (various species of vertebrates). The increased probing rate exhibited by parasitized insect vectors serves to increase transmission of leishmania and typanosomes to their vertebrate hosts.

2.2. Changes in the behavior of parasitized hosts may mitigate the impact of the infection on the host. Behavioral fever is a clear example of this. Parasitized individuals seek out warmer temperatures than unparasitized individuals. If the higher temperatures are more deleterious to the parasites than to the hosts, then such behavior can effectively cure parasitized individuals of their infections. House flies infected with the pathogenic fungus *Entomophthora muscae* can be cured of their infection if exposed to high temperatures, and parasitized flies prefer warmer temperatures than do unparasit-

ized flies. The cool-temperature preferences of bumblebees parasitized by conopid flies results in slower development of the fly larvae and thus increases the bees' longevity.

Reduced activity levels of parasitized individuals may benefit the hosts if it allows them to use more energy to mount a defensive reaction against the parasites.

2.3. Changes in host behavior may benefit neither the host nor the parasite, but simply be a nonadaptive consequence of the pathology of parasitic infection. For parasites that require a living host to reproduce, an increase in host mortality as a result of parasitism will reduce the fitness of both the parasitized host and its parasites.

2.4. Finally, preexisting behavioral variation among hosts may affect an individual's chance of becoming parasitized. Thus, behavioral differences may be a cause rather than a consequence of differences in parasitism among hosts. For instance, many insects exhibit genetically based variation in their feeding and oviposition preferences for various host-plants. It is also known that the rates of parasitism of a given insect species can depend on their host-plant species. For instance, parasitism of the gypsy moth both by nuclear polyhedrosis virus and by egg parasitoids is lower among larvae feeding on pitch pine than among those feeding on oak. Genetic variation for habitat preference is widespread in other groups of animals, and it is likely that this can result in differences among individuals in their exposure to parasites.

3. Regardless of which, if either, party benefits from altered host behavior, one can inquire into the evolutionary origin of changes in host behavior

associated with parasitism. Phylogenetic analysis of a group of related hosts parasitized by a group of related parasites can be used to identify those host–parasite associations in which alterations in host behavior are derived. Experimental infections of each species of host with each parasite could, in principle, be used determine whether the altered behavior results from evolutionary changes in the hosts, the parasites, or both. However, there have as yet been no studies on parasite-induced changes in behavior in a group of related host species whose phylogenetic relationships are known. The increasing use of molecular data to construct phylogenetic trees and appreciation of the need for phylogenetic analysis suggest that such studies will be done in the near future.

III. POPULATION-LEVEL PROCESSES

Parasites can potentially affect both the growth and regulation of populations of their hosts. The net deleterious impact of parasites on a host population depends on two factors: (1) the degree to which the survival and reproduction of parasitized individuals is reduced and (2) the fraction of the population affected by parasites. More specifically, for microparasites, the population-level effect is determined by the virulence of the parasites (reduction in host fitness) and the prevalence of parasitism. For macroparasites, the population-level impact depends on the relationship between host fitness and the number of parasites per host and the statistical distribution of parasites among hosts. These factors will be considered in turn.

A. Parasite Virulence

It had long been believed by many parasitologists that parasites and hosts live in a state of benign coexistence. Parasites should be avirulent, it was thought, because a parasite that killed its host was, in effect, committing suicide. This concept was bolstered by laboratory studies showing that some parasites had little effect on host survival. However, both the theoretical and empirical underpin-nings of this view have recently been called into question.

Because the laboratory environment may be a poor model of the natural world, laboratory studies give little indication of the magnitude of parasite-induced reductions in host fitness in natural populations. Several important sources of mortality, including predation, unfavorable weather conditions, food limitation, and infection by other species of parasites, are generally absent in the laboratory. If parasites only weaken their hosts, this could increase the risk of succumbing to these other factors. In addition, laboratory studies generally measure only a subset of fitness components, often ignoring variables such as ability to acquire a territory, mating success, and number of broods produced per season.

Recent theoretical and empirical studies show that parasites will not necessarily evolve to become avirulent. If parasites and hosts compete for limited resources, increases in the rate of parasite reproduction can only come at the expense of the host, resulting in decreased host survival and reproduction. Under such conditions, a completely benign parasite will garner no resources and thus be unable to reproduce. On the other hand, a parasite that uses all of the available resources, leaving none to the host, may also have a low reproductive rate, if the host's survival is very much diminished. Thus, an optimal, intermediate level of parasite virulence that results in maximum parasite reproduction is expected to evolve.

The mode of parasite transmission can also affect parasite virulence. Theoretical studies show that parasites that are transmitted vertically, that is, from an infected mother to her offspring, should be less virulent than those that are transmitted horizontally (from an infected individual to the offspring of other individuals). Empirical studies on bacteria parasitized by viruses and fig wasps parasitized by nematodes confirm this expectation. Parasites carried from one host to the next by insect vectors (e.g., malaria-causing *Plasmodium*) may also be highly virulent, as incapacitation of the host does not greatly reduce parasite transmission rates.

For some parasites with complex life cycles, transmission to the next host requires death of the present host. Acanthocephalans alter the behavior of their intermediate hosts (invertebrates) in ways that increase rates of predation on these hosts by their definitive hosts (vertebrates). Thus, the success of these parasites is directly related to the degree to which their hosts suffer increased rates of predation.

Finally, parasites can affect host fertility as well as survival. Parasitic castration occurs in a wide variety of hosts, including plants, crustaceans, fishes, echinoderms, mollusks, insects, and mammals. Because they are sterile, such hosts make no more contribution to host population growth than one that is dead. Parasitic castration can actually result in increased host longevity, because the effort involved in reproduction can reduce an animal's longevity.

These considerations lead to the expectation that many, if not most, parasites adversely affect the survival and fertility of their hosts. Thus, most parasites probably have some effect on the growth rates of their host populations.

B. Prevalence and Intensity of Parasitism

The key variable affecting the population dynamics of microparasites is their effective reproductive rate (R), defined as the number of secondary infections resulting from a single infected host. For directly transmitted parasites (i.e., those not dependent on vectors or requiring intermediate hosts), $R = \beta LN$, where β is the transmission rate, L is the expected duration of infectiousness of an infected host (taking into account parasite-induced mortality), and N is the density of susceptible hosts in the population. Spread of these parasites requires that the number of infected hosts increase through time. For this to occur, R must exceed unity, that is, $\beta LN > 1$. This can be rearranged to yield $N > 1/\beta L$ as the conditon necessary for spread of the infection. Because β and L are parameters (not variables), dependent on the host and parasite species but independent of host and parasite numbers, the density of susceptible hosts in the population is the key determinant of whether or not the parasite continues to spread.

This simple formation is important in a number of respects. First, it shows that there is a *threshold density* of hosts (N_T), below which parasites cannot become established. When the parasite first appears, every individual in the host population is susceptible to infection. If the total host density is below N_T, the parasite will fail to spread. If the density exceeds N_T, the prevalence of infection will increase through time, that is, there will be an epidemic.

Second, the existence of a threshold density affects the fate of an infection even after it has become established in a population. Consider a hypothetical epidemic as sketched in Fig. 1. If the host density is above N_T, the microparasite can become established ($R > 1$), as indicated by the rise in the number of new infections per unit time. As more of the hosts become infected or recover with immunity to subsequent infection, the number of susceptible hosts declines. Eventually, the density of susceptible hosts falls below N_T, at which point the number of new infections per unit time begins to decline. The density of susceptible hosts then increases through birth (or aging of the population, if the infection is age specific), immigration, and loss of immunity. If the density of susceptible hosts then surpasses N_T while some individuals are still infectious, a second epidemic can take place. Thus, the incidence of disease may exhibit sustained or damped oscillations. Several features of the host and parasite can make the occurrence of such periodic epidemics more likely, including variation

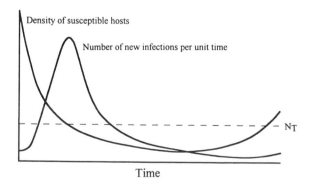

FIGURE I Epidemic of a hypothetical disease. The number of new infections per unit time will increase as long as the density of susceptible hosts remains above threshold density (N_T). A second epidemic may occur if the density of susceptibles surpasses N_T while some hosts from the first epidemic remain infectious.

TABLE III
Stabilizing and Destabilizing Factors for Host–Parasite Population Dynamics

Stabilizing	Destabilizing
Parasite-induced host mortality	Parasite-induced reductions in host fertility
Spatially structured populations	Spatially homogeneous populations
Aggregation of parasites within hosts	Lack of genetic diversity in hosts and parasites
Vertical transmission of parasites	Time delays between parasite reproduction and subsequent infection of new host (e.g., long-lived infective stages)
Density-dependent reductions in parasite reproduction in multiply parasitized hosts	Multiplication of parasites within infected hosts
Host death rate increases faster than linearly with parasite number	

among host age-classes in parasite transmission rates and seasonal variation in rates of contact among host individuals, which entrain the periodicity (see also Table III). Many childhood diseases of humans exhibit distinct periodicity in the occurrence of epidemics, with interepidemic periods ranging from two to six years (Table IV).

If there is a low rate of recruitment of susceptible hosts, then the last infected individual may recover or die before the density of susceptibles exceeds N_T. In this case, the parasite population will die out. Thus, the *critical population density* necessary to sustain an infectious parasite indefinitely is larger than the threshold density required for the initial establishment of a parasite. The measles virus can become established in isolated island populations comprising only a few thousand people. However, a population of 200,000 to 500,000 individuals is required to sustain measles indefinitely. Comparable results have been found in other systems. For example, a minimum density of Japanese beetle larvae in the soil is required for the successful establishment of *Bacillus popilliae,* which is used as a biological control agent of the beetles. Recent experimental work with fungal pathogens of soil-dwelling nematodes has shown that below a certain density of hosts, the parasites do not persist. These considerations suggest that as a host population increases in density (as humans have done), more and more disease-causing parasites can become established and sustained.

Finally, the concept of a threshold density shows how a *vaccination* program can eradicate a disease, even if only a fraction of the host population is immunized. By keeping the density of susceptibles below N_T, the reproductive rate of the parasites is kept below unity. Such vaccination was responsible for the worldwide eradication of smallpox in the 1970s.

The population dynamics of macroparasites are more complicated than those of microparasites, although the concept of threshold population size still applies, because the basic reproductive rate of a macroparasite depends on the density of hosts. The population-level impact of a macroparasite de-

TABLE IV
Periodicity of Some Childhood Diseases of Humans[a]

Infection	Period between epidemics (years)	Time period
Measles	2	England and Wales (1948–1968)
Rubella	3.5	Manchester, U.K. (1916–1983)
Mumps	3	England and Wales (1960–1980)
Chicken pox	2–4	New York City (1928–1972)
Diphtheria	4–6	England and Wales (1897–1979)
Smallpox	5	India (1868–1948)

[a] Adapted with permission from R. M. Anderson and R. M. May (1991). "Infectious Diseases of Humans." Oxford University Press, Oxford, England.

pends on the statistical distribution of parasites among hosts and how host fitness varies with parasite number.

Macroparasites are almost invariably distributed in an aggregated manner among their hosts. A large fraction of the parasite population is often harbored in a relatively small number of hosts. Although the ubiquity and potential importance of macroparasite aggregations have attracted considerable attention, for many species we know little about the specific causes of these aggregations. In general, such aggregations are due either to heterogeneity among hosts in their exposure to the infective stages of parasites or to their susceptibility to parasitism. For instance, the probability that a plant will become infected by a specific strain of a fungal pathogen often depends on the genotype of the plant. If the infective stages of parasites are nonrandomly distributed across the environment, there will be variation among host individuals in their exposure to parasites. Finally, variation among hosts in their present level of parasitism may affect their probabilities of further infection.

Aggregations of macroparasites have important consequences for both the hosts and the parasites. First, such aggregations are likely to play an important role in regulating populations of the parasites, because a substantial fraction of the parasites inhabit heavily infected hosts. Variables dependent on the density of parasites within hosts, including parasite survival and reproduction, host survival, and the intensity of host defensive reactions, can serve to hold parasite populations in check (Table V). The more intense regulation of a parasite population resulting from parasite aggregation can also facilitate the coexistence of several parasite species utilizing the same species of host.

The effect of parasite aggregations on host population dynamics depends on how host fitness changes with parasite number. It is sometimes asserted that parasite aggregation reduces the net impact of macroparasites on populations of their hosts. However, if host fitness declines in a linear fashion with the number of parasites per host, then the population-level impact of the parasites is independent of their statistical distribution among hosts. If host fitness declines in an accelerating fashion with parasite number, then aggregation can increase the net deleterious effect of parasites on a host population. Only if host fitness declines in a decelerating fashion with parasite number does aggregation lessen the impact of parasites on the host population.

The deleterious effects of infection on individual hosts, combined with observed levels of parasitism in natural populations, suggest that parasites may often have a major effect on the abundance of their hosts. The most dramatic examples come from deliberate or accidental introductions of parasites. For example, the deliberate introduction of myxoma virus to Australia resulted in the crash of rabbit populations there. Escape to *Endothia parasitica,* the fungus that causes chestnut blight, nearly caused the extinction of native chestnuts in North America.

Just as disease epidemics can recur in a periodic fashion, regular oscillations in host population size may be driven by parasites. The densities of breeding birds within populations of red grouse in Scotland fluctuate nearly 10-fold with a periodicity of about 5 years. These cycles are believed to be driven by reductions in host fecundity caused by a parasitic nematode. Similarly, many forest insects exhibit dramatic fluctuations in density over periods of 5

TABLE V
Examples of Reductions in Parasite Fitness Dependent on Number of Parasites Per Host

Host species	Parasite species	Fitness component
Cyclops bicuspidatus (crustacean)	*Triaenophorus crassus* (propcercoids)	Parasite body size
Drosophila spp.	*Howardula aoronymphium* (nematode)	Host survival
Xenopus laevis (amphibian)	*Protopolystoma xenopodis* (Monogenea)	Parasite fecundity
Rats	*Hymenolepis diminuta* (cestode)	Parasite body size
Humans	*Ascaris lumbricoides* (nematode)	Parasite fecundity

to 12 years that may be driven by their interactions with viral parasites.

Although parasites often have adverse effects on infected hosts, they may sometimes have no effect on the growth rate of host populations. In many species of vertebrates, only the fittest individuals succeed in mating or in obtaining territories. For instance, in many birds and mammals, there are more individuals than there is available habitat for their territories. In such species, parasite-induced reductions in host fitness may have no effect on the number of breeding pairs and thus no effect on the potential growth rate of the host population. Such *compensatory effects* are thought to be more common among vertebrate than invertebrate hosts.

IV. COMMUNITY-LEVEL EFFECTS

Community-level effects of parasites range from relatively minor shifts in the relative abundance of different host species to wholesale changes in the species composition and trophic structure of a community. It has long been known that slight shifts in environmental conditions can reverse the outcome of interspecific competition. Similarly, if the net deleterious effect of parasites varies among host species, whether through differences in virulence or prevalence and intensity of infection, then the outcome of interspecific competition may depend on whether or not a particular parasite is present in a community. In the 1940s, T. Park demonstrated that the outcome of competition between two species of *Tribolium* flour beetles is determined by whether or not a sporozoan parasite (*Adelina*) is present. When *Adelina* is absent, *T. castaneum* is competitively superior to *T. confusum*. However, when these parasites are present, the outcome of competition is reversed, because *T. confusum* is less severely affected by *Adelina* than is *T. castaneum*. I have found in field-cage experiments that the competitive superiority of *Drosophila putrida* over *Drosophila falleni* is significantly reduced or eliminated by a parasitic nematode. These nematodes completely sterilize females of *D. putrida*, but reduce the fertility of *D. falleni* by only 50%.

In natural populations, the distribution of competing species is occasionally correlated with the distribution of a shared parasite. In the eastern United States, *Drosophila putrida* is much more common than *D. falleni* in areas where parasitic nematodes are absent, whereas the two *Drosophila* species are about equally abundant where the nematodes are found. This pattern suggests that geographical variation in the structure of *Drosophila* communities is affected by these parasites. On the Caribbean Island of St. Maarten, the local distribution of two species of *Anolis* lizards is apparently determined by the distribution of a malarial parasite: the competitively inferior, but malaria-resistant species, *Anolis wattsi,* occurs only where the malaria is present. Similarly, the distribution of two species of macaques in India is correlated with the distribution of a mosquito vector of *Plasmodium knowlesi*. Where the mosquitoes are present, the malaria-tolerant and competitively inferior *Macaca fascicularis* replaces *Macaca mulatta,* which is highly susceptible to *Plasmodium* infections. Finally, the distribution of some cervids in North America may be determined by the parasitic nematode *Parelaphostrongylus tenuis*. These worms are far more pathogenic to moose, caribou, and pronghorn antelopes than to white-tailed deer. Cervids other than white-tailed deer are often unable to persist in areas where these parasites are endemic.

These results show that parasites can actually bring about an increase in the abundance of some host species. Although parasites have adverse effects on the fitness of individual hosts, their presence may actually be of benefit to a host species, if another competing host species is even more severely affected by the parasites.

If parasites affect one of the dominant species within a community, their effects can be far more dramatic, ramifying throughout the community. In such cases, the parasites can be considered *keystone species*. Until recently, chestnut was the dominant tree species in many forests in the eastern United States. However, invasion by the fungus that causes chestnut blight, *Endothia parasitica,* nearly caused the extinction of chestnuts. As a result, the vegetational composition of eastern forests has changed dramatically in this century.

In the 1800s, a *Plasmodium* that causes avian malaria and its mosquito vector became established in Hawaii. The native Hawaiian birds, which are highly susceptible to malaria, are now largely restricted to xeric and high-elevation habitats, where the mosquitoes, and thus malaria, are rare to absent. Birds that have been introduced to Hawaii are far less susceptible to malaria and are therefore abundant in habitats where the native birds have been excluded. Thus, the *Plasmodium* has had a major effect on the distribution of the Hawaiian avifauna.

A microsporidian parasite of the caddis fly *Glossosoma nigrior* invaded several Michigan streams around 1988 and has caused populations of its hosts to collapse. Because these caddisflies are dominant grazers of periphyton, the invading microsporidians will probably have communitywide effects in these streams.

Finally, the introduction of myxoma virus to England in 1953 resulted in the mortality of about 99% of the rabbits in some areas. The loss of the rabbits, which were important grazers, affected the structure of ecological communities in a variety of habitats. For instance, sand dunes became stabilized, because the dune vegetation was no longer subject to overgrazing by rabbits. After the decimation of the rabbits in chalk downlands, the abundance of palatable grasses increased and the general appearance of these habitats changed from lawnlike to meadow. Voles and ants benefited from the increased height of the vegetation.

It is thus apparent that parasites can have dramatic and wide-ranging effects on the structure of ecological communities. However, most of the examples discussed here involve introduced parasites and, therefore, unnatural host–parasite associations. It is possible that the ecological effects of parasites in these cases are atypical, simply because the hosts and parasites have not had a long history of association. The conventional view of parasitologists that hosts and parasites coevolve to a state of benign coexistence implies that the ecological effects of a given parasite should diminish through time. In fact, the myxoma virus that was introduced to control rabbits in Australia has evolved to become less virulent. However, the

APPENDIX

Groups of Organisms Including Parasitic Species (Most Groups Include Nonparasitic Species as Well).
Hosts = Bacteria (b), Fungi (f), Plants (p), Invertebrates (i), Vertebrates (v).

Viruses: all (b, f, p, i, v)

Bacteria: many families,
  including:
  Spirochaetaceae (i, v)
  Spirillaceae (b)
  Enterobacteriaceae (p, i, v)
  Neisseriaceae (v)
  Rhizobiaceae (p)
  Veillonellaceae (v)
  Micrococcaceae (v)
  Streptococcaceae (p, i, v)
  Actinomycetaceae (v)
  Rickettsiaceae (i, v)
  Bartonellaceae (i, v)
  Anaplasmaceae (v)
  Mycoplasmataceae (v)
  Spiroplasmataceae (p, i)

Fungi: parasitic species found
  in all five subdivisions,
  including:
  Mastigomycotina (p, i, v)
  Zygomycotina (f, p, i)
  Ascomycotina (f, p, i, v)
  Basidiomycotina (f, p)
  Deuteromycotina (p, i, v)

Chromophycota
  Genus *Notheia* (p)

Plants
  Convolvulaceae (p)
  Monotropaceae (p)
  Orobranchaceae (p)
  Scrophulariaceae (p)
  Rafflesiaceae (p)
  Loranthaceae (p)

Protozoa
  Phylum Dinoflagellata (i)
  Phylum Zoomastigina (i, v)
  Phylum Sacrcodina (i, v)
  Phylum Sporozoa (i, v)

Phylum Myxozoa (i, v)
Phylum Ciliophora (i, v)

Animals
  Phylum Cnidaria
    Class Hydrozoa (v)
    Class Anthozoa (i)
  Phylum Mesozoa
    Class Rhombozoa (i)
    Class Orthonectida (i)
  Phylum Platyhelminthes
    Class Turbellaria (i)
    Class Monogenea (v)
    Class Trematoda (i, v)
    Class Cestoidea (i, v)

  Phylum Acanthocephala (i, v)

  Phylum Rotifera
    Class Monogononta (p, i)

  Phylum Nematoda
    Class Secernentea (p, i, v)
    Class Adenophora (i, v)

  Phylum Nemertea
    Class Enopla (i)

  Phylum Mollusca
    Class Bivalvia (i, v)
    Class Gastropoda (i)

  Phylum Annelida
    Class Polychaeta (i, v)
    Class Clittelata (i, v)
    Class Hirundinoedea (i, v)

  Phylum Arthropoda
    Class Crustacea (i)
    Class Arachnida (i, v)
    Class Insecta (p, i, v)
    Class Pentastomata (v)

  Phylum Chordata
    Class Agnatha (v)

recent theoretical and empirical studies discussed in the previous section indicate that, at evolutionary equilibrium, parasites may be quite virulent. Thus, parasites may well have dramatic community-level effects, even in communities where they have been present for long periods of evolutionary time.

If parasites occasionally cause the extinction of their hosts, and thus of themselves, their effects may be manifest even in communities where they are no longer present. It remains a challenge to identify how parasites, past and present, have helped shape the structure of present-day ecological communities.

Glossary

Critical population density Minimum host population density required to maintain an infection indefinitely.

Definitive host Host in which sexual reproduction of the parasite occurs.

Epidemic Increase in disease in time and space.

Intensity of parasitism Number of parasites per infected host.

Intermediate host Host in which asexual proliferation of a parasite occurs.

Macroparasite Parasites that do not have direct reproduction within the definitive host, such as most helminths and ectoparasitic arthropods.

Microparasite Parasites that have direct reproduction within their hosts (i.e., they can complete numerous generations of population growth within a singel host), such as viruses, bacteria, some protozoa, and fungi.

Parasite Organism that lives and feeds on another organism (the host), usually reducing the fitness of the host but not killing it outright.

Parasitold Insects whose larvac live and feed within the bodies of other insects, causing the death of the host.

Prevalence of parasitism Fraction of host population infected with a particular parasite.

Threshold population density Minimum host population density required for the initial establishment of an infection.

Vector One of the hosts in the life cycle of a parasite that transmits the parasite to another host species by biting.

Bibliography

Anderson R. M., and May, R. M. (1991). "Infectious Diseases of Humans." Oxford, England: Oxford University Press.

Brooks, D. R., and McClennan, D. A. (1993). "Parascript: Parasites and the Language of Evolution." Washington, D. C.: Smithsonian Institution Press.

Burdon, J. J. (1987). "Diseases and Plant Population Biology." Cambridge, England: Cambridge University Press.

Esch, G. W., and Fernandez, J. C. (1993). "A Functional Biology of Parasitism: Ecological and Evolutionary Implications." London: Chapman & Hall.

Ewald, P. W. (1991). Transmission modes and the evolution of virulence. *Human Nature* **2**, 1–30.

Moore, J., and Gotelli, N. J. (1990). Phylogenetic perspective on the evolution of altered host behaviors: A critical look at the manipulation hypothesis. New York: Taylor and Francis. (C. J. Barnard and J. M. Behnke, eds) pp. 193–233. *In* "Parasitism and Host Behavior."

Price, P. R., Westoby, M., and Rice, B. (1988). Parasite-mediated competition: Some prediction and tests. *Amer. Nat.* **131**, 545–555.

Scott, M. E. (1988). The impact of infection and disease on animal populations: Implications for conservation biology. *Conservation Biol.* **2**, 40–56.

Park and Wilderness Management

James K. Agee
University of Washington

I. Introduction
II. Legal and Political Framework
III. Management of Parks and Wilderness
IV. Challenges for Park and Wilderness Management

Proper management of park and wilderness areas is essentially the management of change. Park and wilderness designations have been applied to ecosystems where the usual management objectives are to perpetuate natural conditions. These ecosystems are not static, and the "natural conditions" include natural disturbances such as forest fires and hurricanes as well as plant and animal communities. Impacts of human activities can detract from these natural conditions unless carefully managed. Regional to global influences on natural areas suggest that these areas are not ecological islands. Ecosystem management is a preferred strategy for preserving parks and wilderness for future generations.

I. INTRODUCTION

Parks and wilderness areas, as discussed here, are generally remote and relatively unspoiled areas set aside to preserve their natural values. The management of these areas is intended to preserve these natural values for future generations while allowing use that is compatible with these objectives. [*See* NATURE PRESERVES.]

The evolution of ecological thought from early in the twentieth century makes this preservation mandate difficult to accomplish, as we now know that "natural conditions" have fluctuated over time. A great variety of past "natural" landscapes have existed, and a similar variety will occur in the future, inevitably altered by anticipated human-caused global changes. The setting of appropriate park and wilderness goals is as difficult as finding means to achieve them.

II. LEGAL AND POLITICAL FRAMEWORK

National Parks and wilderness preservation began formally over a century ago in North America. In 1864, Yosemite Valley was ceded to the state of California as the first formal preserve (it was later to become the heart of Yosemite National Park in 1890), and in 1872, Yellowstone was created as America's first national park. The origins of legal wilderness are less clear, but early in the twentieth century, Forest Service employee Arthur Carhart and Aldo Leopold successfully proposed the first allocations of national forest land specifically for wilderness preservation.

The passage of the National Park Service Organic Act in 1916 provided a mandate for national park management "to conserve the scenery and the natural and historic objects and the wildlife therein, and to provide for the enjoyment of the same in such manner and by such means as will leave them unimpaired for the enjoyment of future generations." Nearly half a century later, the Wilderness Act of 1964 established a National Wilderness Preservation System composed of wilderness areas "where the earth and its community of life are untrammeled by man." Wilderness was further defined as "an area of undeveloped Federal land retaining its primeval character and influence . . . which is protected and managed to preserve its natural values." In 1975, the Eastern Wilderness Act was passed to incorporate many of the smaller natural areas of the eastern United States into the National Wilderness Preservation System, with somewhat less stringent guidelines for inclusion in the system.

Wilderness areas can be found on Forest Service, Bureau of Land Management, Fish and Wildlife Service, and National Park Service lands (Table 1). The backcountry areas of many national parks are also designated wilderness areas, but not all national park land is designated as either wilderness or backcountry. For the purposes of this article, emphasis associated with national parks is placed on their backcountry portions, which have much

TABLE I
Areas of Parks and Wilderness in the United States

Agency or land unit	Year	Land area (ha)
United States Wilderness		
Forest Service, USDA	1992	13,954,183
National Park Service, USDI[a]	1989	15,588,082
Fish and Wildlife Service, USDI	1989	7,827,116
Bureau of Land Management, USDI	1989	189,408
United States Parks and Preserves		
National Park Service, USDI[a]	1990	32,451,815
Worldwide Protected Areas		
IUCN Categories I–V[b]	1985	423,784,398

[a] Wilderness managed by the National Park Service area is a subset of the total National Park Service-managed land shown under Parks and Preserves.

[b] These are designated by the World Conservation Union (IUCN) as I, scientific or nature reserves; II, national parks or equivalent; III, national monuments; IV, managed nature reserves and wildlife sanctuaries; and V, protected landscapes or seascapes.

in common with wilderness areas in terms of management opportunities and constraints.

Park and wilderness lands in America have a dual mandate: first, to preserve natural conditions to the extent possible, and second, to provide for solitude or primitive recreational experiences. The park and wilderness concepts of the United States have influenced international efforts at wilderness designation and management. Because parks and wilderness are cultural as well as natural concepts, it is not surprising that various countries have adapted rather than adopted these concepts to meet their own societal needs. [See CONSERVATION AGREEMENTS, INTERNATIONAL.]

III. MANAGEMENT OF PARKS AND WILDERNESS

Wilderness and park management appears to be a paradox. *Wilderness* and *parks* are intended to be primeval and unaffected by humans, whereas *management* implies human control. Yet management of parks and wilderness is essential to their preservation. The impact of direct use by visitors, the need to understand and perpetuate natural processes, and the need to offset undesirable regional to global influences on park and wilderness areas all argue for some form of management intervention. Management can take a "light on the land" approach, however, and must use the minimum tools, regulation, and enforcement required to achieve objectives.

A. Natural Resources Management

Park and wilderness managers in the early 1900s believed that resources management was resources protection—the protection of things or objects, not natural processes like fire. Fires were bad, killing trees and blackening the landscape; native predators were bad, because they killed desirable animals like deer; and native insects were bad, because they killed green trees. Natural resources management in nature preserves mirrored our ecological understanding of the time: if we protect these objects, a stable system will be the result. A half century would pass before improved ecological understanding would convince us that change is the only constant in nature.

The species of present temperate plant communities have not coevolved in stable communities over millennia but have individually migrated over the landscape in response to changing climates and disturbance regimes. What was sage-steppe 10,000 years ago may be forestland today. This process of ecological change continues, so that the same site may be woodland or grassland in the distant future. At any spot in a park or wilderness, a variety of past natural landscapes have existed due to changing natural conditions, diverging and coalescing over time much like the channels of a braided stream. To take only one cross section in time as "natural," whether today or the time of Columbus, is to deny change, the only constant in park and wilderness management.

The recognition of park and wilderness areas as dynamic ecosystems came relatively abruptly. Wildlife management problems in Yellowstone National Park in the early 1960s caused Secretary of the Interior Stewart Udall to establish a commission to investigate the problems. The commission report, later known as the Leopold Report after its chairman, wildlife scientist A. Starker Leopold, noted that most parks were composed of a mosaic of plant communities maintained by periodic disturbance—fires, floods, and hurricanes. To maintain that vegetation diversity and the wildlife associated with it, park managers would have to manage disturbance, particularly fire. New ecological tools would have to be developed to maintain primitive America for future generations. The old paradigm of "draw a line around it and leave it" was replaced by a new one known as "natural regulation," which focused on management of natural disturbance processes. Fire was generally the most important and most controversial of these processes.

Some natural disturbance processes are relatively easy to manage, as they are beyond management control. There was no need to develop a "let-it-grind" policy for glaciers, or a "let-it-blow" policy for hurricanes. But fire had a long sociocultural history that through a history of burning by Native Americans, could not be separated from its natural history. There was a need for a "let-it-burn" policy that prescribed when and where we would allow fire to roam across the landscape. A quarter century after fire was recognized as a valid landscape process in parks and wilderness, we still debate its merits. [See FIRE ECOLOGY.]

Soon after fire was incorporated into national park management policy, it was gently reintroduced into some parks like Yosemite and Sequoia–Kings Canyon, where light, benign fires had once burned across the forest floor at 5- to 15- year intervals. These fires reduced excessive fuel loads, stimulated reproduction of certain species, encouraged wildflowers to bloom, and lengthened vistas through stately columns of mature, unharmed trees. Naturally occurring lightning fires were allowed to burn in selected areas of the backcountry, where they would be confined to sparse fuels in the partially forested high country. As the success of these programs grew, fire was reintroduced into more difficult situations: chaparral landscapes that naturally burn every few decades at high intensity, or forests with long fire return intervals but typically intense, uncontrollable fires (Fig. 1).

The year 1988 convinced managers that there was a limit to natural process management. That year, almost half of America's first and largest (at least in the lower 48 states) national park, Yellowstone, burned in a combination of naturally caused management fires and human-caused wildfires. The smoke from these fires drifted across America and signalled a new, more conservative era of fire management in natural areas. Before the policy reviews of the fires were complete, the serotinous cones of fire-killed lodgepole pines opened across much of the burned area, raining abundant seed on nutrient-rich ashbeds from which shrubs and wildflowers would also bloom. Yet the resiliency of the Yellowstone ecosystem appears in sharp contrast to the resistance to ever allow another Yellowstone fire episode. [See FOREST STAND REGENERATION, NATURAL AND ARTIFICIAL.]

Fire is still an important part of resource management programs in park and wilderness areas. Monitoring programs have expanded, and prescription criteria (the conditions under which a fire may

FIGURE I Forest fires in subalpine areas are an important natural process in creating long-lasting subalpine meadows and maintaining a balance between meadow and forest. (A) A recent fire in dense subalpine forest of subalpine fir *(Abies lasiocarpa).* (B) The center of a 55-year-old burn, with remaining snags from the fire. (C) The edge of the 55-year-old burn in (B), with higher tree recolonization because of the nearby seed source of unburned forest. (D) A 90-year-old burn at the top of a south-facing slope, still a relatively open landscape after nearly a century.

burn) are more explicit. The use of prescribed, manager-ignited fire may expand to maintain the role of fire as a more conservative natural fire policy is employed. However, this practice within wilderness boundaries is still controversial. Whether or not prescribed fire can adequately substitute for a wild, natural process remains to be seen.

Alien plant species, or those nonnative to parks and wilderness, are becoming more of a problem. Most are encouraged by disturbance, be it human or naturally caused, and many are highly competitive with native species. River channels are a common corridor for such species to invade parks and wilderness. Trails are another set of vectors for such species. Management may include pulling individual plants or individually treating them with biocides; widespread use of pesticides is undesirable in park and wilderness areas.

Fish and wildlife habitat issues have not all been solved by the trend toward natural process management. Taking of fish and wildlife in both parks and wilderness is commonly allowed. Hunting is allowed in some wilderness (such as Forest Service areas in the lower 48 states and most Alaska wilderness, even within some national park units), and fishing is allowed in most all parks and wilderness. Both are incompatible with the basic mission of natural area management, but are allowed because of recreational pressure and historical tradition.

More serious wildlife issues are associated with the wide ranges of larger wilderness animals, such as the grizzly bear. Many migratory animals leave park and wilderness areas each year during their life and may be affected by activities and management policies on these surrounding lands. Inventory of wildlife populations and monitoring the movements of wildlife through techniques such as radio-telemetry are an important part of effective wildlife management in parks and wilderness. The range of a grizzly bear during its life may be 250,000 ha, and the overlapping ranges of a viable population far exceed the existing or potential area of park and wilderness, at least in the lower 48 states. Elk may migrate into high-elevation park and wilderness areas in the summer, and spend the winter on lands outside of the nature preserves, where the land may be managed for very different objectives. Where the land has largely been usurped for livestock grazing or development, elk populations in the wilderness may be reduced below natural levels. Conversely, where elk winter range exists on adjacent lands in largely forested terrain with little winter snow, timer harvesting may provide much better winter range and result in above-natural elk populations on the park and wilderness summer range. Both conditions are undesirable when trying to manage for natural population levels, but in many cases may be unavoidable. The first step toward a solution is to possess adequate information about wildlife numbers and movements. [*See* WILDLIFE MANAGEMENT.]

Wilderness wildlife, if legally threatened or endangered under the Endangered Species Act, may cause species-specific concerns to override a more holistic ecosystem-level management strategy. For example, parks and wilderness areas within the habitats of the northern spotted owl may have to be managed for one successional stage—old growth, preferred by the threatened owl—rather than what nature might provide through periodic stand-replacing forest fires. Although the species is threatened largely because of land management practices outside of parks and wilderness, nature preserves have necessarily become a significant part of the species recovery plan.

Fishery issues have revolved around fish stocking: placement of fish, usually trout for recreational purposes, largely in backcountry lakes that historically did not contain fish. Such lakes are popularly known as barren lakes, although previous to stocking they may have contained healthy populations of other vertebrate and invertebrate animals and a diverse plant life. There has been a trend away from

stocking barren lakes and more recognition of the value of natural aquatic ecosystems in park and wilderness management. Catch-and-release programs are becoming more popular in areas with native populations of fish. Such programs allow maintenance of a quality fishery while avoiding the need to stock nonnative species. [*See* FISH CONSERVATION.]

B. Visitor Management

Visitor experiences, whether the opportunity for solitude, the viewing of wildlife, or a primitive camping or hiking experience, are part of the management mandate of park and wilderness areas. With too many visitors, however, the opportunity for any of all of these experiences may decline for present or future generations. The problems associated with overuse were recognized as early as the 1930s, and visitor management within limits known as local carrying capacities have been in place for decades. Visitor carrying capacity is the maximum sustainable level of use as determined by ecological and social criteria. Carrying capacity varies from area to area, not only because the resiliency of the landscape varies, but because the expectations of visitors to encounter others in trail or camping locations will also vary.

A new model to determine user carrying capacity has been developed by the Forest Service and is called Limits of Acceptable Change (LAC). The LAC approach focuses on defining desirable conditions rather than numbers of users; user limits are defined when desirable conditions are exceeded. The LAC model has four major components, determined through a cooperative process between agency planners and the public. The first is to define acceptable or achievable resource and social conditions. Most park and wilderness areas have distinctive resource characters that should be maintained through time. Each will have a distinctive set of experiential opportunities; for example, a remote area will have more opportunity for solitude than one close to an urban area. Existing wilderness areas experience more than a hundredfold variation in use per unit area.

The second component of LAC is to assess existing conditions versus acceptable ones. Usually, some areas of parks or wilderness receive heavy use and others receive little. Across the park or wilderness, what are the measurable resource conditions, such as trail conditions, water quality, or wildlife populations, compared to those desired? A comparison of measurable social indicators, such as opportunities for solitude, conflicts between users such as horse and foot parties, or noise levels near popular campsites, is another part of the assessment process.

The third LAC component integrates the first two by defining management alternatives to achieve desired conditions. This step includes the implementation of preferred alternatives. The fourth LAC component includes monitoring and evaluating the effectiveness of the chosen alternative. Through the measured natural and social parameters, existing conditions will be shown to move toward or away from desired conditions, and adjustments to management are then implemented.

The management of use may include a number of strategies. For illegal acts, law enforcement is the desirable strategy, but most users are in compliance with law. Careless action, such as littering, may be dealt with by either rule enforcement or education, and unskilled or uninformed actions are usually best dealt with through education. Some effects of use are unavoidable and yet create undesirable physical impact to the site. In these cases, direct management action is required, such as prevention of erosion adjacent to heavily used trails.

The need for these actions is usually limited to a very small percentage of the landscape, but it is the portion of the landscape most used by visitors: trail corridors (Fig. 2), campsites, or featured sites such as lakes or promontories. Some are more susceptible than others to trampling, campfires, trail construction, recreational stock grazing, and water pollution. One of the principles of park and wilderness recreational impact assessment is that initial periods of heavy use usually do the most damage. Therefore, limiting use will normally confine impact to those sites already impacted, rather than result in site recovery. Dispersing use, unless to resistant sites, will usually result in more impact

FIGURE 2 Heavy past use in this park area created numerous parallel trails and eventually forced managers to harden the main path and provide adequate drainage for it. The older paths will recover very slowly without rehabilitation.

to a greater variety of sites. Temporary closures or limiting length of stay are of limited utility in site recovery. Limiting party size and encouraging minimum impact camping (fewer campfires, camping on resistant on resistant sites such as dry meadows, hauling out trash, etc.) can prevent further deterioration of high-use sites.

Rehabilitation of overused trails and sites is labor-intensive but is being applied more widely as successful techniques are developed. These include trail redesign for better water drainage and less soil erosion, or rerouting trails from chronically wet areas to nearby drier routes. Revegetation of trampled sites can be accomplished by adding soil amendments and transplanting shrubs and trees, or planting greenhouse-grown native stock. Proper education and future management can prevent the recurrence of many of the resource problems created by past use.

IV. CHALLENGES FOR PARK AND WILDERNESS MANAGEMENT

Parks and wilderness can no longer be considered ecological islands. They are inextricably tied to neighboring areas for better or for worse. We can no longer rely on natural process management as the sole means of achieving natural area goals because these are not, in a cybernetic or systems analysis context, *closed* systems. The "walls" of park and wilderness boundaries are political realities, but not social or biological barriers.

A. Natural Resources Issues

Consider the effect of global climate change, clearly a human-induced phenomenon, and one that may occur in less than a century, a fraction of the lifetime of many park and wilderness organisms. Giant sequoias can live 2000 years, and most western conifers can live 500 years or more. The range and timing of this human-induced climate change is subject to debate, but most scientists agree that a global warming is likely. In the Pacific Northwest, the seasonal snowline, by some projections, is anticipated to increase from about 1000 to 1500 m. Under such a scenario, most subalpine meadows will become potentially forested sites; the desert sagelands of Oregon will be approaching the Crater Lake rim, now dominated by subalpine forest. Serious ecological problems due to human-induced change are on the horizon, and current management paradigms or models are not sufficient to the task. This horizon, although decades off, requires immediate planning by scientists and managers: we are dealing with organisms that live for centuries to millennia. Other important natural resources issues in parks and wilderness include the effects of regional air pollution, habitat fragmentation around the borders of these areas, and invasion of alien (nonnative) species into these natural ecosystems. [*See* GLOBAL ANTHROPOGENIC INFLUENCES.]

B. Visitor Use Issues

There is little correlation between national population trends and trends in park and wilderness use, even though as national population has increased, total use in parks has also increased. Most backcountry use in parks and wilderness peaked in the 1970–1975 period and has stabilized or declined since in many areas. This trend is not evident for all such areas, but seems to be occurring nationally. Increasing age of the U.S. population, increasing constraints on leisure time, and changes in social preferences and tastes may all contribute to the level

or declining use. From the standpoint of resource protection, this trend may be encouraging, but in terms of the political future of such areas, lack of visitation could develop into lack of a constituency for wilderness designation and perpetuation. To date, there remains a high and continuing interest in wilderness preservation by the American people.

C. Ecosystem Management

We need to place our park and wilderness goals in a regional framework, looking at larger, regional goals for biodiversity and ecosystem sustainability. Park and wilderness areas are a part of the solution, part of a network of lands managed for multiple goals, not just preservation and not just commodity goals. The boundaries of parks and wilderness areas are permeable membranes. In the future, cooperative management approaches are most likely to preserve the values for which these areas were established, given the inevitable changes we foresee in and around these areas.

A four-part ecosystem management strategy is envisioned: first, defining precise objectives, improving and incorporating our knowledge of the natural world and our recognition that humans are a part of the problems and solutions; second, defining components of concern, each of which may have a different boundary overlapping different adjacent lands; third, developing cooperative management strategies; and fourth, monitoring the results of such strategies.

The future of management of the various components of parks and wilderness areas depends on the direction of overall resources policy and coordinated management strategies more than on the technical refinements possible in the various disciplines applied to park and wilderness management. Those choices may not be scientific ones but value judgments within the broad range of those allowable under the National Park Service Organic Act and the Wilderness Act. The changing physical, biological, and cultural environments in and around these truly great nature preserves will force us to more specifically define the values we wish to preserve and actively pursue them with our neighbors, or we are likely to see them erode. As the golden anniversary of the Wilderness Act approaches in the year 2014 and the Organic Act centennial in 2016, we should hopefully be looking back at the wise choices we made in the 1900s.

Glossary

Alien plant Plant that is not native in a natural landscape; also known as an exotic plant.

Backcountry Roadless area of a park or wilderness, generally accessible only by trail.

Ecosystem Any part of the universe chosen as an area of interest, with the line around that area being the ecosystem boundary and anything crossing the boundary being input or output.

Ecosystem management Process of regulating internal ecosystem structure and function, plus inputs and outputs of the system, to achieve socially desirable conditions.

National park Unit of land managed by the National Park Service generally to preserve natural conditions for the enjoyment of present and future generations.

Natural fire Wildland fire ignited by natural causes, such as lightning, and considered a natural process.

Prescribed fire Wildland fire ignited by managers under specified conditions to achieve management objectives.

Radiotelemetry Monitoring technique for wildlife by which an animal is temporarily captured, fitted with a battery-operated radio transmitter, and then released and tracked by triangulating from directional radio signals.

Serotinous Seedbank strategy commonly observed in species of pines (*Pinus* spp.) in which mature seed is retained in resin-coated cones on the trees until a fire burns the forest, melts the resin, and allows the cone to open and spread seed within days after the fire.

Visitor carrying capacity Maximum level of use that an area can sustain given natural and social constraints.

Wilderness Area of undeveloped federal land generally retaining its "primeval" character and influence, and designated as part of the National Wilderness Preservation System.

Bibliography

Agee, J. K., and Johnson, D. R., (1998). "Ecosystem Management for Parks and Wilderness." Seattle: University of Washington Press.

Frome, M., ed. (1985). "Issues in Wilderness Management." Boulder, Colo.: Westview Press.

Hendee, J. C., Stankey, G. H., and Lucas, R. C. (1990). "Wilderness Management," 2nd ed., revised. Golden, Colo.: North American Press.

Lucas, R. C., compiler. (1986). "Proceedings, National Wilderness Research Conference: Issues, State-of-Knowledge, Future Directions," USDA Forest Service General Technical Report INT-220. Ogden, Utah: USDA.

Nash, R. (1982). "Wilderness and the American Mind," 3rd ed. New Haven, Conn.: Yale University Press.

Physiological Ecology of Forest Stands

T. T. Kozlowski

University of California

I. Introduction
II. Competition
III. Succession
IV. Environmental Control of Growth and Development
V. Physiological Control of Growth and Development

The development of forest stands is viewed as a complex process that is regulated by the hereditary potential and environment of trees operating through their physiological processes. Trees in forest stands are subjected to continuous and periodic abiotic and biotic stresses that interactively influence their growth. Plant succession in disturbed forest stands proceeds toward a community of species that have high capacity to tolerate the stresses of competition.

I. INTRODUCTION

Forests are complex and dynamic ecosystems that vary from pure even-aged stands composed of a single species of trees to mixed stands of several species and different age-classes (Fig. 1). The much greater complexity of mixed over pure stands reflects variations among species in growth rate, crown form, phenology, and longevity. In addition to such differences, forest stands exhibit some similarities and change predictably during their development. For example, a developing forest progresses through four sequential stages (Fig. 2): (1) regeneration stage, in which seedlings invade spaces created by tree harvesting or disturbance: (2) thinning stage following canopy closure, when competition among trees is intense and many trees are destined to die: (3) understory initiation stage, during which gaps form in the canopy and the understory is reinvaded by advance regeneration: and (4) steady-state (old-growth) stage, when understory trees have filled spaces released by death of overstory trees. The stand now consists of a mosaic of patches of various sizes and ages. However, the steady-state stage often is not reached because of the long time required to develop to this stage and the high probability of major disturbances before this stage can be achieved.

Rates of tree mortality vary appreciably during stand development. Trees undergo their greatest mortality risk in the ungerminated embryo of the seed and in the young seedling stage. High seedling losses are traceable to lack of seed viability, seed dormancy, injury by fungi, and consumption of seeds by insects and higher animals. If a seed germinates, the young seedling is vulnerable to attacks by various organisms and is sensitive to environmental stresses. Because of their low reserves of carbohydrates and mineral nutrients, mortality of seedlings in the cotyledon stage is particularly high. Even a mild environmental stress can induce death of

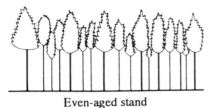

Even-aged stand

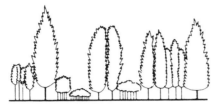

Balanced uneven-aged stand

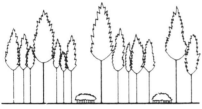

Irregular uneven-aged stand

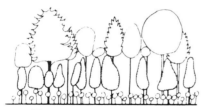

Even-aged stratified mixture

FIGURE I Variations in structure of forest stands. The trees of the top three stands are composed of the same species. The bottom stand is made up of several species. [From D. M. Smith (1986). "The Practice of Silviculture." John Wiley & Sons, New York.]

seedlings in the cotyledon stage. In established forests, mortality of trees is higher in young stands and declines as the trees age. Although mortality decreases exponentially, the annual rate is rather constant, between 1 and 2%. However, when severe environmental stresses (e.g., insect attacks, disease, fire) are superimposed on the persistent stresses induced by plant competition, tree mortality often is greatly increased. In contrast to the thinning stage of stand development, death of canopy trees during the transition and steady-state phases is associated with catastrophic events rather than with deficient resources. [See INSECT INTERACTIONS WITH TREES.]

II. COMPETITION

Trees in stands compete with each other and with shrubs and herbaceous plants for light, water, and mineral nutrients. Hence, competing trees become involved in a severe struggle as shown by growth inhibition and eventual death of many trees. Competition may result in decreased growth of the smallest trees only or of the smaller but different-sized trees. [See FORESTS, COMPETITION AND SUCCESSION.]

Some idea of the stressful effects of plant competition may be gained from studies of development of even-aged stands. In young single-species plantations, for example, the trees are evenly spaced and their crowns are similar in size and shape. As these trees increase in size and their crowns close, some individuals become suppressed and occupy low positions in the canopy; others express dominance and become the largest and most vigorous trees. As the competitive capacity of the smaller trees progressively declines, their rates of shoot growth, cambial growth, and root growth are reduced. Eventually the suppressed trees are most likely to die (Fig. 3).

Foresters classify trees in even-aged stands in four crown classes—dominant, codominant, intermediate, and suppressed (overtopped) trees—depending on their degree of dominance (Fig. 3). The most dominant trees have crowns above the canopy and are larger than average trees in the stand. Codominant and intermediate trees are progressively smaller and more crowded. Suppressed trees do not receive direct light from above or from the sides and their crowns are entirely below the level of the crown cover. [See FOREST CANOPIES.]

Whereas in open-grown trees the crown covers much of the main stem, the crowns of trees in mature stands are restricted to the upper stem. After the crowns of trees close in a developing stand, the lower branches begin to die, causing a progressive decrease in the live-crown ratio (percentage of length of stem with live branches). When the live-crown ratio decreases to a critical low value, tree growth is severely depressed. Wood production declines and is redistributed along the stem. Not only is the annual ring of wood formed in suppressed trees thinner than that in dominant

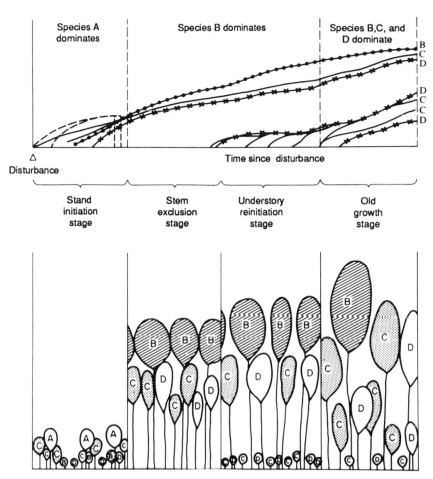

FIGURE 2 Stages of development of a forest stand after a major disturbance. As competition intensifies, the number of trees decreases and the dominant species changes. Species A and C are pioneers with relatively fast early growth. Species B and D, although present early, do not assert themselves in early competition. Eventually, however, species D becomes dominant. [From C. D., Oliver (1981) *For. Ecol. Manage.* **3,** 153–168.]

trees, but very suppressed trees may not form any wood at all in the lower stem. The reduction in stem diameter growth of suppressed trees reflects late seasonal initiation and early cessation of cambial growth, as well as a decrease in the rate of production of wood cells by the cambium. In contrast to the higher sensitivity of diameter growth to the spacing of trees in stands, height growth of canopy trees is much less sensitive to spacing.

III. SUCCESSION

Disturbed forest stands tend to regenerate to a previous conditon. A cleared forest stand, for example, is colonized by the remaining trees in the vicin-

ity. In addition to such *secondary succession* in a forest stand disrupted by major disturbances, *primary succession* occurs in areas of severe erosion, volcanic ash, or rocks. [*See* FOREST STAND REGENERATION, NATURAL AND ARTIFICIAL.]

Succession in forest stands progresses toward a community of tree species that possess high capacity to tolerate the stresses of competition. Hence, shading, drought, and mineral deficiency associated with competition tend to alter the balance of species by eliminating those most susceptible to deficiencies of resources. Success of a competing species also may be influenced by sudden imposition of a catastrophic stress. For example, a pollution episode may increase growth of certain species if they are given a competitive advantage by a rela-

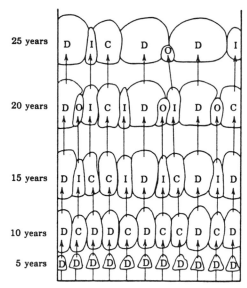

FIGURE 3 Differentiation of trees of a pure even-aged stand into crown classes. D, dominant; C, codominant; I, intermediate; O, overtopped (suppressed). [From D. M. Smith (1986). "The Practice of Silviculture." John Wiley & Sons, New York.]

tively greater inhibitory effect of the pollutants on other species in the stand.

A good example of forest succession is the replacement of pure pine stands by broad-leaved species of trees in the Piedmont Plateau of the southeastern United States. There *Andropogon* grass is dominant on abandoned fields but is succeeded by *Pinus taeda* or *P. echinata* by the fifth year. Within 10 to 15 years, closed stands of these pines have developed. However, pine seedlings do not become established under closed pine stands, whereas several species of broad-leaved trees do. Hence, the understory of a young pine stand is composed of broad-leaved trees such as oaks and hickories and the overstory pines are naturally thinned out as they become overmature in 70 to 80 years.

During uninterrupted succession, a climax forest may eventually evolve with a relatively stable species composition. However, mature forest stands are never completely stable but are maintained in an oscillating steady state by continuous elimination of suppressed and old trees and addition of new ones. The type, frequency, and severity of disturbance determine the composition and successional stage of a forest. Although catastrophic disturbances (e.g., crown fires, hurricanes, tornadoes,

snow avalanches, landslides, soil erosion and deposition, and harvesting of trees) lead to reversion to pioneer stages of succession, mild disturbances (e.g., surface fires, windthrow of some canopy trees, insect and disease attacks that kill only canopy trees, partial cuttings, and lightning strikes) tend to keep a mature forest in a relatively steady state.

A. Colonization of Gaps

Formation and filling of gaps created by disturbances play an important role in succession and maintenance of species diversity in mixed forest stands. Small gaps form frequently; large gaps form less commonly.

I. Temperate Forests

The species that colonize gaps in temperate forests already were present at the time of disturbance as suppressed or overstory trees (of sprouting and suckering species), as buried seeds in the soil, or as seeds newly dispersed into the gap. Small gaps usually close by extension of branches of adjacent trees; larger gaps close by height growth of saplings and sprouts within the gap. Shade-intolerant species (e.g., *Liriodendron tulipifera*) usually grow rapidly in large gaps but not in small ones. Generally they are confined to gaps large enough so they are not closed by growth of branches of adjacent trees or by previously suppressed seedlings.

Large gaps often are occupied by shade-intolerant, early-successional species and later replaced by shade-tolerant species, or species of both groups may arrive at the same time, with the faster-growing, short-lived pioneer species eventually being eliminated. Rates of height growth at higher light intensities characteristically are high for early-successional species and low for late-successional species.

2. Tropical Forests

Creation and filling of gaps are also important in regeneration of tropical forests. Small gaps are created by falling tree branches, and larger gaps by falling trees or groups of trees because of old age, disease, wind, fire, earthquakes, or lightning. Very small gaps are filled by growth of branches of adjacent trees; somewhat larger gaps are filled by accelerated growth of seedlings that were present before the gap formed, and by resprouting. Still larger gaps are filled by plants emerging from seeds present in the soil; very large gaps are colonized by plants growing from seeds brought in by animals and wind after a gap formed.

3. Seed Banks

The importance of seed banks for regeneration of gaps varies with forest type and gap size. In general, deciduous forests support richer and denser seed banks than do conifer stands. Buried seeds in temperate coniferous forests often lack or rapidly lose viability. Early-successional species are much better represented in buried seed reserves than are late-successional species. Regeneration of coniferous forests is commonly linked to disturbances by fire. Regrowth results from interactions among seed banks, serotiny, sprouting, seed dispersal, and climatic conditions. [See SEED BANKS.]

Species represented in the seed banks of temperate deciduous forests differ appreciably from those in the standing vegetation (except in large, recently created gaps or abandoned fields). Both the density and richness of seed banks typically decrease during succession. Hence, seed banks consist largely of shade-intolerant or early-successional species. In the North Carolina Piedmont Plateau, seed banks usually contain most seeds in first-year-old fields, with seed density declining by more than 90% in seed banks of old-growth *Quercus–Carya* stands. The decrease with stand age of species richness and diversity in seed banks has been attributed to changes in canopy species, input of seeds from within and outside the stand, and decay of the seeds that were present in initial banks. Seeds of herbaceous plants usually are absent in seed banks of forests.

Only a few late-successional species depend on seed banks for regeneration. These species also reproduce vegetatively, by growth of their shade-tolerant seedlings and rapid seedling turnover. Species that are poorly represented in seed banks generally exhibit one or more of the following attributes: dispersal by wind, wide germination tolerance, shade tolerance of seedlings, high capacity of roots to penetrate leaf litter, slow growth rate, tolerance of herbivores, and a long vegetative stage.

In lowland tropical forests, the number of viable seeds in banks decreases rapidly after a gap forms because of seed germination and mortality. The importance of seed banks changes as gap size increases. Small gaps regenerate primarily by canopy ingrowth, to a lesser extent by sprouting, and to a small degree by seedling bank and advance regeneration. In progressively larger gaps, the importance of canopy ingrowth decreases rapidly while sprouting and seedling bank and advance regeneration increase. In very large gaps, regeneration by combined seed banks and seed rain predominates. The importance of seed banks for regeneration of tropical forests is modified by the type of disturbance. For example, fire often kills seedlings, sprouts, and seeds in the soil.

IV. ENVIRONMENTAL CONTROL OF GROWTH AND DEVELOPMENT

Forests are subjected to many continuous and periodic environmental stresses that alter the growth of their component species. Important abiotic stresses include shading, drought, flooding, temperature extremes, wind, low soil fertility, and fire. Major biotic stresses include attacks by insects and pathogens, herbivory, and various activities of humans, including production and release of pollutants. Trees in stands are stressed much more than isolated trees because plant competition reduces the availability of light, water, and mineral nutrients to individual trees.

A. Light

The amount of light available to understory trees is low. For example, the light transmitted through a forest canopy commonly is only 10 to 15% of full sunlight in pine stands and less than 1% in some tropical forests. In individual trees the amount of light decreases with increasing depth of the crown. The crowns of some species are so dense that essentially no light reaches their interior.

B. Water

The distribution of forests, their species composition, and productivity are controlled to a large extent by water supply. In the equatorial zone, where the temperature changes only slightly during the year, vegetation types vary from semideserts to rain forests. Where there is more precipitation, the number of woody species increases until deciduous forests predominate. Evergreen forests occur where there is even more rain and a shorter dry season. The wettest portions of the equatorial zone, without a distinct dry season, support lush rain forests. Water supply also affects biomass production. For example, annual net primary productivity varies from as much as 3000 g m^{-2} in regions with abundant and seasonally well distributed rainfall to 250 to 1000 g m^{-2} in semiarid regions, but only 25 to 400 g m^{-2} in arid regions.

Water deficits retard growth of forest stands during all stages of development. Seed germination differs with seedbed characteristics that influence the availability of water. Mineral soils usually are good seedbeds because of their high infiltration capacity, adequate aeration, and close contact between the soil particles and imbibing seeds. Litter is a less suitable medium because it prevents seeds from contacting mineral soil, dries rapidly, impedes penetration by roots, and shades small seedlings. Nondormant seeds must imbibe water in amounts approximating two to three times their dry weight to initiate the metabolic processes essential for germination. Once seeds germinate, the young seedlings require a continuous supply of water, with the needed amount increasing progressively as leaves develop and transpiration increases. Most viable seeds (those with impermeable seed coats are an exception) can absorb enough water for germination from soil at field capacity. Germination capacity decreases rapidly as the soil dries but varies among species.

Following seed germination competition for water becomes progressively severe as emphasized by greater availability of water to the trees remaining after some of the competing trees are removed. Competition for water also occurs between trees and shrubs and between trees and herbaceous plants. Control of water-consuming herbaceous weeds in young forest plantations with herbicides may increase biomass of trees by several hundred percent.

C. Mineral Nutrients

Competition for mineral nutrients is common in both seedlings and older trees. Severe growth reduction of trees often has been attributed to their inability to compete successfully with other woody and herbaceous plants for mineral nutrients. Existence of mineral deficiency in competing forest trees is emphasized by acceleration of tree growth following application of fertilizers to forest stands. However, the amount of growth stimulation by fertilization varies with species and genotype, site quality, age of stand, amount and type of fertilizer applied, and the time of fertilizer application.

D. Complexity of Environmental Control of Growth

Variations in competitive capacity enable one plant, genotype, or species to exclude another from an ecosystem. The mechanisms of exclusion are complex and may involve attributes that (1) deny resources (e.g., light, water, mineral nutrients) to neighboring plants, (2) affect the capacity of one plant to tolerate being denied resources by adjacent plants, and (3) influence a plant's capacity to maximize fecundity when denied resources.

Success or failure of individual trees in a forest stand reflects an integrated response to the influences of different continual and periodic environmental stresses. For several reasons it is difficult to quantify the importance of individual environmental stresses in stand development. Environmental stresses often occur randomly and their impacts are modified by interactions with the others. An example is the interaction between water supply and soil fertility. When the soil dries, ions become less mobile as air replaces water in the pores between soil particles, making the path from the soil to the root surface less direct. Other examples of interaction include those between light and soil moisture, soil temperature and soil moisture, and pollution and soil moisture (pollutants are less harmful during a severe drought because stomata are closed, hence uptake of gaseous pollutants by leaves is reduced). The relative importance of individual environmental stresses on growth and development of trees varies with time and with their severity. For example, the importance of temperature in influencing growth of *Quercus ellipsoidalis* trees in Wisconsin decreases in late summer and the effect of water supply increases progressively as soil dries. A suddenly imposed severe stress such as fire, disease, insect attack, or a pollution episode may dominate over other stresses that previously influenced growth. The effect of one environmental stress on tree growth is influenced not only by prevailing environmental regimes, but also by preceding and following stresses. In the morphogenesis of a plant, early ecological influences often persist and may be expressed in later stages of development. Sensitive periods for preconditioning appear at the time of initiation and formation of buds and seeds and time of growth initiation after a dormant or resting stage.

Another problem in assessing the role of individual stresses on trees is that growth responses to an environmental stress, or to its alleviation, may not be obvious for a long time. For example, in many species of trees a favorable environment during the year of bud formation results in formation of large buds that produce long shoots in the following year. Long lag responses to alleviation of environmental stresses often are shown by negligible increases in diameter growth of residual trees for the first few years after a stand is thinned.

V. PHYSIOLOGICAL CONTROL OF GROWTH AND DEVELOPMENT

The effects of environmental stresses on tree growth and development are mediated through changes in physiological processes, predominantly involving carbohydrate, water, mineral, and hormone relations. The persistent environmental stresses associated with competition (e.g., shading and deficiencies of water and mineral nutrients),

together with additional randomly imposed stresses, set in motion a series of physiological dysfunctions that inhibit tree growth.

There is close physiological interdependency between the growth of roots and shoots. Exposure of tree crowns to an environmental stress adversely affects photosynthesis and hormone synthesis, followed by a decrease in the amounts of carbohydrates and hormones transported to the root system, thereby decreasing root growth. Such inhibition of root growth decreases absorption of water and mineral nutrients and their transport to the crown, further decreasing production and expansion of leaves. In addition, stresses exerted directly on the root system (e.g., insect attacks, root diseases) that lower the capacity of roots for absorption of water and mineral nutrients eventually will inhibit growth of leaves, which subsequently will lead to further reduced root growth, and so on. Hence, both the aboveground and belowground stresses imposed on trees set in motion sequential physiological disturbances and feedback responses that inhibit growth of entire trees.

Both seeds and seedlings of early- and late-successional species exhibit differences in physiological characteristics that account for their establishment and subsequent success or failure as competition intensifies in developing stands. Seedlings of early-successional plants are competitively inferior to those of late-successional plants. Typically the early-successional plants use available resources and grow rapidly. For example, sun-adapted species such as *Rhus glabra* and *Fraxinus americana* grow faster than the shade-adapted *Acer saccharum*.

A. Carbohydrate Relations

More than two-thirds of the dry weight of trees consists of transformed sugars. Hence growth and survival of trees depend fundamentally on synthesis of carbohydrates, their transport to sites of meristematic activity, and assimilation into new tissues. Carbohydrates are the chief constituents of cell walls. Large amounts are oxidized in maintenance and growth respiration. Maintenance respiration is associated with maintaining existing tissues, and growth respiration with synthesis of new tissues. The energy produced by maintenance respiration is variously used in (1) resynthesis of substances that are renewed in metabolic processes (e.g., enzymatic proteins, ribonucleic acids, and membrane lipids), (2) maintenance of required gradients of ions and metabolites, and (3) processes involved in physiological adaptations to changing or stressful environments.

The dry weight of trees represents only a small amount of the photosynthates they produce. This discrepancy results not only because of consumption of plant tissues by fungi and herbivores and by shedding of plant tissues, but also because of depletion of carbohydrates by respiration, which often reduces the daily production of photosynthate by 30 to 60%, and carbohydrate losses by leaching, exudation, and transport to other plants through root grafts and mycorrhizae.

Both currently produced photosynthetic products and stored carbohydrates are used in metabolism and growth of trees. The carbohydrates that accumulate in buds, leaves, branches, stems, and roots play an important role in growth and in cold hardiness, defense, and postponement or prevention of disease of trees. Accumulation of carbohydrate reserves is very sensitive to late-season environmental stresses, and failure to accumulate abundant carbohydrates adversely affects growth of trees in the following year. For example, in addition to their role as a respiratory substrate, stored carbohydrates are mobilized and used in early-season growth of both conifers and broadleaved trees. The dependence of metabolism and growth of trees on reserve carbohydrates is shown by rapid depletion of stored carbohydrates from twigs as buds open and expand into shoots. It has been estimated that stored reserves supply half to two-thirds of the carbohydrates used in growth of shoots and leaves of *Malus* trees very early in the growing season.

Trees with very low carbohydrate reserves are those most likely to die. Very high mortality rates of seedlings in the cotyledon stage of development have been linked to their low amounts of stored carbohydrates. Mortality of planted seedlings also has been attributed to their low carbohydrate reserves. Furthermore, mortality of planted pine seedlings also has been attributed to depletion of carbohydrates during cold storage of seedlings over winter.

As trees become increasingly shaded during stand development, their production of carbohydrates declines as a result of both stomatal closure and inhibition of the photosynthetic process. Stomata may close directly in response to shade or through photosynthetic depletion of CO_2 in the mesophyll, or both may occur at the same time. The effects of shading on photosynthesis are modified by other environmental factors. For example, photoinhibition (reduced quantum efficiency and lowered photosynthetic capacity in plants exposed to very high light intensity) is accentuated by drought. Although both shade-intolerant and shade-tolerant species show photoinhibition, shade-tolerant species generally are affected most by bright light. Thus photosynthesis declines rapidly in understory trees that are suddenly exposed to high light intensity by thinning of forest stands or by gap formation.

Species such as *Tsuga canadensis, Fagus sylvatica,* and *Acer saccharum* are tolerant of shade and thrive in an understory; others like *Populus tremuloides, Liriodendron tulipifera,* and many pines are intolerant of shade and thrive only in the sun. Growth of trees in the shade depends on high capacity to trap the available light, convert it to chemical energy, and allocate a high proportion of the carbohydrate pool into leaf growth. The greater adaptations of shade-tolerant species to low light intensities are associated with different effects of shade on leaf structure and on photosynthetic capacity. A lower dark respiration rate, reduced light compensation point for photosynthesis, and higher quantum efficiency are involved in maintaining the carbon balance of shade-tolerant species even at low light intensities.

To a considerable extent the replacement of shade-intolerant species by shade-tolerant species during forest succession is related to the more efficient photosynthesis of the latter group at low light intensities. Many shade-tolerant species show maximum photosynthetic capacity at light intensities as low as 5 to 10% of full sunlight. In contrast, shade-intolerant species characteristically show maximum photosynthesis at much higher light intensities. For example, the rate of light-saturated photosynthesis of shade-intolerant tropical trees (e.g., *Terminalia superba, Triplochiton scleroxylon*) is approximately twice as high as that of shaded seedlings. In contrast, seedlings of the shade-tolerant *Khaya senegalensis* show little change in the rate of photosynthesis whether grown at high or low light intensities.

B. Water Relations

Development of forest stands depends on adequate hydration of leaf tissue. Water is essential as a constituent of physiologically active cells, as a reagent in photosynthesis and hydrolytic processes, and as a solvent in which solutes move from cell to cell. Water deficits reduce tree growth directly by decreasing cell turgor and indirectly by adversely influencing photosynthesis, mineral nutrition, enzymatic activity, and hormone relations.

Soil water deficits lead to leaf dehydration, which reduces photosynthesis by closing stomata, decreasing efficiency of the photosynthesis process, inhibiting leaf formation and expansion, and inducing leaf abscission. Stomata close as the water content of leaf subsidiary cells is reduced. This creates a water potential gradient between the subsidiary cells and guard cells, causing water to move out of the guard cells. As the guard cells lose turgidity, the stomatal pores close. Stomata generally close during early stages of a drought and remain closed as a drought continues. When droughted plants are irrigated, many stomata may not reopen, at least for a long time, even though leaf turgor is increased.

Some of the decrease in photosynthesis during a drought results from stomatal closure (hence reduced diffusion of CO_2 to chloroplasts in mesophyll cells), and some to adverse effects on the photosynthetic mechanism. Close correlations often are found between stomatal aperture and rates of photosynthesis. Midday reductions in photosynthesis have been attributed to temporary stomatal closure. Some investigators concluded that stomatal inhibition of photosynthesis of droughted plants

is not caused entirely by leaf dehydration. Stomata sometimes close before leaf turgor changes appreciably. Such closure has been attributed to root sensing of soil water deficits and transport of a hormonal signal (presumably in the form of abscisic acid) from the roots to the leaves, leading to stomatal closure.

During a prolonged drought, nonstomatal inhibition of photosynthesis has been attributed to decreases in carboxylating enzymes, electron transfer capacity, and chlorophyll content. However, even in the short term, reduction in the rate of photosynthesis may involve effects on the photosynthetic apparatus as well as on stomatal closure.

C. Mineral Relations

Differences of mineral nutrient contents as well as nutrient imbalances adversely affect growth of forest stands. The importance of mineral nutrients for growth of trees is emphasized by their roles as constituents of plant tissues, coenzymes, osmotic regulators, catalysts in biochemical processes, components of buffer systems, and regulators of membrane permeability.

As a prelude to inhibiting tree growth, mineral deficiencies decrease the rate of photosynthesis, with nitrogen deficiency lowering the rate of photosynthesis more than deficiencies of other nutrients. Both chlorosis and necrosis of leaves may accompany the lowered photosynthetic capacity of mineral-deficient trees. However, the rate of photosynthesis often is lowered even when visible symptoms of deficiency are not present. The rate of photosynthesis in mineral-deficient leaves may be lowered by stomatal closure (hence lowered CO_2 uptake), inhibition of chlorophyll synthesis, decreased capacity for photosynthetic electron transport, lowered activity of photosynthetic enzymes, and increased respiration. In the longer term, total photosynthesis is reduced by a slowing of leaf production and leaf expansion.

D. Hormone Relations

Environmental stresses influence tree growth by gradually altering the amounts and balances of hormonal growth regulators, reflecting differences in hormone synthesis, destruction, and transport. Endogenous hormones, including auxins, gibberellins, cytokinins, abscisic acid, and ethylene, regulate bud dormancy, shoot growth, cambial growth, root growth, and reproductive growth at concentrations much lower than those at which carbohydrates and mineral nutrients affect growth.

Plant hormones influence growth by regulating many processes, including cell division and elongation, membrane permeability, activity of enzymes, and senescence of tissues. Hormonal effects may be achieved by (1) a balance or ratio between different hormones, (2) opposing effects of hormones, (3) alteration of the effective concentration of one hormone by another, and (4) sequential actions of different hormones. Biological effects of hormones commonly are initiated through "second messengers," which may or may not be hormones. For example, exposing plants to a mechanical stress stimulates synthesis of ethylene, which then may become a second messenger.

A common response of plants to environmental stress is a profound change in plant hormone balance. For example, many plants that produce very little ethylene in favorable environments synthesize very large amounts when exposed to drought, flooding, chilling, air pollution, or attack by microorganisms. In fact the amount of stress-produced ethylene often has been used to indicate the degree of stress to which the plants are exposed. Environmentally stressed plants can synthesize more abscisic acid and often less cytokinins than are produced by unstressed plants. A prevailing view is that hormones are the triggers that directly elicit growth reduction in response to environmental stresses. Low availability of resources activates this stress–response system.

E. Allelopathy

In addition to competing for light, water, and mineral nutrients, some herbaceous and woody plants may adversely influence seed germination and growth of neighboring plants by releasing toxic chemicals. Such compounds, called allelochems, include a wide variety of substances that are released by volatilization, leaching, exudation, and decay of plant tissues. A good example of allelopathy is the release of the toxic chemical *juglone* by plants of the genus *Juglans*.

Many laboratory experiments have shown that extracts of a wide variety of plant tissues inhibit seed germination and growth of tree seedlings. It may be difficult to quantify the ecological significance of allelopathy on the basis of such experiments because the concentrations of toxic compounds in the plant extracts often were higher than those to which plants in the field are exposed. Accumulation of allelochems in forests is modified by soil moisture and soil type. Furthermore, allelochems often are destroyed by soil microflora.

Glossary

Allelochems Substances produced by plants that are harmful to nearby seeds or plants.

Allelopathy Production of allelochems by plants.

Cotyledon The first leaf; part of the embryo in the seed.

Field capacity Water content of a soil after downward drainage by gravity has become very slow and the soil water content has become relatively stable. The field capacity varies with soil texture, and is low in sandy soils and high in clay soils.

Genotype Genetic constitution of a plant, which may be latent or expressed (as contrasted with a plant's phenotype).

Guard cells Pairs of epidermal cells surrounding a stomatal pore; opening and closing of stomata is caused by changes in turgidity of the guard cells.

Light compensation point Light intensity at which photosynthetic uptake of carbon dioxide and its release in respiration are equal. At the light compensation point there is no net gas exchange between the leaf and the atmosphere.

Mesophyll Photosynthetically active ground tissue of a leaf, located between the upper and lower epidermis.

Net primary productivity Sum of (1) increase in biomass, including leaves, stems, roots, and reproductive structures, (2) litter production, and (3) the amount of biomass consumed by animals and microbial decomposers.

Phenotype Physical appearance resulting from interaction between the genotype of a plant and its environment.

Photoinhibition Reduction in the rate of photosynthesis as a result of exposure of plants to a very high light intensity.

Photosynthesis Production of carbohydrates by green plants from carbon dioxide and water by using the energy of light.

Serotiny Late opening of cones. In *Pinus contorta* for example, the cones containing viable seeds may remain unopened on the tree for years. When the resinous material on the cones is destroyed by fire, the cone scales open and the seeds are released.

Subsidiary cells Epidermal cells that are associated with guard cells around stomatal pores.

Suckering Reproduction of some plants by sprouts produced from the roots of some species such as poplars.

Turgor (turgidity) Degree of cell distension resulting from a hydrostatic pressure (turgor pressure) exerted against the cell wall.

Water potential Physical term referring to the free energy status of water in the plant or soil as compared to that of free water.

Bibliography

Kimmins, J. P. (1987). "Forest Ecology." New York: Macmillan.

Kozlowski, T. T. (1992). Carbohydrate sources and sinks in woody plants. *Bot. Rev.* **58,** 107–122.

Kozlowski, T. T., Kramer, P. J., and Pallardy, S. G. (1991). "The Physiological Ecology of Woody Plants." San Diego: Academic Press.

Leck, M. A., Parker, T., and Simpson, R. L., eds. (1989). "Ecology of Soil Seed Banks." San Diego: Academic Press.

Mooney, H. A., Winner, W. E., and Pell, E. J., eds. (1991). "Response of Plants to Multiple Stresses." San Diego: Academic Press.

Peet, R. K., and Christensen, N. L. (1987). Competition and tree death. *BioScience* **37,** 586–594.

Pickett, S. T. A., and White, P. S., eds. (1985). "The Ecology of Natural Disturbance and Patch Dynamics." San Diego: Academic Press.

Rice, E. L., (1984). "Allelopathy," 2nd ed. Orlando, Fla.: Academic Press.

Smith, D. M. (1986). "The Practice of Silviculture." New York: John Wiley & Sons.

Waring, R. H., and Schlesinger, W. H. (1985). "Forest Ecosystems: Concepts and Management." Orlando, Fla.: Academic Press.

Plant–Animal Interactions

W. Scott Armbruster
University of Alaska

All animals depend either directly or indirectly on plants (or autotrophic bacteria) for food. Nearly all plants are attacked by plant-eating animals, and most plants depend on animals for certain essential services, including pollination, dispersal of seeds, and sometimes protection from enemies. The ecological significance of plant–animal interactions is unparalleled: plant–animal interactions form the basis of most nonmicrobial food webs in nearly all ecosystems. Coevolution between plants and animals has resulted in some of the most fascinating and elaborate relationships known in nature. The coevolution of plants and animals has also resulted in a vast array of secondary compounds of utility to humans, including drugs, insecticides, antibiotics, spices, and perfumes. Furthermore, the evolutionary results of plant–animal mutalisms have significantly enhanced our aesthetic appreciation of nature, in the form of colorful flowers and palatable fruits. The study of plant–animal interactions has revolutionized our understanding of both community ecology and evolutionary biology. Detailed understanding of relevant plant–animal interactions is of vital importance in conservation: it is critical in the planning and management of biological reserves.

I. INTRODUCTION

Most of the energy entering the earth's ecosystems comes from photosynthetic fixation of carbon by plants. Transfer of energy to other trophic levels occurs by animal consumption of plant products and microbial decomposition. Thus plant–animal interactions play a major role in most community and ecosystem processes. [*See* GLOBAL CARBON CYCLE.]

Animals generally interact with plants by consuming some portion of them. Plants may suffer from the loss, as when their leaves, stems, roots, or seeds are eaten by herbivores. Alternatively, the plant may benefit from the relationship with hungry animals, as when bees or hummingbirds consume nectar and pollinate flowers in the process. Other benefits that plants experience in exchange for providing food (or other plant products) include protection from enemies and dispersal of seeds to better sites for establishment. Sometimes plants turn the tables on animals by "tricking" them into providing a service, even though no compensation is offered, and by catching or digesting animals as a source of mineral nutrients. Most plant–animal interactions can be recognized as predominately (1) antagonistic and exploitative (one organism benefits, one loses), (2) mutualistic (both benefit), or (3) commensalistic (one organism benefits, the other experiences no net effect). Many relationships are intermediate between these extremes and exhibit a mixture of their features. Plant–animal interactions are not static; they can change over evolutionary time. For example, mutualistic relationships have probably repeatedly evolved from antagonistic relationships, and the reverse may have occurred as well. [*See* PLANT ECOPHYSIOLOGY.]

The study of plant–animal relationships has led to major breakthroughs in how we think about adaptation and evolution. Charles Darwin drew on his knowledge of floral form and pollination (especially in orchids, of which he was particularly fond) to muster evidence for his new theory of evolution by means of natural selection. The term "coevolution" was coined in the 1960s and applied to the relationships between butterflies and their host-plants; the concept, however, dates back to Darwin's study of plants and their pollinators. Today, evolutionary biologists frequently use the interactions between plants and animals as study systems for understanding how organisms adapt to their biotic environments.

As noted by Darwin, the relationships between plants and animals provide some of the most dramatic examples of natural selection and adaptation. The close match between the length of orchid nectar spurs and the pollinating moths' proboscides (tongues), the amazing diversity and potency of chemical defenses of plants, and the remarkable, obligate symbiosis between ants and acacias are examples of the dramatic evolutionary consequences of plants and animals interacting.

Three approaches to studying plant–animal interactions have been used. The oldest method is field observation of the natural history and behavior of animals and plants. Field biologists equipped with only binoculars and notebooks have made valuable contributions to our understanding of how plants and animals interact. The second approach is comparative analysis; it is an extension of the observational method and generally depends on field or laboratory observations for raw data. The observational data are analyzed to detect patterns and to infer processes. For example, if two closely related plant species have different-colored flowers and are pollinated by different animal species, it is tempting to attribute the difference in pollinators to differences in flower color and pollinator color preference. Because such patterns could be generated by chance, however, it is necessary to analyze such associations statistically. Statistical analyses range from simple correlation analysis to multilevel analyses of variance and covariance. Most recently, comparative analyses have been conducted in the framework of the phylogenies of the interacting plants and animals. This last method promises to yield new clues to the early history of plant–animal relationships. The third approach is experimental. Some aspect of the interaction between plants and animals is manipulated, and the consequences are observed in the context of predictions derived from an initial hypothesis. The experimental approach has several advantages over comparative approaches, for example, controlling variation and demonstrating likely causality, but has disadvantages in that the results reflect only short-term responses, and the conditions and manipulations may be, by necessity, unrealistic.

II. CHEMICAL MEDIATION OF PLANT–ANIMAL RELATIONSHIPS

Major breakthroughs in our understanding of plant–animal interactions have come about as the result of collaboration between ecologists and organic chemists. Research in chemical ecology, as the field is now known, has shown that most plant–animal interactions are mediated at least partially by organic compounds. These compounds range from sugars and starch consumed by animals for the calories they contain, to toxic secondary compounds that make plant parts unpalatable to herbivores, to other secondary compounds, such as monoterpenes and aromatic fragrances, that are collected by male bees to enhance their sexual success. Clearly plants and animals engage in both chemical warfare and chemical friendships.

Chemical ecologists recognize several basic groups of compounds that affect plant–animal interactions. Primary metabolites are generally taken by animals as food. Sugars, starches, lipids, proteins, and amino acids fall into this category. The compounds may be taken exploitatively by animals (e.g., folivory and seed predation) or mutualistically (e.g., pollination and frugivory/seed dispersal). Other compounds, called secondary metabolites, may play diverse roles. The monoterpene cineole, for example, deters hare feeding in balsam poplars in Alaska, while the same compound attracts and rewards the pollinators of many orchid species in the tropics. Other secondary compounds that serve to attract animals include flavonoid pigments in flowers and fruits and triterpene resins that reward pollinating bees. Secondary compounds that deter attack by herbivores include toxic alkaloids, cyanides, terpenoids, mustard oils, saponins, and tannins. An unusual defense compound is a nonprotein amino acid that mimics a normal amino acid, fooling the consumer's transfer-RNA and leading to the synthesis of inactive proteins and the death of the animal eating the plant parts containing the compound.

The distinction between chemical deterrents and chemical attractants is not always clear. For example, compounds such as mustard oils, which repel some insect herbivores, also attract and stimulate feeding by other herbivores. It is puzzling that plants would produce secondary compounds that are attractive to their enemies. The answer to this puzzle appears to be that the compound originated as a defense against all herbivores, but some specialist herbivores evolved detoxification systems for the toxin. They later evolved to use the special toxin (which was now harmless to them) as a system for recognizing the host-plants to which they were now specialized.

III. PLANT–HERBIVORE RELATIONSHIPS

Among the most pervasive and best-studied interaction between plants and animals is the interaction between plants and herbivores. It is here where we see chemical warfare fully developed by plants, with detoxification countermeasures having evolved by herbivores. Insects are by far the most abundant and diverse of herbivores, although in certain biomes (e.g., tundra, taiga, and tropical savanna) mammals also play very important roles. By convention, the term herbivore refers to animals that exploit plants by feeding on them, in contrast to pollinators, for examples, which also usually eat plant parts or products, but are mutualists. Herbivores include plant exploiters that consume leaves (folivores, including grazers and some browsers), stems (borers and some browsers), roots, flowers (florivores), and seeds (seed predators). Technically, most herbivores, except seed predators, are parasites, because they do not kill their host plants; in contrast, seed eaters usually kill their host-plants (recall that a seed is a whole plant) and are therefore predators.

One of the earliest organizing concepts in this discipline sprang from the observation that host-plant use by butterfly larvae was not randomly distributed among butterfly taxa, but rather organized phylogenetically. For example, whites (Pieridae: Pierinae) as larvae generally feed on members of the closely related mustard and caper families (Brassicaceae and Capparidaceae), whereas members of the monarch subfamily (Nymphalidae: Danainae) as larvae feed mostly on members of the closely related milkweed and periwinkle families (Asclepiadaceae and Apocynaceae), and members of the heliconiine subfamily (Nymphalidae: Heliconiinae) feed almost exclusively on passionflower vines (Passifloraceae). At the level of butterfly genera, the association with plant groups is even tighter. The unifying concept that emerged from these observations was that plants have evolved special chemical defenses (e.g., toxins) in response to herbivore pressure, closely related plants employ the same defensive chemicals, and certain herbivores have evolved specific detoxification systems or other kinds of tolerances that allow them to feed and specialize on plant with those toxins. This relationship has led to specialization on host-plant groups becoming the rule not just in butterflies, but in many groups of insects: specialist feeders are usually more common than generalists.

In contrast, mammalian herbivores are usually more generalized in their feeding. The reason for this is not entirely clear, but it is partly because mammals are much larger than insects and often their host-plants. Mammals usually feed on many plant individuals over the course of their lives; many insects, in contrast, spend their entire lives on a single plant. An example of generalist feeding can be seen in the snowshoe hare in Alaska. They feed preferentially on twigs of willow and birch in winter, but also feed on spruce and a variety of shrubs. In summer they feed on a wide variety of herbs. Moose, deer, and other members of the deer family are generalist browsers, feeding on a variety of woody plants. The same is true of grazers, such as equids and bovids: these animals eat a wide variety of grass and herb species. Recent studies have shown that having a highly mixed diet is actually critical to the survival of most mammalian herbivores. Mammals have generalized detoxification systems that can be overwhelmed by large amounts of any one secondary compound (as a result of eating large amounts of only one plant species). Cows, for example may become fatally ill when they consume excessive quantities of single plant species, whereas they thrive on smaller amounts of the same plant when mixed with other species. Mixing plant species in the diet prevents overloading of the detoxification system. The detoxification systems of most insects, in contrast, are usually adapted to dealing with large quantities of specific compounds and are usually not overwhelmed by the amounts normally ingested.

The literature on how plants defend themselves against animal attack has grown rapidly over the past two decades. It is now known that plants employ a large variety of techniques to defend their leaves stems, roots, flowers, and seeds. Among the most obvious defenses employed by plants are mechanical defense systems. These include spines, thorns, and trichomes (plant hairs) that impale herbivores or impede their feeding activity. Hard seed coats may prevent entry or seed cracking by seed predators. Thick bark helps prevent herbivore access to the cambium. Sticky resins are secreted from leaf glands and protective bud scales, and are exuded from wounds in stems and trunks; they can immobilize insects or clog up their mouthparts. The toughness of leaves can discourage many animals from feeding.

Some plants appear to defend their parts by adaptive timing of their vulnerable stages, a strategy known as phenological defense. In many tree species, for example, most of the population produces few or no seeds in most years and large numbers only in certain years, a pattern called mast fruiting. Mast fruiting appears to result in seed predators being overwhelmed by the abundance of a normally scarce resource and unable to eat all the seeds produced.

As discussed earlier, plants also defend themselves from attack by herbivores by synthesizing and storing special chemical compounds. Apparently not all defense compounds act in the same way. Although many defense compounds are toxic, actually killing herbivores or making them ill, others, such as some tannins, may act by reducing the animals' ability to digest plant tissues and obtain nutrients (especially proteins) from them.

An interesting twist in the chemical ecology of plant–herbivore relationships is that some herbivores deal with toxic defense compounds by sequestering them in parts of their bodies where they do no harm. However, these compounds are still active and provide effective defense for the herbivores themselves. Borrowed toxins (usually sequestered by immatures, but sometimes by adult insects) are a common chemical defense system found in insects. Insects with borrowed toxins are often brightly colored, or aposematic (= warning coloration), and themselves may be mimicked by other highly edible species (Batesian mimicry). The classic mimicry system of the monarch (model) and the viceroy butterfly (mimic) may be an example of this phenomenon; the monarch obtains its defense compounds (cardenolides) from its milkweed host-plant. Recent evidence, however, suggests that the viceroy itself synthesizes toxins and should instead be considered a Müllerian mimic (a distasteful species mimicking another distasteful species).

Recent research on the chemical defense of plants against animals has focused on why some plants are highly defended while others have virtually no chemical defense systems, and how so many kinds of plant defense systems have originated. One unifying hypothesis is that the availability of resources in a plant's environment influences the plant's investment into defenses. This may influence plant defenses over the short term, as plants respond to variation in the environment. For example, defense compounds based on carbon (an element fixed during photosynthesis) tend to increase in plants growing in sunny environments, but decrease in shaded environments (where plants are carbon stressed because of low photosynthetic rates). Over the longer term, plants that grow in resource-limited environments tend to evolve slow growth rates and are slow to replace tissue lost to herbivores. Thus the cost of losing tissue is greater than in fast-growing plants living in high-resource environments. Slow-growing plants have therefore generally evolved higher levels of investment into defense.

Another area of active research concerns the immediate responses of plants to herbivore attack. It has been observed that some plants respond to herbivore attack by producing more defense compounds and/or mobilizing them to the site of attack (short-term induction). Some species respond by producing regrowth that is more highly defended than the original attacked tissue. There is some evidence that the latter process may drive the hare population cycles of the boreal zone, and perhaps the population cycles of other mammalian and insect herbivores as well. These interactions may, in turn, affect rates of decomposition, nutrient cycling, and succession. Thus plant–herbivore interactions may often have major effects on numerous ecosystem processes.

IV. PLANT–POLLINATOR RELATIONSHIPS

The study of plant–pollinator interactions has been organized in two ways. One approach is to focus on the animal partner and look for similarities in the interactions between plants and each group of animals: for example, bee-pollination versus bird-pollination versus bat-pollination. A second approach, one that I find more instructive, focuses on the material or behavioral basis of the interaction. I will employ the second approach in the discussion that follows.

There are two major floral functions in which pollinators assist. One is the male floral function: to get pollen grains to the stigmas of other flowers (typically on other plants). The other is the female floral function: to receive pollen on the stigmas, allowing fertilization to occur and seeds to be produced. Usually pollinators are involved in both functions simultaneously. Flying animals generally make the best pollinators because they can move rapidly from plant to plant; not all pollinators are volant, however.

Attraction of pollinators to flowers, and their subsequent employment as pollinators, has two components. The first task of a flower is to attract the attention of the pollinator and give it recognition cues by which it can remember that the flower is rewarding to visit. This task is met by the advertisement system. Advertisements include bright colors, distinctive petal shapes, and characteristic fragrances. Often a flower will express all three of these common advertisements. The ultimate reason why most pollinators visit flowers, however, is to obtain a reward. This is usually, but not always, some edible substance.

By far the commonest reward is nectar. Although it is widely known that nectar contains sugar and is sweet (it is what honeybees use to make honey), recent research has shown that the chemistry of nectar is quite complex. In addition to sugars, many nectars contain a variety of essential amino acids, and some pollinators depend on nectar as an amino acid source. The sugar constituents of nectar vary from plant species to species. This variability is at least partly related to the apparent food preferences of the primary pollinators. For example, flowers pollinated by long-tongued bees generally contain sucrose-dominated nectar. The same is true of flowers pollinated by moths, butterflies, or hummingbirds. In contrast, flowers pollinated by short-tongued bees, flies, perching birds, or bats generally contain hexose-dominated nectar (glucose and fructose are the main constituents). Other nectar-seeking animals that are sometimes pollinators include several South American and Australian marsupials (e.g., opossums), mice, lemurs, and monkeys. Beetles, wasps, and ants also consume nectar and may serve as pollinators.

The next most common reward consumed by pollinators is pollen. Although all perfect flowers (having both male and female structures) and staminate flowers (having male structures only) have pollen, in many flowers it is partially protected in some way from being consumed. In other flowers, however, pollen is the only reward available and it is offered freely to the pollinators. Animals that visit pollen-reward flowers include female bees, some flies, beetles, and some bats. In the process of eating pollen, the pollinator gets some pollen dusted over its body and carries it to the stigmas of other flowers.

A less common nutritive reward is oil (lipid compounds). Despite its relative uncommonness, it is quite widespread, being employed by plants in the tropical and temperate regions of both Old and New Worlds. More than half a dozen plant families exhibit this syndrome, including Malpighiaceae, Krameriaceae, Cucurbitaceae, Scrophulariaceae, Orchidaceae, Liliaceae, Solanaceae, and Primulaceae. This reward system has apparently evolved independently in each family. Oil rewards are collected by female bees in the Anthophoridae, Melittidae, Ctenoplectridae, and possibly Apidae. The bees mix the collected oil with pollen and feed it to their offspring.

Another nutritive reward system is the presentation of food bodies rich in sugar, starch, lipid, or protein. Pollinators feed on the food bodies instead of the sexual flower parts, flying from flower to flower for food, pollinating them in the process. Beetles are the commonest pollinators of food-body flowers.

Several reward systems are nonnutritive, and these have been elucidated only recently. The commonest nonnutritive pollinator rewards in the neotropics are fragrances, much like those employed as advertisements. Fragrances are collected from flowers by male euglossine bees (Apidae: Euglossini) and are apparently modified biochemically to become sex pheromones to attract females for mating. Initially males euglossines probably exploited flowers employing other reward systems but using fragrance as advertisement. They may sometimes

have proved to be effective pollinators, causing selection for increased fragrance production, cessation of production of the original reward, and specialization on male euglossine pollinators. This system of reward has evolved independently in several plant families, including Orchidaceae, Araceae, Euphorbiaceae, Solanaceae, Gesnericeae, Bignoniaceae, Haemodoraceae, and Palmae.

Another nonnutritive reward system that is fairly well understood entails production of triterpene resins by flowers or extrafloral glands. The resin is collected by female bees in the Apidae and Megachilidae families; they use the resin as a building material in nest construction. Two large genera are known to offer resin rewards: *Dalechampia* (Euphorbiaceae) and *Clusia* (Clusiaceae).

All the systems described here are based on a mutualistic relationship between plant and pollinator. Sometimes, however, the relationship is exploitative, with the plant getting pollinated without offering a reward or the animal obtaining a reward without pollinating flowers. Examples of the former include floral mimicry, in which a nonrewarding flower resembles a rewarding flower (in a general or specific fashion), and carrion flowers, which attract carrion flies which mistakenly lay their eggs on the carrionlike flowers and pollinate them (the flies' larvae starve, however).

Another example is the pseudocopulatory orchids, most of which occur in the Mediterranean region or Australia. These orchids produce flowers that resemble female wasps or bees (or, in a few cases, flies) in appearance, texture, and odor. The male insect is attracted to the flower and attempts to copulate with it. In Australian hammer orchids (*Drakaea*) and many of the Mediterranean bee orchids (*Ophrys*), the pollination system depends on the flowers being produced early in the spring when the males are unable to find the real thing (females generally emerge from hibernation later than males). The flowers are pollinated in the copulation attempt, but the male insect appears not to benefit from the relationship.

Often animals visit flowers and obtain nectar without contacting the sexual structures. Sometimes this is simply the consequence of some geometrical incompatibility (e.g., the animal is too small to touch the stamens or stigmas). In other cases the animal pierces the sepals and/or petals to obtain nectar "illegitimately," bypassing the fertile parts. Such creatures are called nectar thieves and robbers, respectively.

Relationships between plants and their pollinators are commonly asymmetrical. Plants are often dependent on only a few species of animal pollinators, whereas animals generally visit many flower species for their resources. Sometimes flowers are as promiscuous as their pollinators, being pollinated by a wide variety of animals. In other cases, the relationship between plant and pollinator is highly species specific: the flower is pollinated by only one (or a few) animal species and the animal is dependent on that plant alone for critical resources. Although such specificity is unusual, two of the widely known (but not well understood) pollination systems are of this type. These are fig trees and pollinating fig wasps and yuccas and pollinating yucca moths. In both cases the plant is pollinated by one or a few species of insect and the insect obtains virtually all necessary resources from the flowers, seeds, and fruit of the host-plants.

V. PLANTS AND SEED DISPERSERS

Plants are for the most part immobile organisms, yet in order to colonize new habitats, escape enemies, and improve the chances of their offspring's survival, they need to transport their seeds to distant localities. Plants employ wind, water, and animals to disperse their seed to appropriate sites that are some distance from the maternal parent. This section will focus on animal dispersal of seeds.

Animal seed-dispersal systems work much like pollination systems in plants. There are generally both advertisements and rewards. Advertisements include the bright colors and distinctive odors of most fruits. Rewards include fleshy or juicy fruit walls that may contain sugars, starches, lipids, and amino acids. Other rewards include arils (fleshy appendages of the seed coat), containing lipids, carbohydrates, and/or amino acids, and oil-rich elaeosomes (also seed-coat appendages).

Fruits with rewarding walls are eaten by mammals (both flying and nonflying) and birds that pass seeds through their digestive tracts and disperse them in their feces. Some birds eat fleshy-walled fruits with seeds too large to pass through the gut; the fruit and seed is ingested into the crop, where the gizzard strips the flesh from the seed. A short time later and, ideally, some distance away from the parent tree, the cleaned seed is regurgitated and dropped to the ground. This is how wild avocados and many other large-seeded members of the laurel family (Lauraceae) are dispersed. The seed dispersers of these plants in the neotropics are commonly toucans, trogons, quetzals, and other birds with large gapes.

Large, arillate seeds are dispersed in a similar fashion; wild nutmeg and its relatives (Myristicaceae) are good examples of this system. (Humans also ingest nutmeg arils; the nutritive aril of the commercial nutmeg is dried and ground to produce the spice known as mace.) Smaller arillate seeds such as those of *Clusia* (Clusiaceae) may pass through the digestive tract of birds and be deposited in the feces. Elaeosomes are seed-coat appendages (arils in the broad sense) that are generally lipid rich and usually eaten by ants. Ants collect the seeds plus elaeosomes under the parent plant, carry them to their nest, and then clip the seed off the elaeosome just prior to or after taking the food reward into their nest. Why ants carry the relatively heavy seed all the way to their nest prior to separating it from the elaeosome remains a mystery; the benefit for plants obviously accrues from this behavior.

VI. PLANTS AND ANTS

Ant–plant mutualisms are among the better studied of plant–animal interactions. The classic example of mutualistic relationships between plants and ants is the symbiosis between neotropical swollen-thorn acacias and the ant *Pseudomyrmex*. The acacia provides house and balanced diet for the ants: hollow thorns for the ants' nests, sugar-rich nectar from leaf nectaries, and fats and proteins in the form of Beltian bodies (small outgrowths of the leaflets on young leaves). All the needs of an entire colony of ants are provided by the host-plant and the ants spend virtually their whole lives on the plant. The ants in turn are aggressive protectors of the plant, biting and stinging any animal that touches the acacia, and even clearing away competing plants.

Similar species-specific mutualisms between plants and protective ants have evolved independently in the African whistling-thorn acacia, neotropical *Cecropia,* and a variety of other genera throughout the tropics. Looser associations between plants and ants occur throughout the world. Plants provide nectar from extrafloral nectaries and are visited and protected by a variety of local ants. Members of the legume family (Leguminoseae), for example, the vetches (*Vicia*) of temperate Europe and North America, very commonly have extraflora nectaries on their leaves and loose associations with protective ants.

In some areas where ants are less common or less active, they may be replaced as visitors to extrafloral nectaries by predaceous or parasitic wasps (Hymenoptera). These wasps capture or lay eggs on herbivorous insects and probably attack insects near where they feed on nectar, thereby providing a degree of protection for the plant. A shift from ants to parasitic wasps on extrafloral nectaries is seen in highland Costa Rica and Alaska, where ants are less common than in warmer regions. For example, quaking aspen (*Populus tremuloides*) secretes nectar from the bases of its leaves; these extrafloral nectaries are visited by ants in temperate North America. In Alaska this nectar is fed upon by parasitic wasps that commonly attack the aspens' major herbivore, the tortrix moth (*Choristoneura*). The attraction of large numbers of the parasitic wasps to a tree probably reduces the tree's herbivore load.

Not all ants that visit plants are mutualists. The commonest ant visitors are ants that tend aphids (Homoptera: Aphidae). Aphids are small insects that tap sugar-rich phloem sap. Because they ingest more sugar than they can use in their quest for amino acids and other rarer nutrients in phloem sap, aphids dump large amounts of sugar out their anuses. Ants are attracted by this sugar dumping and regularly tend aphids, transporting them and protecting them from predators. Because aphids

inflict serious damage to plants, their protection by ants has a negative effect on plants. A few ants tend lepidopteran larvae in a similar fashion. This has the same negative consequence to the plant.

One group of ants that directly inflicts injury to plants is the leaf-cutting ant (*Atta* and related genera; Hymenoptera: Attini). Leaf cutters are common and important herbivores in neotropical forests; they cut leaves and carry them to their nests, where they macerate the leaves and raise a fungal mutualist on the resulting medium. They then harvest and eat the fruiting bodies of their cultivated fungus. Another group of ants that injure plants directly is the harvester ant (e.g., *Pheidole* and *Pogonomyrmex;* Formicidae: Myrmicinae). These ants are seed predators, gathering seeds, storing them in their nests, and feeding on them. They are especially abundant in drier regions, such as the southwestern United States and Australia.

VII. CARNIVOROUS PLANTS

Occasionally plant exploitation of animals takes the form of carnivory. Carnivory has evolved independently in half a dozen or more different plant lineages. Nearly all carnivorous plants live in areas with low nutrient availability, such as bogs, sandy areas, or in trees as epiphytes. Carnivory has evolved to supplement the mineral nutrient budget of plants. Animals (usually insects) are captured and digested or broken down microbially so that they release the mineral nutrients contained in them. Digested animals do not provide energy to plants as they do to animal predators.

Plants have evolved numerous different mechanisms for capturing animals, although all operate as traps. The simplest are sticky glands on leaf blades that act like fly paper. This mechanism is exhibited in butterworts (*Pinguicula,* Lentibulariaceae). The trap in sundew (*Drosera,* Droseraceae) is similar but the sticky glands are at the end of tentacular processes. Somewhat more elaborate traps are pitfall traps as seen in pitcher plants. Pitcher traps have evolved independently at least three times: in the New World pitcher plants (Sarraceniaceae); *Nepenthes* (Nepenthaceae) of Southeast Asia, Australia, and Madagascar; and the Western Australian pitcher plants (*Cephalotus*, Cephalotaceae). The most elaborate trap mechanisms are exhibited by the Venus flytrap (*Dionaea*, Droseraceae) and bladderworts (*Utricularia*, Lentibulariaceae), in which a rapid movement causes closure of the trap after the animal has entered. Bladderworts are aquatic plants, and their traps differ from those of Venus flytraps in that they are much smaller and occur underwater.

VIII. ORIGINS AND EVOLUTION OF PLANT–ANIMAL INTERACTIONS

Plants and animals have interacted with one another since the beginning of the two life-forms. Higher plants in particular have a long history of interacting closely with animals. For example, the origins and great success of flowering plants (Anthophyta) in the modern landscape appear to be associated with their use of insect pollinators and possibly avian seed dispersers. The most primitive fossil flowering plants appear to have been pollinated by flies or beetles; wind pollination originated later in the evolution of flowering plants.

We are just beginning to understand how new relationships between plants and animals originate and how relationships evolve. Many relationships appear to have originated by modification of preexisting relationships. For example, insect pollination in flowering plants may have originated by modification of the relationship with flowering-feeding herbivores. Floral rewards and advertisments in some plants probably originated as chemical defenses against herbivores, and by chance assumed secondarily attractive functions. Thus the course of evolution of plant–animal relationships apparently has been influenced by a variety of coincidences.

Changes in the nature of plant–animal interactions may be influenced by both chance and natural selection. Clearly, the fine-tuning of mutualistic relationships is the result of coevolution by natural selection, as is the evolution of defense and counterdefense systems in plants and herbivores. Shifts from one pollinator to another, for example, may occur by natural selection as the abundance or effec-

tiveness of pollinators change. Chance, nonadaptive shifts may also occur, however. Simple mutations, for example, in fragrance chemistry, may sometimes result in new pollinators being attracted in place of the original pollinators. This results in a complete shift in pollinators and may lead to the "instantaneous" formation of a new species. The relative frequency of these two kinds of shifts in relationship has not yet been determined, but they both appear to be important.

IX. PLANT–ANIMAL INTERACTIONS AND CONSERVATION BIOLOGY

There is growing concern today about the loss of biotic and genetic diversity that is occurring through human habitat modification (e.g., deforestation) and species extinction. One approach to reducing extinction rates has been to establish nature reserves. Understanding the ecology of plant–animal interactions is of fundamental importance in understanding the causes of extinction and in the design and management of nature reserves. [*See* CONSERVATION PROGRAMS FOR ENDANGERED PLANT SPECIES.]

We now know something (although not nearly enough) about webs of interdependence between plants, their specialist herbivores, their exclusive pollinators, and their seed dispersers. The extinction of a single species may, in turn, cause extinctions of other species (e.g., the first species' obligate mutualists) and other serious, unanticipated disruptions of the ecosystems. Mutualisms and other interactions are themselves potentially endangered entities that deserve our conservation efforts.

Research focusing on plant–animal interactions has much to contribute to conservation and reserve management. When possible, management and conservation decisions should be based on solid understanding of how all the plants and animals in a particular community interact. In this way we can improve our abilities to preserve what remains of the earth's biotic diversity.

Glossary

Antagonism Interaction between organisms in which one or both are harmed in some way. This includes exploitative antagonisms (e.g., herbivory and predation) and nonexploitative antagonisms (e.g., allelopathy).

Anther Saclike portion of the stamen (the fertile male part of a flower) that contains the pollen grains.

Browser Animal (conventionally a mammal) that feeds on stems and/or leaves of woody plants.

Coevolution Evolution of two interacting populations of two species in which one evolves in response to selection generated by the activities of the second population, and this evolutionary change in the first population induces an evolutionary response in the second. Coevolution may be tight, where one population coevolves with one other, or diffuse, where populations of several species interact and coevolve with populations of one or several other species.

Folivory Feeding on leaves.

Frugivory Feeding on fruits.

Generalist herbivore Animal that feeds on several to many unrelated or distantly related species of plants.

Grazer Animal (conventionally a mammal) that feeds on aboveground parts of herbaceous plants, especially graminoids (grasslike plants).

Mutualism Interaction between organisms in which both benefit in some ways.

Nectar Secretion produced by glands (nectaries) in flowers or on leaves or stems. Nectar is an aqueous solution of sugar and sometimes amino acids and other trace substances.

Phylogeny Tree of relationships (by descent) among populations, species, or higher taxa, analogous to a genealogy for human families. Usually the cladistic method is used to estimate or reconstruct phylogenies.

Pollen Grainlike 2- to 3-celled male gametophyte that contains the sperm. Pollen is produced by the anther sacs at the tip of the stamens in the flower.

Pollination Process by which a pollen grain moves from the anther sac of a flower to the stigma of a flower.

Secondary compounds Several diverse groups of compounds that are not obviously involved in primary metabolism (e.g., photosynthesis, respiration) of organisms (usually plants). Secondary compounds were initially thought to be waste products, but are now thought to play roles in plant defense and other functions.

Specialist herbivore Animal that feeds on one or a few closely related species of plants.

Stigma Portion of the pistil on which pollen lands and germinates. Pollen tubes grow from the stigma down the style and convey sperm cells to the ovules, where they can unite with egg cells.

Symbiosis Tight relationship between two organisms in which all or much of their lives are spent in close physical association with one another.

Bibliography

Albert, V. A., Williams, S. E., and Chase, M. W. (1992). Carnivorous plants: Phylogeny and structural evolution. *Science* **257,** 1491–1495.

Armbruster, W. S. (1992). Phylogeny and the evolution of plant–animal interactions. *BioScience* **42,** 12–20.

Futuyma, D. J., and Slatkin, M. (1983). "Coevolution." Sunderland, Mass.: Sinauer.

Gilbert, L. E. (1980). Food web organization and the conservation of neotropical diversity. *In* "Conservation Biology. An Evolutionary–Ecological Perspective." (M. E. Soule and B. A. Wilcox, eds.), pp. 11–33. Sunderland, Mass.: Sinauer.

Gilbert, L. E., and Raven, P. H. (1980). "Coevolution of Animals and Plants," revised edition. Austin: University of Texas Press.

Janzen, D. H. (1966). Coevolution between ants and acacias in Central America. *Evolution* **20,** 249–275.

Meeuse, B., and Morris, S. (1984). "The Sex Life of Flowers." New York: Facts on File.

Proctor, M., and Yeo, P. (1972). "The Pollination of Flowers." New York: Taplinger.

Real, L. (1983). "Pollination Biology." New York: Academic Press.

Willson, M. F. (1992). The ecology of seed dispersal. *In* "Seeds. The Ecology of Regeneration in Plant Communities" (M. Fenner, ed.), pp. 61–85. London: CAB International.

Plant Conservation

R. W. Enser
Rhode Island Natural Heritage Program

I. Introduction
II. Causes of Plant Rarity
III. Inventory
IV. *In Situ* Conservation
V. *Ex Situ* Conservation
VI. Restoration
VII. Conservation Programs

According to recent projections, more than 10% of the estimated 20,000 native plant species in the United States are in danger of extinction. Although most early efforts to preserve endangered species were largely directed to vertebrate animals, increased attention is now being given to the unique values and needs of plant conservation. A primary focus of conservation has been the inventory of regional floras to identify the most imperiled taxa and to protect rare species localities, along with the development of management techniques for the maintenance or restoration of specialized habitats. To further preserve plant diversity, species that are particularly vulnerable are propagated in botanical gardens and preserved in seed banks. Regional plant conservation programs have been effective in prioritizing rare species and formulating the research needs and management schemes that are of particular importance within each region.

I. INTRODUCTION

The decline of plant diversity in the United States is exemplified by the nearly 400 plants already listed as federally threatened or endangered. In addition, approximately 2250 species, subspecies, and varieties of plant taxa are currently being reviewed for possible listing under the Endangered Species Act. Other species previously proposed for listing have now been removed from consideration. Some have been documented as more common than previously assumed, but 97 species are no longer being considered because they are believed to be extinct. [*See* CONSERVATION PROGRAMS FOR ENDANGERED PLANT SPECIES.]

The ecological values of both a diverse flora and fauna are equitable, but the support for conservation within these two biotic groups has traditionally been skewed in favor of animals. Plants do not generally excite the public as much as the more charismatic vertebrates, and science education tends to focus on large mammals or birds in endangered species curricula. However, as outlined here, new directions and ideas are emerging to address the specialized needs of plant conservation.

II. CAUSES OF PLANT RARITY

Most plants inhabit ecological communities that are defined by strict physical characteristics of bedrock and surficial geology, hydrologic regime, soil morphology and chemistry, and microclimate. Most species have also developed specific relationships with other organisms such as pollinators or symbiotic fungi. In general, the plants most in danger of becoming rare are those that exhibit the highest level of ecological sophistication.

Rarity can arise for a variety of reasons. Several species have been identified as rare because they exist as relict populations of ancient lineages. The Florida Torreya (*Torreya taxifolia*), a member of the Yew family, once inhabited large portions of eastern North America. During the last ice age, the distribution of Torreya receded and became localized in western Florida and inexplicably never repopulated its former range following retreat of the glaciers. Conversely, some plant taxa are rare because they have only recently evolved and have not had time to become more widespread. Although plants may be localized owing to evolutionary youth or senescence, in most cases infrequency of populations or individuals has been exacerbated by the detrimental impacts associated with the alteration or destruction of habitats by anthropogenic forces.

The filling and drainage of wetlands to provide agricultural land, transformation of lakeshores and coastal zones into public beaches and cottage communities, and the multiplication of residential developments on well-drained soils of pine barrens and other upland forests are only a few examples of habitat conversion. Although all biota suffer when habitats are degraded, plants are particularly vulnerable because they exhibit more localized distribution patterns. The xeric scrub communities of the sand ridge country in central Florida represent a specialized habitat supporting many rare endemic species, including Scrub Plum (*Prunus genticulata*) and Short-leaved Rosemary (*Conradina brevifolia*). It is estimated that less than 3% remains of the original extent of the scrub community, and the remnants are highly threatened by clearing for citrus groves, residential expansion, and commercial development.

Rarity may also be the result of exploitation for commercial or private gain. Parts of certain plants hold actual or perceived medicinal properties that ascribe a financial worth to their collection. The harvest of American Ginseng (*Panax quinquefolius*), the root of which has been valued at over $300 per pound, is stringently controlled through state and international regulations. Other plants, especially orchids, cacti, bromeliads, and carnivorous plants such as the Venus flytrap (*Dionaea muscipula*), are prized by collectors who do not understand the biological consequences of displacing rare plants from native populations to private gardens.

Plant populations can also suffer degradation through a variety of indirect impacts. Herbicides intended for controlling weeds may be toxic to native plants, and nonspecific insecticides aimed at pests may also reduce or eliminate beneficial pollinators. Exotic plants imported for ornamental or utilitarian use may escape and become invasive over large areas, effectively crowding out native vegetation.

Some plant populations have been reduced because of impacts from native species. A notable example in recent years involves unusually high numbers of White-tailed Deer (*Odocoileus virginianus*), which are the products of successful wildlife management campaigns. A consequence of the increase of this browsing mammal has been the reduction of some populations of herbaceous plants, especially orchids and lilies.

Ultimately, the most debilitating impact to plant populations will be global climate change. Current predictions of the rate of temperature increase suggest that many plant populations will succumb because of an inability to migrate long distances to new habitats in advance of lethal climatic modifications. [*See* GLOBAL ANTHROPOGENIC INFLUENCES.]

To provide for the perpetuation of a high plant diversity, conservationists have identified several avenues of research and management: inventory, *in situ* and *ex situ* conservation, and restoration.

III. INVENTORY

The conservation of plant diversity appears foreboding because a large number of species require attention and the financial burdens of implementing protection are substantial. These restrictions dictate that priority conservation actions be directed to those plants in the greatest risk of extinction. The determination of priority relies heavily on the inventory process.

Data on plant distributions have traditionally been compiled by botanists associated with universities, museums, or biological survey units. During the past two decades, botanical inventory has also been one responsibility of Natural Heritage Programs that exist in the 50 states, and elsewhere in several Canadian provinces and Latin American countries. Initially created as joint ventures between local governments and The Nature Conservancy, a private land conservation group, Natural Heritage Programs function as biodiversity data bases within their respective regions. Program botanists investigate the historical record available through herbarium specimens, published literature, and personal journals to provide a listing of species worthy of concern. This research is augmented by field surveys to assess the vigor of known populations of rare species and also to locate previously undocumented occurrences. Data concerning the location and history of each population are incorporated in a biological/conservation data base (BCD). This system can be assessed to identify important rare species habitats for land protection efforts, or to provide data within established protocols for determining the environmental impacts of land-altering activities.

Because all programs utilize the same BCD format, data can be synthesized to evaluate the status of a species over its entire range. These analyses allow Natural Heritage Program botanists to assign a *Global* or *G rank* for each plant taxon that defines its relative rarity on the planet. Species with G ranks of 1 (G1) are critically imperiled throughout their range and typically have fewer than 6 occurrences in the world, or fewer than 1000 individuals. G2 species have between 6 and 20 occurrences, or fewer than 3000 individuals. Approximately 2400 United States plant species are currently ranked G1 or G2. The relative number of species assigned to each Global rank is illustrated in Fig. 1.

One accomplishment that will greatly benefit the inventory process in the upcoming years is the Flora of North America Project coordinated

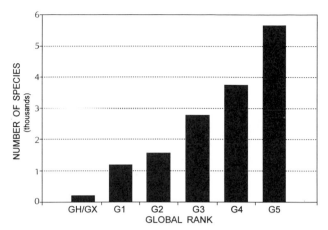

FIGURE 1 Numbers of United States plant species assigned to global ranks. Definitions of ranks are: GH/GX, of historical occurrence (H) or believed to be extinct (X); G1, critically imperiled (1–5 occurrences or fewer than 1000 individuals; G2, imperiled (6–20 occurrences or fewer than 3000 individuals); G3, vulnerable (21–100 occurrences or fewer than 10,000 individuals); G4, apparently secure, G5, demonstrably secure. [Source: The Nature Conservancy and Natural Heritage Network.]

through the Missouri Botanical Garden. This endeavor will result in the publication of a continental flora complete with identification keys, summaries of habitat and geographic range, phenological data, and other pertinent biological information. The product will be a definitive taxonomic reference and source of distributional information that is essential when determining which species require immediate attention for conservation.

Although species prioritization relies significantly on a quantitative assessment of populations, the process may be refined by the consideration of other factors. Other determinants could be the known or potential economic value of a species, its ecological value within the natural community it inhabits, its taxonomic uniqueness, and the number of its populations that are already protected.

IV. *IN SITU* CONSERVATION

The goal and primary method of conserving rare plants is the on-site or *in situ* protection and management of the habitats within which they have evolved. Comprehensive inventory work leads to the identification of vulnerable populations that may be protected through land acquisition by governmental or private conservation concerns or, in cases where acquisition may be inappropriate, landowners may agree to conservation restrictions designed to protect rare species.

In some cases, the safeguarding of populations in preserves, where threats from habitat conversion or other impacts are lessened, may be sufficient to secure the target species; however, most conservation projects must also provide for management to maintain the viability of rare species populations. *In situ* management may be applied to maintain a community at a particular successional stage, alleviate competition from exotic species, and control human interference.

A widely used management scheme has been the use of prescribed burning to restore natural fire disturbance cycles that historically sustained entire vegetation communities such as prairies and pine barrens, or stimulated the reproduction of certain plants. For example, controlled burns have been used at a Nature Conservancy preserve in Virginia to enhance germination of the endangered Peter's Mountain Mallow (*Iliamna corei*), a species requiring the high temperatures of fire for breaking of the seed coat.

Prescribed burning is commonly used to eliminate competing vegetation, but in areas where this may not be a practical approach, mechanical mowing, grazing, or cutting with hand tools may be substituted. A particularly difficult task is the control of exotic plants that can spread aggressively and usurp rare species populations. Autumn Olive (*Elaeagnus angustifolia*) has been widely introduced for ornamental use and as a wildlife food supplement. The avian-dispersed seeds of this shrub enhance its capacity to invade native plant communities, such as grassland types where fire may be used successfully. In other situations, the most effective control has been labor-intensive cutting of individual shrubs followed by an application of herbicide painted directly on the stump to prevent vegetative regrowth.

Human management can be accomplished in many ways. Recreational pursuits that cause deterioration of habitats can be ameliorated by fencing or the rerouting of hiking trails. In extreme cases, personnel may be required for policing sites to prevent the covert collection of plant material.

V. *EX SITU* CONSERVATION

It is not always possible to preserve a plant taxon simply by protecting its surviving occurrences. Some plants exist in so few populations, or at sites with so few individuals, that a single stochastic or random event has the potential of removing a significant portion of a species' genetic diversity. In circumstances where a species continues to decline despite the application of best known management, or in cases where the only known populations are imminently threatened by habitat conversion, an alternative is *ex situ* or off-site conservation.

A principal *ex situ* strategy is collection and storage of the propagules of rare plants in seed banks. In general, seed is collected from as many individuals of a population as possible without jeopardizing its reproductive potential in the wild. A sample of the collected seed is tested for germination and the remainder is dried and stored in airtight containers at $-20°C$. Periodically some of the stored seed is tested to determine its viability over time, and the resulting garden-grown plants may then provide an alternate seed source; however, it is most desirable to replenish seed bank stocks from wild populations to ensure genetic variability. A confined population consisting of a small number of individual plants may suffer declines in genetic health due to inbreeding. Higher levels of genetic variation can be maintained in seed banks by collecting a sample of seed from a large number of individuals if they still exist in wild populations. [*See* SEED BANKS.]

Seed banking is not a viable option for all species. Some have seeds that do not withstand the desiccation or freezing associated with storage; others may survive storage but their requirements for germination and successful reproduction in the greenhouse or garden have not been ascertained. In such cases, live plants are maintained in collections at botanical gardens. These living collections may eventually serve as a backup for wild plants lost to catastrophic events or as a source of genetic material for research. Also, they can be grown in public educational displays to illustrate the realm of botanical diversity. Plants received into living collections are chiefly those grown from seed or propagated from cuttings obtained from wild populations, but they could also be transplants rescued from sites that are imminently threatened with destruction.

VI. RESTORATION

A potential use for plants produced by *ex situ* methodologies is in restoring native populations that have been degraded or lost. This process can occur in several ways. *Augmentation* takes plants propagated from an existing wild population and transplants them back to strengthen the same population. The New England Wildflower Society has successfully grown plants of the endangered Robbin's Cinquefoil (*Potentilla robbinsiana*) using seed collected from the only known population in the White Mountains of New Hampshire. The new plants are being established as subpopulations in nearby locations beyond the scope of hiking traffic, which has impaired historic sites. [*See* RESTORATION ECOLOGY.]

Reintroduction is the use of plants propagated from existing sources to restore a species at a site where it was historically known to occur. This technique has been used in the recovery effort for Pitcher's Thistle (*Cirsium pitcheri*), a threatened species found on the dunes and beaches bordering several of the western Great Lakes. Biologists using seed collected from populations in Michigan and Wisconsin have grown plants for reintroduction at an historic site in Illinois, where the species has not been recorded since 1919.

Finally, an *introduction* of propagated plants may be attempted in an area where the species has not been historically recorded but where the habitat conditions appear compatible. This technique has not been widely attempted and is usually considered a viable option only as a last resort in preventing the extinction of a species.

Many biologists are reluctant to use introduction as a conservation tool because in some attempts the apparently early survival of transplants has been followed by deterioration and/or disappearance of the new populations in succeeding generations. Unfortunately, a premature conclusion of success may be perceived by nonbiologists as proof of the efficacy of simply moving rare species to new sites as a mitigating answer to the destruction of native habitats for development.

A successful introduction requires a complete knowledge of the plant's life history. Initially the strict physical and chemical characteristics of the subject's abiotic environment must be identified, but an understanding of the species' biology is also essential. Some important considerations include the following:

What is the reproductive system?
  Sexual or asexual
  Self-fertilizing or outcrossing
  Annual, biennial, or perennial
  Specificity of pollinating organism
How is the species dispersed?
  Type of seed
  Dispersal agent (wind, water, or animal)
What are the specialized needs for
  germination?
  Passage of seed through animal digestive
    system
  Environmental disturbance (fire, flooding,
    desiccation)
What are the needs for establishment?
  Mycorrhizal fungi
  Hosts for parasitic and saprophytic species

Although introduction has not been widely accepted, it has been suggested as a partial solution to the impacts of global climate change. Theoretically, a species could be transplanted to new sites in latitudes north of its recorded distribution in advance of global warming. This translocation would constitute a form of "jump-starting" new populations before sites within the current range became uninhabitable. Superficially this method appears to be a viable alternative for preserving individual plant species in the face of drastic climatic change, but only species for which the best knowledge is available will be successfully introduced to new sites. This paradigm connotes an artificial selection of species for survival, and emphasizes the urgency of acquiring more data on the ecological requirements of rare species.

VII. CONSERVATION PROGRAMS

The preservation of plant diversity has relied primarily on land conservation organizations and botanical gardens to respectfully tackle *in situ* and *ex situ* conservation. In recent years a new strategy has emerged that emphasizes the integration of the two methodologies along with new initiatives in research, education, and public policy. One organization that has successfully applied this integrated approach is the Center for Plant Conservation (CPC), a national organization headquartered at the Missouri Botanical Garden. The CPC coordinates the work of 25 botanical gardens and arboreta and acts as a clearinghouse for current research on rare plant conservation. The CPC is national in scope and has devoted much of its effort to those parts of the country that are the most botanically diverse and support the largest number of rare species. A report published by the Center in 1988 recognized Hawaii, California, Texas, Florida, and Puerto Rico as collectively containing three-fourths of the 680 species then considered to be the most vulnerable in the United States.

Both the Center for Plant Conservation and the federal government, through provisions of the Endangered Species Act, are primarily focused on nationally endangered species. This approach implies that a species is usually not recognized for special conservation effort until it has declined to such a low population level that emergency action is required. The scope of the plant diversity crisis dictates that national programs must concentrate on those species most critically imperiled, but a large number of additional species (evidenced by the more than 2000 taxa identified for potential federal listing) also demand attention from the conservation community to prevent their decline to endangered levels.

A response to this need has been the organization of regional efforts as exemplified by the New England Plant Conservation Program (NEPCoP).

Formulated in 1991, NEPCoP is a consortium of 65 governmental and private groups that works to prevent the extirpation and promotes the recovery of New England's endangered plants. It is administered by the New England Wildflower Society, which also provides propagation and seed-banking facilities at its headquarters in Framingham, Massachusetts. The NEPCoP regional council, consisting of experts from the six New England states, has focused on developing a regional list of plants in need of conservation. This list includes taxa of global concern (ranked G1, G2, and G3 by The Nature Conservancy/Natural Heritage Program Network) and also taxa of regional concern. This latter group includes plants that are uncommon in New England but widespread outside the region, and those that may be common in some parts of new England but for which particular populations hold special ecological significance. An example is the shrub Inkberry (*Ilex glabra*), which is found throughout southern New England but also occurs as a single disjunct population at a bog in Maine. According to the regional listing criteria, only the Maine population would be recognized for special conservation status.

To facilitate its work, NEPCoP has organized individual task forces to develop conservation strategies within each state for those species designated as highest priority on the regional list. Known populations of priority plants are identified for survey work that consists of monitoring, landowner contact, and/or seed collection. A goal of NEPCoP is to demonstrate that a regional integrated conservation program, developed through the cooperative effort of many organizations, can succeed in conserving rare plants. In this manner, NEPCoP will serve as a model for analogous programs in other regions.

Glossary

Endemic Restricted to a particular region or locality.
Genetic diversity Number and frequency of alleles within a population or species.
Herbarium Collection of dried and mounted plants compiled by field botanists and housed at universities and museums.
Mycorrhizal fungi Fungus complexes that exist symbiotically in or on the roots of many species, including trees,

orchids, and heaths. The fungi obtain food from the roots, and benefit the plant by aiding in the absorption of minerals and water. Some plants are unable to survive without mycorrhizal fungi.

Taxon (plural taxa) Basic unit of classification for plants and animals, which is usually designated as the *species*. Subpopulations of species that are characterized by particular variations from the typical form may be designated as a *subspecies,* or the more commonly used botanical designation of *variety*.

Taxonomic uniqueness Degree of distinction of a particular species. For example, the ginkgo (*Ginkgo biloba*) is not closely related to any living plant and is therefore more taxonomically unique than species of large families, such as the grasses.

Bibliography

Ayensu, E. S., and DeFilipps, R. A. (1978). "Endangered and Threatened Plants of the United States." Washington, D.C.: Smithsonian Institution and World Wildlife Fund.

Falk, D. A., and Holsinger, K. E., eds. (1992). "Genetics and Conservation of Rare Plants." New York: Oxford University Press.

McMahan, L. R. (1990). Propagation and reintroduction of imperiled plants, and the role of botanical gardens and arboreta. *Endangered Species Update* **8,** 4–7.

Morin, N. A., ed. (1993). "Flora of North America, Volume 1: Introduction." New York: Oxford University Press.

Natural Heritage Data Center Network (1993). "Perspectives on Species Imperilment," revised printing. Arlington, Va.: The Nature Conservancy.

New England Wildflower Society (1992). New England Plant Conservation Program. *Wildflower Notes* **7**(1), 1–79.

U.S. Fish & Wildlife Service (1993). Endangered and threatened wildlife and plants; Review of plant taxa for listing as endangered or threatened species—Notice of review. *Fed. Reg.* 50 CFR Part 17.

Plant Ecophysiology

Ludger Kappen

University of Kiel, Germany

I. History of Ecophysiological Research
II. The Meaning of Ecophysiology
III. The Subject of Research
IV. Environmental Factors and Their Relevance to Plant Responses
V. Stress
VI. Water Relations and CO_2 Exchange
VII. Scaling Up from Leaf to Ecosystem
VIII. Methodical Approach: Techniques

Plant ecophysiology or physiological ecology is the physiological approach of investigating plant ecology. Physiological methods are used in the laboratory as well as in the field. Whereas physiologists investigate plants for basic understanding of functions, ecophysiologists measure the functioning of plants under complex and varying environmental conditions and within the continuum of life processes (Fig. 1). Ecophysiology analyzes the nature of the various environmental factors as well as the actual or potential plant responses. These responses also result from adaptation, adjustment, reversible dysfunction, or injury. Ecophysiology attempts to understand the performance of whole plants, plant populations, and even ecosystems. Ecological thinking may stimulate physiologists to study the influence of environmental factors on the level of a cellular membrane or in metabolic patterns or on a molecular basis. Economically applied botany, such as agriculture and forestry, essentially carry out ecophysiological research, however, mainly with the aim of optimizing crop production and yield.

I. HISTORY OF ECOPHYSIOLOGICAL RESEARCH

The core question of ecophysiology—how plants function under certain environmental conditions—may be as old as the interest of humankind in growing plants for its use. For instance, the Greek philosopher Theophrastos (372–287 BC) raised ecophysiological questions in his treatises, which contained questions relevant to modern ecophysiology, such as the influence of wind, water, heat, and soil on the plant. Thus, some of these issues had been addressed long before the term "ecology" was coined by E. Haeckel in 1866, meaning the relations between an organism and its environment. [*See* ECOPHYSIOLOGY.]

Early studies on temperature and water relations (e.g., calculation of transpiration rate of plants in the field) began at the end of the eighteenth century. Scientists, such as H. Dutrochet (1776–1847), J. Sachs (1832–1897), and L. Cockayne (1855–1934), tackled ecophysiological problems, but considered themselves to be physiologists. Among other

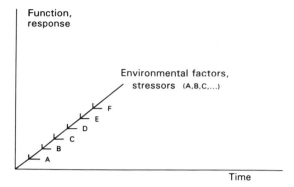

FIGURE I Dependence of plant functions and responses on the various environmental factors (stressors) such as (A) light, (B) temperature, (C) water stress, and (D–F) other factors in the time continuum.

things, they studied plant tolerance to water stress and to extreme temperatures. The plant physiologist E. Stahl (1848–1919) demonstrated the ecological relevance of his results. He is considered one of the pioneers in ecophysiological research in Central Europe. Stimulated by the ideas of his colleague A. F. W. Schimper, who wrote the book "Pflanzengeographie auf physiologischer Grundlage" (Jena, 1898), Stahl investigated, for example, the influence of light on plant leaves, described the difference between sun and shade leaves and the interaction between plants and snails, and recognized the morphology of cacti as a response to drought. As examples of how functional adaptations of plants to environmental factors were realized by scientists during the last 20 years of the nineteenth century, the studies about physiological anatomy of tropical leaves by G. Haberlandt (1854–1945) in southeast Asia and the early ecological field studies of the laboratory-trained botanist G. Volkens (1855–1917) on African desert plants should be mentioned here.

Much of the theoretical background for the physiological investigations of ecological phenomena was reported at the turn of our century in books that mainly focused on explaining the reasons for the existence of plants in certain regions of the globe. One of the first institutes dedicated primarily to ecophysiological work (on desert plants), the Carnegie Institution's Desert Research Laboratory outside Tucson, Arizona, was founded in 1903.

Important input to the understanding of plant ecophysiology in the beginning of this century came from G. Kraus (1841–1915) and R. Geiger (1894–1981), who made pioneering studies on the habitat conditions and microclimate around plants and in the soil. Between 1925 and 1950, modern ecophysiology became established in Europe, Israel, Russia, the United States, and Australia.

Until the 1960s, physiologists and ecophysiologists gathered (from field and laboratory experiments) a broad knowledge about the adaptation of wild and cultivated plants to water stress, heat, cold, and other stresses. Microclimate and plant energy budgets were analyzed, as well as chemical–plant interactions (allelopathy). Publications about CO_2 exchange and water relations of plants from various climatic regions were rare up to 1960. Ecophysiology as a special scientific approach was reviewed for the first time in the *Annual Review of Plant Physiology* in 1957, by W. D. Billings. In that review, "physiological ecology" was placed along an "ecotone between physiology and ecology that are closely interrelated" and the wide scope of this subject matter in biology was emphasized.

The technological progress in the past 30 years enabled ecophysiologists to work with more appropriate and precise methods, particularly for field research. Between 1970 and 1980, the first textbooks about physiological ecology appeared. Within this period, ecophysiological monographs were written and new journals with particular emphasis on ecophysiology appeared.

In the beginning of the 1980s (1981–1983) the first large compendium on "Physiological Plant Ecology" within the new series "Encyclopedia of Plant Physiology" was published. It consists of four volumes (Vols. 12A–D), representing the four columns of modern ecophysiology. The first volume discusses responses to the physical environment (except water), comprising the relations and adaptations to abiotic environmental factors. The second volume deals with water relations and carbon assimilation, the third treats plant responses to chemical and various kinds of biotic environmental factors, and the fourth volume deals with processes in ecosystems such as primary production and mineral cycling in plants and the influence of humans.

II. THE MEANING OF ECOPHYSIOLOGY

Various terms in the literature, such as ecological or physiological plant geography, biophysical plant ecology, physiological ecology, ecological physiology, experimental ecology, and ecophysiology, indicate the interrelationship between physiology and ecology of plants. Other terms, such as physical autecology and biotic autecology, are used only in zoology.

In the last century the meanings of ecology and physiology were not clearly distinguished, notwithstanding Haeckel's efforts, but the term ecology was rarely used. Schimper may be the first who discerned the ecological and physiological aspects, in 1898: "The ecology of plant distribution will succeed in opening out new paths only on condition that it leans closely on experimental physiology."

The meaning of physiological ecology was frequently based on that of autecology, a term coined in 1896 (its *antonym* is synecology). Autecology focuses on the relationships between an individual plant and its environment and does not seek the relation to higher integrated systems (populations or ecosystems). But autecology has a more limited scope than ecophysiology because it omits interactions with other species as biotic factors and structural elements of the environment.

In 1909, B. E. Livingston, one of the early staff members of the Carnegie Institution's Desert Research Laboratory, said that "physiological ecology [studies] factors which determine the occurrence and behaviour of plants growing under uncontrolled conditions" and thus emphasized a principal difference from plant physiology. In the 1950s, some researchers evaded the conflicting term physiology by using the term experimental ecology or later environmental biology.

The term ecophysiology appeared in the literature in 1960 (F. E. Eckardt, 1923–1990: ecophysiologie) and came into use in 1962 at a symposium held at Montpellier, France. It was understood not as "a confines between ecology and physiology but as the analysis of the functional relationships among organisms and between organisms and their environment." This definition meets the understanding of ecology by Haeckel (1866), but in a wider sense. At present, there is still no final consensus about the meaning of ecophysiology. It is a young discipline with a varying scope that can be as wide as that of ecology. We may see it as a "discipline within plant ecology that is concerned fundamentally with the physiology of plants as it is modified by fluctuating external influence" or as a "discipline that is not a science in isolation but an integral part of the whole of ecology." The broad and integrating scope of ecophysiology is widely accepted among scientists: on one hand, as a study of growth, biological processes, and reproduction within plant populations in natural and controlled environments, and on the other, as an objective to explain plant performance, survival, and distribution in physiological, biophysical, biochemical, and molecular terms. It is essential to state that the goal of ecophysiology is ecological but the methodology uses physiological information. Thus, we can equally use the term physiological ecology.

III. THE SUBJECT OF RESEARCH

The interaction between genetic composition and environmental influences during evolutionary history has created a great diversity of physiological and morphological properties in plants. It demonstrates to the ecophysiologist the range of potential and actual response patterns, at species level but also in subspecies and ecotypes.

In ecological terms a major division can be made between vascular and nonvascular plants. Because of a primitive organization, the ecological performance of nonvascular plants results directly from physiological properties. In vascular plants, physiology is more diversified, that is, different metabolic pathways and tolerance mechanisms are present. The adaptive direction of physiological properties in plants may be very obvious or be obscure or ecologically unexplainable. Structural and morphological properties strongly determine the ecological performance of vascular plants.

Plant size is genetically controlled in principle but can be largely modified by environmental conditions ("miniplants," a consequence of water stress, intoxification, etc.). Plant response is the sum of the various responses of the plant organs. The many different shapes of leaves and stems and their coverage by scales, hairs, or spines can have strong ecophysiological relevance. Ecophysiologists mainly consider responses of vegetative structures. Crop scientists, on the other hand, investigate the extent to which reproductive organs react to changes of environmental conditions.

The concept of plant life-forms describes the plant shape and the way plants can persist in their habitats. Therophytes are herbaceous taxa that finish their life cycle during one season. Geophytes are perennials with respect to their subterranean storage organs, but with shoots only during a favorable season. Hemicryptophytes are herbaceous perennials with a permanent presence of sprouts and buds above the soil surface. Chamaephytes and phanerophytes are small and large woody species (such as shrubs and trees), respectively, that permanently maintain their leaves as evergreens or shed leaves if deciduous. The hydrophytes and helophytes in aquatic bodies and swamps can be taken as modifications of geophytes or therophytes. These life-forms can be further subdivided according to their shape and growth strategies (e.g., rosettes, cushions, climbers, and creepers). The life-form determines the large-scale way in which these plants are potentially exposed to the environment (open stands, dense canopies, etc.).

The way plants show a special adaptation to habitat conditions can be described by means of ecological terms such as aerophytes (epiphytes are plants attached to another terrestrial plant without physiological contact to it) or the many types of edaphic specialists (e.g., calcicolous plants on limestone or halophytes growing on saline soils), phreatophytes (plants capable of extracting a stable water source), or finally the great number of phenotypes that function as xerophytes (adapted to drought in soil and air, as illustrated in Fig. 2).

Habitat specialists exemplify how plants are adapted to certain environmental factors. It is no wonder that ecophysiological research was primarily focused on regions with prevailing extreme environmental conditions: arid and polar regions, high mountain habitats, or raised bogs. Physiological and anatomical properties of the many ecologically specialized plants challenge our understanding of adaptations to the environment.

For instance, do plants with different CO_2 assimilation patterns (C_3 and C_4 plants) grow in the same environment of a grassland? Another puzzling problem arises with species of the same genus of the family Bromeliaceae that fix CO_2 in different ways (C_3 or CAM plant) and sit side by side on the same branch of a tropical tree. Finally, it is interesting to explain the strategies of plants with two different types of leaf morphology equally adapted to the Mediterranean climatic regions, for example, the leathery leafed (sclerophyllous) and the soft leafed (mallacophyllous) plants.

Symbiotic systems, mainly resulting from relationships between photosynthetic (autotrophic) and nonphotosynthetic (heterotrophic) organisms such as lichens, or nitrogen–fixing microorganisms with vascular plants, or the hemiparasites of vascular plants (mistletoe type), also raise the question of ecophysiological adaptation. A further challenge to ecophysiologists is the investigation of integrated responses by aggregates of the same or different life-forms of plants (canopy, community, ecosystem) to their inner and outer environment. At this level, a new dimension of organismic response comes into play.

IV. ENVIRONMENTAL FACTORS AND THEIR RELEVANCE TO PLANT RESPONSES

To define the environmental influence on a plant or vegetation, a multifactorial system has to be considered. Among the major influential parameters, primary and secondary factors can be defined. Primary factors are the components of radiative energy (UV, visible light, IR, and thermal energy), water, mechanical factors (wind, hydrostatic pressure), and chemical agents and gases. Secondary factors are those that result from an interaction between the primary factors and other media forming soil, contour of landscape, biotic factors, canopy structure, and fire. Both kinds of factors exist

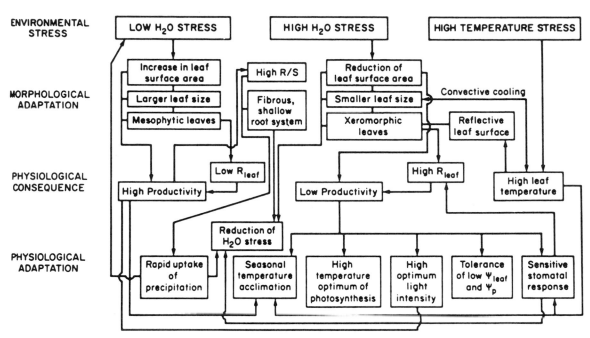

FIGURE 2 Plant stress–response at various levels (physiological adaptation, physiological consequence, and morphological adaptation), exemplified by multiple stress interactions in the evolution of drought-deciduous desert shrubs. [From P. W. Rundel, (1991). *In* "Response of Plants to Multiple Stresses" (H. A. Mooney *et al.*, eds.). Academic Press, San Diego, Calif.]

in the atmosphere as well as in the hydrosphere (Fig. 3). To investigate the role of these factors, ecophysiologists have to distinguish between direct and indirect action, and also between immediate effects and aftereffects. Considerable input to our knowledge comes from physics, meteorology, chemistry, hydrology, and the geosciences. Ecophysiological investigations with emphasis on the quality of these factors were also referred to as biophysical ecology or environmental physiology.

Action and interaction of the first three primary factors form the microclimate, and all primary factors together form the microenvironment. The spatial dimension of the microenvironment or microclimate always depends on the size and shape of the organism under consideration or the part of vegetation in which physiological responses are investigated. Ecophysiologists describe and quantify these factors and correlate them with aftereffects and plant responses recorded at the same time.

A. Light

Ultraviolet radiation forms a major factor in high alpine and polar regions and its effect may increase due to the present widening of the ozone hole. Although injury was rarely recorded for land plants under present conditions, ecophysiologists must be aware of deleterious effects of enhanced UV-B radiation.

Visible light is relevant to plant functions in various ways: photoperiodism, for example, differs with geographical latitude, season, and the diurnal course and has caused specifically adapted plants. During the course of the day the light quality varies greatly, particularly the red/far-red ratio influences growth, morphogenetic responses, and seed germination. The physiological and morphological strategies of sun and shade plants as plastic or species-specific responses to open and shaded habitats are a matter of ecophysiological research. The photosynthetically active radiation (spectral energy between 380 and 710 nm) is the energy source for photosynthetic production. But sudden incidence of excessive light intensity causes disturbance or injury to the photosynthetic apparatus of plants (photoinhibition) in various climatic regions and affects photosynthetic productivity. On the other hand, sun beams reaching plant leaves in the shade of the canopy for periods of seconds or minutes

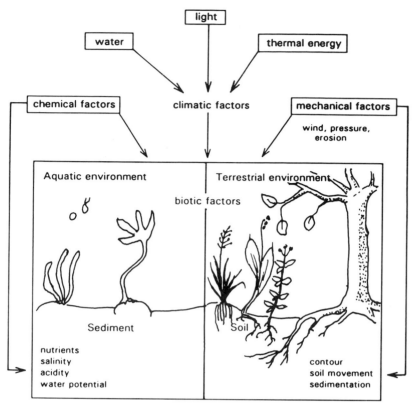

FIGURE 3   The microenvironment of terrestrial and aquatic plants influenced by primary factors (microclimate, chemical, and mechanical factors) and secondary environmental factors (contour, soil, sediment, their chemical and mechanical properties, and the biotic factor).

(lightflecks) cause an enhancement of the photosynthetic rate, which is maintained as an aftereffect for several minutes.

B. Thermal Energy

Physiological processes in plants are for the most part directly governed by the thermal conditions of the environment, because plants have very limited temperature control. Air temperature is usually not indicative of the conditions in the plants because leaves, stems, and roots may have different temperatures. Energy budgets of plant organs are calculated by accounting for radiant heat (acquired from various sources of the environment) and latent heat (as a function of evapo-transpiration).

Plants have a species-specific temperature range that matches their environment. However, their physiological optimum is rarely reached in their natural habitats. Physiological processes in plants can be adjusted within certain temperature ranges and show seasonal variations. Temperature adaptations of plants have been extensively investigated and related to habitat conditions and distribution area.

C. Water

Water has many roles to play in the ecophysiology of plants. It is a medium in which processes take place and nutrients are transported. It is also an environmental factor that exists in various physical states, as a liquid, ice, or vapor. Consequently, water has effects by its quantity (hydration, turgidity, relative water content) and by its physicochemical potential (water potential, osmotic potential).

Water uptake (namely, by roots, in some plants by leaves, and in poikilohydrous plants through the whole surface), water conduction, and transpiratory water loss are driven by water- and osmotic potential gradients in the plants and by soil- and atmospheric water relations. An example is the water vapor pressure deficit (VPD) between air and leaf. Ecophysiologists established the model of a soil–plant–atmosphere continuum (SPAC) of water flow along potential gradients. A wide range of ecophysiological research deals with how roots acquire water from the soil, and with the water relations in the shoot and their dependence on the atmospheric water potential. [See RHIZOSPHERE ECOPHYSIOLOGY.]

Dew and mist have particular relevance to plants in arid regions. They may serve as a water source and as a precipitate on plants and soil surfaces. Poikilohydrous plants (such as many lichens and aerial algae) can also be activated by water vapor uptake.

The role of soil- and plant water deficit has been investigated particularly in arid and semiarid regions. In arid regions, seeds respond in a very complicated manner to soil moistening. These seeds have distinct threshold levels for swelling and for the activation of the embryo. Several plant species are able to produce two differently responding seed types or propagules. In annual plants, the size can be a direct function of the water availability from the soil. [See SEED DISPERSAL, GERMINATION AND FLOWERING STRATEGIES OF DESERT PLANTS.]

Ice has mostly a deleterious effect on plants. However, it may also function as an insulator against heat loss of plant organs, and ecophysiologists are aware of plant activity under snow, particularly in late winter.

D. Mechanical Factors

Pressure by wind, sand and snow blast, snow pack, ice, or other material causes deformation of plants and reversible or irreversible damage. Of particular interest is wind as a convective force. It controls the laminary boundary layer on plant and canopy surfaces and thus greatly influences the heat and water flux from and to the plant. The boundary layer resistance is highest under calm conditions.

E. Chemical Factors

Components of the air such as oxygen, CO_2, and nitrogen are essential for plants, whereas other gases such as SO_2, NO_2, HF, and other pollutants, which occur mostly in aerosols, are toxic. Oxygen is a product of plant photosynthesis and reached its present atmospheric concentration only since the Paleozoic (earlier than 250 million years ago). This means that it was an air pollutant for the earlier dominating anaerobic microorganisms. CO_2 may turn out to be a pollutant in the future if its concentration increases in the presently expected manner. Doubling of the CO_2 concentration stimulates plant productivity according to short-term experiments. In the long term, the effect is not clear. At the very least the carbon/nitrogen ratio in plant litter was shown to be increased, which may have consequences to microbial leaf destruction. The effects of concomitant warming in the atmosphere are a continuing focus of ecophysiological research.

The deleterious effect of SO_2 and other toxic gaseous components is well established by physiological experiments. However, ecophysiological investigations are difficult to perform, although great technical efforts have been made. The aerodynamic conditions in test chambers can hardly simulate the natural situation, and we know too little about the long-term effects of low gaseous concentrations and of a mixture of toxic substances in the atmosphere.

F. Ionic Relations

Regarding ionic relations, plants must deal with the availability and balance of nutrients and essential trace elements, and the presence of toxic ions. Nutrient availability depends on the mineral composition of the soil. Ecophysiologists have investigated the adaptation of plants to acidic and alkaline soils and why certain species are edaphic specialists (e.g., on limestone, gypsum, serpentine, heavy metal soils, saline soils). Plants living on saline soils (halophytes) along seacoasts and in arid regions have to cope with the salt load that acts as an osmotic agent or is toxic to their cellular functions. Most halophytes not only control the salt load in the tissues, but also use it for uptake of water against low soil water potentials, and some grow better under saline conditions than in nonsaline soils. In addition, extensive ecophysiological research is focused on the complex situation that causes forest decline. In some cases the effects of increased soil acidity blocked the availability of magnesium for trees, and bleaching of the photosynthetic organs was detected. Another problem is the ionic imbalance caused by inadvertent and anthropogenic nitrogen fertilization distributed through air and water transport.

G. Soil

As a secondary factor, soil has relevance to plant physiological functions because of its structural properties. Clay soils and sandy soils have different water-soaking capacities; in arid regions, clay soils are more hostile to plants than sandy soils. An important factor for root respiration and mineral availability is the aeration of the soil. Compression of soil has dramatic effects on water relations and nutrition of plants. The water that concentrates under stones in desert soils can be exploited by plant roots. A temporal phenomenon is that both soil and vegetation pass through successional stages due to the changes of nutrient availability in the soil at different stages of soil maturity. [*See* Soil Ecosystems.]

H. Landscape Pattern and Contour

Landscape patterns and contours are formed by surface erosion. The vegetation composition varies strongly with compass direction, exposure, and steepness, all of which influence water support to plants as well as soil properties and microclimate. Plant growth and shape are affected by mobility of the slope material and snow accumulation, run-off, and flooding.

I. Biotic Factors

Biotic factors show great diversity. Plants may act as competitors for light, water, and nutrients. Furthermore, many plant species release chemicals to the soil or atmosphere that inhibit other plants. Animals not only reduce phytomass by browsing and grazing, they also induce chemical responses of the plant, causing impalatability or poisoning. Many of the so-called secondary plant products have been identified as chemical defenses against insects or fungi. On the other hand, plants can be attacked by various kinds of pathogenic microorganisms and fungi that alter their physiological function, such as control of water relations. Ecophysiologists also investigate the beneficial role of trees and other plants as shelter for the understory, as phorophyte for epiphytes, or as host for vascular parasites and epiphytes.

J. Fire

Fire as a natural ecological factor plays an important role in Mediterranean-type and other sclerophyllous vegetation. Trees and shrubs show various adaptations, such as insulating bark, quick regeneration of leaves, inflammable oils, lignotubers, and the dependence of fruit-opening and flower-setting on fire heat. Other beneficial effects of fires, particularly in habitats with poor soils, are that plant ash is a fertilizer and that plant-inhibitive chemicals from other plants are decomposed. [*See* Fire Ecology.]

As a whole, each of these factors acts within a range between highest (the physiological optimum) and lowest (the physiological pessimum) benefit to the plants. They may control the presence and function of the plant or they may become stressors (Fig. 4).

V. STRESS

The extent to which an environmental factor is a stressor depends strictly on the response of the plant. Stress can produce reversible or irreversible physiological changes in the plant. Stress resistance can be tested by experimentally exposing a plant to a stressor at various intensities. If the stress is irreversible, the plant has reached the limit of its resistance. Resistance to environmental stressors may result either from stress avoidance, which

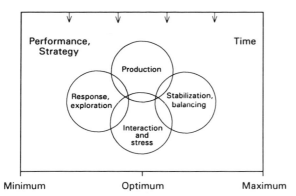

FIGURE 4 The four main types of plant response to single or multiple environmental factors (stressors) with respect to the physiological and ecological status of the plant (as a result of the response) living in optimum or pessimum conditions.

means that the plant excludes the stress (e.g., by morphological properties), or from tolerance, which means that the stress is effective but the plant survives. Stress tolerance can result from avoidance and/or tolerance of stress at different organismic levels (Fig. 5). The plant can also repair stress effects. The intensity of a stress that irreversibly affects vitality or is lethal determines the potential tolerance limit of a plant.

The physiological stress tolerance is genetically determined and varies among species and ecotypes. It is also dependent on the developmental stage of the plant and varies within the plant (roots, stems, leaves, etc.). Tolerance to heat or to freezing varies

in most perennial plants during the seasonal course. The effect of heat depends on the duration of heat exposure, whereas low-temperature stress acts in different ways. Many subtropical and tropical plants are killed by time-dependent temperature stress between 8° and 0°C (chilling sensitivity). Most annual plants and those from regions with very mild winters die as soon as ice forms in the tissues. Freezing-tolerant plants are not affected by extracellular ice formation and depend mainly on tolerance of dehydration of the cells. Whereas plants are able to avoid natural heat strain by anatomical properties and by transpiratory cooling, they can hardly avoid low-temperature stress. The investigation of heat and cold resistance of plants is a means to explain their adaptation to their habitats and the range of their geographical distribution.

Waterlogging is stressful because it causes injury to plants by anoxia (oxygen deficit in soils). Water stress is usually understood as drought stress; it affects the plants by lowering its water and osmotic potential and the water content. In vascular plants the critical (i.e., sublethal) water loss of the tissues is a measure of desiccation tolerance. It varies with season as does thermotolerance. A number of nonvascular plant species and seeds are able to survive extended periods of extremely high water loss. Plants have developed an enormous diversity of

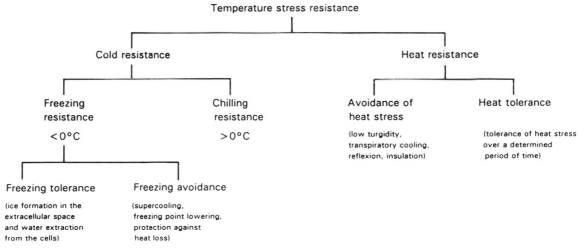

FIGURE 5 Plant response to environmental stress, here exemplified by plant response to temperature stress. The components of cold and heat resistance are shown. [After J. Levitt (1980). "Response of Plants to Environmental Stresses," Vol. I. Academic Press, New York.]

ways to avoid drought stress: morphological properties, life-forms, reproductive strategies, water storage, and metabolic pathways that diminish water loss through the stomata. Figure 6 shows an example of how plants respond to stress over time.

Chemical stress acts mostly by the toxicity of the substance at the cellular level. However, saline (e.g., NaCl) stress also acts by decreasing the osmotic potential and consequently causes water stress. By their capacity of physiological acclimation to some stressors, plants are able to attain a temporary higher tolerance level (hardiness). Adaptation to these stresses is the result of evolutionary processes of the plant taxon. Plants are exposed permanently to various stressors.

Multiple stress means the action and interaction of different stressors in the plant, such as heat, drought (see Fig. 2), and saline stress. Another combination is strong light and low temperatures or water stress (eventually causing photoinhibition). Synergistic effects of air pollutants (SO_2, NO_x) cause altered nutrient balances in trees and reduce carbon transport into the roots. Chronic limitation of mineral nutrient supply affects the

capability of water uptake and carbon assimilation, and loss of the competitive status of the stress-sensitive plants is expected. According to the concept of limiting factors, the deficiency of one nutrient can exclude a plant species from a habitat. The composition of fertilizers for agricultural production takes this into account. However, in their natural environment wild plants may react in a more flexible manner. A great diversity of strategies and temporal commutation of developmental phases allow plants to compensate for nutrient deficiencies and other stresses.

VI. WATER RELATIONS AND CO$_2$ EXCHANGE

Much of the ecophysiological literature deals with the various ways and strategies by which plants acquire water from their environment and with the economy of plant–water relations. Nonvascular plants and a few vascular plant species have no or very limited means of controlling their water relations. These so-called poikilohydrous plants

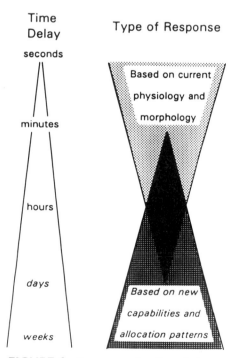

Time Delay	Type of Response	Processes for Response to Water Stress
seconds	Based on current physiology and morphology	Increase in apoplastic water tension
		Decrease in cell turgor
minutes		Root water potential decreases
		The hormone abscisic acid moves to guard cells
		Stomatal aperture decreases
		Transpiration slows
hours		Carbon is allocated to accumulation of osmoticum
		Polysomes disaggregate, new mRNA's transcribed
		Abscisic acid synthesis increases
		Synthesis, accumulation of osmotic agents
days		*Tissues with smaller-sized cells formed*
	Based on new capabilities and allocation patterns	*Increased root growth*
weeks		*Drought tolerant condition*

FIGURE 6 Sequence and scaling of plant responses to stress (here water stress or drought) as processes of gaining drought tolerance. [From D. P. Geiger and G. C. Servaites (1991). *In* "Responses of Plants to Multiple Stresses" (H. A. Mooney *et al.*, eds.). Academic Press, San Diego, Calif.]

(algae, lichens, mosses, and a few fern and phanerogam species) are able to switch from anabiotic to an active stage. In the anabiotic stage they are extremely tolerant to water and thermal stress. Their opportunistic strategy allows many of them to be present in all regions of the globe, even where vascular plants cannot exist (e.g., on the Antarctic continent).

As a result of their anatomical organization, most vascular plants are able to maintain their water relations at a level that allows metabolic activity against a much lower water potential in the air (homoiohydrous plants). In fact, plants cannot preserve water in a closed system for maintaining their turgor. They depend rather on a turnover of water through the plant, which is equally necessary for nutrient uptake. Aquatic macrophytes depend on special transport mechanisms.

Carbon assimilation cannot be considered separately from plant–water relations. Stomata play a key role because they form the conduit where CO_2 diffuses into the leaf and water vapor diffuses out (transpiration). Ecophysiologists study stomatal function and conductivity. Stomatal aperture depends directly on external factors, mainly on irradiance, and in many plant species on the water vapor pressure difference between leaf and air. Aperture is also controlled by the CO_2 concentration in the substomatal leaf region and by the plant's internal water relations. The latter are mediated primarily by the hormone abscisic acid, even for long distances such as from root to leaf. Stomatal transpiration can be as effective as 60% of the evaporation of an open water surface. Stomatal aperture varies during the diurnal course as a result of feedback control. The amount of released water is up to 1000 times larger than the CO_2 intake measured at the same time. [See PLANTS AND HIGH CO_2.]

The economy of water, measured as water use efficiency (WUE) of the photosynthetic production, is highest if carbon is assimilated at the lowest possible cost by water loss. It is hypothesized that plants tend to optimize the balance between CO_2 intake and water loss. This is particularly relevant under arid conditions, where plants have to struggle between water stress and CO_2 demand by stomatal regulation. In moister climates, plants may also be below optimum because they have to exploit nutrients from poor soils by increasing the water flux from soil. As a consequence, the economic ratio of water use efficiency is exceeded. Ecophysiologists recently agreed that wild plants do not function to maximize their carbon gain as is intended with crop plants, because other factors interact according to the carbon allocation, growth strategy, and stress resistance of the plant.

Light, CO_2 concentration, and temperature control assimilation at the chloroplast level. Ecophysiologists measure carbon assimilation mainly at the leaf or whole-plant level as net photosynthesis, the difference between the amount of gross CO_2 assimilation and that of simultaneous photorespiratory CO_2 release. The species-specific photosynthetic performance is determined by its dependency on the photon flux density. Of ecological interest is the light compensation point (LCP) of net photosynthesis (the threshold for positive CO_2 balance) and the light saturation (LS) of net photosynthesis (Fig. 7). In the majority of plants (C_3 plant type), light saturation of net photosynthesis is limited by the ambient CO_2 concentration and the capacity of CO_2 carboxylation. Photosynthetic rates at LCP and LS differ not only by species and ecotype but also within plant organs, for example, between sun-exposed and shaded leaves. Shade leaves usually have low photosynthetic rates, which reflects their adjustment to the environment. Leaf LCP also varies as a function of nocturnal respiratory loss. Taking into account the entire plant (particularly that with a large nonphotosynthetic phytomass), all leaves have to balance a great amount of respiratory carbon loss and their "effective LCP" may reach a very high irradiance level.

The temperature dependency of net photosynthesis is determined by a lower and an upper temperature compensation point and the optimum (see Fig. 7). The temperature range of net photosynthesis may vary with season and is indicative of the adaptation of plant taxa and ecotypes to warm or cold environments. Carbon dioxide uptake in vascular plant leaves ceases around $-6°C$, as a consequence of ice formation in the tissues, however, nonvascular plants such as lichens continue to photosynthesize at temperatures near $-20°C$.

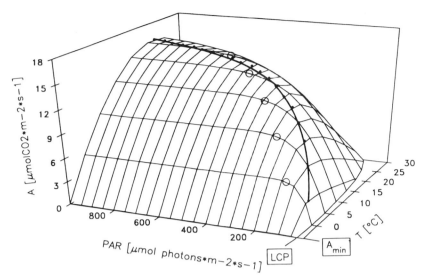

FIGURE 7　Net photosynthesis (A) of leaves of an alder tree as a function of temperature (T) and photosynthetic active radiation (PAR). LCP = light compensation point; ○ = light saturation level; A_{min} = lower temperature compensation point; solid top line = optimum temperature for net photosynthesis. [After C. Eschenbach (1993), unpublished.]

Dark respiration is strictly temperature-dependent but shows species- and organ-specific differences. It measures the metabolic costs for maintenance, growth, and stress response (imbalance and repair) of the living phytomass, and reduces the carbon budget. The capacity of plants for compensating excessive respiratory losses has major ecophysiological relevance.

Those plant taxa that developed CO_2 concentration mechanisms at their metabolic sites have attracted the special attention to ecophysiologists. C_4 plants have a stronger affinity to CO_2 and prove to be more productive than C_3 plants because they do not release photorespiratory CO_2 and have a much better water use efficiency due to their greater enzymatic affinity to CO_2. This may lead to the simple conclusion that, if nutrients are not the limiting factor, C_4 plants are generally superior to C_3 plants. The higher efficiency of C_4 metabolism allows plants to grow quickly and to reach a large size (maize, sugarcane). C_4 plants are therophytes or perennial herbs with high resprouting activity. But taking into account that C_4 plants have a higher light and heat demand than C_3 plants, they are (with a few exceptions) successful and highly competitive only in hot, moist, or semiarid climates and not in temperate and cold regions. They are also inferior

to C_3 plants in dense stands and in forests. This demonstrates the risk of extrapolating from only physiological capacity to whole-plant performance in its native environment.

Plants with crassulacean acid metabolism (CAM), which perform carbon-assimilation and storage as malate during the night and photosynthetic metabolism during the day, reveal an extremely high water use efficiency (e.g., Crassulaceae, Cacti, Mesembryanthemaceae). They metabolize CO_2 at the costs of only cuticular water loss during the day and assimilate CO_2 under moderate water stress during the nighttime. In contrast to C_4 plants, CAM plants grow very slowly, which is a result of their lower amount of diurnal carbon gain, respiratory losses, and slow nutrient acquisition. Many CAM plants are able to switch between CAM and C_3 metabolism according to habitat water relations. CAM is in many cases combined with a high water storage capacity of the tissues (succulence). The obviously successful performance of CAM plants in environments with scattered but regular rainfall is widely studied and demonstrates their adaptation to arid regions. However, recent attention was directed to those CAM plants that also exist in moist habitats or even in water. At very damp sites CAM guarantees survival to epiphytic

Bromeliaceae as it provides regulated CO_2 uptake at times of low transpiration and possible prevention of photoinhibition by internal CO_2 circulation. In the aquatic environment, CO_2 can be limiting for plants during daytime owing to warming and CO_2 depletion by other plants. Here a CAM plant may fill a gap in that it assimilates carbon at night. In this context, the recently discovered CO_2-concentrating mechanisms (although metabolically different from CAM and C_4) in cyanobacteria and many microalgae have to be considered as adaptive responses to rapid changes in the availability of dissolved inorganic carbon. The concept of a carbon-concentrating mechanism presents new directions for ecophysiological research in studying environmental constraints that lead to the evolution of modified carbon pathways in plants.

The performance of a plant in its habitat depends largely on how it uses its assimilates. This can be illustrated by a cost–benefit analysis. Part of the carbon is allocated for acclimation, which restores and maintains homeostasis under environmental stress. Other assimilates are allocated to various plant organs for strengthening the efficiency of functioning. Partitioning of carbon into supporting stems may allow for a strategy of quickly merging into higher canopy levels where the plant will gain more light. Formation of more leaves increases the photosynthetic capacity and competitive strength of the plant. Carbon reserves in storage organs prepare a plant for survival of stress and regeneration (e.g., drought, fire). Carbon investment to root growth enhances the capacity of water and nutrient exploitation (e.g., in poor soils). The role of nitrogen for carbon production becomes clearer by understanding allocation processes (Fig. 8). Nitrogen supply, up to a certain threshold, increases CO_2 assimilation rates. Above this threshold the rate is saturated, but more leaves are formed and nitrogen may be stored. The enormous biomass production and competitive capacity of nitrophytes result from stimulated root growth and recycling of nitrogen from outshaded into newly formed leaves. It is obvious that growth of a plant cannot be simply related to the photosynthetic rate of its leaves. Primary production of a plant or a population of plants is therefore best estimated by calculating biomass increment over time.

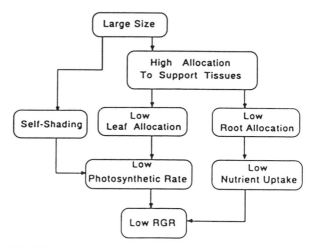

FIGURE 8 The role of allocation of carbon for the growth rate/plant size relation. In this case, allocation was low in most compartments and allowed large size (stem elongation) even at low growth rate. [From F. S. Chapin III (1993). *In* "Scaling Physiological Processes. Leaf to Globe" (J. R. Ehleringer and C. B. Field, eds.). Academic Press, San Diego, Calif.]

Phytomass production is the result of an environmentally controlled complex interaction between carbon and nutrient fluxes in the plants. Ecophysiology not only attempts to investigate the physiological processes and control mechanisms, but also to understand their environmental control to explain how plants grow in their habitat. Thus, conceptual models (top–down models linking submodels) may bridge the gaps of our ignorance and provide predictions about growth responses. These models may be validated by future experimental approaches designed to analyze complex processes at higher integrated levels (Fig. 9). It is unproven to what extent biomass partitioning can be considered as a growth optimization process. The mechanism that regulates constant internal C/N pools has to be investigated. How can a controlled partitioning schedule be explained that ensures maximal growth for the prevailing environmental conditions?

VII. SCALING UP FROM LEAF TO ECOSYSTEM

Measured response of small herbaceous plants and leaves to environmental factors can be modeled and predictions can be made by calculating the model

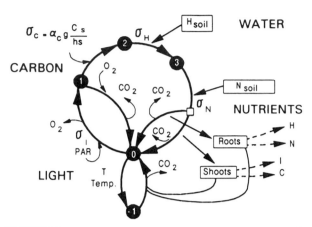

FIGURE 9 Shoot and root growth in relation to carbon assimilation and maintenance respiration (CO_2 release). The model depicts cyclic flow from node to node, determined by light (PAR), O_2, CO_2, soil water (H), soil nutrients (N), and temperature (T). Node 0 to 1, light absorption; 1 to 0 photorespiration; 1 to 2 CO_2 uptake; 2 to 3, water uptake; 3 to 0, nutrient uptake; 0 to −1, maintenance respiration. The fluxes from the main cycle (lower right of the figure) describes the allocation fractions for C (carbon), H (water), N (nutrient), I (light energy). [From P. J. H. Sharpe and E. J. Rykiel, Jr. (1991). *In* "Responses of Plants to Multiple Stresses" (H. A. Mooney *et al.*, eds.). Academic Press, San Diego, Calif.]

with the environmental variables over a defined time series. The difficulty increases if responses of large and canopy-forming plants with an extended life span are to be modeled.

In a canopy of a crop population or of a tree crown, new structures and variables must be considered. A plant canopy has its own boundary layer and varying and inhomogeneous internal light, temperature, and moisture conditions. These conditions can be measured by means of micrometeorological methods. The measurements of leaf area index (LAI) or leaf area density (LAD) provide the basis for estimating light interception in the canopy and the area-related potential of CO_2 and H_2O exchange. The combination of environmental factors determines the formation of structural and functional partitioning in plants. The foliage of a shrub or tree can be considered as populations of differently reacting single leaves (reflectiveness, thickness, sun leaves, shade leaves, etc.). We envisage here the linking between ecophysiology and population biology. [*See* FOREST CANOPIES.]

There is in principle no difference whether ecophysiologists study a single plant or whole canopies composed of different growth forms. A simple example may explain this, the photosynthetic performance and water relations of a lichen thallus may be understood like those of a plant leaf, although the lichen is a symbiotic system of a fungal meshwork and a population of unicellular algae or cyanobacteria. Net photosynthesis of a lichen results from the CO_2 balance between algal net photosynthesis and fungal respiration. It is also a system response if a lichen reacts to strong water uptake shown by a decrease of its photosynthetic rate, because swelling of the surrounding fungal mass increases the resistance for gas diffusion to the algal symbiont.

Considering the plant canopy of a forest or a hedgerow, questions are directed to the ecological constraints for understory plants, for example, the growth control by light reflection from other plants or the consequences of production and allocation strategies of different shrub and tree species that result in ramification pattern and canopy architecture. Also relevant to this aspect are the contributions of epiphytes or of seasonally developing understory plants, because they all respond to the canopy structure and contribute to an overall performance of the canopy. A preliminary, but reductionistic approach is to consider the leaves of a canopy as a "super-leaf" having its average stomatal conductance and saturation deficit. This approach may provide critical tools for scaling ecophysiological concepts at the level of plant canopies or even ecosystems. So-called plant–environment models use leaf-level parameters to infer canopy-level information about net photosynthesis and water budget.

Micrometeorological measurements of fluxes of water, momentum heat, and CO_2 (e.g., by using the Eddy correlation from canopies to the atmosphere) are used by ecophysiologists as parameters for their models to understand overall functions such as transpiration and CO_2 exchange of canopies. Another available tool for investigating photosynthetic light absorption—perhaps as a predictor of net primary production—is remote sensing.

Values gained on this highly integrated level automatically comprise contributions from animals, microorganisms, and soil. These components are connected through carbon and nitrogen fluxes, which, as a whole, can be understood as physiological functions of an ecosystem. Modern isotope techniques can trace pathways of processes and provide an integrated aspect of the long-term behavior of plants. Overall measurements yield important information about broad-scaled influences of air pollutants, climatic changes, raised CO_2 in the atmosphere, and forest decline (Fig. 10).

The question simply formulated by W. D. Billings of "why plants function in their environment the way they do" can be seen as a trigger for investigating the whole complex of functioning under natural conditions as individuals or by forming a canopy or as the primary producer layer of an ecosystem. Finally, ecophysiology contributes to answering the question of how ecosystems operate in a changing biosphere. Answers to these questions will eventually provide us with tools to manipulate and manage plant strategies, for example, to increase crop or wood production. On the other hand, these answers can be an essential basis for estimating the influence of air pollution and CO_2 and for guidelines in nature conservation and landscape management.

VIII. METHODICAL APPROACH: TECHNIQUES

Understanding of ecophysiology has highly profited from physical, and technological findings, and from modern chemical analytics. Not only their new techniques and methodology but also their modern theoretical principles are applied to explain ecophysiological processes. Examples of theoretical concepts are the understanding of water vapor and heat transfer from leaves, the role of the boundary layer, and energy budgets. Water relations are understood by means of the thermodynamic principle of water potential and stomatal control. Gas and water exchange processes are described by electric circuit analogs (potential gradients, conductivity, resistance, capacitance, etc.). Economic principles were introduced for considering energy or carbon balances or cost–benefit relations. Stable isotopes are used to evaluate fixation and metabolic discrimination of carbon and nitrogen. Conceptual, empirical, and mechanistic models are widely applied.

Technical innovations have provided better measuring equipment, precision, and control applications. Much of the technical design was constructed in cooperation between researchers and engineers, and only a limited number of instruments was produced. But with increasing public interest in environmental problems and ecological processes an increasing demand for scientific research created a market for commercially available instruments for ecophysiological measurements.

Many components of the physiological status of plants have to be analyzed in the laboratory, such as enzyme activities (e.g., CAM, flooding response), the presence of hormones (water stress response), pool sizes of metabolites (stress response, acclimation), the nutritional state, and salt content. Similarly, ecophysiologists have to bring harvested plant material to the laboratory for measurements of the osmotic potential, carbon/nitrogen relations, or biomass and shoot/root ratios. Many specific experimental designs must be performed in the laboratory, such as acclimation experiments with whole plants in conditioned chambers or phytotrons, the analysis of the freezing and heat tolerance in a series of controlled chambers, and the measurement of the CO_2 and H_2O exchange as steady-state response in controlled environments (e.g., by means of infrared gas analyzers, or IRGA).

Moreover, ecophysiology is concerned with environmental conditions and performance of plants in the field. The technical and theoretical progress in micrometeorology allowed the establishment of more accurate methods of describing environmental components such as evaporation, precipitation, wind profile, and atmospheric turbulence. Of particular relevance are the methods measuring water vapor pressure in the air on the basis of dew point determination or by means of electric capacity sensors. The amount of light that activates photosynthesis is measured as photon flux density, which is an appropriate measuring unit for the conversion of radiant energy into chemical energy for the assimilation of CO_2. In plant canopies, spectral reflectance of foliage and light transmission and dis-

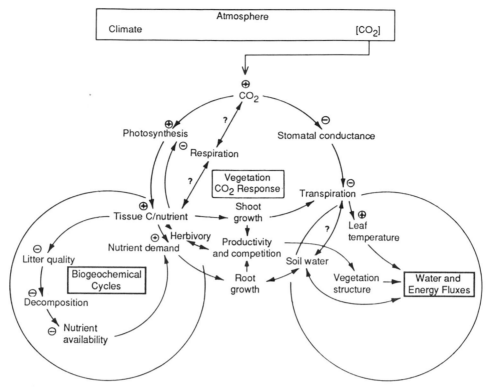

FIGURE 10 Ecophysiology at the level of an ecosystem: "ecosystem physiology" with respect to the influence of (increased) CO_2. Ecosystem responses by biogeochemical cycles, water and energy fluxes, and vegetation response and their interactions are shown in the model. +, positive; −, negative response. [After H. A. Mooney (1992). *In* "Global Change Report No. 21" (W. L. Steffen *et al.*, eds.). I.G.B.P. Secretariat, The Royal Swedish Academy of Sciences, Stockholm.]

persion are measured by radiation sensors or by photographic methods.

The difficult investigations of soil and plant moisture status variables are carried out by tensiometric, psychrometric, manometric, or resistance measurements or by means of neutron probes. A great advance in ecophysiological research has been the Scholander pressure chamber, which measures the tension of water in the conducting elements of plant organs (xylem water potential). Thermoelectric methods are applied to measure the speed of water flow in trees.

Experimentation in the field requires mobility of the instrumentation as well as robustness and weatherproof construction of sensors and recorders. Sensors should be small and should not influence the plant (microsensors). Data should be recorded continuously, illustrating the fluctuations of the environmental conditions. The usually weak signals have to be electronically amplified.

Modern sensors and recorders run on low voltage supply and can operate with battery or solar power generators. Mobile field labs have been installed in all regions on earth, including ice-free sites in Antarctica. They comprise micrometeorological equipment with air and plant sensors, CO_2 and H_2O analyzing equipment with conditioned plant chambers of various sizes, portable pressure chambers, and various other kinds of sensors.

Miniaturization of electronic devices, datalogging equipment, and sensors provide increasing flexibility of application in field experiments. Plant CO_2 exchange and transpiration can be measured with portable porometers. The fluorescence of chlorophyll as a measure of photosynthetic activity and indicator of stress impact to the photosynthetic apparatus can now be detected by means of small portable computerized instruments in the field under ambient conditions.

Furthermore, in the past 20 years the vigorous development of ecophysiological research has prompted the establishment of new monograph series such as "Ecological Studies" (Springer-Verlag, 1970), and "Physiological Ecology" (Academic Press, 1971), and journals like *Advances in Ecological Research* (1962), *Photosynthetica* (1967), *Oecologia* (1971), *Environmental and Experimental Botany* (1976, formerly *Radiation Botany*), *Plant, Cell and Environment* (1978), *Agricultural and Forestry Meteorology* (1984, formerly *Agricultural Meteorology*), and *Functional Ecology* (1987).

Glossary

Allocation, partitioning Investment of metabolites (mainly C and N compounds) into discrete plant organs (stem, root, fruit, etc.).

Apoplastic water Water from intercellular space and cell wall but not from the protoplast.

Biotic Living organism as a factor.

C_3 plant Plant that fixes CO_2 solely by means of the enzyme ribulose bisphosphate carboxylase and initially produces a three-carbon compound. Most frequent in forests and extratropical vegetation.

C_4 plant Plant that fixes CO_2 additionally by means of a more efficient enzyme phosphoenol pyruvate carboxylase and initially produces a four-carbon compound in the cells of the normal leaf tissue. This four-carbon compound (malate or aspartate) forms the substrate for subsequent carboxylation in larger, specialized cells around the conductive tissue. These plants use more energy for fixation of one molecule of CO_2 than the C_3 plants. They occur mainly in tropical and subtropical regions.

CAM plant Plant characterized by fleshy stem and leaves that fixes CO_2 by means of the enzyme phosphoenol pyruvate carboxylase mostly at night. During the subsequent daytime, the four-carbon compound (malate) forms the substrate for photosynthetic CO_2 carboxylation. This temporal separation between CO_2 fixation by night and photosynthetic CO_2 assimilation by day prevents excessive water loss through the stomata on hot and dry days. CAM plants typically grow in hot deserts with scattered but regular rainfall.

Compensation point Point where the photosynthetic rate passes the zero line; it is reached at a threshold irradiance level or at certain temperatures. At the compensation point the CO_2 uptake balances the photorespiratory CO_2 release.

Dew point Any temperature that causes the precipitation of water from moist air; it indicates water saturation in the air.

Eddy correlation method Method of estimating the fluxes (of water vapor) above a canopy. Parameters are the instantaneous vertical velocity and the instantaneous concentration (of water) at that level measured in close temporal sequence.

Epiphyte Plant (vascular or nonvascular) that lives on another plant (shrub, tree) without harmful effect.

Hydration Degree of water imbibition.

In situ measurement Measurement on the intact plant in its natural environment.

Laminary boundary layer Layer of air influenced by a body; it increases with body surface size and with decreasing air movement (wind).

Leaf area index (density) Surface area of the foliage of plants per square meter (cubic meter) related to one square meter of ground surface.

Lignotuber Mostly voluminous lignified (woody) part of the stem basis of a shrub (or tree), well-insulated against fire heat (frequent in Australia).

Metabolite Product of metabolic processes.

Nitrophyte Plant that positively responds to nitrogen compounds in the soil. Nitrophytes are highly productive and grow fast.

Nonvascular plant Plant also called a thallophyte, and lacking water-conducting elements and water loss protection.

Optimum/pessimum Conditions allowing the highest rates of activity and the conditions limiting activity to the lowest rate respectively.

Phorophyte Tree or shrub that carries epiphytes.

Plant litter Plant material dropped to the ground (mostly dead).

Poikilohydrous Pertaining to plants with extremely labile water relations. Being land plants, they change between active (water-soaked) and anabiotic (dehydrated) states and are highly desiccation tolerant.

Species diversity Measure of the natural composition of a number of plant species.

Turgidity Pressure-related stability due to water content in the tissue or cell.

Vascular plant Plant with roots and shoots and containing conducting elements for transport of water and other substances. They are protected against water loss by an epidermis with cuticular and wax layers. Their water release is controlled by the stomatal conductance.

Vegetative structures Leaves, stems, branches, and roots, but not flowers, fruits, or seeds.

Bibliography

Billings, W. D. (1985). The historical development of physiological plant ecology. *In* "Physiological Ecology of North American Plant Communities" (B. F. Chabot and H. A. Mooney, eds.), pp. 1–15. New York/London: Chapman & Hall.

Ehleringer, J. R., and Field, C. B., eds. (1993). "Scaling Physiological Processes. Leaf to Globe." San Diego, Calif: Academic Press.

Evans, J. R., Caemmerer, S., and Adams, W. W., III, eds. (1988). "Ecology of Photosynthesis in Sun and Shade." Melbourne, Australia: CSIRO.

Jones, H. G. (1992). "Plants and Microclimate. A Quantitative Approach to Environmental Plant Physiology." New York: Cambridge University Press.

Kappen, L. (1988). Ecophysiological relationships in different climatic regions. *In* "CRC Handbook of Lichenology" (M. Galun, ed.), Vol. II, pp. 37–100. Boca Raton, Fla.: CRC Press.

Lange, O. L., Nobel, P. S., Osmond, C. B., and Ziegler, H., eds. (1981–1983). "Encyclopedia of Plant Physiology, New Series Vol. 12A–D: Physiological Plant Ecology, I–IV." Berlin/Heidelberg/New York: Springer-Verlag.

Larcher, W. (1991). "Physiological Plant Ecology." Berlin/Heidelberg/New York: Springer-Verlag.

Levitt, J. (1980). "Response of Plants to Environmental Stresses. Vol. I. Chilling, Freezing, and High Temperature Stress." New York: Academic Press.

McIntosh, R. P. (1987). "The Background of Ecology. Concept and Theory." New York: Cambridge University Press.

Mooney, H. A. (1991). Plant physiological ecology—Determinants of progress. *Funct. Ecol.* **5,** 127–135.

Mooney, H. A., Winner, W. E., and Pell, E. J., eds. (1991). "Response of Plants to Multiple Stress." San Diego, Calif: Academic Press.

Nobel, P. S. (1983). "Biophysical Plant Physiology and Ecology." San Francisco: Freeman.

Pearcy, R. W., Ehleringer, J. R., Mooney, H. A., and Rundel, P. W., eds. (1989). "Plant Physiological Ecology." London: Chapman & Hall.

Schulze, E.-D., and Zwölfer, H., eds. (1987). "Potentials and Limitations of Ecosystem Analysis" Ecological Studies 61. Berlin/Heidelberg/New York: Springer-Verlag.

Plants and High CO$_2$

Jelte Rozema
Vrije Universiteit, Amsterdam

I. INTRODUCTION

The burning of fossil fuel releases carbon dioxide to the atmosphere. The rising concentration of CO$_2$ in the Earth's atmosphere is a well documented and generally recognized part of current global changes caused by human activities. The atmospheric concentration has risen continuously from 270 ppm before the Industrial Revolution (1870) to more than 355 ppm at present (1993) (Fig. 1). By the end of the next century, the atmospheric CO$_2$ concentration is expected to increase up to 500–600 ppm. A yearly recurrent sinusoidal pattern in atmospheric CO$_2$ concentration appears to reflect seasonal and CO$_2$ fixation activity of terrestrial vegetation in the northern hemisphere. Plant growth depends to a large extent on the outcome of net photosynthesis, whereby through utilization of photosynthetically active (solar) radiation (PAR), atmospheric CO$_2$ is fixed and, in combination with absorbed water, carbohydrates and oxygen are produced:

$$\mathrm{CO_2 + H_2O} \xrightarrow{\text{PAR}} \underset{\text{(carbohydrates)}}{\mathrm{CH_2O}} + \mathrm{O_2}$$

The great majority of terrestrial angiosperms use atmospheric CO$_2$ as the sole carbon source for photosynthetic carboxylation rather than soil-derived CO$_2$. One of the few exceptions is the vascular land plant *Stylites andicola,* which lacks stomata. This land plant absorbs CO$_2$ through its roots for photosynthetic fixation. Submerged angiosperms (helophytes) appear to use sediment-derived CO$_2$ in addition to an efficient internal recycling of respiratory CO$_2$. Within the plant, CO$_2$ is transported from the root system to the sites of carboxylation in the leaves via air tissue (aerenchyma). Although CO$_2$ concentrations in lacunae in aerial parts of the emergent wetland grass species common reed (*Phragmites australis*) may accumulate to 5000 ppm, this plant exploits only atmospheric CO$_2$ for photosynthetic carboxylation. [*See* GLOBAL ANTHROPOGENIC INFLUENCES.]

II. METHODOLOGY OF CO$_2$ ENRICHMENT

To describe and analyze the response of plants to elevated atmospheric CO$_2$, a careful experimental

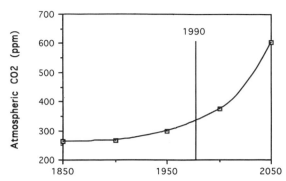

FIGURE 1 Time course of the atmospheric carbon dioxide concentration from preindustrial time (1850) to the present day (1990) and predicted atmospheric CO$_2$ into the future. The preindustrial atmospheric CO$_2$ concentration is projected to double over the next 60 years as a result of continued CO$_2$ emission from burning of fossil fuels. Many experimental CO$_2$ enrichment studies use 350 ppm CO$_2$ as ambient, control treatment and 700 ppm as elevated CO$_2$ treatment.

approach is needed. Ideally the experimental treatment should relate only to the increase of atmospheric CO$_2$ to a desired level, while all other relevant environmental conditions remain constant (controlled environment) or follow natural diurnal and seasonal variation of environmental conditions (see later discussion of OTC and FACE).

Controlled environment facilities, including leaf cuvettes, whole-plant growth chambers, growth cabinets, climate rooms, and greenhouse compartments, are enclosed spaces with control of light, air and soil temperature, relative humidity, and soil nutrient and soil water status. These climatized controlled environments may be called Soil–Plant–Atmosphere Research (SPAR) facilities. The majority of published experimental data on the responses of crop plants and native plant species to elevated CO$_2$ refer to studies performed in controlled environments.

CO$_2$ enrichment generally stimulates photosynthesis and growth of many C$_3$ plant species. This stimulation of net photosynthesis may be a short-term effect, which gradually decreases. It has been hypothesized that this photosynthetic acclimation is caused by the limited space and nutrient availability in the pots used to cultivate experimental plants. Most CO$_2$ enrichment studies in natural plant communities appear not to reveal such photosynthetic acclimation. Reduction of net photosynthesis of plants grown in small pots in a controlled environment is sometimes referred to as the pot-size effect and may be regarded as a methodological artifact. Yet the possibility of photosynthetic acclimation brings to focus that the long-term effect of CO$_2$ enrichment depends on the strength of sources and sinks of photosynthetic products. End product inhibition represents a feedback mechanism affecting the rate of net photosynthesis. [*See* GLOBAL CARBON CYCLE.]

A. Closed Controlled Environments

Greenhouses are frames covered with a screen consisting of PAR-transparent glazing such as (UV-B absorbing) glass or synthetic polymer material (polycarbonate, acrylate). Heaters and cooling systems are used in controlled environments for temperature control often in combination with control of the air humidity. The level of photosynthetically active radiation in greenhouse compartments and other controlled environments is often (but not necessarily) considerably lower than ambient levels. Artificial light may be used to enhance the light level and for control of photoperiod.

Control of the CO$_2$ concentration of the air inside greenhouse compartments is often easy and reliable. Greenhouse compartments with control of CO$_2$, temperature, and air humidity therefore allow relevant experimental studies of plant responses to elevated CO$_2$ in interaction with other environmental conditions such as light, temperature, air humidity, nutrients, and soil water. Many early CO$_2$ enrichment studies refer to closed controlled environment (CCE) experiments. Despite the limitations, the CCE represents an important tool for plant ecophysiologists to unravel plant responses to atmospheric CO$_2$ enrichment.

The controlled environment in a greenhouse may differ markedly from the outside environment, in that environmental conditions (light, temperature, and humidity) are held constant, while outdoors these abiotic factors follow a natural and seasonal pattern. With glazing of greenhouses consisting of glass, indoor plants can be grown in an environment without (solar) UV-B, as glass absorbs solar UV-B radiation. Currently UV-B-transparent glazing is used for greenhouses. In horticultural practice, transmission of solar UV-B radiation may affect both quantitatively and qualitatively the greenhouse crops. This may improve the quality of horticural crops. Also the CCE is usually milder than outside. Extrapolation of greenhouse CO$_2$ enrichment studies to the field situation should always be considered with care.

B. Open-Top Chambers

Besides closed controlled environments, so-called open-top chambers (OTC) (Fig. 2) represent an important tool in experimental CO$_2$ enrichment studies. OTCs may be cylindrical or polygonal. Air taken from a height of 2.50 m above the soil surface is mixed with CO$_2$ from a CO$_2$ gas cylinder.

FIGURE 2 An open-top chamber (OTC) for atmospheric CO$_2$ enrichment. Ambient air is taken in through the vertical PVC tube (chimney) and enriched with CO$_2$. The desired inside CO$_2$ level is controlled and monitored with an automatic mass flow control system and an infrared gas CO$_2$ analyzer. The glazing of the OTC consists of Acrylate, which transmits 90% of incoming solar UV-B radiation. There is no marked temperature variation along a vertical gradient from the bottom to the open top. The crop grown inside is wheat (*Triticum aestivum*) cultivated in pots with nutrient-enriched garden soil. Doubling of ambient atmospheric CO$_2$ (350 ppm) to 700 ppm CO$_2$ led to a 50% increase of wheat plant biomass.

This CO$_2$ supply can be controlled with a mass flow controller. This air is mixed with additional CO$_2$, blown through OTC, and leaves the chamber via the open top. Air is sampled from the chamber and led to an infrared gas CO$_2$ analyzer. These CO$_2$ measurements are coupled to the mass flow control system directing the CO$_2$ supply to the air inlet.

OTCs are applied worldwide as a tool for studies of plant responses to experimental CO$_2$ enrichment. The open top permits capture of natural solar radiation, although part of solar radiation may reflect to the vertical walls. UV-B-transmitting glazing can be applied to avoid the exclusion of solar UV-B. Light, temperature, and air humidity follow, to a large extent, natural diurnal and seasonal variation of these environmental conditions. In well-designed and -operated OTCs, increase of temperature and change of air humidity can be kept to a minimum. From this point of view, plant responses obtained in OTCs may be extrapolated to "real-world ecosystems" more easily than those from controlled environment studies.

Yet, inside the OTC, temperature and humidity may differ from the outside environment. Condensation occurs on the glazing of OTCs, reflecting these temperature differences, which temporarily alters the radiation climate inside the chamber. Wind and turbulence inside the chamber will also differ from outside conditions. Like closed controlled environments, conditions inside open-top chambers, although closer to those in the natural environments, are often milder than in the outside world.

When an OTC is closed for a short period, comparison of the CO$_2$ concentration of the air entering and leaving the chamber permits calculation of the net rate of photosynthesis and respiration (in the dark) of all the plant parts enclosed. Thus, OTC studies permit study of the responses of a single leaf, a whole plant, or even a plant canopy to atmospheric CO$_2$ enrichment.

C. Free Air CO$_2$ Enrichment

The technique of Free Air CO$_2$ Enrichment (FACE) represents a "real-world" approach to realistic experimental analysis of the effects of global atmospheric CO$_2$ enrichment on natural ecosystems, agroecosystems, and other human-made and -controlled ecosystems. The major advantage is the absence of an enclosure around the plant canopy. The FACE methodology uses a circular system of pipes for the provision of CO_2 for CO_2 enrichment of the ambient air of the area enclosed by the pipeline circuit. CO_2 gas released from the pipeline distribution system is diluted and mixed by horizontal advection and by vertical turbulent eddy diffusion. Ideally, desired elevated atmospheric CO_2 concentrations are reached without considerable temporal and spatial variation in CO_2 levels.

There are some advantages of the FACE-type CO_2 enrichment technique over indoor and outdoor controlled environment chambers:

1. The solar radiation environment is not reduced. Levels of photosynthetically active radiation (400–700 nm) of 2000 μEinstein $m^{-2} sec^{-1}$ may be exceeded, whereas in indoor controlled environments much lower PAR levels occur.
2. No absorption of solar UV-B, as occurs with the glazing material of most OTCs and in most greenhouses.
3. No occurrence of unnatural flow of wind, turbulence, and chamber-derived micrometeorological patterns of temperature, humidity, and infrared energy exchange.
4. No change of natural soil water characteristics as with OTCs, where the vertical glazing material and the top of the OTC markedly alter natural precipitation.

Generally speaking, application of the FACE technique avoids the undesired "chamber effect" of CCEs and OTCs. In field CO_2 enrichment studies, plants grown inside OTCs appeared to show delayed senescence. Such chamber effects hamper long-term CO_2 enrichment studies with OTCs.

However, FACE facilities tend to be large-scale facilities; circular FACE systems in arable land in Arizona (U.S.A.), planted with wheat, have a diameter of 22 m. Eight replicate arrays of these FACE rings have been realized. OTCs tend to be smaller (1- to 5-m diameter, 1-4 m in height),

which generally allows more replications. The advantages and disadvantages of the various CO_2 enrichment methods discussed here are listed in Table I.

III. PLANT RESPONSES TO ATMOSPHERIC CO_2 ENRICHMENT

A. High CO_2 and Photosynthesis

Enrichment of atmospheric CO_2 stimulates photosynthesis, particularly in C_3 plants. In C_3 plants, the first molecule formed as a result of CO_2 fixation is the C_3 compound phosphoglyceric acid. Atmospheric CO_2 that enters the plant by diffusion through the stomata is fixed in mesophyll cells by combining with ribulose-1,5-biphosphate. In the presence of oxygen, oxygenation of ribulose-1,5-biphosphate occurs, which is called photorespiration. In C_3 plants, 20–50% of the CO_2 photosynthetically fixed is immediately lost by photorespiration. At increased atmospheric CO_2 levels, the ratio of CO_2 and O_2 is increased at the site of carboxylation, and photorespiration is thereby depressed. As a result, doubling of atmospheric CO_2 from 350 to 700 ppm may lead to a 40% increase of net photosynthesis.

C_4 plants have a different CO_2 acceptor, phosphoenolpyruvate (PEP), with a high affinity for CO_2. When PEP fixes a CO_2 molecule, a C_4 compound such as oxaloacetate is formed in the mesophyll cells. Via intermediate compounds (malate and aspartate), and after transport, the C_4 compounds in the chloroplasts of bundle sheath cells release CO_2, which is then fixed as in C_3 plants, by ribulose-1,5-biphosphate to form photosynthates. The two-step fixation of atmospheric CO_2 in C_4 plants leads to an internal accumulation of CO_2, causing a high CO_2 to O_2 ratio at the site of carboxylation in the chloroplasts of bundle sheath cells. Photorespiration is therefore absent or very low in C_4 plants. Because of the high affinity of the CO_2 fixation for phosphoenolpyruvate catalyzed by the enzyme phosphoenolpyruvate carboxylase, the net photosynthesis of C_4 plants is already saturated at a relatively low concentration of atmospheric CO_2:

TABLE I
Comparison of CO$_2$ Enrichment Methods

CO$_2$ enrichment method	Advantages	Disadvantages
Closed controlled environments (CCE)	Small, permits study of interactions of elevated CO$_2$ with PAR, temperature, and relative humidity; many replications are possible.	Extrapolation of results to natural ecosystems is difficult, PAR level often less than ambient solar radiation level, and artifacts (pot size, photosynthetic acclimation) may interfere with physiological analysis of plant responses; plant size limitations; only short-term experiments.
Open-top chambers (OTCs)	Permits field studies of crops and natural vegetation, with natural ambient PAR levels; many replications are possible; closed configuration of OTC allows study of carbon balance of whole-plant or whole-crop canopy system.	Chamber effect may occur due to altered radiation, wind, and humidity inside OTC.
Free air CO$_2$ enrichment (FACE)	Very close to natural environment (temperature, PAR, humidity) and large area for cultivated or natural plants; permits long-term (years) experiments; allows study of effects of CO$_2$ enrichment on agroecosystems and natural ecosystems and consequences of ecosystem changes (e.g., transpiration, leaf and canopy temperature) on microclimatological parameters (evapotranspiration).	Limited replication and high costs in particular for tall crops and natural vegetation (shrubs and trees); spatial and temporal gradients in atmospheric CO$_2$.

about 350 ppm CO$_2$. As a result, a further atmospheric CO$_2$ enrichment does not cause an increase of net photosynthesis in C$_4$ plants (Fig. 3).

Examples of C$_3$ plants are sugar beet (*Beta vulgaris*), rice (*Oryza sativa*), potato (*Lycopersicon esculentum*), wheat (*Triticum aestivum*), and soybean (*Glycine max*). The C$_4$ CO$_2$ fixation pathway occurs in crops like corn (*Zea mays*), sugarcane (*Saccharum officinarum*), Bermuda grass (*Cynodon dactylon*), and weed plants like amaranth (*Amaranthus retroflexus*). Native C$_3$ plants prevail in the temperate and cool climate zones on earth, whereas C$_4$ plants inhabit subtropical and tropical areas. Along a north–south gradient in the United States, the number of C$_3$ plant species gradually decreases and the C$_4$ plant species become more abundant.

B. High CO$_2$ and Water Use

The cuticle on the epidermis of terrestrial plants forms a barrier for undesired loss of water from the plant tissue to the surrounding atmosphere. Control of opening and closure of stomata situated in the epidermis optimizes water use and gain of atmospheric CO$_2$. In a low-CO$_2$ environment,

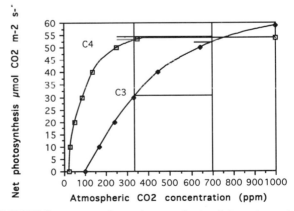

FIGURE 3 Curves of net photosynthesis of C$_3$ and C$_4$ plant species. The vertical lines at 350 and 700 ppm indicate the 1993 atmospheric CO$_2$ concentration and the doubled atmospheric CO$_2$ concentration projected in the next century. The increase of net photosynthesis of C$_3$ and C$_4$ plants as a result of doubling of atmospheric CO$_2$ is indicated.

wide-open stomata could facilitate diffusional CO_2 supply. Atmospheric CO_2 enrichment allows land plants to narrow the opening of stomata, thus reducing transpirational water loss (Table II). As a result, the water use efficiency, that is, the ratio of carbon dioxide fixed to water transpired, increases. Leaf temperature may rise with decreased transpirational water loss. Plants in a high-CO_2 atmosphere show improved water use efficiency. This implies that plants become more resistant to drought and water stress. From an agricultural point of view, C_3 crops are likely to grow better in a high-CO_2 atmosphere, because of increased net photosynthesis *and* reduced transpiration.

C. Atmospheric CO_2 Enrichment and Crop Plants

Because of great economical and agricultural importance and the direct impact on world food production, numerous studies of the effects of elevated CO_2 have been conducted on crop species.

C_3 crop plants generally show a marked increase in biomass accumulation and harvestable yield. The C_4 crop plants like sorghum and corn show much less biomass and yield increase if any (Table II and

III). This difference in responsiveness of C_3 and C_4 plants to elevated atmospheric CO_2 can be related to differences of the CO_2-fixation pathways in these plant groups (see Section III,A).

Leaf stomatal transpiration decreases in both C_3 and C_4 crop plants. This may lead to reduced water consumption by crops and less crop growth in dry areas. However, this prediction is uncertain because elevated CO_2 may lead not only to reduced transpiration but possibly to increased leaf area.

The majority of the available data indicate that elevated atmospheric CO_2 has a positive effect on crop biomass and crop yield. Belowground processes of agroecosystems are less well understood, because of lack of appropriate and accessible methods.

D. Atmospheric CO_2 Enrichment and Wild Plant Species

Studies of the impact of atmospheric CO_2 enrichment on plants from natural ecosystems are far less numerous. Crops and agricultural cultivars have been selected for a high growth rate and biomass allocation to harvestable plant structures. Crops grow in relatively favorable environments with a good supply of water and nutrients and without high weed competition. Wild plant species undergo natural selection and need survival and defense ad-

TABLE II

A Qualitative Survey of the Effects of Increased Atmospheric CO_2 on C_3 and C_4 Plant Growth[a]

Characteristics	C_3	C_4
Photosynthesis	+	0
Growth	+	0
Number of shoots	+	0
Nitrogen concentration of tissue	−	0
Ratio of carbon/nitrogen tissue concentration	+	0
Senescence	−	0
Tillering	+	−
Transpiration	−	−
Water use efficiency	+	+
Leaf temperature	+	+
Leaf water potential	+	+
Stomatal density	−	−
Stomatal index	−	−

[a] +, positive effect; 0, no effect; −, negative effect.

TABLE III

Response of C_3 and C_4 Crop Plants to Doubling of Atmospheric CO_2, Expressed as Percentage Change at 700 ppm CO_2 Compared with Control (350 ppm)[a]

Crop plant	Crop parameter		
	Biomass	Yield	Transpiration
C_3			
Barley	+ +	+ + +	−
Wheat	+ +	+ +	−
Rice	+ +	+	−
Cotton	+ + +	+ + +	−
Soybean	+ +	+ +	−
C_4			
Sorghum	+	−	−
Corn	0	0	−

[a] +, increase; −, decrease; 0, no change; +, 0–25% increase; + +, 25–50% increase; + + +, >50% increase.

aptations to a great variety of environmental stresses. Plants from natural ecosystems may therefore respond differently than crop systems to elevated CO_2. Most crops are harvested after one growing season and long-term CO_2 enrichment studies are scarce.

CO_2 enrichment studies on plants from natural communities confirm to a large extent the findings reported for crops. There may be seasonal and year-to-year variation in the plant responses to elevated CO_2, dependent on rainfall, temperature, and the occurrence of extreme events such as severe frost or storms. Photosynthetic acclimation, that is, reduced stimulation of photosynthesis in response to elevated CO_2, has been reported for crop plants but not for CO_2 enrichment studies of natural salt marsh ecosystems in the United States. The strength of sinks for photosynthates may be greater in natural plant communities than in many (annual) agroecosystems.

IV. PLANTS AND HIGH ATMOSPHERIC CO₂: INTERACTIONS WITH OTHER ENVIRONMENTAL CONDITIONS

Plant growth is not only limited by the content of atmospheric CO_2. In addition, many other environmental conditions may affect plant growth, for example, water, temperature, light, nutrients, salinity, and air pollution (Table IV).

A. Water

Partial closure of stomata promotes water conservation and elevated atmospheric CO_2 tends to alleviate water and salinity stress. This allows plants and crops to grow better under drier conditions. Such partial closure of stomata from increased atmospheric CO_2 may result in reduced damage from air pollution, which enters plants through the stomata. Generally, water stress has been repeatedly reported to be ameliorated by elevated atmospheric CO_2.

TABLE IV

Plant Responses to Atmospheric CO_2 Enrichment: Interactions with Other Environmental Conditions

CO_2 × (soil) water availability	Plants need less water in high CO_2, transpiration rate is reduced.
CO_2 × temperature	Extremely high or low temperatures limit source or sink strength and reduce response to elevated CO_2.
CO_2 × light	In high light, C_3 plants are more responsive to CO_2 enrichment, yet there is a (small) response in low light.
CO_2 × mineral nutrients	In nitrogen deficiency, response of C_3 species to elevated CO_2 is generally less.
CO_2 × salinity	CO_2 enrichment may partially alleviate growth reduction induced by increased salinity.
CO_2 × air pollution	Elevated CO_2 causes decreased stomatal conductance, which reduces effects of air pollutants.

B. Temperature

With a projected doubling of the preindustrial atmospheric concentration of CO_2 to about 560 ppm in the next century, there will be a 1.5–4.0°C increase in global warming. This increase of the mean global temperature will favor the growth of many C_4 plants, rather than C_3 plants. For C_4 plants, the temperature optimum for various physiological and biochemical processes, among them photosynthesis, is relatively high, that is, between 25° and 40°C. For C_3 plants, this optimum temperature is generally about 5–10°C lower. Thus the consequences of the greenhouse effect—atmospheric carbon dioxide enrichment combined with a simultaneous increase of the mean global temperature—will be different for C_3 and C_4 plants. C_3 plants are markedly favored by CO_2 enrichment, but not so by global warming. Carboxylation of C_4 plants is not stimulated by elevated CO_2 but is promoted by temperature increase. The prediction of the response of C_3 and C_4 plant communities to an increase of CO_2 and temperature is therefore rather complex. While the optimum temperature for C_4 species is higher than for C_3 species at the current atmospheric CO_2 concentration, this difference becomes smaller when the CO$_2$ concentration increases. At high CO$_2$ the C$_3$ species behave more like C$_4$ species, including a higher optimum temperature. The relative advantage for C$_4$ species with temperature increase may therefore be smaller when CO$_2$ has also increased.

The temperature dependence of the response of plants to elevated CO$_2$ becomes even more complex when low temperatures (below 18.5°C) are considered. Wheat (*Triticum aestivum*) may show 30% increased growth at elevated CO$_2$ at 25°C, but at 15°C no such increase occurs. At low temperature, reduced or blocked translocation of photosynthates may inhibit photosynthesis and growth. Growth of other plants and crops is even reduced under elevated CO$_2$ and low temperature.

C. Light and Nutrients

Light and atmospheric CO$_2$ are two major factors driving the photosynthesis of plants. Generally the highest percentages of growth increase at elevated CO$_2$ are obtained when light intensity is saturating the rate of photosynthesis. This saturating light intensity may differ between crops. Plant responses to CO$_2$ enrichment will depend on levels of irradiance. Although plant growth is often stimulated by atmospheric CO$_2$ enrichment, sufficient supply and availability of nutrients are required. Reduced growth stimulation at elevated CO$_2$ appears under conditions of inorganic nitrogen limitation. It is remarkable that many species still show a significant response when N is limiting. This is possibly because CO$_2$ increases the Nitrogen Use Efficiency. When P is limiting, the effect of CO$_2$ on growth appears to be much smaller.

D. Salinity

Atmospheric CO$_2$ enrichment may (partially) alleviate salt stress, which implies that salt tolerance of crops and wild plant species is increased with elevated CO$_2$. This effect is probably linked with increased water use efficiency of plants cultivated at elevated CO$_2$. Decreased loss of transpirational water at elevated CO$_2$ may be related to a decrease of salt uptake. Also, less negative values of the

water potential at elevated CO$_2$ have been reported, which may lead to an increased turgor pressure and increased leaf expansion. Alternatively, additional supplies of photosynthates at elevated CO$_2$ may reduce the loss of carbohydrates due to respiration linked with salt tolerance.

E. Air Pollution

Stomata show partial closure at elevated CO$_2$, which infers the reduction of damage to plants by air pollution. Reduced air pollution damage by O$_3$, SO$_2$, and NO$_x$ at elevated CO$_2$ has been reported for several crops. On the other hand stomata may open more wide in polluted air, allowing increased fixation of atmospheric CO$_2$. This response of increased photosynthetic CO$_2$ fixation with air pollution as well as with fertilizer application has been suggested as the "missing sink" in the global carbon budget. This missing sink is the result of the fact that global sources of CO$_2$ (fossil fuel burning, deforestation), exceed global sinks of CO$_2$ (atmospheric increase and ocean uptake of CO$_2$).

V. PLANT RESPONSES TO HISTORICAL CHANGES IN ATMOSPHERIC CO$_2$

Effects of the current increase of atmospheric CO$_2$ can be studied by exposing plants to a control level of ambient atmospheric CO$_2$, for example, 350 ppm, and an elevated level, 700 ppm. Plant biomass increase, gas exchange, water relations, and other physiological parameters can then be measured. In this way the responses of plants, communities, or ecosystems to future enhanced levels of atmospheric CO$_2$ can be predicted.

In addition, there is evidence from the past that terrestrial plants responded to the gradual increase of atmospheric CO$_2$. First, analysis of plant material in herbaria revealed that stomatal density and plant nitrogen have decreased over the past centuries. Second, natural sites exist all over the world where, for long periods, CO$_2$ springs or CO$_2$ vents naturally release CO$_2$ to the atmosphere. Plants occurring in the vicinity of these natural CO$_2$

springs show long-term adaptations to a high-CO_2 environment.

F. I. Woodward showed that the stomatal density (stomatal number per unit of leaf area) of leaves of some British plant species declined over the last 200 years by comparing plant material dating from 1750 to the present day. Stomatal density measurements were also made on leaves of Egyptian olive (*Olea europaea*) originating from King Tutankhamen's tomb dating from 1327 B.C. The stomatal density measurements were compared with values from Egyptian *O. europaea* from pre-332 B.C. and A.D. 1818, 1978, 1991. The results demonstrated that the stromatal density of plants over this time scale was reduced from about $700 \, \text{mm}^{-2}$ (1327 B.C.) to about $500 \, \text{mm}^{-2}$ as the atmospheric CO_2 concentration increased.

Stomatal density represents the number of stomata per unit leaf area. Because leaf expansion and leaf area of plants may vary dependent on water availability and solar radiation, stomatal density will not always relate only to changes of atmospheric CO_2. The stomatal index, that is, the proportion of epidermal cells that are stomata, does not depend on the size of epidermal cells. Therefore the stomatal index will be a more appropriate parameter to test plant responses to historic changes in atmospheric CO_2 than stomatal density.

Further evidence in favor of plant responses to CO_2 enrichment during the last two centuries comes from the fact that leaf nitrogen of collected plant material from 1750 to the present day decreased. This is in accordance with findings of experimental CO_2 enrichment studies of plants. The ecophysiological significance of a reduction of stomatal density and index lies in a reduction of the rate of transpiration.

VI. PLANT RESPONSES TO NATURALLY CO₂ ENRICHED SITES

Thermal springs and gas vents emit gases. In some cases the released gas is composed largely of carbon dioxide (92.5%) and small amounts of nitrogen (5%), methane (2.5%), and traces of oxygen (0.005%). In the proximity of the CO_2 springs, the concentration of carbon dioxide in the atmosphere may easily reach 3000–5000 ppm. Animals have been found dead near CO_2 vents as a result of asphyxiation. Carbon dioxide is generally a significant component of volcanic fumaroles. A CO_2 concentration of about 20% (20.000 ppm) will have lethal effects on animals and humans. There is evidence that the death of citizens of Pompeii after the eruption of the Vesuvius volcano in A.D. 79 was caused by lethal toxicity related to high atmospheric CO_2 concentrations. The evidence comes from the postures of the bodies, which is a characteristic of CO_2 intoxication effects. More recently, natural massive emissions of CO_2 from two lakes in Cameroon, Lake Monoum and Lake Nyos, in 1984 and 1986 killed at least 1800 local inhabitants.

Naturally CO_2 enriched locations offer the opportunity to study long-term adaptations of plants and plant communities to a high-CO_2 world. Plants growing near such sites of CO_2 discharge demonstrated reductions in stomatal density and stomatal conductance. Also decreased leaf tissue nitrogen and increased starch concentrations have been observed in plant individuals growing at naturally CO_2 enriched sites.

However, the aforementioned morphological and physiological characteristics of plants from CO_2 springs may also relate to effects of phytotoxic trace gases. Therefore ecophysiological studies of plant adaptations to the high-CO_2 environment of CO_2 vents should be accompanied by proper experimental CO_2 enrichment research.

VII. CONCLUSIONS

Carbon dioxide is gradually increasing in the atmosphere. A doubling of preindustrial atmospheric CO_2 levels is likely to occur during the next century, which will directly affect terrestrial plants.

Many CO_2 enrichment experiments, conducted in greenhouse compartments, open-top chambers, or Free Air CO_2 Enrichment arrays, have demonstrated increased growth and yield, increased net photosynthesis, reduced transpiration and stomatal conductance, and increased water use efficiency. Growth stimulation by elevated atmospheric CO_2

may range from 30 to 50% for C_3 crops. C_4 crops and C_4 wild plant species do not show such a growth stimulation.

Increased water use efficiency of plants in a high-CO_2 world will permit wild plants and crops to grow better under arid and semiarid conditions. Similarly, partial closure of stomata under elevated atmospheric CO_2 may lead to increased (less negative) values of the water potential of plants. The rising atmospheric CO_2 concentration has been demonstrated to enhance the salt tolerance of crop plants and wild plant species by 5–15%. The doubling of atmospheric CO_2 will be accompanied by 1.5–4.0°C in global warming. Because of differences in the optimum temperature for physiological processes, many C_4 plant species will be favored by global warming.

The response of C_3 and C_4 plant communities to combined elevated atmospheric CO_2 and global warming is difficult to predict. Mounting evidence from plant responses to historical increase in atmospheric CO_2 supports the results of (short-term) CO_2 enrichment experiments. Currently, large-scale and "real-world" Free Air CO_2 Enrichment experiments in the United States and Europe will help us understand the direct effects of elevated CO_2 and interactions with nutrients, water, and other environmental conditions.

Glossary

C_3, C_4 photosynthesis In C_3 plants the initial step in CO_2 fixation is the reaction of CO_2 with the C_5 compound ribulose-1,5-biphosphate, catalyzed by the enzyme ribulose-1,5-biphosphate carboxylase-oxygenase (Rubisco). An unstable C_6 compound is thus formed, producing two C_3 molecules: 3-phosphoglyceric acid. In C_4 plants an alternative acceptor, phosphoenol-pyruvate with a high affinity for CO_2, fixes CO_2 to form a C_4 molecule, such as oxaloacetate.

CCE Closed Controlled Environment.
FACE Free Air CO_2 Enrichment.
OTC Open-Top Chamber.
PAR Photosynthetically Active Radiation has a wavelength range from 400 to 700 nm. Plants absorb most photosynthetically active radiation and the energy involved is used for photosynthesis.
Photorespiration Oxidation of the C_5 compound ribulose-1,5-biphosphate, occurring in the light and catalyzed by Rubisco, the same enzyme catalyzing the CO_2 fixation reaction in C_3 plants. Photorespiration competes against carboxylation (photosynthesis) for the C_5 substrate. Photorespiration is high in C_3 plants at normal ambient CO_2 concentration and low or negligible in C_4 plants.
Rubisco Rubulose-1,5-biphosphate carboxylase/oxygenase, a key enzyme in photosynthesis.
Stomatal density Number of stomata per unit leaf area.
Stomatal index Proportion of epidermal cells which are stomata.

Bibliography

Beerling, D. J., and Chaloner, W. G. (1993). Stomatal density responses of Egyptian *Olea europaea* L. leaves to CO_2 changes in 1327 B.C. *Ann. Botany* **71**, 431–435.

Goudriaan, J., van Keulen, H., and Van Laar, H. H., eds. (1990). "The Greenhouse Effect and Primary Productivity in European Agro-ecosystems." Wageningen, The Netherlands: Pudoc.

Hendrey, G. R., ed. (1992). "FACE: Free-Air CO_2 Enrichment for Plant Research in the Field," Critical Reviews in Plant Sciences, Vol. 11. Boca Raton, Fla.: CRC Press.

Houghton, J. T., Jenkins, G. J., and Ephraums, J. J., eds. (1990). "Climate Change: The IPCC Scientific Assessment." Cambridge, England: Cambridge University Press.

Rozema, J., Lambers, H., van de Geijn, S. C., and Cambridge, M. L., eds. (1993). "CO_2 and Biosphere." Dordrecht, The Netherlands: Kluwer Academic Publishers.

Strain, B. R., and Cure, J. D., eds. (1985). "Direct Effects of Increasing Carbon Dioxide on Vegetation." Washington, D.C.: U.S. Dept. of Energy.

Woodward, F. I. (1993). Plant responses to past concentrations of CO_2. *In* "CO_2 and Biosphere" (J. Rozema, H. Lambers, S. C. van de Geyn and M. L. Cambridge, eds.). Dordrecht, The Netherlands: Kluwer Academic Publishers.

Plant Sources of Natural Drugs and Compounds

Olivier Potterat and Kurt Hostettmann

University of Lausanne

For a long time, plants were the only source of drug available to humans. Even today, 25% of all prescriptions in the United States contain substances derived from higher plants. Moreover, virtually each pharmaceutical class of drug includes a natural prototype. Plants have provided and will continue to provide not only directly usable drugs but also a great variety of chemical compounds that can be used as starting points for the synthesis of analogues having improved pharmacological properties. Only a small fraction of the existing plant species have been investigated up to now. Thus, the plant kingdom represents an untapped reservoir of valuable chemical compounds still to be discovered.

I. INTRODUCTION

The use of plants as a source of relief from illness is as old as humankind itself. Herbal remedies were for centuries the only cures available. Even today medicinal plants represent for the majority of the world's population practically the only source of drugs. In industrialized countries, substances derived from higher plants still constitute about 25% of prescribed medicines, and some 120 plant-derived compounds obtained from about 90 plant species are currently used in modern therapy. Although emphasis shifted away from plant-derived drugs after 1945 with the tremendous development of synthetic pharmaceutical chemistry and microbial fermentation, major pharmaceutical companies have shown a renewed interest over the last decade in higher plants as a source for new lead structures.

II. PLANT-DERIVED DRUGS IN MODERN MEDICINE

Modern medicinal plant research emerged in the early nineteenth century with the development of organic chemistry, biology, and medicine. Plants known to possess medicinal or toxic properties have been systematically investigated and numerous pharmacologically active compounds, mostly alkaloids, have been isolated and characterized, several of which are still of great value in modern medicine. If a few plant metabolites are nowadays produced commercially by chemical synthesis (e.g., caffeine, ephedrine, emetine, papaverine), the great majority are still extracted and purified directly from plants. The structures of some therapeutically and economically important plant-derived drugs are depicted in Fig. 1.

Morphine and narcotine were isolated in 1816 from opium, the dried latex obtained after incision of unripe poppy capsules (*Papaver somniferum*, Papaveraceae), followed a few years later by codeine, narceine, and thebaine. Morphine and codeine are still used as analgesics and antitussives, respectively. Morphine has additionally served as a model for the development of many powerful synthetic analgesics such as methadone. Almost at the same time (1820), quinine was discovered in cinchona bark (*Cinchona pubescens,* syn. *C. succirubra,* Rubiaceae). Until the Second World War, this alkaloid was one of the only antiparasitic agents effective against malaria. It was used as template for the development of potent synthetic antimalarials such as chloroquine. Digitaline, a crystalline mixture of cardiac glycosides, was obtained in 1866 from foxglove (*Digitalis purpurea,* Scrophulariaceae). The cardiac glycosides, the first of which were isolated in the early 1930s, are of enormous value for the treatment of atrial fibrillation and congestive heart failure. Digitoxin, gitoxin, digoxin, lanato-

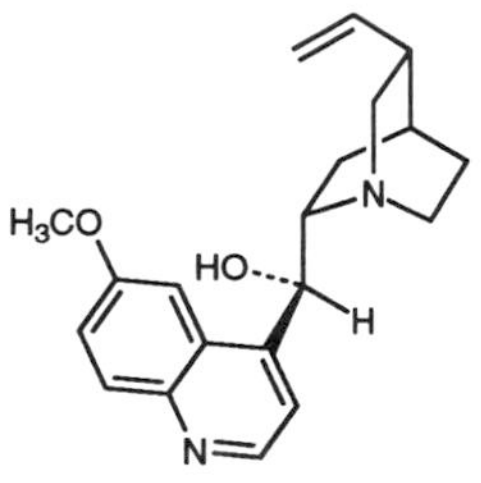

Quinine

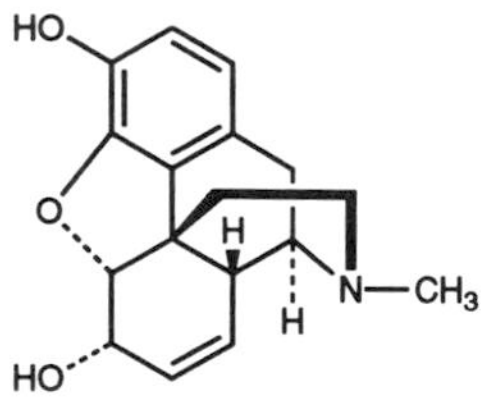

Morphine

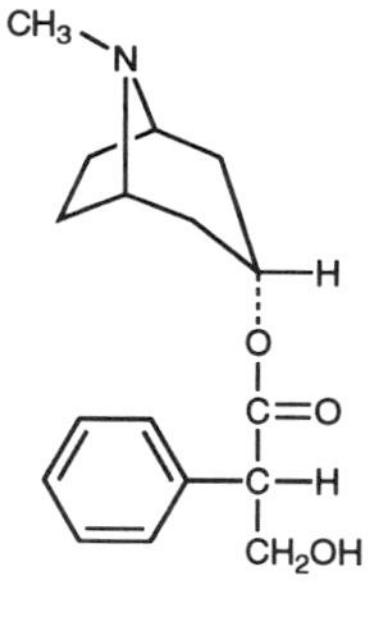

Atropine

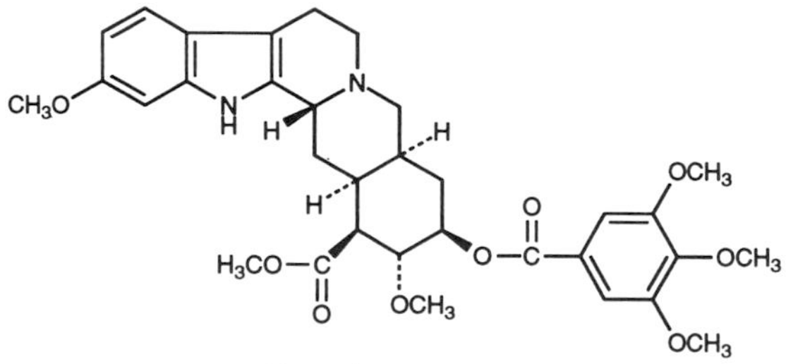

Reserpine

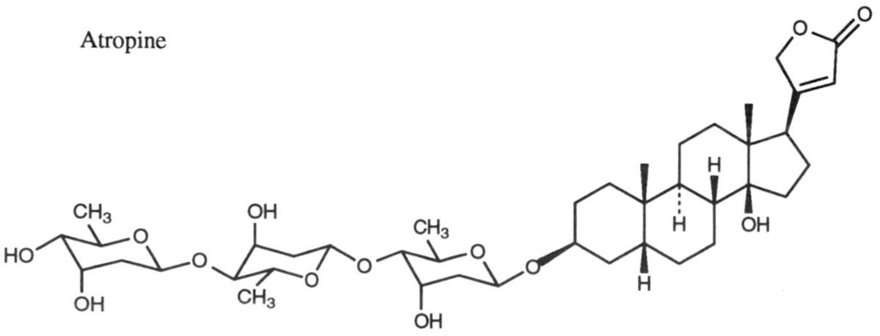

Digitoxin

FIGURE 1 Some therapeutically and economically important plant-derived drugs.

side C, and some related glycosides are still extracted from the leaves of *Digitalis purpurea* and *D. lanata.*

Other significant medicines of plant origin include pilocarpine, a cholinergic alkaloid from *Pilocarpus jaborandi* (Rutaceae) used for treating glaucoma; the antihypertensive and psychotropic drug reserpine from the Indian snakeroot, *Rauwolfia serpentina* (Apocynaceae); colchicine, isolated from *Colchicum autumnale* (Liliaceae), which is used to relieve gout; the anticholinergic tropane alkaloids atropine, hyoscyamine, and scopolamine from *Atropa belladona, Datura stramonium, Hyoscyamus niger,* and other solanaceous species; and the anticancer agents vinblastine and vincristine from the Madagascar periwinkle (*Catharanthus roseus,* Apocynaceae).

In addition to the pharmacologically active plant secondary metabolites that have found direct medicinal application in an unmodified state (e.g., vinblastine and vincristine), many other biologically active plant compounds have served as *leads* or *model compounds* (templates) for synthesis or semisynthesis of novel drugs. Dicoumarol, isolated from fermented sweet clover (*Melilotus officinalis,* Fabaceae) as the causative agent of "sweet clover disease" in animals, was the prototype structure for the discovery of the anticoagulants of the 4-hydroxycoumarin type (ethylbiscoumacetate, phenprocoumon). The alkaloid vasicine from the Ayurvedic medicinal plant *Adhatoda vasica* (Acanthaceae) served as template for the secretolytic drug bromhexin. Khellin, a furanochromone from *Ammi visnaga* (Apiaceae), a plant used in Egyptian traditional medicine as a remedy for bronchial asthma and cardiac irregularities, provided the lead for the discovery of the antiasthmatic drug cromoglycate and the antiarrhythmic agent amiodarone.

The alkaloid cocaine from *Erythroxylon coca* (Erythroxylaceae) was the starting point for the development of local anesthetics, such as lidocaine, procaine, and benzocaine. Last but not least, the isolation of salicin, a simple phenolic glucoside from the bark of various *Salix* spp. (Salicaceae) known for their analgesic properties, led later to the discovery and commercialization of the synthetic drug aspirin (acetylsalicylic acid).

A third way in which plants may contribute to the development of drugs is by providing building blocks for the synthesis of pharmacologically active substances with complex structures. The steroidal saponin diosgenin purified from yam species (*Dioscorea* spp., Dioscoreaceae) is used as starting material for the synthesis of oral contraceptives. The related glycoside hecogenin from sisal (*Agave sisalana,* Agavaceae) is suitable for the manufacture of corticosteroids.

III. SEARCH FOR NEW DRUGS

Among the approximately 250,000 known plant species, only a small percentage have been phytochemically investigated, and the fraction submitted to biological or pharmacological screening is even smaller. For example, even though the American National Cancer Institute had screened over 35,000 species of higher plants for antitumor activity by 1986 and is currently undertaking the acquisition and testing of some 20,000 tropical species, these plants still have to be considered "uninvestigated" with respect to any other biological activity. The plant kingdom thus represents a largely unexplored reservoir of pharmacologically valuable compounds to be discovered. Obviously, not every one of the 250,000 different species can be submitted to a complete panel of pharmacological and biological assays. The selection of the plants to be studied is therefore a crucial factor for the ultimate success of the investigation. Besides random collection of plant material, targeted collection based on consideration of chemotaxonomic relationships and the exploitation of ethnobotanical information is currently carried out. Ethnopharmacological screens taking into account the empirical knowledge of traditional medicine appear to be the most likely to yield pharmacologically active compounds.

Despite the emergence of powerful techniques, developing a clinical drug from a plant remains a formidable challenge. The process that leads from a plant to a pharmacologically active constituent and further to a useful drug is a very long interdisciplinary task taking up to 20 years or more and requiring the collaboration of biologists, chemists, biochemists, pharmacists, and physicians. This process, which is summarized in Fig. 2, can be divided into the following stages:

- collection, proper botanical identification, and drying of the plant material;
- preparation of appropriate extracts and preliminary chromatographic analysis by thin-layer chromatography (TLC) and high-performance liquid chromatography (HPLC);
- biological and pharmacological screening of crude extracts;
- isolation of the active constituents; several consecutive chromatographic steps are required; each fraction has to be biologically tested to localize the active constituent(s) during the isolation process (activity-guided isolation);
- elucidation of structure by chemical and physicochemical methods, including mass spectrometry (MS) and nuclear magnetic resonance (NMR);
- partial and/or total synthesis;
- preparation of derivatives/analogues for the investigation of structure–activity relationships; development of more active and/or less toxic analogues;

- large-scale isolation for pharmacological testing, toxicological studies and clinical trials;
- development of appropriate pharmaceutical preparations.

At each stage of this very long multidisciplinary procedure numerous hurdles have to be overcome, a lot of which are not restricted to plant-derived drug research. Dealing with plants, however, results in additional drawbacks that have discouraged many pharmaceutical companies in the past from investigating plant-derived compounds. While microbial metabolites can be produced in any desired quantity, a continuous supply of plant material in adequate amounts for drug development and commercialization represents a major problem (e.g., see the difficulties encountered by the development of the new anticancer drug taxol). Higher plants and trees in particular grow slowly; they are submitted to seasonal changes and numerous other extrinsic factors that may be difficult or impossible to control. In addition, for tropical plants or rare species, political and ecological issues have to be dealt with. Plant cell culture on an industrial scale, similar to microbial fermentation, would have the potential to circumvent these problems. However, because of the slow growth of plant cells, their sensitivity to mechanical stress, and an insufficient understanding of regulatory mechanisms involved in the expression of secondary metabolites, plant cell culture will not be an alternative in the near future.

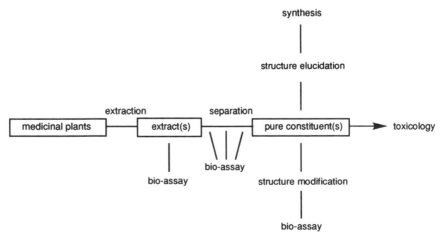

FIGURE 2 Procedure for obtaining the active principles from plants.

Large-scale isolation of plant constituents can also be problematic. Plant extracts are highly complex mixtures containing hundreds or thousands of different compounds. In contrast to microbial metabolites usually produced after optimization in high yields, biologically active plant constituents are often present in very small amounts (e.g., 5 tons of fresh material are required to isolate 1 g of vincristine or vinblastine). Consequently, isolation of plant constituents on an industrial scale can be a tedious and expensive process requiring sophisticated chromatographic methods.

IV. BIOASSAYS AND PHARMACOLOGICAL MODELS

Bioassays constitute the fundamental element of a drug discovery program. In the screening phase, identification of promising plants with the desired biological activity greatly depends on the reliability of the test system used. Bioassays play a crucial role during the isolation process: they serve as a guide for the identification and isolation of compounds showing activity. Thus, all fractions are biologically evaluated and those continuing to exhibit activity are carried through further isolation and purification until pure active principles are obtained. This strategy, known as "activity-guided isolation," is assumed to afford the pure active constituent(s). There are unfortunately many obstacles that can seriously hamper this approach. First, it is possible that a broad range of structurally diverse compounds contribute to the overall pharmacological activity of a plant extract and synergistic effects between these active principles may exist. This is particularly the case for medicinal plants possessing less specific activities. Typical examples include extensively studied plants such as *Panax ginseng* (Araliaceae). Second, there is an urgent need for more appropriate pharmacological models. Existing assays are quite often not reliably predictive for clinical efficacy. Third, for a number of common diseases of unknown or multifactorial origin (e.g., arteriosclerosis), no suitable pharmacological models have yet been developed.

The discovery of promising plant extracts and the subsequent activity-guided isolation place specific requirements on the bioassays to be used. Bioassays must be simple, inexpensive, and rapid to be compatible with a large number of samples, which include extracts from the screening phase and all fractions obtained during the isolation procedure. They must be sensitive enough to detect active principles that are generally present only in small concentrations in crude extracts. Their selectivity should be such that the number of false positives is reasonably small. Poor solubility of extracts under test conditions is quite common. False negative or false positive results are therefore a much more serious problem than when dealing with pure compounds.

The targets for bioassays can be classified into five major groups: (1) lower organisms, such as microorganisms, insects, crustaceans, and molluscs; (2) isolated subcellular systems, such as enzymes, receptors, and organelles; (3) cultured cells of human or animal origin; (4) isolated organs of vertebrates; and (5) whole animals. With a deeper understanding of cell biology and molecular pharmacology, mechanism-based assays using subcellular systems have become increasingly important. Because of their selectivity and sensitivity, combined with good reproducibility and high sample throughput, this type of assay is given preference for large screening programs in industry or in collaborative ventures. However, mechanism-based assays will not detect compounds with unknown mechanisms of action; nonspecific interactions such as enzyme inhibition by tannins will lead to false positive results.

Most of the bioassays require specialized facilities for cell culture and laboratory animal handling and the know-how of a biochemist, biologist, or pharmacologist. Bioactivity-guided fractionation has to be carried out in cooperation with research groups possessing the appropriate infrastructure. Even in a good collaborative setting, the waiting times for the results may be weeks or months, thus slowing down the isolation process considerably. Moreover, in the past many pharmacologists have been quite reluctant about the idea of testing crude extracts in their assays. A number of natural product laboratories have therefore been looking out for inexpensive, simple "bench-top" bioassays that can be carried out by nonspecialized personnel. Over the last few years, a number of such assays have emerged, some of which are now routinely used by phytochemical laboratories. Lower organisms have in particular been put forward as alternative screening models for the discovery of new anticancer agents. Sea urchin eggs and the unicellular green alga *Micrasterias denticulata* have been used to detect antimitotic activity. The crustacean *Artemia salina* (brine shrimp) has been proposed as a low-cost substitute for cytotoxicity assays. The predictivity of some of these simple substitution assays is still a matter of controversy.

V. SOME BIOACTIVE PLANT CONSTITUENTS DISCOVERED DURING THE PAST 30 YEARS

A. Antagonist of Platelet Activating Factor

Platelet activating factor (PAF, 1-*O*-alkyl-2-acetyl-*syn*-glycero-3-phosphocholine) is involved in a broad range of biological processes that have not yet been fully explored. It is a potent mediator of anaphylaxis and inflammation, and is implicated in shock, graft rejection, ovulation and ovoimplantation, pulmonary hyperreactivity, and acute ischemia. There is also growing evidence that PAF is capable of modulating the immune response.

Numerous nonspecific PAF antagonists have been reported, whereas specific antagonists of synthetic origin are a growing group of structurally heterogeneous compounds. However, two of the largest and most extensively studied groups of specific antagonists are of plant origin, namely, the ginkgolides from the maidenhair tree and the kadsurenone-related lignans from the Chinese climbing plant "haifenteng."

The ginkgo or maidenhair tree (*Ginkgo biloba* L.) is the sole survivor of the family Ginkgoaceae that flourished during the Cretaceous period, 80 million years ago. Originating from southwestern China, this "fossil tree" has been introduced as an orna-

mental plant in Europe and in North America. The maidenhair tree is an old Chinese medicinal plant and is still part of the modern Chinese pharmacopoeia as an antiasthmatic and antitussive drug. In Europe, the first pharmaceutical specialities containing ginkgo extracts were commercialized in the 1960s. Indications for these preparations are various disorders of peripheral blood circulation and adjuvant treatment of the sequels of cerebral ischemia. The clinical efficacy of ginkgo extracts was for a long time ascribed to the phenolic constituents (flavonol glycosides, biflavonoids). Only a few years ago it was discovered that extracts of ginkgo specifically antagonize platelet aggregation induced by platelet activating factor. Four compounds responsible for the inhibitory effect were subsequently isolated and identified as ginkgolides A (**1**), B (**2**), C (**3**), and M (**4**). Interestingly, these compounds, which are structurally unique diterpenes, were already known for a long time but they were considered devoid of any pharmacological activity.

As a consequence of the implications of PAF in numerous physiological and pathological processes, the pharmacological profile of the ginkgolides has been extensively studied. Besides a specific inhibition of PAF-induced platelet aggregation, ginkgolide B antagonizes thrombus formation *in vivo* and also produces thrombolysis. Ginkgolides exert a protective effect against bronchoconstriction induced in asthmatic patients. A synergistic effect of ginkgolide B with immuno-

1 $R_1 = OH; R_2 = R_3 = H$, ginkgolide A

2 $R_1 = R_2 = OH; R_3 = H$, ginkgolide B

3 $R_1 = R_2 = R_3 = OH$, ginkgolide C

4 $R_1 = H; R_2 = R_3 = OH$, ginkgolide M

suppressive drugs has been observed in the suppression of cell-mediated graft rejection. In various models of cerebral ischemia, ginkgolides reduced hypoxic damage as well as postischemic lesions. Clinical trials confirmed the PAF antagonistic effect in humans. In asthmatic patients, protection against pollen- and cold-induced asthma was observed. Ginkgolide B is currently being clinically evaluated for its efficacy in the treatment of shock, stroke, graft rejection, organ preservation, and extracorporal circulation (hemodialysis).

The neolignan kadsurenone (5) was isolated from the Chinese medicinal plant "haifenteng" (*Piper futokadsura* Sieb., Piperaceae) as a potent and specific inhibitor of PAF-induced platelet aggregation. Herbal preparations of the plant are used for the general relief of bronchoasthma and stiffness, inflammation, and pain in rheumatic conditions. The correlation between empirical use in traditional medicine and the pharmacological properties is quite remarkable. Kadsurenone is present at very small concentrations in the plant but is a potent inhibitor of the binding of PAF to its receptor. Kadsurenone antagonizes various PAF-induced phenomena, such as effects on heart and on skin permeability. It also partially counteracts endotoxin shock. Compared with the ginkgolides, kadsurenone has a relatively short action. Several structurally related lignans, such as kadsurin A, kadsurin B, and piperenone, were also isolated from *Piper futokadsura*. Their weak activity compared with kadsurenone emphasizes the high degree of structural specificity required.

B. Cardiovascular Agents

In the search for novel plant-derived cardiovascular drugs, Indian scientists discovered the blood pressure-lowering and antispasmodic effect of *Coleus forskolii* Briq. (Lamiaceae), a plant used since ancient times in East Indian folk medicine. The main active principle, forskolin (6), is a potent stimulator of adenylate cyclase activity. It shows vasodilatatory properties and a positive inotropic action on the heart, lowers the intraocular pressure, and inhibits platelet aggregation. The toxicity of forskolin is low. Clinical studies have so far focused on cardiovascular and bronchospasmolytic effects and on the treatment of glaucoma. Additionally, tests carried out on animals have revealed a potential antimetastatic activity.

C. Anticancer Agents from Higher Plants

Efforts to obtain anticancer agents from higher plants go back more than four decades. As mentioned earlier, a screening program initiated at the National Cancer Institute (NCI) of the United States in 1956 screened more than 35,000 plant species for antineoplastic activity up to 1986. Similar, though less extensive, programs were conducted in several other countries, notably in the People's Republic of China and in France. Recently, the NCI has been developing a new approach to antitumor activity screening using a broad panel of human tumor cell lines. Contracts for the collection of 20,000 plant species from Africa, Central and South America, and Southeast Asia have been awarded.

The first clinically useful compounds to be isolated from a plant were the *Catharanthus* alkaloids. The antineoplastic properties of the Madagascar periwinkle (*Catharanthus roseus* G. Don., Apocyna-

5 kadsurenone

6 forskolin

7 $R_1 = R_2 = CH_3$; $R_3 = Ac$, vinblastine

8 $R_1 = CHO$; $R_2 = CH_3$; $R_3 = Ac$, vincristine

9 $R_1 = CH_3$; $R_2 = NH_2$; $R_3 = H$, vindesine

10 5'-nor-anhydrovinblastine

ceae) were independently discovered by Canadian and American research teams. Combined efforts led to the isolation and structure elucidation of the active bis-indole alkaloids vinblastine (**7**), vincristine (**8**), leurosine, and leurosidine in the late 1950s and the early 1960s. Two of them, vinblastine and vincristine, have been developed as commercial drugs and are still among the most important chemotherapeutic agents in use. These compounds are antimitotics inhibiting cell growth, at least in part, by disrupting microtubules, causing the dissolution of cell mitotic spindles and the arrest of cells at metaphase. The two compounds differ in their clinical utility and toxicity. The major use of vinblastine is the treatment of patients with Hodgkin's disease, non-Hodgkin's lymphomas, and renal, testicular, head, and neck cancer. Vincristine is widely used in combination with other anticancer agents in the treatment of acute lymphotic leukemia in childhood, and for certain lymphomas and sarcomas, small-cell lung cancer, cervical and breast cancer. Bone-marrow toxicity is associated with vinblastine, whereas vincristine has a more pronounced neurotoxicity. Numerous structural variants were synthesized to find compounds with lower toxicity and/or a different specificity. Among them, vindesine (**9**) (desacetyl vinblastine amide) and the derivative 5'-nor-anhydrovinblastine (**10**) are currently used as anticancer drugs. 5'-Nor-anhydrovinblastine possesses a broader anticancer activity and a lower neurotoxicity than the other *Catharanthus* alkaloids.

Podophyllin, the resin obtained by alcoholic extraction from rhizomes of *Podophyllum peltatum* L. and *P. emodii* Wallich (Berberidaceae), demonstrated antimitotic activity and antitumor properties in experimental cancer in animals. Podophyllotoxin (**11**), an aryltetralin lignan, was isolated as the main active principle in the late 1940s. Further investigation of the resin in the 1950s resulted in the isolation and characterization of several related lignans, such as α-peltatin, β-peltatin, demethylpodophyllotoxin, deoxypodophyllotoxin, and the corresponding glucosides. However, early clinical results with podophyllotoxins were disappointing; for example, α-peltatin, which was active against several transplantable rodent tumors, showed severe nonspecific side effects and no significant response in humans. Systematic structural modification led to the first therapeutically useful derivatives podophyllinic acid ethylhydrazide (SP-I) and podophyllotoxin benzylidene-β-D-glucopyranoside (SP-G). Some derivatives of 4'-demethylepipodophyllotoxin also exhibited promising antitumor activity. Two of them, etoposide

11 podophyllotoxin

12 R = CH₃   etoposide

13 R = [thiophene structure]   teniposide

(**12**) and teniposide (**13**), were developed as anticancer drugs. In contrast to the spindle poison podophyllotoxin, which causes arrest of the cell in the metaphase, etoposide and teniposide do not affect microtubule assembly but prevent cells from entering mitosis. Etoposide has proven to be useful in the treatment of patients with small-cell lung cancer, testicular cancer, Kaposi's sarcoma, lymphoma, and leukemia. Teniposide is mainly used to treat acute lymphatic leukemia neuroblastoma in children, as well as non-Hodgkin's lymphoma and brain tumors in children.

The diterpenoid taxol (**14**) was discovered as part of an NCI-sponsored screening program. Isolated for the first time from the stem bark of the Pacific yew *Taxus brevifolia* Nutt. (Taxaceae) in the late 1960s, this compound has since been found in other *Taxus* species such as the European yew *T.*

baccata L. Taxol possesses a unique mode of action. In contrast to other antimitotic drugs, taxol enhances both the rate and yield of microtubule assembly, which leads to the formation of abnormal arrays or "bundles" of tubulin. In clinical trials, responses in patients with colon, lung, and ovarian cancer were obtained. The major obstacle in the development of taxol to the clinical stages has been its very limited supply from natural sources. The compound occurs almost exclusively in the stem bark and at low concentration. More than 10,000 trees had to be cut down to provide adequate amounts required for clinical trials. *Taxus* species are among the slowest-growing trees in the world and with the current demand for taxol it has been estimated that yew populations worldwide would soon become extinct. On the other hand, although some progress has been made, the total synthesis of taxol is likely to remain impractical on an industrial scale. An economically and technically realistic approach to overcome the supply problem has been devised by French researchers at the Centre National de la Recherche Scientifique (CNRS) with the partial synthesis of taxol from 10-deacetylbaccatin III (**15**), a more accessible congener found in the leaves. At the same time, the semisynthesis

15 10-deacetylbaccatin III

14 taxol

16 taxotère

of taxol has given access to a series of structural analogues. Among them, taxotère (**16**), which has reportedly a greater chemotherapeutic potential than taxol itself, is now being clinically tested. Taxol received final approval from the U.S. Food and Drug Administration (FDA) for the treatment of drug-refractory ovarian cancer in December 1992.

The monoterpenoid–alkaloid camptothecin (**17**) was isolated from the Chinese ornamental tree *Camptotheca acuminata* Decne (Nyssaceae) in the late 1960s. However, despite potent antitumor properties, the clinical success remained moderate owing to toxic effects and poor solubility. Interest for camptothecin was recently revived with the discovery of its novel mode of action: camptothecin inhibits DNA-topoisomerase I, an enzyme involved in DNA replication, thus inducing single-strand breaks of cellular DNA. Considerable efforts have since been made to find more active and/or less toxic structural analogues. Clinical trials with the two derivatives topotecan and irinotecan are currently under way. Topotecan has already shown responses against colon and nonsquamous cell lung cancer. Irinotecan exhibits anticancer activity against lung, colorectal, ovarian, and cervical cancer, as well as non–Hodgkin's lymphomas.

The pyridocarbazole alkaloid ellipticine (**18**) was first isolated in 1959 from the leaves of *Ochrosia elliptica* Labill. (Apocynaceae). Subsequently, this compound was also shown to occur in many other Apocynaceae belonging to the genera *Aspidosperma, Tabernaemontana,* and *Strychnos.* Its inhibitory properties on the growth of various experimental tumors were discovered in the late 1960s, but toxicity problems impeded its further development. Recently, French researchers from the CNRS, in collaboration with a pharmaceutical company, developed the highly

18 ellipticine

active and less toxic derivative 2-*N*-methyl-9-hydroxyellipticinium acetate (elliptinium), which exhibits very promising activity for the treatment of breast cancer metastases, kidney sarcoma, and hepatoma.

Harringtonine (**19**) was isolated from the bark of the evergreen *Cephalotaxus harringtonia* (Knight ex Forb.) K. Koch (Cephalotaxaceae). Harringtonine and other related esters such as homoharringtonine were subsequently found in other *Cephalotaxus* species. Following the demonstration of anticancer activity in preclinical studies at the NCI, harringtonine and, later, homoharringtonine were selected for preclinical development in 1969. The limited supply of *Cephalotaxus* bark forced a halt to the development in the United States for some years, however. The reestablishment of relations with the People's Republic of China (the genus *Cephalotaxus* is indigenous to China) solved the supply problem. In clinical trials conducted in China and the United States, homoharringtonine yielded responses in both myeloid and lymphoblastic leukemia.

D. Antimalarial Agents

Malaria is still the most important tropical disease. The number of clinical cases is estimated to be

17 camptothecin

19 harringtonine

around 200 million annually and more than 2 million people, mainly infants and young children, die from it yearly. Malaria is caused by parasitic protozoa of the genus *Plasmodium*. The disease is transmitted by the bite of infected female *Anopheles* mosquitoes, causing sporozoites to enter the human blood circulatory system. Because of increasing resistance of *Plasmodium* strains to currently used drugs, there is an urgent need for new antimalarial agents. Screening programs aimed at new antimalarial drugs of plant origin were initiated soon after World War II. These attempts were encouraged by the fact that many plants are used in tropical traditional medicine for the treatment of malaria.

The most promising compound discovered so far is artemisinin (qinghaosu, **20**), which was isolated in 1972 by Chinese scientists from qinghao (*Artemisia annua* L., Asteraceae), a plant used for over 2000 years in China as a febrifuge and for the treatment of malaria. Artemisinin is a sesquiterpene lactone containing an endoperoxide group. It represents a completely new chemical class of antimalarial compounds and shows a high activity against resistant *Plasmodium* strains. Because of the highly lipophilic nature of artemisinin, however, there are some problems with its administration as a drug. A series of derivatives including ethers, aromatic and aliphatic esters, and carbonates have therefore been synthesized. Among the most active of these compounds are artemether (**21**), arteether (**22**), and sodium artesuanate (**23**). Artemether has been developed as a drug by the Chinese. At the same time, Rhone Poulenc Rorer have commercialized artemether under the trade name Paluther for oral, intramuscular, and intravenous use. These products have been licensed as medicines in several African countries, Brazil, China, Burma, and Thailand. Arteether is currently being investigated by the World Health Organization and the Walter Reed Army Institute of Research for use in the clinic. Artemisinin and its derivatives have been used for antimalarial chemotherapy in more than 1.5 million patients.

Plants of the family Simaroubaceae are pantropically used in traditional medicine for the treatment of malaria. Several species have also shown antimalarial activity *in vitro*. Detailed investigations revealed that quassinoids, a unique class of bitter-tasting terpenes, were responsible for the antimalarial properties. Glaucarubinone (**24**) isolated from *Simarouba amara* Aubl. was particularly potent both *in vitro* and *in vivo* against *Plasmodium berghei*. Brucein B (**25**) and brusatol (**26**) isolated from *Brucea javanica* (L.) Merr. were also very effective. However, the therapeutic use of quassinoids would be affected by the problem of the high toxicity of these compounds.

20 artemisinin

21 R = CH₃, artemether

22 R = CH₂CH₃, arteether

23 R = COCH₂CH₂COONa, sodium artesuanate

24 glaucarubinone

25 R = CH₃, brucein B

26 R = CH=C(CH₃)₂, brusatol

At the same time that the potential of other medicinal plants was being realized, there has been resurgence of interest in quinine, now recognized as an indispensable and effective drug in the treatment of *P. falciparum* malaria resistant to 4-aminoisoquinolines and antifolate. Chloroquine-resistant infections can also be treated with quinine in combination with other drugs. Other *Cinchona* alkaloids are likewise effective in malaria treatment. Quinidine is reported to be more potent than quinine, whereas cinchonine shows greater activity than both against several strains of *P. falciparum*.

E. AIDS-Antiviral Agents

In part because of the complexity of the techniques involved, there have been only a limited number of antiviral screening programs of plant extracts in the last few decades. Nevertheless, some antiviral agents having a good chemotherapeutic index have been isolated and characterized. A series of 3-methoxyflavones isolated from *Euphorbia* species (Euphorbiaceae) appear to be promising leads for the development of antirhinovirus drugs. The triterpene saponin glycyrrhizic acid showed a potent antiviral activity *in vitro* against herpes simplex, varicella-zoster, and human immunodeficiency viruses. Whether these compounds have any clinical potential remains to be established.

The identification of the human immunodeficiency virus (HIV) as the causative agent of AIDS has strongly stimulated the search for novel antiviral agents. A large-scale screening program aimed at the discovery of new HIV inhibitors of natural or synthetic origin is currently under way at the American National Cancer Institute. This research has already resulted in the discovery of several active natural products that appear to be promising lead compounds for the development of anti-HIV drugs. The first plant-derived compounds that emerged from this screening were hypericin (**27**) and castanospermine (**28**). The anthracenic derivative hypericin occurs in certain species of *Hypericum* (Guttiferae), such as St. John's Wort (*H. perforatum* L.). Hypericin exhibits antiretroviral activity both *in vitro* and *in vivo*. Clinical trials are currently under way and successful long-term treatment of HIV

27 hypericin

28 castanospermine

patients has already been reported. The tetrahydroxy indolizidine alkaloid castanospermine occurs in the toxic chestnutlike seeds of the evergreen Australian tree *Castanospermum australe* A. Cunn et Fras. (Fabaceae). Castanospermine blocks HIV replication by inhibiting the enzymes α-glucosidases I and II, which are of key importance in the production of glycoproteins contained within the envelope of the virus. Without the proper glycosylation coating, the virus is unable to spread to uninfested cells. However, rather high toxicity may impede further development of castanospermine.

Recently, new lead compounds that are candidates for preclinical development have been discovered at the NCI. The naphthalene-tetrahydroisoquinoline alkaloid michellamine B (**29**) was isolated from the tropical liana *Ancistrocladus korupensis* (Ancistrocladaceae), a previously undescribed species. Michellamine B demonstrated a broad-range anti-HIV activity against diverse virus strains and is considered by the NCI as a high-priority candidate for preclinical studies. The phorbol ester prostratin (**30**) was obtained from the bark of the small tree *Homalanthus nutans* (Euphorbiaceae), a plant used by Samoan traditional healers for the treatment of various diseases. Bark extracts revealed HIV antiviral properties and prostratin was identified after activity-guided fractionation as

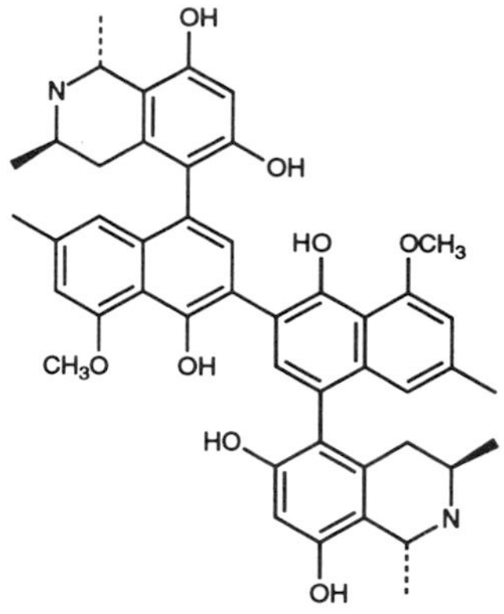

29 michellamine B

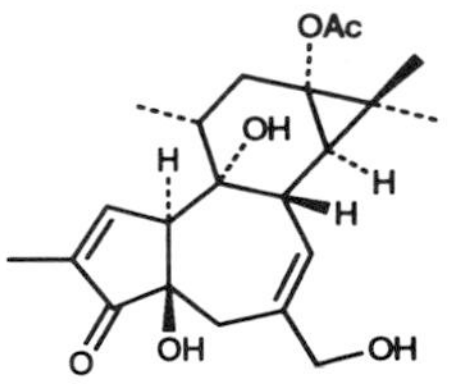

30 prostratin

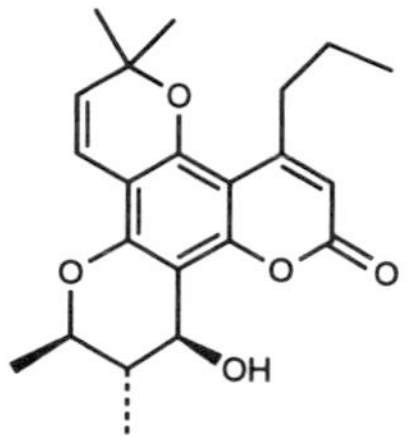

31 calanolide A

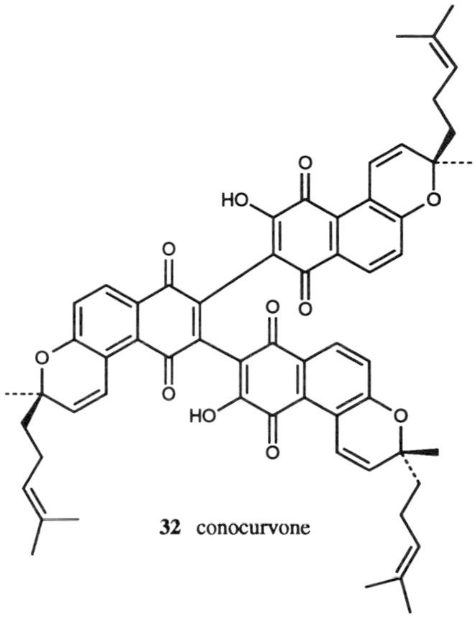

32 conocurvone

the active principle. This compound was already known but its antiviral properties were not investigated. Interestingly, prostratin is devoid of the irritant and tumor-promoting properties typical of phorbol esters and has instead a potent antitumor-promoting activity.

The prenylated coumarin calanolide A (**31**), which was isolated from *Calophyllum lanigerum* (Guttiferae), completely blocks the replication of the HIV-1 virus by inhibiting the enzyme reverse transcriptase. Chemical synthesis of calanolide A has been recently reported and the serious supply problems which have so far hampered the further development of this highly interesting substance are likely to be avoided. The latest HIV-active compound discovered at the NCI is conocurvone (**32**). This unique naphthoquinone trimer was recently found in the endemic Australian shrub *Conospermum* sp. (Proteaceae). Conocurvone is the most potent anti-HIV substance identified so far by the NCI natural product screening program. Detailed biological investigations are currently under way to determine its mechanism of action.

F. Molluscicides

In tropical countries, eradication of disease vectors is a direct contribution to medicine. Schistosomiasis, commonly known as bilharzia, is a parasitic disease caused by threadworms of the genus *Schistosoma* and is endemic throughout South America, Africa, and the Far East. It affects more than 250 million people in over 76 countries. The reproductive cycle of schistosomes involves a stage implicating aquatic snails, in which the parasite multiplies into cercariae. These cercariae, after leaving the snails, can penetrate the skin of humans who come into contact with contaminated water. Chemotherapy with orally administered synthetic antischistosomial drugs is the best method to cure infected people. However, this approach is rather expensive and in many cases the possibility of reinfection exists. An alternative approach is to destroy the intermediate host, thus interrupting the life cycle of the parasite and preventing its further development.

Molluscicides therefore represent crucial factors in integrated programs to control schistosomiasis. At the present time, Bayluscide (2,5'-dichloro-4'-nitrosalicylanilide) is the only molluscicide recommended by the World Health Organization. However, synthetic compounds are expensive, tend to be general biocides, and may result in deleterious long-term effects in the environment. In contrast, the use of locally available plants with molluscicidal properties is a simple, inexpensive, and appropriate technology for the control of the snail vector. Over 1100 plant species have been screened for molluscicidal activity and some very potent, structurally diverse, plant molluscicides have been characterized. Among them, triterpene saponins are molluscicides of special interest because of their favorable water solubility and their widespread distribution in nature. One valuable saponin-containing plant is *Phytolacca dodecandra* l'Hérit (Phytolaccaceae), the berries of which contain the highly potent triterpenoid saponins lemmatoxin (**33**), lemmatoxin C (**34**), and oleanoglycotoxin (**35**). Because of their strong molluscicidal properties, aqueous extracts and slurries of *P. dodecandra* have been the subject of field trials in Ethiopia. Promising results have also been obtained with two saponin-containing Leguminosae species, *Swartzia madagascariensis* Desv. and *Tetrapleura tetraptera* Taub.

33   R₁ = H;   R₂ = Gal,   lemmatoxin
34   R₁ = Glc²-Rha;   R₂ = H,   lemmatoxin C
35   R₁ = Glc;   R₂ = H,   oleanoglycotoxin

VI. CONCLUSION

Plants have provided and will continue to provide humans with useful drugs. Natural products not only have significant medicinal properties in their own right, but also function as very important lead compounds. The limited selection of bioactive plant constituents and therapeutical applications described here—by no means an exhaustive review—gives a general idea of the extraordinary potential of higher plants for the discovery of substances with valuable pharmacological properties. It also brings out the numerous difficulties associated with this research. Expansion of the screening programs, made possible by the increasing automation of testing procedures, and introduction of new bioassays will without doubt result in the discovery of a great number of valuable new drugs from plants. These promising prospects are tempered, however, by the rapid depletion of biological diversity. Tropical rain forests contain about two-thirds of all existing plant species and a great number of undiscovered species. Their destruction will lead to the extinction of numerous plants before they have been investigated or even described. (M. J. Balick from the New York Botanical Garden estimated that 50–60% of all existing species could become extinct in the next ten years!) Considerable efforts to preserve the biological diversity and effective screening programs are urgently required to preserve, investigate, and develop the irreplaceable global heritage constituted by plants and living organisms. [*See* BIODIVERSITY.]

Glossary

Activity-guided isolation (or activity-guided fractionation) Separation process coupled with a bioassay; the bioassay is used as a guide for the identification and selective isolation of the active principle(s).
Bioassay Biological test system used to detect and follow a desired activity.
Chromatography Large group of sophisticated separation methods based on the differential physicochemical properties of molecules (polarity, charge, solubility, etc.).

Ethnobotany Botanical investigation and systematics of plants used in traditional medicine.

Ethnopharmacology Pharmacological studies on plant, animal, or mineral preparations used in traditional medicine.

Lead compound (lead, template) Compound used as starting point or model for the synthetic or semisynthetic preparation of analogues assumed to have improved pharmacological properties.

Plant extract Mixture of substances obtained by treatment of fresh or dried plant material with water or organic solvents, and subsequent removal of the solvent.

Semisynthesis Partial chemical synthesis of a compound using a natural substance as starting material or building block.

Bibliography

Borman, S. (1991). Scientists mobilize to increase supply of anticancer drug taxol. *Chemical & Engineering News,* September 2, 11–18.

Chadwick, D. J., and Marsh, J., eds. (1990). "Bioactive Compounds from Plants," Ciba Foundation Symposium 154. Chichester, England: John Wiley & Sons.

Cordell, G. A. (1993). The discovery of plant anticancer agents. *Chemistry & Industry,* 841–844.

Hamburger, M., Marston, A., and Hostettmann, K. (1991). Search for new drugs of plant origin. *Adv. Drug Res.* **20,**167–215.

Hostettmann, K., ed. (1991). Assays for bioactivity. *In* "Methods in Plant Biochemistry" London: Academic Press.

Hostettmann, K., ed. (1995). "Phytochemistry of Plants Used in Traditional Medicine," Proceedings of the Phytochemical Society of Europe, Vol. 37. Oxford: Oxford University Press in press.

Kinghorn, A. D., and Balandrin, M. F., eds. (1993). "Human Medicinal Agents from Plants," ACS Symposium Series 534. Washington, D.C.: American Chemical Society.

Nicolaou, K. C., Dai, W.-M., and Guy, R. K. (1994). "Chemistry and biology of taxol." *Angew. Chem. Int. Ed. Engl.* **33,** 15–44.

Pollution Impacts on Marine Biology

T. R. Parsons
University of British Columbia

I. Introduction
II. Point Source Pollutants
III. Hydrospheric Pollutants

Pollution impacts on marine biology are most noticeable in local situations, such as where a factory, sewage plant, or oil spill may have caused obvious signs of stress on marine plant and animal communities. These impacts are generally the most easily remedied through appropriate measures for the removal of the pollutant. Other impacts on the marine biota involve the pan-oceanic contamination of the seas by small amounts of anthropogenic substances, such as plastics and pesticides. Sometimes the impact of these pollutants is not well understood in terms of their effect on global ecology. In addition, their presence in the world's oceans is much more difficult to control. However, a number of United Nations international agencies have made some progress in alleviating some of the worst cases, such as the dumping of oil at sea when cleaning tankers and restrictions on the use of chlorinated hydrocarbons.

I. INTRODUCTION

The Intergovernmental Oceanographic Commission gives the following definition of marine pollution: "Marine pollution is the introduction by man, directly or indirectly, of substances or energy into the marine environment (including estuaries), resulting in such deleterious effects as: harm to living resources; hazards to human health; hindrance to marine activities including fishing; impairing the quality (or) use of seawater; and reduction of amenities."

The number of pollutants that have entered the marine environment is very large and new ones are added every day as a result of an expanding and ever-changing industrial economy. Because the volume of the oceans is more than 10 times greater than the volume of the land-mass, it is not surprising that the oceans have served as a sink for many natural pollutants (e.g., from volcanic activity) as well as, more recently, for anthropogenic substances. The "cleansing" role of the oceans in this respect has gone on throughout the geological time of our planet. Many substances that now enter the oceans as a result of human activity are so diluted by the volume of the oceans that they may not cause any damage to the marine biota. In spite of this, the cleansing action of the ocean through attenuation, biodegradation, and deposition in sediments is not inexhaustible. Two general categories of pollutants that are of particular concern today are point source pollutants, which have become concentrated near particular coastlines or in local seas, and hydrospheric pollutants, which have often been carried by the atmosphere to be deposited in all the world's oceans. The point source pollutants visually have had the greatest impact on the marine biota. The role of hydrospheric pollutants is less well understood because they are generally present in very small amounts and, though they may not have acute effects on the marine biota, often no one knows if there are long-term chronic effects to some sensitive organisms.

II. POINT SOURCE POLLUTANTS

Point source pollutants are derived from industrial or domestic outfalls, from rivers entering the sea, or from the accidental spillage of a pollutant from a vessel or an industrial complex. There are numerous examples, and the types of pollutants involved are almost infinite in number. It is only possible to consider here those that are currently believed to be among the most harmful to the marine biota.

A. Petroleum Hydrocarbons

The most photogenic form of marine pollution is an oil spill (Fig. 1). A partial list of some of the largest oil spills together with some important smaller spills is given in Table I. The largest marine oil spill occurred during the Gulf War when approximately 1 million tons of crude oil were deliberately released into the sea. Historically, however, it was the spillage of oil from the tanker *Torrey Canyon* that first drew attention to the devastating effects of oil spills. The animals most severely affected in all such oil spills are marine birds and, in the case of the *Torrey Canyon* spill, approximately 30,000 marine birds were killed. About 10 years later in almost the same location in the English Channel, the *Amoco Cadiz* ran aground on the coast of Brittany. This spill was in an area of intensive shellfish fisheries and it is estimated that together with a further 10,000 to 20,000 bird mortalities, approximately 300 km of shoreline were covered with various amounts of oil, resulting in the elimination of at least 30% of the marine benthic fauna, including commercial species of clams and mussels as well as noncommercial fauna such as marine worms and many small crustaceans. Seaweeds were also killed, which in turn removed much of the natural cover for inshore animals. The affected areas of both of these oil spills took 5 to 10 years

to recover to a relatively pristine state, although species that reproduce slowly, such as bird populations, may take 50 years to recover. The sequence of biological restoration that follows an oil spill often involves bizarre colonization by large amounts of green seaweeds together with one or two species of marine worms; later, as predators and other species are carried by currents to the decimated areas, a greater diversity of organisms occur, until most species return to their former level. The exceptions to this scenario of biological recovery are a few species that may have been at the northern or southern limit of their geographic distribution and for which there is no immediate opportunity for recolonization from adult individuals in the same area.

Some of the largest spills in Table I have been less damaging than some of the smaller spills. For example, the 1974 spill from a storage tank in the Inland Sea of Japan resulted in a massive loss of marine plants and animals that were being grown at aquaculture sites. The *Exxon Valdez* oil spill, although relatively small compared to others in Table I, was one of the most damaging to wildlife, partly because of its location on the coast of Alaska, which is an area abounding in marine wildlife, and partly because the oil was carried by strong near-shore currents for over 1000 km along the coast. On the other hand, the spill of 30,000 tons of oil from the *Argo Merchant* is believed to have caused relatively little damage because the oil stayed at sea and was decomposed, evaporated, and sunk without becoming stranded in a sensitive marine habitat, such as near bird colonies. In summary, oil spills have had their greatest impact on birds and benthic organisms; far less to negligible effects are found on most pelagic organisms, including fish and mammals, because they can swim away from the oil. However, fur-bearing marine mammals can be severely affected, as was the case in the *Exxon Valdez* spill when several thousand sea otters died.

B. Sewage and Industrial Outfalls

The disposal of human and industrial waste from large cities is usually accomplished through a sew-

FIGURE I An aerial view of oil from the *Amoco Cadiz* stranding on the beach at Roscoff, France. Note that while oil covers most of the foreshore, beach areas in the lee of islands and bays are relatively free of oil. (Photo compliments of P. Lasserre.)

TABLE I
A Partial List of Some Major Oil Spills in the Oceans

Location	Source	Approximate amount spilled (tons)	Date
Arabian Gulf	Gulf War	1,000,000	1990–91
Gulf of Mexico	Oil well	440,000	1981
Coast of Brittany, France	Tanker, *Amoco Cadiz*	220,000	1978
Coast of Cornwall, England	Tanker, *Torrey Canyon*	117,000	1967
Japan Inland Sea	Storage tank	8–40,000	1974
Coast of Alaska, U.S.	Tanker, *Exxon Valdez*	30,000	1989
North Sea	Oilwell, *Ekofisk Bravo*	15,000	1979
Northwest Atlantic	Tanker, *Argo Merchant*	30,000	1976

age system that, in most cases, combines the industrial wastes with domestic sewage. Because many large cites are located along the coasts, the final disposal site for the combined anthropogenic effluents is often a large pipe that runs directly into a shallow coastal area. This raises problems for the indigenous biota because the effluent has many forms of toxicity, including heavy metals, pesticides from gardens, organics from industry (including petroleum), PCBs, detergents, and occasional high levels of highly lethal compounds used in some industries (e.g, cyanide). In addition to these toxic substances, domestic sewage will introduce pathogenic bacteria and viruses, pharmaceuticals, and nutrients, particularly organic nitrogen and phosphorus, which can lead to eutrophiciation of local waters. [*See* ECOTOXICOLOGY.]

A sewage outfall usually creates a zone of anoxic water and sediment in the immediate vicinity of the pipe, where the only marine inhabitants are anaerobic bacteria. Just outside this zone of extreme degradation, one or two animals may have become adapted to the highly polluted conditions and grow in great abundance. For example, the polychaete *Capitella* is a benthic worm that is found in great abundance near marine outfalls. Still farther from the mouth of a marine outfall, larger numbers of species are present, some of which are dependent on the nutrient enrichment of the sewage and others occur as part of the natural marine biota of the area. Because of the highly variable conditions in this region, the species diversity is usually greatest at this point, being composed of both opportunistic and natural species. Farther still offshore, the number of opportunistic species declines and the habitat appears essentially normal with respect to diversity and abundance of species. The distance covered by the zonation described here could be up to 1 or 2 km from the end of a major outfall entering into open-ocean waters; in estuaries, bays, and lagoons, the biological effects of an outfall will extend farther owing to restricted exchange of water with the sea.

Pathogenic bacteria, such as some species of *Salmonella,* may survive for 1 to 2 weeks in seawater depending on local conditions. The presence of both pathogenic bacteria and viruses is usually monitored by measuring the numbers of a common fecal bacteria, *Escherichia coli.* Although this is not a pathogen, its presence at concentrations of 100 to 1000 per 100 ml of seawater indicates that the water is dangerous for swimming. In addition, pathogenic bacteria can be taken up by shellfish, such as clams, mussels, and oysters, and the harvest of these animals near an outfall is usually forbidden.

Heavy metals in organisms living near a sewage outfall may be higher than in individuals of the same species living at some distance from the source of pollution. This is illustrated for commercial crabs taken from the Fraser River estuary, which was the site of a major outfall from the city of Vancouver, British Columbia, Canada. The amount of mercury in crabs that lived close to the outfall was about ten times that in the same species living 10 to 50 km away. The levels reported in Table II for the Sturgeon Bank crabs are above the

TABLE II

Mercury in Muscle of 2-Year-Old Crabs (*Cancer magister*)
Collected near an Outfall, 10 km and 50 km Distant[a]

Location	Sample size	Mean Hg content (ppm ± S.D.)
Sturgeon Bank (outfall location)	8	2.4 ± 0.64
Roberts Bank (10 km distant)	5	0.42 ± 0.09
Cowichan Bay (50 km distant)	7	0.22 ± 0.02

[a] Data collected in 1972 before the installation of a new outfall for the city of Vancouver, B.C. Canada. [From Parsons *et al.*, (1972). *J. Fish. Res. Bd. Canada* **30**(7), 1014–1016.]

permitted value for commercial sale of fish produce, which in most countries is 1 ppm. The crabs, which were high in mercury, did not show any ill effects, although in other cases, high metals may be carcinogenic. Other metals that were elevated in all animals living in the same area as the Sturgeon Bank crabs included silver, copper, zinc, and cadmium.

Another aspect of sewage disposal is that it is often necessary to transport sewage sludge to an offshore dumpsite where it is believed that it will have minimum impact on marine biota. Off the city of New York, various dumpsites have been located at greater and greater distance from the coastline. These include a municipal dumpsite located only 20 km off the coast in less than 50 m of water and a 2.4-km-deep site located 160 km off the coast. In all cases it is found that the smothering of benthic animals, the accumulation of toxins, and the turbidity plume from such underwater disposal, have a measurable effect on the local biota.

C. Coastal Industries

Historically, perhaps the worst case of coastal marine pollution occurred in the 1970s in Minamata, Japan. The inhabitants of this village consumed local fish and shellfish that had become contaminated with mercury from a plastics factory located nearby. By January of 1975 there were 798 verified cases of mercury poisoning among humans (known as the "Minamata disease"). Production at the factory had been stopped in 1968, but the full extent of the damage to the human population and the marine environment was not fully documented until a number of years later.

A second notable case of industrial poisoning was the occurrence of DDT (dichlorodiphenyltrichloroethane) in the coastal areas off Los Angeles in the 1970s. At that time, approximately two-thirds of the world supply of this insecticide was from one company located near Los Angeles. Some of the DDT was released into the city's sewage system and, over a period of 20 years, this pesticide entered the ocean food chain and had effects on the biota that could be detected for more than 100 km along and around the coast. On the basis of concentrations of DDT and its derivatives found in various animals, it is believed that the effects of this pesticide included loss of reproduction by California sea lions (*Zalophus californianus*), loss of successful breeding of the brown pelican (*Pelecanus occidentalis*) due to egg shell thinning, death of seagulls and cormorants in the Los Angeles zoo, and accumulations of DDT in lower trophic levels, particularly the lantern fish (mycotophids). Similar effects of DDT, particularly with respect to aquatic bird populations, have also been found in eastern North American and European species.

Throughout the world, various mining activities in coastal areas cause different forms of pollution of the marine habitat. These activities range from dredging beaches for gravel to the use of coastal areas as dumpsites for mine tailings. The latter activity is usually the most devastating because of the enormous volume of tailings involved. On Bougainville Island in Papua New Guinea, for example, a large copper mine disposes of about 80,000 tons of tailings per day. The tailings are carried a short distance down a river to be deposited in Empress Augusta Bay, where a large delta has begun to form from the new sediment. The two main biological impacts of this activity are destruction of the natural estuary, which was an important habitat for local fish, and smothering of the nearshore benthos with a thick layer of ground rock. After such mines close down, it is usually found that although the overall ecology has been altered, a new ecology becomes established based on the soft sediment of the mine tailings, which remain, partly submerged and partly as a beach. Heavy metals, often still associated with the tailings, may cause increased concentrations in the local biota, but usually not to any level that is biologically damaging to the organisms. This is because all organisms have some limited natural defense against heavy metals in the form of a protein (metallothionein) that detoxifies small traces of heavy metal pollutants.

Paper production is another industry that often locates in coastal environments. On the coastline of British Columbia there are 13 pulp mills. In the Baltic Sea they are also a major source of marine pollution. Pulp mills contribute large amounts of soluble organic carbon to the marine environment, which results in a high biological oxygen demand (BOD). By careful control of the rate of effluent addition to seawater, the problem of the pulp mill's BOD can usually be managed to keep local waters at least partially oxygenated. A greater problem is the accumulation of organic solids, such as bark and wood fibers, in the benthic community. Unless the site of the pulp mill is in an area that is very well flushed, the accumulated organic debris invariably gives rise to an anoxic sediment covered with a white growth of bacteria known as *Beggiatoa* mat. High levels of the heavy metals zinc, mercury, and cadmium have been commonly associated with organisms living close to a pulp mill outfall, although recently mercury has no longer been used as an antifungal agent in the pulp mill process. However, other antifungal agents composed of a variety of chlorophenols are now used to preserve wood, particularly by the lumber industry for the shipping of unseasoned wood overseas. Extremely toxic derivatives of organochlorinated compounds, known as dibenzo-*p*-dioxins (commonly referred to as dioxins), have been found in traces in coastal organisms associated with local drainage from the lumber industry. Other sources of these compounds are not always evident as they can be formed from pesticides, by incineration of wastes, and in various chlorination processes. Because dioxins are one of the most toxic compounds found in the marine environment, the permitted level in fish for human consumption (usually in the 20 ng/kg range) is much lower than that for other substances.

Radioactive isotopes have entered the marine habitat from several sources, including bomb testing, losses of nuclear submarines, ocean dumping of radioactive wastes, nuclear power stations, and nuclear fuel processing plants. Because of their radioactivity, isotopes from all of these sources are easily traced in the marine environment. At the Windscale nuclear fuel processing plant in the northeast corner of the Irish Sea, levels of cesium-137 can be detected in seawater that are up to 2000 times higher than levels found in the open ocean. However, according to the International Atomic Energy Agency (IAEA), the level of radioactive substances in the marine biota associated with such point source emissions is still below that estimated to cause somatic and genetic damage to organisms. In the light of further studies on radioactive substances in the oceans, it may be necessary to question the currently accepted safety level, which for human exposures is usually taken to be about 100 mrem/yr.

Power plants require several 100,000 m³ of cooling water per hour. When these plants are located near the coast, they use seawater, with the result that discharge waters may be heated to between 1 and 5°C above ambient temperatures. In some areas where water is very cold, this may be beneficial to the growth of organisms (e.g., for aquaculture). However, in tropical areas the effect is usually not beneficial. In the 1960s, a power plant operating in Biscayne Bay of southern Florida discharged water that was 5°C above the ambient temperature. In the immediate vicinity of the discharge, algae and sea grass disappeared and were replaced by a blue-green algal mat; in an adjacent zone that was 3°C warmer, some algae disappeared and some remained; at about 1°C of heating, the habitat was essentially normal. However, in addition to discharging heated water, power plants may also add chlorine to the water to control fouling organisms within the power plant. The presence of as little as 0.1 ppm of chlorine can begin to affect respiration and reproductive rates of marine fauna, particularly at elevated temperatures.

Marinas and harbors generate a variety of pollutants ranging from sewage and petroleum to very specific compounds used as antifouling paints on the hulls of both small and large vessels. In the latter category, it was recently discovered that the addition of tributylin (TBT) to marine paints gave them extraordinary antifouling properties. As a result of this discovery, many boats were painted with TBT during the 1980s. Unfortunately TBT is slowly leached from the boats into the surrounding seawater. TBT levels of 0.1 to 100 μg/liter have been found in the waters of harbors and marinas; this is the same approximate level that has been found by experimentation to cause 50% mortality among larval stages of oysters and mussels, as well as causing growth inhibition in diatoms. The use of TBT paints has now been banned in a number of countries.

D. Estuaries and Enclosed Seas

Rivers entering the sea are generally point sources of pollution because, in addition to being the sites of many of the impacts discussed earlier (e.g., harbors and sewage outfalls), they also carry pollutants derived from the coastal hinterland. If the hinterland is industrial, then the pollutants will be heavy metals, hydrocarbons, and other organics; if it is agricultural, the pollutants will be nutrients, herbicides, and pesticides. Further, if the river enters a relatively enclosed sea, such as the Baltic, Mediterranean, or Black Sea, the attenuating power of the sea will be greatly reduced because of restricted exchange of water with the oceans; in such cases, the effects of pollutants are more pronounced.

A major problem in determining the biological effects of pollutants on marine organisms in estuaries is that rapidly changing temperature and salinity in the estuarine environment can alter the concentration at which a pollutant becomes toxic. All plants and animals have a certain physiological ability to withstand the effects of small amounts of pollutants. The level that animals can withstand is usually quoted as the LC_{50}, which is the concentration at which 50% of the population of that organism will survive over a given time period (e.g., within 96 hr) of constant exposure to that concentration of pollutant. In the highly changeable environment of estuaries, however, plants and animals can be placed under added stress by very wide changes in temperature and osmotic pressure (salinity), which fluctuate with each tidal cycle. Figure

2 illustrates this problem; in this figure, the small estuarine crustacean *Corophium* could survive a chromium concentration of 16 mg/liter for much longer (>300 hr) at low temperatures and high salinities compared to their survival time (<40 hr) at high temperatures and low salinity. Thus the problem of managing estuaries with respect to the biological impact to pollutants is more difficult than open-ocean systems, where there is generally a narrow range of temperature and salinity.

The North Sea receives point source pollutants from six large rivers and numerous smaller ones. The largest river is the Rhine, which has a mean discharge of over 2500 m^3/sec and drains an area inhabited by 41 million people. The whole region is one of heavy industrialization and intensive agriculture. Therefore, it is not surprising that nearly

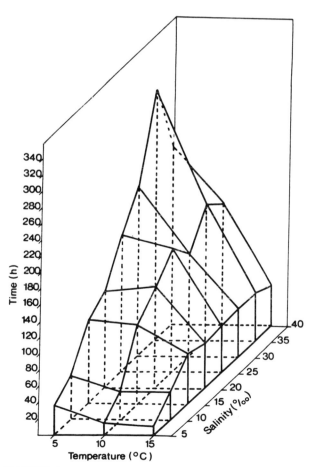

FIGURE 2 The effect of temperature and salinity on median survival time (hr) of an amphipod (*Corophium volutator*) at a chromium concentration of 16 mg/l. (Used by permission from Bryant *et al.*, (1984). *Mar. Ecol. Prog. Ser.* **20,** 137.)

all the rivers entering the North Sea carry many pollutants. In many cases the fate of these pollutants is unknown. For example, there are hundreds of different chlorinated hydrocarbons in the estuarine waters of the North Sea, but to study the fate of each compound and its effect on the marine biota would be an impossible task. Rather, it is only possible to trace the distribution of a few of the more abundant, or more toxic, compounds and then to record what the total effect of all compounds has been on the marine biota.

The distribution of one widely used pesticide, lindane (used to control soil insects and household pests), is shown in Fig. 3. Relatively high concentrations are found over about three-quarters of the area and the source is clearly coastal, where it is found in its highest concentrations in estuaries. The quantity of lindane in hermit crabs, which are found throughout the area, has also been measured and found to be as high as 50 ng/g in coastal areas, decreasing to less than 15 ng/g away from the point sources of pollution.

Not all the pollutants in the North Sea follow a pattern associated with rivers. Some materials are airborne and others may be dumped at some distance off the coast. One program, which has since been abandoned, involved the burning of PCBs (polychlorinated biphenyls) from electrical transformers. A specially designed vessel was used for the high-temperature burn that was necessary to

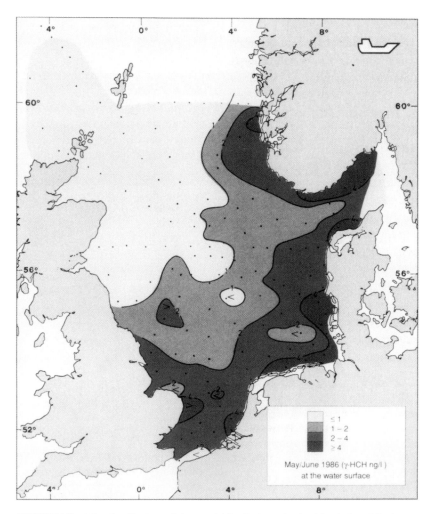

FIGURE 3 The distribution of the pesticide, lindane, in the North Sea. Maximum concentrations are shown to occur in coastal waters. (Used by permission from H. Huhnerfuss, Alfred-Wegener-Institut fur Polar- und Meeresforschung, Bremerhaven.)

destroy this compound. The burn site for all such incinerations at sea was unfortunately located over a major fishing area in the North Sea. A by-product of the incineration of chlorinated hydrocarbons and other garbage is the toxic compound hexachlorobenzene (HCB). It was later found that this compound was accumulating in the sediments around the burn site and in the biota. The distribution of the compound depended largely on the prevailing wind, subsurface currents, and ecosystem interactions.

As a result of the many pollutants entering the North Sea, scientists have seen an increase in the incidence of diseased fish (e.g., skin tumors generally associated with heavy metals) and changes in the overall ecology, as illustrated in Fig. 4. This figure is a composite drawing of scientifically documented changes that have occurred on a shallow portion of the North Sea, known as Dogger Bank. It illustrates that during a period of about 30 years, the bottom community in this area has undergone a typical pollution stress in that a highly diverse, natural community has been replaced by a community dominated by a few species of worms.

Another area where rivers have contributed to a marked change in estuaries and offshore waters is the Black Sea. This sea has an area of 420,000 km^2 but twelve major rivers draining into the sea come from a combined area of over 2×10^6 km^2 with an annual river flow of 350 km^3. The sea has a very limited exchange with the Mediterranean, which means that any substance introduced into the Black Sea is likely to remain there for a long time. Annually the substances entering the Black Sea add up to 2000 tons of cadmium, 70 tons of mercury, 6300 tons of lead, and 45,000 tons of detergent. As a result of these and many other additions, some of the biological changes in the Black Sea make it perhaps one of the most severely impacted local marine habitat.

For example, Yu.P. Zaitsev of the Ukranian Academy of Sciences has described anomalous mutants among planktonic organisms; disappearance of many planktonic species; concentrations of DDT derivatives in the eggs of white pelicans similar to those found in pelicans off the coast of Florida, where reproductive failure was encountered; disappearance of the brown algae *Cystoseira barbata;* reduction in the number of commercial species of fish from 26 to 5; and a decline in the size of a red seaweed bed (*Phyllophora*) from approximately 10^7 tons in 1950 to 10^5 tons in 1990. In addition to these changes, there has been a seasonal extension of areas of hypoxic conditions in the Black Sea over a 20-year period, as shown in Fig. 5. The list of changes also includes the introduction of a new species of jellyfish that has now become one of the dominant predators of plankton and larval fish.

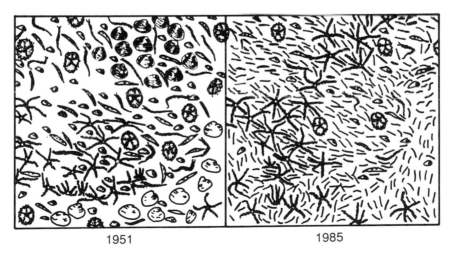

1951   1985

FIGURE 4 A comparison of bottom fauna on Dogger Bank in the North Sea as recorded in 1951 and 1985. In this 30-year period the fauna has changed from a diverse community to one largely populated by marine worms. (Adapted by permission from I. Kroncke, Biolische Anstalt Helgoland, Hamburg.)

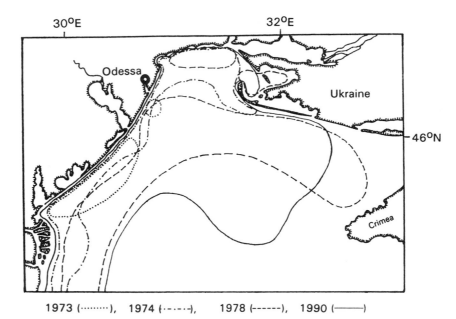

30°E   32°E

Odessa

Ukraine

46°N

Crimea

1973 (·········), 1974 (·–··–·), 1978 (------), 1990 (———)

FIGURE 5 The development of hypoxic areas on the northwestern shelf of the Black Sea, 1973–1990. The hypoxia develops in the late summer and causes the death of most benthic fish and invertebrates; the area of this mortality has been steadily increasing during the period shown. (Used by permission from Y. P. Zaitsev, UAS, Ukraine.)

This event was not associated with estuarine pollution but occurred by the introduction of the species in the ballast waters of commercial vessels.

The Baltic is another enclosed sea that receives pollutants from a great many rivers. Of all the large seas, the Baltic is in fact the least saline and parts of it (e.g., the Gulf of Bothnia) are more similar to a subarctic lake than the deep ocean. Perhaps for this reason, Swedish scientists studied nutrient loading (eutrophication) to compare the level of eutrophication in this large sea with the same effects that have caused many problems in large lakes (e.g., the Great Lakes). Table III is a summary of

the estimated total input of nitrogen and phosphorus to the Baltic, including material transported by the atmosphere and nitrogen fixed by blue-green algae, which tend to grow when there is an imbalance in the nitrogen to phosphorus ratio. The nutrient loading reflects an increase in phosphorus by a factor of eight and nitrogen by about four compared to estimates made prior to 1900. In Fig. 6, the phosphorus loading is then compared to the ratio of the mean depth to water residence time (a measure of flushing used in limnology); the Baltic appears to be presently at a state just above "permissible" and just below "excessive" in terms of caus-

TABLE III

Estimated Total Nitrogen and Phosphorus Inputs to the Baltic Sea in 1985 and before the Twentieth Century When Man's Influence Was Still Comparatively Small[a]

Time		Rivers	Atmosphere	Nitrogen fixation	Total (tons/yr)
Prior to 1900	P	6,800	2,800	–	9,6000
	N	150,000	83,000	67,000	300,000
1985	P	51,600	5,500	–	77,000
	N	640,500	322,000	134,000	1,189,700

[a] From Larsson *et al.*, (1985). *Ambio* **14**(1) 9–14.

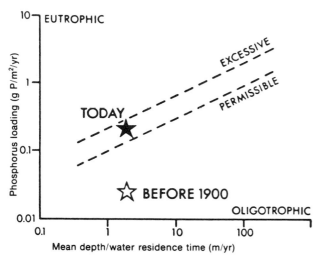

FIGURE 6 The trophic status of lakes is illustrated in order to compare the present level of nutrient enrichment with the level before 1900. The bottom scale is the ratio of depth-to-water residence time. It indicates that deep water bodies with little input of water remain the most oligotrophic. However, as the phosphorus loading is increased (left-hand scale) there is a point reached in all water bodies at which the phosphorus load is "permissible," in retaining the normal ecology of the water body, compared to that which is "excessive" in causing eutrophication. In 1985 the Baltic Sea reached a point that was approaching the "excessive" nutrient loading. [From U. Larsson *et al.*, (1985). *Ambio* **14**, 9.]

ing damage to the whole biota from eutrophication alone. Since these results were published, much attention has been given to the removal of nutrients, particularly phosphorus, from various sources entering the Baltic. [*See* EUTROPHICATION.]

III. Hydrospheric Pollutants

Many pollutants that are carried by the atmosphere tend to be brought back to the earth's surface by the rain. Because 70% of the earth's surface is ocean, some of these substances end up being distributed very widely in the oceans of the world. In addition, other pollutants (e.g., petroleum oil and plastics) have been so widely distributed by shipping that they too have become ubiquitous in the marine environment. Ocean dumping is a third source of pan-oceanic pollution. The extent to which material dumped off the coast of the continents or farther out at sea in much deeper water is carried by ocean currents into other parts of the world's oceans is not well known. Even though

pollutants from these three sources may eventually be diluted by the ocean waters to levels of parts per billion or less, the fact that they can accumulate in the sea is of increasing concern in the management of the oceans.

A. Ocean Plastics

Ocean plastics range in size from large objects, such as monofilament nylon fish nets abandoned during storms, to pellets and fragments less than a millimeter in diameter. The obvious hazards of large objects range from the entanglement of marine organisms in nets to the consumption of plastic bags by leatherback turtles that mistake such objects for their more usual diet of jellyfish. Plastic pellets can also be mistaken as food items by birds feeding on fish eggs and plankton. Plastics consumed by marine animals may cause the direct blockage of the digestive tract or reduced assimilation of food and hence a gradual weakening of the animal.

Figure 7 shows the percentage of juvenile male fur seals found entangled in netting during the commercial hunting of this species. This percentage has declined from a high in the 1970s to a level in 1990 that is one-third of the highest level previously observed. At one time it was estimated that entanglement was causing a 1% decline in the fur seal population per year; however, more recent data indicate that the greatest effect on this population is caused by the pollock fishery, which has removed much of their food supply. Entanglement of birds in monofilament fish netting may have caused the death of several hundred thousand seabirds per year. However, with the recent international control of the use of such nets, the earlier levels of mortality will not be sustained; thus there is probably no lasting effect on seabird populations that often number in the millions.

In a study on the ingestion of plastics by two species of shearwaters (*Puffinus* spp.) in the North Pacific from about 40° to 50°N, it was found that in a sample size of 450 birds, more than 80% contained some plastics in their stomachs. Species differences were seen in the size of plastics ingested, but essentially both species appeared to be mistak-

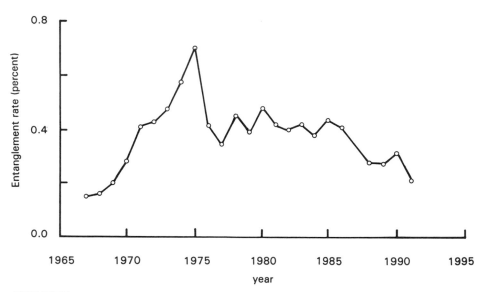

FIGURE 7 The percentage of juvenile male seals found entangled with fish netting in the commercial harvest on St. Paul Island, Alaska. (Used by permission from C. W. Fowler, NOAA, Seattle.)

ing floating plastic debris for small shrimps and fish on which the species normally feed.

In summary, the problem with the hydrospheric occurrence of plastics in the sea is that while it is known that large amounts of plastics are released into the marine environment, in many cases it is still very difficult to accurately gauge the long-term impact of this form of pollution.

B. Chlorinated Hydrocarbons

During the 1960s, DDT was widely used in tropical countries to decrease the incidence of mosquito-borne diseases, particularly malaria; it was also used in agriculture. The method of application was to spray from the air, which resulted in some of the DDT (or its derivatives) being carried out to sea, where it was eventually deposited in the ocean. The extent to which DDT entered Indo-Pacific and Antarctic waters is shown in Fig. 8. DDT residues of 1 to 2 mg/kg wet weight have been found in some fish, and even in Antarctic birds there are reports of 0.1 to 1 mg/kg of DDT in fatty tissue, where it tends to concentrate. The long-term effects of such concentrations are not known. However, in studies on much more severe contamination with DDT in coastal areas, the normal breeding cycle of pelicans did not resume until their

prey fish had less than 0.1 to 1 mg/kg wet weight of DDT. Thus some levels of chlorinated hydrocarbons may be near the level at which they might influence the food chain of the oceans. In general, however, values of 0.1 mg/kg or less do not appear at present to be harmful.

A second group of chlorinated hydrocarbons are the PCBs, which have been widely used in industry since the late 1920s. These compounds have found their way around the world largely through careless disposal systems; only quite recently was a program introduced to destroy these compounds by neutralizing or burning them. The extent to which PCBs have entered the marine environment is shown in Fig. 9, which shows the amount of PCB found in zooplankton in the Atlantic Ocean.

C. Mercury and Other Heavy Metals

At the time of the mercury poisoning of a large number of people in Minamata, the mercury content of many of the fish products consumed daily was in excess of 10 mg/kg wet weight. Concern was then expressed regarding the mercury level of other fish products in the world and what was to be considered a safe level for consumption. From a large number of analyses performed in the 1970s it was found that certain species of commercial fish,

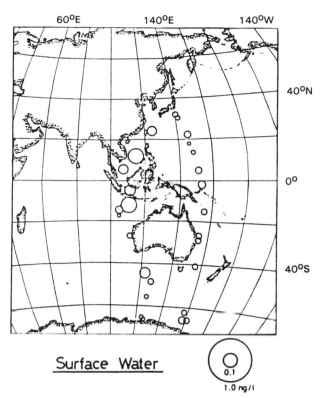

Surface Water

FIGURE 8 The distribution of DDT and its derivatives in the surface waters of Southeast Asia and the Antarctic. [From S. Tanabe et al., (1982). *J. Oceanogr. Soc. Japan* **38**, 137.]

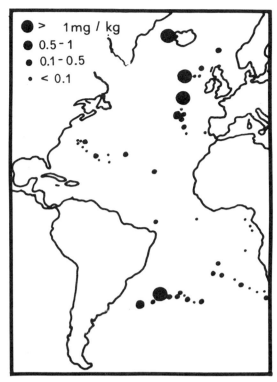

FIGURE 9 The PCB content of zooplankton in the Atlantic Ocean. [From Harvey *et al.*, (1974). *J. Mar. Res.* **38**, 103.]

notably large tuna and swordfish, had mercury concentrations in the range of 0.5 to 1.5 mg/kg wet weight. At the time these were considered to be very high levels and there was a temporary ban on the sale of both fish if the level of mercury was found to be in excess of 0.5 mg/kg. However, in searching for a global source of mercury pollution that could be related to these levels in larger, older fish, it was generally established that the global production of mercury was 20,000 tons per year and that since this was an insignificant fraction of the 70 million tons of mercury dissolved in the oceans, the source of the mercury in tuna and swordfish was considered natural and not likely to increase owing to anthropogenic activity. The second problem that needed to be resolved was whether the natural background level of mercury in these fish was dangerous, or whether some tolerance levels could be established. The solution to this problem was that at a concentration of 1 mg/kg wet weight there was a fivefold safety factor

for an average consumption of fish of 200 g/day. Because this rate of fish consumption is itself very high for tuna and swordfish, it can generally be assumed that there is no problem in the average use of these fish products by humans.

Two other heavy metals that have shown some anomalies in their oceanic distribution are lead and cadmium. Lead is far less toxic than either mercury or cadmium but its use in gasoline additives has caused measurable increases in local sea areas of 10 times the oceanic average of about 2 ng/liter. However, toxicity values are generally regarded to be in excess of 100 ng/liter. Thus, although the source of increased lead levels in some surface waters has been clearly established as being due to gasoline additivies, the levels are still 10 times lower than those considered toxic; the banning of these additives is likely to cause a reduction in present surface concentrations of lead. Cadmium has been found to occur in quite high levels in a number of marine organisms, but its presence, like mercury, appears to be due to natural enrichment, particularly from upwelled water, and not from anthropogenic sources.

D. Petroleum Hydrocarbons

Petroleum hydrocarbons enter the oceans from a large number of sources, including natural seeps, rivers, the atmosphere, and oil tankers. Although oil tankers are not the largest contributors to the total petroleum carbon flux, the persistent routing of these vessels from Arabia to Europe and the Far East leaves a trail of petroleum carbon tars that can be directly attributed to this heavy maritime traffic. Table IV shows that the heaviest concentrations of residual lumps floating at the ocean surface are found in European waters (i.e., the Mediterranean, the Sargasso Sea, and the Gulf Stream) and in the Pacific Northwest off the coast of Japan. The concentration of tar lumps is particularly heavy in the Mediterranean because there is little chance for them to escape from this semienclosed sea. Since the 1970s, the quantity of tar released by tanker traffic has decreased following a recommendation by the International Maritime Consultative Organization that tankers should no longer be allowed to clean their tanks at sea.

The widespread occurrence of tar lumps, which range in size from a few millimeters to several centimeters, is not a particularly serious threat to the marine biota. In fact many of these lumps are colonized by seaweeds and small animals such as barnacles that use them as floating platforms from which they feed and grow. Tar lumps that become entangled in a bird's feathers do cause a problem because they change the insulating properties of the feathers and any water that can seep into the bird's underfeathers will eventually cause the death of that animal. Because petroleum oils are poorly soluble in seawater, the levels of dissolved hydrocarbon in the oceans are extremely low, although still easily measured. Near the surface, values of around 5 ppb can be encountered, whereas below the surface values are less than 1 ppb. From experiments on the toxic effects of hydrocarbons it is generally reported that toxicity, as manifested by abnormal growth, mortality of small crustaceans, and changes in algal photosynthesis, occurs at hydrocarbon levels above 1 or 2 ppm. Thus in general, the levels found in the hydrosphere are well below those that are presently known to cause acute toxicity to marine organisms. However, in cases where large oil spills occur at sea, it can be documented that the much higher concentrations of oils found in the immediate vicinity of such spills will affect marine life all the way from direct mortality of birds to disorientation of small crustaceans, possibly due to the narcotic effects of some hydrocarbons or to interference in water-soluble pheromones that may be released by some animals, particularly during mating season. The use of most detergents to clean up such spills is usually more devastating to the marine biota than the original hydrocarbons because such compounds upset the hydrophobic properties of an animal's integument.

An approximate summary of the amount of petroleum hydrocarbon entering the world's oceans per year is given in Table V. From these data it will be seen that the largest amount of oil comes from coastal runoff. At times, this can be so severe that drainage water from a sudden storm in the area of Los Angeles can dump a year's accumulation of road oil into the local ocean environment.

TABLE IV
Distribution of Floating Tar Lumps in Different Oceans[a]

	Average values (> 25 estimates, mg/m^2)
Mediterranean	23.0
Kuroshio Current (Japan)	2.1
South Pacific	0.0003
North East Pacific (Canada)	0.03
North East Atlantic (Europe)	6.0
Sargasso Sea	9.3

[a] Data from D. R. Green, University of British Columbia.

E. Radioactive Materials

During the period of atomic bomb testing in the atmosphere (ca. 1950 to 1970), large amounts of radioactive materials were released into the atmo-

TABLE V

Estimates of the Amount of Petroleum Hydrocarbons (Crude Oils and Refined Products) Entering the Oceans

Source	Petroleum flux (10^6 tons/yr)
Natural seeps (including offshore production)	0.6
Tankers (including accidents, terminal operations, bilge water)	1.3
Atmosphere	0.6
Coastal runoff (including rivers, urban runoff, industrial wastes)	2.5
Total	5.0

sphere where they were carried into the oceans. Many of these materials decayed rapidly within 1 or 2 years, but a few isotopes of cesium, strontium, and particularly plutonium decay slowly. Over time these isotopes have accumulated in ocean sediments so that very small amounts (about 2.0 mCi/km^2) can be detected. In turn, the sedimentary animals then tend to accumulate some of these isotopes in very small quantities. These values are even lower than those discussed with respect to point source emissions of radioactive materials and the marine biota on a global basis is not believed to be threatened by the accumulated fallout from bomb testing. Since this period of history is hopefully behind us, there is no immediate threat of radioisotopes spreading on a pan-oceanic scale unless the vast amounts of plutonium stored by the industrialized countries are somehow allowed to escape into the atmosphere and oceans such as what may have occurred through the loss of nuclear submarines or through careless disposal of nuclear engines in the Arctic by the former Soviet Union.

F. The Carbon Dioxide Problem

Carbon dioxide is not a pollutant in the ocean. However, the continuing increase in atmospheric carbon dioxide, which could double within the next 50 years, may cause atmospheric warming of 1°C to 4°C. This in turn could cause a partial melting of the polar ice caps, and a rise in sea temperature. The former effect will have a profound impact on coastal marine communities

such as mangrove forests and coral reefs, which may not be able to "grow up" with the inundation. The latter effect could greatly affect fisheries by changing migration routes (e.g., salmon prefer cool water) and by affecting the reproductive success of fish through changes in larval survival. The problem of carbon dioxide accumulation in the atmosphere has a strong connection with the sea in that about half of the carbon dioxide already produced by industrial activity is believed to have been taken up by the sea. The world's oceans already contain some 60 times as much carbon dioxide as is found in the world's atmosphere. However, the total capacity for the sea to continue to absorb a large fraction of the increased atmospheric carbon dioxide is not at present known. [See GREENHOUSE GASES IN THE EARTH'S ATMOSPHERE.]

Glossary

Biological oxygen demand Oxygen debt in a sample of seawater incubated in the dark for 5 days under standard conditions.

Blue-green algae Class of algae, the Cyanophyceae, that can metabolize atmospheric nitrogen.

Eutrophication Condition of enrichment with nutrients, usually leading to large algal blooms.

Hypoxic Level of oxygen sufficiently low to cause stress and eventual death among many organisms.

LC$_{50}$ Concentration of a pollutant that causes 50% of a population of organisms to die in a specified time period (usually 96 hr).

Metallothionein Low molecular-weight protein that is produced by organisms to detoxify heavy metals in the body.

Minamata disease Disease first named after residents of Minamata, Japan, who were found to be suffering from mercury poisoning.

Species diversity Number of species found in a habitat; usually represented as an index, such as the number of species divided by the logarithm of the total number of individuals.

Bibliography

Guthrie, D. (1988). Castoff plastic debris. *Oceanus* **31**(3), 29–36.

Goldberg, E. D. (1976). "The Health of the Oceans." Paris: UNESCO Press.

Hollister, C. D., ed. (1990). Ocean disposal reconsidered. *Oceanus* **33**(2), 1–96.

McLusky, D. S. (1989). Estuarine pollution. *In* "The Estuarine Ecosystem," 2nd ed., pp. 133–176. New York: Chapman Hall.

Waldichuk, M. (1989). The state of pollution in the marine environment. *Mar. Pollut. Bull.* **20**(12), 598–602.

Zaitsev, Yu.P. (1992). Recent changes in the trophic structure of the Black Sea. *Fish. Oceanogr.* **1**(2), 180–189.

Pollution of Urban Wet Weather Discharges

Ghassan Chebbo and Agnes Saget
CERGRENE, France

What exactly are urban wet weather discharges? The answer to this question lies where an urban sewer network meets the receiving waters, usually surface waters. There are three types of wet weather discharge: at the exit of a sewage treatment plant, by a combined sewer overflow, or the outlet of a separate storm sewer network. The heavy flows of the last two categories of discharges (combined sewer overflows and stormwater discharges) can have considerable negative effects and are not nearly as clean as they were considered several decades ago.

For this reason, researchers and specialists have studied the pollution of urban wet weather discharges in an attempt to shed some light on the following points:

- Knowledge of the processes of pollution formation and transfer.
- Understanding mechanisms linked to the negative effects of this pollution in receiving waters.
- Assessment of pollutant flows and comparison with those of other recognized sources of pollution.

- Study of the main characteristics of this pollution.
- A study of possible improvements in sewer network design and of the means of pollution control to be used.

I. INTRODUCTION

Since the beginning of modern town planning, stormwater has been looked at from the hydraulics point of view, namely, the quickest possible evacuation of runoff water, so as to protect the population against floods. This hydraulics objective has consequently led to large-scale discharge into the receiving waters (river, lake, or sea) of runoff water that is to a greater or lesser extent, depending on the type of sewer network, mixed with sewage.

Over the past 30 years, this quantitative approach has been complemented by a recognition of the problem of stormwater quality and of the seriousness of the effects produced by its discharge into the receiving waters. At the present time, there is no longer any doubt that urban stormwater runoff is a considerable source of pollution that must be treated according to the desired quality objectives of the receiving waters in each country.

To meet this new concern for the protection of the receiving waters, many studies devoted to pollution of urban stormwater have been carried out, based in many cases on experimental campaigns that were extensive, entailing a large investment in money and labor. The results have provided some basic knowledge, which we present in this article.

II. ORIGIN OF POLLUTION AND ITS TRANSFER WITHIN AN URBAN SYSTEM

Periods of dry weather between rainfall events result in towns getting dirty. Pollution accumulates on urban surfaces in a solid form, during periods of greater or lesser duration, and therefore in greater or lesser quantities. Moreover, the atmosphere is also polluted by emissions of a natural or human origin. Therefore, from the moment it is formed, rain is contaminated. As rain reaches the surface, it washes off a greater or lesser part of the products deposited there and picks up various types of pollution. The water, which is already very loaded when it enters the sewer network, continues to accumulate pollution by mobilizing the various products that have accumulated within the network, and by mixing with the water that is stagnating or flowing there. A thorough understanding of the different steps of this process is indispensable to clearly identifying the causes of the pollution of urban wet weather discharges and, consequently, the possible points of action.

A. The Pollution of Rainwater Reaching Urban Surfaces

The formation of rain is dependent on atmospheric pollution. Without the presence of atmospheric dust, water droplets would not be able to fuse together to reach a sufficient mass to reach the ground. It is therefore normal—contrary to popular belief—that rain is unclean!

Until the recent past the Earth's atmosphere was fed with various particles and aerosols from widespread natural causes: wind, fires, and volcanic eruptions. For the last century and a half, humans have been assisting nature in this task.

Overall, it is generally estimated that 15 to 25% of the pollution contained in runoff water can be directly imputed to pollution found in rain. This proportion can be even greater for some pollutants like, for example, heavy metals. Moreover, the very acid nature of some rain (pH < 4.5) is likely to increase its ability to remove the materials over which it runs off. [*See* ACID RAIN.]

B. Deposition and Accumulation of Pollutants on Urban Surfaces during Dry Weather

The contaminants that are deposited during dry weather come from many sources. They may be endogenous, that is, come from the town itself, or exogenous, that is, come from very long distances away before being deposited on urban ground. In the same way, pollutants produced by the town can leave it and be deposited at considerable distances.

Except in exceptional cases, when the pollution can come in large measure from outside the town through atmospheric transportation, it appears that an appraisal of the flow of pollutants imported and those exported generally shows that urban sources are larger. The following typology is proposed for these sources.

1. Automobile Traffic

Vehicles are one of the main direct sources of a large number of pollutants. This is particularly the case for hydrocarbons (oils and gasoline), lead (gasoline), rubber (wearing out of car tires), and different metals coming from the wearing out of tires (zinc, cadmium, copper) and metal parts (titanium, chromium, aluminum).

Automobile traffic also requires the use of sand and deicing salt (sodium chloride and calcium chloride), which often contains various additives (chromates, cyanides, and dissolving agents containing ethylene glycol).

2. Soil Erosion and Construction and Repair Works

Soil erosion during dry weather, either by the wind or by the mechanical action of vehicle wheels, is an important source of suspended solids (SSs). This pollution is generally mineral and inert, however, in some cases it can contain active agents (e.g., tarmacadam). The mass of pollutants accumulated can be greatly increased by the presence of construction and repair works.

3. Industry

Industry's part in the pollution of urban surfaces is obviously variable, for it depends on the types of activity and their location in relation to the town. As we emphasized earlier, atmospheric transportation can entrain some pollutants over very long distances. Industry is generally considered to be the source of metals (lead, cadmium, and zinc), some gasoline residues, and many organic micropollutants (solvents in particular).

4. Solid Waste

The generic term "waste" designates everything that humans throw away, voluntarily or involuntarily, onto the ground and street surface. This waste has many forms. Because "street cleaning" is of greater or lesser efficiency, some waste is washed off by runoff water. Even more serious are occasional accidental or deliberate discharge (motor oil, cleaning of market places) into stormwater or combined sewer networks. To this waste is added the excrement of various animals, domestic or wild, whose proliferation can cause serious problems.

5. Vegetation

Urban vegetation produces large masses of carbon compounds that are biodegradable to a greater or lesser extent (in particular dead leaves in autumn, flowers, seeds, pollen, and twigs). This form of pollution is also an indirect source of nitrogen compounds and phosphates (fertilizers), as well as organochlorinated products (pesticides and herbicides).

C. Accumulation and Wash-Off Functions of Pollutants before Entering the Sewer Network

As soon as the quantity of rain on the ground accumulates beyond several millimeters, or even several tenths of a millimeter, runoff begins, which causes erosion and the wash-off of mobilized particles. The capacity of rain to erode permeable ground and to mobilize pollutants accumulated on impermeable surfaces would seem to depend on its granulometry (i.e., the size and kinetic energy of the drops) and on its instantaneous intensity over very short periods of time, the majority of the particles being detached under the direct impact of raindrops. The absence of measurement of such intensities or of granulometric spectra, accompanied by concomitant measurements of pollutant concentration in runoff water, makes these hypotheses debatable.

For its part, the capacity of the rain to wash pollutants into a sewer network depends mainly on the flows that it is capable of generating, in particular, in gutters, which seem to supply a notable part of the pollution deposited during a dry weather period. The two main, pluviometric parameters that can explain pollutant flows (or at least pluviometric parameters, which are the most often quoted!) are the total depth of the water fallen during the rainstorm and the maximum intensity over a short period (let us say of less than 5 to 6 minutes).

Generally, it is considered that the mass of mobilizable pollutants is finite. To represent the increase—in terms of time, over a scale of several days—of the pollutant masses that have accumulated and that can be mobilized, some authors suggest increasing linear functions, the majority of which indicate that there is a rapid tendency toward an asymptote that could be reached from 10 days onward.

D. Transfer and Evolution within the Sewer Network

1. Various Contributions

In networks that are strictly separate (which are, at least for stormwater sewer networks, more often an ideal than a reality), the pollution losses or gains can only be linked to sedimentation or the erosion of particles deposited during preceding rainstorms or after intrusion through drain openings during dry weather, accidentally or not. In a separate storm sewer network, all water comes from the surface. In combined sewers (or not very strictly separated sewers), two other types of gains should be taken into account: domestic or industrial sewage received by the sewer network during runoff and the erosion of deposits accumulated in the sewer network. The former can be assessed quite directly using knowledge of the corresponding flows and the quality of water in dry weather (taking into consideration possible decantations occurring in dry weather and not in wet weather).

2. A Description of Transfer Functions

Just as for pollution washed off on the surface, dissolved matter is in fact transported by the water and does not require, a priori, special treatment in models. The originality of the models springs from how they try to take into consideration particulate matter. We shall limit ourselves, therefore, to these aspects in the following brief description. Hydraulic phenomena will not be discussed here.

Solid transportation has been studied for a long time in rivers and, of course, these techniques have been carried over to other media. However, these transfers have sometimes been carried out too quickly, without a real consideration of the specific characteristics of the sewer networks (e.g., particles of a very small size, with a certain cohesion, "dry weather" suspensions different from those of "wet weather," form of the sewers, transitory character of the flow, and so on). The difficulty of data collection in a sewer network and the relatively recent interest of scientists in a "small-scale" understanding of how they function have not yet provided a completely satisfactory formulation of these phenomena. Besides, many methods are used in these models. These methods are distinguished from each other, on the one hand, by how the solids are described, and on the other hand, by the modeling of their transportation, especially by the consideration (or not) of different types of transportation (suspension and bed loading). Sewer network mod-

els generally neglect bed load transportation (i.e., transportation occurring from slipping or saltation on the sewer bed).

III. EXTENT OF POLLUTION IN URBAN WET WEATHER DISCHARGES

Many measurement campaigns have been carried out all over the world to obtain a description of urban wet weather pollution. The results have (1) provided orders of magnitude for pollution masses discharged during wet weather, downstream of the urban sewer network, and (2) allowed us to identify the parameters that are most harmful to receiving waters.

A. Overall Orders of Magnitude

Table I shows ranges and average values of ratios of masses annually discharged at the outlet of seven French urban drainage areas, a surface of several ten to several hundreds of hectares, and drained by a separate storm sewer network. The symbols correspond to the following parameters:

SSs: total solids in suspension
COD: chemical oxygen demand
BOD$_5$: biological oxygen demand in 5 days
HC: total hydrocarbons
Pb: lead

These values show that even at this point the masses to be considered are quite large. By examining the annual average concentrations, comparisons can be made of the results obtained from the outlet of separate storm sewer networks and combined sewer networks (Table II). We can see that generally, the wet weather effluent of a combined sewer network is more loaded than that of a separate sewer network, but that the differences from one site to another override this difference in nature.

B. Parameters to be Prioritized

Although our knowledge of the negative effects of urban wet weather discharges is still limited, it is

TABLE I

Ranges and Average Values of the Annual Masses Transported, in Wet Weather, at the Outlet of Seven French Urban Drainage Basins That Are Drained by a Strictly Separate Network

Estimated pollution masses[a]	Statistics	SSs	COD	BOD₅	HC	Pb
M1	Minimum	16,130	9,980	1,520	264	12.3
	Average	28,495	19,726	3,035	645	27.9
	Maximum	41,230	58,964	8,335	1,030	52
M2	Minimum	176	55	9.1	1.4	0.19
	Average	491	290	47	7.7	0.42
	Maximum	957	478	86	19	1
M3	Minimum	503	235	39	4.1	0.6
	Average	1,033	608	98	17.0	0.88
	Maximum	2,278	1,076	206	35.0	1.8
M4	Minimum	0.8	0.4	0.07	0.0063	0.0009
	Average	1.7	1.0	0.17	0.025	0.0015
	Maximum	3.5	1.7	0.33	0.05	0.0027
M5	Minimum	2.7	1.3	0.25	—	0.0009
	Average	5.2	2.8	0.42	—	0.0035
	Maximum	8.2	4	0.57	—	0.006
M6	Minimum	182	83	13	1.5	0.1
	Average	286	184	28	5.1	0.27
	Maximum	456	339	48	9.3	0.47

[a] M1: annual pollution mass (kg); M2: annual pollution mass (kg) per hectare (1 hectare = 10^4 m²); M3: annual pollution mass (kg) per impermeabilized hectare (1 hectare = 10^4 m²); M4: annual pollution mass (kg) per impermeabilized hectare (1 hectare = 10^4 m²) and per millimeter (1 mm = 10^{-3} m) of rain; M5: annual pollution mass (kg) per inhabitant; M6: annual pollution mass (mg) per liter (1 liter = 10^{-3} m³) of water discharged.

possible to draw up an inventory of the risks linked to this kind of discharge. This list leads us to distinguish the parameters that are, a priori, likely to play an important role from the point of view of impact on the receiving waters. A comparison of the loads contained in urban wet weather discharge with the loads engendered by urban dry weather discharge, or by runoff on agricultural or "natural" surfaces, allows us to refine our approach.

I. Negative Effects

The most frequently quoted effects are linked to the discharge of:

- *organic matter* (oxygen depletion and its consequences on fish mortality, smells, etc.) "overall" indicators: COD, BOD₅

- *solids* (Solids exert in the first place their own special impact: they contribute to colmation of the receiving water bed and cause turbidity. Moreover, they ensure, as we shall describe in greater detail, a role as a pollutant vector. Consequently, they contribute to the prolongation of impacts, especially concerning oxygen consumption and the release of toxicants.) "overall" indicators: SSs

- *toxicants* (acute mortality and long-term effects of the degradation of flora and fauna, in the river and the estuary) indicators: hydrocarbons, (total hydrocarbons, polyaromatic hydrocarbons, etc.) and organic (polychlorinated biphenyls) and metallic (lead, zinc, cadmium, etc.) micropollutants

TABLE II

Average Annual Concentrations of Urban Wet Weather Discharge (in mg/liter; 1 mg/liter = 10^{-3} kg/m³)

Type of sewer network	Statistics[a]	SSs	COD	BOD₅	HC	Pb
Storm sewer	C_{min}	182	83	13	1.5	0.1
	C_{av}	286	184	28	5.1	0.27
	C_{max}	456	339	48	9.3	0.47
Combined	C_{min}	232	172	41	4.1	0.16
	C_{av}	338	280	76	5.5	0.29
	C_{max}	545	425	113	9.2	0.43

[a] C_{min}: minimal concentration; C_{av}: average concentration; C_{max}: maximal concentration.

Let us note that the hydrocarbons, if they are in sufficient quantity to form a film, will limit water–air interface exchanges and will contribute, in this way, to a low level of oxygenation of the water.

- *pathogenic germs or viruses* (water rendered unfit for some uses such as swimming, oyster farming, etc.)
  indicators: total coliform bacteria, streptococci fecalis, and so on
- *nutrients* (eutrophication and its consequences) indicators: nitrogen and phosphor under different forms (total Kjehldahl nitrogen, ammonia, total phosphor, phosphates, etc.)
- *floating bodies* (visual pollution)

For an analysis of a pollutant impact, two "quantitative" elements are important:

- *dilution of discharge:* This can be interpreted by different indicators, for example, a flow ratio in the case of a river and a volume ratio in the case of still water.
- *the time scale of the degradation processes occurring in the receiving waters.*

This leads us to integrate the flows discharged on different time scales, so as to assess, pertinently, the impact of each pollutant with regard to its different possible effects.

The various pollutant parameters of urban wet weather discharge can cause three major types of effect:

- *Shock effects,* linked mainly to rapid oxygen depletion, as well as to immediate toxic effects. Although they are called "shock effects," these phenomena seem to occur, for the majority of receiving waters, over a period of several hours to several days. We shall therefore consider a first approach where, in this case, the pollutant flow to be taken into consideration is the mass discharged during a rainfall event.
- *Cumulative effects* over long periods, especially concerning toxicants, solids, and nutrients. In this case the characteristic flow would be, for example, the annual mass discharged.
- *Stress effects,* of which little is known. They concern the deterioration of the recovery capacities of the receiving waters owing to the discharge of rather large quantities too often. The indicator taken into consideration here is the frequency of discharge.

2. Comparison with Urban Dry Weather Discharge

There are many comparisons of urban wet weather discharge and dry weather discharge. To show what urban discharge in wet weather represents in annual masses, we begin by presenting an assessment concerning the discharge of a hypothetical town of 10,000 inhabitants with a surface area of 167 hectares (1 hectare = 10^4 m²), with 30% impermeabilization. The results appear in Table III. The following were observed:

- the extent of the quantities of SSs, above all in the combined sewer networks. In

TABLE III

Annual Masses [in tons (10^3 kg)] Discharged by the Sewage Treatment Station, the Combined Sewer Overflows, and the Separate Sewer Network of a Theoretical Town [10,000 Inhabitants, 167 ha (1 ha = 10^4 m²), 30% Impermeabilization]

Pollution parameters	Sewage treatment plant discharge	Combined sewer network	Separate sewer network
SSs	10–17	40–200	25–100
COD	30–50	40–130	10–50
BOD₅	10–17	15–30	2.5–10

consideration of the order of magnitude of the micropollutant contents of SSs in urban wet weather discharge, the consequences on the annual toxicant quantities are to be noted.

- that BOD_5 and COD are of the same order of magnitude in the dry weather discharge from the sewage treatment plant and the wet weather discharge at the outlet of sewer networks.

Other studies have taken nutrients into account. They seem to indicate, however, that nutrients occur in sewage in greater quantities than in urban stormwater runoff.

The ratio between the masses discharged per rainfall event and the quantities discharged in one day after treatment is even more remarkable. Table IV compares the masses transported by the most polluted rainfall events (return period of 6 months to a year) to pollution during dry weather, before and after treatment, produced in 1 day, per impermeabilized hectare. We take into consideration the same type of land use as that of the foregoing hypothetical town. An impermeabilized hectare therefore corresponds to 3.33 hectares of real surface and houses 200 inhabitants. The quantities of lead, hydrocarbons, suspended solids, and chemical oxygen demand are especially high.

3. Comparison with Rural Runoff Pollution

Table V compares for several pollutants the annual loads (in kilograms) per hectare due to urban and agricultural runoff. Two things in particular were found: the metal content of urban runoff is clearly greater and, as far as SSs are concerned, the situation is a great deal more variable, because of the variability of pollution flow in rural areas [annual load washed off by runoff varies between 100 kg/ha (1 kg/ha = 10^{-4} kg/m^2) in a forest and 40,000 kg/ha in a recently deforested area]. According to Table III, it is possible to calculate that the runoff in an urban drainage area with 30% impermeabilization transports an annual average of between 150 kg/ha (minimum for a separate sewer network) and 1200 kg/ha (maximum for a combined sewer network). This order of magnitude is comparable to that obtained for rural areas that cannot be easily eroded.

It should be noted, however, that even if their quantities are equivalent, the SSs eroded in a rural and urban area can present very different characteristics! Finally, as far as nutrients are concerned, the quantities produced per unit of surface area are of the same order of magnitude in the two media.

4. Conclusion on the Pertinent Parameters

In consideration of these comparisons, the first things that should be examined, per year as well as per rainfall event, are the loads discharged for (1) SSs, in particular because they are, as we shall see later, a vector for other pollution; (2) COD and BOD_5, especially because of shock effects such as oxygen depletion; and (3) heavy metals, organic micropollutants, and hydrocarbons, and more particularly their cumulative effect.

TABLE IV

Comparison of the Pollution Masses (kg) Transported per Impermeabilized Hectare (1 hectare = 10^4 m^2) by Dry Weather Wastewater in 1 Day and by Wet Weather Wastewater from a Rainfall Event Occurring with a Return Period of 6 Months to 1 Year

Pollution parameters	Sewage before treatment (1 day of discharge)	Sewage after treatment (1 day of discharge)	High-pollutant rain (return period of 0.5 to 1 year)
SSs	7	0.6–1	65
COD	14	1.6–2.7	40
BOD_5	7	0.6–1	6.5
HC	<0.05	—	0.7
Pb	<0.002	—	0.04

TABLE V

Comparison of the Annual Pollution Masses (kg) per Hectare (1 hectare = 10^4 m^2) Due to the Urban and Agricultural Runoff for Some Pollutants

Type of drainage area	SSs	Total phosphor	Nitrogen	Pb	Zn
Agricultural area	3–5600	0.02–9.1	0.6–42	0.002–0.08	0.005–0.3
Urban area	50–2300	0.1–4.1	1.9–14	0.06–7	0.02–12

IV. CHARACTERISTICS OF THE POLLUTION IN URBAN WET WEATHER DISCHARGES

Recent research programs have tried to obtain a clearer idea of the characteristics of the pollution in urban wet weather discharges. This research has been used to assess the orders of magnitude and the variation ranges of parameters, demonstrating the distribution in urban wet weather discharges of dissolved pollution and pollution fixed onto suspended solids and the size, specific mass, and settling velocity of SSs in stormwater effluent.

A. Particulate Pollution and Dissolved Pollution

Table VI shows the distribution between the proportion of urban wet weather pollution fixed onto solids (size greater than 0.45 μm) and the proportion dissolved, downstream of different types of

TABLE VI

Particulate Pollution (Fixed onto Solid Particles) as a Percentage of the Total Pollution (Particulate + Dissolved)

Type of sewer network	COD	BOD$_5$	TKN	HC	Pb
Stormwater	84–89	77–95	57–82	86	79–96
Combined sewer overflow	88	83	48	—	99
Combined sewer	83–92	91	70–80	82–99	>99

sewer network (strictly stormwater, combined, combined sewer overflows). It seems clear in the three cases presented that the pollution is mainly fixed onto solids (except, perhaps, for nitrogen), even as far as hydrocarbons are concerned. This is even more the case for micropollutants: 93% of polychlorinated biphenyls, 90% of polyaromatic hydrocarbons, 97% of benzopyrene, and 85% of fluoranthene are fixed onto solids in suspension. Attention to the physical characteristics of solids in suspension in urban wet weather discharges is therefore indispensable.

B. Characteristics of Solids in Urban Wet Weather Discharges

1. Granulometry

Fine particles represent the majority of the solids transported (mainly in suspension) in sewer networks during rainfall events. The median diameter is generally between 25 and 44 μm and the percentage in mass of the particles less than 100 μm is on the order of 66 to 85%. This point is important, because road-sweeping procedures are often very inefficient against particles smaller than 100 μm.

2. Specific Masses

For combined sewers, solids are less dense (1700 to 2200 kg/m^3) than those in separate sewers (>2200 kg/m^3), which reveals that there is a large specific quantity of deposits in networks (and sewage flowing at the time of runoff) that is added to the pollution transported in combined sewer networks during wet weather. However, for a given type of network (combined or separate), the specific mass of the particles seems to be homogeneous from one site to another.

3. Settling Velocities

The settling velocities of particles transported in suspension during wet weather are generally very high: median settling velocity V_{50}, a threshold such that 50% of the mass of particles fall slower, equals 4 to 11 m/h (1 m/h = 1/3600 m/s). V_{90} indicates that 90% of the particles fall slower (corresponding therefore to a decantation efficiency of 10%), and it is weaker in a combined sewer (45 to 90 m/h) than in a stormwater sewer (50 to 120 m/h), but remaining the same order of magnitude. However, V_{10} (representing a decantation efficiency of 90%) is highly variable, which has strong implications. The reference values for the design of decantation works intended for depollution should be close to the indicators V_{10} or V_{20}, and the amplitude of their variations will have direct repercussions on those of the planned dimensions. Table VII summarizes the ranges of variation observed for the values V_{20}, V_{50}, and V_{80}.

V. MEANS OF POLLUTION CONTROL

A. Curative Actions

1. Effluent Decantation

Because the majority of pollution is transported by solid particles, and because they are easy to decant,

TABLE VII

Settling Velocity [in m/h (1 m/h = 1/3600 m/s)] of the Suspended Solids in Urban Wet Weather Discharges

Type of sewer network	$V_{20}{}^a$	$V_{50}{}^b$	$V_{80}{}^c$
Stormwater	0.73–2.4	5.5–9.0	22–55
Combined	0.06–1.3	3.7–11.0	23–53

^{a} V_{20}, threshold such that 20% of the mass of particles fall slower.

^{b} V_{50}, threshold such that 50% of the mass of particles fall slower.

^{c} V_{80}, threshold such that 80% of the mass of particles fall slower.

one of the most effective techniques in the fight against pollution is decantation works. These decantation works can be of several types:

- a strictly storage-decantation works of a rather large size, where an effort is also made to optimize the throughflow by using geometry and hydraulics advantageously;
- a strictly in-line water treatment works, where the admissible flow is optimized by limiting the vertical distance that must be covered by particles to be decanted, and where the flow is suitably distributed between elementary units that make up the work; and
- a combination of the two preceding types (water treated on-stream for a module of maximum flow corresponding to a return period of some months, for example, and storage-decantation for a part of the excess volume).

The efficiency of the decantation works has been confirmed by experiments carried out in real size on retention basins of various types (separate, combined, combined sewer overflows), where there was a comparison between the pollutant masses entering and those leaving after a stay of several hours. The reduction of pollutant loads was on average 86% for SSs, 82% for the COD, 80% for the BOD_5, and 80% for Pb.

If decantation is an effective means of combating pollution, it is nonetheless difficult to implement. In fact the volumes and the flows of water involved are considerable and it is necessary to take this into consideration for their treatment. Above all, this requires a definition of the criteria for assessing the reduction of negative effects on receiving waters.

a. Selection of Criteria for Assessing the Reduction of Negative Effects on Receiving Waters

To design treatment capacities, let us examine three criteria that have already been mentioned:

- pollutant masses discharged over long periods (e.g., an annual cycle), which are significant with respect to cumulative effects;

- pollutant masses discharged during the most loaded rainfall events, which are significant regarding immediate shock effects; and
- the frequency of residual discharge, which is of a greater or lesser size.

b. Estimation of the Volumes and Flows to be Treated

The data gathered downstream from four French drainage areas that are strictly separate, with several tens of hectares of surface, have given the following results:

- the efficient treatment of annual pollution (cumulative effects) requires volumes of 100 m³ per impermeabilized hectare (ha. imp.) (the case of strictly storage-decantation works) and flows of 15 to 20 liters/s/ha. imp. (1 liter = 10^{-3} m³) (the case of in-line treatment works);
- to act against the most critical events (shock effects), volumes of 200 m³/ha. imp. and flows of 75 liters/s/ha. imp. are required; and
- the reduction of the frequency of the most critical events is satisfactory with volumes of 200 m³/ha. imp. and flows of 40 liters/s/ ha. imp.

It is important to remember that these results can be strictly applied only to strictly stormwater drainage areas of several tens of hectares. Studies of the same nature are under way on combined sewer systems, and/or on drainage areas of greater size, to analyze to what extent they can be generalized. However, it is doubtful that the orders of magnitude of the volumes and the flows to be treated can be significantly reduced.

2. Possibilities of Using Sewage Treatment Plants

Water transported during wet weather in sewer networks has a large impact on the running and the efficiency of activated sludge treatment plants and provokes disruptions that may appear as:

- large peaks of concentration in SSs in treated water;

- a slight lowering in efficiency in the treatment of organic pollution;
- a strong decrease, even a halt, in nitrification, with ammonium peaks occurring of several mg/liter on exit;
- a slightly deferred fall in denitrification; and
- storage of sludge in the clarifier tank and possibly a leakage of this sludge into the receiving waters.

It is also possible that rainfall events have longer-term consequences, especially these provoking changes in the equilibrium of bacterial populations.

The quality of the water treated is drastically altered during wet weather. Present management consists, most frequently, in avoiding as much as possible sludge leakages, without really trying to optimize the treatment. On the other hand, during wet weather, the receiving water receives all the direct discharge from stormwater overflows upstream of the station.

It is therefore indispensable to address two issues. The first is to have an overall vision of the system composed of the drainage area, the network, and the sewage treatment plant. In fact, the seeking of a balance between the different techniques available for the treatment of stormwater over its whole journey is essential, with the main concern of minimizing all the discharge from a drainage area into the receiving waters. The second is to go from a static or average vision of the running of plants to a dynamic vision of the process, as much in dry weather as in wet weather. This should lead to large modifications in the design and the management of treatment plants. Three main questions therefore need to be asked:

- How to improve the running and the management of the existing treatment plants during wet weather?
- How to design and size future sewage treatment plants so that they meet new requirements under the best conditions during wet weather?
- In the overall management of stormwater on a drainage area scale, how much of this water is

it necessary and reasonable to treat in sewage treatment plants?

3. Direct Action on the Receiving Waters

The last type of curative action involves acting not on the discharge but on the receiving water itself. The risk coming from the meeting of a vulnerable element (the receiving water) and a disruptive element (urban wet weather discharge) can be reduced by reducing the vulnerability of the receiving water. It is important to remember that shock effects were caused by massive urban discharge due to storms, as well as the chronic lack of dissolved oxygen upstream of the discharge points. Different strategies, which are certainly complementary, can be envisaged to decrease these shock effects:

- either the instantaneous oxygen demand is limited during wet weather by a policy of decreasing pollutant discharge
- or the quantity of dissolved oxygen available in the receiving waters is increased [this solution can itself be implemented using different procedures, mechanical (stirring the water, oxygen insufflation, the injection of water oversaturated in oxygen) or hydraulic (water falls at navigation dams)];
- or the consequence of the shock effects on fauna is limited by instituting protected areas where the fish can take refuge in the case of "brutal" pollution (gravel pits, parallel canals).
- or the aggressed receiving water is restored by alevinage campaigns, by cleaning up around the immediate area and downstream of the overflows, and by gathering up floating objects.

B. Preventive Actions

1. Alternatives to Sewer Networks

Alternatives to sewer networks consist of replacing the all-tunnel solutions by solutions aiming at a local management of runoff water. Two principles can be used: water storage and its infiltration. These works should be installed as far upstream as possible.

In practice, these two principles are often used simultaneously. It is thus possible to describe an "alternative" work as a storage work, which can be empty or full, installed on private property, integrated into architectural or urban design, and, possibly, likely to empty itself by infiltration.

For example, it is possible to use ground surfaces with a reservoir structure. In this case, the body of the surface is used as a storage works; the water is introduced either directly through the surface (permeable surface) or through drains; and the water is evacuated either toward the ground by infiltration or toward the traditional sewer network.

Knowing the good rate of sedimentation of runoff water, it is easy to imagine that this type of technique can limit pollution discharged into the receiving waters. Measurements carried out over the last few years in fact show large levels of depollution.

The main problem caused by this type of technique concerns the running and maintenance of these works. Two main questions require a satisfactory answer: How to clean them? What to do with the extremely polluted sludge and dust that is recovered?

2. Actions Upstream of the Sewer Network

One last form of preventive action consists of acting upstream of the sewer network. Actions of this type are certainly difficult to implement because they can neither be decided nor applied by sanitation engineers. Sometimes they even require political decisions taken at a national level, as well as support from the public. However, this type of solution is undeniably the future path and must be promoted from now on.

Moving from the local to the more general, the following actions can be emphasized:

- Modification of the local street-cleaning practices (greater frequency of cleaning, a combination of sweeping, hoovering, and washing).
- Modification of local household garbage storage and collection practices, in particular to improve the water tightness of containers.

- Use of road surfaces and paint that do not contain toxicants.
- Lesser and more careful use of road salts.
- Control of use in towns (generally by the municipality) of fertilizers, herbicides, and other plant products.
- Improvement in the efficiency of depollution systems on industrial producers of smoke (in particular, urban heating and household garbage incineration installations).
- Implementation of incitements or regulations aiming at encouraging industrialists to improve their storage areas.
- Promotion of public transport.
- Promotion of lead-free gasoline.
- Improvement in vehicles, targeting in particular a decrease in pollutant release.

This panoramic view would be incomplete if it was limited to the actions to be taken in town. It is absolutely indispensable that we stop segmenting the water cycle. There is no town water and field water. Ground-water, just like surface water, is a complex and unique system. Any localized action on this system is doomed to failure. Only an overall approach to problems and a seeking of overall solutions to be implemented are likely to allow future generations to have a well-preserved aquatic heritage. The management of urban wet weather discharge is part of the problem, as are the management of urban dry weather discharge, that of rural discharge, and that of industrial discharge.

Glossary

Biological oxygen demand in 5 days (BOD₅) Quantity of oxygen expressed in mg/liter (1 mg/liter = 10^{-3} kg/m^3) required for the destruction or deterioration of organic matter in a body of water with the assistance of microorganisms, which develop there, over a period of 5 days at a constant temperature of 20°C.

Chemical oxygen demand (COD) Quantity of oxygen expressed in mg/liter (1 mg/liter = 10^{-3} kg/m^3) that is consumed by oxidizable matter present in the water.

Combined sewer network Sewer network where domestic and industrial sewage and stormwater are concentrated within one sewer. During storms, the excess water that cannot be conveyed to the sewage treatment plant is discharged directly into the receiving waters through stormwater outlets.

Combined sewer overflows (CSO) Water discharged through stormwater outlets from a combined system.

Drainage area (or watershed, or catchment) Boundary defining a surface where all precipitation drains to a common point.

Event mean concentration (EMC) Total mass of pollutant in runoff divided by the total volume of runoff.

Granulometry Distribution of particles according to their size.

Separate sewage network or separate sanitary sewers Network that collects only domestic and industrial sewage.

Separate storm sewer network (or stormwater sewer network) Network that collects only stormwater.

Settling velocity Speed of a particle when it settles in water measured in calm conditions.

Specific mass Mass per unit of volume.

Suspended solids (SSs) Overall indicator expressed in mg/liter (1 mg/liter = 10^{-3} kg/m^3) of the solid content in a body of water.

Total Kjeldahl nitrogen (TKN) Nitrogen present in organic and ammoniacal forms, excluding nitrous and nitric forms (NO_2 and NO_3).

Water pollution And condition of a body of water that reflects unacceptable water quality, usually due to human influences.

Bibliography

Chebbo, G., and Bachoc, A. (1992). Characterization of suspended solids in urban wet weather discharge. *Water Sci. Tech.* **25**(8), 171–179.

Ellis, J. B. (1989). "Urban Discharge and Receiving Water Quality Impacts." New York: Pergamon Press.

Saget, A., Chebbo, G., and Bachoc, A. (1992). Evaluation of urban runoff volume to be treated at the outlet of separately sewered catchments. *Water Sci. Tech.* **25**(8), 225–232.

Service Technique de l'Environnement et les Agences de l'Eau (1993). "Guide Technique des Bassins de Retenue d'Eaux Pluviales." Paris: Editions Technique et Documentation–Lavoisier.

Valiron, F., and Tabuchi, J.-P. (1992). "Maîtrise de la Pollution Urbaine par Temps de Pluie." Paris: Editions Technique et Documentation–Lavoisier.

Population Regulation

Charles J. Krebs
University of British Columbia

Two fundamental observations can be made about populations of any plant or animal. The first is that abundance varies from place to place. There are some "good" habitats where the species is, on the average, common and some "poor" habitats where it is, on the average, rare. The second observation is that no population goes on increasing without limit, and the problem is to find out what prevents unlimited increase in low- and high-density populations. This is the problem of *population regulation*. What prevents population growth? This apparently simple question underlies many ecological problems from the conservation of declining bird species to the exploitation of forests and fisheries.

I. INTRODUCTION

Prolonged controversies have arisen over the problems of population regulation. Before 1900 many authors, Malthus and Darwin included, had noted that no population goes on increasing without limit, that there are many agents of destruction that reduce the population. It was not, however, until the 20th century that an attempt was made to analyze these facts more formally. The stimulus for this came primarily from economic entomologists, who had to deal with insect pests, including both introduced and native species. Most of the ideas that exist on population regulation can be traced to entomologists.

Figure 1 illustrates the kind of observations made long ago by Darwin and Malthus. On the Serengeti Plains of East Africa the wildebeest population increased from the mid-1950s to the mid-1970s and then stabilized. Over this same time period the zebra population has changed very little in size. The two questions of population regulation can be applied to this example: why did the wildebeest population stop growing around 1975, and what has prevented the zebra population from increasing during this same time period? The basic principles of how these kinds of questions of population regulation can be answered is shown best by a simple graphical model.

II. A SIMPLE MODEL OF POPULATION REGULATION

If populations do not increase without limit, what stops them? This question can be answered hypo-

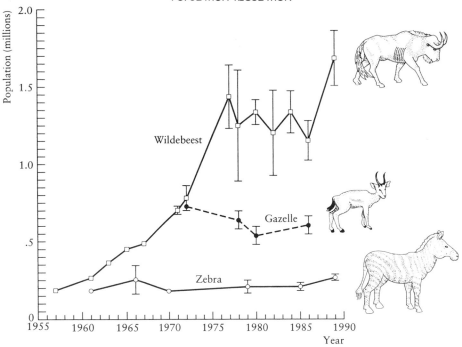

FIGURE 1 Population changes in migratory ungulates in the Serengeti Plains of East Africa. Estimates were obtained by aerial census. Vertical lines are error estimates (one standard error). [Data are from Sinclair and Norton-Griffiths (1982) and Dublin *et al.* (1990).]

thetically with a model. A population in a closed system with no immigration will increase until it reaches an equilibrium point at which

Birth rate per capita = death rate per capita

Figure 2 illustrates three possible ways in which this equilibrium may be defined. As population density goes up, birth rates[1] may fall, death rates may rise, or both changes may occur. To determine the equilibrium population size for any field population, only the curves shown in Fig. 2 need to be determined. Note that this simple model in no way depends on the shapes of the curves, provided that they are smoothly rising or falling shapes. In particular, these curves do not need to be straight lines. [*See* EQUILIBRIUM AND NONEQUILIBRIUM CONCEPTS IN ECOLOGICAL MODELS.]

A few terms need to be introduced to describe the concepts shown in Fig. 2. The death rate per

capita is said to be *density dependent* if it increases as density increases (Figs. 2a and 2c). Similarly, the birth rate per capita is density dependent if it falls as density rises (Figs. 2a and 2b). Another possibility is that the birth or death rates do not change as density rises; such rates are called *density-independent*.

Note that Fig. 2 does not include all logical possibilities. Birth rates might, in fact, increase as population density rises or death rates might decrease. Such rates are called inversely density dependent because they are the opposite of directly density-dependent rates. Inversely density-dependent rates can never lead to an equilibrium density; hence they are not shown in Fig. 2. Figure 2 can be formalized into the first principle of natural regulation: No closed population stops increasing unless either the per capita birth rate or death rate is density dependent. As a population grows in size, some type of negative feedback must occur, affecting the birth or death rates.

This simple model can be extended to the case of two populations that differ in equilibrium density to answer the question of why abundance var-

---

[1] In all discussions of population regulation, birth rates always mean per capita birth rates, and death rates always mean per capita death rates.

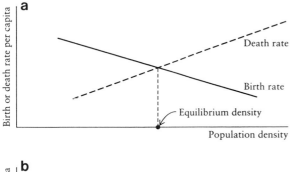

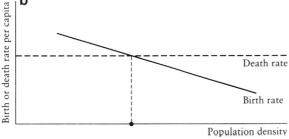

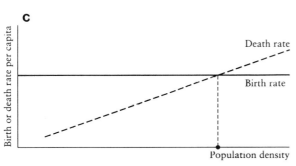

FIGURE 2 Simple graphical model to illustrate how equilibrium population density may be determined. Population density comes to an equilibrium only when the per capita birth rate equals the per capita death rate, and this is possible only if birth or death rates are density dependent.

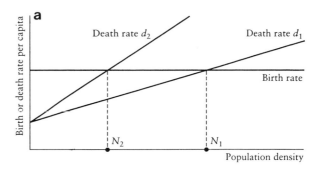

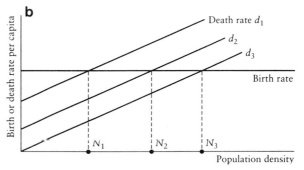

FIGURE 3 Simple graphical model to illustrate how two populations may differ in average abundance. (a) The two populations differ in the amount of density-dependent mortality. (b) The populations differ in the amount of density-independent mortality. Dotted lines mark the equilibrium population densities.

ies from place to place (Fig. 3). Consider first the simple case of populations with a constant (density-independent) birth rate. Equilibrium densities vary for two reasons: (1) Either the slope of the mortality curve changes (Fig. 3a) or (2) the general position of the mortality curve is raised or lowered (Fig. 3b). In case 1, the density-dependent rate is changed because the slopes of the lines differ, but in case 2, only the density-independent mortality rate is changed. This graphic model derives the second principle of natural regulation: Differences between two populations in equilibrium density can be caused by variation in either density-dependent or density-independent per capita rates of birth and

death. This principle seems simple: It states that anything that alters birth or death rates can affect equilibrium density. Yet this principle was in fact denied by many population ecologists for 40 years. This historical controversy is reviewed in the next section.

III. HISTORICAL PERSPECTIVE

The *balance of nature* has been a background assumption in natural history since the time of the early Greeks and underlies much of the thinking about population regulation. The simple idea of early naturalists was that the numbers of plants and animals were fixed and in equilibrium and that observed deviations from equilibrium, such as the locust plagues described in the Bible, were the result of a punishment sent by divine powers. Only after Darwin's time did biologists try to specify how a balance of nature was achieved and how it might be restored in areas where it was upset.

A considerable amount of activity around the turn of the 20th century centered around attempts to control insect pests by the introduction of parasites. In 1911, Howard and Fiske, two economic entomologists with the U.S. Department of Agriculture, studied the parasites of two introduced moths, the gypsy moth and the brown-tail moth, in an attempt to control the damage these defoliators were doing to New England trees. Howard and Fiske believed that each insect species was in a state of balance so that it maintained a constant density if averaged over many years. For this balance to exist, they argued, there must be, among all the factors that restrict the insect's multiplication, one or more *facultative agents* that exert a relatively more severe restraint when the population increases. They argued that only a very few factors, such as insect parasitism, were truly facultative. [*See* PARASITISM, ECOLOGY.]

Furthermore, Howard and Fiske said that a large proportion of the controlling factors, such as destruction by storms, high temperatures, and other climatic conditions, should be classed as *catastrophic* since they are wholly independent in their action of whether the insect is rare or abundant. For example, a storm that kills 10 out of 50 caterpillars on a tree would undoubtedly have destroyed 20 if 100 had been there or 100 if 500 had been there. Thus the average percentage of destruction (the per capita death rate) remains the same no matter what the abundance of the insect.

Finally, Howard and Fiske noted[2] that other agencies, such as birds and other predators, work in a radically different manner. These agents maintain constant populations from year to year and destroy a constant number of prey. Consequently, when the prey species increases, the predators will destroy a smaller and smaller percentage of the prey (i.e., they work in a manner that is the opposite of facultative agents). Howard and Fiske did not give factors of this type a distinct name. [*See* BIRD COMMUNITIES.]

These two entomologists concluded that a natural balance can be maintained only through the operation of facultative agencies that destroy a greater proportion of individuals as the insect in question increases in abundance. Facultative agencies thus cause the per capita death rate to increase with prey density, as in Figs. 2a and 2c. Howard and Fiske believed that *insect parasitoids* were the most effective of the facultative agencies; *disease* operated only rarely, when densities got very high; and *starvation* was the ultimate facultative agency, which almost never operated.

Howard and Fiske were the prototypes of the *biotic school* of population regulation, which proposed that biotic agents, principally predators and parasitoids, were the main agents of population regulation.

Meanwhile, another school of thought, the *climate school,* was in the process of formation. F. S. Bodenheimer in 1928 was one of the first to suggest that the population density of insects is regulated primarily by the effects of weather on both development and survival. Bodenheimer was impressed by all the work done in the 1920s on the environmental physiology of insects, showing, for example, how low temperatures affect the rate of egg laying and speed of development. He was also impressed by the fact that weather was responsible for the largest part of the mortality of insects, often 85 to 90% of the insects in their early stages were killed by weather factors.

In 1931 the Russian ecologist B. P. Uvarov published a larger paper, "Insects and Climate," in which he reviewed the effects of climatic factors on growth, fertility, and mortality of insects. He emphasized the correlation between population fluctuations of insects and the weather, and he regarded these weather factors as the prime agents that control populations. Uvarov questioned the idea that all populations are in a stable equilibrium in nature and emphasized the instability of field populations. [*See* INSECT DEMOGRAPHY.]

Three important ideas were expressed by the early climate school: (1) Insect population parameters are strongly affected by the weather, (2) insect outbreaks could be correlated with the weather, and (3) insect population fluctuations were emphasized, not stability.

It is important to realize here that all this controversy was over insect populations and their regulation; work on vertebrate populations had hardly

---

[2] From a historical vantage point this is now deemed incorrect.

begun by 1930, and there had been no work on the populations of other invertebrates or plants.

In 1933 the *Journal of Animal Ecology* published a supplement entitled "The Balance of Animal Populations" by A. J. Nicholson, an Australian economic entomologist. Nicholson was interested in the parasite–host system of insects, and he teamed up with mathematician V. Bailey to construct a model of this system. Nicholson expanded his ideas on the parasite–host system to cover all interactions between animals. According to Nicholson, the controlling factor was always *competition*: competition for food, competition for a place to live, or the competition of predators or parasites. Nicholson's theory was predominantly a biotic one, and he is usually considered the cornerstone of the biotic school.

Nicholson's viewpoint was given much stronger emphasis by Harry Smith in 1935, who considered the problem of population regulation in some detail. He pointed out first of all that populations are characterized both by stability and by continual change. Population densities are continually changing, but their values tend to vary about a characteristic density. This characteristic density itself may vary. Smith compared a population to the sea, the surface of which is paradoxically a universal point for altitude measurements but which is continuously being changed by tides and waves. Thus Smith reaffirmed Nicholson's ideas on balance.

Different species of animals tend to have different average densities and the same species will have different average densities in different environments. The variations about the average density are stable because there is always a tendency to return to the average density (i.e., populations seldom become extinct or increase to infinity). This is what is loosely termed the "balance of nature."

The equilibrium position, or average density, may itself change with time. This is what causes the economic entomologists so much trouble. The equilibrium position of an introduced pest may be so high that constant damage occurs to crop plants. Smith then set out to analyze the factors that determine the equilibrium position or average density. He pointed out that the number of injurious insects is very small relative to the total number of insects and that we must study both common and rare species if we hope to understand the reasons for the abundance of species.

Smith recognized the distinction Howard and Fiske made between facultative and catastrophic agencies, and he renamed these density-dependent mortality factors and density-independent mortality factors. The average density of a population, Smith concluded, can never be determined by density-independent factors. Only if the death rate line has a slope (i.e., a density-dependent component) can the population reach equilibrium. Thus only density-dependent mortality factors can determine the equilibrium density of a population.

He went on to point out that the density-dependent factors are mainly *biotic* in nature (disease, competition, predation) and that the density-independent factors are mainly physical, or *abiotic*, factors, mainly climate. But, Smith pointed out, it should not be concluded from this that the average population densities of species are never determined by climate. Climate, he states, may act as a density-dependent factor under some circumstances. As an example, he suggests the case of *protective refuges*: If there are only so many of these to go around and all the unprotected individuals are killed by climate, this climatic mortality would be density dependent.

Smith thus restated the main points of Nicholson, adding the terms density-independent and density-dependent, and stated, in contrast to Nicholson, that climate might act as a density-dependent factor in some cases. By this time, then, the main tenets of the biotic school had been crystallized: the idea of balance in nature, that this balance was produced by density-dependent factors, and that these factors were usually biotic agents, such as predators and diseases.

In 1954 Andrewartha and Birch, two Australian zoologists who completely disagreed with Nicholson's ideas, published an important book "The Distribution and Abundance of Animals" which attacked the ideas of the biotic school. They revived in their book a highly modified version of the climatic school's ideas. Andrewartha and Birch concentrated on the individual organism and based their whole approach on this question: What are the factors that influence the animal's chance to survive and multiply? Given this question, they proceeded to classify environmental factors.

First, they rejected the distinction Howard and Fiske and others made between the physical (abiotic) and biotic factors. For example, food and shelter may sometimes be biotic, sometimes abiotic. Hence this distinction does not help much in classifying the environment.

Second, they rejected the classification of the environment based on density-dependent factors and density-independent factors. This distinction was rejected because they believed that there is no component of the environment whose influence is likely to be independent of the density of the population (i.e., all factors are density dependent). Here they are striking at a key principle of the biotic school. For example, they say, consider the action of frost. Between a large population and a small population there may be genetic differences in cold hardiness and, in addition, the places where the insects live may differ with respect to the degree of protection from frost. Thus large populations may be forced to occupy marginal habitats and so suffer more from frost. They conclude that density-independent factors do not exist, so there is no need to attach any special importance to density-dependent factors in classifying the effects of the environment on a population.

How then can one classify environmental factors? Andrewartha and Birch suggested that the environment may be divided into four components:

1. Weather
2. Food
3. Other animals and pathogens
4. A place in which to live

Andrewartha and Birch were concerned only with animal abundance, so plants appear principally as food in this classification. These components of the environment are nonoverlapping and they may be subdivided if necessary, for example, into other animals of the same species and other animals of different species. Together these four components and the interactions between them completely describe the environment of any animal.

Consequently, said Andrewartha and Birch, for any given species it must be asked which of the four components of environment affect the animal's chances to survive and multiply. Once this question is answered, the reasons for the animal's distribution and abundance in nature can be determined. Andrewartha and Birch presented a general theory of the numbers of animals in natural populations. First, they stated that one cannot use expressions like "balance," "steady states," "equilibrium densities," or "ultimate limits" because there is no empirical way of giving a meaning to these words. Second, they noted that one must take account of the fact that all animals are distributed patchily in nature, never uniformly. These "patches," or *local populations,* are the basic component with which they deal.

According to Andrewartha and Birch, the numbers of animals in a natural population may be limited in three ways: (1) by limited supply of material resources, such as food, places in which to make nests, and so on; (2) by inaccessibility of these material resources relative to the animal's capacities for dispersal and searching; and (3) by shortage of time when the rate of increase (r) is positive. Of these three ways, they believe that the last is probably the most important in nature with the first probably the least important. Regarding the third case, the fluctuations in the rate of increase may be caused by weather, predators, or any of the components of the environment.

Andrewartha and Birch were principally concerned with insect populations as their field experience was with insects occupying the severe desert and semidesert areas of Australia. Their main contribution to ecology has been to reemphasize the importance of getting *empirical data* on the problems of population regulation. They continually raised the ever-bothersome question: How can a particular idea be tested in real populations?

The two main schools of population regulation—the biotic school and the climate school—have concentrated on the role of the *extrinsic* factors in control: food supply, natural enemies, weather, diseases, and shelter. Many of these theories tend to assume that the individuals that make up the population are all identical, like atoms or marbles. This neglect of the importance of individual differences in population regulation has been challenged by a group of workers in diverse fields who have proposed theories of self-regulation. Their rallying point has been a search for *intrinsic* changes in populations, changes that might be important in population control.

Two basic types of changes can occur in individuals, *phenotypic* and *genotypic,* and the proponents of self-regulatory mechanisms differ in what importance they attach to each of these basic types. Of course, no matter what mechanism is operating, it must have been evolved in the species concerned. Consequently, these theories of self-regulation all become concerned with evolutionary arguments.

In 1955 Dennis Chitty, working at Oxford, presented the fundamental premise underlying all ideas on self-regulatory mechanisms. Suppose, Chitty argued, that a population is observed at two times, 1 and 2, and that at time 2 the death rate (D_2) is higher than the death rate at time 1 (D_1). This death rate is the result of the interaction of the organisms (O) with their mortality factors (M). The problem now is to determine why D_2 is greater than D_1. The first hypothesis to be explored is that on both occasions the biological properties of the organisms are identical. In this case, there must be a difference between the mortality factors at the two times. In other words, there might have been more predators or parasites or the weather might have been less favorable at time 2. Some population changes can certainly be explained in this manner, but in other cases this method has failed to turn up the right clues. This matter must be looked at from another angle.

Consider, Chitty continued, the possibility that the environmental conditions are much the same at all times, that there is no real difference between the mortality factors at times 1 and 2. In this case, any change in the death rate must be because of a change in the nature of the organisms, a change such that they become less resistant to their normal mortality factors. For example, the animals might die in cold weather at time 2, weather they might have survived at time 1.

These ideas can be summarized as:

	First hypothesis		Second hypothesis	
Time	1	2	1	2
Death rate	$D_1 <$	D_2	$D_1 <$	D_2
Organisms	$O_1 =$	O_2	$O_1 \neq$	O_2
Environment	$M_1 \neq$	M_2	$M_1 =$	M_2

Changes in the individual organisms in the population may be physiological or behavioral, and they may be phenotypic changes or genotypic changes.

The first hypothesis describes the classic approach to population regulation, used, for example, by both the biotic school and the climate school. The second hypothesis describes an ideal self-regulatory approach to population regulation. It is unlikely in nature that this second situation would occur in such a pure form but it is more likely that some mixture of these two situations would be found in self-regulatory populations. Note that the concept of density dependence becomes ambiguous under the second hypothesis. The idea that the environment can be subdivided into density-dependent and density-independent factors has meaning only insofar as the properties of the individuals in the population are constant. Self-regulatory systems have added an additional degree of freedom to the system, the individual with variable properties.

Variation among the individuals in a population may be either genetically based or environmentally induced. In 1931 the British geneticist E. B. Ford was one of the first to point out the possible importance of genetic changes in population regulation. He suggested that natural selection is relaxed during population increases, with the result that variability increases within the population and that many inferior genotypes survive. When conditions return to normal, these inferior individuals are eliminated through increased natural selection, causing the population to decline and at the same time reducing variability within the population. Thus, Ford argued, a population increase inevitably paves the way for population decline.

From a study of population fluctuations in small rodents, Chitty set up the general hypothesis (now called the Chitty hypothesis) that all species are capable of regulating their own population densities without destroying the renewable resources of their environment or requiring enemies or bad weather to keep them from doing so. Not all populations of a given species will necessarily be self-regulated, and the mechanisms evolved will be adapted only to a restricted range of environments. The species may well live in poor habitats where this mechanism seldom if ever comes into effect.

The actual mechanisms by which self-regulation can be achieved in natural populations involve some form of mutual interference between individuals or intraspecific hostility in general. Mechanisms of self-regulation do not require genetic changes and may be entirely phenotypic. This hypothesis can be applied only to species that show mutual interference or spacing behavior. The most important environmental factor for such populations is other organisms of the same species.

The problem of self-regulation has been approached from another angle by V. C. Wynne-Edwards, a British ecologist whose major work has been on birds. Wynne-Edwards began his analysis with the observation that most animals have highly effective mechanisms of movement. If we look in nature, we will usually find that organisms concentrate at places of abundant resources and avoid unfavorable areas. This is the first point to note—that animals are dispersed in close relation to their essential resources.

The essential resource most critical to animals is clearly *food,* Wynne-Edwards observed. Of course, many other requirements must be met before a species can survive in an area, but food is almost always the critical factor that ultimately limits population density in a given habitat. The food resource must then be studied as the key to understanding population control.

Wynne-Edwards suggested that some artificial and harmless type of competition has evolved in many species as a buffer mechanism to stop population growth at a level below that imposed by food exhaustion. The best example of this kind of buffer mechanism is the territorial system of birds. The territories that birds defend so fiercely are just a parcel of ground, but the possession of a territory eliminates competition for food since the owner and its dependents enjoy undisputed feeding rights on that area. Provided that the size of the territory varies with the productivity of the habitat, a perfect illustration of this model exists: Population density is controlled by territoriality, which ensures that the food supply will not be exhausted.

The margins of a species' range will probably not show this self-regulation, Wynne-Edwards stated. Physical factors will predominate in these harsh environments, and hence attention should be concentrated on the more typical parts of the range, where self-regulation is the usual situation. Also, a few species will ultimately fail to be limited by food, and these will not fit into the scheme of Wynne-Edwards.

IV. A MODERN SYNTHESIS

A great deal of controversy in ecology exists over the concepts of population regulation and the areas of agreement and disagreement need to be highlighted.

Because the definition of terms has always plagued discussions about population regulation, two confusing terms need to be defined: *limitation* and *regulation.* A factor is defined as limiting if a change in the factor produces a change in average or equilibrium density. For example, a disease may be a limiting factor for a deer population if deer abundance is higher when the disease is absent. With regulation, a factor is defined to be regulating if the percentage mortality caused by the factor increases with population density.[3] For example, a disease may be a regulating factor only if it causes a higher fraction of losses as deer density goes up. Most experimental manipulations of populations involve studies of limitation.

The simple model of population regulation shown in Fig. 2 is critically focused on the concept of equilibrium, and the first question to be asked is whether natural populations can be equilibrial

---

[3] Alternatively, if the reproductive rate is reduced as the population rises.

systems. Recent work on ecological stability has given a more comprehensive view of the factors that affect stability (Fig. 4). There is no reason to expect all populations to show stable equilibria. There are two sources of instability in populations: strong environmental fluctuations in weather and biotic interactions. Examples of predator–prey interactions that are unstable have already been mentioned. Time lags can also affect population stability. Real world populations are expected to fall along the whole spectrum from showing stable, equilibrial dynamics to unstable, nonequilibrial dynamics. The simple model shown in Fig. 2 is difficult to detect in a real population that shows unstable dynamics.

The spatial scale can also be important in considerations of stability. If a very small population on a small area is studied, it may fluctuate widely and even go extinct. A large population on a large study area may at the same time be stable in density. The important concept here is that of local populations linked together through dispersal into *metapopulations* (Fig. 5). To study population regulation, one must know if a population is subdivided and how the patches are linked. Ensembles of randomly fluctuating subpopulations, loosely linked by dis-

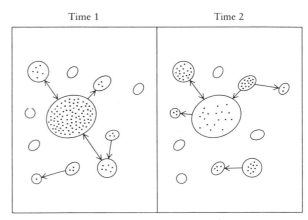

FIGURE 5 Hypothetical metapopulation dynamics. Closed circles represent habitat patches, dots represent individual plants or animals. Arrows indicate dispersal between patches. Over time the regional metapopulation changes less than each local population.

persal, will persist if irruptions at some sites occur at the same time as extinctions at other sites. The result can be that at a regional level the population appears stable while the individual subpopulations fluctuate greatly.

A third complication for the analysis of population regulation is that real world populations rarely show smooth curves like those in Fig. 2. A more usual observation is of a cloud of points, such that

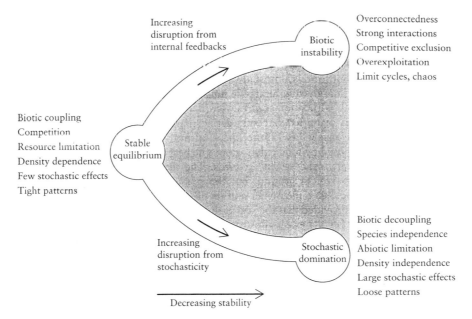

FIGURE 4 Ecological systems in a scale from stable to unstable. Both biotic instability, caused by internal feedbacks, and stochastic domination, caused by strong environmental fluctuations, can result in instability.

density dependence is either "vague" or absent. Figure 6 illustrates the type of density-dependent relationships that might be found in the real world. It may be very difficult to find density-dependent relationships in natural populations.

How then is it possible to study population regulation in view of all these problems? If a population does not continue to increase, it is axiomatic that births, deaths, or movements must change at high density. The first step is to ask which of these parameters changes with population density. Does reproductive rate decline at high density or does mortality increase (or both)? If mortality increases, does this fall more heavily on younger or on older animals, on males or on females? These patterns of changing reproduction and mortality with population density can then be analyzed to see if they occur in a variety of populations. This should be the first step to understanding population regulation in animals.

The second step is to determine the reason for the changes in reproduction or mortality. Determining the cause of death of plants or animals in natural population is not always simple. If a fox or a bat has rabies, a fatal disease, the cause of death is clear. If a caterpillar has a tachinid parasite, it is certain to die from this parasitization. But as more complex cases are examined, decisions about causes of death are not clear. If a moose had inadequate winter food and the snow is deep, it may be killed by wolves. Is predation the cause of death? The answer would be yes, but only in the immediate sense. Malnutrition and deep snow have increased the probability of being killed. Because many components of the environment can affect one another, mortality can be *compensatory*. The idea of compensatory mortality is one of the most important concepts needed to understand population regulation. At the two extremes, mortality may be *additive* or it may be completely compensatory. How can these be distinguished?

Additive mortality assumes the agriculture model of population arithmetic. If a farmer keeps sheep, and one sheep is killed by a coyote, the farmer's flock is smaller by one. Deaths are additive so in order to measure their total effect on a population, deaths are simply added up. But in natural populations, when there are several causes of death, the arithmetic is not so simple. Consider, for analogy, a sheep population in which winter food is limiting so that starvation will kill many individuals by the end of winter. If a coyote kills one of these sheep, it may be doomed to die anyway from starvation. The number of sheep left at the end of winter will be the same, regardless of whether predation occurs or not (in this hypothetical scenario). In this case predation mortality is not additive but is compensatory and simple arithmetic does not work. Figure 7 illustrates how additive and compensatory effects can be recognized.

Compensatory mortality is the reason behind many ecological anomalies that puzzle the average person. If a pest is killed, it will not necessarily become less abundant. BIOLOGICAL CONTROL discusses this question of biological pest control. If a grouse is shot or a fish is caught, their

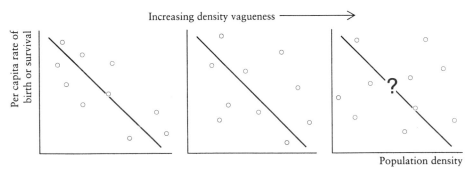

FIGURE 6 Types of density-dependent relationships likely to be found in real world populations. An increasing scatter of points makes it difficult to determine if there is an equilibrium point or where it is.

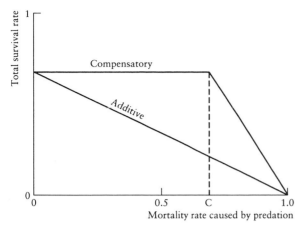

FIGURE 7 Additive and compensatory mortality for losses due to predation. The additive hypothesis predicts that for an increase in predation mortality, total survival decreases an equal amount. The compensatory hypothesis predicts that, below a threshold (C), an increase in predation losses has no effect on total survival. This model can be applied to any mortality agent: predation, disease, starvation, or hunting.

numbers may not necessarily fall. Compensatory mortality has practical consequences when it occurs. [*See* WILDLIFE MANAGEMENT.]

In natural populations, mortality agents are rarely completely additive or completely compensatory. To determine if a particular cause of mortality is compensatory, an experiment in which total losses are measured with and without the particular cause of death must be done. Few of these experiments have been done, and there is an unfortunate tendency to assume all losses are additive in natural populations.

If birth rates change with population density, it is often difficult to pin down the factors that cause reproduction to change. Food supply or nutrient availability may cause birth rates to change but these effects can be overridden or modified by weather or social interactions in natural populations.

Given this background, how can these key questions of population regulation be studied?

V. TWO APPROACHES TO STUDYING REGULATION

There are two competing paradigms about how best to study population regulation. *Key factor anal-*

ysis is a method of analyzing populations through the preparation of life tables and a retrospective analysis of year-to-year changes in mortality and reproduction. *Experimental manipulations* form a second method of analyzing population changes. The advantages and disadvantages of these two approaches are considered next.

A. Key Factor Analysis

In 1957 R. F. Morris developed key factor analysis as a technique for determining the cause of population outbreaks in the spruce budworm, which periodically defoliates large areas of balsam fir forests in eastern Canada. This method was improved by George Varley and G. R. Gradwell in 1960.

Key factor analysis begins with a series of life tables of the type shown in Table I. The life table data are most easily obtained for organisms with one discrete generation per year. The life cycle is broken down into a series of stages (eggs, larvae, pupae, adults) on which a sequence of mortality factors operate. Each drop in numbers in the life table are defined:

$$k = \log (N_s) - \log (N_e),$$

where k is the instantaneous mortality coefficient, N_s is the number of individuals starting the stage, and N_e is the number of individuals ending the stage. For example, 83.0 winter moth larvae entered the pupal stage in 1955 (Table I), and of these, 54.6 were killed by pupal predators (shrews, mice, beetles) during late summer, which reduced the population to 28.4 per m². Thus

$$k_5 = \text{instantaneous mortality coefficient for pupal predation}$$
$$= \log (83.0) - \log (28.4) = 0.47$$

These calculations are done in logarithms to preserve the additivities of the mortality factors. Thus generation mortality K can be defined as

$$K = k_1 + k_2 + k_3 + k_4 + k_5 + \dots.$$

TABLE I

Life Table for the Winter Moth in Wytham Woods, near Oxford, England, for 1955–1956[a]

	Percentage of previous stage killed	No. killed (per m²)	No. alive (per m²)	Log no. alive (per m²)	k value
Adult stage					
Females climbing trees, 1955			**4.39**		
Egg stage					
Females × 150			658.0	2.82	
Larval stage					$0.84 = k_1$
Full-grown larvae	86.9	551.6	**96.4**	1.98	
Attacked by *Cyzenis*	6.7	**6.2**	90.2	1.95	$0.03 = k_2$
Attacked by other parasites	2.3	**2.6**	87.6	1.94	$0.01 = k_3$
Infected by microsporidian	4.5	**4.6**	83.0	1.92	$0.02 = k_4$
Pupal stage					$0.47 = k_5$
Killed by predators	66.1	54.6	28.4	1.45	
Killed by *Cratichneumon*	46.3	**13.4**	15.0	1.18	$0.27 = k_6$
Adult stage					
Females climbing trees, 1956			**7.5**		

Note. The figures in bold are those actually measured. The rest of the life table is derived from these.
[a] After Varley *et al.* (1973).

Key factor analysis assumes that all mortality factors are additive and it ignores compensatory mortality. For the sample data in Table I,

$$K = \log (658) - \log (15) = 1.64$$
$$\text{(No. eggs)} \quad \text{(No. adults of both sexes)}$$

which is identical to

$$K = 0.84 + 0.03 + 0.01 + 0.02 + 0.47 + 0.27 = 1.64.$$

Given a series of life tables like Table I over several years, it is possible to proceed to step 2 of key factor analysis. Figure 8 illustrates this step. It is now possible to ask an important question: What causes the population to change in density from year to year? Simple visual inspection shows that k_1 (winter disappearance) in Fig. 8 is the *key factor* causing population fluctuations. A key factor is defined as the component of the life table that causes the major fluctuations in population size. There is an implication in this definition that key factors could be used to predict population trends.

Finally, the k values are used to answer a second important question: Which mortality factors are density dependent and thus might stop population increase? By plotting the k values against the population density of the life cycle stage on which they operate, density dependence can be estimated. Figure 9 shows these data for the winter moth, and Fig. 10 shows the idealized types of curves that can arise from this type of key factor analysis. Note that the key factor does not need to be density dependent and need not be involved in population regulation. In this example for the winter moth, winter disappearance is the key factor but pupal predation is the major density-dependent factor.

Key factor analysis has been widely applied to insect populations but has some important limitations. It cannot be applied to organisms with overlapping generations, like birds and mammals. Mortality factors may be difficult to separate into discrete effects that operate in a linear sequence and do not overlap and are completely additive.

Finally, density dependence may be difficult to detect if the equilibrium density (Fig. 3) varies greatly from year to year. Nevertheless, key factor analysis has provided a reliable quantitative framework for some populations against which the problems of natural regulation can be discussed.

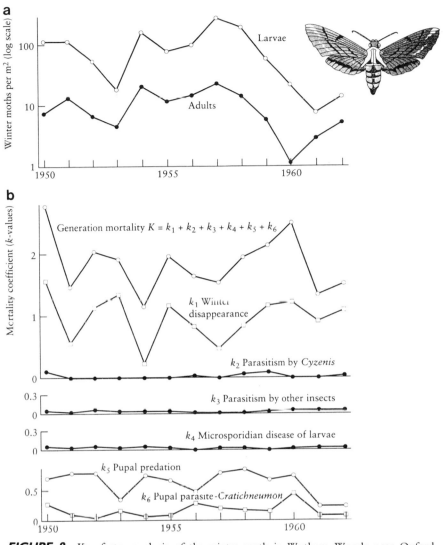

FIGURE 8 Key factor analysis of the winter moth in Wytham Woods near Oxford, 1950–1962. (a) Winter moth population fluctuations for larvae and adults. (b) Changes in mortality, expressed as k values, for the six mortality factors shown in Table I. The biggest contribution to change in the generation mortality K comes from changes in k_1, winter disappearance, which is the key factor for this population.

B. Experimental Analysis

An alternative approach to population regulation is to ask the empirical question, What factors limit population density during a particular study? This approach does not utilize the density-dependent paradigm because density dependence is often impossible to demonstrate with field data. Instead, *limiting factors* should be identified and studied experimentally. A population may be held down by one or more limiting factors, and these factors can be recognized empirically by a manipulation adding to or reducing the relevant factor. If it is suspected that food is limiting a population, the food supply can be increased to see if population size increases accordingly. Alternatively, changes in population density and the supposed limiting factors can be observed over several years to see if they vary together. Observations of this type, however, always provide weaker evidence than experimental manipulations.

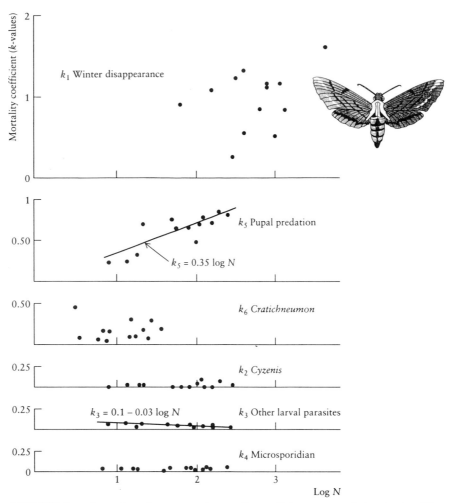

FIGURE 9 Relationship of winter moth mortality coefficients to population density. The k values for the different mortalities are plotted against the population densities of the stage on which they acted: k_1 and k_2 are density independent and are quite variable, k_2 and k_4 are density independent but are constant, k_3 is inversely density dependent, and k_5 is strongly density dependent. Compare these data with the idealized curves in Fig. 10.

The experimental approach uses the most direct and empirical techniques for answering the two central questions of regulation: what determines average abundance and what stops population growth? If parasites are believed to reduce the average abundance of pheasants, the parasite loads can be increased or reduced and the changes can be observed in pheasant numbers. If food shortage is believed to stop population growth in cabbage aphids, then the food resources can be manipulated up or down and the aphid population growth measured. It is important to realize that more than one factor may be involved in population limitation. Both parasite levels and food supplies may affect

average abundance, and once it is established that one factor is significant, it should not be assumed that only one factor is involved.

Often it is not possible to manipulate a suspected factor, either because it is too expensive (e.g., weather) or because it is not possible biologically or politically. Experimental analysis can be carried out without manipulations if the hypothesis is set up and a prediction is made about what can be observed. For example, if it is postulated that cold spring weather stops population growth in spruce budworm, spring weather and population trends can be measured in several sites in several years to test this prediction.

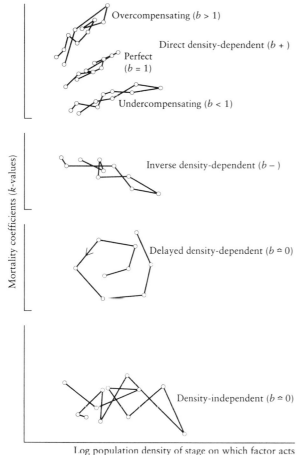

FIGURE 10 Idealized forms of the relationship possible between *k* values determined from key factor analysis and population density. The points are connected in a time sequence. *b* is the slope of the regression line. Compare with Fig. 9.

Experimental analysis looks forward and is oriented toward hypothesis testing about mechanisms of regulation whereas key factor analysis looks backward and is confined to a descriptive analysis of a population. Theoretically, both methods should converge to provide an understanding about population regulation.

VI. PLANT POPULATION REGULATION

Because most plants are modular organisms, population regulation in plants must be discussed as the regulation of biomass instead of numbers. Plant ecologists have not usually addressed the problem of population regulation in the same way as have animal ecologists, but the same principles can be applied. As a plant population increases in numbers and biomass, either reproduction or survival will be reduced by a shortage of nutrients, water, or light, by herbivore damage, by parasites and diseases, or by a shortage of space. Because plants are typically fixed in one location, competition for light or nutrients is often implicated in population regulation. This competition has been described by the $-3/2$ power rule (also called Yoda's law or the self-thinning rule).

The self-thinning rule describes the relationship between individual plant size and density in even-aged populations. Mortality, or "thinning," from competition within the population is postulated to fit a theoretical line with a slope of $-3/2$:

$$\log \overline{m} = -3/2 \, (\log N) + K,$$

where *m* is the average plant weight (grams), *N* is the plant density (ind./m^2), and *K* is a constant. This line has been suggested as an ecological law that applies both within one plant species and between different plant species. Figure 11 illustrates

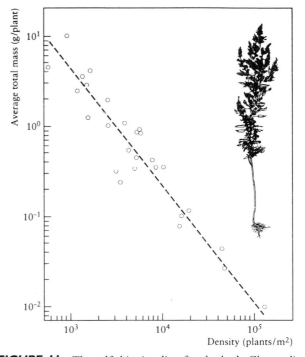

FIGURE 11 The self-thinning line for the herb *Chenopodium album*. The slope of this line is -1.33, close to the theoretical $-3/2$ of the self-thinning rule. Populations started to either side of this line would be expected to move to the line and then reach equilibrium along the line.

the $-3/2$ power rule. The self-thinning rule highlights the tradeoffs that can occur in organisms with plastic growth, such that the size of an individual can become smaller as density increases.

Recent evaluations of the self-thinning rule have found many exceptions to it. Of 63 data sets for particular plants, for example, only 24 fitted the $-3/2$ predicted slope of the self-thinning line. For gymnosperm trees, more shade-tolerant tree species had more shallow slopes than the predicted value. These results argue against a single, quantitative thinning law for all plants. The slope of the thinning line is variable, but this provides further insight into species differences under strong competition for light and nutrients.

VII. EVOLUTIONARY IMPLICATIONS OF POPULATION REGULATION

How are systems of population regulation affected by evolutionary changes? In many of these population interactions, evolutionary changes operate very slowly and are difficult to detect. But recent work in ecological genetics has shown that evolutionary changes may occur very rapidly, so that the evolutionary time scale approaches the ecological time scale. Natural selection may thus impinge upon population regulation in some organisms.

Many changes in abundance can be attributed to changes in extrinsic factors, such as weather, disease, or predation. But some changes in abundance are the result of changes in the genetic properties of the organisms in a population. Such evolutionary changes are produced by natural selection. David Pimentel from Cornell University has suggested that population regulation has its foundation in the process of evolution.

A simple model will illustrate the type of systemic changes that could be produced by natural selection. Consider a two-species system of one plant and one herbivore, and, to make the model simple, only one gene on one chromosome in the plant will be focused on. The hypothetical gene has a major effect on (1) the ability of the plant to survive in its environment and (2) the palatability of the plant to the herbivore. Two different alleles

(A and a) occur at the hypothetical gene locus, and the properties of the genotypes are as follows:

	Genotype of plant		
	AA	*Aa*	*aa*
Ability of plant to survive	Good	Poor	Very poor
Palatability to herbivores	High	Low	Very low

Thus plants of genotype *AA* are able to survive very well but attract many herbivores because they are desirable food. Each plant genotype can support only a limited number of herbivores before it is killed by overgrazing. Finally, it is assumed that the reproductive rate of the herbivore will be affected by the genotype of plant on which it lives, so that highly palatable plants are best for herbivore reproduction.

This simple model is a variant of the discrete generation predator–prey model, in which the plant is the prey and the herbivore is the predator. The major change in the model is that genetic variation is allowed within the plant population. Figure 12 illustrates one pattern of equilibrium for a hypothetical system starting with 150 herbivores and a plant population with genotype frequencies of 0.36 *AA,* 0.50 *Aa,* and 0.14 *aa.* The system stabilizes under some initial conditions (as in this example) but with other assumptions may be unstable.

Pimentel cataloged some spectacular examples of genetic changes of this type that played a role in population regulation. For example, the Hessian

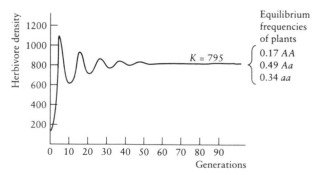

FIGURE 12 Determination of average herbivore density resulting from the interaction of a plant and a herbivore through natural selection. Starting conditions are given in the text.

fly population was reduced drastically in Kansas after 1942 when resistant varieties of wheat were introduced. The herbivore population of Hessian flies was significantly reduced by changing the genetic makeup of the wheat plant. Another example is the myxomatosis–rabbit interaction in Australia. The European rabbit was introduced into Australia in 1859 and increased to very high densities within 20 years. After World War II, an attempt was made to reduce rabbit numbers by releasing a virus disease from South America, myxomatosis. The myxoma virus was highly lethal to European rabbits, killing over 99% of individuals infected. Figure 13 shows the precipitous crash in rabbit numbers that followed the introduction of myxomatosis in 1951.

Since myxomatosis was introduced into Australia in 1951, evolution has been going on in both the virus and the rabbit. The virus has become attenuated so that it kills fewer and fewer rabbits and takes longer to cause death. Since mosquitoes are a major vector of the disease, the exposure time before death is critical to viral spread. Table II summarizes changes that have occurred in the virus. These data are obtained by testing standard laboratory rabbits against the virus so they measure viral changes while holding rabbit susceptibility constant. Since 1951 less virulent grades of virus have replaced more virulent grades in field populations.

Rabbits have also become more resistant to the virus (Fig. 14). By challenging wild rabbits with a constant laboratory virus source, it has been shown that natural selection has produced a growing resistance of rabbits to this introduced disease.

The evolution of resistance to the virus in rabbits is easily explained by selection operating at the individual level: rabbits that are more resistant leave more offspring. It is more difficult to explain the evolution of reduced virulence in the virus. Virulence in a virus is related to fitness because more virulent viruses make more copies of themselves. But if more virulent viruses kill rabbits more quickly, there will be less time available for transmission of the virus through mosquitoes or fleas. The result for the myxoma virus is group selection operating to reduce virulence to a moderate level. Group selection occurred because less virulent viral colonies are favored over more virulent viral colonies because they take longer to kill the host rabbit (Table II). Host–parasite systems may be ideal candidates for group selection along these lines.

It is not known if the rabbit–myxoma system has reached a stable equilibrium or whether continuing evolution will allow the rabbit population to recover to its former levels (Fig. 13). There is some evidence that the rabbit–myxomatosis interaction in Britain is changing toward more resistant rabbits and more virulent viruses; the population size of rabbits in Britain seems to be increasing again.

Pimentel's concept of genetic changes involves interspecific interactions and suggests some possible implications of these interactions to the determination of average abundance. This concept emphasizes the role of evolution in questions of average abundance, and in doing so it serves as a warning to the continual introduction of new species into ecological communities of distant areas.

Self-regulatory populations present yet another problem in evolutionary ecology. How does a population evolve the machinery to be self-regulating? Self-regulation is clearly a desirable adaptation for any population that has the potentiality of destroying its resources. The problem is that it is an adaptation that is favorable for the population, not necessarily for individuals. Darwinian natural selection

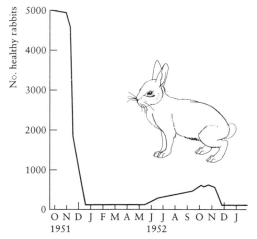

FIGURE 13 Population crash of the European rabbit (*Oryctolagus cuniculus*) at Lake Urana, New South Wales, after the virus disease myxomatosis was introduced in 1951. Numbers of healthy rabbits were counted on standardized transects.

TABLE II

Virulence of Field Myxoma Virus Types in Rabbits in Australia[a]

Grade of severity	Virulence type grade					
	I	II	IIIA	IIIB	IV	V
Mean survival times of rabbits (days)	<13	14–16	17–22	23–28	29–50	—
Case mortality rate (%)	>99	95–99	90–95	70–90	50–70	<50
Australia						
1950–1951	100	—	—	—	—	—
1958–1959	0	25.0	29	27	14	5
1963–1964	0	0.3	26	33	31	9

[a] After Fenner and Myers (1978).

is individual selection. How can population adaptations arise?

The answer is simple, by *group selection.* Just as natural selection can operate at the level of the individual organism—individuals that are more fit leave more descendants in the next generation—it can also operate at the level of the group. Groups that have an adaptation may avoid extinction. Group selection was invoked by Wynne-Edwards to explain his theory of self-regulation. Most workers, however, reject group selection as a possible mechanism of evolution and try to explain all adaptations on the basis of individual selection.

Group selection is typically invoked as an evolutionary force when some adaptation is good for the population but bad for an individual. In these cases, group selection pushes up the frequency of

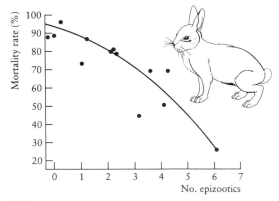

FIGURE 14 Mortality rates of wild European rabbits from the Lake Urana region of southeastern Australia after exposure to several epizootics of myxomatosis. Mortality was measured after a challenge infection with a strain of myxoma virus of grade III virulence.

a trait, while individual selection pushes it down. For group selection to work, whole populations (groups) must become extinct while others survive. Since more individuals die than groups become extinct, individual selection is always much stronger than group selection. Obviously, if group selection is pushing the same way as individual selection, the trait will be favored at both levels and no problem arises.

So the critical question becomes whether self-regulatory adaptations are good for populations (agreed) and good for individuals as well (questioned). If natural selection operating on individuals can favor mechanisms of self-regulation, the problem is solved, and group selection can recede into the background. How might selection favor self-regulation?

The answer is very similar to that used to explain the evolution of competitive ability. Natural selection favors individuals that increase their fitness by means of higher reproduction or lower mortality, but it also favors individuals that reduce the fitness of their neighbors by any technique of interference competition. Fitness is relative, and the mistake of many evolutionary ecologists was to assume that organisms were trapped in an upward spiral of ever-increasing fitness through ever-increasing reproductive rates. Thus self-regulatory mechanisms can be explained most easily by individual selection operating on mechanisms of interference competition within the species.

One mechanism of self-regulation is emigration of organisms from optimal to suboptimal environments. Emigration would always appear to be disadvantageous to the individuals involved (since many of them die in the process of moving), and thus, Robert MacArthur argued in 1972, emigration cannot evolve by individual selection. This argument is not correct. Individual selection can favor self-regulation by emigration because not all individuals have equal access to resources in natural populations and there is both spatial and temporal variability in natural environments. Thus there is no need to invoke group selection to explain self-regulation.

VIII. SUMMARY

Populations of plants and animals do not increase without limits but show more or less restricted fluctuations. Two general questions may be raised for all populations: (1) What stops population growth and (2) what determines average abundance?

Three general theories answer these two questions by focusing on the interactions between the population and the environmental factors of weather, food, shelter, and enemies (predators, parasites, diseases). The biotic school suggests that density-dependent factors are critical in preventing population increase and in determining average abundance. Natural enemies are postulated to be the main density-dependent factors in many populations. The climate school emphasizes the role of weather factors affecting population size and suggests that weather may act as a density-dependent control. In contrast, the self-regulation school focuses on events going on within a population, on individual differences in behavior and physiology. The general premise of this school is that abundance may change because the quality of individuals changes. A population increase may be stopped by a deterioration in the quality of individuals as density rises instead of by a change in environmental factors. Average abundance may be altered by genetic changes in populations. Quality and quantity are both important aspects of populations.

Population regulation theory has focused on equilibrium conditions, and many ecologists now emphasize nonequilibrium concepts and ask what factors reduce stability for populations. The spatial scale of a study affects conclusions about stability, and if a population is subdivided into local populations stability may be increased for the entire population. Mortality agents may cause additive or compensatory losses in populations. Additive losses may limit or regulate population density, but compensatory losses may be irrelevant to both limitation and regulation.

The theories of population regulation are not mutually exclusive but overlap, and a synthesis of several approaches may be most useful in attempts to answer practical questions. The natural regulation of populations is a critical area of theoretical ecology because it is central to many questions of community ecology and because it has enormous practical consequences for conservation, pest control, and harvesting.

Glossary

Additive mortality Losses to a population caused by one factor that accumulate in effect so that they are perfectly correlated with total mortality (Fig. 7).

Compensatory mortality Mortality caused by one factor that varies independently of total mortality (Fig. 7).

Density-dependent factors Factors that increase in per capita (percentage) effect as a population increases in density (Fig. 2).

Density-independent factors Factors that show no relationship between per capita effect and population density.

Metapopulations A series of local populations that may operate independently but are linked together by dispersal movements (Fig. 5).

Population limitation Ecological processes that affect the average density of a population, so that adding or subtracting the process causes the population density to increase or drop. Processes that limit density may also regulate density (Fig. 2).

Population regulation Ecological processes that stop populations from increasing in density and thereby provide negative feedback to population growth (Fig. 2).

Self-regulation Population regulation achieved by a deterioration in the physiological, behavioral, or genetic quality of individuals as population density increases.

Bibliography

Deangelis, D. L., and Waterhouse, J. C. (1987). Equilibrium and nonequilibrium concepts in ecological models. *Ecol. Monogr.* **57**, 1–21.

Hassell, M. P. (1985). Insect natural enemies as regulating factors. *J. Anim. Ecol.* **54,** 323–334.

Krebs, C. J. (1994). "Ecology: The Experimental Analysis of Distribution and Abundance," Chapters 11–19. New York: Harper Collins Publishers.

McNamara, J. M., and Houston, A. I. (1987). Starvation and predation as factors limiting population size. *Ecology* **68,** 1515–1519.

Sinclair, A. R. E. (1989). Population regulation in animals. *In* "Ecological Concepts" (J. M. Cherrett, ed.), pp. 197–241. Oxford: Blackwell Scientific Publications.

Strong, D. R. (1986). Density-vague population change. *Trends Ecol. Evol.* **1,** 39–42.

Watson, A., and Moss, R. (1970). Dominance, spacing behaviour and aggression in relation to population limitation in vertebrates. *In* "Animal Populations in Relation to Their Food Resources" (A. Watson, ed.), pp. 167–218. Oxford: Blackwell Scientific Publications.

Population Viability Analysis

Craig M. Pease
University of Texas

Russell Lande
University of Oregon

A population viability analysis (PVA) assesses the current status of a species and systematically determines the conservation measures needed to prevent its extinction. The analysis should determine current trends in population size, identify deterministic thresholds for population persistence, and quantify the risks of extinction from stochastic fluctuations in the environment and in the genetic composition of a population. Large-scale habitat destruction and alteration are the primary threats facing most currently endangered species. Any single PVA should be part of an iterative cycle that involves use of the PVA to generate hypotheses concerning what conservation measures are necessary, monitoring the population and its habitat to test these hypotheses during their implementation, and reconsideration of the PVA in light of information on the effectiveness of the conservation measures recommended and of new biological data.

I. INTRODUCTION

Many characteristics of a population determine its risk of extinction, including population size, den-sity, inbreeding levels, genetic variability in traits underlying fitness, and habitat abundance, quality, and configuration. To design a conservation program that prevents extinction, one must understand the impact of each of these variables. [See EVOLUTION AND EXTINCTION.]

The purely empirical approach to these questions is fraught with difficulties. Directly manipulating the causes of extinction usually is not feasible because the time scales involved may be long and because replicate populations often do not exist in which to study different factors affecting the risk of population extinction. The available empirical studies of extinction include notable observational data on natural populations of plants, amphibians, birds, and mammals and experimental studies of various plant and invertebrate species and domesticated animals.

However, important parameters affecting extinction risks can be readily measured using existing methodologies. Examples include dispersal distances, fecundity and survivorship schedules, the genetic variability of traits underlying fitness, and the magnitude of environmentally induced variation in fecundity. Instead of directly manipu-

Copyright © 1995 by Academic Press, Inc.
All rights of reproduction in any form reserved.

lating the causes of extinction, the more productive and prevalent approach is to construct mathematical models predicting extinction probabilities from these more readily measured parameters. This article focuses on such models; it does not critically review the empirical literature on extinction, nor does it review methods for measuring key population parameters. Although the following does provide a broad framework for undertaking PVAs, there is great diversity between species in aspects of their biology that control the chance of extinction; hence, in many cases the following general models will have to be modified or extended before being applied to specific species.

II. THE GOAL OF POPULATION PERSISTENCE

The goal of all PVAs is to determine the steps needed to prevent a particular population from going extinct in the foreseeable future. Any PVA should explicitly identify and justify the quantitative definition of persistence it chooses. Many authors take a risk analysis perspective, defining a population with a high chance of persistence as one that has, say, a 99% probability of persistence over 100 years.

Though the choice of time span and probability level may be influenced by practical considerations, there are biological aspects of these decisions. The time span chosen should be long in comparison to the length of relevant ecological time lags, including the average life span of individuals, the characteristic return time of catastrophes, and the regeneration time for disturbed habitat. One could also reasonably argue that the probability of extinction should be the same order of magnitude as that implied by background species extinction rates measured from the fossil record. But this is rarely if ever done, in part because the prospects for endangered species are so poor in comparison to the average life span of species.

III. DETERMINISTIC FACTORS

The most immediate and direct threats to a population arise from deterministic downward trends in its size and deterministic degradation of other key parameters for persistence, such as genetic variability and habitat quality and abundance. A decreasing population has zero probability of persistence if its decline is not halted. Moreover, purely deterministic analyses of a population's biology often predict the existence of various thresholds for population persistence, implying a high probability of persistence above the threshold and a low probability of persistence below it. To prevent extinction, it is necessary to maintain key population parameters well above any such threshold.

Unfortunately, many PVAs initially concentrate on the threat of extinction due to stochastic factors, underemphasizing the critical importance of deterministic trends in population size and other key population parameters. The first steps of a PVA should include documenting current trends in population size and identifying deterministic thresholds for population persistence.

A. Censuses and Leslie Matrices

The most direct method for determining the trend in population size is to obtain two or more estimates of population size at different times. Because exhaustive censuses are often impractical or impossible, it is often advantageous to exploit a mark–recapture experimental design in order to estimate population sizes and growth rate. Alternatively, when age (or stage)-specific fecundity and survivorship data are available, these can be analyzed using a Leslie (or Lefkovitch) matrix to estimate the intrinsic rate of natural increase of a population at a particular time.

B. Habitat Destruction and Alteration

The ultimate cause of a decline in population size is usually habitat destruction and degradation. Moreover, data on habitat abundance and quality are often readily obtained by remote sensing or by carefully designed ground surveys of available habitat. The indirect assessment of population status afforded by habitat trend data is often an important adjunct or substitute for population status estimates obtained from census, survivorship, and fecundity data, inasmuch as these more direct measures are often costly, inaccurate, or simply unavailable.

Very little of the original habitat remains for many endangered species. For example, the grizzly bear currently occupies only 2% of its former range in the lower 48 states in the United States, in large part because of widespread destruction of the wilderness habitat on which it depends. Habitat destruction is ultimately caused by human population growth and resource use, and it is the single most important cause of deterministic downward trends in the abundance of species. When habitat destruction or alteration is not implicated in a species decline, the ultimate problem is often overharvesting or introduced predators, competitors, or parasites.

C. Thresholds for Population Persistence

A threshold for population persistence arises when there is a critical value of population density, habitat abundance, quality or configuration, or another key population parameter, below which a population inevitably becomes extinct. Such thresholds are widespread, and it is incorrect to assume, for example, that abundance of a species and abundance of its habitat are directly proportional. Simply halting a decline in habitat quality or population size is insufficient, by itself, to save a population that is already below a threshold.

Table I summarizes the thresholds discussed in the following paragraphs. These thresholds arise from two distinct classes of models, (1) those that predict the overall population growth rate ("Single Spatial Locality" and "Spatially Distributed Population" in Table I) and (2) those that predict the percentage of suitable habitat occupied ("Metapopulation Considerations" in Table I). A model that implies a threshold for population persistence predicts one of these two quantities as a function of parameters such as mean-square dispersal distance or rate of environmental movement. A parameter or combination of parameters that makes the population growth rate or percentage of habitat occupied equal to zero is a threshold for population persistence. Thus the formulas in Table I for population

growth rate and percentage of habitat occupied can be used to solve for the various thresholds discussed next.

The *Allee effect* is a threshold in organism density that arises because social interactions, such as finding a mate, become more difficult as the population becomes less dense. Such effects are of particular importance for vagile, sparse species whose numbers have been greatly reduced by humans.

Edge effects occur when the fecundity and/or survivorship of individuals living on the edge of a habitat patch is reduced in comparison to those of individuals living in the patch interior. For example, passerines living in deciduous forest patches can suffer increased nest predation and brood parasitism up to several hundred meters from an edge, and trees within a hundred meters of tropical habitat edges occupy different light, wind, and moisture regimes.

A *critical patch size* arises from the balance between population growth within a patch of suitable habitat and dispersal into inhospitable habitat outside of the patch. This is a consequence of the fact that the perimeter of a circular habitat patch increases disproportionately as habitat area decreases, causing an increasing proportion of the population to disperse into inhospitable habitat. More complex models of this sort give rise to a *critical rate of environmental movement;* when the spatial location of an organism's habitat is moving, the population may go extinct because dispersal into inhospitable areas occurs too rapidly and/or genetic adaptation to the novel environment occurs too slowly to permit a positive population growth rate. Such environmental movement could arise from global warming or a biotic invasion. A population confined to an island, park, or isolated habitat patch may be unable to track its environment beyond the boundaries.

Metapopulation models of local extinction and recolonization predict the existence of a *critical density of suitable habitat*. As the density of suitable habitat patches decreases, recolonization becomes less frequent and a decreasing proportion of the patches are occupied. At the critical density, the proportion of patches occupied at equilibrium drops to zero; essentially, recolonization of vacant

TABLE I

Some Factors Giving Rise to Thresholds for Population Persistence

Biological consideration	Population growth rate or proportion habitat occupied	Notes
	One locality	
Edge effects	$r - \dfrac{2\Delta r_x X}{R}$	r = maximum possible population growth rate Δr_x = difference of population growth rates in interior and edge R = radius of a circular patch X = edge width
Balance of dispersal/ population growth[a]	$r - \dfrac{\sigma^2 \pi^2}{2L^2}$	σ^2 = diffusion coefficient (variance of individual dispersal distance per generation) L = length of one-dimensional patch of suitable habitat. See Okubo for generalizations, including density dependence and two spatial dimensions.
Maintain genetic variation for selection[b,c]	$r - \dfrac{k^2}{2\gamma^2 G^2}$	k = rate of change of optimal phenotype γ = strength of stabilizing selection G = additive genetic variance
Maintain genetic variation for selection[b,c]	$r - \dfrac{\left(\dfrac{\gamma}{4}\right)A^2\omega^2}{\omega^2 + \gamma^2 G^2}$	ω = frequency of cyclic fluctuations in location of optimal phenotype A = amplitude of these fluctuations
Genetic and phenotypic variation can create maladaptation[b,c]	$r - \dfrac{\gamma}{2}P - \left(\dfrac{\gamma}{2}\right)\sigma_\theta^2\left[1 + \dfrac{\gamma}{2}G\right]$	σ_θ^2 = coefficient that determines the extent of the random fluctuations in the location of the optimum phenotype P = phenotypic variance
Allee effects[d]	Population decline if it falls below p_c	p_c = the critical population density
	Spatially distributed population	
Balance of dispersal/ population growth[e]	$r - \dfrac{1}{2}\left[\dfrac{\sigma^2}{w_{11}(1 - \rho^2)}\right]^{\frac{1}{2}}$	$[w_{11}(1 - \rho^2)]^{1/2}$ = environmental tolerance σ^2 = as above
Critical rate of environmental movement[e]	$-\dfrac{\nu^2}{2\sigma^2}$	ν = rate of movement of optimum phenotype across the landscape
Maintain genetic variation for selection[e]	$+\dfrac{\rho^2 G}{2w_{22}(1 - \rho^2)}$	$[w_{22}(1 - \rho^2)]^{1/2}$ = niche breadth ρ = correlation between environmental position and optimum character value
	Metapopulation considerations	
Extinction/recolonization of habitat patches[f]	$\hat{p} = 1 - \dfrac{e}{m}$	p = proportion of habitat occupied e = extinction rate m = recolonization rate
Extinction/recolonization of territories[g]	$\hat{p} = 1 - \dfrac{(1 - k)}{h}$	k = demographic potential h = proportion of territories in the landscape that are suitable habitat

[a] Okubo, A. (1980). "Diffusion and Ecological Problems." Springer-Verlag, Berlin.

[b] Lynch, M., and Lande, R. (1992). Pages 243–250. *In* "Biotic Interactions and Global Change." Sinauer Associates, Sunderland, MA.

[c] Lande, R. (1993). The meaning of quantitative genetic variation in evolution and conservation. Pages 27–40. *In* "Biodiversity in Managed Landscapes." (R. Szaro, ed.) Oxford University Press.

[d] Dennis, M. (1989). Allee effects: population growth, critical density and the chance of extinction. *Natural Resources Model.* **3**, 481–538.

[e] Pease, C. M., Lande, R., and Bull, J. J. (1989). A model of population growth, dispersal and evolution in a changing environment. *Ecology* **70**, 1657–1664.

[f] Levins, R. (1970). Extinction. Pages 77–107. *In* "Lectures on Mathematics in the Life Sciences," Vol. 2. The American Mathematical Society, Providence, Rhode Island.

[g] Lande, R. (1987). Extinction thresholds in demographic models of territorial populations. *Am. Nat.* **130**, 624–635.

habitat patches occurs too slowly to overcome local extinction. A special case is the *critical territory density* that arises in species that are territorial or that require small, discrete habitat patches. In this case, the habitat patch is occupied by a single mated pair, and the proportion of juveniles that successfully disperse to unoccupied patches declines as the habitat patches become more sparse. This is an important consideration in the conservation of the northern spotted owl, an old-growth species found in the Pacific Northwest of the United States.

For metapopulations maintained by a balance of local extinction and colonization, not all of the suitable habitat patches will be occupied at any time. Protection of suitable unoccupied habitat may be crucial to the persistence of such a species by providing areas for future colonization.

D. Thresholds Involving Genetic Variation

It is useful to distinguish two distinct effects of environmental and genetic variation on population persistence. First, if the variation is maintained at a roughly constant value, it may be treated as another parameter in analyses of population growth rate and percentage occupied habitat. Thus various models (Table I) give rise to *critical values of genetic variance* needed to track changing environments. Second, environmental and genetic variability implies random fluctuations (random genetic drift) or adverse trends (inbreeding depression) in traits strongly correlated with fitness. Random fluctuations can drive a population to extinction even when the average population growth rate and percentage of habitat occupied are both positive. These considerations are discussed in Section IV and Section V,A.

IV. STOCHASTIC ENVIRONMENTAL FACTORS

Many, if not all, currently endangered species became so because of deterministic processes that have drastically reduced their population size. But even if conservation steps successfully halt these deterministic processes, and the average growth rate of the population below carrying capacity is positive, there is no guarantee that the population will persist.

One must consider the pattern of variation in fitness itself and in traits that are major components of fitness, such as fecundity, survivorship, and dispersal rates. The following discussion of variation in such traits is based on the starting point of elementary quantitative genetics: the phenotype is determined by both genetic and environmental factors. Importantly, it is impossible to make any generalizations regarding whether increasing this variation will have positive or negative consequences on population persistence. For example, increasing genetic variability has a positive impact on persistence under many, but not all, circumstances, while increasing environmental variability in the phenotype often has a negative impact.

Many types of environmental variation in demographic parameters have been recognized in the literature. Fundamentally, these differ with respect to the temporal and spatial extent of the variation. The following discussion first outlines the types of variation in a population occupying a single spatial location, and then considers the types of environmental variation that arise in species with multiple spatially distributed populations.

A. Demographic Stochasticity

Demographic stochasticity is random variation among individuals of a single population in fecundity or survivorship. It is an important cause of extinction only in small populations; in fact, the mean time to extinction due to demographic stochasticity of a population of size K scales as $\exp(2K\bar{r}/V_1)/K$ as K becomes large, where $\bar{r} > 0$ is the average growth rate of the population when it is below K, and V_1 is the variance in individual fitness (per unit time). For example, a population of 100 individuals is 10-fold larger than a population of 10 individuals, yet its mean time to extinction due to demographic stochasticity is about 800 times less than that of the smaller population (assuming $\bar{r} = 0.05$ and $V_1 = 1.0$).

B. Environmental Stochasticity

Temporal variation in survivorship or fecundity affecting all individuals in a population is called environmental stochasticity. Many authors make a distinction between different rates and magnitudes of environmental variation. Infrequent events that affect a large proportion of the population are called catastrophes (e.g., hurricanes, floods, etc.), while the term environmental stochasticity is often reserved for variation that occurs more frequently and affects only a small proportion of the population (e.g., annual variation in rainfall that causes variation in fecundity). The mean time to extinction due to environmental stochasticity scales as K^c when $\bar{r} > 0$, where $c = 2\bar{r}/V_e - 1$, and V_e is the environmental variance in r. When $\bar{r} > 0$, the mean time to extinction from random catastrophes also scales as a power function of carrying capacity, with the power decreasing as the magnitude and frequency of the catastrophes become larger in relation to the average population growth rate in normal years. These scaling relations imply that environmental stochasticity and random catastrophes will generally be more important causes of extinction than demographic stochasticity in large populations.

C. Spatial Variation

It is widely recognized that a broad spatial distribution can ameliorate some of the risk of extinction due to environmental stochasticity. This is so because the environmental variations at distinct sites will often not be perfectly correlated, allowing the population to spread its risk. Although few quantitative results are available, maintaining an endangered species across a broad geographic range has considerable merit.

Demographic and environmental stochasticity refer to types of variation in a single population. In many practical situations there are a wide variety of other considerations, including spatial variation in fecundity and survivorship, and spatial variation that changes through time (e.g., annual changes in location of favored foods, or a disturbance regime or plant succession shifting the location of favored habitat).

V. GENETIC FACTORS

It is relatively straightforward to determine what effect a given magnitude of environmentally caused variation in fecundity and survivorship has on the probability of extinction of a small population. When the environmental variation is not autocorrelated, as most theoretical models assume, the deviation from average fecundity and survivorship a population experiences in one year will not influence the deviation experienced in future years. As the magnitude of demographic or environmental stochasticity increases, the probability of a run of bad years sufficient to cause extinction increases.

The variation between individuals caused by genetic factors has a much different impact on probability of persistence than does environmental variation. In a small population, sampling accidents during reproduction cause random changes in gene frequencies (random genetic drift) which tends to reduce genetic variation in a population. When an allele becomes fixed or is lost, it remains so for all future generations, at least in the absence of mutation or migration. This autocorrelation between the genetic structure of the population in successive generations greatly complicates the analysis of the relation between sampling of genetic variation and extinction. Random genetic drift increases the probability of population extinction because deleterious recessive mutations become expressed in homozygous form and because there is less variation for adaptive evolution.

A. Inbreeding Depression

Inbreeding depression is the reduction in fecundity and/or survivorship that occurs when a formerly large outbred population is rapidly reduced in size, becoming inbred and hence homozygous, and causing the unmasking of lethal and partially lethal recessives as well as loss of any heterozygote advantage. Little information exists concerning the level of inbreeding that a population can sustain without going extinct, although it is known that the intense inbreeding (i.e., continued self-fertilization or brother–sister mating) of formerly outbred species often results in the extinction of roughly 90% of the lines within 10 to 20 generations. The endangered Florida panther, which now numbers only 30 to 50 adults, is a good example of genetic management concerns. This animal is showing signs of past inbreeding, including increased incidence of male sterility and congenital heart defects. A population viability panel recently recommended immediate introduction of several adolescent females from the nearest abundant population (in Texas) to revitalize the gene pool.

To determine the size at which a formerly large outbred population needs to be maintained to limit the impact of inbreeding depression on extinction, the following indirect argument is useful. Both theoretical models and empirical data support the idea that when a formerly outbred population goes through a single bottleneck with inbreeding coefficient $f \ll 1$, the mean fitness is reduced approximately by the proportion fB in the generation immediately after the bottleneck, where B, the average number of lethal equivalents in the genome, is of order one for many laboratory and domestic animals and captive endangered species. Thus limiting the increase in f to a few percent per generation, as would be achieved with an effective population of 50 in a diploid species, should usually limit the rate of reduction in fitness to at most a few percent per generation.

Determining the cumulative impact of inbreeding on a population sustained at a small size over a number of generations is more complicated and involves understanding how natural selection purges recessive deleterious alleles from the population when the inbreeding coefficient increases gradually. This consideration potentially causes the fitness impact of sustained inbreeding to be much less than fB.

B. Adaptive Evolution in a Changing Environment

Genetic variation generally allows a population to adapt more rapidly to environmental changes, although this does not invariably promote persistence (Table I). When the environmental change being tracked is directional or cyclic, increasing the genetic variance decreases the difference between

TABLE II
Management Threats to Population Viability

Numerous problems arise when human institutions attempt to implement the biological recommendations of a PVA

The conservation goals are not explicitly stated

There are conflicting goals

The conservation strategies for achieving these goals are not explicitly stated

No method for gathering data to evaluate the effectiveness of the conservation strategies, or the method is ineffective

Long time lags between improved data and analyses, and appropriate revisions in conservation strategies

Ineffectiveness communication between those gathering new data and those responsible for devising and implementing improved strategies

Failure to achieve a goal leads to its redefinition instead of adoption of a more effective conservation strategy

Continued habitat destruction while improved scientific data are gathered and analyzed, and policy studies are completed

the population growth rate of a perfectly adapted population and the population growth rate actually achieved. However, when the environmental changes are unpredictable (vary randomly with no autocorrelation) about some mean, increasing genetic variation actually decreases the population growth rate.

If the assumption that the population is confined to a single spatial location is relaxed, and the population is allowed to track its environment across space, then genetic variability increases the population growth rate, while environmental movement decreases it. The critical rate of environmental movement referred to earlier is obtained by solving for the rate of environmental movement that makes the intrinsic growth rate of the population zero (Table I).

Determining the population size necessary to maintain a given level of genetic variation is a complicated issue. The answer depends on whether the trait is determined by a single locus or varies quantitatively, and the nature of the phenotypic distribution of mutants and of selection against variant alleles. A population of roughly 500 individuals should theoretically maintain typical levels of genetic variation (roughly equal to environmental variation in the phenotype) for many quantitative traits, although much larger sizes may be necessary to maintain molecular heterozygosity or variation at particular loci (e.g., for major disease or pesticide resistance alleles).

There are numerous examples of genetic mismanagement in forestry and fisheries where population health and/or abundance declined because restocking programs used a nonlocally-adapted source population or failed to deal with loss of fitness from inbreeding depression or adaptation to artificial breeding conditions.

VI. COMPREHENSIVE APPROACHES

Whatever framework a PVA uses, the standard approach is to consider separately, rather than in combination, each factor that could lead to a population's extinction. This approach is justified on the grounds that considering several factors simultaneously generally leads to intractable complications. It is nevertheless true that interactions between various factors considered separately could lead to radically different extinction dynamics than those predicted by considering each factor alone. An analysis that included both edge effects and environmental stochasticity could, for example, conceivably produce a different result from that produced by separate analyses of edge effects and of environmental stochasticity. Several researchers are currently developing models that allow one to make predictions about extinction that simultaneously consider several risk factors; this research focuses both on the general analytic theory and on numerical applications to particular species. Another active area of research involves combining remote-sensing data with spatially explicit demographic simulations for particular species of concern.

The inherent uncertainty in predictions about extinction (see the following discussion), the long time lags in ecological systems, and the possibility of unknown interactions all conspire against the predictions of the most thorough PVAs. Aggressively protecting a species' habitat provides some measure of protection against these unknowns, inasmuch as habitat protection involves preserving time lags and interactions that are currently part of the species' ecology. Consequently, a PVA should

involve both the factor-by-factor consideration of causes of extinction developed earlier, as well as habitat protection per se.

VII. ECONOMIC, SOCIAL, AND POLITICAL CONSIDERATIONS

Diverse human institutions are responsible for managing rare species, including government agencies, private ranches, corporations, individuals, and indigenous peoples. The potential for bureaucratic inertia can be truly staggering, as in the Yellowstone ecosystem of the western United States where over 10 government agencies, each with its own goals and constituencies, manage various aspects of an ecosystem occupied by a remnant grizzly bear population. Similarly, bureaucratic conflicts delayed efforts to restore the Everglades ecosystem in south Florida, where extensive habitat alteration has reduced wading bird populations to less than 10% of their historical abundance. Even if management is coordinated by a single agency, there is no guarantee it will be effective; Table II enumerates some frequent problems.

An additional important class of economic and political threats to population viability arises not from the agency or individual directly responsible for management, but from various outside economic or political forces which may be in direct conflict with the goal of population persistence. The issue is often sustainability. The conservation recommendations of PVAs typically require sustained use of natural resources. Cost–benefit analyses often demonstrate a broad range of positive long-term benefits from implementing the recommendations of a PVA that compensate for the short-term problems inherent in shifting away from nonsustainable natural resource uses. These benefits can include increased tourism, increased property values, and protection of small-scale extractive natural resource uses.

VIII. CHANGE, UNCERTAINTY, AND THE SCIENTIFIC METHOD

Conservation plans for rare species must account for change at a number of levels, including continual refinements in our knowledge of the species' biology, changes in the species' environment such as those brought on by global warming or habitat degradation, and even changes in the magnitude and nature of human factors such as tourism.

Even without these changes there would always be uncertainty about the precise consequences of implementing a PVA's recommendations, regardless of how thorough the analysis. There is uncertainty (1) that arises from stochastic variation in environmental and genetic factors, (2) in our knowledge of critical parameters describing the species' biology, and (3) in the choice of what empirical or mathematical model the probability of persistence is estimated from. Many PVAs concentrate heavily on the first type of uncertainty, evaluating, for example, the probability that a population of a given size will persist in the face of environmental stochasticity, while discounting or even overlooking the latter two types of uncertainty, although they can be readily quantified using standard statistical methods and sensitivity analyses.

The scientific method is the only reliable approach for ensuring progress toward a goal in the face of change and uncertainty. Since the PVA is the central scientific document underlying a conservation plan, it is essential that it adhere to the basic tenets of the scientific method. The conclusions of a PVA serve dual roles; each is simultaneously a conservation recommendation and a scientific hypothesis.

In undertaking a PVA, it is useful to first analyze the species' biology to determine what biological characteristics it needs to survive and only then to secondly and separately consider what economic, social, and political steps are necessary to implement these recommendations. Of course the latter analysis may show that certain of the biological recommendations are infeasible, in which case the biological PVA will have to be reconsidered so as to determine how to best reduce the probability of extinction, given the constraints imposed by human institutions. Ideally, this would result in an iterative process that would lead to identification of an effective conservation strategy. Two common problems that often prevent this iterative process

from occurring are failure to explicitly identify what assumptions are being made about constraints imposed by humans and failure to recognize that a PVA is never final, but instead is one step in an iterative loop.

The northern spotted owl (*Strix occidentalis caurina*) in the Pacific Northwest of the United States has been the subject of several cycles of increasingly sophisticated and realistic PVAs since 1987, including both demographic estimates of current population growth rate and predicted effects of habitat fragmentation on future population growth. These culminated in state-of-the-art statistical analysis of demographic data from extensive mark–recapture field observations, and detailed spatially explicit computer models of habitat and population dynamics. Results of the PVAs were important in the 1990 decision to list the subspecies as threatened under the U. S. Endangered Species Act and for the development and implementation of increasingly comprehensive conservation plans, which now include sustainable management of the entire old-growth forest ecosystem on which the owl and numerous other species, as well as many aspects of the regional economy, depend. [*See* VIRGIN FORESTS AND ENDANGERED SPECIES, NORTHERN SPOTTED OWL AND MT. GRAHAM RED SQUIRREL.]

Glossary

Adaptive management Natural resource management that involves explicitly setting goals, using raw data and its analysis to identify conservation strategies for achieving these goals, monitoring the implementation of these strategies, and using the monitoring information to first identify and then implement increasingly effective strategies.

Allee effect A critical density below which a population declines to extinction. Its value is often set by social factors, such as the increased difficulty of finding a mate at low densities.

Demographic stochasticity Random variation in individual fitness in a population.

Environmental stochasticity Environmentally caused temporal variation in fitness. At any point in time, this source of variation causes the same deviation from mean fitness in all individuals.

Inbreeding depression The reduction in fitness that occurs when a formerly outbred population is rapidly reduced in size. It is caused primarily by increased levels of homozygosity of deleterious recessive mutations.

Minimum viable population The threshold population size for persistence predicted by deterministic models, or the population size necessary to achieve a certain probability of persistence during a given span of time in stochastic models.

Threshold for population persistence The value of any population parameter that makes the population growth rate or the proportion of habitat occupied equal to zero.

Bibliography

Boyce, M. (1992). Population viability analysis. *Annu. Rev. Ecol. System.* **23**, 481–506.

Burnham, K. P., Anderson, D. R., and White, G. C. (1994). Estimation of vital rates of the northern spotted owl. National Biological Survey, Colorado.

Gilpin, M. E., and Hanski, I., eds. (1991). "Metapopulation Dynamics: Empirical and Theoretical Investigations." London: Academic Press. [Reprinted from the Biological Journal of the Linnean Society, Vol. 42, Numbers 1&2, 1992.]

Gilpin, M. E., and Soulé, M. E. (1986). Minimum viable populations: Processes of species extinction. Pages 19–34. *In* "Conservation Biology: The Science of Scarcity and Diversity" (M. E. Soulé ed.). Sinauer Associates, Sunderland, MA.

Hedrick, P. W., and Miller, P. S. (1992). Conservation genetics: Techniques and fundamentals. *Ecol. Appl.* **2**, 30–46.

Kareiva, P. M., Kingsolver, J. G., and Huey, R. B., eds. (1993). "Biotic Interactions and Global Change." Sinauer Associates, Sunderland, MA.

Lande, R. (1993). Risks of population extinction from demographic and environmental stochasticity, and random catastrophes. *Am. Nat.* 142: 911–927.

Lande, R. (1993). The meaning of quantitative genetic variation in evolution and conservation. Pages 27–40. *In* "Biodiversity in Managed Landscape" (R. Szaro, ed.) Oxford University Press.

Lande, R., and Barrowclough, G. F. (1987). Effective population size, genetic variation, and their use in population management. Pages 87–123. *In* Viable Populations for Conservation (M. E. Soulé, ed.) Cambridge University Press, New York.

McKelvey, K., Noon, B. R., and Lamberson, R. H. (1993). Conservation planning for species occupying fragmented landscapes: The case of the northern spotted owl. Pages 424–450. *In* "Biotic Interactions and Global Change" (P. M. Kareiva, J. G. Kingsolver, and R. B. Huey, eds.). Sinauer Associates, Sunderland, MA.

Myer, N. (1987). The extinction spasm impending: Synergisms at work. *Conservation Biol.* **1**, 14–21.

Pease, C. M., Lande, R., and Bull, J. J. (1989). A model of population growth, dispersal and evolution in a changing environment. *Ecology* **70**, 1657–1664.

Simberloff, D. (1988). The contribution of population and community biology to conservation science. *Annu. Rev. Ecol. System.* **19**, 473–511.

Walters, C. (1986). "Adaptive Management of Renewable Resources." Macmillan, New York.

Range Ecology and Grazing

John F. Vallentine
Brigham Young University

I. Ecological Perspectives
II. Management Philosophies and Misconceptions
III. Principles of Grazing Management
IV. Effects of Grazing
V. Environmental Enhancement by Grazing
VI. Grazing and Land Use Planning

Grazing in most natural ecosystems is as much a part of the system as is the need for forage by grazing animals. Lack of grazing in a community that has evolved under grazing should thus be considered a disturbance factor. The objectives of rangeland management should often give priority to a metastable ecosystem or disclimax vegetation that is productive and resilient instead of attempting to maintain or achieve the hypothetical climax conditions. The management of livestock and big game grazing is a tool that can be used not only to harvest the grazing resource on rangelands but also to improve it and for environmental enhancement generally. Rangeland foraging by livestock is both economically and culturally significant in all parts of the world. The future demand for animal grazing must be incorporated into land use planning, both on public as well as private rangelands, in order to meet the future needs of society.

I. ECOLOGICAL PERSPECTIVES

Ecosystems are assemblages of living organisms in association with their physical and chemical environment; within each rangeland ecosystem com-

plex interrelationships and interdependencies exist among its various components. Grazing management involves the manipulation of grazing animals within rangeland ecosystems to accomplish desired results in terms of animal, plant, land, or economic responses. The *grazier* is the person who manages the grazing animals, i.e., the *grazers* (including *browsers*). Grazing of the standing forage on rangeland and pasture is the counterpart of machine processing of harvested forage crops, except that the grazing animal is the consumer, the modifier, and the converter of the standing crop as well as the harvester. [*See* FORAGING STRATEGIES.)

Both the grazing process and the managerial efforts to manipulate it are influenced by a common set of ecological principles. Opportunities to enhance the efficiency—primarily energy efficiency—within the soil–forage–ungulate herbivore complex common to all grazed ecosystems, whether natural or developed, exist in three principal areas:

1. Increase the conversion of radiant solar energy by photosynthesis into usable chemical energy forms during growth of the forage plants, i.e., by enhancing the quantity and quality of forage produced on the site.

2. Increase the consumption by grazing animals (both livestock and big game) of chemical energy fixed by forage plants through optimal management of grazing and reducing forage waste and nonproductive consumption. (In the absence of herbivores this energy is transferred directly into mulch and litter following plant senescence.)

3. Increase the conversion of the energy ingested by the grazing animal into products directly usable by humans through improved animal genetics, nutrition, and health.

Grazing management is intended to minimize the detrimental consequences of intrinsic ecological constraints on plant and animal production within each of the three areas just listed. Nevertheless, in grazed range ecosystems, less than 1% of solar energy is typically captured in primary production; less than 20% of the total primary production is consumed by livestock; and only about 10% of the consumed energy is converted into animal products, i.e., secondary production. Since these constraints are universally present and defy even the best managerial strategies, grazing management must work within instead of attempting to circumvent these limitations. It is readily concluded that the most important concerns in grazing rangeland ecosystems should be the stability of the range ecosystem and its productivity with respect to the desired products which include forage for livestock and wildlife but also water, timber, recreation, and other uses.

A dilemma encountered in grazed ecosystems is that severe grazing (high levels and frequency of defoliation) ensure that available production is efficiently harvested. However, severe grazing eventually reduces primary production by minimizing the leaf area of desirable forage plants for the subsequent capture of solar energy. Alternatively, lenient grazing maximizes primary production, but a greatly enlarged portion of the primary production evades secondary production through herbivore consumption and is merely reduced by fire, insects, rodents, and decomposers in the ecosystem. The grazing intensity which maximizes sustainable animal production per unit area is that which optimizes the processes of solar energy capture, harvest

efficiency, and conversion efficiency within the grazed ecosystem.

Grazing in most natural ecosystems is as much a part of the system as is the need for forage by grazing animals. Most native rangelands evolved under animals grazing plants and plants tolerating grazing, i.e., the evolution of the herbivores and edible plants was simultaneous. The selective grazing pressures over thousands of years favored plants that developed resistance to grazing, browsing, and trampling. Lack of grazing in a community that has evolved under grazing should thus be considered a disturbance factor.

It is generally concluded that associated grazing pressures played a greater role in the evolution of the Great Plains grasslands and prairies than in the sagebrush steppe ecosystems of the Great Basin. Nevertheless, the exclusion of livestock over periods of many years from desert and semidesert rangelands has commonly had little beneficial effects on the vegetation. The removal of all grazing—by livestock and/or native ungulate herbivores—often results in the development of plant communities greatly different from that which originally developed under grazing. [*See* FORAGING BY UNGULATE HERBIVORES.]

II. MANAGEMENT PHILOSOPHIES AND MISCONCEPTIONS

Different philosophies prevail as to the optimal orientation that management and manipulation of range ecosystems should take in regards to climax or pristine conditions. Where the objectives of management and the desirability of climax coincide, such as in the case of most native perennial grasslands grazed by cattle or bison, there is minimal conflict between the range condition classes—these generally rank from poor to excellent based on similarity to climax vegetation/soil conditions—and the objectives of rangeland management. But where unusable vegetation (i.e., for grazing) makes up a greater proportion of the climax vegetation, the relationship between management objectives and the range condition concept weakens.

To manage for a less than excellent range condition appears incomprehensible, but it is in reality often the desired management goal for either domestic livestock or big game herbivores. Multiple uses of range ecosystems (e.g., non-game wildlife habitat, camping, and hiking as well as herbivore grazing) may best be provided when a site's vegetation is very different from its natural potential composition. For example, an optimal deer habitat is usually provided on a site when shrubs and forbs as well as grasses are abundant. But if the site's natural potential were either almost solely perennial grasses or a monolithic stand of trees and judged in relation to climax, it would be given a top condition rating, even though it failed to support the desired plant species in the desired relative amounts.

The conclusion seems to follow that the objectives of wildlands management should often give priority to a metastable ecosystem or disclimax vegetation that is productive and resilient instead of attempting to maintain or achieve the hypothetical climax conditions. Forage plant species are not superior or inferior because of their place on a successional scale in relation to climax but instead differ in production characteristics and must finally be judged on their ability to support animal production and soil stability. The philosophy that native climax vegetation is always optimal may, in fact, lead to a plateauing of rangeland ecosystem productivity and limit productivity to that level. Targeting rangeland research solely to climax is not always optimal, even though viewed by some as heretical to do otherwise.

Classifying the condition of range ecosystems solely in relation to climax when using the terms "excellent," "good," "fair," and "poor" can be misleading and may lead to misinterpreting the results of managment. This nomenclature suggests erroneously that "excellent" is always the managment objective sought when in fact "good" or even "fair" may be more beneficial. A change in descriptive terms might better describe a site's existing vegetation relative to its natural potential while not subjecting it to bias, leading to the more appropriate terms of "potential natural community," "late seral," "mid seral," and "early seral" in describing the status of range ecosystems.

The traditional concept of single equilibrium communities that progress steadily toward or away from climax depending on grazing pressure does not seem to apply to many arid and semiarid ecosystems. Examples of alternative steady states, abrupt thresholds, and discontinuous and irreversible transitions are becoming increasingly abundant for both succession and retrogression. When one group of plants has been displaced by another as a result of altered climate–grazing–fire–human interactions, the new assemblage of plants may be long-lived and persistent, despite the application of optimal grazing management practices or even the exclusion of grazing. [See FIRE ECOLOGY.]

Ecological processes associated with grazing have probably not changed appreciably since the initial appearance of grasses and grazers in the fossil record of time periods long past. However, a rapidly expanding human population, escalating degradation of natural resources, and increasing socioeconomic pressures have all increased the complexity associated with the management of grazed ecosystems. Political–judicial–legislative constraints often become serious, particularly on public lands. Single-purpose laws, executive orders, and legal attitudes toward problem solving in range ecosystems can thwart their scientific management and enhancement. Political pressures and law suits by special interest groups often result primarily only in red tape, delays, and diversion of finances.

Social constraints to rangeland ecosystem development are interrelated with legal and political constraints. Extremes in conservation and environmental thought continue to surface. While ecological principles, particularly plant competition and succession, must necessarily remain the basis of ecosystem manipulation, simplistic notions of "turning it over to mother nature," "getting the livestock off the range," and "management by legislation" continue to surface. Simplistic "natural" approaches, to the exclusion of man-directed accelerated tools, are frequently slow by themselves and of uncertain direction and results.

Social constraints are not infrequently the result of misunderstanding climax communities or their import. One example in the West is mistaking juniper as the true climax species on many sites, when in fact it is only an invader there. When actively invading a sagebrush–grass community, principally a result of the absence of fire, juniper is capable of converting it into a closed juniper community to the exclusion of most native and desirable introduced plant species.

Livestock grazing—and sometimes even native herbivore grazing—is often equated only with improper, destructive grazing and deterioration of the environment. The penchant for describing the bad effects of overgrazing often far overshadows descriptions of successful grazing programs and the good results from proper grazing. Early rangeland research was largely involved with studying the impacts of livestock grazing on rangeland vegetation when relatively little control of grazing was applied. Even today, scientists sometimes knowingly but probably more often unknowingly, report comparison of the impact of "no livestock grazing" with "livestock grazing," when often all that was compared was severe livestock use much beyond the pale of proper use but with no qualification made as to this aspect.

The western public lands in the United States are necessarily managed to provide sustained production of multiple, renewable natural resources, including water, recreation, lumber, firewood, open space, and forage for wild and domestic animals. However, in recent years, livestock grazing on public lands has become embroiled in a controversy fueled by misinformation. Public lands were severely abused in the late 1800s and early 1900s because of improper livestock grazing. This was brought on, in no small part, by the lack of knowledge existing at that time, even in the scientific community, about what comprises optimal herbivory. It also resulted from ill-conceived land laws that provided inadequate land areas for economic ranch units and no means for grazing control on the remaining unallocated public lands; both factors directly promoted overgrazing.

In some locations, soil and vegetation are still recovering from these past abuses. Also, improved livestock grazing management is needed even today on many privately owned and public grazing lands. In general, public as well as privately owned rangelands are currently in the best condition that

they have ever been in this century, and the improvement is continuing.

III. PRINCIPLES OF GRAZING MANAGEMENT

Common to the management of all grazing lands must be forage plant considerations such as plant growth requirements, providing for plant vigor and reproduction, defoliation and other animal impacts, and seasonality and fluctuations in forage production. But equally high in priority are animal considerations, including animal performance, animal behavior, nutrient intake levels, forage quality relative to animal needs, and forage palatability/animal preference.

The principles of grazing management remain the same regardless of the kind of grazing land, i.e., (1) optimal stocking rate, (2) optimal season(s) of grazing use, (3) optimal grazing system, (4) optimal kind or mix of animal species, and (5) optimal grazing distribution. Nevertheless, their application and the relative emphasis on cultural treatments (weed control, reseeding, fertilization, other renovative practices such as pitting and furrowing, and even waterspreading and irrigation) may vary greatly depending on the kind of grazing land, management objectives, and the economic implications.

Ecological principles provide the basis of managing both cropland pasture and rangelands; all must be viewed as managed ecosystems and involve overcoming limitations inherent in the interrelationships of climate, soil, plants, and harvesting. Opportunities exist to manipulate the forage resource to better fit the specific kind or class of grazing animal, but another approach is to manipulate the grazing animal (species of animal, state or production, and calendar events in the animal production cycle) to better fit the forage resource. A third and even better alternative is the integration and simultaneous application of both approaches.

Grazing management practices designed for improving cropland pasture can generally target the maximization of animal production in the short term; such pastures are mostly limited to growing season utilization during immature growth stages, are generally characterized by prolonged growth periods, are often comprised of a limited number of forage species relatively tolerant of grazing, and can generally justify occasional total renovation or reestablishment. In contrast, rangelands are generally comprised of more complex plant–species mixtures and the plants have more limited growth periods, have slower and less reliable regrowth, are often grazed during vegetation dormancy, and must be maintained in productive condition for the long term. On most rangelands the option of periodic site restoration is seldom practical or economical and is often impossible as well.

The animal unit month (AUM) is the basic quantitative measure that permits comparison of the amount of grazing needed with the amount available and in achieving a balance between them. Grazing capacity is necessarily limited to that portion of the standing crop that is available and acceptable to the grazing animal, that provides for animal maintenance and appropriate production levels, and that can be harvested by the grazing animals without damage to individual plants or the forage stand.

An inventory or estimate of how much forage is or will become available for grazing is the basis of projecting how many animals can be grazed and for how long. Grazing capacity is determined by a complex of plant production and utilization factors, and no estimate of grazing capacity can be realistic without considering how the grazing land will be grazed and to what extent cultural treatments will be applied. However, monitoring of forage production and utilization throughout the grazing period provides the basis of making further adjustments in stocking rates as the grazing period progresses.

One of the best documented but poorly appreciated characteristic of the growth of forage plants is the rapid turnover of tissue that takes place. The standing crop is dynamic, with new herbage being added and existing herbage simultaneously disappearing. Forage production into the standing crop does not enter a static storage situation. Dry matter flows continuously and rapidly through the standing crop of graminoids, forbs, and shrubs as new leaves are continually produced and old ones die. Herbage that is not consumed by grazing eventu-

ally dies, decays, and disappears from the standing crop. A major problem faced by grazing managers is annual and seasonal fluctuations in forage production resulting from climate. The challenge is to keep from overgrazing and damaging the forage resources in low production years or seasons but still realistically fully utilize the forage produced in high production periods. Seasonal and annual deviations from average grazing capacity must be incorporated into both the grazing inventory and the grazing plan.

The dispersion of grazing animals and associated forage utilization within a management unit or area is one of the most important facets of proper grazing management. The goal is to obtain the maximum safe use over as wide an area as possible without causing serious damage to any portion within it. Grazing distribution has a direct effect on realizable grazing capacity. If grazing animals are not kept well distributed, the areas grazed too heavily as well as those grazed too lightly expand in size while those areas receiving optimal use become smaller. Many grazing units that appear overstocked may merely be suffering from lack of uniform use over the full grazing unit, leading to the potential for localized overgrazing when the unit as a whole is actually understocked.

In determining when to harvest the standing forage crop with grazing animals, one must consider plant factors, physical site factors, animal factors, and economic and management factors. While some range forage stands can be utilized any season of the year, many are adapted to grazing only when grazing is confined to a specific season of the year. Thus, the excess grazing capacity on an annual basis can be associated with a serious deficiency during one or more seasons of the year. Achieving a seasonal balance between forage production and forage needs by both domestic and wild ungulate herbivores is a key aspect of the grazing plan.

IV. EFFECTS OF GRAZING

Grazing animals affect plant communities in several interrelated ways, including plant defoliation, nutrient removal and redistribution through excreta, and mechanical impacts on soil and plants through trampling. Each of these effects, in turn, interacts on the productivity and welfare of the grazing animal. The short-term or immediate effects of grazing on a plant can (1) be detrimental, i.e., reduce plant vigor or even kill it; (2) have no apparent beneficial or negative effect on it; or (3) be beneficial, i.e., increase plant vigor, plant size, or growth rate.

Grazing by large herbivores in the short run often is of little importance in the process of vegetation change, unless grazing is so excessive that the grazed plants cannot restore themselves. The long-term effects of grazing largely depend not only on the adaptation of the plant to local environmental factors, but also on the relative effects of grazing on associated plants and plant species. The extent to which a plant is under competitive pressure from other plants will determine, in large part, its tolerance of defoliation and trampling by grazing animals. Intensive competition by surrounding plants may be more suppressive of foliage yield than severe defoliation, but the two in combination are additive in the negative effects on yield and survivability.

A. Defoliation by Grazing

Defoliation refers to the removal of leaves from a plant; the term is generally extended to include edible inflorescences, stems, and twigs. Although primary attention here is directed to defoliation by grazing and browsing ungulates, defoliation can also result from insects and rodents, fire, hail, frost, mowing, and contact herbicides. All forage plants can be safely and properly defoliated to some degree; the challenge in grazing management is to apply the maximum safe level that will still maintain optimum, sustained plant productivity. The component parts of optimum defoliation are: (1) how much (intensity of defoliation); (2) when (seasonality of defoliation); and (3) how often and over how long (frequency and duration of defoliation).

The response of a forage plant to defoliation is lessened and plant repair and regrowth accelerated if it (1) has adequate carbohydrate storage reserves to meet immediate needs, (2) has adequate leaf area remaining to quickly resume photosynthesis, (3) has available meristematic tissue and new tiller buds to replace removed parts, (4) has adequate root area and viability to support aboveground growth, and (5) is in a phenological/morphological stage that permits a return to active growth. However, each of these must be considered in light of the current environmental conditions and the constraints that they place upon the forage plant in tolerating and making rapid recovery following defoliation. Different forage plant species and even different cultivars of the same species have different mechanisms for avoiding or tolerating defoliation.

Although grazing can have beneficial effects on forage plants, as discussed in Section V, overgrazing (excessive levels of defoliation over time) can have negative effects on the vegetation. The vigor of grazed plants may be severely reduced if defoliation is severe and prolonged, and distorted growth patterns in the plants may result, i.e., grasses reduced to decumbent forms and shrubs become hedged. Other negative effects of overgrazing include reducing the plant root system, reducing leaf area and photosynthetic capacity and thus yield, and delaying growth response of the key forage species even when temperature and soil moisture are optimal. The cumulative effects of overgrazing often result in the key plant species being replaced by less desirable or even worthless plants, and even to the invasion of brush, poisionous plants, and noxious, alien weeds, i.e., "green migrants." Excessive reductions in the amount of vegetation through herbivory may seriously reduce the amounts of organic biomass covering the soil, thereby increasing the probability of wind and water erosion.

Defoliation may adversely—less commonly beneficially—affect reproduction by seed in existing forage stands but often has minimal lasting effect. Perennial grass-dominated range communities at higher levels of succession have minimal seedling recruitment needs or opportunities. On excellent condition range, plant mortality is so low and competitive dominance is so complete that seedings of climax grass species are rarely found. Where rangelands are in low condition, recruitment of additional individuals is typically desirable but is relatively slow if from seed alone, unless the vegetation exists as an open community or is made so by cultural treatment. Major obstacles to recruitment, particularly from seed, are weather variables and seed depredation as well as plant competition. Although reproduction from seed is mandatory for annual plants and is an important survival strategy for perennials, vegetative reproduction by perennial plants or rangeland is often faster, more efficient, and more reliable.

B. Physical Effects of Grazing on Plants

Clipping, shearing, and mowing are commonly used to simulate the effects of grazing on plants. However, the eating and trampling action of grazing affects forage plants in several ways not simulated by simple defoliation alone. Most of these effects are deleterious to the plants affected but will be tolerated unless they become excessive. The eating action involves biting, pulling, and breaking off plant parts at random heights. This often results in pulling unpalatable plant parts from the plant that are subsequently discarded instead of being ingested. Stolons, shallow rhizomes, and even entire plants, particularly seedlings, may be pulled out of the ground if they are not well rooted. For this reason, new forage plant seedings must be allowed to become firmly rooted before being grazed.

Another impact of grazing not applied by clipping is the trampling and treading of both the plants and the soil by the hooves of grazing animals, resulting in some plants being crushed, severed, or bruised. Trampling losses of forage, resulting in a direct addition to the mulch component, may become excessive in dense forage stands and may be a major factor contributing to utilization inefficiency. However, trampling damage and loss of forage on arid and semiarid rangelands are generally much less than on mesic sites.

Additional mechanical effects of foraging on woody plants include intentional or inadvertent breaking off of limbs and bark wounding by rubbing, foraging, or hoof impact. The transport of seed and other plant propagules by grazing animals, either internally through the digestive systems or externally by temporary attachment to hair, fleece, or hooves, is always an ecological factor affecting a perennial forage stand; but the consequences range from favorable to deleterious depending on the plant species and the site being affected. Also, the covering or otherwise fouling of vegetation with feces must be considered in grazing management, primarily in dense forage plant stands.

C. Physical Effects of Grazing on Soil

The treading of soil by grazing animals has the potential of being deleterious to soil in the following ways: (1) compacting the soil, (2) penetrating and disrupting the soil surface, (3) reducing infiltration, (4) displacing soil vertically on steep slopes, (5) developing animal trails, and (6) increasing erosion. All grazing land receives treading to a greater or lesser extent as a natural consequence of grazing. The interaction of many site, soil, weather, and vegetation factors determines the severity of hoof action on the soil and the rate of recovery therefrom; the effects range from inconsequential, or less commonly beneficial, to destructive.

Excessive grazing by either livestock or big game can result in removing protective vegetation from a watershed, resulting in (1) increasing the impact of raindrops, (2) decreasing soil organic matter and soil aggregates, (3) increasing surface soil crusting, and (4) decreasing water infiltration rates and increasing overland flow. Soil treading generally has a much greater impact on wet, heavy soils, and deep treading can result in disrupting soil structure and soil surface; the shearing of foliage, growing points, and even roots; and the deposition of mud on the herbage. Unfortunately, many watershed grazing studies historically have compared only heavy or uncontrolled grazing with no grazing, thus implying that livestock grazing is necessarily synonymous with deleterious effects. However, recent watershed research data strongly suggest that the watershed condition can be maintained or improved under moderate grazing intensity and spatial distribution of grazing. The application of advanced grazing management techniques can mostly minimize or prevent the physical impacts of grazing on soils, even on riparian sites, or limit their effects to the short term by providing opportunities for rapid recovery later.

V. ENVIRONMENTAL ENHANCEMENT BY GRAZING

Grazing is a tool that can be used not only to harvest the grazing resource but also to improve it; grazing can be used for environmental enhancement. In recognizing grazing animals as positive environmental manipulators when properly managed, the Society for Range Management in 1985 issued the following tenet: "The Society further recognizes the value of livestock as a management tool to bring about desired trends or changes in certain plant communities to improve forage production, water conditions, wildlife habitat, recreational and aesthetic quality, and other tangible or intangible products."

Both common use (grazing two or more species of grazing animals in common) and biological control of undesirable plants by grazing are based on selective grazing. Common use increases the actual grazing capacity by grazing a mix of animal species with different dietary and/or terrain preferences to properly and uniformly harvest all potential forage plants. In contrast to common use, biological control by grazing is applied to achieve a desired directional succession in vegetative composition. It is effective when the right combination of grazing animal species, season and system of grazing, and stocking rate results in heavy selective grazing of undesirable or less desirable plants to the competitive advantage of the favored plants.

Certain levels and combinations of grazing increase overall plant species diversity by decreasing the capacity of competitive dominants to exclude other plant species and by creating gaps available for occupation by the other species. With moderate levels of stress, including that provided by grazing, species resistant to stress and species susceptible to stress are both able to survive and reproduce, resulting in maximum species diversity. While this phenomenon of increased diversity occurs at moderate levels of disturbance, the plant diversity is typically lowered by high frequencies or intensities of grazing disturbance. Thus, prolonged, high levels of grazing stress can be expected to lower plant diversity even below that found under prolonged nongrazing.

Prescribed grazing positively influences the desirable components of the standing crop by delaying maturation and holding plants in a vegetative, forage-producing state; by stimulating growth or regrowth by a pruning effect; by accelerating nutrient recycling in the ecosystem and making some nutrients more available for plant use; and by enhancing the nutritive value of the available herbage by increasing the new growth:old growth ratio.

In high biomass-yielding ecosystems, grazing provides an important role in removing or preventing excessive thatch buildup, thereby enhancing the attractiveness of the standing crop to grazing animals. Also, grazing reduces the excess accumulations of standing dead vegetation and mulch that may chemically and physically inhibit new growth and delay soil warming in cool-climate areas. Grazing in dense forage stands can maintain optimal rather than excessive leaf area index in order to make more effective use of available sunlight in photosynthesis. Grazing can also play a positive role in removing excessive foliage when plants are entering a harsh cold period; reducing transpiring surfaces when plants are going into a drought period may also be beneficial in reducing water stress, increasing stomatal conductance, and conserving soil moisture.

Soil treading by grazing animals may have beneficial effects in planting naturally or artificially broadcasted seed. Benefits may be derived from the effect of hoof impact working the seed into the soil surface and compacting the soil around the seed. Simultaneous grazing at this time may further benefit the establishment of the seeded species by reducing competition from the residual vegetation by defoliating them. Nevertheless, a near consensus has been reached by range scientists that intentional, accelerated trampling on a frequent basis by grazing animals cannot benefit most rangeland ecosystems regardless of the grazing system followed.

VI. GRAZING AND LAND USE PLANNING

Rangeland foraging by livestock is both economically and culturally significant in all parts of the world. Grazing lands play a prominent role in the conservation ethics comprising grassland agriculture, a land management system emphasizing cultivated forage crops, pasture, and rangelands for livestock production and soil stability. The development of grazing resources provides a means of utilizing otherwise mostly idle or nonproductive land. The widespread practice of permitting abandoned or unproductive cropland and other devastated but potentially productive sites to go without improvement can only be considered as a waste of natural resources; many such lands require treatment through improvement or establishment of vegetative cover for site stabilization and conservation use as grazing lands.

On a worldwide basis, rangeland is the largest land resource, encompassing about 50% of the land area of the earth. Additional acreage, possibly 5 to 10%, comprises cropland pasture and permanent pasture. Approximately half (50.6%) the land area of the United States is grazing land. Total rangeland, both open and forested, constitutes 41.9% of the U.S. total land area, while pasture, excluding rangeland, constitutes 8.7%.

Grazing lands, including rangelands, contribute an estimated 40% of the feed consumed by livestock in the United States, harvested forages contribute about 20%, and concentrate feed, including grains and protein supplements, contribute the remaining 40%. The proportion of total feed in the United States obtained from grazing varies with different kinds of livestock: sheep and goats, 80%; beef cattle, 74%; horses, 51%; dairy cattle other than lactating cows, 43%; and milking cows, 18%. Feed for maintenance of breeding herds of beef cattle, sheep, and goats and the production of their offspring comes primarily from grazing lands; rangelands, together with forestlands with a forage understory, are the largest and most productive habitats for big game animals.

The future demand for animal grazing must be incorporated into land use planning, both on public as well as private rangelands. Grazed forage consumption in the United States for 1985 has been estimated at 431.2 million AUMs; it is projected to increase to 483.9 million AUMs by 2000 and to 637.2 million AUMs by 2030. Livestock grazing on public lands play a significant role in U.S. cattle and sheep production, makes a significant contribution to western U.S. economy, and is one of the multiple uses mandated on federal lands administered by the Bureau of Land Management (BLM) and the U.S. Forest Service, (USFS). In the 11 western states, about 12% of the total forage consumed by livestock comes from public rangelands, with an individual state high of 49% in Nevada.

Great strides have been made in improving federal grazing lands in the United States since the turn of the century as livestock grazing control became a reality, thereby permitting the application of optimal grazing techniques and practices. Summations made by the Society for Range Management in 1989 reveal that 89% of the USFS lands and 79% of BLM lands are either improving or stable in ecological condition. Similarly, 76% of the USFS lands described as suitable for livestock grazing have a satisfactory livestock forage resource value rating.

Another finding is that big game populations on public lands are now increasing in the presence of regulated livestock grazing. The renewed resolve of public land managers, big game representatives, and ranchers to improve and properly manage public grazing lands and coordinate big game and livestock grazing have been highly rewarding. Not only are big game ungulate herbivores an important part of the natural environment in many parts of the United States including the West, they also provide indirect benefits to local economies and direct income to agricultural firms through game ranching enterprises.

The benefits of livestock grazing, particularly on range but also on cropland pasture, are very significant from the standpoint of national, regional, and local economies. These society benefits include (1) low fossil fuel expenditure per pound of livestock weight gain, (2) reduction or elimination of need for commercial fertilizers, and (3) the production of red meat and fiber for export with an improved balance of trade. Additional benefits include increased rural income at the agricultural firm level but also in local communities through secondary and third level money turnover. In releasing feed grains for human consumption and/or export, each AUM of grazing is equivalent to 8 bushels of corn; rangeland grazing capacity provides animal protein necessary for meeting the nutritional needs of people around the world.

Glossary

Animal unit month (AUM) (1) The potential forage intake (i.e., animal demand) of one animal unit for a period of 1 month (30 days); (2) amount of carrying capacity in terms of forage required by an animal unit for 1 month (30 days), based generally on 750 pounds of dry matter consumed or less frequently on 540 therms of digestible energy.

Browse (n) Leaf and twig growth of shrubs, woody vines, and trees available and acceptable for animal consumption; (v) to consume browse (syn. browsing).

Ecosystem An assemblage of living organisms together with their physical and chemical environment, forming an interacting system and inhabiting an identifiable space.

Forage That part of vegetation that is available and acceptable for animal consumption, whether considered for grazing or mechanical harvesting; includes herbaceous plants in mostly whole plant form and is generally extended to include browse.

Grazing The act of consuming a standing forage crop by ungulate herbivores; often expanded to include browsing.

Grazing capacity (1) Total number of animal unit months (AUMs) produced and available for grazing from a pasture unit, a grazing allotment, the total ranch, or other specified land area; (2) the maximum stocking rate possible without inducing damage to the soil, vegetation, or related resources or deleteriously affecting grazing animal response.

Grazing management The manipulation of grazing animals to accomplish desired results in terms of animal, plant, land, or economic responses.

Range condition The present state of vegetation of a range site in relation to the potential natural (or climax) plant community for the site based on kinds, proportion, and amounts of plants present; suggests current productivity relative to natural productivity potential.

Range ecology The study of the interrelationships of organisms with their environment, with particular application to rangelands.

Rangelands Uncultivated grasslands, shrublands, or forested lands with an herbaceous and/or shrub understory producing forage for grazing or browsing; may be native or seeded.

Range management The art and science of planning and directing the development, maintenance, and use of rangelands to obtain optimum, sustained returns based on the objectives of land ownership and on the needs and desires of society.

Bibliography

Forbes, T. D. A. (1988). Researching the plant-animal interface: The investigation of ingestive behavior in grazing animals. *J. Anim. Sci.* **66**(9), 2369–2379.

Gee, C. K., Joyce, L. A., and Madsen, A. G. (1992). "Factors Affecting the Demand for Grazed Forage in the United States." USDA, For. Serv. Gen. Tech. Rep. RM-210.

Heady, H. F., and Child, R. D. (1994). "Rangeland Ecology and Management." Boulder, CO: Westview Press.

Heitschmidt, R. D., and Stuth, J. W., eds. (1991). "Grazing Management: An Ecological Perspective." Portland, OR: Timber Press.

Holechek, J. L., Pieper, R. D., and Herbel, C. H. (1989). "Range Management Principles and Practices." Englewood Cliffs, NJ: Prentice Hall.

Kamil, A. C., Krebs, J. R., and Pulliam, H. R., eds. (1987). "Foraging Behavior." New York: Plenum Press.

Mosley, J. C., Smith, E. L., and Ogden, P. R. (1990). "Seven Popular Myths about Livestock Grazing on Public Lands." Moscow, ID: Idaho Forest, Wildlife, and Range Experiment Station.

National Research Council/National Academy of Science (1984). "Developing Strategies for Rangeland Management." Boulder, CO: Westview Press.

Society for Range Management, Public Affairs Commission (1989). "Assessment of Rangeland Condition and Trend of the United States." Denver, CO: Society for Range Management.

Stoddart, L. A., Smith, A. D., and Box, T. W. (1975). "Range Management," 3rd Ed. New York: McGraw-Hill.

Vallentine, J. F. (1990). "Grazing Management." San Diego: Academic Press.

Restoration Ecology

John Cairns, Jr.
Virginia Polytechnic Institute and State University

Restoration ecology is the field devoted to repairing damaged ecosystems to as close an approximation of their predisturbance condition as possible. Because each ecosystem is the product of a sequence of climatological and biological events unlikely to be repeated in precisely the same sequence, restoration to precise predisturbance structural and functional conditions is unlikely; however, substantial ecological improvement in both structural and functional attributes is quite possible given standard restoration practices and is rapidly improving. Ecological restoration is necessary because human society has maintained low prices by externalizing costs in metals, energy, and food production. This is accomplished by destroying "ecological capital" such as old growth forests, wetlands, and aquatic ecosystems (just a few examples). If material goods were produced with less environmental impact, they might be more expensive to the manufacturer but not necessarily to society since environmental damage would be reduced. Society's "frontier mentality" held that natural systems were available in endless supply. In addition, people fond of the amenities of natural systems could move to a new, undamaged area when the local one had been damaged. This mind-set is equivalent to the Mad Hatter's tea party in "Alice in Wonderland:" instead of washing cups, simply move around the table to a set of unused cups. Of course, ultimately, all the unused cups will be used, and the teacups will have to be washed or restored (i.e., restored to an acceptable condition). The Mad Hatter's tea party in the environment has now reached this point—the population continues growing and relatively undamaged ecosystems are becoming scarcer and scarcer; future generations will be unable to have the relationship with natural systems that earlier generations enjoyed.

I. WHAT IS RESTORATION ECOLOGY?

Restoration ecology is a new field of science devoted to returning damaged ecosystems to a condition that is structurally and functionally similar to the predisturbance state. Since the healing process requires significant time and since each ecosystem is a product of a sequence of climatic and biological events unlikely to be repeated precisely, exact replication is improbable. In fact, with the present global loss of species and habitat, precise replication may not be possible for another reason: loss of component species. Sometimes component species can be obtained from other undamaged areas; however, then an ethical question arises: "Should a pristine (or less damaged) area be put at risk to obtain recolonizing species for a damaged ecosystem?" Removal of species for use elsewhere might have damaging effects on the source ecosystem and might not successfully colonize the damaged area, resulting in biotic impoverishment in two areas instead of one. Thus, if ecological restoration is widely practiced, preserves should be established in each ecoregion to serve as sources of recolonizing species. This would help ensure that the remaining, essentially undamaged, systems are left to serve as restoration models, especially when, as is often the case, precise records of ecological structure and function are not available for a damaged ecosystem following an oil spill or, equally commonly, when incremental damage has occurred over a long period of time. [*See* ECOLOGICAL RESTORATION.]

Some may be concerned "that restoration offers a license to kill" (W. Jordan, as quoted in *The Scientist,* September 28, 1992, p. 15). That is, if restoration is viewed by society as a simple task, there will be no incentives to preserve and protect the remaining intact ecosystems. However, now that realistic costs of ecological restoration are more widely appreciated, casual ecosystem destruction is less likely. [*See* NATURE PRESERVES.]

II. RATE OF ENVIRONMENTAL DESTRUCTION INCLUDING GLOBAL LOSS OF SPECIES

Ecological restoration is necessary because global landscapes have undergone tremendous alterations and devastation. In the United States, over 90% of the prairies, wetlands, and virgin forests have been altered by various types of development. In one sense, attention to ecological restoration has its genesis in continued ecological destruction. If the human population continues to grow at its present rate, as is highly probable over the next half century, and ecological destruction continues at its present rate over the same period of time, the world will have over twice its present population and practically no mature ecosystems that are unaffected by human activities. Even some types of ecological preserves, such as national parks, are feeling the pressure of too many visitors, with consequent altering of ecological balances. Ecological restoration is a way of partially replacing lost ecological values. However, large ecosystems can suffer severe deleterious effects overnight (as was the case with the Exxon Valdez oil spill in Prince William Sound, Alaska, or the larger spill off the Shetland Islands), even though the recovery time may be years or decades. In many cases, recovery or restoration to the predisturbance condition is highly improbable. Even if restoration efforts are successful, initiating restoration immediately following ecological damage will not ensure a no net loss of healthy, robust ecosystems because damaging ecosystems almost invariably takes less time than restoring them. [*See* GLOBAL ANTHROPOGENIC INFLUENCES.]

At present, regulations in the United States and some other countries are intended to replace "in kind" ecosystems unavoidably damaged by construction of public amenities such as airports and other large projects. That is, constructing a comparable ecosystem or restoring a damaged ecosystem at some other location is an attempt to achieve no net loss. This is called mitigation. In a wetlands mitigation case near Gunnison, Colorado, constructed wetlands intended to replace lost natural wetlands were not associated with the hydrologic cycle in the same pattern that natural wetlands were. As a consequence, the wetlands species planted in the constructed wetlands died and were replaced by upland vegetation. Thus, although the intent was to replace in kind, the goal was not realized. In another instance near Gunnison, Colorado, an elk wintering ground was lost when a dam was built and the wintering ground was flooded.

A replacement elk wintering ground was not established until nearly a decade later, and this was not comparable to the lost wintering ground, although, conceivably, it may ultimately become more similar. Thus, mitigative efforts suffer from two major drawbacks despite their lofty intent: (1) matching a high quality ecosystem through construction or restoration is rarely possible, and (2) even if the match is possible, it does not occur instantaneously and, thus, the ecosystem services are lost for some extended period of time.

III. MAINTENANCE OF ECOSYSTEM SERVICES

Widespread societal support for ecological restoration may be intimately associated with ecosystem services. Illustrative ecosystem services include maintenance of the gaseous balance in the atmosphere and improvement of surface freshwater quality, food, recreation, etc. These services are often unappreciated at present and ignored because they are free. However, such services should not be taken for granted, as deteriorating water quality around the world has shown. A major problem is determining how much ecosystem degradation can occur before these services are excessively impaired or irretrievably damaged. Nowhere is public misunderstanding of how science works more evident than in ecological degradation. Science operates primarily by disproving hypotheses and, if a substantial amount of time lapses in which a hypothesis has been checked by others and has not been disproved, it gains acceptance and becomes popularly viewed as a "fact." Consequently, when people assert that scientific evidence is not available to show that ecological restoration is needed or that half the earth's species are lost because their habitat has disappeared, they are correct. This does not mean that the hypothesis is wrong because there is no "proof"; it merely means that no robust scientific evidence bears on this hypothesis. If the hypothesis is restated as "environmental destruction can continue indefinitely at its present rate with no adverse effects on human society," which is the unstated hypothesis being offered by criticizing the

other, there is no scientific "proof" of this either. If the second hypothesis is validated by experiment (namely, by destroying most of the earth's ecosystems and seeing what happens), human choices in the future may have been restricted to a severe degree that, in retrospect, will prove unacceptable. Thus, "learning by doing" in respect to the restoration of ecosystems is a better choice, so that, when more robust information is available decades from now, a more enlightened choice of options can be made.

IV. TIMING OF RESTORATION

Restoration is usually less expensive if carried out relatively soon after ecological damage. For example, the demonstration project that restored a portion of the Kissimmee river in Florida was successful because the wetlands, when reflooded, still retained seeds, soil characteristics, and other attributes of wetlands so that little reintroduction of species was necessary. Evidently, if a wetland is disconnected from the larger hydrologic cycle (as many wetlands that adjoined the Kissimmee were), hydric soil (a typical wetland soil) characteristics are retained for a number of years. When soils do become altered, atypical wetland plants adapted to hydric soils cannot readily become established even when they are returned to the damaged area. If restoration activities are delayed, a disturbed ecosystem will reach a new equilibrium appropriate to the changed conditions. This can be seen in urbanized areas when a city lot is left vacant following demolition of a house. This new equilibrium, often characterized by undesirable species, is usually not highly desirable or even acceptable. If a new array of weed species becomes established, eradicating them will be necessary. However, if, following damage, desirable species are successfully introduced, the entire project is much less costly.

V. SELECTION OF RESTORATION OPTIONS

Figure 1 shows four basic options for restoration (restoration to predisturbance condition, rehabili-

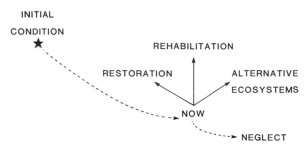

FIGURE 1 Basic options for restoration. [From Magnuson *et al.* (1980).]

tation, created alternative ecosystems, neglect). It is unlikely that any large-scale restoration will succeed or survive, once established, without strong societal support. This means that the public and its representatives must become far more environmentally literate than they now are.

Restoration to predisturbance condition is usually not a viable option despite its attractiveness because each ecosystem is the result of a sequence of climatic and biological events unlikely to occur again. Additionally, information on the precise predisturbance condition is usually not sufficiently detailed and, even on those rare occasions when such information is available, some of the critical species may have been lost.

Rehabilitation is restoring some, but not all, of the ecological attributes that are lost. It is not clear whether only attributes of societal interest can be restored without an ongoing subsidy (e.g., a fish hatchery, nutrients, etc.), but it appears highly unlikely. A fully restored system should be self-maintaining, but a rehabilitated one will probably not be. Regrettably, most ecosystems require considerable attention during the restoration process and probably significant attention on a continual basis thereafter. The world is simply too crowded with humans and too intent on development to believe that ecosystems, once restored, can be left unattended.

Created alternative ecosystems, that is, ecosystems not identical to those originally present before ecological destruction, should be a viable option in many cases. In California, which has lost 91% of its wetlands, any opportunity to create an ecologically functional wetland should not be lost. Additionally, in areas highly vulnerable to oil spills and other ecological disturbances, an ecosystem more resistant to disturbance might well be considered in place of a very fragile ecosystem.

The American public is eager for immediate action following any sizable environmental catastrophe. This may not always be wise. *Neglect,* while appearing irresponsible and callous, may often permit natural recovery under certain circumstances. For example, following an oil spill, dispersants (more toxic to some organisms than oil) are frequently utilized as part of the clean-up process. Other methods include steam cleaning of rocks, suction to remove oil (which removes not only the oil but the organisms associated with the oil), and, alternatively, power spraying to remove oil from rocks. All of these and other similar activities may be a greater threat to the surviving organisms than the residual oil.

VI. SELECTION OF RESTORATION GOALS

If the human population reaches 10 billion or more by the year 2050 and the present rate of environmental destruction continues, ecosystems will number less and people will number more. In addition to the fact that most people want natural systems for recreation, persuasive evidence indicates a deep-rooted psychological need for natural systems as well. Even for those who are content to live in highly urbanized areas and never see a natural system except on television, the world's ecosystems still provide a variety of services (such as regulating the gaseous balance in the atmosphere that affects both micro- and macroclimate). The precise amount of ecosystem services per capita that is minimal or optimal is not known; however, it would be better to wait until science determines these levels with more robust evidence than to find them out accidentally by having some unfavorable, and perhaps irreversible, change occur. Only a few options exist.

1. Do nothing and see what happens, which could also apply to toxic chemicals, AIDS, and the budget deficit.

2. Restore at a rate less than the rate of damage. This would have the same ultimate result but gives more time to explore options, develop the science of ecological restoration further, and become more environmentally literate.

3. Require that all ecosystems damaged either accidentally or deliberately (e.g., surface mining) have restoration occur elsewhere so that there is a no net loss of ecosystem quality and ecosystem services. It is important to remember that ecosystems can be damaged at a rate far greater than the rate of destruction. Even this would mean far less natural area per capita and far fewer ecosystem services per capita if the population continues to grow.

4. Restore at a rate exceeding destruction, which would still result in less ecosystem services per capita unless the human population is stabilized.

VII. DEVELOPING A NATIONAL RESTORATION STRATEGY

Scientists have been understandably reluctant to get involved in political and social decisions involving national strategy and policy because certain difficulties exist.

1. Scientific decisions may be based on scientific evidence and probabilistic determinations, whereas political and social decisions depend heavily on value judgments that are less easily communicated and more difficult to validate or verify.

2. Scientists are accustomed to reserving judgment until a persuasive weight of evidence is available, whereas political decisions must often be made in time frames that do not permit developing an adequate evidence base.

3. Scientists should be as objective and dispassionate as possible, whereas influencing strategy and policy decisions may require a degree of advocacy that makes them uncomfortable.

4. Scientists are accustomed to communicating with their peers and can assume a substantial shared base of knowledge, common terminology, and professional standards; communicating with those outside of their profession or discipline may require

scientists to condense information to an unacceptable level or simplify matters to a degree that makes these scientists uncomfortable.

5. The peer-review process, although not without difficulty, has served science well; however, the review process in the political arena is one with which most scientists are unfamiliar and find unacceptable when they are well acquainted with it.

6. The sums of money involved in resolving large societal problems (e.g., restoration of hazardous waste sites) might have a corrupting influence on scientific judgment. This is, of course, a gross oversimplification of a very complex situation, and more information on this subject is abundant in the professional and popular literature.

Despite these misgivings, scientists should become involved in the development of restoration strategy and policy. One persuasive reason is that most restoration projects transcend the capabilities of a single discipline and, thus, represent integrated science or integrated problem solving instead of the reductionist science with which most scientists are most comfortable. All landscape-level restoration projects that are properly designed are carried out by teams involving a variety of disciplines. As a consequence, the unifying theme is most likely to be a strategic or policy matter, although with good scientific support. Finally, large-scale restoration projects have substantial temporal and geographic dimensions in addition to the diversity of detail already mentioned and, as a consequence, are unlikely to succeed without general public understanding and support. The Guanacaste dry forest restoration in Costa Rica, conceived and implemented by Professor Daniel Janzen of the University of Pennsylvania, represents a world-class ecological design coupled with scrupulous attention to development of understanding at the top administrative levels in that country as well as understanding by an array of age groups of citizenry of the area. In a very real sense, humans are regarded as an integral part of the ecosystem and, if their actions can be modified by a generally accepted strategy and policy so as to enhance ecosystem integrity without compromising the scientific soundness of the ecological procedures, both society and science will benefit.

In his budget message for 1991, President Bush stated, "Today, a consensus is emerging in our society: investments in maintaining and restoring the health of the environment can now be seen as responsible investments for the future." This presidential statement implies that a national restoration strategy needs to be developed to meet changing societal demands on natural resources. Viewed from a social strategy standpoint, restoration is not warranted simply because it recreates a historic condition (although many would support this as a sole goal), but, at a more general level, ecological restoration is justified whenever restored ecosystem processes serve socially desired goals for human or ecological functions and services. The ecological services aspect has, regrettably, not received the attention in the past that it has deserved. In many cases, societal goals may only be achieved by restoring the natural processes of ecosystems and, in many other cases, using the natural processes may be the least costly, albeit not the only, way of achieving these goals (e.g., maintenance of water quality).

During the pre- and postelection debates of 1992, there was considerable sensitivity to the levels of federal and state expenditures. It would be easy to interpret a call for a national restoration strategy as an endorsement for large increases in federal spending on a new program. This is not the intent, although it would be a viable option if our federal financial situation were more robust. It is a *sine qua non* that restoration of damaged ecosystems may, in many instances, be extremely expensive, particularly where hazardous waste sites or dams are to be decommissioned or other major physical or chemical alterations are needed as part of the restoration process. If the dams or hazardous waste sites were created by the federal government, federal funds will certainly have to be used in the restoration process. If the sites resulted from activities of private industry or government at the state or some lower level of government, the institution or individual causing the problem should be required to bear the restoration costs. Many restoration programs should focus on the interests, opportunities,

and cost-sharing possibilities provided by private land owners, local governments, and states. Trout, Unlimited, for example, has restored the fishery of many damaged trout streams. Similarly, many wetland restoration efforts have been initiated by duck hunting clubs, land trusts, waterfront renewal organizations, and other regional or state agencies.

A damaged ecosystem may be restored at a comparatively modest cost if undertaken opportunistically as part of a broader plan. For example, it is often prohibitively expensive to clear a riparian zone of houses and to reestablish natural vegetation and contours under normal circumstances if the sole goal is restoring a natural system. However, restoration is almost invariably more financially attractive and feasible after a severe flood event in which houses and other structures have been destroyed or irreparably damaged and landowners wish to move from the area and flood insurance benefits, etc. are available to pay much of the cost of relocation. Stated differently, if govenment subsidies for restoring the housing are not particularly attractive, relocation may be the option of choice even if insurance payments for relocation are minimal. Many such opportunities existed following the hurricane that devastated South Florida in 1992.

A national restoration strategy for both aquatic and terrestrial systems should have the following components:

1. explicitly stated national restoration goals,
2. well-reasoned principles for determining priorities and facilitating decisions,
3. a major policy and program redesign for federal agencies emphasizing integrated environmental management including restoration, and
4. innovation in financing and use of land and water markets along the lines of the illustrative examples just given.

VIII. DEVELOPING A NATIONAL RESTORATION POLICY

Of course, it is easy to state that the United States is, as a nation, in favor of restoring damaged ecosystems. However, it becomes an entirely different matter to spell out how much restoration will be achieved and in what time intervals.

It is extremely important that ecological damage assessment not be confused with ecological restoration. For example, over $50 million has been spent thus far on the Exxon Valdez oil spill for assessing environmental damage. Another figure that is reasonably current is the AVX settlement in New Bedford Harbor of $66 million, presented in a lump sum for both clean-up of hazardous materials and ecological restoration. The Superfund site clean-up has thus far funneled virtually all the funding into assessment instead of ecological restoration or even, in most cases, clean-up of hazardous materials. Ecological restoration must not be given a bad name by co-mingling funds for assessment, clean-up of toxic materials, and legal fees with true ecological restoration costs. When setting target goals, it is important to distinguish between these activities and allocate money specifically to each category.

A. National Restoration Goals

Both species and habitat are being lost at a historically unprecedented rate. Restoration ecology is such a new field that there is neither time nor financing to generate the necessary information base for skillful ecological restoration unless we learn by doing. Thus, setting national restoration goals should include a combination of repairing damage already done together with generating the information base necessary to make ecological restoration more precise and reduce the uncertainty of the outcome when certain actions are taken. Practitioners of ecological restoration have made marvelous improvements in ecological conditions following damage merely by following common sense procedures that incorporate existing knowledge. Consequently, the outcome will almost certainly be markedly superior to the damaged condition, even though all of the processes involved may not be well understood. This is not uncommon. The steam engine came first and the theory of why it worked came second. Two illustrative national restoration goals follow.

1. A national restoration policy should be directed toward broad-based and measurable goals. The United States has a large number of ecoregions. Just as it would be irrational to give precisely the same advice to farmers in Florida and South Dakota, so also must the ecological strategies be modified from one ecoregion to another. Similarly, if different methods and procedures are to be tried, there must be more than one restoration project in each ecoregion. Finally, abundant circumstantial evidence indicates that large-scale restoration projects are more likely to be self-maintaining (can be left unmanaged without disastrous results), and, therefore, the scale of the restoration activity should be as large as possible since long-term maintenance costs should be avoided whenever possible. The most explicitly stated restoration goals are in the National Research Council's 1992 book, "Restoration of Aquatic Ecosystems: Science, Technology, and Public Policy" (National Academy Press, Washington, D.C.).

2. A national ecosystem assessment process should be established to monitor the achievement of the nation's terrestrial and aquatic ecosystem restoration goals. Stated goals must be followed by evaluation and assessment.

B. Principles for Priority Setting and Decision Making

1. Policies and programs for ecosystem restoration should emphasize a landscape perspective.

Attempting to solve environmental problems in isolation from each other, or having one governmental agency ignore what others are doing, simply does not work. Integrative management is required and will only be successful if the larger system in which each event occurs is considered. In short, a landscape perspective is necessary both in ecological management and in ecological restoration.

2. Restoration policies and individual restoration projects must be designed and executed following the principles of adaptive planning and management.

Droughts, floods, invasions of exotic species, and the like may cause problems in restoration projects extending over multiple-year spans. Therefore, adaptive management is far preferable to adhering rigidly to long-term plans.

3. Evaluation and ranking of restoration alternatives should be based on an assessment of opportunity cost rather than traditional benefit-cost analysis.

Details of this principle are given in the National Research Council's book (1992).

C. Policy and Program Redesign for Federal Agencies

1. A uniform definition of *restoration* should be incorporated into all appropriate federal legislation and other federal documents.

A major problem in legislation is that the word *restoration* is not uniformly defined. This tends to confuse the public as well as those trying to communicate from one agency to another.

2. A national restoration strategy should allocate leadership to the federal government for landscape restorations of national significance but should rely on other nonfederal and federal units of government to coordinate restoration programs in various regions.

3. The federal government should initiate an interagency and intergovernmental process to develop a unified national strategy for ecosystem restoration.

4. The development of a unified national program for ecosystem restoration should be facilitated, and then maintained, under the leadership of a single, responsible federal organizational unit.

5. Current and proposed federal programs should take full advantage of all opportunities for ecosystem restoration, including derelict or abandoned lands, farm lands no longer used for that purpose, abandoned surface mines, and the like.

IX. SOURCES OF INFORMATION

Until relatively recently, other sources of information on ecological restoration were difficult to lo-

cate except in major university libraries. However, the situation is now much improved. For example, the December 1992 issue of *Omni* magazine has a very nice article by Kathleen Stein, and the December issue of *Geotimes* has an extended review of restoring aquatic systems by Mary Beck Desmond. A half-hour program on Daniel Janzen's restoration of the Guanacaste dry forest was shown on public television. The National Research Council's book (1992) costs less than $40 (plus $3 shipping) from the National Academy Press and should be available in most major libraries and to small libraries on inter-library loan. The book "Rehabilitating Damaged Ecosystems" (1989, CRC Press, Boca Raton, Florida) has a number of examples of terrestrial restoration as does a forthcoming book "Rehabilitating Damaged Ecosystems, Second Edition" (Lewis Publishers, Chelsea, Michigan). Journals such as *Restoration and Management Notes* (University of Wisconsin) and *Restoration Ecology* (Blackwell Scientific Publications, Inc.) are dedicated to this subject. Additionally, special issues of professional journals (such as Vol. **13**(3), 185–292, 1991 of the *Environmental Professional*) are entirely devoted to the restoration of ecosystems. There are also many restoration projects in the United States and elsewhere in the world which either are regularly open to visitors or can be viewed by persons willing to make a modest effort to obtain permission. Most citizens who have participated in a restoration project are justifiably proud of their efforts and enjoy showing visitors what they have done. There are a number of case histories for aquatic ecosystems from a wide variety of locations, including the Kissimmee River in Florida, Hackensack Meadowlands (in New Jersey near New York City), the Powell River Project in southwestern Virginia, and so forth. Such areas can often be identified by writing to the land grant university in each state. One of my graduate students and I, with the help of a number of graduate students and local citizens and industry, are designing the restoration of a public landfill associated with Explore Park near Roanoke, Virginia. The Explore Park is readily accessible from the Blue Ridge Parkway and, when the project is underway, a guided tour of restoration efforts will be provided for visitors willing to make the short trip from the parkway. John Berger has produced a splendid guide to restoration projects in the San Francisco Bay area but, regrettably, there are as yet no comparable books for other regions of the United States. Organizations such as Trout, Unlimited; the Sierra Club; the Audubon Society; the American Fisheries Society; the Society for Ecological Restoration; and the Nature Conservancy will almost always know of illustrative case histories. If not, they will know how to get the information. A librarian doing a computer search for these books can use a variety of key words, including *restoration, rehabilitation, reclamation, ecological repair,* and even *cleanup.* There are even books such as "Creating Freshwater Wetlands" (Lewis Publishers) and "Restoring the Nation's Marine Environment" (Maryland Sea Grant Program, 1992) that can be located in a search by using a word such as *restoring.* As is the case of most new fields (or even well-established fields), the beginner often finds the terminology confusing; regrettably, different professions and different organizations use the same word with quite different meanings. Nevertheless, even mild persistence and modest skills will result in obtaining literature on the subject.

X. ILLUSTRATIVE CASE HISTORIES

The beauty of most case histories of ecological restoration is that they demonstrate not only factors of scientific interest but also the vast array of value judgments that human society makes. One of the interesting aspects of studying case histories comes in tracking the evolution of societal values and judgments related to natural ecosystems. In fact, ecological restoration has been receiving greater attention, not only because the rate of destruction has accelerated but also because societal values are changing.

Another interesting aspect of case histories of ecological destruction and restoration is the frequent finding that individual decisions may be viewed as having insignificant environmental impacts in the aggregate, but, over time, they almost invariably do. For example, forests may disappear

one tree at a time to provide cooking and heating fuel for households that cannot afford other sources.

Another aspect of isolated, incremental decision making is the fact that each individual and each organization may act as if it is "the only flower facing the sun." For example, at the federal level, construction practices espoused by the Department of Transportation or agricultural practices espoused by the Department of Agriculture or practices of the Bureau of Reclamation (e.g., permitting overgrazing) may increase siltation in surface aquatic systems, particularly rivers. At the same time, the Corps of Engineers is charged with keeping channels clear and, thus, must dredge the silt that partly results from these policies or those of other agencies. This is one of the reasons why the U.S. Congress and the executive branch have been seriously considering a single United States federal agency charged with the responsibility of all environmental problems to which other agencies would be responsible. Restoring a river is not reasonable if some federal agencies continue in the practices that damaged the river.

Successful ecological restoration requires a landscape perspective. For example, ecological restoration of the Chesapeake Bay is useless unless steps are taken to control environmental pollution not only in the bay area itself but in the entire drainage basin of the rivers (e.g., Susquehannah) that discharges into the Chesapeake Bay. Regrettably, the present age of specialization usually rewards reductionist science and has only recently begun to reward integrative science. As a consequence, attention has been focused primarily on components of systems instead of on entire systems. In higher education, the more advanced the degree, the more likely that the person with that degree will be specialized. Even undergraduates are rarely exposed to simultaneous mixtures of engineering, ecology, economics, and the like, despite the fact that these disciplines are interactive components of all of earth's societal systems. Similarly, organizations are accustomed to playing a very explicitly stated role at the federal level assigned by the government by means of line item budgets. Integrative science takes effort, and rarely are even token funds allocated for this, although occasionally agencies are exhorted to interact with each other. Naturally, organizations pay most attention to the line items with the biggest budgets despite the fact that society would almost certainly benefit from an integrative approach that would not have one agency creating problems for another. An example outside the area of ecological restoration would be the adverse effects of tobacco smoke; both first- and second-hand effects are well documented, but the federal government has more vigorously enforced reducing exposure to asbestos than it has reduction to first- and second-hand tobacco smoke.

Improvement in both the educational system to include integrative approaches and in the larger social system would facilitate environmental restoration where and when needed. The following examples are not in the order of importance but, rather, have been selected to illustrate different aspects of ecological restoration.

A. Tidal Thames River in London, England

Damage to the Thames estuary can arguably be traced back as far as the occupation of that part of England by the Roman Empire. A splendid summary, primarily of the political and social events that resulted in the pollution and subsequent restoration of the Thames estuary, is recounted by A. L. H. Gameson and A. Wheeler in a chapter in "Recovery and Restoration of Damaged Ecosystems" (University Press of Virginia, Charlottesville, 1977). As early as 1620, the Bishop of London expressed the hope that the river might be clean. Tobias Smollett wrote in Humphry Clinker "If I would drink water, I must quaff the mawkish contents of an open aqueduct, exposed to all manner of defilement; or swallow that which comes from the River Thames impregnated with all the filth of London and West Minster. . . ." The decline of the Thames as a fishery is thought to have begun around 1830, and, by the 1950s, the only fish able to survive in the most polluted sections were eels. Since 1955, the pollution load from London's sewage discharges has been progressively reduced, and other discharges, including upstream

waters, have been reduced as well. The reason this case history is so interesting is that it was carried out during a period when ecological restoration was in its infancy, when waste treatment technologies were far inferior to those available today. Furthermore, the overall economic benefits and social amenities were remarkable. The river smelled better, benefitting both tourism on the river itself and in the adjacent areas. The resource provided amenities such as recreational and commercial fishing that had been absent for many years. There was no indication that any organization was forced into bankruptcy (that was not already inevitably going to become bankrupt for other reasons) as a result of this cleanup. There is persuasive evidence that the benefits were sufficiently well understood by human society to cause no regrets about the ecological repair. In this case history, the term *ecological recovery* seems to be a better descriptor than *ecological restoration,* although neither is inappropriate. In fact, very little ecological management of the ecosystem itself was necessary. The pollutional stresses were merely substantively reduced, and nature took over the healing process. Ecological recovery is not inappropriate because that was not the intent of the pollution reduction.

B. Rocky Mountain Biological Laboratory near Crested Butte, Colorado

Rocky Mountain Biological Laboratory is one of the most productive biological field stations in the world (it also offers academic courses for students). Such well-known research investigators as Paul Ehrlich (Stanford), John Harte (University of California at Berkeley), David Allen (University of Michigan), and Barbara Peckarsky (Cornell) carry out research there, and these names are only a fraction of a long list of research investigators who publish in the most prestigious national and international ecological journals. In addition, the number of investigators wishing to carry out ecological research exceeds the space available. What makes these two facts so fascinating is that all the research is carried out in an area that was ecologically trashed by mining activities roughly 100 years ago. This indicates that ecological recovery from the activities of human society is indeed possible since the results of these investigations pass rigorous peer review in top scientific journals and more researchers wish to work there. An equally fascinating aspect of the Rocky Mountain Laboratory ecological recovery is that the present ecosystem, although suitable for the most distinguished ecological research, does not resemble the predisturbance ecosystem in some aspects. This provides the important message that naturally recovered or restored ecosystems may not closely resemble the predisturbance ecosystem but, nevertheless, can be extremely useful to human society and to theoretical ecologists.

The University of Michigan Biological Station at Douglas Lake is another site that was badly damaged around the turn of the century, in this case, by extensive clear-cutting of old growth forests. Research investigators also gather at this station from all over the world and publish substantial numbers of papers in prestigious scholarly journals. In 1986, I decided to write to 61 directors of field stations from a list of more than 200. A substantial number of the respondents indicated that the two biological field stations just mentioned are not unique examples but might even well represent the majority of field stations with regard to prior ecological damage. Despite the small sample, four conclusions can be drawn from this survey (covered in more detail in Chapter 38 of the book "Biodiversity," National Academy Press, 1988): (1) damaged ecosystems in temperate regions can recover rather rapidly (in geologic time) so that the biological diversity is substantively improved; (2) simple management practices, even benign neglect, often result in a vastly improved ecological condition; (3) ecological research can be carried out on formerly damaged ecosystems, indicating that the processes in the recovered or restored ecosystems are ecologically interesting and comparable in many respects to those in undisturbed ecosystems; and (4) the quality of these ecosystems relative to most of the surrounding countryside is sufficiently superior to justify such designations as nature reserves or preserves and experimental ecological preserves.

C. Biocultural Restoration of the Guanacaste in Costa Rica

Tropical rain forests have deservedly received much attention in the news media. However, tropical dry forest is arguably the most endangered of the once widespread habitat types in meso-America, as well as in the remainder of the tropical world. Today, only 0.08% of the original 550,000 km^2 of dry forest on the Pacific side of Central America is in preserves. Professor Daniel H. Janzen (University of Pennsylvania), the prime mover in the restoration of the Guanacaste tree habitat, uses the word *biocultural* for this activity. This emphasizes that attention must be given to the sociological aspects of integrating this important ecosystem into Costa Rican society if the effort is to endure. Of course, sound ecological science is essential, but people interact with this important habitat continually and, of course, were responsible for its destruction. Therefore, ignoring the cultural aspects almost ensures that the restoration effort will be unsuccessful and, even if it were successful, would probably not endure for sustainable, long-term use because human society would not play an appropriate stewardship role. This is a particularly interesting case history for other reasons as well: (1) ecological restoration in a developing country poses a different set of management problems than restoration efforts in developed countries, (2) it documents that persons in developed countries more affluent than the average person in Costa Rican society have taken a global view with regard to ecological restoration and are helping substantively, (3) the project is simultaneously of enormous practical benefit to human society and to theoretical ecology, (4) abundant evidence has already been generated that citizens of Costa Rica (ranging from the president to local villagers and representing a wide variety of age groups) can understand why the relationship with natural ecosystems should be improved and how this can be done, and (5) the combination of applied and theoretical research appears to have enhanced Janzen's professional career. More information on the Guanacaste biocultural restoration can be found in Volume II of "Rehabilitating Damaged Ecosystems" (1988,

CRC Press, Boca Raton, Florida). Requests to Dr. Janzen himself for information on this project should be accompanied by a contribution to the Guanacaste biorestoration effort.

XI. CHOICES FACING HUMAN SOCIETY

In 1992, a joint statement entitled "Population Growth, Resource Consumption, and a Sustainable World" was released by the officers of the Royal Society of London and the U.S. National Academy of Sciences. This brief report focuses attention on the 1991 report on world populations produced by the United Nations Population Fund (UNFPA), which states that population growth is occurring more rapidly than was forecast in the UNFPA report of 1984. The global population is expected to rise from 5.4 billion in 1991 to 10 billion in 2050. The report states that this rapid rise *may be unavoidable*. Considerably larger increases in human population must be expected if fertility rates do not stabilize at the replacement level of about 2.1 children per woman and, of course, if war or disease does not cause dramatic reduction in population size.

The rapidly growing human population and the pressure to develop economies, especially in developing countries, are leading to substantial increased damage to the local environment. This damage is the result of many factors, including direct pollution from energy use, other industrial activities, and activities such as forest clearing and inappropriate agricultural practices. Less developed countries face overwhelming challenges in coping with their environmental and resource problems alone; although, in some areas where natural resources are cherished (e.g., Bhutan), population growth does not inevitably imply serious deterioration in the quality of the environment.

Human society must adjust to three fundamental and environmental laws: (1) pay now or later, (2) no species is exempt, and (3) "solving" one problem in isolation from other problems will not work.

The following assumptions appear to be valid when considering human society's relationship with natural systems.

1. Ecosystem services (e.g., maintaining atmospheric gas balance and water quality) are important to the survival of human society in its present form.

2. The continuing destruction of ecosystems and the concurrent rapid increase in human population size are serious threats to the delivery of ecosystem services.

3. The rapid global loss of species intimately associated with the loss of habitat will also impair the delivery of ecosystem services.

4. Healing of damaged ecosystems will enhance both the quality and quantity of ecosystem services.

Until these presently plausible assumptions are proven incorrect, it seems prudent for human society to consider the various policy options in its relationship with natural ecosystems. Some illustrative options follow.

1. Continue ecosystem destruction and population growth at their present rates and see what happens before any action is taken. This would almost certainly mean dramatic erosion of ecosystem services per capita.

2. Adopt a no-net-loss of ecosystem services, which means that a balance between destruction and repair must be achieved at all times. Destruction of an ecosystem may take hours or months, but restoration may take years. Exercising this option with continued human population growth would mean a rapid per capita decline of ecosystem services but not as rapid as in Option 1.

3. Exceed a no-net-loss of ecosystem services, which means implementing a policy of healing or repairing damaged ecosystems at a more rapid rate than the rate of destruction. This would still mean a loss of ecosystem services per capita given the predicted rate of human population increase and the rate at which ecosystem services could be restored, but would still provide more ecosystem services per capita than either Options 1 or 2.

4. Stabilize human population growth and exercise restoration Option 2, which would maintain the status quo of ecosystem services per capita.

5. Stabilize human population growth and restore ecosystems at a greater rate than damage,

which would improve ecosystem services per capita.

Doubtless, readers can modify or supplement the information earlier taken from the Royal Society and the National Academy of Sciences. Neither group, by any stretch of the imagination, can be considered fanatic environmentalists, but they represent the best in mainstream science and their conclusions are based on a rigorous examination of available evidence. As Louis Pasteur notes, "Change favors the prepared mind." Therefore, even though the future always remains uncertain, one certain thing is that it becomes the present. The point has been reached in human population growth and power where the cumulative actions and decisions of individuals and nations can no longer be considered in isolation from each other. This is especially true when humans have the power to destroy within their lifetimes the remaining, relatively unimpaired ecosystems on the planet. In a relationship this important to each individual as well as to the future of human society, informed discussions are essential. This author remains optimistic about what can be done but is not yet optimistic about what will be done.

Glossary

Amenities Ecological features, traits, or characteristics that add to the physical or material comfort of human societies.

Damage assessment An accounting of the magnitude, physical extent, and types of damage suffered by an ecosystem.

Ecological capital Stock of ecological goods and facilities devoted to providing ecosystem services.

Ecosystem services Valuable functions ecosystems provide free of charge to human societies, including maintenance of atmospheric gases; regulation of the hydrologic cycle; provision of potable water, fertile soil, wood, fish, and other consumable products; processing of wastes; pollination of crops; and a genetic library from which people have domesticated crops and are now designing new foodstuffs.

Integrated environmental management Coordinated control, direction, or influence of all human activities in a defined environmental system to achieve and balance the broadest possible range of short- and long-term objectives.

Mitigation Actions taken to avoid, reduce, or compensate for the effects of environmental damage, including activities that restore, enhance, create, or replace damaged ecosystems.

No net loss A policy by which the total amount of some habitat type is not decreased though individual units may change.

Reclamation Putting a natural resource to a new or altered use.

Rehabilitation Improvements to a natural resource that return it to a good condition but not the condition prior to disturbance.

Restoration Return of an ecosystem to a close approximation of its condition prior to disturbance.

Bibliography

Cairns, J., Jr. (1989). "Rehabilitating Damaged Ecosystems." Boca Raton, FL: CRC Press.

Cairns, J., Jr. (1989). Restoring damaged ecosystems: Is predisturbance condition a viable option? *Environ. Profess.* **11,** 152–159.

Cairns, J., Jr. (1995). "Rehabilitating Damaged Ecosystems, Second Edition." Chelsea, MI: Lewis Publishers.

Cairns, J., Jr., Dickson, K. L., and Herricks, E. E., eds. (1977). "Recovery and Restoration of Damaged Ecosystems." Charlottesville, VA: Univ. Press of Virginia.

Ehrlich, P. R., and Ehrlich, A. H. (1991). "Healing the Planet: Strategies for Solving the Environmental Crisis." Reading, MA: Addison-Wesley Publishing Co.

National Research Council. (1992). "Restoration of Aquatic Ecosystems: Science, Technology, and Public Policy." Washington, D.C: National Academy Press.

Magnuson, J. J., Regier, H. A., Christie, W. J., and Sonzogni, W. C. (1980). To rehabilitate and restore Great Lakes ecosystems. *In* "The Recovery Process in Damaged Ecosystems" (J. Cairns, Jr., ed.), pp. 95–112. Ann Arbor, MI: Ann Arbor Science Publishers Inc.

Wilson, E. O., ed. (1988). "Biodiversity." Washington, D.C: National Academy Press.

Rhizosphere Ecophysiology

E. Paterson, E. A. S. Rattray, and K. Killham

University of Aberdeen

I. Introduction
II. Ecophysiology of Plant–Microbe Interactions
III. Ecophysiology of Microbe–Microbe Interactions

The rhizosphere, the zone of soil influenced by the plant, is a biologically complex and distinct microhabitat within the terrestrial ecosystem. The characteristic ecophysiology of the rhizosphere, manifested in the interactions between rhizosphere populations and between these populations and plants, is a consequence of the physiological attributes of microorganisms which are adapted to inhabit this region. Rhizosphere ecophysiology encompasses interactions between plants and their microbial pathogens and mutualistic partners. The distinct biological composition of the rhizosphere is partly the result of the chemical and physical conditions imposed by the plant, but is also because of the competition between microbial populations. The understanding of these ecophysiological interactions has practical implications in plant nutrition and health through the manipulation of rhizosphere microbial populations.

I. INTRODUCTION

The term "rhizosphere" was first used by Hiltner (1904) to describe the specific interaction between bacteria and legume roots. This definition has now widened to encompass the region of influence exerted by roots generally on biological, chemical, and physical processes in soil. As a distinct ecological microhabitat, which is of considerable importance in relation to agricultural productivity and terrestrial nutrient cycling, an understanding of the rhizosphere and the ecophysiological interactions occurring therein is necessary. This article concentrates on interactions between plants and microorganisms (bacteria, fungi, and protozoa) in the rhizosphere and outlines some of the most important processes occurring in this biologically complex habitat. [*See* Nutrient Cycling in Forests; Soil Ecosystems.]

II. ECOPHYSIOLOGY OF PLANT–MICROBE INTERACTIONS

A. Nutrient Dynamics

1. Rhizodeposition and Microbial Nutrition

The rhizosphere effect in soil is generated by the continual loss of carbon from roots, amounting to up to 40% of the carbon fixed by the plants. Rhizodeposition includes all carbon losses and can be classified into four components: exudates, secretions, lysates, and gases. Water-soluble exudates are released from the root cells through passive leaking and are compounds of low molecular weight, such as sugars, amino acids, hormones, and vitamins. Insoluble exudates are also released from roots, mainly as insoluble polysaccharide material termed mucigel and mucilage. Secretions may also include low molecular weight compounds, but may require metabolically active processes for their release. Secreted substances may also include polymeric carbohydrates and enzymes. Lysates result from the autolysis of older cells and include cell wall material, sloughed whole cells, and whole roots. Gases (e.g., ethylene and CO_2) are also lost from roots through diffusion. Results from ^{14}C tracer experiments suggest that 25% of root-derived material may be of low molecular weight (e.g., sugars, amino acids, and other organic acids), whereas 15% of root-derived carbon is associated with the more recalcitrant high molecular weight compounds. In addition to the root-derived carbon losses already described, rhizodeposition also includes the carbon utilized immediately by the microbial biomass and the subsequent evolution of CO_2 through microbial respiration (see Fig. 1).

The root system is highly dynamic with turnover representing a large proportion of rhizodeposition. The rate of root turnover is highly dependent on plant and soil type. In forests, for example, between 30 and 90% of fine roots are lost and replaced annually. Mineralization of such recalcitrant material requires saprotrophic degradation, to release more readily available forms of carbon for the microbial biomass. Another source of macrorhizodeposition is root cap cells, of which up to 5000 a day may be lost from maize. These cells may contain mixed substrates for microbial mineralization, as root cap cells are generally imbedded in polysaccharides. Seminal roots of young wheat plants can contain a large proportion of dead cortical cells, which lyse to release their cellular contents. In terms of sources of substrate for the microbial population within the rhizosphere, root exudates and lysates will be of more importance in creating a physiologically diverse rhizosphere population.

The composition of rhizodeposition varies depending on plant type, phenological stage, and soil abiotic factors (e.g., temperature, pH, moisture). Substrate composition, even between closely re-

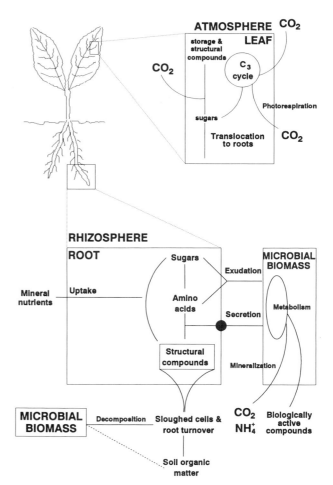

FIGURE I Rhizosphere processes driven by photosynthetic assimilation and the subsequent distribution of carbon.

lated plants may differ, resulting in very different growth rates of closely related bacteria. The nature of the root-derived carbon source and the state of root degradation and senescence will affect its availability as nutrients for the microorganisms. Particular components of rhizodeposition may therefore stimulate specific populations of microorganisms, for example, some bacteria are capable of degrading root mucilage, while zoospores of pathogens are attracted by soluble exudates. Bacterial numbers are generally highest at cell junctions on the root, possibly indicating regions of enhanced exudation. Rhizodeposition therefore provides the driving force for microbial nutrition and subsequent activity in the rhizosphere. This in turn may benefit the plant, as some microbial metabolites can act as plant growth promoters or soil-stabilizing agents, resulting in improved soil structure and water availability, both contributing to enhanced plant growth.

When nutrients are in short supply, both the microorganisms and the root will compete for these nutrients. Using the principles described earlier, Monod kinetics have been used to develop simplistic model describing rhizosphere population dynamics. Such models (e.g., Newman and Watson, 1977) have been used to predict the abundance of the microbial population in relation to the distance out from the root and the amounts of soluble root exudate.

2. Nitrogen Dynamics and the Relationship between Microbial Mineralization/Immobilization and Competition for N between Plants and the Rhizosphere Population

The C:N ratio of the microbially available rhizo-deposition and of available soil organic matter determines whether mineral nitrogen (either added as fertilizer or previously mineralized in the soil) will be immobilized by rhizosphere microorganisms or taken up by plant roots. This competition for N also depends on the physiology of the rhizosphere microbes. In particular, substrate assimilation efficiency and the cellular C:N ratio determine the competitive sink strength of these microbes and, in particular, whether mineralization or immobilization of N will occur.

The concentration and form of available N are critical in controlling the competition for this pool. The effect of concentration is largely a function of comparative Michaelis Menten kinetics for absorption by both roots and microorganisms, although such kinetics are a considerable oversimplification. The kinetics of root N uptake vary depending on the mycorrhizal status of the root and the extent of suberization, with absorption being highest at or near the apical tips of young, unsuberized roots. Superimposed on this, however, is the form of available N. Most plant roots and microbes readily take up NH_4^+ and NO_3^- (there is often a preference for NH_4^+, although NO_3^- tends to be more freely available). In some cases, however, the nitrate reductase enzyme is either absent or very

slowly induced. Competition for mineral N between roots and microorganisms may, therefore, also be influenced by the rate of nitrification (the microbial, dominantly chemoautotrophic oxidation of NH_4^+ to NO_3^-).

3. Plant-Induced Specificity of Rhizosphere Microbial Populations

As outlined in the introduction, the plant creates a specific physical and chemical environment in rhizosphere soil immediately surrounding the root. As a consequence, the plant exerts selective pressures on the soil microbial population for organisms physiologically adapted to the physical and chemical conditions imposed by the plant. This effect of plants on rhizosphere populations is of general applicability, although the physical extent of influence in soil and the magnitude of selection pressure vary depending on soil conditions and plant species.

In addition to the general influence which all plants have on soil microbial populations, the inhibition or promotion of particular components of the rhizosphere population through root secretion/exudation of physiologically active compounds has been demonstrated. Indeed, from an evolutionary perspective, such mechanisms of plant-mediated control of rhizosphere populations would be expected as means of influencing the activities of microorganisms beneficial or detrimental to plant growth and productivity.

The presence of compounds in root exudates of pine inhibitory to nitrification of ammonium mediated by *Nitrosomonas* has been found and is in part responsible for soils under established forest stands having low nitrifying activity. The use of ammonium as opposed to nitrate, by plants as a source of nitrogen, is energetically more efficient. Therefore, the production of compounds to prevent microbial transformation of ammonium to nitrate may be beneficial with respect to plant nutrition.

Rice plants produce the enzyme catalase in root exudates under anaerobic soil conditions, present with paddy-field cultivation. This may function to promote oxidation of phytoinhibitory sulfide ions, through bacterial transformation, which is inhibited in the presence of autotoxic peroxides which accumulate in soil in the absence of the catalase enzyme.

Chemoattractants produced by roots may influence the composition of the rhizosphere microbial population. Compounds released from roots such as simple sugars and amino acids may induce accumulation of a broad spectrum of motile organisms. In addition, chemotaxis to secondary metabolites released by roots may result in the accumulation of specific organisms, for example, attraction of rhizobia to flavones released by legume roots and *Agrobacterium tumefaciens* to acetosyringone released from lesions. These compounds further reduce population diversity through their relatively broad antibacterial activity. Roots also produce inhibitory compounds such as amino acid analogs which select for physiologically adapted rhizosphere microorganisms.

4. Microbial Stimulation of Root Exudation

Rhizosphere microbial activity is stimulated by readily assimilated components of rhizodeposition, primarily root exudation, as discussed in previous sections. In turn, rhizosphere microorganisms tend to influence root function, be they invasive of the root or noninvasive.

Inoculation of root growth media with microorganisms generally increases root exudation. Presumably, microbial metabolites contribute to the carbon efflux, assuming the microorganisms are noninvasive. The number of microbial metabolites which affect root function is very considerable, with a multiplicity of possible mechanisms of action. Such modes of action vary from being beneficial to detrimental to plant function.

As well as the production of metabolites, root infection and rhizosphere colonization also affect root exudation through mechanisms such as damage to cell function (if pathogenic microorganisms are involved) to stimulation of root function (if beneficial microorganisms are involved).

Use of Carbon-14 tracer methodology enables the scale of microbial stimulation of root exudation to be determined, since it facilitates discrimination between plant and microbially derived carbonflow. However, it may be that of equal, if not greater, ecological significance relates to microbially induced changes in the composition rather than to the quantity of root exudation and other forms of rhizodeposition.

5. Enhanced Plant Nutrition through Solubilization of Mineral Nutrients in the Rhizosphere

Increased plant nutrition can occur through beneficial plant–microbe interactions in the rhizosphere. For example, the development of mycorrhizal fungi on plant root systems brings about physiological changes of benefit to the plant, especially the enhancement of phosphorus nutrition under phosphorus-deficient conditions (through increasing the absorbing surface area of the plant roots). Since phosphate in most soils is normally organically bound, phosphatase and phytase activity of the microbial population in the rhizosphere is important in its solubilization. Many soil and rhizosphere bacteria produce acid phosphatase enzymes that solubilize phosphorus from mineral sources. Phosphorus is thus more available to the plant. Plant growth can be enhanced when phosphatase-producing bacteria are coinoculated with VA mycorrhizal fungi.

Mycorrhizal roots have the ability to produce ligands to chelate ions in the rhizosphere. Plant roots and rhizosphere microorganisms can increase the solubility of phosphorus by producing metabolites which release phosphorus from insoluble phosphataes by the formation of soluble complexes with metal ions. One such group of metabolites are siderophores. Many Gram-negative aerobic bacteria produce siderophores to increase iron availability for the microorganism. Mycorrhizal fungi play an important role in host iron nutrition, especially in high pH soils where iron solubility is limiting. The plant roots can then utilize the ferric iron from microbial siderophores.

Under low redox potential conditions, fermentation metabolites such as acetic acid can accumulate, causing the pH of the soil solution to drop and the solubilization of some metal ions (e.g., Fe^{2+} and Mn^{2+}). These mineral nutrients therefore are made potentially available to plants.

6. Plant/Fungal Symbiosis: Mycorrhizas

Mycorrhizas are formed by association between the roots of a plant and a fungus. By far the majority of vascular plants are involved in these associations which are of paramount ecophysiological importance.

There are a number of different forms of mycorrhizal association (endomycorrhizas, where the fungal tissue is dominantly within the root of the plant, include the vesicular arbuscular, arbutoid, ericoid, and orchid mycorrhizas, and ectomycorrhizas where the fungal tissue sheaths the roots of mainly tree hosts), although in many cases the fungus involved (the mycobiont) enhances the plant uptake of nutrients (major, minor, and trace) from the soil. Mycorrhizas are also generally associated with increased rootlet size and longevity, often with root protection from soilborne pathogens, and sometimes with enhanced drought resistance.

The main mechanism of enhanced nutrient acquisition by mycorrhizal plants is the increased soil volume exploited by the extramatrical mycelium of the mycobiont. As a result, nutrients can be transported from beyond the narrow nutrient depletion zone that surrounds nonmycorrhizal roots. In addition, the fungal mycelium can often absorb nutrients at lower solution concentrations than an equivalent surface area of plant root alone. Thus, the kinetics of nutrient uptake differ between mycorrhizal and nonmycorrhizal roots.

A further ecophysiological advantage of the mycorrhizal association lies in the ability to access nutrients in an organic form that would normally be unavailable. There is also evidence that some mycorrhizal plants are connected via hyphal strands and that this link can provide a means of nutrient transfer within plant communities.

The preceding dicussion on the ecophysiology of mycorrhizas focused on the combined association. It must also be remembered, however, that the mycorrhizal fungus is provided with a habitat free of competition and of particular ecophysiological significance, with a steady supply of photosynthate carbon (the exceptions to this are the orchid and monotropoid mycorrhizas where carbon flow is away from the fungus to the host) for growth/energy requirements.

7. Plant/Rhizobial Symbiosis

The ecophysiology of the legume Rhizobium association is probably better documented than that of any other plant–microbial interaction. This is largely because of the ecological as well as economic significance of this N_2-fixing association which accounts, at the global level, for 140×10^6 tons of nitrogen annually (this is more than the combined input to soil from artificial fertilizers and atmospheric deposition).

Nitrogen fixation is an energy demanding process and this is provided by sunlight via legume photosynthesis. Photoassimilated carbon is harnessed via the TCA cycle of the rhizobia in the legume nodule. Rhizobia are therefore a carbon drain in legume hosts, although the plant obtains fixed nitrogen in the form of amino acid-N in return. A cost of nitrogen fixation of 6.5 g C per g N has been estimated, although this does not tend to translate into a loss in plant yield as plants are rarely close to their upper levels of productivity where a carbon drain would offer a significant burden.

The flow of carbon to the legume nodule represents almost as great a flow as that to the entire remainder of the legume root system. Of the carbon diverted to the nodule, about 15% is stored in the nodule, 40% is respired into the atmosphere, and the remainder is retranslocated (in the xylem) to the aboveground tissue of the plant.

Catalysis of the fixation of atmospheric N by Rhizobium is carried out by the enzyme nitrogenase. The enzyme is highly sensitive to oxygen inactivation and so a key ecophysiological feature of the legume–rhizobial symbiosis is the regulation of oxygen diffusion in the legume nodule. Oxygen concentrations in the nodule are maintained at less than $0.1 \ \mu M$ by means of a membraneous sac consisting of the pigment leghemoglobin with strong oxygen control characteristics.

Although this section has concentrated on the ecophysiology of the combined legume–Rhizobium associated, it should be pointed out that, because rhizobia are also free-living soil bacteria, an important ecophysiological aspect of their ecology relates to their stimulation by rhizosphere carbon flow in the soil. This stimulation need not necessarily involve legumes, as high populations of *Rhizobium* can also be maintained by other "nurse" plants. However, only legumes provide the additional chemical signals which initiate root infection/nodulation and hence establish N_2 fixation.

8. Nonsymbiotic N₂ Fixation in the Rhizosphere

Nitrogen-fixing bacterial systems loosely associated with the roots of plants are sometimes referred to as rhizocoenoses and the bacteria involved include *Azotobacter, Beijerinckia,* and *Azospirillum.*

The major ecophysiological determinant of nonsymbiotic N_2 fixers in the rhizosphere is the supply of carbon from the plant roots. This is because free-living N_2 fixers, with the exception of cyanobacteria, are heterotrophs and the rate of N_2 fixation is carbon limited. There is a very high demand for carbon from the inefficient nitrogenase enzyme system (partly because only, about half of the electron flow to nitrogenase is transferred to N_2, the remainder being lost to H_2 evolution) so that for each gram of glucose used as a substrate by *Azotobacter,* somewhat less than 20 mg of N_2 is fixed. This very high demand for substrate carbon demonstrates why the rhizosphere is the main site for nonsymbiotic N_2 fixation in soil.

A simple calculation can be made of the likely ecological and agronomic significance of nonsymbiotic N_2 fixation based on the ecophysiological considerations of the carbon demand for nitrogenase activity in free-living systems and the carbon supply from plant roots. Using this approach, if it is assumed that all the root-derived carbon flow is exploited by free-living N_2 fixers such as *Azotobacter,* then the maximum likely rate of N_2 fixation would only be 10 kg N ha^{-1} $year^{-1}$ for a temperate arable crop. Clearly, since N_2 fixers have to compete with other heterotrophs for rhizosphere carbon flow, then the actual likely rate of fixation will be much lower, and of little importance per unit area, compared to symbiotic fixation.

B. Plant–Pathogen Interactions

I. Host-Locating Mechanisms

Location of a suitable host by a pathogen is the primary interaction in the pathogenic process. This may be passive (on the part of the pathogen), as propagules are stimulated as the root system expands during plant growth to exploit increasing volumes of soil. Host roots are able to promote the growth of pathogenic organisms and, in some instances, this involves a highly host-specific stimulation of inactive or dormant propagules. *Sclerotium cepivorum,* the white rot pathogen of onion, survives in soil as dormant hyphal masses (sclerotia) and only germinates in the presence of its host. This is a response to sulfoxides released by the host and metabolized by indigenous soil bacteria, producing volatile alkylsulfides which are the active compounds in stimulating the dormant fungal propagules.

Tactic responses of motile organisms are important mechanisms of host location over short distances in soil. Motile organisms are known to have tactic responses to a number of stimuli associated with plant roots. Ecophysiologically, the most important of these in relation to plant roots are chemotaxis and electrotaxis, which are responses to chemical and electrical gradients, respectively. Motile bacteria and fungal zoospores have been found *in vitro* to be positively attracted to a wide range of low molecular weight solutes known to be released from growing roots. The electrotaxis of zoospores in response to electrical fields associated with apical growth and root lesions has been demonstrated and has also been suggested as a mechanism of host location by pathogens.

2. Host Recognition

Once in contact with the host, initial binding to the host surface may be mediated by specific interactions between surface structures of the host and pathogen. The most studied interaction of this type involves recognition of specific saccharide structures by bacterial glycoproteins (lectins). The initial binding of cells to plant surfaces is important in facilitating more permanent binding, for example, through production of extracellular polysaccharide and cellulose microfibrils.

Lectin-mediated attachment is a chemically specific process, in that lectins recognize specific saccharide configurations. However, the biologically specificity of these interactions is determined by the frequency and distribution of these binding sites in plant tissues. The binding of Enterobacteriaceace and fluorescent pseudomonads to roots is mediated by lectins present on extracellular projections called fimbriae. This binding is relatively nonhost specific, in that the lectin recognizes a saccharide structure common to surfaces of root hairs of many plant species. As such, the binding can be considered tissue specific and may be an ecophysiolgoical adaptation to promote occupancy of a microbially profitable niche.

There is evidence for host-specific binding mediated by lectins for both bacterial and fungal plant pathogens, and interference with this process (chemically or biologically) has been suggested as a potential means of disease control.

3. Host Defense Mechanisms

The host plant has many methods of defense against invasion by potential pathogens. Such defense mechanisms include both increased physical protection, through structural changes in the plant cell wall and through physiological changes. The physical structure of the plant cell wall may change by an increased lignin content or by forming additional cross-links between the structural β-carbohydrates which increases resistance to degradation. Increased biological protection can result from changes in the host's metabolism, seen as an increase in the permeability of the plasmalemma or a rise in the respiration rate. Changes in enzyme levels may also occur. For example, plant chitinases and β-1,3-glucanases are induced following pathogen invasion and may serve a defense function since chitin is a major component of the fungal cell wall and some plant chitinases have lysozyme-like activity against bacteria. Phytoalexins, which are absent in healthy plants, accumulate in plants after exposure to pathogenic microorganisms. Death of the plant cells, resulting from a hypersensitive response, may also serve as a defense mechanism, isolating potential pathogens from the remaining live cells.

Induced resistance, in which the host defense mechanisms recognize and respond to a harmless mimic, thus enabling the plant to respond when invasion by a pathogen occurs, may be a means of biocontrol. Tomatoes, for example, can be protected from the wilt fungus *Fusarium oxysporum* by dipping the roots into a culture of nonpathogenic *Fusarium*.

The production of organic compounds such as phenolics, tannins, and agglutins may serve as host defense mechanisms. Phenolics and tannins are able to precipitate infective viruses or potentially damaging enzymes in host plant tissue. Agglutins serve as a general defense mechanism, nonspecifically agglutinating microbial cells in the rhizosphere, thus preventing infection.

4. Microbial Invasion Mechanisms

The major determinant in root disease as a result of pathogen invasion is rhizodeposition, either directly by influencing pathogen population dynamics or indirectly by affecting the general microbial population, through nutrient availability. For biocontrol to be successful, the mechanisms pathogens employ to invade host plants should be elucidated. There are different types of pathogens, which have different survival and invasive strategies. Some pathogens may survive as dormant propagules (e.g., spores, resistant chlamydospores, or sclerotia). These types of pathogens germinate, grow, and infect as the opportunity arises. Pathogens, such as *Rhizoctonia*, live saprotrophically on soil organic matter between their attacks on live hosts. This type of pathogen tends to have a high competitive saprophytic ability.

Pathogen life cycles are basically composed of four stages. Infection, which involves germination of the spore or other propagule after host location and recognition has occurred. These processes are affected by many factors (e.g., weather conditions, nutrient status of the host, and the initial recognition process). *Gaeumannomyces graminis*, the causative agent of "take-all" in wheat, requires nitrogen for germination, hyphal growth, and penetration. The pathogen produces enzymes for the degradation of proteins, pectins, and cellulases and initially infects root hairs. The second stage involves nutrient acquisition and growth of the pathogen, during which the host may deter the pathogen. The feeding stage for *G. graminis* involves mycelial increases in the form of clumps and extensive root hair colonization. This stage is followed by dispersal from the initial site of infection, which for *G. graminis* is the spread of spores to infect seminal root axes.

The last stage involves the survival of the pathogen, for example, as a spore or a sclerotium.

III. ECOPHYSIOLOGY OF MICROBE–MICROBE INTERACTIONS

A. Activity of Protozoa in the Rhizosphere

Protozoa typically constitute 1–6% of the soil microbial biomass and are the dominant microfaunal predators of bacteria in most soils. Protozoa become progressively more numerous around the root after emergence as a consequence of the associated increase in bacterial numbers during this period. The protozoan predation of bacteria is important in releasing ammonium into the rhizosphere as a by-product of the digestion of bacterial cells. Such mineralization may be an important source of inorganic nitrogen to the plant, particularly where soil temperature permits the proliferation of protozoan populations.

The susceptibility of bacterial cells to predation is influenced by spatial location in the rhizosphere, the water status of the soil, and the physiological condition of the cells. Protozoa are large in comparison to bacteria, and therefore bacteria present in pores inaccessible to protozoa are physically protected from predation. Physical protection may also occur as a result of discontinuity of water-filled pathways, necessary for protozoan motility. Such discontinuity increases as a soil dries and is particularly prevalent in the rhizosphere which experiences diurnal variation in water status resulting from uptake of water by roots, driven by transpiration. The bacterial production of extracellular polysaccharides may provide protection from predation, particularly when this is associated with attachment to soil surfaces and growth in microcolonies. Thus, protozoan predation imposes a strong population pressure in the rhizosphere, which may be selective in action. The ecophysiological importance of predation to the functioning of the rhizosphere habitat is an area which is little understood and warrants future research.

B. Microbial Production of Inhibitory Compounds

The production of inhibitory compounds by rhizosphere microorganisms is an important aspect of the ecophysiology of the plant–soil system, as the substances present in the rhizosphere determine its community structure.

The range of inhibitory microbial metabolites involved in microbe–microbe interactions in the rhizosphere includes nonspecific compounds such as phytotoxins and compounds which act specifically such as antibiotics, ionophores, lectins, agglutins, and lytic enzymes. Aliphatic acids, phenolic acids, hydrogen sulfide, and cyanide are all phytotoxic substances produced in the rhizosphere. Phenolic acids are generally more phytotoxic than aliphatic acids and may play an important role against plant pathogens. Hydrogen sulfide can be produced by obligate anaerobes under low redox potentials. This is more likely in flooded soils, resulting in inhibiting respiration and the cytochrome oxidase systems of plant roots and microorganisms. The production of cyanides (CN-) by microorganisms may be toxic to plant roots. Many plant roots can produce cyanides, and some fungi are pathogenic to such roots because of their ability to detoxify the cyanides. The microbial production of cyanide may be a significant factor in biological control mechanisms. For example, *Pseudomonas fluorescens* produces cyanide which is toxic to *Thielaviopisis basicola* (the causative agent of black rot of tobacco).

Ionophores have been extensively studied, especially siderophores which chelate iron. The production of siderophores is indirectly inhibitory to plant pathogens. The involvement of siderophores in biocontrol of pathogens is discussed in the following section.

The most extensively studied of the microbial metabolites in soil is the production of antibiotics for commercial and medical purposes. Many antibiotics have been produced *in vitro*, but surprisingly few have been detected *in vivo*. One of the proven commercial success of biocontrol agents is the antibiotic agrocin, active against the pathogen *Agrobacterium tumefaciens*. Many of the antibiotics characterized in biocontrol of plant diseases are antifungal in nature. For example, patulin, produced by *Penicillium patulu,* is biologically active against damping-off diseases in tomatoes. Antibiotic production is further discussed in the section on biocontrol.

Lectin binding appears to be involved as a mechanism for the biocontrol of plant pathogens. Specific binding and coiling by the antagonist *Trichoderma* to the pathogen (*Rhizoctonia*) are followed by penetration resulting from the production of lytic enzymes, such as chitinases and glucanases.

C. Biological Control of Plant Pathogens

The control of soilborne plant pests in the rhizosphere is associated with a range of soil bacteria and fungi. The most effective of these biocontrol agents are able to readily exploit root carbon inflow and colonize the rhizosphere (rhizosphere competent). In the case of bacteria, the term "rhizobacteria" is used to describe such aggressive colonizers of the rhizosphere.

Many of the mechanisms involved in biological control in the rhizosphere relate to the different ecophysiologies of the agents involved. Some strains of *Arthrobacter,* for example, parasitize plant pathogenic fungi by producing enzymes such as chitinases which lyse the fungal cell wall. On the other hand, control of the "take-all" fungus *Gaeumannomyces graminis* by fluorescent pseudomonads is related to antibiotic suppression of the pathogen. The pseudomonad control of damping-off fungi associated with seedlings may well be due to competition for infection sites through a chemotactic response by the pseudomonad cells. Control by *Trichoderma harzianum* of a broad range of plant pathogenic fungi appears to be linked to the production of alkyl pyrones, volatiles which act as paramorphagens to assist in the penetration of the pathogen by the antagonist. The "mycoparasitism" of nematodes provides a strong means of biocontrol—the fungus *Harposporium,* for example, parasitizes nematodes both externally and internally through germination of adsorbed conidia.

Although parasitism, ammensalism, and competition are often the dominant mechanisms of biocontrol in the rhizosphere, other ecophysiological strategies may also be employed by antagonists. The production of iron-chelating "siderophores," for example, may reduce the availability of iron to rhizosphere plant pathogens. In this context, siderophores are metabolites produced that enable the antagonist to take up ferric iron (Fe^{3+}) as an organic complex that is available through recognition by specific membrane acceptors. [*See* PARASITISM, ECOLOGY.]

It should be pointed out that it may well be that improved plant growth due to the activity of rhizobacteria and rhizosphere-colonizing fungi may often be associated with ecophysiological strategies other than those related to biological control. Because of this, the term plant growth-promoting rhizobacteria is used when any beneficial effects on plant growth result from any bacteria. The beneficial effect may result from the production of plant growth-promoting substances as readily as from a means of biocontrol.

Glossary

Antagonist An organism exerting a detrimental effect on another organism.

Chemotaxis Ability of microorganisms to migrate toward a range of chemical attractants.

Competitive saprophytic activity Ability of a pathogen to compete with saprotrophs during a stage in its life cycle when it is not being pathogenic.

Immobilization Assimilation by organisms of elements through conversion from an inorganic to an organic form.

Mineralization Conversion of elements in an organic form to an inorganic form.

Pathogen An organism with the ability to cause a disease in a plant, animal, or microorganism.

Rhizodeposition Term describing all carbon lost from plant roots.

Rhizosphere competence Ability of an organism to aggressively colonize the rhizosphere.

Sclerotium Resting structure of fungi, comprising a mass of hyphae, surrounded by a protective waterproof layer of thickened hyphal walls.

Symbiosis Living together of two organisms, regardless of whether the relationship is beneficial or detrimental.

Bibliography

Curl, E. A., and Truelove, B. (1986). "The Rhizosphere." New York: Springer-Verlag.

Killham, K. (1994). "Soil Ecology." Cambridge: Cambridge University Press.

Lynch, J. M. (1983). "Soil Biotechnology." Oxford: Blackwell Scientific Publications.

Lynch, J. M. (1990). "The Rhizosphere." Chichester: Wiley.

Metting, F. B., Jr. (1993). "Soil Microbial Ecology." New York: Marcel Dekker.

Read, D. J., Francis, R., and Finlay, R. D. (1985). *In* "Ecological Interactions in Soil: Plants, Microbes and Animals." (A. H. Fitter, ed.), pp. 193–217. London: Blackwell.

River Ecology

J. V. Ward

Colorado State University

I. Introduction
II. Aquatic Biota
III. Interactive Pathways
IV. Biodiversity

River ecology is the study of running waters from a perspective that focuses on the assemblages of aquatic organisms inhabiting riverine environments, their adaptations to habitat conditions, and their interactions. Unidirectional flow, the primary environmental attribute of running waters, confers several advantages to organisms adapted to living in current. A comprehensive understanding of river ecology must also consider influences from the surrounding landscape and from headwater tributaries. Given the dynamic and highly interactive nature of river systems, human-induced disruption of flow and alterations in linkages between rivers and adjacent environments may profoundly affect river ecology.

I. INTRODUCTION

Ecological conditions in rivers integrate the influences of climate and geology, which determine vegetation and soil development in the catchment, all of which interact to determine channel evolu-

tion, flow and temperature regimes, sediment transport, substrate and current, and the chemical milieu of the aqueous medium. The interactions of these and other environmental factors ultimately structure the biotic communities inhabiting running waters.

The term "river" connotes large running water segments in the lower reaches of a drainage network, but such riverine environments should not be viewed in isolation. River ecology cannot be fully understood without the consideration of headwater streams or floodplain dynamics. Therefore, to achieve a comprehensive perspective of river ecology it is necessary to examine not only habitat conditions and aquatic biota within the channel, but also longitudinal resource gradients and lateral exchanges with floodplain/riparian systems and groundwater aquifers. [*See* LIMNOLOGY, INLAND AQUATIC ECOSYSTEMS.]

II. AQUATIC BIOTA

A. Habitats and Communities

The main biotic communities of running waters are shown in Fig. 1. The pleuston community consists of organisms that reside at the air–water interface. Water striders and whirligig beetles use the

---

Some of the material contained herein is extracted from the article "River Ecosystems" by J. V. Ward published in the *Encyclopedia of Earth System Science,* copyright © 1992 by Academic Press, Inc.

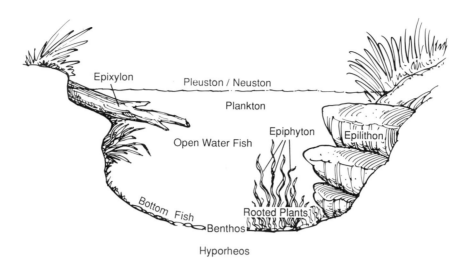

FIGURE I The main biological communities of running waters.

water surface for support as they prey on terrestrial arthropods trapped in the surface film. Immature mosquitoes hang from the underside of the air–water interface with their respiratory horns piercing the surface film. Mayflies used the surface film for support as they shed their nymphal shucks to attain the winged terrestrial stage. Free-floating plants such as duckweed and water hyacinth are also part of the pleuston community. The surface film concentrates lipoproteins and other organic molecules and is also inhabited by a microscopic assemblage of organisms designated as "neuston." Although a few surface film animals are specially adapted to rapidly moving water, the best developed pleuston and neuston communities occur in slow-moving reaches and quiet backwaters where lake and pond forms find suitable conditions.

Three primary communities—biofilm, benthos, and rooted plants—are associated with the bottom–water interface of rivers. The biofilm community includes sessile organisms attached to solid surfaces such as rocks (epilithon), submerged wood (epixylon), or the leaves of aquatic plants (epiphyton). This attached community consists of a diverse assemblage of microorganisms (bacteria, fungi, algae, protozoans) and micrometazoans such as rotifers. In addition, attached plants, such as bryophytes (liverworts and mosses), may be included. Sessile animals, such as bryozoans and sponges, are also part of the biofilm community. The biofilm

community plays an important role in the uptake of dissolved organic matter (DOM), thereby making DOM available to animals that feed on the biofilm. The photosynthetic components of the biofilm are primary producers that synthesize food from inorganic carbon while simultaneously producing oxygen. The biofilm provides habitat structure, as well as food, for a variety of small motile animals (e.g., oligochaetes, nematodes).

The rooted aquatic plant community consists of angiosperms (flowering plants) that maintain position by anchoring themselves in bottom sediment. A few species of aquatic angiosperms root in gravel, but most require sand or mud bottoms and are absent from high-gradient erosional reaches. Although waterfowl and some tropical fishes feed on the living tissues of angiosperms, few aquatic herbivores rely on higher aquatic plants as a food resource. Chemical defenses against herbivory may account for their general unpalatability. The major contributions of higher aquatic plants to the food base of running waters occur via detrital pathways following death and by increasing the substrata available for colonization of epiphytic algae while alive. [*See* AQUATIC WEEDS.]

The term benthos refers to the macroinvertebrates that burrow in bottom sediments or move upon the various substrata. Included are aquatic insects, various worms and crustaceans, water mites, clams, and snails. Some are filter feeders, using nets or hair fringes to strain particles from the current. Others feed on the biofilm or the organic matter deposited on or within the sediment; yet others are predators. Most members of the benthic community are subject to predation by other macroinvertebrates or fishes. Stream fishes are normally not considered part of the benthos, although some of them are intimately tied to benthic habitats.

A special type of benthic community, called hyporheos, penetrates deeply within the stream bed, inhabiting the interstitial spaces between sediment grains. The hyporheic zone is an ecotone or transition zone between the surface water of the channel and the groundwater aquifer. The fauna consists of two major elements: (1) members of the surface benthos that temporarily migrate some distance into the substratum, and (2) specialized groundwater forms that rarely, if ever, occur in the surficial stream bed. The hyporheic zone serves as a refuge for the surface benthos, offering shelter from floods, drought, and temperature extremes, and providing suitable and predictable conditions for immobile stages such as eggs, pupae, and diapausing larvae. The hyporheic zone affords protection from large predators and contains a faunal reserve capable of recolonizing the surface benthos should the latter be depleted by adverse conditions.

The open water column of rivers provides habitat for two remaining communities, plankton and fishes. The term plankton refers to small organisms suspended in the water column that are nonmotile or such weak swimmers as to be at the mercy of water currents, even in lakes. In fact, the plankton found in running waters originate in standing water bodies and consist of those species that can also survive and reproduce under riverine conditions. There is no evidence for a truly riverine plankton community distinct from the plankton of lakes. Bacteria, diatoms, rotifers, and the smaller planktonic crustaceans commonly occur in rivers. High-gradient headwater streams are devoid of plankton unless supplied by an upstream lake or drainage from adjacent marshes; even then, the plankton rapidly disappears downstream. Only the lower reaches of large, slowly flowing rivers develop persistent reproducing populations of plankton.

Even though freshwater bodies occupy only a tiny fraction of the earth's surface, 42% of the 20,000 or so species of fishes are found in fresh waters, and many of those occur in running waters. Many are migratory, spending only part of their lives in rivers. Not all riverine fishes are adapted to swim against the full force of the current. Some are closely associated with the benthic habitat, residing between and under rocks, for example. In tropical rivers a diverse assemblage of crevice-dwelling fishes inhabit the hollows of submerged wood. Truly pelagic (open water) fishes occur in the lower reaches of large rivers where many are planktivorous. Riverine fishes exhibit a diverse array of feeding habits. Some are strictly predaceous, herbivorous, or planktivorous, but most species are opportunistic and utilize a variety of food types that typically includes a range of invertebrates, both aquatic and terrestrial.

B. Adaptations to Running Waters

Current, by continuously replenishing needed substances (e.g., oxygen, food, nutrients) and carrying away waste products, reduces the energy expended by organisms for those functions. Current also prevents the establishment of a depletion zone around organisms, thereby increasing the "effective physiological concentration" of essential ions and enhancing their uptake from the dilute medium in which most riverine organisms are immersed. To take advantage of the benefits conferred by running waters, riverine organisms had to evolve ways to maintain their position against unidirectional water movement.

Many species seek current refugia under stones, in the dead-water zones behind boulders, within plant beds and algal mats, or in crevices within wood or bottom sediments. Invertebrates and fishes of high-gradient streams may possess morphological and behavioral adaptations that operate in concert to allow them to avoid the full force of the current. For example, small size, flattening of the body, or an elongate body shape, when associated with negative phototactic behavior, result in organisms seeking crevices where current is reduced. Such adaptations may serve more than one function in the same or different species. Flattening of the body, which serves as a current adaptation by allowing animals to seek crevices or remain within the laminar sublayer, also makes animals less available or conspicuous to predators.

Most fishes of rapid streams exhibit one of two constrasting morphologies. Strong swimmers such as salmon are circular in cross section and have a fusiform (torpedo-like) body shape to minimize frictional resistance. Bottom dwellers such as sculpin are ventrally flattened fishes that do not depend on strong swimming ability. Bottom-dwelling fishes of rapid streams exhibit numerous adaptations in addition to flattening. The pectoral fins, for example, are shaped as reverse hydrofoils; when the leading edge faces the current the forces generated press these large fins against the bottom. The swim bladders of bottom dwellers tend to be reduced, which reduces buoyancy.

In naturally turbid rivers, cutaneous sense organisms of the lateral line system may be especially well developed in bottom fishes and some species also have barbels to detect prey. Electric fishes (mormyroids of Africa, gymnotoids of South America) use distortions in their electric fields to navigate and to detect food items. The signals emitted are species specific and are also used in schooling and courtship.

Many running water organisms possess mechanisms to cling to the substream. Algae may attach themselves to solid surfaces with gelatinous secretions or stalks. Filamentous algae have basal cells modified as hold-fasts that remain attached and viable through the winter when the filaments die back. Mosses use rhizoids for attachment and the special attached angiosperms found on tropical waterfalls have modified roots that serve as hold-fasts. Stream animals also exhibit a variety of clinging mechanisms, including hooks, suckers, friction pads, and silky secretions. Black fly larvae spin a safety line of silk that allows them to regain their position, should they be accidently dislodged or forced to release their grip to avoid a predator. Some fishes of torrential streams, especially in the tropics, have modified mouth structures or fins that form suction discs.

Unidirectional flow may be exploited by riverine organisms of low mobility. Invertebrate "drift,"

the downstream transport of benthic animals by current, is a universal phenomenon in running waters. Some river ecologists view drift as an active (volitional) process of habitat search/dispersal instead of merely passive (accidental) entrainment of the benthos by current. The majority of benthic invertebrates exhibit nocturnal drift maxima which, according to the "risk or predation" hypothesis, is an adaptive response to the presence of drift-feeding fishes. Such fishes are size selective visual predators, which would account for ontogenetic shifts toward nocturnal drifting as nymphs mature and become more vulnerable to predation. In historically fishless streams, mayfly nymphs of all sizes may exhibit aperiodic drift patterns, thereby providing additional support for the risk of predation hypothesis.

Drift compensation mechanisms involve upstream-directed oviposition flights by aquatic insects as well as upstream migration within the water. Downstream drift, combined with mechanisms to recolonize upper reaches, enables benthic animals to exploit different habitats in different life stages, avoid desiccation or other adverse conditions, and optimize resource utilization.

III. INTERACTIVE PATHWAYS

To understand the interactive nature of river ecosystems, it is helpful to use the four-dimensional framework portrayed in Fig. 2. This figure empha-

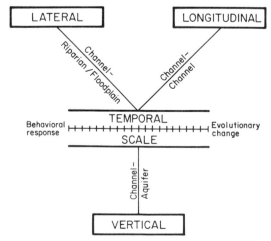

FIGURE 2 Conceptualization of the four-dimensional nature of river ecosystems. [From J. V. Ward (1989). *J. North Am. Benthol. Soc.* **8**, 2–8.]

sizes that rivers integrate interactions along three spatial dimensions, each of which operates over a range of time scales.

A. Upstream–Downstream Patterns and Processes

Profound changes occur along the longitudinal dimension extending from the headwaters to the river's mouth. The headwaters are typically cool and are continuously saturated with oxygen; high gradients and heterogeneous substrata dominated by gravel, cobble, and boulders result in turbulent currents. The term "rhithral" is used to describe upland streams with the just-mentioned combination of characteristics. It is in rhithral habitats that running water organisms attain their most distinctive character. In the cool, permanently well-oxygenated environment there has been little selective pressure to evolve adaptations to cope with low oxygen conditions. Consequently, many animals have high respiratory rates with little or no ability to regulate oxygen consumption or to tolerate even temporary oxygen deficits. Some species are so dependent on current that they succumb when placed in still water, even if the water is fully saturated with oxygen. Such organisms are unable to tolerate conditions in the lower reaches of rivers, the "potamal."

In the potamal environment, water temperatures exceed 20°C for extended periods, dissolved oxygen deficits occur at times, turbidity is higher, turbulence and gradient are lower, and the substratum is finer and more homogeneous than in the rhithral. Organisms common to the potamal do not necessarily exhibit special adaptations to running waters; many of the plants and animals also occur in the lakes and ponds of the region. Plankton is present, but consists merely of the lake species that can tolerate riverine conditions and the same is generally true for rooted aquatic plants. The fish fauna of floodplain rivers, discussed in the next section, best exemplify special adaptations to potamal conditions.

The River Continuum Concept (RCC), developed by stream ecologists in North America, views longitudinal changes not as zones, but as resource gradients along which the biota are predictably structured. According to the original RCC model, headwaters are shaded by a canopy of terrestrial vegetation. Little aquatic plant production can occur in such light-limited headwaters. The vast majority of aquatic production is based on terrestrial leaf litter. When terrestrial leaves enter the stream they are colonized by bacteria and fungi. The microbial community makes the leaves more palatable to "shredders," large-particle detritivores. Shredders convert the leaf litter into finer organic particles that are fed upon by the "collectors," or fine-particle detritivores. Some predators are also present, but shredders and collectors dominate the benthic fauna in forested headwaters. Animals that feed on biofilm, the "scrapers," are rare because low light and nutrient levels limit the production of attached algae.

As tributaries enter, the stream becomes larger and eventually in the mid-reaches becomes wide enough that the canopy opens. The combination of light, nutrients, and patches of fine substratum in the mid-reaches provides suitable conditions for the development of rooted aquatic plants. In addition, the surfaces of rocks, wood, and the higher plants are colonized by biofilm, especially attached algae. No longer is terrestrial leaf litter the main energy resource. Shredders decline in abundance and scrapers become prominent. Fine organic particles, some of which are imported from the headwaters, remain abundant in the mid-reaches and so do the collectors which feed upon them.

In the lower reaches of rivers, fine detritus and plankton are the primary organic resources. Light attenuation, because of greater depth and turbidity, and the paucity of stable substrata limit the production of rooted aquatic plants and attached algae. Collectors, therefore, dominate the fauna of the main channel.

Is the RCC universally applicable to undisturbed deciduous forest rivers as described earlier? Can the predictions of the RCC be modified to fit rivers of different types? Such questions cannot be answered with certainty at this time. One problem is that few rivers remain undisturbed. The serial discontinuity concept uses the RCC as a basis for predicting the effects of disrupting the longitudinal resource gradient by dams. For example, impounding the headwaters has little or no influence on the already high water clarity, whereas a dam placed on a turbid river increases water clarity resulting in major changes in ecological conditions downstream.

Many riverine fishes undertake extensive migrations within river systems (potamodromy), often moving hundreds of kilometers to spawn in small tributary streams. Other fishes, such as salmon, mature in the ocean, but spawn in the headwater stream where they were born (anadromy). Longitudinal migrations serve to disperse fish stocks, expose eggs and larvae to lower predation pressures, partition the habitat between species and life stages, and lead to a more complete use of food resources. The headwater streams where anadromous species spawn tend to be very low in nutrients and are of generally low fertility, yet salmon production from such streams may be high. Using stable isotopes as tracers, researchers have shown that nearly all of the biomass of returning Pacific salmon was elaborated in the sea. Migrating salmon, therefore, transport significant amounts of nutrients and organic matter to the headwaters; following spawning their decaying carcasses initiate a food chain leading to young salmon of the next generation. Because migrations play such an important role, dams or other interruptions along the longitudinal dimension may have major effects on the ecology and productivity of river ecosystems.

River regulation by dams may also disrupt interactions between rivers and their estuaries. For example, the eutrophication potential of fertile estuaries may not be realized because of light limitation resulting from the highly turbid water supplied by the river. The clarifying effect of an impoundment on a river's lower reaches may, therefore, tip the scales and result in eutrophication problems in the estuary. The reproduction of many plants on the estuary and delta relies on the natural cycle of flooding and exposure which is suppressed and altered in time by river regulation. Human alteration of these natural cycles threatens the important role of estuaries as nursery grounds for marine fishes and invertebrates.

B. The Land–Water Interface

The lateral dimension includes interactions between the channel and adjacent riparian and flood-plain systems. In addition to supplying leaf litter, riparian vegetation may also influence light and temperature regimes, control bank erosion and sediment routing, and provide cover for fishes. Even in substantially altered drainage basins, some protection of the integrity of land–water interactions may be achieved by retaining "buffer strips," stream-side corridors of riparian vegetation from which activities such as grazing and logging are excluded.

Interactions along the lateral dimension are perhaps best exemplified by "flood rivers," riverine reaches that have extensive fringing floodplains and predictable annual flooding. At the beginning of the flood season, rising water in the river channel overspills the banks and inundates the floodplain. At the height of the flood, the entire floodplain surface is under water and the ponds, lakes, and creeks on the floodplain are no longer isolated water bodies. The floodplain is rapidly transformed into a highly productive aquatic habitat, fertilized by riverborne nutrients and decaying organic matter elaborated during the dry phase. Aquatic plants and their attached epiphytes grow rapidly. Plankton, initially diluted by the rising water, develop dense populations. The inundated floodplain forest provides additional food for aquatic animals as fruits, plant litter, and terrestrial arthropods fall into the water. Fishes from floodplain water bodies and fishes from the river channel move onto the floodplain surface. "Flood-dependent" fishes rely on lateral migrations between the river channel and the floodplain to meet their ecological requirements. Such species use the channel as a dry season refuge and as a longitudinal migration route. The flood season is the major growth period and spawning is highly synchronized with annual inundation of the floodplain, assuring that adequate food and shelter are available for young fishes. As flood waters subside, adult and juvenile fishes move to the river channel. Those that wait too long are trapped in floodplain depressions that will dry or become deoxygenated. Additional losses occur from predators that congregate in the mouths of drainage channels.

In much of the world, floodplains have been drained for agricultural development, rivers have been channelized, artificial levees have been constructed to prevent river overspill, and the natural flood regime has been altered by flow regulation. These attempts to isolate rivers from their floodplains result in major disruptions of the lateral transfers of energy and matter upon which the ecological integrity of the system depends and interfere with the lateral (and longitudinal) migrations of fishes. Intact riverine–floodplains systems are in fact the best flood control measures, presuming that one recognizes the floodplain as an integral component of river systems.

C. The Surface–Subsurface Interface

Just as floodplains are extensions of river ecosystems, so are groundwater aquifers. The strength of channel–aquifer interactions varies along river courses, being weak in canyon segments and strong in gravel bed rivers with well-developed alluvial floodplains. The groundwater within porous alluvial gravel deposits is hydraulically connected to the water in the river; water levels recorded in wells several kilometers from the river may closely track water level changes in the channel. In alluvial rivers of low fertility, within-channel production may be substantially enhanced during baseflow when groundwater containing higher nutrient levels enters the river. Conversely, excess nutrients in groundwater from agricultural land are substantially reduced before entering the river during subsurface passage beneath the riparian zone.

Many members of the surface benthos, as mentioned earlier, penetrate a short distance into the hyporheic zone. A few riverine species undergo extensive subterranean migrations as part of their normal life histories. Stoneflies of several species spend their entire nymphal lives deep within the alluvial gravels of a Rocky Mountain river, moving up to 2 km away from the river channel to complete their growth as part of the subterranean food web. Eventually, they must return to the river, emerge as winged adults, mate along the shoreline, and deposit eggs in the river. How they find their way back to the river and how

such subterranean migrations evolved remain a mystery.

IV. BIODIVERSITY

Natural river ecosystems are characterized by a high level of heterogeneity that is manifest across a range of spatial and temporal scales. High biodiversity is attributable, in large part, to the high habitat diversity and dynamic nature of rivers. This is especially true if the floodplain is considered an integral component of the riverine environment.

The three spatial dimensions in Fig. 2 each form environmental gradients along which species assemblages change, thereby increasing biodiversity. For example, dramatic species replacements occur from the headwaters to the lower reaches of river systems.

As rivers migrate across their alluvial plains they maintain high levels of habitat diversity by forming new habitats and altering extant habitats, resulting in a variety of aquatic and riparian environments in different stages of succession. Many riverine animals require access to a variety of habitat types. For instance, spawning may occur in the headwaters, feeding may require access to inundated floodplains, juveniles may occupy different floodplain habitats than adults, and the winter or the dry phase may be spent within the river channel.

Riverine organisms face a multitude of threats to biodiversity. Superimposed on relatively well-known problems such as pollution, overharvest, and introductions of exotic species are more subtle threats associated with human-induced reductions in the natural dynamics and interactive nature of the riverine environment. For example, river training structures constrain channel migration, which truncates succession; dams block longitudinal migration pathways; and dredging, flow regulation, and levee construction isolate the river from its flood plain.

According to this perspective, the ecological integrity of river systems depends on sustaining, rather than suppressing, environmental heterogeneity. To be successful, river restoration efforts must reestablish interactive pathways and reconstitute a semblance of the natural dynamics.

Glossary

Alluvium Sediment deposited by running waters.
Aquifer Alluvium or porous rock that is saturated with water.
Benthos Aquatic animals associated with bottom habitats.
Biotic community Collectively, all of the species populations inhabiting a given environment.
Drift Downstream transport of benthic organisms by current.
Flood plain Flat valley of alluvium formed by riverine erosional/depositional processes. Inundated by flood waters about once per year, on the average, in free-flowing river systems.
Hyporheos Animal community living in the interstitial spaces between sediment particles in river beds.
Laminar sublayer Microzone of nonturbulent water immediately above the river bed.
Plankton Small organisms suspended in open water with little or no directional swimming ability.
Pleuston/neuston Macroscopic/microscopic organisms associated with the surface film.
Riparian vegetation Corridor of shoreline vegetation along water courses, including floodplain plant communities.

Succession Gradual changes in biotic community structure following a disturbance.

Bibliography

Allan, J. D., and Flecker, A. S. (1993). Biodiversity conservation in running waters. *BioScience* **43,** 32–43.

Boon, P. J., Calow, P., and Petts, G. E., eds. (1992). "River Conservation and Management." Chichester: Wiley.

Calow, P., and Petts, G. E., eds. (1992). "The Rivers Handbook: Hydrological and Ecological Principles." Oxford: Blackwell.

Cushing, C. E., Cummins, K. W., and Minshall, G. W., eds. (1994). "River and Stream Ecosystems." Amsterdam: Elsevier (in press).

Flecker, A. S. (1992). Fish predation and the evolution of invertebrate drift periodicity: Evidence from Neotropical streams. *Ecology* **73,** 438–448.

Gregory, S. V., Swanson, F. J., McKee, W. A., and Cummins, K. W. (1991). An ecosystem perspective of riparian zones. *BioScience* **41,** 540–551.

Schlosser, I. J. (1991). Stream fish ecology: A landscape perspective. *BioScience* **41,** 704–712.

Stanford, J. A., and Ward, J. V. (1993). An ecosystem perspective of alluvial rivers: Connectivity and the hyporheic corridor. *J. North Am. Bentholog. Soc.* **12,** 48–60.

Ward, J. V. (1994). The structure and dynamics of lotic ecosystems. In "Limnology Now: A Paradigm of Planetary Problems" (R. Margalef, ed.), pp. 195–218. Amsterdam: Elsevier.

Ward, J. V. (1994). Ecology of alpine streams. *Freshwat. Bio.* **32,** 277–294.

Sahara

J. W. Lloyd
University of Birmingham, United Kingdom

I. Introduction
II. Climate and Surface Water
III. Groundwater
IV. Biology
V. Conclusion

The Sahara is the principal desert area of Africa, occurring approximately between latitudes 17° and 30°N and longitudes 10° and 33°E. The precipitation across the area is very sparse and very irregular in time, space, and amount. For convenience the zone within the annual average precipitation 20-mm isohyte is considered as the Sahara, although this isohyte is very poorly defined and the concept of annual precipitation can be misleading for deserts. Moisture within the Sahara is provided by the very intermittent precipitation, dew, and localized groundwater discharge. Apart from the latter, the moisture amounts are extremely small, but together with groundwater contributions, a surprisingly diverse flora and fauna are present locally.

I. INTRODUCTION

Together with parts of the Gobi and Atacama deserts, the Sahara is one of the driest areas in the world. Although a well-known geographical area, it is not particularly well defined, in part because of the paucity of precipitation data and because the desert is a dynamic entity that is currently expanding.

For the purpose of this discussion the Sahara is defined as that area occurring within the 20-mm isohyte in Fig. 1, also embracing the very localized higher rainfall mountainous zones that occur within that isohyte. The Nile valley is excluded as it is an anomalous imported water source, representative of the Sahal.

II. CLIMATE AND SURFACE WATER

The presence of a vast arid area (approximately 2.5×10^6 km^2) such as the Sahara is clearly not the result of purely local climatic controls. Dessication commenced in the area in the late Cainozoic associated with the general northerly movement of the African plate. This movement, together with the development of the easterly jet stream, a steep temperature gradient between the North Pole and the equator, and a cooling of the ocean surface, established aridifying conditions, although a pattern of climatic ossilations ensured. The periodicity of the oscillation is uncertain; however, a long-term progressive aridity has been maintained with interspersed short, slightly wetter periods. The last onset of aridity occurred about 20–30000 years B.P. (before present) with slightly wetter periods recorded at about 12–10000, 5–6000, and 2–3000

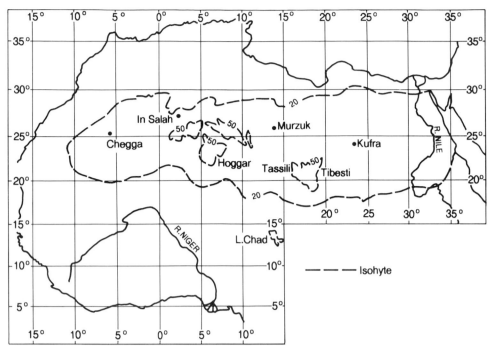

FIGURE I Location of the Sahara desert defined as occurring within the <20-mm isohyte.

years B.P. Savannah conditions occurred in the central Sahara during the latest wet period with elephant remains recorded in the Tebesti (see Fig. 1), showing that even quite recently a diverse and prolific biological environment existed. Aspects of the diversity of this environment are shown particularly by the animals featured in Saharan Neolithic rock paintings. [See DESERTIFICATION, CAUSES AND PROCESSES.]

The Sahara is the most continental of the hot deserts and is removed from major oceanic influences. Currently the desert lies below the northern descending limb of the north equatorial Hadley Cell which continually feeds low moisture air into the north-easterly trade winds moving toward the equator. The winds are unable to acquire moisture over the North African landmass, aridity is therefore accentuated, and a climate is created of extremely low precipitation, punctuated infrequently by localized high intensity storms. [See DESERTS.]

Elevation affects rainfall amounts when storms do occur so that centrally in the Sahara, where mountains are present, isolated areas experience precipitation above the norm for the general desert area of less than 20 mm per annum.

A whole year's precipitation may occur in 1 day, resulting in high rainfall intensity that can occasionally cause sheet or wadi runoff. This usually rapidly evaporates or recharges but does provide a desert moisture source, albeit intermittent and of extremely variable distribution.

Coupled with the low rainfall are very high radiation receipts and very high temperatures. Although temperatures are high, diurnal temperature changes can be large, up to 20°C, and despite the generally limited humidity in the near ground surface air, can cause the deposition of dew. This forms briefly at about dawn and rapidly evaporates. Although a small moisture feature of the desert, it is nevertheless locally of biological importance.

Clearly, the current atmospheric conditions do provide some moisture input of biological importance in the form of precipitation, runoff, or dew; however, this input does little to promote extensive thriving floral and faunal populations.

III. GROUNDWATER

Geologically the Sahara is underlain by large regional basins that are comparatively undisturbed structurally and contain thick strata of sandstones and limestones which form huge regional aquifers. These aquifers were recharged substantially prior to the last main onset of aridity, principally along the margins of the central mountainous spine running from the Hoggar to Tibesti, although some recharge undoubtedly occurred in the more recent wetter periods. Because of the very long groundwater flow paths involved and other hydrogeological factors, vast amounts of good quality groundwater still exist in the aquifers. Estimates indicate that nearly 10% of the world's fresh groundwater lies under the Sahara.

Most of the groundwaters are very old, some dating at least about 0.5 million years. These are usually too deep to affect the biological conditions of the desert, although there are suggestions that they may contain sulfate-reducing bacteria. However, locally, particularly beneath wadi courses and at internal (within the landmass as opposed to coastal) groundwater discharge points, old groundwaters occur at very shallow depths or emerge at the surface to form oases of sebkhas (Fig. 2). In such areas a range of vegetation is present together with indigenous and migratory fauna, and it is these localized groundwater sources that provide the principal moisture input to the Sahara.

IV. BIOLOGY

A. Flora

The flora of the Sahara, in common with other desert flora, exhibits remarkable capacities to maintain effective water balances under extreme and impersistent drought conditions. Such plants are generally termed xerophytes. [*See* SEED DISPERSAL, GERMINATION AND FLOWERING STRATEGIES OF DESERT PLANTS.]

The most primitive forms of flora are lichens and algae that occur extensively and depend principally upon rainfall and dew when other moisture sources are not available. Such flora can withstand dessication for long periods by remaining dormant and becoming metabolically active when water becomes available.

Ephemeral plants that have a life cycle restricted only to wet periods also typify the Sahara. These normally have extensive shallow-depth roots and prolific seed production. The root structure results in widely spaced individual plant growth with marked dying off and dwarfing during desiccating periods. Quézel (1965) notes therophyte associations of such plants comprising *Leysera leyseroides*—*Trigonella anguina* and *Lotus glinoides*—*Matthiola livida* in the Hoggar and *Astragalus eremophilus* and *Bidens pilosa* in Tibesti.

In addition to lichen and algae, perennial plants are present, utilizing moisture retained in succulent leaves or stems. Many such plants are closely associated with sebkha areas and attribute their succulence to internal salt concentration. Succulent species are *Euphorbia* and *Caralluma* while examples of plants with succulent leaves or stems are *Nitraria, Haloxylon, Anabasis,* and *Salsola.*

The xerophytes mentioned earlier are small plant forms that can occur throughout the Sahara in any advantageous microtopographic setting conducive

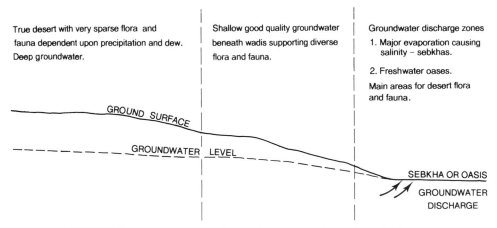

FIGURE 2   General concept of groundwater control of Saharan biology.

to moisture accural. Regionally, such moisture is from the atmospheric source. However, where this moisture is supplemented by shallow groundwater within the reach of desert plant roots or where groundwater is the dominant source of moisture, different floral species predominate. While moisture availability is the controlling feature in these areas the floral speciation is also linked to the soil types present, which in turn are associated with past and present hydrological factors.

In areas close to incised wadi beds (see Fig. 2) where depths to groundwater become shallow, the presence of a higher overall moisture content in the ground is identified by increases in the density of small plants such as *Panicum turgidum*, *Pergularia tomentosa*, and *Retema raetam*. As moisture availability increases, *Cornulaca monacantha*, *Atriplex halimus*, *Zygophyllum album*, and *Traganuum mudatum* appear. The latter two species are usually associated with some soil salinity.

Within the sandy gravelly wadi bed spreads the increased moisture levels support progressively more and larger plants. *Calligonum* sp. occur, although these may also characterize some sand dune areas. Arboreous vegetation occurs typically with *Acacia* spp. (*raddian*) while representative herbs and shrubs are *Panicum turgidum*, *Pennisetum dichotomum*, *Pergularia tomentosa*, *Pituranthus* sp., and others.

Where groundwater is very shallow or emerges at the surface (see Fig. 2), vegetation species are closely associated with water quality. In the shallow groundwater, salinity concentrations may be higher than at depth, but these waters are of good quality compared to those that are actively evaporated at emergence on sebkha surfaces.

Adjacent to sebkhas in the better water quality areas, *Tamarix* sp. and *Phoenix dactylifera*, with occasional other palms, are common. As a result of the presence of moisture, coupled with wind deflation and plant growth, soil stabilization in these areas gives rise to phytogenic hillocks often covered by *Tamarix* sp. and *Nitaria retusa*. Many species, however, can contribute to hillocks, including geophytes with rhizomes, tussock-forming hemicryptophytes, and various chamaephytes.

Tamarix sp. are also present on the highly saline sebkhas and can be associated with *Aeluropus lago-poides*, *Phragmites communis*, *Juncus maritimus*, and others. Locally, where the upward emergence of groundwater flow exceeds evaporation, isolated saline lakes can occur that support such hydrophytic vegetation as *Cyperus laevigatus*, *Typha australis*, *Phragmites* sp., etc.

In parts of the Sahara, notably the western desert of Egypt, oases form lakes with reasonably good quality water. These are normally surrounded by date palm groves which may form a monospecific plant community. *Phoenix dactylifera* is a common species. Such groves, however, do merge into mixed communities to include plants such as *Imperata cilyndrica*, *Zygophyllum album*, and many others. In many oases, irrigation is practiced so that cereals, alfalfa, and some fruit trees occur and couch grass (*Cynodon dactylon*) is often introduced.

B. Fauna

As with any other environment the faunal habitat is inextricably linked with the floral cover. Similar to the vegetation the faunal populations are more prevalent where the highest moisture levels occur in the desert, namely groundwater discharge zones.

I. Nonmammals

Of the most primitive life forms, some species of Protozoa are recorded surviving on dew and precipitation in favored localities, particularly in the mountainous zones. They are encysted except when there is moisture.

Various forms of arthropods are thought to be relatively common, but data are sparse (Lewis, 1984). Crustacea (principally woodlice) such as *Porcellio* sp. are reported together with *Agabiformius obtusus* and *Hemilepistus reaumuri*. *Porcellio* sp. occur mainly in the oases and wetter areas, although *P. simulator* is recorded in the Hoggar. *H. reaumuri* is a species very resistant to aridity and is commonly seen on the desert surface during the cooler periods of the day. Millipedes such as *Polyxenus lagurus* also occur.

Insects in the Sahara are manifold, varying considerably in population density from low in dry periods to very high in wetter periods. As with some of the other anthropoda, their presence is most discernable in the cooler dark hours when humidity is relatively higher. Many of the species have adapted to live at a shallow depth below the ground surface and have developed a very low permeability body cuticle cover that minimizes transpiration rates.

Lepismatidae occur commonly, as do cockroaches, ants, and termites. Termites are often associated with *Tamarix*. As with most other parts of the world, flies (*Musca domestic*) seem to appear from nowhere while sand flies (Psychodidae) and crane flies (Tipulidae) also feature locally. Mosquitos are usually present in sabkha and oasis areas in the winter. Beetles of the Tenebrionidae and Scarabaeidae families are present most prevalently close to oasis areas, where goats and camals provide dung for their nourishment, although they are not totally dependent upon dung.

Perhaps the most universally known of the North African insects is the desert locust (*Schistocerca gregaria*) and the African migratory locust (*Locusta migratoria*). However, the eggs of the locust require a higher humidity than normally found extensively in the Sahara so the pest does not widely inhabit the true desert, but ravages the Sahel margins.

The arachnid fauna comprises scorpions and various species of spider, although the latter are not numerous. Of the scorpions, *Leiurus quinquestriatus* and *Buthus occitanus* are among the most common species with *Androctonus australis* and *Scorpio maurus* also present. The spiders normally inhabit wet areas but the species are not well documented away from these areas. *Heteropoda walckenaeri* and *Xysticus* sp. and *Plesiippus* sp. are recorded in the desert.

While frogs, such as *Rana occipitalis* (green frog), are amphibians found in the wetter mountainous areas, the most visible of the vertebrates that occur are reptiles, notably lizards and snakes. Examples include *Stenodactylus* sp., *Acanthodactylus* sp., *Agama* sp., and the monitor *Varanus griseus griseus*, which is seen in the Hoggar and elsewhere. *Psammophis* spp. are probably the most common of the snakes. Lizard and snake numbers are the most numerous where the insects that comprise their main food source are frequent.

In some parts of the Sahara, even away from oases or the wetter areas, early mornings are characterized by the presence of larks (e.g., *Ammonanes* sp.). Closer to moisture sources and vegetation cover indigenous warblers, finches, and sand grouse are found. In the mountainous areas a large range of breeding birds is present, including various predatory species, while in the oases the desert breeding species are augmented by doves, sparrows, ravens, etc. By far the greatest number of birds seen in the Sahara are migratory species that traverse the desert southwards during August–October and northwards during March–May. These are added to by birds that may move periodically into the desert from the margins during wetter periods.

2. Mammals

Rodents are by far the most successful of the small mammals, represented by gerbils and jirds. Surprisingly, hedgehogs occur but shrews are rare, as are bats. Of the Carnivores, three species of fox are recorded, and although a relatively large range of small mammal species is present the harshness of the environment does not permit a plentiful food supply so that populations are small. Mobility is a feature of many of the species, which is a function of the low density populations, but also ensures their survival.

The situation with large mammals is unclear. Since the advent of high-powered rifles most of the large mammal population has been killed and it is uncertain as to what actually existed during this century. Antelope (*Addax nasomaculatus*) and gazelle (e.g., *Gazella dorcas*) can still be seen in the desert. Newby (1984) and others record aoudad (*Capra lervia*) and cheetah (*Acinonyx jubatus*) in the mountainous areas, but it is uncertain how many still exist.

V. CONCLUSION

Despite an extremely harsh environment the Sahara supports a surprisingly large and diverse biology, although the current knowledge of this biology is limited. Of the three moisture sources that occur, precipitation, dew, and groundwater, it is the latter which is the dominant and focal source for the floral and faunal species.

Glossary

Wadi Arabic term for valley.

Aquifer Rock strata that can transmit groundwater sufficiently to provide an exploitable resource.

Recharge Water from the ground surface that enters the groundwater flow system; can be from precipitation or surface runoff.

Oasis Isolated "fresh" surface water body in a desert.

Sebkha Groundwater discharge area where discharging water is subject to evaporation so that both groundwater and soil are saline.

Xerophyte Flora that exist under impersistent moisture availability conditions.

Bibliography

Fauré, H. (1967). Lacs Quarternaires du Sahara. Int. Ass. Theoretical and Applied Limnology (SIL). Communications No. 17, 131–146.

Lewis, J. G. E. (1984). Woodlice and Myriapods. *In* "Sahara Desert" (J. L. Cloudsley-Thompson, ed.), pp. 115–127. Oxford: Pergamon Press.

Newby, J. E. (1984). Large Mammals. *In* "Sahara Desert" (J. L. Cloudsley-Thompson, ed.), pp. 277–290. Oxford: Pergamon Press.

Quezel, P. (1965). La Végétation du Sahara. Du Tchad à la Mauritanie. Stuttgart: Gustav Fischer Verlag.

Sahel, West Africa

Sharon E. Nicholson

Florida State University

The Sahel is a densely populated, semiarid region of West Africa bordering the Sahara desert. Its highly diverse inhabitants have a long history in the region, its fabled cities having been tremendous trade centers many centuries ago. The Sahelian cultures have recently been threatened by political upheaval and environmental calamities, such as severe and sustained drought, and the region's future is uncertain.

I. INTRODUCTION

The Sahel is a narrow, semiarid expanse of grassland, shrubs, and small, thorny trees which stretches some 5000 km across northern Africa from the Atlantic Ocean eastward nearly to the Red Sea. It includes much of the countries of Mauritania, Senegal, Mali, Niger, Chad, the Sudan and northern fringes of Nigeria and Burkina Faso. Extending from approximately 14°N to 18°N (Fig. 1), the Sahel is an ecological zone of climate and vegetation which represents a transition between the desert and the more humid savanna to the south. In fact, the term derives from an Arabic word meaning the "fringe" or "shore" of the desert. The

Sahel is historically important because centuries ago it was the site of numerous prosperous cities and empires taking part in trans-Saharan trade. Today it is the home of millions of people representing many diverse ethnic groups, cultures, and languages. Major cities in the Sahel include Dakar and St. Louis (Senegal), Niamey (Niger), Tombouctoo (Mali), Ndjamena (Chad), and Khartoum (Sudan).

The Sahel region has become well known in more modern times because of the devastating droughts which it has experienced. Conditions were catastrophic in the early 1970s, with herds decimated, millions of people displaced to refugee camps and cities, and perhaps 100,000 deaths from famine. The drought affected a much larger region, including most of the wetter savanna to the south.

A precise location of the Sahel is hard to specify because the features which distinguish it, climate and vegetation, form a continuum of rapid change from north to south, as described in Sections II and III. Also, both of these environmental characteristics change over time, even within decades. Because of this and the extension of the recent droughts to the wetter savanna further south, the term Sahel is commonly (but mistakenly) applied to the entire arid/semiarid region from the southern Sahara to

261

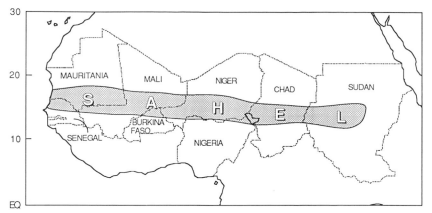

FIGURE I Location of the West African Sahel.

the forest margin at about 10°N latitude. This entire region is described in further sections because the regions are nearly inseparable geographically and culturally.

The region's geography is closely linked with its climate. Climatic constraints have been a major force in the development of vegetation, soils, lifestyles, and agriculture. Heat is ever present, and water is seasonally scarce; its availability is erratic in time and space. Over centuries life has adapted to such conditions, but in recent times politics and drought have waged war with tradition and the region is undergoing tremendous hardships fraught with continual change.

II. CLIMATE

The Sahelian climate is characterized by high temperatures throughout the year and by a brief rainy season occurring during the "high sun" or summer season of the year. The alternation between the few months of rainfall and the dryness during the rest of year is linked to the shift of the large-scale general atmospheric circulation features which control the climate and weather of West Africa. This shift is illustrated by conditions of August and January (Fig. 2).

During January the subtropical high pressure cell of the North Atlantic is displaced far enough southward that it influences much of northern Africa. The characteristics of high pressure cells, sinking and divergent air motion and temperature inver-

sions, produce marked aridity. On the equatorward side of the high are northeast trade winds, called the Harmattan over northern Africa. This dry Harmattan flow prevails in the Sahel from October to May or June. South of the equator is another quasi-permanent subtropical high in the South Atlantic, with southeasterly trades on its equatorward flank. When these winds cross the equator, they become southwesterly; they form a humid current of air often referred to as a monsoon.

During the dry months, rainfall is rare but it does occasionally occur. One example is the "heug" weather of coastal areas, such as parts of Mauritania. Troughs in the upper atmosphere may bring several days of continuous light rain in December or January. Similar situations, coupled with mid-latitude weather systems, can bring up to 25 mm to the central Sahel during almost any month. Such light and persistent rain can be highly efficient in producing soil moisture, so that even this meager rainfall can in some cases be exploited for agriculture.

During August, the general circulation features are displaced toward the northern hemisphere. The subtropical high moves north and circulation over northern Africa is dominated by a heat low over the west central Sahara and the southwesterly monsoon flow to the south of it. Thus the humid air penetrates far inland.

The Harmattan and the southwesterly monsoon meet in a region called the Intertropical Convergence Zone (ITCZ). Where airstreams converge, air is forced to rise; since the air in them is warm, humid, and unstable, this leads to cloud formation

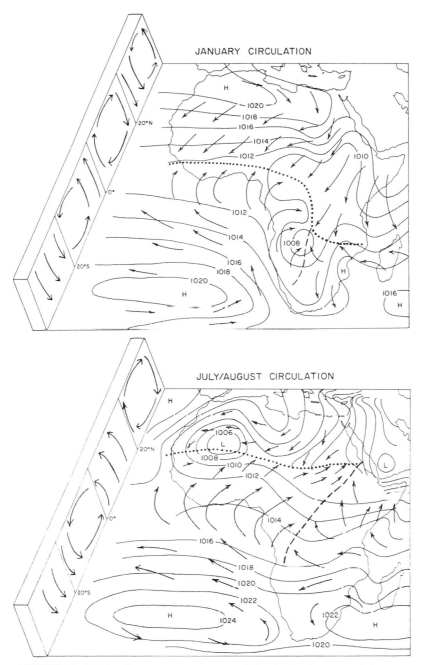

FIGURE 2 Wind and pressure patterns over Africa during January and August.

and convective activity. Thus, rainfall is linked to the ITCZ. Its seasonal migration also represents the seasonal migration of the tropical rains. It moves north and south with the sun, bringing rains to the Sahel during the "high sun" season of northern hemisphere summer.

Although the seasonal cycle of rainfall is linked to the ITCZ, most rainfall is not confined to local thunderstorms imbedded in it. Instead, there are organized disturbances, called cloud clusters and easterly waves, which traverse the Sahel from east to west. These systems, linked to a jet stream in the mid-troposphere, bring most of the rainfall.

In the Sahel and southward to the Guinea coast of the Atlantic, the amount of rainfall received during the year and the length of the rainy season are both determined by the length of time the ITCZ is the dominant circulation feature. Thus, both the amount of rainfall and the length of the season increase from north to south (Fig. 3). The region traditionally considered the Sahel receives, in the annual mean, about 100 to 200 mm where it gives way to desert in the north and its southern limit receives 500 to 600 mm annually. The length of the rainy season ranges from about 2 months in the north to about 3 months in the south. Throughout the region, the month of maximum rainfall is August.

The amount of rainfall received in much of the Sahel, 500 to 600 mm, is about the same as received in much of the Great Plains of the United States, where agriculture prospers. In the Sahel, however, the rains are considerably less effective for several reasons. One is the concentration in a relatively brief season; this severely restricts the length of the growing season. Second, most is received in intense showers, so that much of the water is lost to runoff and relatively little enters the soil and becomes available to plants. The rainfall is also erratic in both time and space. The few showers that do occur are often confined to areas on the order of tens of kilometers in size, so that in a given season the distribution of rainfall is quite patchy.

This spottiness also contributes to a large variation in rainfall from year to year, although other factors play a role. Thus, the rains are unreliable

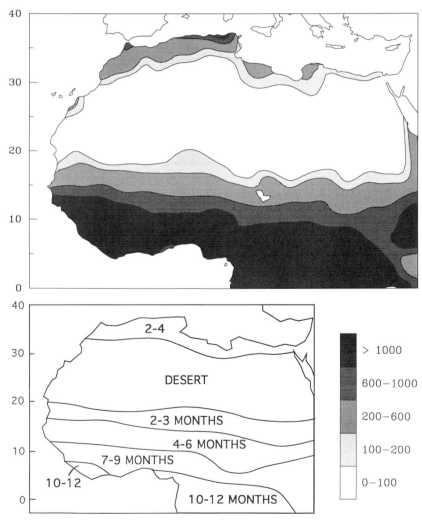

FIGURE 3 Mean annual rainfall and length of the rainy season over West Africa.

and droughts are frequent. At Niamey (Niger), where the mean annual rainfall is 559 mm, annual totals have varied between 281 and 938 mm; the variability of annual totals is 25%. At drier St. Louis (Senegal), with 352 mm in the mean, annual totals have varied between 94 and 943 mm; interannual variability is 45%. Such variable conditions pose problems for agriculture. The seasonal distribution is also unreliable, another difficulty to which agriculture must adapt. However, during most drought years, rainfall is below normal primarily during the months of July to September, even in areas south of the Sahel, where the rainy season is considerably longer.

The rainwater in the Sahel is lost not only to rapid runoff, but also to evaporation. Because of the Sahel's location in low latitudes and its relatively small cloud cover most of the year, incoming solar radiation is strong year round. It is reduced somewhat during the rainy season because of increased cloud cover. The incoming radiation largely determines the potential evapotranspiration, i.e., the amount of water which could be evaporated if the land surface were continually wet. This quantity determines "water need" or "water demand" and, together with precipitation, determines the overall arid or humid character of a climate.

Potential evapotranspiration along the northern and southern fringes of the Sahel is given in Fig. 4, together with monthly rainfall. It is about

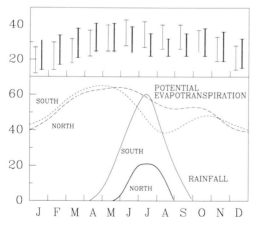

FIGURE 4 Typical values of potential evapotranspiration and rainfall January to December along the northern and southern margins of the Sahel.

2000 mm per year or greater in the north and on the order of 1700 mm per year in the south. Rainfall exceeds potential evapotranspiration usually at most 1 to 2 months of the year in the southern Sahel and not at all in the north. Thus, the Sahel has a prevailing semiarid character, with little water available for vegetation or crop growth.

Because of high insolation, temperatures are high year round in the Sahel. The warmest month varies within the region, ranging from April to July; it is always prior to the onset of the high-sun rainy season. During this month, daily mean temperatures average about 32 to 35°C throughout the zone. In January, the coldest month nearly everywhere, temperatures average about 20°C along the northern margin of the Sahel and 22 to 25°C in the south. Thus, the annual temperature range is only about 10 to 15°C.

The temperature contrast between day and night is equally extreme. The mean daily maximum is about 27 or 28°C in January and 40 to 42 °C in the warmest month. The mean nightly minimum is about 14 to 17°C and 27 to 29°C during these same months. The dryness of the soil, relative lack of cloud cover, and sparseness of vegetation lead to these daily extremes. During the rainy season daily mean temperatures are about 5°C cooler but daily maximum temperatures average about 10° cooler. Temperatures are also considerably lower near the coast, and the diurnal and annual range more moderate.

To enhance human comfort, the population deals with these extremes in many ways. One is the open flowing style of the clothing and, particularly along the desert fringe, its reflective white color. Another is the architecture. The thick walls of the adobe dwellings, common throughout the region, keep out the heat by day but retain it at night, minimizing the diurnal fluctuations and maximizing the time period of bearable temperatures. Activity is reduced to a minimum during the hot afternoon hours.

This does little to protect the population from other climatic hardships: dust storms, floods, and droughts. The dust is churned up by the vigorous harmattan winds at the end of the wet season, reaching up as high as 5 km. It is often visible as a reddish cast to the cumulus clouds occurring early in the rainy season. The dust layer is most intense in the mid-troposphere, where it is transported westward by the African easterly jet stream. Dust from the Sahel is carried as far as the southeastern United States, the Caribbean, and South America. Dust is scavenged from the atmosphere by intense storms early in or just before the rainy season. The sudden onset of a storm, marked by strong gusty winds, can transform bright sunlight to an eerie, dusk-like red haze within minutes.

The floods the region experiences are promoted by the intensity of the rainfall, the impenetrability of much of the soil (see Section III), and the lack of permanent drainage channels in much of the region. Most of the season's rainfall is concentrated in a few intense storms, which often bring as much as 100 to 200 mm a day to the region. A few tens of millimeters suffices to produce intense streams of runoff and flash floods. These commonly destroy houses, bridges, and roadways.

The other extreme, drought, is just as common, but droughts last longer and are generally more devastating. The most extreme recent drought years in the Sahel have been 1972 and 1973 and 1982 to 1984; however, rainfall has been below the long-term mean every year since 1970 (Fig. 5). This remarkable period of aridity is in stark contrast to the period of favorable rainfall which prevailed throughout the 1950s and ended abruptly in 1960. Rainfall in the Sahel has decreased by a factor of two since the 1950s (Table I). These are not localized drought episodes, but instead tend to affect the entire Sahelian zone, as well as the areas well to the north and south of it.

The characteristics of such dry episodes, their extreme magnitude, duration and spatial extent,

TABLE I
Mean Annual Rainfall (mm)

	1950–59	1970–84
Bilma	20	9
Atbara	92	54
Nouakchott	172	51
Khartoum	178	116
Agadez	210	97
Timbuktoo	241	147
Nema	381	210
Dakar	609	308
Banjul	1409	791

and their abrupt onset, all exacerbate the problem of living with drought. Such droughts have plagued the region for centuries. A similar drought of nearly two decades occurred during the 1820s and 1830s. Another decimated the Sahelian population during the 1730s to 1750s.

Overall the region has experienced many longer-term fluctuations of rainfall, in response to which the prosperity of the region has waxed and waned (Fig. 6). One humid episode occurred around the 8th through 13th centuries, a period when Sahelian empires thrived (see Section V). In the 16th through 18th centuries, the climate was generally wetter and the environment was more humid than today. Perhaps coincidentally or as a result, these were periods of renewed prosperity of the large Sahelian states. The decline of these wetter conditions occurred around 1800, but for a few decades at the end of the 19th century a more favorable climate again prevailed and the Sahel flourished. The lakes expanded; the annual river floods, which provide water for agriculture, were continually good; and the harvests were bountiful. Conditions were so good in the Niger bend that Tombouctoo became the "bread basket" of West Africa, exporting wheat to surrounding regions. These conditions quickly gave way to the drier ones of the 20th century toward 1900.

III. REGIONAL GEOGRAPHY

A. Land

The terrain of the Sahel is relatively flat; the only large topographic feature is the Jebel Marra in the

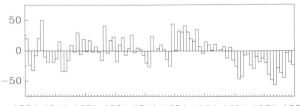

FIGURE 5 Annual rainfall in the Sahelian zone from 1901 to 1990, expressed as a percentage of departure from the long-term mean.

FIGURE 6 The Dogon people are recognized as some of Africa's foremost traditional artists. The doors on many of the local houses and graineries are intricately carved pieces of wood.

western portion of the Sudan Republic. In most of the region, relief is slight in both height and extent. Stabilized sand dunes less than 20 m in height are found in the north. There are also banks emerging from flood plains and a system of low plateaux in the central Sahel. In a few areas there are remnants of old mountains, usually not exceeding a few hundred meters in height. Isolated inselbergs (small, mountain-like rocky massifs) are common.

B. Water Resources

Although the region is semiarid and drought-prone, the area's rivers provide some environmental security (Fig. 7). The two major ones, the Senegal and the Niger, originate in more humid regions to the south and provide perennial flow and expansive floods after the rainy season. The flood waters provide areas of cultivation and pasture. The area's largest lake, Lake Chad, is also fed by water from further south, via the Logone and Chari rivers. During most of this century, the lake spread out over nearly 25,000 km^2, but it has been nearly desiccated since the early 1980s as a result of continuous drought. Other lakes in the region, such as Faguibine in Mali, also dried out because of the drought. The rivers, their volume of flow and extent of floods, also change from year to year with the region's rainfall. The land covered by the Senegal's flood, for example, is about twice as great during wet years as during dry years.

C. Soils

The soils of the Sahel are products of its recent geologic past, which includes both periods of extreme aridity and more humid conditions with more extensive rivers and lakes within the last 20,000 years. The soils can be divided into primarily three types. One is the sandy aeolian soils, which are deep and often uniform over vast expanses of the northern Sahel; they are remnants of arid periods. These soils are the most widespread, covering about 50% of the region, but are generally confined to areas where mean annual rainfall does not exceed about 500 mm. There are also iron-rich soils, many with bare stretches of hard, rocky and inpervious sandstone or laterite (hardpan) crusts (Fig. 8). These usually have a loamy texture, are quite heterogeneous, and are prone to intense runoff. Water collects on these soils in spotty, temporary pools or on flood plains. In some areas there are also fluvial or lacustrine soils, produced from sediments of former lakes and streams or formed more recently in the inland Niger Delta region. These are often sandy on the surface with clay layers below, so drainage can be poor. The fossil soils tend to be impoverished, but the recent fluvial and lacustrine soils are generally quite fertile. [*See* SOIL ECO-SYSTEMS.]

D. Vegetation

The vegetation of West Africa is determined by a combination of factors related to both climate and soil, essentially exhibiting the same longitudinal zonation as rainfall (Fig. 9) and an equally rapid

FIGURE 7 The Sahel's rivers have helped the regions agriculture and commerce to prosper by providing fish, transportation and, in many regions, natural irrigation systems. Shown above is the confluence of two major rivers, the Niger and the Bani, near Mopti (Mali). The contrast in the water indicates the confluence.

north–south transition among desert, savanna, and forest. The structure of the vegetation also changes progressively from north to south. The further south, the taller the vegetation, the greater the proportion of woody species (trees, shrubs, bushes), and the higher amount of ground cover.

Further north along the desert's edge is open grass vegetation, mixed in some areas with occassional woody species of small trees or shrubs. Grasses are perennials, generally not taller than 80cm, and annuals; woody species are often thorny, like the typical Acacia. This is the Sahel proper. Here the surface is to a large extent bare soil and the vegetation tends to be clustered in sites of favorable conditions of soil or runoff, creating a mosaic pattern.

This pattern is carried to extremes in one widespread vegetation formation, which is called "brousse tigrée" (tiger bush) because, viewed from above, it resembles stripes on a tiger (Fig. 10). In the tiger bush, vegetation grows in dense stands, perhaps 200 m in length and 20 m wide, separated by bare soil; the bare patches are about twice the size of the green clusters. The causes of this landscape are not really known, although several theories exist based on features of the soil or surface relief.

South of the Sahel is a region referred to as the Sudan (or Soudan), a term derived from Arabic and meaning "land of the Blacks." There, the low grass cover gives way to savanna grasslands with a taller and more continuous cover. The grasses here tend to be perennials. Much of the savanna is also covered with scattered bushes and trees within the grassland. Mean annual rainfall ranges from around 600 mm at the transition to the Sahel to 1000 mm in the southern margin of the Soudan. The savanna woodland is further south, a region in which trees become the dominant vegetation and which forms the transition to the forests nearer the coast.

IV. ANIMALS

Animals typical of the savanna—elephants, giraffe, diverse species of antelopes, zebra, and so

FIGURE 8 Dogon people of Mali occupy villages scattered along steep cliffs in the Bandiagara plateau region. The Dogon are primarily an agricultural people who have inhabited the region for many centuries. They practice irrigation by collecting water in pools, which they fill with fertile soil.

on—once roamed the Sahel. In recent years these have become rare, victims both to climatic fluctuations and to the impact of humans in the region. Herds, which were common in the region as recently as the 19th century, have been described in the journals of numerous explorers. However, as recently as a few decades ago, elephants and other animals of wetter environments could be seen as

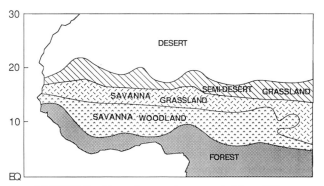

FIGURE 9 Map of vegetation zones in West Africa.

far north as southern Mauritania, where the mean annual rainfall is on the order of a few hundred millimeters. Now such animals are scarce even in the protected game parks of the region.

Domesticated animals are now the region's main inhabitants. Millions of goats, sheep, zebu cattle, and camels are herded by the peoples of the Sahel. Their numbers increased steadily in mid-century. In Mali, for example, the number of cattle rose from about 3.5 to 5.5 million between 1961 and 1971. However, in most Sahelian countries the drought years of the 1970s virtually halved the number of animals.

V. HUMAN INHABITANTS

The peoples of the Sahel are diverse racially and culturally. Most inhabitants are Negroid, but many are of mixed race, especially the nomads of the

FIGURE 10   Much of the Sahel is covered by "tiger bush," a vegetation formation characterized by meandering stripes of dense shrubs separated by large sectors of bare ground.

northern fringe of the region. Many are descended from Arabic peoples who have lived and traded in the region since the 7th century. The largest groups include the Hausa, Djerma, Fulani (or Peul), Malinké, and Soninké. This short list belies the complexity of culture and language in the region. In the Sahel and regions just to the south, hundreds of local languages are spoken. In most of the region, the lingua franca is French, a legacy of a century or more of colonial domination. Many religions are practiced, but Islam was widely adopted in the trading centers of the Sahel between the 11th and 15th centuries.

West Africa, especially the Sahel, was home to many ancient cities and empires. A number of substantial towns existed as early as Roman times, perhaps as early as 100 BC, near the current cities of Djenne in Mali or Kano and Zaria in northern Nigeria. The Arab conquest of North Africa in the 7th century brought renewed vigor to the Sahelian region and spawned a period of urban growth. Towns like Djenne, Gao, and Tombouctoo emerged in the heart of the trade network linking east and west and bridging the desert, becoming populous trade centers (Fig. 11).

Later, beginning probably between the 8th and 10th centuries, powerful and expansive organized states appeared along the northern fringe of the Sahel. The Ghana empire, between the bends of the Niger and Senegal Rivers, linked the gold fields to the west with the rest of North and West Africa. Gao, in the Niger Bend, was sustained by the river, which provided fish, water for irrigation, and silt for the fertility of the soil. The empires of Kanem and Bornu arose east and west of Lake Chad. All thrived on the caravan trade, which carried millet, salt, cotton, slaves, gold, textiles, and other merchandise across the Sahara.

After these empires declined, new states replaced them in later centuries, first the Mali and Kanuri states and later the Songhai and a revived Kanuri state in nearby locations around the Niger Bend and near Lake Chad. They prospered until around the 16th or 17th centuries; the Kanuri was so successful that it spread as far north as the Fezzan in the northern Sahara.

VI. AGRICULTURE AND ECONOMY

Livelihood in the Sahel is dictated by environmental constraints (Fig. 12). Most of the Sahelian population practices some form of agriculture, although

FIGURE II The ancient city of Djenne, Mali, was the heart of trans-Saharan trade many centuries ago. This local resident wears a head covering in the traditional Muslim fashion.

fishing is also important near rivers and coasts. Except in the wetter south, subsistence farming is the rule. Agricultural practices fall into mainly three classes: nomadic pastoralists, who roam the driest regions with their herds of animals; transhumant pastoralists, who seek a more southerly dry-season base and move their animals to more northern pastures during the rainy season; and sedentary farmers who cultivate crops and tend their herds from a settlement year round. Throughout the Sahel, land use is mainly pastoral, with cropping re-

stricted to relatively small and dispersed areas. To the south, in the Soudanian savanna, cultivation becomes extensive.

The nomads' way of life is particularly well adapted to the nature of the dry environment of the Sahel. Their mobility allows them to take advantage of the very patchy nature of the rainfall in the region, moving their herds to the locations where the best rains fall. Their continual movement disperses use of the meager resources over large areas, thus preserving the environment by min-

FIGURE 12   The typical Sahelian village is a cluster of small houses and graineries with mud-baked walls and thatched roofs, often surrounded by fields of millet or sorghum.

imizing their impact in a given locale. Their diversified herds of goats, sheep, zebu cattle, and camels allow them to optimally exploit the environment, since each favors a somewhat different habitat, and to reduce the risk imposed by the ever-present threat of drought, disease, or other calamities.

The seminomadic herdsmen (the transhumant pastoralists) move their animals northward to the drier regions at the start of the rains. They do so to escape the biting flies in the Soudan zone and benefit from the annual grasses of the Sahel, which are more nutritious than the perennial grasses of the Soudan. The animals graze for 2 or 3 months until the rainy season ends and drinking water becomes scarce. During the dry season, they retreat to the wetter south, moving their herds to lakes and rivers or to brouse on the stubble of farm fields. In some cases they establish temporary villages, planting some crops, such as cereals.

In the southern Sahel, rainfall is sufficient to support rain-fed cultivation. Agricultural methods are simple and basic, with little use of technology, special varieties of crops, or fertilizer. Land fertility is restored by letting fields lie fallow for 1 or more years. Like climate, the crop regions tend to be zonally arranged because as rainfall increases southward, so does the length of the growing season. It increases from 1 month or less in the desert fringe to 3 months in the south where the Sahel gradually gives way to Soudanian savanna. However, the boundaries between crop zones are diffuse since soil and local moisture are as important as rainfall in determining them.

Millet is the major crop in regions with less than 400 mm rainfall, especially on sandy soils, but cowpeas can also be grown in relatively dry areas. Further south sorghum is grown, especially on clay soils, as are millet, cowpeas, groundnut, and beans. In extensive areas of irrigation along the rivers and lakes, rice is cultivated. In the savanna where annual rainfall exceeds 800 mm, cotton and maize are grown, in addition to the other crops. Groundnut and cotton are cash crops, the rest are subsistence crops. Nomadic pastoralists keep sheep, goats, and zebu cattle for milk and meat which provide both food and exchangeable goods for market (Fig. 13). Milk and butter fat are exchanged for millet and sorghum grown by sedentary farmers. The farmers are clustered in small villages, around which there are frequently overgrazed regions on the order of tens of kilometers.

FIGURE 13 Trade has been an important part of Sahelian life for centuries. Merchants come to regional markets, such as this one in Baleyara, Niger, to sell and trade animals, goods, and agricultural products.

The Sahel is a human ecotone between the nomadic and seminomadic pastoralists and the sedentary farmers, a loci of constant interaction among these mutually dependent groups. Traditionally, the nomads provided the others with salt from the desert and manufactured products from North Africa. The pastoralists of the steppe supplied cattle and horses, which are difficult to breed in the humid savanna. The sedentary farmers of the savanna supplied millet and other food products, as well as cola nuts and gold from the forest frontier further south.

The symbiotic reciprocity among these peoples helped them survive a relatively harsh environment and sustained them through long and recurrent periods of drought. The nomads brought their herds to the farmers when water was scarce; the animals grazed on the stubble of fields and fertilized the surrounding soil. During sustained droughts the nomads become laborers in the villages and towns.

The nomads and farmers, together with the townspeople, also played an essential role in the trade system which had been the mainstay of the West African economy for centuries. Merchants distributed goods, farmers provided the caravans with goods, and the nomads transported animals across the desert, acted as guides, and gave protection to travelers (Fig. 14).

These interactions have integrated many cultural elements among the various peoples, providing some homogeneity to rather diverse groups. Each group has been integral to the survival and lifestyle of the others. Their cooperation formed a security system which, unfortunately, has been broken down by the establishment of international boundaries and the imposition of colonial rule throughout West Africa.

VII. CHANGE IN THE SAHEL

Some of the most profound—and most abrupt—changes in the Sahel were evoked by the imposition of colonial rule. With the exception of Liberia, all of West Africa was colonialized by Britain, France, and other European countries during the 19th century. In the Sahel, France was the dominant power. The colonial governments discouraged nomadic ways and promoted urbanization. Urban systems were restructured to serve the inter-

FIGURE 14 In the markets of Niger, a common local product is natron, a sodium-carbonate mixture which is fed to grazing animals to assist digestion. When water is left to evaporate in pools, the sodium-rich residue is collected and baked in long tubes, as shown above.

FIGURE 15 Wood and wood products are important for fuel and construction, but in the Sahel trees are relatively scarce and overexploitation of the woodlands may have serious consequences.

ests of the colonialists. They drew West Africa into the international market economy, at the expense of disrupting the traditional subsistence agriculture by favoring cash crops, such as peanuts.

With a few exceptions, the foreign governments retained power until around 1960. With independence came new changes. The colonialists had been in power over many generations, long enough to destroy the traditional organizational structures and economic systems which helped the region to prosper in earlier centuries. Coupled with independence was a rapid demise of the favorable climatic conditions which had prevailed for most of the previous century.

By the late 1960s, a steady rainfall decline had led to a drought which became devastating in the 1970s. Drought and dry conditions were likewise the rule in the 1980s. Much of the regional development by the colonial powers had occurred during a period of more favorable environment. These systems were ill-equipped to deal with the disaster, and traditional coping strategies were no longer viable. At the same time, both animal and human populations in the Sahel were rapidly growing as a result of improved medical and veterinary care.

The combination of drought, increased population, and the breakdown of traditional ways of life induced many hardships in the region. Among these are a dramatic increase in urban population, a breakdown of the trade system, and a near destruction of the nomadic way of life. Another is severe environmental degradation. Overuse of the land around villages and water holes, fuel wood gathering, poor land management, and increasing numbers of grazing animals have reduced land productivity, a process referred to as "desertification" (Fig. 15). At the same time, foreign aid is being offered to many countries, a double-edged sword when strategies and technologies inappropriate to the Sahelian environment are introduced. Thus, the present and future are rough and uncertain ones, times of rapid change for many of the Sahelian countries.

Glossary

Brousse tigrée English translation: "tiger bush," a common vegetation type in West Africa in which vegetation grows in dense clusters separated by vast areas of bare, often unpenetrable soil; from the air, the formation resembles stripes on a tiger.

Harmattan Dry northeasterly winds which blow southward off the Sahara towards the Guinea Coast of West Africa during the low sun seasons.

Intertropical Convergence Zone (ITCZ) Region where the trade winds of the two hemispheres converge, producing rising motion, clouds, and intense rainfall. It migrates with the seasons, following the sun.

Potential evapotranspiration Ideally, or hypothetically, the maximum amount of evapotranspiration which can be sustained by a plant canopy or soil surface with continuous supply of water.

Soudan Ecological zone south of the Sahel in which climate is semiarid to subhumid and the vegetation is a savanna woodland.

Transhumant pastoralist Nomadic herdsmen.

Bibliography

Grove, A. T. (1985). "The Niger and Its Neighbours: Environmental History and Hydrobiology, Human Use and Health Hazards of the Major West African Rivers." Rotterdam: Balkema.

Lewis, L. A., and Berry, L. (1988). "African Environments and Resources." Boston: Unwin Hyman.

Nicholson, S. E. (1995). "Dryland Climatology." New York: Oxford (in press).

Rognon, P. (1989). "Biographie d'un Desert." Paris: Plon.

Seed Banks

Mary Allessio Leck

Rider University

I. Introduction
II. Methods
III. Seed Bank and Vegetation Dynamics
IV. Seed Banks
V. Applied Studies: Agriculture, Conservation, Reclamation, and Mitigation
VI. Future Studies

The seed bank usually refers to viable seeds and seed-like fruits, such as achenes or caryopses, that are on or in the soil. However, in some situations, including chaparrel or other Mediterranean vegetation types, seeds may be stored above ground in serotinous cones or fruits that require fire for opening. The seed reservoir, derived from seeds produced at the site or dispersed into it from elsewhere and with components present for varying lengths of time, may contain a variety of species, genotypes, and/or phenotypes. These can provide considerable flexibility for vegetation response.

I. INTRODUCTION

Coastal California in summer is brown, but in winter, after suitable rainfall, the hills are a glorious orange from large populations of California poppy (*Eschscholzia californica*). A photograph of the Tehachapi Mountains, also in California, shows a hillside awash with orange, blue, yellow, and a touch of green. Deserts can display these spectacular seasonal palettes of color. Plants seemingly spring from nowhere.

Similar renewal also occurred during World War I; red poppies (*Papaver dubium*) grew from trenches, vehicle tracks, and graves of the Somme (France) and other battlefields. Poppies had not been observed there since about the time of the Franco-Prussian War in 1870 when fields had last been cultivated. This red poppy has become a symbol of rememberance on Memorial Day in the United States and on similar holidays elsewhere.

These are especially colorful examples of vegetation developing from seeds in soil, but gardeners, farmers, and other observers have long known that weeds can dominate crops when not carefully tended and that seeds in soil may persist for decades. A two-page article published in the first volume of the *Journal of the Linnean Society* (London) in 1857 by J. Salter may be the first written account to document the presence of seeds in soil. He noted that harbor mud contained the viable seeds of agricultural and wild species, including several rare ones that grew at some distance from Poole Harbor (England); the mud flora, in fact, was distinct from that surrounding the mud quay. However, despite its later date, an observation by Charles Darwin in "Origin of Species" published in 1859 is more widely cited. He had 537 seedlings emerge from three tablespoons of mud obtained from the edge of a pond. In 1860 Henry David Thoreau, in an essay entitled "Succession of Forest Trees" (*New York Tribune*, 6 October), reported on plants found in the cellar of a house, dating to at least 1703, that had been razed in 1859. A number of species were found growing in the cellar that he had not found before, that were rare, or that had been cultivated at least 50 years before. Also, in the 19th century, the first detailed seed bank studies were carried out in German forests where seeds were discovered at three depths and where species characteristic of agricultural fields or grasslands were found in soils of young forests of known age, but only seeds of woody species were found under old forests.

In the 20th century many early studies of soil seed banks involved agricultural lands. Only in recent years, since 1970, have natural communities been studied extensively; for some, such as arctic and alpine tundra, studies date from 1980. Recent interest in seed banks parallels the growing understanding of the importance of regeneration to the functioning of plant populations and to the structure of plant communities.

Over the years seed bank studies have explored many topics, including succession, species diversity, and zonation patterns. Seed bank studies are commonly incorporated into population and community studies. It is increasingly apparent that seed banks are a critical life history stage of many plant species and of the communities in which they occur.

II. METHODS

Soil seed banks may be evaluated by separating seeds from soil and examining them directly or by germination. Regardless of the method, however, consideration must first be given to selecting an appropriate sampler, number and size of samples, how samples are to be collected and stored, treatment of samples, and, if the germination method is used, germination requirements.

Collection of soil samples has employed a variety of implements, including bulb planters, golf cup corer, Ekman dredges, soil corers, Russian peat sampler, trowels, shovels, PVC thin-walled pipe, augers, knives, and hand samples. Soil and habitat characteristics ultimately determine the appropriateness of the sampler.

The number and size of samples should reliably document both size and composition of the seed bank. Published studies suggest that adequate surface area for a sample should be between 50 and 200 cm^2 with a combined area of ~1000 cm^2. The optimal surface area may vary with habitat, with low diversity habitats being adequately evaluated with smaller sized individual samples (e.g., 50 cm^2). The size of the sample (surface area and depth) may be, depending on the objective(s) of the study, influenced by the size of the seeds that may be encountered. A large number of small samples appears more reliable than a few large ones for the same total volume of soil sampled. In special circumstances, e.g., where particular species are of interest or where seeds are aggregated, the number of samples required may be greater. The optimal size and number of samples may be determined from a preliminary study that determines the impact of increasing numbers or size of samples on size and composition of the seed bank being studied.

Once collected, soil samples should be handled in such a way as to minimize contamination, loss of seeds, and/or loss of viability. If seed banks are to be determined by direct examination, samples may be dried and stored until processed. If the germination method is used, every effort must be taken to prevent loss of viability and induction of dormancy. These problems can be minimized by collecting samples soon after natural afterripening has occurred and placing them immediately into the greenhouse or germinator. If it is necessary to collect seeds at nonoptimal times, afterripening and germinability may be artificially developed by low or high temperature storage depending on habitat and/or species present. Even when direct examination is used, it may be necessary to correct for nonviable seeds when they substantially bias estimates; then storage to maintain viability is necessary. In some studies, freezing samples until they are processed may be warranted. Conditions and duration of storage may be especially significant if nondormant, recalcitrant, or short-lived seeds are present.

For the germination method, some understanding of the habitat is necessary. For example, to elicit germination of all species in a wetland seed bank, it is often necessary to incorporate inundation into the experimental design because some species require aeration while others require hypoxic conditions. Wetting and drying cycles coupled with high temperature may be necessary for certain agricultural weeds. Some fire-prone communities have species that germinate better with charcoal. [*See* FIRE ECOLOGY.]

The timing of soil collection or the number of times soil is collected may be important in documenting the transient/persistent nature of the seed bank. Collecting samples after field germination but before seed dispersal would preclude demonstrating the importance of transient seed banks. Alternatively, the use of exclosures made of weed cloth or other suitable material, which prevent incorporation of seeds into the soil, would provide evidence regarding transient seed banks. Use of exclosures, however, may be complicated by frost heaving or by the presence of animals that transport seeds.

Ultimately, the location of sample collection points depends on how variable the habitat is and on its spatial characteristics. The distribution of seeds in the soil profile also varies. It may be necessary to take samples to a 25-cm depth in cultivated fields but only to a centimeter or so in desert soils.

The primary advantage of direct examination of samples is that results are available as soon as all samples are processed. This may be especially desirable, for example, where it is necessary to predict weed contamination of agricultural soils. Direct examination may also be faster than seedling emergence when seeds are large and dormant. When seeds are very small, this method is tedious. Similarly, processing samples that are high in organic material may require considerable time both for treating and sieving samples and for examining them. A serious limitation for some soils is that viability may be as low as 6–36%. Among the advantages of estimating seed banks using germina-

tion are that seedlings are often easier to identify than seeds and, of course, each seedling represents a viable seed.

Studies comparing the reliability of these methods are few. For 57 Danish fields studied by Hans Jensen, manual separation estimated an average of 135,000 seeds m^{-2}. In contrast, seedling emergence provided an estimate of 19,000 m^{-2}; however, if the dominant species, toad rush (*Juncus bufonius*), was excluded, this estimate was 75% of that recovered by separation.

Katherine Gross, using soil from an annually plowed field, compared isolation by flotation, elutriation, and seedling emergence following cold stratification. The flotation method, which has the disadvantage of requiring different salt concentrations for separation of different species, also results in considerable sample loss during multiple washings and transfers of samples. It is also more expensive and time consuming. Cold stratification yielded the greatest number of species, but elutriation yielded the largest number of seeds per sample. Differences among methods may be because of a lack of suitable germination cues or the difficulty in distinguishing species with similar seeds or seeds that are similar to the sandy soil retained by the sieves. A disadvantage of elutriation was the poor recovery of very small seeds (0.022 mg/seed = 17%; 0.060 mg/seed = 87%).

Despite these problems the elutriation method may be preferable for studies that focus on seed bank dynamics or distribution of only a portion of the species present. Also, for a population study a large number of locations or depths can be sampled and, especially if the seeds are large and/or easily distinguished, samples can be counted rapidly. Elutriated seeds may also be used to determine viability (using tetrazolium stains) or isozyme frequency and potential growth rate for evolutionary and genetic studies.

Descriptive models (e.g., Fig. 1) provide a visualization of seed bank and vegetation patterns and they attempt to account for the important processes driving the system. Such models are useful summaries and provide ways to interpret observations. However, the addition of mathematics enhances the ability to ask questions about relationships

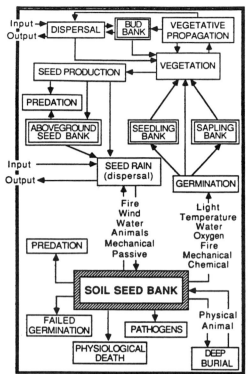

FIGURE I Model of seed bank and vegetation dynamics. [Reproduced with permission from R. L. Simpson, M. A. Leck, and V. T. Parker (1989). *In* "Ecology of Soil Seed Banks" (M. A. Leck, V. T. Parker, and R. L. Simpson, eds), pp. 3–8. Academic Press, San Diego.]

within the model and they may be used to inform research design. Such analysis may provide insights lacking in the original data or may emphasize critical areas requiring study. Models can be used to determine the essential features of an experimental system and then to make predictions about the system. Thus, predictions can be made about how the behavior of the seed bank, vegetation dynamics, etc. change as the environment changes. Then actual experiments can check such predictions. Models are also important tools in exploring questions that cannot be easily answered experimentally. For example, they can be used to explore long-term trends in populations (e.g., over 1000 years), the impact of environment on a seed or seed bank characteristic (e.g., seed size or persistence), or the impact of seed longevity or deterioration on seed bank size.

Because demands of habitats and the nature of the hypotheses vary, it is not feasible to have com-pletely standardized methods. However, data should be collected and presented in such a way as to allow comparisons.

III. SEED BANK AND VEGETATION DYNAMICS

The relationships between the seed bank and vegetation dynamics may be considered on several levels. These include direct and indirect influences on the seed bank itself, population level differences, as well as similarities and differences between the seed bank and the vegetation.

Input into the seed bank is determined by the seed rain. Although dominated by local dispersal, seed rain input may occur by a variety of vectors (Fig. 1) of which animals, water, or wind may provide long distance transport. For a given species, more than one vector may be involved. Losses from the seed bank may be due to germination or to predation, pathogens, deep burial, redispersal, and seed death. [*See* SEED DISPERSAL, GERMINATION AND FLOWERING STRATEGIES OF DESERT PLANTS.]

These inputs and losses determine the number of seeds in the soil as well as the species composition of the seed bank. Furthermore, other life history processes, such as establishment success, pollination, and predation, indirectly influence these parameters and seed bank dynamics.

For some species, vegetative propagation (e.g., bulbs or tubers) may give rise to a bud bank, which, unlike the seed bank, contains only established genotypes. For still other species and in certain habitats it is advantageous to store the genetic reserve in aboveground seed banks or in seedling or sapling banks. For example, Mediterranean dry climate shrub communities may store seeds, retained in serotinous cones or fruits, in the plant canopy. Serotiny is common in certain Australian and African families such as Proteaceae, Myricaceae, and Cupressaceae and in the California chaparrel it is common in Pinaceae. In the New Jersey Pinelands where aboveground seed storage also occurs, pitch pine (*Pinus rigida*) has greater serotiny in the pygmy plains areas where fires are more frequent. In still other habitats, notably tropical and temperate deciduous forests, seedling or sapling banks, which are ecologically equivalent to seed banks and bud banks, provide regeneration potential. Seedlings and saplings are at less risk than the large seasonal seeds and, because they are present throughout the growing season, can exploit gaps or other suitable environments whenever they occur. The relative advantage of these ecological alternatives to soil seed banks may vary spatially and temporally in response to variable climatic or disturbance factors.

Examing the relationship of seed banks and vegetation at the population level provides other insights regarding seed bank and vegetation dynamics. Reliance on a seed bank varies with species; for instance, it may be nonexistent as is the case with a tundra saxifrage (*Saxifraga cernua*) and some sweetflag (*Acorus calamus*) populations that are incapble of producing seeds or mangroves (e.g., *Avicennia germinans*) that usually germinate before dispersal. For other species, such as reed grass (*Phragmites australis*) and cattails (*Typha* spp.), it is of little consequence because once seed dispersal and establishment have occurred, reproduction in a site is vegetative. Still other species are fugitive species whose survival is dependent on long-lived seed banks; examples include beak rush (*Rynchospora inundata*) that establishes only during periods of low water or weed species that establish following disturbance.

Importance of the seed bank may vary geographically. The seed bank of New Jersey populations of jewelweed (*Impatiens capensis*) appear to be transient with spring depletion, while a Nebraska population appears to have a sizable number of persistent seeds. Another example, an African savanna grass (*Digitaria milanjiana*), has low rain ecotypes that are dormant and high rainfall ones that are not.

In species with seed polymorphism, one morph may have greater longevity than others. A California vernal pool species, *Limnanthes douglassii* vars. *rosea* and *nivea* studied by Herbert Baker, germinates only when there is favorable fall moisture and has many populations with both hermaphrodite (flowers are both female and male) and male-sterile plants. Although the male-sterile genotype is not as common, seeds containing it have superior longevity.

Often the composition of the seed bank is not similar to that of the vegetation (Fig. 2). Sometimes one (or very few) species makes up more than 50% of the seed bank while the vegetation has more diversity. This lack of similarity frequently is related to the successional stage and the relative seed longevity of species characteristic of the early stages. Lack of similarity is also related to the disturbance regime, shade intolerance, and dispersability. In habitats such as temperate and tropical forests, undisturbed wetlands, and permanent grasslands where disturbance is minimal, seeds do not build up in the soil. In other communities, such as chaparrel, agricultural fields, and disturbed wetlands where disturbance occurs at either unpredictable or predictable intervals, seed banks accumulate and may be quite large (Table I). When the disturbance regime is predictable e.g., tidal freshwater wetlands (daily tides) or temperate forests (winter), the seed bank has a large transient component. When the disturbance regime is unpredictable e.g., prairie glacial ponds, temporary ponds, and agricultural or successional fields, persistent species are important in the large seed banks that develop. For species such as pin cherry (*Prunus pensylvanica*)

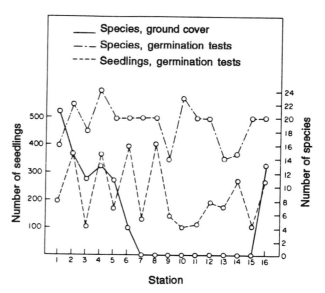

FIGURE 2 Seedlings and species recorded in the vegetation and from soil samples obtained beneath various stages in old field succession in Massachusetts. The 16 stations represent a sequence from a 1-year field (Station 1) to an 80-year-old white pine dominated forest. [Reproduced with permission from R. B. Livingston and M. A. Allessio (1968). *Bull. Torrey Bot. Club* **95,** 58–69.]

TABLE I
Size and Diversity of Seed Banks Representing a Variety of Selected Communities[a]

Location	Vegetation type	Seed bank size $(m^{-2})^b$	Species diversity[b]	Comments
	Tundra	0–3,367	0–13	Range, all sites
United States/Eagle	*Dryas* and Heath	85–3,142	8	Xeric site
Summit, Alaska	Snowbed vegetation	310–2,717	13	Mesic site
	Carex moss	80–2,802	2	Hydric site
	Coniferous forests	0–3,400	0–19	
United States/Colorado	Old growth *Abies lasiocarpa*	53		
	Picea engelmannii	3		
Canada/Alberta	*Pinus contorta*	393–707		
British Columbia	*Pseudotsuga* and *Tsuga heterphylla*	>1000	19	
	Temperate deciduous forest	27–13,650	3–41	Range
Japan	0-year *Mallotus-Aralia*	628	26	
	5-year *Mallotus-Aralia*	1,613	30	
	75-year *Rhododendron-Pinetum*	3,180	38	
	Photinio-Castanopsietum (climax)	850	33	
United States/Massachusetts	1-year horseweed	2,497	17	
	2-year goldenrod	4,564	23	
	8-year *Andropogon*	1,259	19	
	5-year white pine	4,510	25	
	25-year white pine	5,016	21	
	37-year white pine	1,399	21	
	42-year white pine	1,560	14	
	47-year white pine	1,346	20	
	80-year white pine	3,348	20	
	Tropical mature forests	2–3,616	2–79	Range, all sites
		60–2,340	21–79	Range, <100 m elevation
Panama/Barro Colorado	Semideciduous	318	54	25 m
	Semideciduous	742	48	25 m
Australia/Queensland	Evergreen	519	64	60 m
Ghana/Nueng	Wet deciduous	163	22	60 m
Shai	Forest outlier in grassland	107	22	120 m
Obdoben	Dry semideciduous	696	43	120 m
Kade	Moist semideciduous	623	29	180 m
Akosombo	Southern marginal	384	38	340 m
Atewa	Upland evergreen	45	17	780 m
	Tropical secondary growth	48–29,974	4–67	Range, all sites
		196–29,974	4–58	≤100 m
Mexico/Los Tuxlas	Regrowth	862–2,672	4–11	~100 m
Belize/Central Farm	Pasture and fields	6,488	54	
Senegal	Crop field	5,029–7,183	36–38	
Papua New Guinea/Gogal	Just clear cut	4,100	43	≤100 m
Valley	Regrowth forest	757	43	≤100 m
Australia/Queensland	Regrowth forest	1,302–1,440	35–58	<70 m
	Grasslands	406–31,344	7–62+	
United States/Kansas	Short grass prairie	761	16	
	Mixed grass prairie	406	21	
	Medium grass prairie	287	12	
Australia/Tasmania	*Poa gunnii* grassland	2,204	13	
Japan	*Zoysia* grassland	23,430	19	
	Deserts (Arizona)	4,000–63,000	3–10	Range; comparison of microhabitats at 1 site
United States	Sonoran Desert	~7,500		Hot desert (March)
	Great Basin Desert	~8,300		Cold desert (May and September)
	Wetlands	0–410,000[c]	0–108[d]	Range

continues

Continued

Location	Vegetation type	Seed bank size $(m^{-2})^b$	Species diversity[b]	Comments
United States	Tidal marshes			
	Fresh water	1,620–41,010	35–53	
	Salt	63–1,375	9–17	
	Nontidal marshes			
	Fresh water	696–255,000	29–50	
	Brackish	70–8,256	24–34	
	Salt	50–20,182	3–4	
	Other freshwater wetlands	0–410,000	0–108	
	Agricultural lands	250–130,300		Range
	Crop vegetables	250–24,330		+ herbicide
United Kingdom	Cereals	8,329–73,350		− herbicide
	Cereals	17,712		+ herbicide
Germany	Cereals	43,778		− herbicide; no weed control
		87–21,000		Seeds in soil (96–100%)
	Chaparrel (woody dominants)	15–2,133		Canopy seed bank
		5,385–320,000		4% viability
United States/California	*Adenostoma fasiculata*	2,000–21,000		100%
	Arctostaphylos glandulosa	8,422		3%
	Ceanothus leucodermis	87		96%
Australia	*Banksia ornata*	1,002		100%
	B. leptophylla	2,133		63%
	B. menziesii	67		3%
South Africa	*Acacia longifolia*	2,901		99%
	Protea eximia	45		100%

[a] Data are from M. A. Leck, V. T. Parker, and R. L. Simpson, eds. (1989). *In* "Ecology of Soil Seed Banks." Academic Press, San Diego.

[b] Data are not strictly comparable; soil depth and area varied with study.

[c] Data from Welling, C. H. and Becker, R. L. (1990). Aquatic Botany 38:303–309.

[d] Data from Kirkman, L. K. and Sharitz, R. R. (1994). Ecological Applications 4: 177–188.

that are dependent on unpredictable disturbance (e.g., fire or hurricanes) for gap formation, reliance on a seed bank is balanced by the capacity for long distance dispersal (e.g., by birds) because the interval between disturbances may be greater than the longevity on the seeds.

IV. SEED BANKS

A. Types

The nature of soil seed banks varies with population, species, community, or habitat type, as well as spatially and temporally. To determine seed bank type and the importance of the persistent seed bank to vegetation dynamics, it is necessary to evaluate the relative importance of the persistent fraction. This may be accomplished by knowing

two of the three following criteria: (1) seasonal changes in the seed bank, (2) seasonal timing of dispersal, and (3) whether seasonal or facultative dormancy is present.

For those species dependent on seed banks, J. P. Grime and K. Thompson found that persistence in soil varies among species so that a community often has more than one pattern (Fig. 3). Figure 3 describes four common seed bank types based on a study of temperate herbaceous species from 10 habitats. Type I species, summer transients, include annual and perennial grasses of dry and disturbed habitats whose seeds germinate completely during the growing season so that the soil is depleted of seeds during the winter. Type II species, the winter transients, include annual and perennial herbs that produce seeds one growing season and then germinate completely the following spring, similarly depleting the seed bank. Type III species include an-

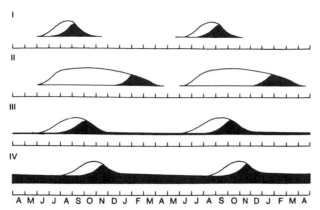

FIGURE 3 Seed bank types: species having summer transient seed banks (I); species with winter transient seed banks (II); and species with persistent seed banks (III and IV). The shaded areas indicate seeds capable of germination if supplied with the appropriate germination conditions. The unshaded areas indicated seeds that are viable but not capable of germinating. [Reproduced with permission from K. Thompson and J. P. Grime (1979). *J. Ecol.* **67**, 893–921.]

nual and perennial herbs that germinate mostly in autumn, but maintain a small seed bank throughout the year. Type IV species, which include other annual and perennial herbs and shrubs, have a large persistent seed bank.

The general applicability of temporal terrestrial (herbaceous) seed bank types requires additional study. For other habitats strict analogy may not be possible. Tropical seed bank strategies, summarized by Nancy Garwood, are more diverse (Fig. 4) and are related to continuous reproduction throughout the year. Transient seed banks (Fig. 4A) are produced by species of all regeneration strategies, including weedy species, short-lived pioneers, long-lived pioneers, and primary species. Primary species germinate and establish in the shaded understory of undisturbed forests. Some primary species form seedling banks (Fig. 4B). Persistent seed banks (Fig. 4C), produced by weedy and short-lived pioneer species, have long-lived seeds with facultative dormancy that are dispersed for short or long periods. In some cases seasonal dormancy may be imposed (Fig. 4D). Pseudo-persistent seed banks (Fig 4E) may also be produced by weedy and short-lived pioneer species. These seeds are short lived, nondormant, and, because they are dispersed continuously, have a continuous soil seed component. If dispersal is frequent, but

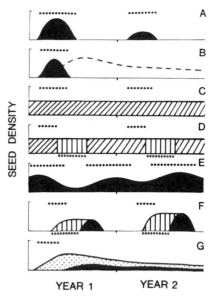

FIGURE 4 Seed bank strategies of tropical species (A) Transient seed bank. (B) Transient seed bank replaced by seedling bank. (C) Persistent seed bank. (D) Persistent seed bank with period of seasonal dormancy. (E) Pseudo-persistent seed bank of fluctuating size. (F) Seasonal transient seed bank. (G) Delayed transient seed bank. Fruiting periods are noted by asterisks, dry seasons by small open circles, seedling banks by dashes, germinable seeds without dormancy that must germinate or die by black areas, seeds with seasonal dormancy by vertical lines, seeds with facultative dormancy under forest canopy by slanted lines, and seeds with delayed germination by stippling. [Reproduced with permission from N. C. Garwood (1989). *In* "Ecology of Soil Seed Banks" (M. A. Leck, V. T. Parker, and R. L. Simpson, eds.). pp. 149–209. Academic Press, San Diego.]

not continuous, the size of the seed bank fluctuates. Seasonal transient seed banks (Fig. 4F), as with transient seed banks, may occur in any regeneration strategy. Seeds are seasonally dormant, with intermediate longevity, and are dispersed over short or long periods. Delayed transient seed banks (Fig. 4G), found primarily in long-lived pioneer or primary species, have seeds with delayed germination which is not associated with seasonally adverse conditions. Some seeds may remain in the soil for 1–2 years. All of these strategies may occur simultaneously. While one strategy may be most important, dominance may change seasonally.

B. Characteristics of Seeds

Certain seed characteristics appear to be associated with particular seed bank strategies. Summer transient seeds (Type I, Fig. 3) are large and/or have elongated structures such as awns; require little afterripening and lack dormancy; are able to germinate over a wide range of temperatures, which for some species may be as low as 5°C; and are able to germinate in either light or continuous dark. If moisture and temperature are appropriate, species with summer transient seed banks germinate soon after dispersal. These large, variably shaped seeds are less likely to be deeply buried. In addition, large seeds provide ample reserves for the rapid establishment of a large seedling capable of surviving winter conditions. For some species, such as *Salix* spp, which have short-lived transient seed banks, the small seeds are dispersed early in the growing season, allowing time for seedlings to establish before winter.

In contrast to the summer transient species, winter transient ones (Type II) typically have a chilling requirement prior to germination. Seeds are also relatively large, not deeply buried, germinate equally well in light or in dark provided they are afterripened, and germinate well at low temperatures as would be experienced in early spring. Early spring germination, at the beginning of the growing season, allows these species to exploit seasonally available gaps.

The majority of species with persistent seed banks (Types III and IV) have very small seeds. The temperature range for germination is relatively narrow and tends to be inhibited at cold temperatures. Often germination of some seeds is inhibited by continuous dark. These features facilitate burial and prevent germination in the soil. Some may be found at considerable depth (100 cm in a marsh). Among large-seeded species with persistent seed banks, physical dormancy (Table II) prevents germination and improves the probability of burial.

Dormancy mechanisms within a seed bank type vary. For example, giant ragweed (*Ambrosia trifida*), jewelweed, and arrow arum (*Peltandra virginica*), all high marsh species found in a tidal freshwater marsh, are Type II species. In the case of arrow arum, the fruit coat, late fall dispersal, and low temperatures prevent germination of nondormant seeds. Both giant ragweed and jewelweed require 3–4 months at low temperature for afterripening

to occur. All three species germinate equally well in light or in dark. In addition, giant ragweed tolerates drying, but seeds of the other two do not.

Species responses to a major environmental factor also vary within a habitat. Among wetland species, oxgen requirements differ; under prolonged hypoxia (3–4 months), jewelweed seeds lose viability; burmarigold (*Bidens laevis*) tolerates hypoxic conditions but requires adequate aeration for germination; canary reed grass (*Phalaris arundinacea*) can germinate under hypoxic conditions but root growth is impaired; and water plantain (*Alisma subcordatum*) and pickerel weed (*Pontederia cordata*) require hypoxic conditions.

Seed characteristics including size but also surfaces (whether smooth or tuberculate, having wings or other appendages, or mucilage), influence dispersal, anchorage, and incorporation into the soil and/or availability of moisture. One of the most remarkable examples of this is *Blepharis perisca* from the Negev Desert studied by Y. Gutterman and his colleagues. This plant's seed coat has unicellular hairs. When seeds fall with their flat sides touching moist soil, the hairs hydrate and swell, raising the seeds to an angle of 30–50° relative to the soil surface, bringing the micropyle into contact with the soil. All this occurs within a few minutes. The ends of the hairs then become mucilaginous so that as they dry, they join seed and soil, helping with anchorage of the seedling and root penetration. When the hairs are experimentally removed and even if seeds are properly oriented relative to the soil, root penetration is poor. Other species may have flat or concave surfaces, polymorphic seeds, or other seed characteristics that represent adaptations to microenvironments that are heterogeneous.

C. Seedling Characteristics

The relationship of seed bank dynamics to vegetation pattern is dependent on recruitment from the seed bank and establishment of seedlings. In some habitats the vegetation itself influences germination because it may alter light quality, especially red : far red ratios, light intensity, and/or soil temperature fluctuations. Vegetation may also compete for re-

TABLE II
Dormancy Types of Mature Seeds[a]

Types	Causes of dormancy	Embryo characteristics
Physiological	Physiological-inhibiting mechanism of germination in the embryo	Fully developed
Physical	Seed coat impermeable to water	Fully developed; nondormant
Combinational	Impermeable seed coat; physiological-inhibiting mechanism of germination in the embryo	Fully developed; dormant
Morphological	Underdeveloped embryo	Underdeveloped; dormant
Morphophysiological	Underdeveloped embryo; physiological-inhibiting mechanism of germination in the embryo	Underdeveloped dormant

[a] Reproduced with permission from J. M. Baskin and C. C., Baskin, (1989). *In* "Ecology of Soil Seed Banks" (M. A. Leck, V. T. Parker, and R. L. Simpson, eds.), pp. 53–56. Academic Press, San Diego.

sources and establish the availability of gaps. Furthermore, litter affects light and temperature, and constitutes a mechanical barrier to small seedlings and may be the source of germination inhibitors. Recruitment is also influenced by conditions such as soil topography, moisture, and seed depth. Seed size, shape, and dormancy breaking requirements interact with the heterogeneous environment.

The rate of emergence of seedlings is a function of seed size and burial depth. In arid habitats where drought avoidance is of particular advantage, seedlings from larger seeds can grow more quickly and/or shoots can emerge from greater depths. Trade-offs often exist. Large seeds are less likely to be buried, but seedlings from buried seeds have better moisture regimes; moreover, when deeply buried, only large seeds are likely to establish seedlings. Survival in closed vegetation is related to seed size. In addition, seed size influences seedling emergence and growth in competitive situations. Seedlings of large-seeded biennials, for example, are better able to exploit available resources or survive adverse conditions such as shade.

Seed size affects the competitive ability of a seedling as it determines the area of cotyledons. The high shoot/root ratios of seedlings from large seeds suggest that the priority is for light gathering instead of nutrient absorption.

Other aspects of seedling morphology may also be involved in establishment. In species with hypogeal germination, seedlings achieve height through thick litter or above surrounding vegetation without photosynthesizing because resources are present within the large cotyledons. Epigeal germination is associated with small seeds and hypogeal with large seeds. Hypogeal seedlings tend to be shade tolerant whereas epigeal seedlings are light demanding.

Seedling death may result from herbivory, but few studies have documented this seedling establishment problem. If the growing point is uninjured by the grazer, the seedling may survive. Also, if grazed seedlings are unshaded, axillary buds may grow and seedlings may survive, but if shaded by neighboring plants, buds do not develop before resources are depleted and the seedlings die. Seedling size and morphology affect the susceptibility to predation as does the grazing technique of the herbivore. Shepard's purse (*Capsella bursa-pastoris*) seedlings are not killed when grazed by slugs (*Agriolimax caruanae*), but those of speargrass (*Poa annua*) are.

D. Relationship of Seed Bank Type with Habitat

Seed banks reflect the action of establishment environments and local selection on plant life histories. They integrate past and present selective pressures on seeds and seedlings, and, in turn, influence life history characteristics of species as well as community dynamics.

Species that have transient seed banks appear to be adapted to predictable stable habitats where seedling establishment gaps occur regularly. Persistent seed banks, on the other hand, provide for population recovery following disturbances which may be unpredictable and decades apart. Where habitat changes are profound, significant changes may occur in community seed banks. In Britain, Type III species, capable of dispersing dormant seeds while conditions are not favorable for germination, are increasing in abundance associated with increasingly disruptive land use patterns. Type II species are declining. These changes reflect the loss in native broad-leaved deciduous woods where Type II seed banks predominate among shrub, tree, and herbaceous species. Divergence in seed banks can be due to changes in physiology or longevity related to site characteristics or to disturbance or stress gradients.

Thus, some preliminary comments can be made regarding seed bank types and habitats. Questions regarding constancy of seed bank type over a biome and changes with latitude or with increasing frequency of disturbance must await further studies.

E. Seed Bank Processes

Seed bank processes include seed production, predation, dispersal, and dormancy. Changes in relative importance determine differences observed in seed banks and, accordingly, in the dynamics of populations at a given site.

I. Seed Production

The importance of seed production varies with regenerative strategy. Seeds of species with persistent seed banks accumulate as long as there are reproductive individuals in the vegetation, often forming large seed banks (Table I). Their seeds may be dormant in the soil for decades until disturbance provides the environmental cues required for germination. Transient seed banks require annual renewal (Fig. 3). For species dependent on vegetative reproduction, seed (or bud) dispersal is necessary to colonize a site, but then seed production is not important.

Seed production is affected by many factors. Pollination failure or lack of a suitable pollinator may preclude seed formation. Light availability, competition for nutrients or water, temperature, and plant predators may determine the photosynthate available for seed formation. Ovule and seed predators may limit seed production, and post disperal predation may prevent the accumulation of seeds in soil. Sequestering seeds in aboveground seed banks by species with serotinous cones or fruits may, in fact, be a means of avoiding seed predation. In addition to environmentally influenced limitations, ovule losses may be related to lethal genetic combinations in pollen or ovules.

2. Predation

Seeds of many plants are consumed by a wide range of animals. They are primary food sources for many insects, birds, rodents, and primates, which often can act as agents for dispersal. The impact of predation varies among plant species and among communities, but descriptive data are uncommon. The obvious impact of predators on seed bank dynamics is on the availability of seeds. This impact may occur at all stages in the life cycle.

Predators themselves may influence seed and seed bank characteristics. By affecting morphology, species composition and spatial distribution, dormancy and longevity, and density, predators influence the persistent/transient nature of the seed bank. The evolutionary impact of predators on plant and seed traits is the result of predator–seed interactions operating in a variable and changing environment and requiring constant adjustment.

Several generalizations regarding the impact of predation have been made by Svata Louda. Predators may influence seed bank strategy because they select seeds differentially; by finding larger seeds they create pressures that select for persistent seed bank characteristics (e.g., small seed size and hard seed coats). The impact of predation may vary along the environmental gradient occupied by the species and with the heritability of differences that deter the predator (e.g., production of toxic chemicals). Some species, such as fugitive ones that depend on seeds for colonization and persistence in habitats that are moderately disturbed, may be at risk depending on the demography of the seed predators. Sometimes impact may result in a compensatory response. For example, defoliation/defloration may increase seed weight and coat thickness, yet may increase vulnerability.

3. Dispersal

At any site, establishment of a seed bank is ultimately determined by dispersal that may be accomplished by a host of vectors (Fig. 1). However, the frequency of a particular dispersal agent varies with vegetation type. In early successional stages, wind-dispersed seeds are dominant. Later, when there is more structural complexity, bird dispersal becomes important. In temperate deciduous forests, the dominant species are wind dispersed (e.g., maple, *Acer*) or have seeds taken by cache-hoarding mammals or birds (e.g., beech, *Fagus*). In tropical forests the proportion of species with wind, bird, mammal, or other dispersal agents varies with the life form; lianas, for example, are more likely to be wind dispersed than trees while trees are more likely than shrubs. The dispersal agent also varies with forest type; dry tropical forests have more wind dispersal than moist or wet forests.

Seed morphology is often related to the dispersal agent. Seeds or fruits adapted to external transport by birds or other animals often have burrs, hooks, or sticky material. However, small seeds or seedlings with no obvious morphological adaptations may be transported, adhering to feet, boots, and even tires. In general, seeds adapted to wind dispersal are made more buoyant by wings or other structures which reduce the rate a seed falls to the ground. Seeds of many species from a wide range of temperate and tropical communities rely on ants to carry them to a favorable germination site. Such seeds usually have a small mound of nutritive tissue (e.g., the elaisome) upon which the ants feed, leav-

ing an undamaged seed in the ant nest where soil is richer. Dispersal by other animals may also provide seeds with improved microsites. For example, cottontail rabbits (*Sylvilagus floridanus*) provide an alternate means of dispersal as well as a nutrient-enhanced microsite for achenes of various knotweeds (*Polygonum* spp.), which are primarily dispersed by water. Seeds successfully dispersed by rabbits are smaller ones that survive mastication; the undamaged, thick lignified fruit coat protects seeds from digestion.

In addition to dispersal, some animals may actually be responsible for burying seeds in the soil. Jays, for example, bury acorns at depths suitable for germination and establishment. Seeds may also be hoarded below ground by rodents such as squirrels, mice, and kangaroo rats.

For species with more than one type of seed, the dispersal pattern of each type may differ; in amphicarpous species, seeds produced above ground are more widely dispersed than subterranean ones. In addition, particular vectors like ants are likely to ensure seed burial where the seeds may be protected from fire and surface-feeding seed predators. Moreover, the dispersal agent itself may alter seed bank dynamics. For some tropical species, passage through a disperser may kill the larvae of seed predators. In addition, sometimes the dormancy mechanism is altered. For example, when seeds of *Cecropia obtusifolia* pass through spider monkeys and tayras (a mustelid), far-red light inhibition of germination is lost; this does not occur if bats are the dispersal agent.

4. Longevity

Longevity of seeds in the seed bank, as noted earlier, varies from transitory, with extreme examples being viviparous species such as mangroves that germinate prior to dispersal, to persistent species with seeds capable of remaining viable for decades. Longevity of seeds in soil is often difficult to evaluate. For example, the presence of species in the seed bank but their absence in vegetation of known age provides only circumstantial evidence for longevity (Fig. 2). Seeds may be dispersed considerable distances by wind, water, or animals to sites unsuitable for immediate germination where earthworms, capable of making burrows reaching 210–240 cm below ground, may effectively work them into the soil. Such seed movement may obviate carbon-14 dating of nearby materials. Radiocarbon dating of seeds themselves, however, may provide the best evidence of longevity. This method was used with the large seeds of the sacred lotus (*Nelumbo nucifera*) from an ancient Manchurian lake bed; they were found to be between 466 and 705 years old at the time of germination.

Experimental evidence for seed longevity is provided by burial studies where seeds are mixed with soil or sand in bottles or other containers. One such study, involving 20 species, found only 3 species capable of germinating after 100 years: moth mullein (*Verbascum blatteria;* 42% germination), common mullein (*Verbascum thapsus;* 2%), and dwarf mallow (*Malva rotundifolia;* 2%). Evening primrose (*Oenothera biennis*) and curly dock (*Rumex crispus*) survived 80 years. In another study most crop species (22 of 26) had negligble survival after more than 1 year. Where seed additions to soil are prevented, persistent seed banks of arable weed seeds decline exponentially over a period of 5 years. Trends, however, may be confused by hard-seededness; longevity of dry hard seeds has a different physiological basis than imbibed soft seeds. Under experimental conditions longevity is often enhanced at a lower moisture content, but under natural conditions seeds may experience intermittent wetting.

Differences in seed characteristics involving the responses to drying hamper comparisons of the great differences in seed longevity. The majority of crops and many noncultivated species have orthodox seeds. Recalcitrance is found in a number of large-seeded woody species, including cacao (*Theobroma cacao*) and chestnuts (*Castanea*). Intermediate categories may exist for a limited number of species that are injured by low temperature or additional dessication. Examples include coffee (*Coffea arabica*) and possibly beech (*Fagus sylvatica*).

Why some species survive longer in soil despite hydration levels that are too high for satisfactory storage in the laboratory or for agricultural use is still a mystery. Many factors appear to contribute to the aging of seeds, including loss of membrane,

ribosome and DNA integrity, and cellular compartmentation and enzyme activity. Little is known about such changes under natural field conditions. Longevity may be constrained by several factors, including genetic effects that may vary with parent or population, environment during maturation, degree of seed maturity, and mechanical injury.

Among species with persistent seed banks, longevity often exceeds a decade. However, available evidence shows that recruitment is from the youngest cohort of seeds or primarily during the first year after sowing. Such observations suggest that older seeds contribute little to a recruited population.

5. Dormancy

Seeds that are long lived in soil generally are one of two kinds. Hard-seeded ones fail to imbibe until the seed coat is altered; the others survive while partially or fully imbibed but their metabolism is controlled. The two strategies are not independent of one another; combinations of these and other dormancy mechanisms (Table II) occur. Dormancy varies among offspring of one individual or between geographically distinct populations. In hard-seeded species (e.g., mallow, *Malva* spp.), for example, the coat integrity of seeds in soil has been observed to change over 12 years. Seeds that survive for long periods while imbibed do so by remaining dormant; their germinability is controlled by either internal or external factors. Burial alters light, temperature, and ambient atmosphere and may lead to enhanced dormancy involving development of a light requirement for germination.

Seeds with all types of dormancy (Table II) can be found in seed banks. However, in temperate regions most species have physiological dormancy with physical dormancy being second in importance. Seeds with physiological dormancy exhibit annual dormancy–nondormancy cycles, but patterns may vary (Fig. 5). In addition, when nondormant seeds are brought to the surface, changes in germination status and germination requirements interact with each other and with the soil surface environment to control the timing of germination. For seeds with physical dormancy, germination

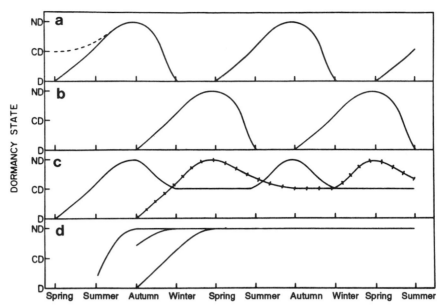

FIGURE 5 Dormancy patterns of seeds with physiological dormancy. (a) Obligate winter annual with annual dormancy/nondormancy cycle. Freshly matured seeds in some species are dormant (———) while others are conditionally dormant (- - -). (b) Spring-germinating summer annual with annual dormancy/nondormancy cycle. (c) Facultative winter annual (———) and spring- and summer-germinating summer annual (—+—+—+—) with annual conditional dormancy/nondormancy cycle. (d) Perennials with no changes in dormancy state after seeds come out of dormancy. [Reproduced with permission from J. M. Baskin and C. C. Baskin (1989). *In* "Ecology of Soil Seed Banks" (M. A. Leck, V. T. Parker, and R. L. Simpson, eds.), pp. 53–66. Academic Press, San Diego.]

depends on the seed coat becoming permeable in a specific area. For example, in some mallows the cells in the chalazal region of the seed pull apart, while in others the cells around a plug break down; in some legumes water entrance through the site of entry (the strophiole) depends on the disruption of the strophiolar plug by high temperatures. Softening of other hard seeds may also be accomplished by high temperatures, which occur during fire or in summertime.

In any habitat the seed bank is composed of species which collectively demonstrate a variety of germination strategies. Knowledge of the habitat may provide clues as to germination behavior of seeds in the seed bank. This is especially true if seeds with physiological dormancy are present. If the habitat has a predictable period of stress, seeds may exhibit annual dormancy/nondormancy cycles with seeds being dormant during the period of stress.

V. APPLIED STUDIES: AGRICULTURE, CONSERVATION, RECLAMATION, AND MITIGATION

Vegetation management practices, which involve seed banks, have widely divergent goals. In some managed habitats the goal is to elimate species (e.g., agricultural weeds), while in others it is to increase numbers (e.g., rare species or those desired for vegetation restoration).

Viable weed seeds in soil may exceed 130,000 m^{-2} (Table I). Cultivation is used to reduce numbers. However, timing relative to dormancy cycle and frequency of cultivation influences effectiveness. Prevention of seed input, whether by crop rotation, fallowing, or other crop management methods, such as cultivation and herbicide use, effectively reduces the seed bank in 1–4 years. Interestingly, because they do not affect all species in the same way, herbicides can alter the weed flora dramatically. Long-time herbicide use can result in great increases in resistant species. In England where modern agricultural techniques have caused some weed species to become rare, an effort is being made to maintain some traditionally farmed areas to prevent their extirpation.

Reestablishment of vegetation, whether for reclamation (e.g., in areas of erosion or strip mining), mitigation (e.g., creating wetland to compensate for destruction by road building or other human disturbance), or other purposes, requires certain information. It is necessary to know whether a seed bank is present, its species composition and density, and the conditions necessary for recruitment. Sometimes, if past experience has established its presence, a preliminary seed bank study is not essential. Otherwise such a study provides data indicating whether desirable or undesirable species may become established or whether desirable species are missing. Based on the few sets of data available, greenhouse results may accurately predict which genera would be present, but not their abundance in the field.

Seed banks are routinely used by wetland managers in the American Midwest to select for a particular type of vegetation. Depending on the drawdown (low water) regime, open water with submerged species, emergent vegetation, or mud flat annuals may be favored. Also, in areas where natural vegetation has been destroyed, cultivated or pasture soil may contain relict seed banks of the native species, but few studies have examined the usefulness of such seed banks in reestablishing natural vegetation.

On unvegetated sites, such as mine spoil or wetland mitigation areas, use of donor seed banks may provide an inexpensive way to reestablish vegetation. The soil, containing the seed bank, is removed, stored, and then put into the area to be revegetated. The new vegetation will not be identical to the local mature vegetation because seeds of some native species may be absent or lose viability rapidly. Before beginning a regeneration project, it is necessary to consider time of year the soil should be collected, depth (no more than 25 cm), storage conditions, and conditions for

recruitment of both desirable and undesirable species.

Seed banks may be used for conservation. Once it is determined which species are rare or endangered either locally or globally, management goals, based on site constraints, can be worked out. The choice of goals may involve concensus of a variety of interest groups. The simplest scenario would be to protect the site from human interferences. But often some manipulation of the disturbance agent is necessary. Perpetuation of some prairie, savanna, heathland, or pinelands species requires fire, while cycles of drawdown and flooding are needed in wetlands to maintain floristic diversity, including rare species. For some species, particularly those of late successional or closed habitats, seed banks may not be important conservation tools because persistent seed banks are not formed.

Models to predict change of both natural and management-created vegetation, although useful, require information that is often unavailable. Data about seed longevity, germination requirements, seed responses to different environmental conditions, seedling establishment requirements, and adult mortality need to be obtained. Ultimately for precise predictions for vegetation management, all aspects of desired species life histories would be required.

VI. FUTURE STUDIES

The role that seed banks play in vegetation dynamics can best be determined by studies that incorporate empirical data, experimental manipulation, and modeling. Authors of previous studies have noted many areas in which understanding is lacking. What follow are some questions raised by their attempts to make seed bank ecology predictive.

Few studies compare methods. Understanding the limitations of particular methods would improve our ability to make comparisons among vegetation types, communities, or habitats.

In some vegetation types few published seed bank studies exist. Additional studies are needed before the significance of seed banks to vegetation dynamics can be determined. Integration of seed bank patterns into vegetation studies would allow comparisons among and within sites. Questions that have been raised include: Do species of comparable successional stages have similar seed bank dynamics? How do the effects of edges, contrasting landscapes, and other aspects of biogeography relate to seed bank characteristics? How variable are seed banks within a biome? Do generalizations extend across biomes? How do physiological aspects of germination ecology affect dynamics at the community level? How adaptive are seed banks? What is the relationship between seed bank type and regenerative strategy in different habitats? What is the relationship between seed bank type and other reproductive and dispersal characteristics? Is selection for a seed bank strategy influenced by whether the population is increasing or decreasing in size? What are the genetic implications of differential mortality among species, populations, and cohorts in a given site or among sites with differences in a given environmental parameter (e.g., fire)? How do interactions between genetic variability in dormancy and spatial varability in the microenvironment affect seed banks? How does seed aging affect growth, reproduction, and competitive ability? What are the relationships of seed characteristics, seed bank strategies, and seedling adaptations?

Seed bank processes also require careful study. Little is known about germination death because of difficulties in monitoring seeds and seedlings in soil. Studies are needed for tropical species to determine the relative importance of seed bank and seed rain in regeneration and how this changes with forest type, gap size, and/or age. In the tropics and elsewhere little is known about the fate of seeds in soil or how seed physiology changes during transport by one or another dispersal agent or during burial. Not much is known about the relationships among dispersal, dispersal vector, and incorporation (burial?) into the seed bank. The relative importance of dispersal vectors for a species needs to be examined. There is scant information regarding the impact on seed banks of seed predation pressure as it varies in selectivity and intensity in space and time.

Another area of investigation involves testing models. The reliability and accuracy of models for predicting the future compostion of vegetation have rarely been tested. In addition, models need to be used to help explain the possible mechanisms for ecological patterns in seed banks. Interaction between empirical and theoretical studies will enhance insights regarding seed bank dynamics.

Much has been accomplished during the past 2 decades. Understanding of the role of seed banks in community dynamics has progressed to the point where studies often include, in addition to estimates of buried seed densities obtained two or more times during the year, measurement of seed input (seed production and/or seed rain), and determination of the relative importance of colonization by seeds and vegetative propagation. Ultimately, the significance of seed banks depends on integration of such ecological studies with germination physiology and with seedling morphology and ecosphysiology.

Acknowledgments

Many of the ideas presented here are not original. I am especially indebted to works by C. and J. Baskin, P. B. Cavers, M. Fenner, N. Garwood, J. P. Grime, K. L. Gross, P. A. Keddy, S. Louda, M. McDonnell, S. T. A. Pickett, H. A. Roberts, K. Thompson, and D. L. Venable.

Glossary

Afterripening Physiological and biochemical processes involved in the loss of dormancy by a seed.

Chalaza Area of the ovule (seed) where the stalk connecting the ovule to the placenta is (was) fused.

Dormancy Condition of arrested growth in which seeds (or other plant parts) do not begin to grow when supplied with optimal conditions for growth (water, temperature, and oxygen). Dormancy ensures that germination is delayed until favorable conditions for growth and establishment in the field are met.

Dynamics Processes, such as germination, growth, death, and replacement, that relate to the dynamic nature of seed banks and vegetation and emphasize the changes that occur either spatially or temporally.

Eleutriation Method of washing soil samples using air to agitate samples suspended in water, and then sieving through a series of three sieves to collect particles of different sizes.

Epigeal During germination cotyledons appear above ground and usually become functional photosynthetic leaves that contribute to seedling establishment.

Establishment Growth of a seedling to the point of independence from seed reserves.

Fruit A seed-containing ripened ovary of a flower and any flower parts that develop with it, such as achenes, caryopses, berries, and capsules.

Hypogeal During germination cotyledons remain at or below ground level, usually within the seed coat, and do not become photosynthetic. Seedlings rely on nutrient reserves with the cotyledons.

Hypoxia Condition involving low oxygen supply.

Life history Stages of the life cycle of an organism, e.g., seed, seedling, juvenile, adult, and reproductive adult.

Micropyle In the ovules of seed plants, the opening through which the pollen tube usually enters.

Orthodox seeds Can be dried to a low moisture content without damage, and longevity increases with a decrease in seed storage moisture content and temperature in a predictable way.

Recalcitrant seeds Do not survive dessication and are killed if the seed water potential drops below ~ −1.5 to −5.0 MPa.

Recruitment Processes by which individuals are incorporated into a community.

Regeneration strategy Includes vegetative expansion, seasonal regeneration (including Type I and II seed banks, see text), persistent seed banks (including Types III and IV seed banks), numerous widely distributed seeds, and persistent juveniles.

Seed The sexually produced reproductive unit of cone- and flower-bearing plants (gymnosperms and angiosperms) having a seed coat, an embryo with one or more cotyledons (seed leaves), and shoot and root growing points. Typically there is a stored food reserve located either in the endosperm or cotyledons.

Serotiny/serotinous Fruits or cones having thick resistant tissues that often require fire for opening and seed dispersal.

Stratification Cold moist treatment used to afterripen dormant seeds.

Succession Orderly changes in vegetation composition following a disturbance. Old field succession, where agricultural fields are abandoned, involves changes from weedy annuals, to perennials, and then to woody species.

Bibliography

Fenner, M. (1985). "Seed Ecology." London: Chapman and Hall.

Fenner, M., ed. (1992). "Seeds: The Ecology of Regeneration in Plant Communities." Wallington, UK: C·A·B· International.

Gross, K. L. (1990). A comparison of methods for estimating seed numbers in the soil. *J. Ecol.* **78,** 1079–1093.

Leck, M. A., Parker, V. T., and Simpson, R. L. (1989). "Ecology of Soil Seed Banks." San Diego: Academic Press.

Leck, M. A., and Simpson, R. L. (1993). Seeds and seedlings of the Hamilton Marshes, a Delaware River tidal freshwater wetland. *Proc. Acad. Nat. Sci. Philadelphia* **144,** 267–281.

Mayer, A. M., and Poljakoff-Mayber, A. (1989). "The Germination of Seeds," 4th Ed. Oxford: Pergamon Press.

Murray, D. R. (1985). "Seed Dispersal." Sydney: Academic Press.

Priestley, D. A. (1986). "Seed Aging: Implications for Seed Storage and Persistence in the Soil." Ithaca: Comstock Publishing Associates.

Simspon, G. L. (1990). "Seed Dormancy in Grasses." Cambridge: Cambridge Univ. Press.

Vyvey, Q. (1988). Bibliographic review on buried viable seeds in the soil. *Excerpta Bot. Sect. B* **26,** 311–320.

Vyvey, Q. (1989). Bibliographic review on buried viable seeds in the soil. *Excerpta Bot. Sect. B* **27,** 1–52.

Seed Dispersal, Germination, and Flowering Strategies of Desert Plants

Yitzchak Gutterman

Ben-Gurion University of the Negev, Israel

The critical stages in the life cycle for the survival of annual plants are seed dispersal, germination, and flowering, especially under extreme desert conditions. In deserts with hot and dry summers that receive only winter rains, such as the Negev Desert highlands of Israel, the beginning as well as the end of the growing season and the amount and distribution of rain are unpredictable.

The seed is the most resistant to environmental conditions whereas the seedling is the most vulnerable stage. Therefore, germination at the right time and place of only a portion of the seed population has a very high survival value for these species inhabiting the extreme deserts.

In the Negev the majority of mature seeds are consumed, mainly by ants. Therefore, at least two extreme directions of seed dispersal strategies have developed: the "escape" and the "protection" strategies. At least two extreme strategies have also developed in the germination of these species: the "opportunistic" strategy and the "cautious" strategy. Different species may have different combinations of these strategies. The genotypic strategies increase the fitness of a plant species to its habitat whereas the phenotypic strategies spread the risk. On a single plant, seeds can mature with different germinability. The time of "readiness for germination" of a seed is dependent on its genotypic as well as phenotypic influences including events from maturation to the time of germination.

The majority of the annual plant species tested have a facultative long-day response for flowering. This means that the later seed germination occurs in the season, the longer the daylength and the shorter the time until flowers appear, giving the plant the chance to complete its life cycle and produce mature seeds before the summer, even when seedlings appear after a late rain.

I. INTRODUCTION: FACTORS AFFECTING PLANTS OF HOT DESERTS WITH ONLY WINTER RAINS

A. Environmental Factors

Very hot and dry summers and small, unpredictable amounts of rain in the cool winter, as well as the unpredictable distribution of rain, are typical of many deserts. Thus, for plants such as winter annuals, geophytes, and others the length of the growing season with rains and suitable temperatures is also unpredictable and can be very different from one year to the other. [*See* DESERTS.]

The Negev Desert highlands of Israel are situated in the transition zone between the Saharo–Arabian and Irano–Turanian phytogeographic regions at the south western part of the Irano–Turanian and the northern part of the Saharo–Arabian regions. Sede Boker is situated in the center of the Negev where the environmental factors are as follows:

1. The rainy season may start any time from October in one year or December, or even January, in another. The last rain may appear in March, April, or even May. This means that the appearance of the first rains as well as the last rains can differ by 3–4 months. The number of days with rain during the rainy season may be from 10 in one year to 43 in another. The length of the season between the first and last rains can be from 104 days in one year to 230 in another. The amounts of rain can vary from 30 to 165 mm with very big fluctuations from one season to another. In some years the daily maximum rain can be nearly the same as all the rains during the whole season of another year (Fig. 1).

2. Temperatures during the coldest months of December or January range from the daily average of a minimum of 5°C to a maximum of 15°C. The maximum temperatures of 20–38°C occur during July and August.

3. Evaporation daily maximum rates from free water space during the hot and dry summer (June/July) may reach more than 15 mm compared to an almost 0-mm daily minimum during December/January. Monthly evaporation rates can be from 60 mm in December/January to about 300 mm during June and July.

4. The relative humidity monthly average at midday is the lowest in June (27%) and the highest in December (53%).

5. Dew. The number of hours with dew per month is 160–170 in January, September, and December and the yearly minimum is in April with about 50 hr.

B. Biotic Factors

1. Depressions

Porcupine diggings and other depressions in which seeds and organic matter accumulate by wind during summer and by runoff water during winter are favorable habitats. In loess soil, runoff water develops after only 3 mm per hour of precipitation occurs. Therefore, depressions are very important microhabitats for germination and seedling establishment of annuals as well as other plants inhabiting the desert.

Porcupine diggings in some areas cover more than 2.5 diggings per 1 m^2. In other areas the whole soil surface is covered with diggings in different stages of cover over. Porcupines dig and consume subterranean storage organs of geophytes and hemicryptophytes. The depth of each digging depends on the depth from which these organs have been dug out and the resultant digging can be from 7–25 cm deep, about 30 cm long and 10 cm wide. These diggings act as wind traps in which dry particles of plants, dispersal units, and seeds that are moved by the wind along the soil surface become trapped and accumulate during summer. During winter, runoff water may accumulate and as much as six times the amount of water can penetrate the soil in the digging in comparison to the surrounding area. The bottom of the digging remains moist much longer which gives more seeds the chance to germinate after one rain event, even those seeds that need a longer period of wetting for germination. It has been found that during the 10–25 years of cover over of a porcupine digging, depending on the topography and geomorphology, a succession of annuals develops. There is an increase from one year to another in plant biomass, number of plants, and number of species as well as the seed yield per unit area with a peak when the digging reaches between 50 and 60% of cover over. During the first years, the majority of plant species whose seeds are dispersed by wind appear in the diggings. Later on more species whose seeds are dispersed by rain or runoff water appear and their numbers increase in time when the numbers of other species decrease. When full cover over of the digging is reached, species whose seeds are dispersed by rain remain where the digging was situated.

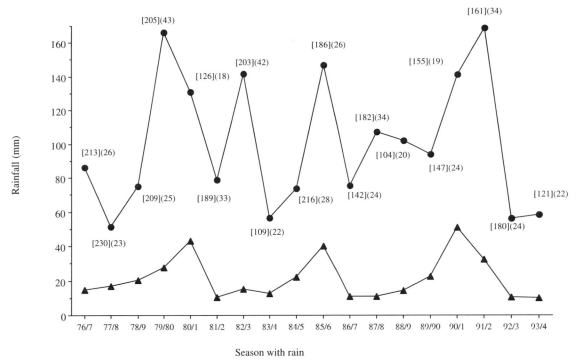

FIGURE I Seasonal total and maximum daily rainfall for the years 1976–1994 at Sede Boker. Seasonal Total (—●—); daily maximum (—▲—); number of days between first and last rain [..]; number of days with rain (..). [After Zangvil and Druian (1983) and Unit of Meteorology, Jacob Blaustein Institute for Desert Research; adapted from Gutterman (1993)]

2. Allelopathy and Inter- and Intraspecific Competition

Competition affecting seed germination and seedling establishment has been found in *Artemisia sieberi* (− *A. herba-alba*) (Asteraceae). This shrub is one of the dominant plant species of the Irano–Turanian region of the Negev Desert highlands. Compounds from these plants inhibit germination and seedling development even as far as 40–80 cm from the center of the shrub. The lower the percentage of cover of *A. sieberi* shrubs in an area of natural plant populations on a hill slope, the more seedlings of annual plants germinate and survive. A similar effect was also found during the development of seedlings of *A. sieberi* after they germinated in very great numbers (up to 159 seedlings per 1 m²) in 1963/1964. As seedlings increased in size, less seedlings remained until they developed into shrubs in 1975 with a distance of 40 to 80 cm between each shrub and only about five plants remained per 1 m².

Intraspecific competition was also found in 1964/1965 when 709 plants of the facultative perennial plant *Diplotaxis harra* (Brassicaceae) remained from the previous winter. None of the 1020 seedlings that appeared during the following winter survived in this area of 16 m².

In the annual plant *Salsola inermis* (Chenopodiaceae), the more seedlings that appeared beneath the dead mother plant the fewer plants survived to produce seeds and less seeds were produced.

3. Seed Consumption

In deserts, seeds are a very important part of the diet of desert creatures. Therefore, there are years in which the majority of seeds that have been produced are consumed.

Because of the very scarce food material in deserts it is typical that such areas will be overgrazed by grass-eating insects and animals from the natural wildlife as well as sheep, goats, camels, etc. The result is that in many habitats at the end of summer most of the area is bare soil without any parts of plant material left of the winter annuals of the previous year. There is also very high pressure from seed collectors and consumers. In some years more than 70 to even 96% of the seeds produced and dispersed is harvested, mainly by ants but also by rodents and other animals, birds, etc.

Under such stress of consumption two main extreme directions have evolved in seed dispersal of desert plants: (A) "escape" or (B) "protection" strategies.

Seed dispersal, germination, and flowering strategies are the most important for the survival of the species in the life cycle of annual plants inhabiting the extreme deserts. These strategies include: (1) the seed dispersal strategies that may "protect" the seeds or avoid their massive consumption by "escape"; (2) genotypic as well as phenotypic regulation strategies affecting germination at the proper time of only a portion of the seeds of a plant species after one rain event; and (3) flowering at the proper time, which enables the plants to produce new seeds before the dry and hot summer.

II. SEED DISPERSAL STRATEGIES

A. "Escape" Strategies of Seed Dispersal

In many plant species inhabiting deserts, "escape" strategies of seed dispersal have developed. These species typically produce large numbers of very small dust-like seeds that after maturation are dispersed at the beginning of summer into depressions or soil splits and may be covered by dust and larger soil particles where they avoid massive consumption by "escape."

In many habitats of the Negev Desert, plant species with a combination of an "escape" strategy of seed dispersal and an "opportunistic" strategy of germination appear very frequently as a dominant plant in the annual plant populations. These plants, such as *Schismus arabicus* (Poaceae), produce very small (0.5×0.7 mm; 0.07 mg) caryopses (seeds) that resemble sand particles. They may mature in very large numbers of up to 10,000 seeds per 1 m². A small portion of their seeds also germinate in large numbers even after as little rain as 10 mm.

Does the success of this plant depend on the very great numbers of seeds that have the chance to germinate? Even if the majority are not successful, there are still enough seedlings to survive, become a dominant species, and year after year produce tremendous numbers of seeds. Is this "shot in the dark" strategy the best way to survive under such unpredictable desert conditions? If so, the greater chance of success of a species as a dominant plant depends on the higher seed numbers of the plant species and the higher numbers of germinating seeds on more rain occasions.

In contrast, the most sophisticated strategies of seed protection and "cautious" dispersal and germination can be seen in the rare species *Blepharis* spp. (Fig. 2).

B. Protected and Delayed Seed Dispersal (Synaptospermy)

In contrast to the escape strategy, in the opposite direction of evolution of seed dispersal strategies that have been developed are the "protection" strategies. Seeds of these plant species are protected from the time of maturation until the following rainy season (synaptospermy), or for many years, by the woody parts of the dead mother plants. This occurs in many of the plants whose seeds are dispersed by rain (ombrohydrochoric plants). The dispersal by rain of a small portion of the "seed banks" that are situated in the dead mother plants when they are able to germinate is a very important strategy for the survival of these species in areas of the extreme deserts where ants consume a high percentage of the seeds.

Blepharis spp. (Acanthaceae) is a rare plant that is found only in limited habitats. This plant produces relatively large seeds in small numbers, which are well protected.

Blepharis spp. has one of the most sophisticated "protection" strategies of seed dispersal: Seed dispersal is delayed until a suitable time for germination when enough rain has moistened the soil. Only then are some seeds dispersed and they germinate within a very short time. The very "cautious" strategy of seed dispersal and rapid seed germination of this species is dependent on the double safety mechanism of seed dispersal which acts through two stages. First is the opening by wetting of the upper and lower sepals which enclose the dry and woody capsule containing two seeds. The opening of the sepals to more than 5 mm takes an average of 30 min. Inflorescences that have been in the field for many years open more quickly than those that flowered the previous season, which open only 65 min after the continuous wetting by drops of rain. The second stage begins when the rain continues after more than 5 mm separates the sepals because then the "lock-area" that is located on the upper part of the capsule may start to be wetted by raindrops. Here again the time of wetting for about a third of the capsules to explode and disperse their seeds is about 30 min. The older inflorescences disperse their seeds earlier whereas those of the previous season disperse later, after approximately 40 min. The explosion of the capsule occurs when the tissue of the lock area is weakened by the continuous wetting and when the tension that exists between the two halves of the dry tissue of the capsule is able to overcome the strength of the lock area.

In the majority of capsules (about 70%), water also penetrates the hydrochastic tissues of the capsule that are situated along the peripheral zone of the septum and thus when wet release the tension. This is the reason why capsules that did not explode after 30–95 min will not do so even after a long period in water (24 hr). After redrying, this capsule is able to explode and disperse its seeds after the weakening of the "lock-area" by a period of wetting.

The capsules explode and thus disperse their seeds up to a few meters from the shrub if the wetting is caused by rain. In plants situated in wadis the dispersal occurs by floodwater where seeds are dispersed into the flood. A mucilaginous layer develops from the seed integument as well as from the multicellular hairs that cover the seed. A layer of mucilaginous gel a few millimeters thick surrounds the seed and prevents enough oxygen from penetrating the embryo. This prevents germination until the excess water drains, leaving the seed situated on the wet soil surface in the wadi to which it will adhere with its mucilaginous hairs. Germination will occur within a very short time (a few hours) (Fig. 2).

In other habitats, during a long rain the dry seeds that are exploded from the capsules will absorb water when they fall onto the wet soil. The multicellular mucilaginous hairs will open, glue the seed to the soil surface, and bring the micropilar zone in contact with the wet soil. Germination starts immediately and within 24 hr of the first wetting the root length could reach as long as 5 cm (Fig. 2).

The timing of germination in this plant is dependent on the double safety mechanisms of seed dispersal; also, if there is an excess of water, the dispersal mechanism reacts during a flood. In this case the mucilaginous layer not only inhibits germination and prevents damage to the roots, but the flood also disperses the seeds a long distance along the banks of the wadi.

In this plant the amount and distribution of rain regulate the time of dispersal and germination to when enough water is available for seedling establishment and development. Despite the good seed protection and very sophisticated "cautious" strategies of dispersal and germination, why are these plants rare? Is it because of the very few seeds that mature on these plants and the very rare occasions in the extreme deserts in which this mechanism could react?

Asteriscus hierochunticus (= *A. pygmaeus*) (Asteraceae) is another example of an annual plant that inhabits the hottest and driest desert areas, such as the southern- and eastern-facing slopes on the hills near Sede Boker. It is also fairly common in the Judean desert, the Jordan valley, Dead Sea, and Arava valleys of Israel. This plant has a Saharo–Arabian distribution.

The main stems, which are only a few millimeters long, terminate in an inflorescence that has a diameter of ca. 16 mm. In this size inflorescence about 200 achenes (seeds) are arranged in 11 whorls. On lateral stems smaller inflorescences develop of about 6 mm in diameter, containing ca. 50 achenes arranged in about 5 whorls. The dry inflorescence (capitulum, head) is surrounded and covered by 11–18 woody bracts that are arranged in 2 whorls. This small woody dry plant containing the "seed bank" in the capitulum is situated very close to the soil surface and remains in place for many years. The bracts are opened when wet by the swelling of the cohesion tissue which is situated in the basal plate of the capitulum as well as of the hydrochastic

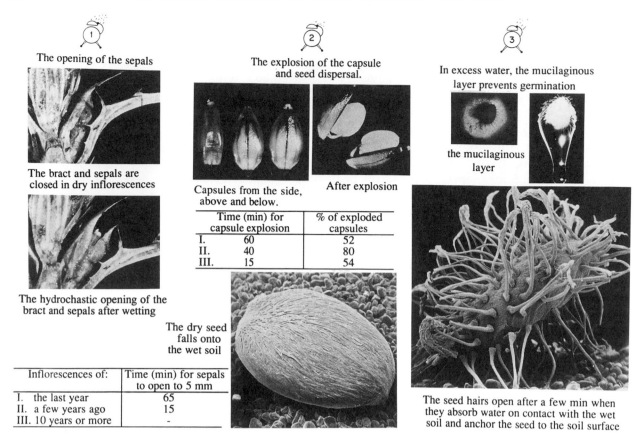

The opening of the sepals

The bract and sepals are closed in dry inflorescences

The hydrochastic opening of the bract and sepals after wetting

The explosion of the capsule and seed dispersal.

Capsules from the side, above and below.

After explosion

Time (min) for capsule explosion	% of exploded capsules
I. 60	52
II. 40	80
III. 15	54

The dry seed falls onto the wet soil

In excess water, the mucilaginous layer prevents germination

the mucilaginous layer

The seed hairs open after a few min when they absorb water on contact with the wet soil and anchor the seed to the soil surface

Inflorescences of:	Time (min) for sepals to open to 5 mm
I. the last year	65
II. a few years ago	15
III. 10 years or more	-

FIGURE 2 The three "water clocks" of seed dispersal of *Blepharis* spp.: (1) the opening of the sepals by wetting; (2) the explosion of the capsule and seed dispersal; and (3) the prevention of germination if there is excess water, which enables these seeds to be dispersed along wadis by the flood (see text). The bract and sepals are closed in dry inflorescences and open after wetting. Bracts of the previous season's inflorescences (I) open more slowly than inflorescences from earlier years (II). The weathered inflorescences (III) open after the shortest time to a distance of more than 5 mm between the upper parts of the sepals. This opening enables drops of rain to wet the "lock area" at the terminal tip of the capsule. The longer the time from seed maturation (I, II, III), the shorter the time of wetting needed for the capsules to explode and disperse its seeds. Only a portion of the capsules explode on one occasion. The dry seed falls onto the wet soil and germinates. After 50 min a radicle geotropic bend may be seen. In an excess of water the layer of mucilage that forms around the seed prevents germination until the excess water disappears. The seed hairs open after a few minutes on contact with the wet soil and anchor the seed to the soil. [After Gutterman *et al.* (1967); Gutterman and Witztum (1977); Witztum *et al.* (1969).]

cohesion tissue situated in the base and the middle of the bracts. The bracts are opened after a short time even during small amounts of rain when only a few drops wet the capitulum. The capitulum bracts close again when dry. The opening and closing of the capitulum bracts cause some achenes from the most peripheral whorl to become disconnected and the rains that cause the opening of the bracts also cause some of the disconnected achenes to drop out of the capsule. These achenes may be dispersed by wind because of their open hair (pappus) or they may adhere by these hairs to the soil surface near the dead mother plant which is the "seed bank" of

this species. If there is a heavy rain and runoff water develops on the slope, seeds could also be carried by the runoff water along runnels of the hill slope or into depressions and later germinate there. Germination takes place after about 10 days of wetting. Whorl after whorl of achenes are dispersed over the years, 1–10 achenes per year. This process can take more than 20 years.

There are other plants whose seeds are dispersed by rain in which the seeds change one shelter for another. When the mucilaginous seeds and/or very small seeds of *Aizoon hispanicum, Mesembryanthemum nodiflorum* (Aizoaceae), *Plantago coronopus*

(Plantaginaceae), *Carrichtera annua* (Brassicaceae), and others are dispersed by rain, they adhere to the soil surface until they germinate. During rainfall the activity of animals, birds, and insects is low but increases after the rain. At this time loose seeds are collected in great numbers.

Both strategies, the "protected" synaptospermic seeds as well as the "escape" strategies in other plants, are very important in preventing seed eaters from consuming all the seeds of the seed bank.

There are also other synaptospermic plant species whose seeds are dispersed by wind. In perennial plants such as some geophytes, there is a combination of a delay of flowering, seed maturation, and seed dispersal which shortens the time that the seeds remain in the dry open capsules in the inflorescences during summer and autumn. At least some of these seeds are dispersed during the following winter rain storms when they may germinate.

In the perennial shrub *A. sieberi* there is a delay in flowering to the following winter (December). The mucilaginous seeds mature in winter and when they are dispersed they are glued by their mucilage to the soil surface by rain and/or dew where they remain connected to the soil and are thus protected from consumption by ants and other granivors until they germinate.

III. GERMINATION STRATEGIES

In the germination process the seed, which is the most resistant stage to extreme environmental conditions, becomes a seedling, which is the most sensitive stage. Therefore, the development of strategies that ensure germination at the right time of the year and season and in the right place is very important for the survival of the species.

In trees and shrubs such as *Dalea spinosa, Olneya tesota,* and *Cercidium aculeatum,* inhabiting wadi beds of the Mojava and Colorado deserts of North America, F. Went found that their seeds germinate only after a very heavy flood. These seeds, with very hard seed coats, can only germinate after scarification by a very heavy flood carrying sand and gravel. This very simple mechanism ensures germi-

nation of these species only at the right place at the right time.

There are also some species in the Negev Desert highlands which have developed mechanisms which ensure that seed germination and dispersal only take place after a suitable amount of rainfall. An example is *Blepharis* spp. (see Section II,B).

A. Genotypic Germination Strategies and the Chance of Seedling Survival

During evolution the development of seed dispersal and germination mechanisms has taken various directions.

1. Plant species with "escape" strategies of seed dispersal and "opportunistic" strategies of seed germination. These produce very tiny dust-like seeds in very great numbers in a good year and often have escape strategies of seed dispersal. Seeds are dispersed immediately after maturation into splits in the soil and thereby avoid massive consumption by ants and granivors. Small portions of seeds germinate even after a rainfall of less than 10 mm which gives no chance for seedling survival unless followed by further rainfalls.

2. Plant species with "protection" strategies of seed dispersal and "cautious" germination strategies produce relatively larger seeds in small numbers which are protected by the woody parts of the dead mother plant. The seeds of different plant species either germinate *in situ* or are dispersed by rain, runoff water, floods, or wind. They germinate only after amounts of rain that are enough for further development of the seedlings and, in other plants species, even for the completion of the life cycle by seed production.

3. Plant species with both "opportunistic" and "cautious" strategies of germination have been found in some amphicarpic species. The aerial seeds of the same species have the more "opportunistic" strategies of germination whereas the subterranean seeds have the more "cautious" strategies.

4. Plant species with heteroblastic seeds in their dispersal units. There are also plant species with such mechanisms that allow only a portion of the seeds of one dispersal unit to germinate after one

rain event, even under optimal conditions, such as *Aegilops geniculata* (see Section III,B,2a).

5. Plant species with slow or fast germinating seeds, the ecological importance. Plant species whose seeds germinate after a short time of wetting belong to species with "opportunistic" strategies of germination. An example is *Salsola kali* (Chenopodaceae) which germinates after only 29 min of wetting in contrast to seeds that require long periods of wetting of more than 15 days to germinate. These seeds belong to species with the "cautious" strategies of seed germination. Examples are seeds of *A. sieberi* which require at least 16 days of wetting on the soil surface because they also need light for germination. For this reason mass germination may occur only once in 30–50 years as in 1963/1964 in the Negev Desert highlands when for more than 16 days the soil surface was wetted continuously by rains. This phenomenon of mass germination of *A. sieberi* following such unique conditions for germination is very rare in extreme deserts. The result was that such an event renewed the population of this dominant shrub and the new seedlings that survived replaced the old shrubs that had disappeared over the years. Such events in the desert cause "age groups" in the perennial plant populations. "Age groups" have also been found in the Negev Desert in *Zygophyllum dumosum* (Zygophyllaceae), another dominant shrub, at elevations of up to 600 m.

In conclusion, in such unpredictable conditions as those that exist in the Negev Desert, a strategy like "shooting in the dark" seems to be more successful in plants that have a combination of "escape" dispersal strategies and "opportunistic" strategies of germination of very small seeds in very large numbers (see Sections II, A and B). Plant species with the most sophisticated "protection" and "cautious" strategies of seed dispersal and germination, which produce fewer seeds such as *Blepharis* spp., are much less successful.

6. Seasonal Seed Dormancy Cycles. Terminal seeds of *M. nodiflorum* harvested near Jericho were found in 1980–1981 to reach a high percentage of germination in petri dishes in laboratory conditions, at constant temperatures from 15 to 35°C. This occurred during the winter season when plants germinate in their natural habitats. However, germination in the laboratory in constant temperatures reached low percentages during summer when, in their natural habitat, temperatures are high and there is no rain.

Baskin found annual seed dormancy cycles in two winter desert annuals, *Erigonum abertianum* (Polygonaceae) and *Eriastrum diffusum* (Polemoniaceae), collected and buried in the Chihuahuan Desert near Portan, Arizona. Germination in winter was low but in the autumn it was high in petri dishes at 20/10°C. This annual dormancy may prevent the germination of seeds of these plant species after an unexpected out-of-season rain in their natural habitat.

B. Phenotypic Influences and the Spreading of Germination in Time

In addition to the genotypic influences that increase the fitness of the species to their habitats, there are also phenotypic maternal and environmental influences that affect the germinability of the seeds. Maternal influences such as position and plant age effects or environmental influences during seed development and maturation, such as daylength, light, temperature, water stress, etc., affect seed germination. The fate of the further generation or generations is dependent, at least to a certain degree, on factors involved when the seeds are still on the mother plant before or during seed maturation. Phenotypic influences also occur during seed storage and seed wetting.

I. Maternal Influences Affecting Seed Germinability

a. Position Affecting Seed Germination

The position of the inflorescences on the plant affects seed germination in some plant species. In others the position of the seeds in the dispersal unit, capitulum, or capsule, as well as the position of the seeds from which the plants originated, have an influence on seed germinability.

i. Position of the Inflorescences on the Plant   In amphicarpic plants (plants with aerial as well as subterranean inflorescences), such as *Gymnarrhena micrantha* (Asteraceae) and *Emex spinosa* (Polygonaceae), the aerial achenes or propagules are dispersed and are much smaller than the subterranean ones.

In *G. micrantha* the aerial achenes are very small and they weigh 0.37 mg. They have hairs (papii), and after maturation at the beginning of the summer they are protected by the inflorescence and achene bracts. These bracts open when wetted during the rain of the following rainy season. The papii are also opened when wetted by rain and the achenes are dispersed by wind. The subterranean achenes are much heavier, weighing 6.5 mg. They are protected by the woody parts of the dead mother plant below the soil surface, have undeveloped papii, and germinate *in situ*. There are big differences in the environmental conditions needed for germination of the aerial and subterranean achenes.

In *E. spinosa* the aerial propagules weigh 24 to 2 mg, depending on the position. The terminals are the smallest with the relatively largest spines. The subterraneans are much larger (74 mg) and have the smallest spines. The woody propagules remain connected to the woody subterranean parts of the dead mother plant and germinate *in situ*. The leachate from aerial propagules contains water-soluble germination inhibitors whereas the subterranean propagules do not.

The percentage of germination of the subterraneans is much lower than the aerials from the same season of maturation. This means that the subterraneans will spread their germination over more years. Together with their lower numbers these factors reduce competition. The aerials germinate better as more water-soluble germination inhibitors are washed away by rains. These inhibitors act as a "rain gauge" or "rain clock" in which the more rain that wets the soil, the more propagules will germinate. Propagules that are trapped in depressions where runoff water accumulates germinate earlier because of the runoff accumulation and the faster dilution of the germination inhibitors.

ii. Position of the Capsule on the Plant Canopy
Large peripheral and small central capsules occur on *Glottiphyllum linguiforme* (Aizoaceae) plants inhabiting the Karoo Desert in South Africa. The large capsules of the peripheral part of the plant contain about 200 seeds per capsule. Their seeds germinate in petri dishes at 25°C after 18 days to 80% in light and 70% in dark. Only a few seeds from the small central capsules, which contain about 125 seeds per capsule, germinate at the same time and in the same conditions. When adhered to wet soil, about 20% of the seeds from the peripheral capsules germinate but none of the seeds do from the central capsules. The peripheral capsules, which are easily separated from the plant canopy, are possibly dispersed by wind or floods whereas the central capsules remain below the canopy and are covered by mounds. These are probably the local seed bank which may supply seedlings to replace the mother plant.

iii. Position of Achenes
The position of achenes in capitulum whorls and their different germination has already been mentioned in *A. pygmaeus* (see Section II,B).

iv. Position of Flowers
The position of flowers and the dimorphism affecting germination were found in *Salicornia europaea* and *Atriplex dimorphostegia* (Chenopodiaceae).

v. Position of the Seeds in the Dispersal Unit
The dramatic position effect that has an influence on more than two generations is the influence of the position of the different grains (caryopses) in a spike, which is the dispersal unit of *Aegilops geniculata* (= *A. ovata*) (Poaceae). For instance, in a spike containing three spikelets and five grains (Table I[a]), the lowest grain (a_1) germinates the best after the shortest time and the single grain of the uppermost spikelet (c) germinates to the lowest percentage after the longest time. Grain b_2 has an intermediate behavior (Table I[b]). The hulls of spikelet A contain less germination inhibitors than spikelet B whereas C has the most (Table I[c]). Not only are there differences between the different grains in the same dispersal units, but they are also different according to the grains from which the mother plant originated. If the germinability of grain a_1 that was harvested from a mother plant that originated from

TABLE I

a	b		c	d	
3-spikelet (A, B & C) dispersal unit and position of caryopses (a₁-c)	Spikelet order	Caryopses % germination from each spikelet	Germination % of *Lactuca sativa* in leachate from hulls of *A. geniculata*	Order of caryopses from which mother plants developed	Germination % of a₁ caryopses matured on different plants from different orders of caryopses
	A	88±4	48±3	a₁	84.4
	B	72±7	36±4	b₂	55.0
	C	10±6	29±2	c	21.2
	Control (water)	-	94±2	-	-

^{a} Schematic drawing of a three-spikelet (A,B, and C) dispersal unit of *Aegilops geniculata* (= *A. ovata*). The position of caryopses a₁, a₂, b₁, b₂, and c is shown.

^{b} The germination (% ± SE) after 5 days of imbibition of caryopses from spikelets A, B, and C sown in soil at 20–30°C. The spikelets were separated from three-spikelet dispersal units.

^{c} The germination (% ± SE) of *Lactuca sativa* achenes in light at 26°C imbibed in leachate from hulls of spikelets A, B, and C of *A. geniculata* dispersal units of three spikelets.

^{d} The influence of the order of caryopses from which the mother plants develop on the germination percentages of caryopses order a₁ matured under long days and temperatures of 15/10°C. After 24 hr of imbibition at 15°C in light.

grain a₁ was compared with the germinability of grain a₁ from a mother plant that originated from grain b₂ or c, their germination would be very different after 24 hr at 15°C in light: 84, 55, and 21% respectively (Table I^{d}).

Position effects on "seed" germination found in *Pteranthus dichotomus* (Caryophyllaceae) were studied by Evenari. In a dispersal unit with seven "seeds" the four terminals germinate the best, the lowest one germinates the worst, and the two intermediates have intermediate germination.

vi. *Position of the Seeds in the Fruit*

The position of the seeds in the capsules of *Mesembryanthemum nodiflorum* (Aizoaceae) affect seed germination. Each capsule contains about 30 seeds. According to the time of wetting needed for seed separation, there are three groups of about 10 seeds in each. In experiments carried out by collecting each group of seeds separately from many capsules and wetting each group separately in petri dishes, the following results were observed: Seeds from the group situated at the uppermost part of the

capsule (the terminals) separate and disperse after 15 min of wetting. Eight years after seed maturation they germinate to about 60% after 12 days of wetting. Seeds from the lowest group are dispersed only after 320 min of wetting and their germination in the same time and conditions of wetting was only 1%. Seeds from the middle group dispersed after 200 min and germinated to 5.5%.

The position effect enables the same genotype (mother plant) to produce seeds with differing germinability. The seeds will be "ready for germination" on different rain occasions which is very important for spreading the risk and preventing all the seeds from germinating at the same time, even under optimal conditions. Such a strategy has a great survival value, especially under the unpredictable appearance of rains in the extreme deserts.

b. Plant Age and Seed Germinability

In addition to the daylength effect (see later) the age of plants affects the degree of seed coat color and impermeability and thereby their seed germinability.

In *Trigonella arabica* and *Ononis sicula* (Fabaceae), under long days older plants also produce green and brown seeds which is typical mainly to seed maturation under short days in young plants. The younger plants under long days produce seeds with yellow, well-developed, impermeable seed coats and the seeds germinate only when the seed coat is damaged or after a suitable treatment.

2. Phenotypic Environmental Effects

a. Daylength

Quantitative long days, quantitative short days, long days, or short days affect seed germinability during seed maturation and are found in seeds of different plant species. Daylength affects the seed coat development, color, and degree of impermeability to water during the last 8 days of seed maturation.

i. Short- or Long-Day Effects
These effects have been found in *O. sicula* and *T. arabica*. These desert annual plants produce, under long days, yellow seeds with seed coats that are impermeable to water. Under short days, brown seeds develop on which the seed coats are poorly developed, and these seeds swell immediately on contact with water. The green seeds that also develop under short days have an intermediate seed coat development as well as an intermediate effect of seed imbibition.

In *T. arabica* the influence of the daylength is not a direct effect. In covered fruit under short days or long days, materials that move from the leaves affect the typical development, color, and impermeability to water of the seed coat according to the daylength during the last stage of seed maturation.

ii. Quantitative Long-Day Effects
These effects are found in *Polypogon monspeliensis* (Poaceae). Seeds (grains) that matured on plants treated with artificial longer days had a higher percentage of germination at 25°C in continuous light than seeds matured under a shorter daylength.

The same was found in *Carrichtera annua* (Brassicaceae). Seeds were matured in 8, 13, and 20 hr in covered fruit during the last stage of seed maturation, which affected seed germinability. The longer the daylength to which the mother plant was exposed, the higher the percentage of seed germination after 144 hr of wetting at 25°C.

Quantitative daylength effects have also been found in *Cucumis prophetarum* (Cucubitaceae), a perennial plant inhabiting desert oases near the Dead Sea. Ripe turgid fruits were harvested and stored under different daylengths of 8, 11, 13, and 15 hr for 9 days. The seeds were then separated and wetted. More seeds from the longer days germinated after 7 days of wetting.

iii. Quantitative Short-Day Effects
This type of effect is found in *Portulaca oleraceae* (Portulacaceae). Plants were transferred before the last 8 days of seed maturation, which is the critical time for the daylength effect, from 16-hr long days to shorter days of 13 or 8 hr. Seed germination was higher when the daylength was shorter.

In conclusion, under artificial daylength the critical time that affects seed germinability is the last period of 7–15 days of seed maturation. The ecological importance of such daylength influences, if they have the same effect in natural conditions, could be the ability of a single plant to produce different seeds with differing germinability according to the daylength during the last 8 days of maturation of each of the fruits, even along one branch. These differences are very important for the survival of plants under desert conditions by spreading seed germination over time.

b. Seed Maturation under Natural Conditions and the Difference in Their Germinability

i. Seasonal Effects
Winter- or summer-matured seeds of Aizoaceae shrubs from South Africa, which flower twice a year and which have been introduced to the Introduction Garden at Sede Boker, differ in their germinability. When these summer- and winter-matured seeds were wetted at different temperatures, different relative amounts of germination were observed. From this, seeds matured on the same shrubs during different seasons display different germinability.

ii. Dates of Achene Maturation and Germinability Achenes of *Lactuca serriola* (Asteraceae) collected during the summer on different dates from July to October from a natural plant population in the Negev Desert highlands near Sede Boker differ in their germinability. These achenes, which were collected during different months, germinated to different relative amounts at different temperatures. The "seeds" from the different maturation dates germinate to different levels when they are wetted for the same time at the same temperature.

Quinlivan found in *Trifolium subterraneum* (Papilloniaceae) seeds matured in Australia in spring under natural daylength, show that the longer the growing season in which seeds matured under longer days, the larger the proportion of hard seeds produced. Similar day length effects were found under artificial daylength in *O. sicula, T. arabica,* and others (see Section III,B,2).

c. Altitude

Altitudinal effects were found by Dorne in *Chenopodium bonus-henricus* (Chenopodiaceae). Seeds collected from plants situated at an elevation of 600 m had higher percentages of germination than seeds harvested from plants from an elevation of 2600 m. The seed coat of seeds from the higher altitude were thicker and contained more polyphenols which inhibited their germination.

d. Light Wavelength during Seed Maturation Affects Seed Light Sensitivity

In seeds surrounded by maternal green tissues still containing high quantities of chlorophyll during the last stage of maturation and dehydration, most of the phytochrome is in the inactive Pr form. Therefore, these seeds require a light stimulus to germinate in the dark. Seeds matured below a plant's green canopy may be affected in the same way.

Treatments of far-red light during seed maturation also have an influence on the seed light sensitivity and the percentage of seed germination in light or dark.

Seeds from mature turgid fruits of *Cucumis prophetarum* that had been stored under far-red or red filters showed differences in the percentage of the active phytochrome (Pfr) and dark germination. Under far-red light more photoreversible phytochrome accumulated, but almost all this phytochrome was found in the inactive (Pr) stage and after 50 hr at 25°C in the dark no germination occurred. After storage of such fruits under red light, much less photoreversible phytochrome accumulated but most of it was in the active (Pfr) stage. These seeds germinated to 100% after 50 hr of wetting at 25°C in the dark.

The opposite results in germination occurred when seeds germinated in light after the storage of such fruits under red and far-red filters. There was a much lower germination in seeds harvested from fruit stored under red light than under far-red light. Under far-red light more photoreversible phytochrome accumulates, which in light turns into the active Pfr form of the phytochrome.

e. Light and Temperatures Affecting Germination

In *A. dimorphostegia* (Chenopodiaceae) and *Hyosyamos desertorum* (Solanaceae), seed germination is inhibited by light at low temperatures. *L. serriola* seeds are not light sensitive at low temperatures but are at higher temperatures. *C. annua* seeds are light sensitive when wetted in the range of temperatures from 10 to 25°C as well as after imbibition at supraoptimal temperatures of 30–40°C.

f. Seeds with No Responses to Light

Seeds of *Blepharis* spp., for example, have no light sensitivity in the whole range of temperatures suitable for germination.

g. Temperatures Affecting Seed Germinability

i. Temperatures during Maturation Maturation of *A. geniculata* grains at temperatures of 15/10°C has a large influence, as mentioned, on the germinability of the a_1 grains, depending from which grain the mother plant originated. In mother plants originating from a_1, a_1 grains germinated to 84%, from b_2 to 55%, and from c to only 21%

within the first 24 hr. At high temperatures of 28/22°C during maturation, all the a_1 caryopses germinated to 100% within 24 hr (Table I).

ii. Temperatures during Storage In the Australian arid zones, Quinlivan found that after a spring with overgrazing the germinability of seeds of subterranean clover (*T. subterraneum*) and blue lupin (*Lupinus verius*) that remained on the bare soil surface was affected by the day and night temperature fluctuations during summer. The higher fluctuation of 75/15°C for 6 months affects the softening of nearly all the hard seeds. When there is no overgrazing and the soil remains covered during summer the fluctuation of day/night temperatures is reduced. The lower the day/night temperature fluctuation during summer the higher the percentage of hard seeds into which water cannot penetrate to the embryo, thus preventing germination.

In the Negev Desert highlands, measurements showed that the day/night temperature fluctuations between the soil surface and different depths decreased from the soil surface to a depth of about 30 cm where the daily fluctuation of temperature almost disappears. On the bare soil during summer the soil surface daily maximum temperature is even higher than 55°C. Similar results were also found in New South Wales, Australia. Day/night temperature fluctuations of 60/15°C were found in northern West Australia.

The majority of seeds of desert plants were found to be situated either above the soil surface on the dead mother plant or in the soil surface to a depth of about 2.5 cm. The activity of birds and animals such as porcupines and gazelles affect soil turnover in which seeds could be trapped and covered even up to depths of 25 cm. Seed cover could also occur by soil erosion affected by runoff water or floods. Different environmental conditions during summer affect seeds at different soil depths, which has an influence on the "readiness for germination" of the seeds at the following rainy season.

Hordeum spontaneum and *S. arabicus* (Poaceae) grains need to be stored dry for at least 70 days at high temperatures above 35°C to be "ready for germination" the following winter. If grains are artificially stored at low temperatures, they are not capable of germinating the following winter.

In eight out of nine annual plant species from the Mojavan and Sonoran deserts, Capon and Van Asdall found that dry storage of 1–5 weeks at 50°C increased seed germination to very high percentages compared with storage at 20°C when these species reached only 0–16% germination after 4–8 weeks.

3. Environmental Factors Affecting Germination during the Time of Wetting

According to Fritz Went, water and temperature are the main environmental factors regulating the time of germination in plants in deserts that receive rains in winter and summer, such as some desert areas of the southwest of North America. In low temperatures, winter annuals germinate and in higher temperatures summer annuals appear.

a. Water

i. Minimum Amount of Water and Germination Germination is the most critical stage of the annual life cycle of desert annuals. Germination at the right time is most important for the survival of the seedlings. As mentioned earlier, at least two extreme directions of development occurred during evolution. One is the "opportunistic" strategies in which portions of the very large numbers of very small seeds of a plant species such as *Schismus arabicus* (Poaceae) will germinate even after less than about 10 mm of winter rain. Thus the seedlings have no chance to survive if this rain is not followed by other rains.

In contrast, there are plant species that have "cautious" strategies of germination. An example is *Blepharis* spp. in which the time of seed germination is dependent on the time of seed dispersal. Seed dispersal occurs only after a long rain or flood. In other plant species such as *Erodium hirtum* (Geraniceae), germination will occur only after a rain of more than 25 mm, which may wet the soil to a depth of about 25 cm in the loess soils of the Negev Desert highlands. In this case all the seedlings that had been observed survived.

In the deserts of the southwestern United States, F. Went and colleagues discovered that some species germinate after only 10–15 mm of rain whereas others only after 30–43 mm. In the Australian deserts, Mott and Groves found that winter-germinating plant species of the Asteraceae require 15–20 mm of rain and germinate within 2–5 days, whereas seeds of summer-germinating grasses germinate after more than 25 mm of rain within 24 hr.

ii. Germination Inhibitors as Rain Gauges

Areas near the Dead Sea where very high salinity accumulates on the soil surface during the summer are inhabited by species such as *M. nodiflorum*. In 1972 mass germination occurred only after 11 rains when the salinity on the surface had been washed down to a certain depth in the soil. The seeds of *M. nodiflorum* plants inhabiting these areas with high soil salinity do not germinate in water containing more than 1% NaCl. Here the amount of water that washes away the salts regulates the time of germination and acts as a "rain gauge."

Other "rain gauges" are found in *A. geniculata* and in many other plant species in which water-soluble germination inhibitors accumulate in the seeds and in other parts of the dispersal units. In a spike with three spikelets, the five different grains as well as the three different spikelets contain different amounts of germination inhibitors. Therefore, the different grains germinate after suitable amounts of rain during different years, which spreads the risk and prevents competition between the seeds germinated from one dispersal unit (Table I).

Chemicals found in plant litter that contain germination inhibitors may also react on seed germination as "rain gauges" or prevent germination beneath the canopy of these plants, which also prevents competition (allelopathy). In this system mass germination occurs only after a fire.

b. Temperatures during Wetting

i. High Temperatures

Seeds of plant species such as *Bergeranthus scapiger* (Aizoaceae) inhabiting the summer rain area of the Karoo desert in South Africa germinate and develop after summer rains. They may be wetted at least for the first few hours, in temperatures as high as 45°C without any damage. The period of wetting of these seeds in high temperatures may even accelerate seed germination when they are wetted later at their optimal temperatures.

The opposite is true in seeds of plants that germinate in winter. These seeds germinate in relatively low temperatures, and periods of high temperatures at the beginning of their wetting inhibits their germination later at their optimal temperatures. This effect could prevent winter-germinating species from germinating in summer. Examples are *Cheiridopsis* spp. (Aizoaceae) from South Africa and *C. annua* and *L. scariola* from the Negev Desert.

In *L. serriola* this thermoinhibition may not appear if after wetting of the achenes in high salinity and high temperatures they are transferred to their optimal temperature after the salinity has been washed away.

ii. Germination in a Wide Range of Temperatures

The germination of seeds of some plant species, such as *Blepharis* spp., is regulated by water only and they may germinate in a range of temperatures from 8 to 40°C in the light or in the dark. In other species, temperature regulation has an influence on their distribution, even on different hill slopes.

iii. Temperatures for Germination and Habitats

The Saharo–Arabian plant *Helianthemum ventosum* (Cistaceae) germinates in high temperatures and inhabits south-facing slopes in the Negev Desert highlands at an elevation of 400–500 m. In contrast, the Irano–Turanian *H. vesicarium* does not germinate in temperatures higher than 30°C and inhabits north-facing slopes in this area.

There are also other plants in which a correlation has been found between their habitat and the temperatures they require for germination. A germination study of three members of the Liliaceae, which inhabit the Negev Desert highlands near Sede Boker, found that *Bellevalia desertorum* is the most thermophilous plant of these species and inhabits south-facing slopes. Its seeds germinate better and faster than the other two species (*B. eigii* and *Tulipa systola*). In all temperatures between 10 and 25°C, germination starts within 7 days. In contrast, *T. systola,* which in this area inhabits north-facing slopes, did not germinate at 25°C even after 56 days of wetting in light, at 20°C they only started to germinate after 28 days, and at 10 and 15°C they started to germinate after 10 or 14 days of wetting. *B. eigii,* which inhabits the loessial wadi beds, has an intermediate temperature requirement between these other two species.

4. Light

a. Light-Sensitive Seeds to the Whole Scale of the Visible Spectrum

The light-sensitive seeds of *Artemesia mono-sperma* (Asteraceae) inhabiting sandy deserts have a higher percentage of germination in light of any wavelength of the visible spectrum than in dark. The sunlight spectrum changes according to the depth of the sand. The deeper the sand the longer the wavelength that can penetrate. If seeds of *A. mono-sperma* are too deep in the sand they will not germinate because of lack of light. But if they are too near the sand surface they will not germinate because the soil water content is too low for the long period of wetting (10 days) required for germination, as was found by Koller, Sachs, and Negbi.

b. Dark-Germinating Seeds

Seeds of plant species inhabiting areas covered by sand, such as *Pancratum maritimum* (Amaryllidaceae), *Zygophyllum coccineum* (Zygophyllaceae), *Calligonum comosum* (Polygonaceae), *Cakile maritima* (Brassicaceae), *Aster tripolium* (Asteraceae), and others, need to be covered by sand before they can germinate. These plants are inhibited by light at all temperatures.

In *P. maritimum* seeds, Keren and Evenari found that the lower the light intensity, the higher their germination after 14 days of wetting. Under 1800 lux only 13% of the seeds germinated. In 52% of this light intensity, 30% germinated and in 26% light intensity, 54% germinated whereas 96% of the seeds germinated in complete darkness.

IV. FLOWERING STRATEGIES OF SOME DESERT PLANTS

A. Flowering Strategies and the Life Span of Some Annual Desert Plants

The regulation of the time of flowering may ensure the production of mature seeds before the dry and hot summer in many winter annuals inhabiting deserts such as the Negev, which receives only winter rains. This includes even plants that germinate after a late rain in the growing season.

There are plant species with different responses to environmental factors such as soil water availability, daylength, and temperature as regulators of the date of flowering. Annual species that have been studied fall into three main groups: (a) Facultative long-day (FLD) plants, with earlier flowering under longer days; (b) day neutral (DN) plant species that flower at the same age in any daylength; and (c) facultative short-day (FSD) plants.

I. Facultative Long-Day Plants

In facultative long-day plants the longer the daylength the earlier the flowering and the shorter the life span. This enables the plants to complete their life cycle and produce mature seeds before the dry hot summer, even if they germinate late in the season. The earlier the germination the more leaves develop and the greater the seed yield.

According to F. Went, the majority of Californian annual plants studied are facultative long-day plants. Of 15 annual plant species studied, 11 are FLD plants, 3 seem to be DN, and only 1 is bimodal.

Evenari and Gutterman found that of the 21 plant species studied from the desert annuals inhabiting the Negev Desert, 12 are FLD plants. Only 3 are DN under water stress conditions, whereas under artificial greenhouse conditions they are also FLD plants. Six others are intermediates, FLD, or DN/FLD plants.

From these two studies it seems that under such desert conditions the FLD flowering strategy is more successful for species survival. Depending on the time of germination after the proper amount of rain (which is unpredictable), the plants orientate themselves according to the daylength to the proper life cycle, enabling them to produce at least a small number of seeds before the plant dries out at the beginning of summer.

In the facultative long-day plant *T. arabica* (Fabaceae), a very large difference was found in the number of leaves and the total biomass as well as the time of flower bud appearance. When the plants grew in 9-hr short days they flowered after 94 days, the plant biomass was 11.2 g of dry weight, and the number of leaves on a plant with about 12 lateral branches was more than 150. Under long days of 18 hr the plant developed only the main stem without any lateral branches. Flowering occurred after about 10 leaves and 31 days. The dry biomass weight of such a plant was only about 0.2 g.

When plants were grown under 8, 12–14.5, and 20 hr per day, the longer the daylength the shorter the time to flower bud appearance. Only under short days of 8–9 hr was there a typical appearance of lateral branches that did not occur under 12-hr days and longer. The time of flowering was gradual.

T. stellata (Fabaceae) flowered under short days of 8 hr after 77 days in the greenhouse and 71 days in outdoor conditions. The number of leaves at flower bud appearance was 54 and 24, respectively. Under long days of 20 hr per day, in which the plants received daylight for 8 hr and additional low light intensity for 12 hr, plants flowered in greenhouse conditions after 37 days and in outdoor conditions after about 60 days. In both treatments the number of leaves until flower bud appearance was 5.9 and 5.5, respectively.

S. arabicus (Poaceae) showed similar results to those of *T. arabica* and *T. stellata*. This plant inhabits the same area of the Negev Desert highlands but is a member of a different family. The number of leaves as well as the time to anthesis (time of the opening of flowers and extension of the anther outside the flower) is 88 and 77 days, respectively, under short days in greenhouse conditions and outdoors. Under long days it is 62 and 59 days, respectively. The number of leaves in this plant is dependent on the number of tillers (lateral branches) (Fig. 3).

2. Day Neutral Plants

Plant species that flower in any daylength after a relatively short time are called day neutral. If there are suitable conditions for growth, such as water availability in the soil, they will develop more lateral branches and flowers and produce more seeds.

One such plant is *C. annua* in which, under greenhouse conditions with low light intensity and no water stress, this plant responds as a FLD plant. But, at the same time, in outdoor conditions of natural temperatures, humidity, and water stress between one irrigation and the next, these plants showed DN behavior for the first flower bud appearance:

i. In both short days of 8 hr or long days of 16 hr, six leaves appeared before the first visible flower bud.
ii. In both, the first flowers appeared after 45 and 43 days, respectively.
iii. The age of the plant at the time of death was 136 and 140 days, respectively.

Under greenhouse conditions, in short days the number of leaves before the first visible flower bud was 17.4 and the age at the time of first flower appearance was 66 days. Under long days the number of leaves was 5.3 and the time of first flowering was 31 days.

In field observations in the Negev Desert highlands near Sede Boker (Avdat) in 1961, seeds of *C. annua* germinated at the beginning of March, which is late in the season. By 21 April the plants had already flowered, and by the end of May they had finished their life cycle, produced mature seeds, and dried out (Fig. 3). In the same area in 1964, seeds germinated at the beginning of January, the plants flowered on February 22, and finished their life cycle by producing mature seeds and drying out at the beginning of May.

As can be seen, the time from germination to first flower was almost the same and within a very short time (about 40 days) whether the germination began late or early in the season. But there was a

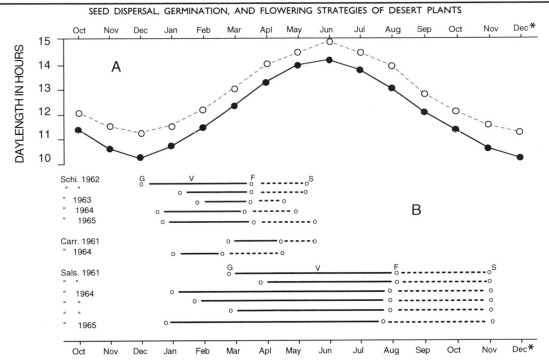

FIGURE 3 (A) Hours of daylength from sunrise to sunset (—) or photoperiodic daylength in hours (from morning light intensity of 5 lux to the same intensity in the evening) (---). (B) Date of germination (G), first appearance of flowers (F), first appearance of seeds (S), and length of vegetative period (V) for *Schismus arabicus* (Schi.), *Carrichtera annua* (Carr.), and *Salsola inermis* (Sals.) as observed in Avdat in various years. The asterisk indicates the 21st of each month. [After Gutterman (1993).]

very large difference in these two years between the time of first flowering to seed maturity and the death of the plant. In 1961 the main rains stopped at the end of January and only a little rain appeared at the end of April (5.8 mm). The time between the first flower appearance and the death of the plant was only about 30 days, whereas in 1964 the time from flowering to the death of the plant was about 65 days. This was because in that year from the time of flowering there were 7 days of rain until April 18 with about 26 mm of rain. This enabled the plants to develop more flowers and more seeds for a longer time on lateral branches (Fig. 3).

Similar responses for flowering and seed maturation have also been observed in *Reboudia pinnata* (Brassicaceae) and the amphicarpic plant *Gymnarrhena micrantha*.

3. Short-Day Facultative Plants and High Temperatures for Flowering

Salsola inermis (Chenopodiaceae) is a biseasonal annual plant species that inhabits the Saharo–Arabian

and also parts of the Irano–Turanian region. The seeds germinate during the rainy season in winter and the plant remains in a rosette form of 14 or 18 small and hairy leaves about 1 cm tall. Only at the beginning of the summer, when the winter annuals produce mature seeds and dry, do the *S. inermis* plants start to elongate their branches and develop lateral branches. They flourish during the hottest and driest season of the year and flower in the autumn under a combination of high temperatures and short days. Under artificially high daytime temperatures of 35–40°C in the greenhouse and under natural short days, *S. inermis* plants start to flower in the spring (April) and not as under natural conditions, in the autumn (Fig. 3).

In an artificial short daylength of 8 hr light per day, in the greenhouse or in outdoor conditions, flowers appear after 108 and 117 days, respectively. Under long days of 20 hr, the first flowers appear after 211 days in the greenhouse and 232 days outside.

The time of flowering also has a large influence on the age of the plant at death. Under short days in the greenhouse and outdoors, death occurs after 172 and 190 days, respectively, whereas under long days it occurs after 351 and 235 days, respectively.

P. oleracea (Portulaceae) inhabits some deserts that receive summer rains. It is one of the most common weeds in summer-irrigated crops in very large areas of the world. From 11 to 15 hr, the longer the daylength the later the flowering and the more leaves appear before the first flower bud develops. Under 11, 13, and 15 hr of light per day, flower buds appear after 10, 14, and 19 days and after the development of 3, 7, and 10 leaves, respectively.

B. The Flowering Time of Perennials

I. Flowering of Desert Shrubs

Only a few perennial shrubs whose daylength response for flowering have been studied.

It is interesting to note that one of the dominant shrubs, *Artemisia sieberi*, has a facultative short-day effect in seedlings of the first year and a short-day effect in adult plants which flower, as already mentioned, during the period with the shortest days of the year, in December. Flower buds already appear in May/June.

The other common and dominant plant on south- and east-facing hillslopes is *Zygophillum dumosum*, a Saharo–Arabian plant, which is a day neutral plant for flowering. If water is available by artificial irrigation, flowers appear throughout the year. In natural conditions flowers appear only in March/April.

Hammada scoparia (Chenopodiaceae) is an Irano–Turanian and Saharo–Arabian plant that is a very important and dominant shrub of the Negev Desert highlands. It appears mainly in deep loess soil in foothills, wadis, and flat areas. This plant has a long-day effect on flowering. Flowers appear in June/July under natural conditions.

Blepharis spp. are facultative perennial or annual plants and day neutral for flowering, as has been found in plant species and ecotypes inhabiting the deserts of Israel and Sinai. The flowering time of seven species and ecotypes showed no differences in the time of flowering of all the plants tested.

Two of the *Blepharis* species and some ecotypes were tested under 8-hr short days or 20-hr long days when seeds were sown on May 7, 1969. No difference was found between the age of the plant at first flower bud appearance or first flower between the plants under short days or long days. These plants are day neutral for flowering.

If germination occurred before the winter, three of the species and ecotypes tested had a much longer vegetative period until the appearance and opening of the first flower (time of anthesis), an average of 201 to 234 days. In comparison with plants whose seeds germinated before the summer (on April 28), the time of anthesis was only between 107 and 116 days.

In *Blepharis* spp., which inhabits extreme deserts, temperature and water stress are the regulating factors of developing and flowering and not the daylength. This is also true of the dispersal and germination strategies of these plant species, as mentioned earlier. Only the amount of water is the regulating system and, after dispersal, seeds germinate very quickly in a very large range of temperatures, from 8 to 40°C, in light or dark.

2. Flowering of Desert Hemicryptophytes

Hemicryptophytes are perennial plants which renew their growth in winter from buds situated on stems near the soil surface.

One of the most common hemicryptophytes in the Negev Desert highlands is *Erodium hirtum* (Geraniaceae), a Saharo–Arabian plant that has daylength-independent flowering. Plants in artificially irrigated plots may flower throughout the year whereas under natural conditions plants flower from February to May. When there is a reduction in soil moisture the plants enter into a less active stage in which all the upper parts die out. Only the roots with their storage organs and part of the stems with buds near the soil surface remain during summer.

An annual rhythm of leaf appearance followed by flowering has been found in *Scorzonera paposa* (Asteraceae), which has a West Irano–Turanian extending into East Mediterranean distribution. This plant is a hemicryptophyte in which, even under dry sand in laboratory conditions, small leaves, 5 mm long, appear before the rainy season. If these plants are kept dry, without any irrigation, after a while the small leaves dry out and the main storage root remains without leaves until the beginning of the next rainy season. This phenomenon repeats itself for 5 years if the storage organs are kept in dry sand in laboratory conditions. After this time the storage organs dry out and no more leaves appear.

Under natural conditions, after the rains wet the soil in winter the plant develops a canopy of leaves and then flowers. In a year with less rains, only a few flowers appear in populations of this plant. At the beginning of summer it enters a less active stage in which, after mature seeds have been produced, all the upper parts of the plant dry out. The seeds are later dispersed by the wind. Even under artificial irrigation, *S. paposa* plants enter the less active stage at the beginning of the summer until the beginning of the next growing season. This contrasts with *E. hirtum* in which plants could continue to flower throughout the hot and dry summer if water is available in the soil by artificial irrigation.

3. Synanthous-Leaved and Hysteranthous Geophytes

Only a few studies have been done on synanthous-leaved and hysteranthous geophytes inhabiting the Negev Desert highlands.

a. Flowering of Hysteranthous Geophytes

Colchicum tunicatum (Liliacea), with a West Irano–Turanian distribution, is one of the most common hysteranthous geophytes found in the Negev Desert highlands, in all the main habitats such as hill slopes and flat loess areas. They have an opposite response to daylength according to the temperatures and flower in autumn. At this time of year very hot days may alternate with colder days.

The annual replacement corms that are situated 15 cm below the soil surface respond to daylength. When the temperature at the depth of the corm is about 15°C the plant will behave like a FLD plant. The longer the daylength the earlier the flowering and the higher the percentage of plants of a popula-

tion that flower. The opposite is seen during September/October when the temperature is about 30°C. Under such temperatures these plants respond as FSD plants in which under short days flowering is earlier and the percentage of flowering is higher.

There are only a few species in the Negev Desert that flower at this time of the year and it is possible that a combination of temperature and daylength also regulates the appearance of the insects that pollinate these flowers. It is possible that there is a certain mutuality between the regulating mechanisms of flowering and the emergence of their pollinating insects from the "resting" stage.

Sternbergia clusiana (Amaryllidaceae) is another species of hysteranthous geophytes that populate the Negev Desert highlands in favorable desert microhabitats receiving runoff water. Their distribution is in the Mediterranean and West Irano–Turanian area. The bulbs of these plants are situated about 15 cm below the soil surface and the plants do not flower until the maximum daily temperature is below 25°C. These plants have a FLD response for flowering. Under a constant artificial daylength of 18 hr, flowering occurs much earlier and a higher percentage of the plant population flowers compared to the group of plants that receive 9 hr of daylength. An intermediate effect of flower appearance was found under natural daylength. During these experiments the natural daylength was shortened from 12.5 to ca. 10 hr per day.

b. Flowering of Synanthous-Leaved Geophytes

The desert tulip, *Tulipa systola* (Liliaceae), with a West Irano–Turanian distribution, is very common in the Negev Desert highlands. This plant has one of the most "cautious" step-by-step strategies of flower development and numbers of mature seeds produced.

Most of the bulbs that are above ca. 8 g in fresh weight will develop a flower bud during summer in addition to three to four leaves. After the rains of the winter season, three leaves will appear. Depending on the amount and distribution of rain, this plant is able to stop at different stages such as the development of the flower bud, flower development, pollen distribution, and flower producing a fruit with mature seeds. The number of seeds in a fruit (capsule) is also dependent on water availability. In a year with less precipitation and less favorable rain distribution, about 200–300 seeds will develop in a capsule. In a year with more precipitation at the right time and proper distribution, up to 1200 seeds will develop per capsule. This very "cautious" step-by-step strategy of reserve allocation into reproductive organs is controlled by the history of the plant as well as the water conditions during the flowering period. If the dry weight of the bulb is less than 3.5 g and fresh weight is less than ca. 8 g, the plant will not develop a flower bud during summer. During the following winter the plant will develop only one leaf but will not flower.

Bellevalia desertorum, with West Irano–Turanian and East Saharo–Arabian distribution, and *B. eigii,* with Saharo–Arabian distribution (Liliaceae), are two other synanthous-leaved geophytes found in the Negev Desert highlands.

In the area near Sede Boker, *B. desertorum* inhabits the south- and east-facing hillslopes which are the more arid microhabitats in this area, as well as the flat areas and wide loessial wadi beds in open areas. The bulb is situated at a depth of only 5–7 cm. *B. eigii* inhabits the wadis and flat loess areas which receive more water. The bulb is situated much deeper than that of *B. desertorum,* up to 20 or even 25 cm below the soil surface.

In both plant species, roots start to develop only when the rain wets the soil at the level where the root initials are situated below the bulb. Leaves develop only after roots start to develop. The different depths and habitats of these two plants have a great influence on the time of flowering as well as the strategy of storage allocation. In *B. desertorum,* early rains wet the soil at the depth of the root initials that start to develop. Therefore, the leaves appear earlier, the plant flowers earlier, and mature seeds are produced much earlier. This is because the soil dries out at the depth of the bulbs of *B. desertorum* earlier than at the deeper level of the *B. eigii* bulb. In *B. eigii* the soil is wetted to the depth of the root initials after more rain, therefore, leaves appear later as well as flowers.

B. desertorum is very "opportunistic" in its strategy of root, leaf, and flower appearance and very "cautious" in its strategy of assimilants and reserve allocation for the reproductive organs. This plant does not reduce its bulb weight to produce seeds.

B. eigii has a much less "cautious" strategy of storage allocation. It is "cautious" in the appearance of roots, leaves, and flowers but more "opportunistic" for the assimilates and storage allocation of its reproductive effects. In this plant, even in a year with low precipitation, more seeds will be produced even if the bulb loses a certain percentage of its weight at the beginning of the season, as found by B. Boeken.

In conclusion, the flowering strategies of the different perennial plant species studied are affected by different environmental factors such as the soil water availability, the soil water availability and temperature, daylength and temperatures, or an annual rhythm.

The proper time of flowering of both annuals and perennials may also ensure pollination of plants by the appropriate insects.

V. CONCLUDING REMARKS

The critical stages in the life cycle for survival of annual plants are seed dispersal, germination, and flowering, especially under extreme desert conditions. In deserts with hot, dry summers that receive only winter rains, such as the Negev Desert highlands of Israel, the beginning as well as the end of the growing season and the amount and distribution of rain are unpredictable. In this area the majority of seeds that mature are consumed, mainly by ants. Under such environmental and consumption pressure, mainly for annuals that are renewed each season from seeds, different strategies of seed dispersal, germination, and flowering have evolved.

Two extreme main directions of seed dispersal strategies have developed: escape and protection. The "escape" strategy consists of plant species producing very large numbers of very small seeds.

These are dispersed at the beginning of the summer into soil cracks and may be covered by dust. In this way most seeds avoid consumption. In contrast, the "protection" strategy in other plant species produces relatively large seeds in small numbers. Seeds are protected by the woody dead mother plant, from the time of maturation at the beginning of the summer until the following rainy season or seasons when they may germinate. The seeds of many of these plant species are dispersed by rain.

At least two extreme strategies have also developed in the germination of these species. The first one is the "opportunistic" strategy in which seeds germinate after only about 10 mm of rain when seedlings have no chance to survive if more rains do not follow within a few days. This strategy is typical of species producing very small seeds in great numbers. Different groups of such plant species have seeds that require a very short time of wetting in order to germinate, e.g., 29 min, 4–6 hr, or 1 to a few days. The opposite of the opportunistic strategy is the "cautious" strategy in which seeds of plant species germinate only after more than 25 mm of rain wets the loessial soil to a depth of about 25 cm. This strategy is typical of annuals producing relatively small numbers of larger seeds. Some of them require a few hours for germination whereas others require a much longer time of 10–16 days or more. In other plant species there is a combination of a delay of seed dispersal by rain and rapid germination as in *Blepharis* spp.

The seed is the most resistant to environmental conditions whereas the seedling is the most vulnerable. Therefore, germination at the right time and place of only a portion of the seed population has a very high survival value for these species inhabiting the extreme deserts.

In addition to the genotypic strategies mentioned, phenotypic influences are also involved. The genotypic strategies increase the fitness of a plant to its habitat and the phenotypic influences spread the risk. On a single plant, seeds with differing germinability ensure that only a percentage of seeds are "ready for germination" in a particular suitable rain. This depends on the history of each seed such as the position of the seed in the dispersal unit or fruit, daylength, temperatures, light and other environmental factors during seed maturation, and seed storage as well as the conditions of seed wetting. Only a portion of the seed population of a plant species will germinate after one rain even under optimal conditions.

One of the flowering strategies of the majority of annual plant species tested is the facultative long-day response for flowering. This means that the later seed germination occurs in the season, the longer the daylength and the shorter the time until flowers appear. This strategy enables even plants that germinate after a late rain to mature seeds before the hot and dry summer. Plants that germinate early in the season may produce large numbers of mature seeds and the late-germinating plants produce less. There are also day neutral plants for flowering in which the main factor is the soil moisture. Others are facultative short-day plants that flower in autumn when the days are shorter under high temperatures.

The flowering strategies of the different perennial plant species studied are affected mainly by the soil water availability in combination with temperatures or an annual rhythm. The hysteranthous geophytes depend only on daylength and temperatures.

Glossary

Seed dispersal strategies Include seed size and number, shape, and structure; also affected by the fruit and inflorescence shape and structure and environmental factors such as wind, rain, etc.

Genotypic influences Typical to each species according to the environmental pressures during evolution which increase the fitness of a species to its habitat.

Germination strategies Include genotypic and phenotypic influences on seeds and/or dispersal units such as accumulation of germination inhibitors, degree of seed coat permeability, length of afterripening, dormancy, etc.; affected by environmental factors which include amounts of rain, length of wetting, temperature, light, salinity, etc.

Flowering strategies Environmental factors such as daylength, water availability, and temperatures or annual rhythm affecting flowering.

Phenotypic influences Include position and maternal as well as environmental effects during seed maturation such as daylength, temperatures, light, etc., also include environmental factors from the time of seed maturation until the time of wetting and germination. All these may spread germination in time.

Seed "readiness for germination" Includes genotypic and phenotypic effects, depending on the history of each seed, which regulate the germination of each seed during and after each rain event.

Bibliography

Chapman, G. P., ed. (1992). "Desertified Grasslands: Their Biology and Management." London: Academic Press.

Evenari, M., Shanan, L., and Tadmor, N. (1982). "The Negev: The Challenge of a Desert," 2nd Ed. Cambridge: Harvard Univ. Press.

Fenner, M., ed. (1992). "Seeds: The Ecology of Regeneration in Plant Communities." CAB International, UK.

Gutterman, Y. (1993). Seed germination in desert plants. Adaptations of Desert Organisms. Berlin/Heidelberg/New York: Springer.

Gutterman, Y. (1994). Strategies of seed dispersal and germination in plants inhabiting deserts. *Botanical Rev.* **60**, 373–425.

Halevy, A. H., ed. (1989). "Handbook of Flowering," Vol. VI. Boca Raton, FL: CRC Press.

Sex Ratio

John H. Werren
University of Rochester

H. Charles Godfray
Imperial College at Silwood Park, United Kingdom

I. Introduction
II. Sex Ratios and Sex Determination
III. Basic Sex Ratio Theory
IV. Population Structure
V. Resource Quality and Parental Health
VI. Genetic Conflict

There is a considerable diversity of sex ratios and sex determination mechanisms found in nature. The selective factors that shape sex ratio evolution include population structure, differences in fitness of males and females dependent on parental health or resource quality, and genetic conflict between genes with different inheritance patterns. Studies of sex ratio evolution have revealed a variety of fascinating adaptations, and future studies promise to be useful for elucidating the relative roles of natural selection, constraints to adaptation, and other processes in evolution.

I. INTRODUCTION

In species with separate sexes, the sex ratio is the proportion of male or female progeny produced. In hermaphroditic species, sex ratio refers to the proportion of male gametes (sperm or pollen) versus female gametes (eggs or ovules) produced. Because ratios are difficult to handle arithmetically, the sex ratio is normally expressed as the proportion of one sex instead of as a true ratio. It is thus a number between zero and one with an unbiased sex ratio equalling one-half.

Sex ratio can be determined for different age classes. The proportion of males or females at conception is the *primary sex ratio,* at the time of birth it is the *secondary sex ratio,* and at adulthood it is the *tertiary sex ratio.* The *operational sex ratio* is the proportion of sexually active males or females in a population.

Sex ratios have significant effects on the ecology and evolution of organisms. Operational sex ratios can strongly influence mating systems and reproductive strategies. In this regard, sexual selection (differential selection between the sexes) and the operational sex ratio interact. For instance, in some species sexual selection leads to increased male mortality because of male–male aggression. The resulting biased operational sex ratio can alter female behavior and promote the evolution of polygyny due to the scarcity of males. Sex ratios also affect the genetics of populations. For instance, biased sex ratios reduce the effective population size, and therefore genetic variability is lost at a higher rate because of genetic drift. This is especially important in endangered species because populations are small. The ability of some organisms to control their sex ratio is a major aspect of their biology, and may have played a role in such fundamental events as the evolution of complex insect societies.

Finally, conflicting selective pressures acting on different "sex ratio genes" may play a significant role in the evolution of sex-determining mechanisms. [*See* POPULATION REGULATION.]

This article focuses on the evolution of primary and secondary sex ratios. The following sections will consider the relationship between sex ratios and sex determination, review basic sex ratio theory, and then discuss the circumstances under which sex ratio adaptations are expected to evolve and the conditions under which "genetic conflict" over sex ratio will arise.

II. SEX RATIOS AND SEX DETERMINATION

There is a great deal of diversity in the primary sex ratios produced by organisms. Evolution of the primary sex ratio is therefore the focus of a considerable amount of empirical and theoretical research. This area of research is intimately related to the evolution of sex-determining mechanisms, the genetic, biochemical, and physiological mechanisms that determine the sex of an individual. Different forms of reproduction occur. In *dioecious* (or *gonochoristic*) species the sexes are separate. Dioecy is common in animals, but is relatively uncommon in plants. The underlying genetic mechanisms for sex determination are highly variable. Some species show morphologically distinct sex chromosomes with the *heterogametic sex* having two distinct sex chromosome types (e.g. X and Y) and the *homogametic sex* having two morphologically similar sex chromosomes (e.g., X and X). In mammals, many vertebrates, and some insects (e.g., fruit flies), males are the heterogametic sex, whereas in birds, snakes, and butterflies females are the heterogametic sex. Sex chromosomes appear to have evolved numerous times independently in different animal and plant taxa. Even in those species with male heterogamety (e.g., humans, mice, and fruit flies), the underlying genetic basis for sex determination can be very different. Many other species do not have morphologically distinct sex chromosomes. Nevertheless, one sex can be heterozygous at a sex locus or loci, and the other sex homozygous.

Several major taxa have a form of sex determination called *haplodiploidy*, in which males develop from unfertilized haploid eggs whereas females develop from fertilized diploid eggs. This occurs in ants, bees, wasps, thrips, pinworms, rotifers, some mites, certain scale insects and their relatives, and a few beetles. Haplodiploid organisms, particularly the parasitic wasps (which lay their eggs within or on other insects), have been used extensively in empirical studies of sex ratio evolution. These organisms have a level of control over the sex ratio among their progeny because females store sperm from matings in a spermathecal organ, from which sperm can be released to produce a fertilized (female) egg or withheld to produce an unfertilized (male) egg. Haplodiploid species are generally believed to most clearly illustrate adaptive control over the sex ratio (see the following discussion), although scientific debate occurs over the precision with which natural selection models can predict details of sex ratio behavior by haplodiploids.

Not all organisms have separate sexes. In *hermaphrodites*, both male function (pollen or sperm production) and female function (ovule or egg production) occur in the same individual. *Simultaneous hermaphrodites* have both male and female function at the same time. Some simultaneous hermaphrodites are capable of self-fertilization whereas others have evolved mechanisms to avoid selfing, such as the self-incompatibility systems found in numerous hermaphroditic plant species. *Sequential hermaphrodites* can be either sex, but are only one sex at any point in time. Examples of sequential hermaphrodites include pandalid shrimp, some molluscs, and a number of fish groups including coral reef wrasses and parrotfish.

In some organisms, sex determination is strongly influenced by environmental factors such as temperature, light, or even the social environment. This phenomenon is termed *environmental sex determination* (ESD). ESD is found in some representatives of turtles (e.g., snapping turtles), other reptiles, fish (e.g., wrasses), and invertebrates (e.g., some shrimp and relatives).

Parthenogenetic organisms reproduce without fertilization. Haplodiploidy represents a form of parthenogenesis in which unfertilized eggs result in males; sexual reproduction is necessary to produce females. In female parthenogenesis (called *thelytoky*), females produce female progeny without fertilization. Female parthenogenesis occurs sporadically in a wide range of animal and plant taxa, but is relatively uncommon. Parthenogenetic species usually have closely related sexual species, suggesting that female parthenogenesis is typically an "evolutionary dead end." Some female parthenogens require courtship or even copulation with males to stimulate egg maturation and development. However, the genetic contribution of the male is not utilized. Female parthenogenesis is caused by associated bacterial parasites in the reproductive tissues of some insect species, such as the parasitoid wasps *Trichogramma deion* and *Encarsia formosa*.

III. BASIC SEX RATIO THEORY

The incredible diversity in sex determination mechanisms and sex ratios found in nature clearly indicates that these traits are evolutionarily labile and suggests that they have been molded by natural selection. An important question is then, "To what extent are primary sex ratios adaptive, and how does natural selection (and other evolutionary processes) shape them?"

In the 19th century, Darwin pondered this question, and remarked on it being a problem best left to future generations. A general explanation for sex ratio selection was provided by the population geneticist Ronald Fisher in 1930. He considered how selection would act on the primary sex ratio in a large randomly mating population, in which there was genetic variability for the sex ratio produced by parents. Fisher concluded (and later theoretical modeling supports the conclusion) that such a population would evolve to "equal investment" in male and female progeny. Investment is measured in terms of the resource that limits reproduction. For example, if available calories was what limited the number of eggs produced, then this would be the limiting resource. For most organisms, male and female offspring are equally costly for the parent to produce, and equal investment will therefore lead to an unbiased primary sex ratio of 1/2 proportion of males.

To understand why this is so, consider a population in which the (primary) sex ratio is female biased. Those parents producing a greater proportion of sons will have more grandchildren because of the reduced competition for mates faced by their sons. In other words, sons will be a better conduit for passing genes to future generations. As a result the genes for greater male production will increase in subsequent generations, leading to a greater proportion of males in the population. Conversely, if the sex ratio is male biased the reverse occurs and females are the more valuable sex. Natural selection then favors the production of more females. Only when the sex ratio is unbiased and males and females are equally valuable (in terms of grandchildren production) does natural selection cease to act on the sex ratio. An unbiased sex ratio is thus evolutionarily stable. It is apparent from this explanation that sex ratio selection is frequency dependent (i.e., selection changes with frequency of a trait). Parents producing the rarer sex have a higher fitness, defined in terms of number of grandchildren produced. An equilibrium sex ratio [sometimes called the *evolutionary stable strategy* (ESS)] is achieved at 1/2.

It is important to understand that differential mortality of the sexes after parental care does not alter the equilibrium primary sex ratio. Suppose that only 1/10 males survive to sexual maturity; the remaining males will achieve 10 times the reproductive success and this will exactly compensate the mother for the increased mortality among her sons. The same argument applies if females suffer greater mortality than males. If females had only a 1/10 survival rate, then the surviving females would have 10 times the proportional representation among the eggs produced in the next generation, balancing the effects of mortality. Hence, even if adult sex ratios are highly biased toward one sex due to increased mortality of that sex, selection will not favor shifting the primary sex ratio from 1/2.

According to Fisher's theory, a biased sex ratio will be found if one sex is more costly for the parent to produce than the other. Such a situation is most likely to arise in species in which the parent provides more resources to offspring of one sex. Fisher's equal investment rule then predicts that evolutionary stability is achieved when a parent allocates equal resources to offspring of each sex. Thus if sons cost more than daughters to produce, the sex ratio should be biased toward females. However, for most organisms, sons and daughters are equally costly.

Perhaps the best evidence from natural populations for Fisher's theory comes from haplodiploid solitary bees and wasps which have different costs for producing sons and daughters. These organisms lay eggs in burrows that they provision with the carcasses of captured insects. In different species, males or females are the larger sex and require more resources (carcasses). In support of the equal investment rule, the sex ratio is typically biased toward the cheaper sex.

In many animals, including most mammals, nearly equal numbers of male and female offspring are produced because of the segregation of sex chromosomes during gamete production. One-half of all male sperm carries an X chromosome and the other half a Y. At first sight, the sex determination mechanism would appear to make Fisher's argument irrelevant. If sex ratios are fixed by the sex-determining mechanism, then they are presumably not free to evolve by natural selection. However, it is clear from the discussion earlier that sex-determining mechanisms also evolve, and Fisher's theory offers an explanation for the evolution of genetic mechanisms that cause equal sex ratios. The presence of a chromosomal sex determination mechanism does, however, limit the ability to adjust the sex ratio to the extent found in organisms with other sex-determining mechanisms. Recent laboratory population experiments support the conclusion that 1/2 sex ratios will evolve in randomly mating populations when genetic variability for the sex ratio is present and both sexes are equally costly to produce.

A large number of species have biased sex ratios and life histories that do not conform to the simplifying assumptions of Fisher's theory. Exceptions to Fisher's theory fall into three major categories, species with (a) structured populations, (b) variation in parental health or resource quality, and (c) genetic conflict due to variation in the inheritance pattern of genes controlling sex ratio.

IV. POPULATION STRUCTURE

Fisher's theory assumes that mating occurs randomly within the population, and in particular that members of a family do not compete with each other for access to mates. In 1967, William Hamilton determined that under circumstances where brothers compete with each other for access to mates, that female-biased sex ratios will evolve. This effect is called *local mate competition*, because sibling competition for mates is most likely to occur in populations subdivided into small family groups where mating occurs prior to dispersal. Under these circumstances inbreeding is also likely, but it is important to realize that competition among brothers for mates is sufficient to cause the evolution of female-biased sex ratios, whether or not inbreeding occurs. Assuming maternal control over the sex ratio, a general formula for the evolutionarily stable sex ratio (proportion of males = x), when mating occurs in local groups composed of N families and G is the ratio of the genetic relatedness of the mother to her daughters versus her sons, is

$$X = [N - 1]/[1 + G]N.$$

The ESS sex ratio is female biased when N is small and converges on $1/2$ when N is large. Under circumstances of strict sibling mating ($N = 1$), the mathematical solution for this formula is zero proportion sons; however, the biological expectation in this case is for a female to produce only enough sons to inseminate her daughters. Inbreeding itself (as opposed to local mate competition) can alter G in certain genetic systems; for example, in haplodiploids inbreeding leads to a greater relative relatedness of mothers to their daughters.

Female-biased sex ratios are found in a variety of insects and mites that experience sibling mating and local mate competition. Some species have the ability to facultatively adjust their sex ratio in response to the level of local mate competition likely to be encountered by their progeny. The mechanisms by which this is achieved are unclear, but cues associated with the presence of other females appear to be involved. Although the general pattern of sex ratio changes (proportion females decreasing toward 1/2 with increasing number of families) conforms to expectations, considerable variation in the sex ratio response occurs. Fig wasp females vary their sex ratio with the number of other laying females within the fig, where mating among the progeny occurs. Different species vary in the average number of laying females per fig, and in any particular species the best quantitative match between observed and theoretically expected sex ratio occurs in figs that have the number of laying females most commonly encountered in that species. In other words, they are best adapted for the most frequently encountered situations. The parasitic wasp *Nasonia vitripennis* parasitizes patches of fly pupae, where mating also occurs after emergence of the wasp progeny. Females increase the proportion of sons with increasing number of other laying females, and also adjust the sex ratio on individual hosts based on whether a prior parasitization has occurred. This species has considerable genetic variability for the sex ratio response, due to both nuclear and cytoplasmically inherited genes.

Local mate competition can occur in a wide range of other circumstances. For instance, the protozoan parasites that cause malaria mate within the guts of mosquitoes and the sex ratio of these parasites is female biased. Plants that have self-fertilization within unopened flowers typically show reduced pollen production.

Other forms of within-family competition have also been invoked to explain biased sex ratios. In some mammals, one sex disperses away from the natal territory while the other remains behind to compete with its parent and same-sex siblings for a territory. The sex ratio of several primates is biased toward the dispersing sex and this has been interpreted as due to local resource competition. In some birds that have two broods in a year, male progeny from the first brood assist in the rearing of the second brood. Helping is the inverse of competition and theory predicts that the sex ratio in the first brood should be biased in favor of the sex that helps. Evidence for this exists in red-cockaded woodpeckers. In these last examples, it is unclear how the parent manipulates the sex ratio since these organisms have chromosomal sex-determinating mechanisms.

Temporal structure to populations can also favor sex ratio adjustment. For example, some insects have highly seasonal life history patterns; individuals born in the summer may reproduce in autumn but are unable to overwinter. In contrast, individuals born in the autumn overwinter and are able to reproduce during both summer and autumn of the following year. This situation creates a partial overlap between summer and autumn generations. When autumn-produced progeny of one sex (e.g., females) have higher adult survival between the summer and autumn reproductive periods of the next year, then natural selection will favor parents to produce a bias toward that sex in the autumn generation and a bias toward the other sex in the summer generation.

Sex ratio changes consistent with this seasonal sex ratio theory are observed in solitary wasps and bees, which have the haplodiploid sex determination that provides control over the sex ratio. It has been proposed that a tendency to produce a female bias later in the season may have predisposed some solitary bees and wasps to evolve toward the complex insect societies found among bees and wasps. The reasoning is as follows. In developing the theory of *kin selection*, William Hamilton noted that haplodiploid females are genetically related more to their sisters than to their own progeny. This higher genetic relatedness can favor haplodiploid females to remain in the natal nest and rear sisters in preference of rearing their own progeny, if other necessary conditions are also met. However, females are actually less related to their brothers than to their own progeny. Therefore, a sex ratio biased toward sisters in the autumn, after the first generation has emerged, would favor the first generation females becoming workers instead of dispersing to rear their own progeny.

V. RESOURCE QUALITY AND PARENTAL HEALTH

An implicit assumption of Fisher's argument is that all females are equally good at producing male and female offspring and that females do not differ in their opportunities for producing young of either sex. In 1973, Trivers and Willard reasoned that female mammals of good condition in species with strong male dominance hierarchies may be selected to produce sons. This is expected to evolve when females in good condition are likely to produce good quality sons which have a great advantage in the strongly hierarchical male society. Females in poor condition would tend to produce daughters. There is evidence for this hypothesis from deer species and several primates, although again mechanisms by which this is achieved in species with chromosomal sex determination have not been explored.

The Trivers and Willard hypothesis was extended in 1979 by Charnov to a broad range of situations where females differed in their ability to raise offspring of each sex. Many parasitic wasps lay single eggs in host insects that provide the complete nutrition for their developing young. The size of the host strongly influences the eventual size of the adult wasp. In these organisms it is the female that gains more from being large than a male because production of eggs is more size dependent than male mating success. When this is true, then theory predicts that female wasps should lay male eggs in small hosts and female eggs in large hosts. In fact, this pattern is widespread in parasitic wasps and was first noted in the 19th century.

The same "variable quality" model can be applied to sequential hermaphrodites. In this case, the "decisions" concern whether to be male or female first and when to change sex. In general, the sex that benefits the greatest from being large (in terms of fitness) is expected to be the later sex. In sequential hermaphrodites in which males do not compete by dominance or aggression and female egg production is correlated with size, such as in pandalid shrimp, individuals reproduce first as males and later as females. In contrast, the Pacific cleaner wrasse has a mating system in which a single male typically controls mating access to a harem of females. The largest individual is male, but if that individual is removed, the largest female then transforms into a reproductive male.

VI. GENETIC CONFLICT

The theories just described implicitly assume that the sex ratio is determined by nuclear genes that are inherited equally through both sexes. However, some genetic elements have asymmetric inheritance through males versus females and are therefore selected to skew the sex ratio. For example, cytoplasmically inherited genes are almost universally inherited through eggs (and hence females) but not through sperm. This asymmetry strongly selects for cytoplasmic genes that bias the sex ratio toward female production. Cytoplasmic genes are found in mitochondria and symbiotic and parasitic microorganisms that are transmitted through eggs but not sperm. Cytoplasmic sex ratio distorters are common and include protozoans and bacteria that kill male embryos in flies, mosquitoes, wasps, and beetles or that convert genetic males into females in isopods and amphipods, bacteria that cause females to reproduce via parthenogenesis or to produce all-female families in wasps, and mitochondria that cause male sterility in plants.

A conflict in the direction of selection for a phenotype that occurs between different genetic elements within an organism is called *genetic conflict* (or intragenomic conflict). Because cytoplasmic genes are selected to produce all-female sex ratios, whereas genes in the nucleus are generally selected to produce a balanced sex ratio (exceptions noted earlier), genetic conflict over the sex ratio is inevitable.

There is also genetic conflict between genes on the sex chromosomes and the other nuclear genes. In many species with XY males and XX females, including mosquitoes, fruit flies, and lemmings, variants of the X have evolved that are transmitted to nearly 100% of the sperm of XY males instead of the expected 50%. This is called *meiotic drive*. Because such chromosomes skew the sex ratio, selection acts on unlinked nuclear genes to suppress the meiotic drive chromosomes, and suppressors are found in a number of meiotic drive systems.

Genetic conflict over sex ratios may play a role in the active evolution of sex-determining mechanisms. For example, the isopod *Armadillidium vulgare* typically has females with heterogametic sex chromosomes, whereas males are homogametic. This species also harbors a cytoplasmic bacterium that converts genetic males into functional females by suppression of a male-determining gland. When the infection becomes very common in a population, there appears to be selection for repressor genes, thus modifying the underlying sex-determining system. One possible consequence is conversion of sex determination from female heterogamety to male heterogamety due to the effects of the repressor genes. The general importance of genetic conflict in sex determination evolution is still unknown.

Genetic conflict over sex ratios can even occur between individuals. For instance, within haplodiploid social insects (such as ants, bees, and wasps), the queen is generally selected to produce a 1/2 sex ratio among the reproductives of the colony, whereas workers, which rear their sisters and brothers, are selected to produce female-biased sex ratios—up to 3/4 sisters when the mother only mates once. There is increasing evidence that such conflict is important within social insect colonies.

Glossary

Genetic conflict (or intragenomic conflict) Conflicting selective pressures acting on different genetic elements in an organism for a particular trait.

Haplodiploidy A form of sex determination in which males develop from unfertilized (haploid) eggs whereas females develop from fertilized (diploid) eggs.

Hermaphroditism A sexual form in which both male function (pollen or sperm production) and female function (ovule or egg production) occur in the same individual.

Local mate competition Competition for mates among related males that occurs in populations subdivided into small family groups where mating occurs prior to dispersal.

Operational sex ratio Proportion of sexually active males and females in a population.

Polygyny A mating system in which one male mates with multiple females.

Primary sex ratio Sex ratio (proportion of males or females) at the time of conception.

Secondary sex ratio Sex ratio at birth.

Tertiary sex ratio Sex ratio at adulthood.

Bibliography

Charnov, E. L. (1982). "The Theory of Sex Allocation." Princeton: Princeton Univ. Press.

Godfray, H. C. J. (1993). "Parasitoids: Behavioral and Evolutionary Ecology." Princeton: Princeton Univ. Press.

Werren, J. H. (1987). Labile sex ratios in wasps and bees. *Bioscience* **37**, 498–506.

Shifting Cultivation

P. S. Ramakrishnan
Jawaharlal Nehru University, New Delhi

I. Introduction
II. Direct Impacts of a Shortened Agricultural Cycle
III. Ecological and Socioeconomic Consequences
IV. Alternates to Shifting Agriculture

I. INTRODUCTION

The forest farmer in the tropics has managed his traditional *shifting agriculture* ("slash-and-burn agriculture") for centuries, with optimum yield on long-term considerations instead of trying to maximize production on a short-term basis. In doing so, the small-scale perturbations ensured enhanced *biological diversity* in the forest, along with multispecies agroecosystem biodiversity. Increasing population pressure and declining land through degradation have resulted in a rapid shortening of the *agricultural cycle,* leading to (1) a drastic reduction in yield to the farmer; (2) a reduced relative agroecosystem and landscape stability (resistance and resilience) in the face of perturbation, leading to social disruption, (3) a decline in biodiversity due to weed takeover, biological invasion, and eventual site desertification; and (4) substantial CO_2 emitted into the atmosphere because of more frequent and extensive burns. All attempts made so far in finding an alternate to shifting cultivation have had little or no impact on the farmer, lacking a holistic approach in dealing with the complex issues. In this context, an evaluation of ecological impacts and finding acceptable but a sustainable solution(s) to the problem become critical.

II. DIRECT IMPACTS OF A SHORTENED AGRICULTURAL CYCLE

A reduced cycle length (5 years or less) in many parts of the world has had many impacts: (1) a drastic decline in soil fertility under short agricultural cycles imposed on the same site over a period of time (Table I); (2) although farmers deal with declines in soil fertility in different ways, a minimum of 10 to 15 years are required for fallow regrowth in order to recoup most of the soil fertility lost during the cropping phase; (3) this loss in soil fertility is accentuated by heavy hydrological losses due to runoff and leaching during frequent perturbations; (4) since herbaceous weed vegetation alone develops during the first 5–10 years of fallow regrowth, the weed potential gets intensified, leading to an arrested succession; and (5) biological invasion by exotic weeds is promoted (e.g., takeover by *Eupatorium* spp. and *Mikania micrantha* in the Asian tropics).

III. ECOLOGICAL AND SOCIOECONOMIC CONSEQUENCES

A. Biodiversity

Approximately 1–2% of the total area of tropical forests is currently being cleared each year; this area

TABLE I
Changes in Soil Fertility Due to Shifting Cultivation[a]

Location	Total N (g/m²)	Available P (g/m²)	Exchangeable cations (m·eq-m⁻²) K	Ca	Mg
West sumatra					
Mature forest	161	1.30	120	70	360
Young fallow[b]	137	0.42	80	250	60
Old fallow[c]	110	0.24	80	30	100
Nigeria					
Mature forest	488	1.40	—[d]	—	—
Young fallow	193	0.25	—	—	—
Old fallow	289	0.75	—	—	—
Northeast India					
Mature forest	—	18.0	1.00[e]	0.95	1.0[e]
Young fallow	978	0.30	0.70	2.5	2.9
Old fallow	1225	0.80	0.20	2.1	1.4

[a] Adapted from Aweto (*J. Ecol* **69,** 609, 1981) and Protor (1989).
[b] Up to 3 years.
[c] Ten to 15 years.
[d] No data.
[e] g/mol/m⁻².

is about the size of the United Kingdom. These rates are driven by intense population pressures, and they are likely to continue. These conversions, along with weed invasion into land once covered by forests and site desertification, have already had an adverse impact on biodiversity. Therefore, a holistic approach toward tropical forest conservation and management could reverse this trend. [*See* DEFORESTATION.]

In situ conservation of agricultural crop biodiversity in the landscape has become critical because of large-scale adoptions of modern agriculture involving only a few selected cultivars under monocropping. In fact, in traditional agroecosystems such as shifting agriculture, high biodiversity exists, often comparable to that of the natural ecosystems and indeed occasionally exceeding it. Under short cycles, biodiversity does drop to 5–6 species per plot compared with up to 40 or more species under longer cycles. A qualitative drop in biodiversity could also exist. Lowered soil fertility levels, leading to a shift in crop mixtures toward more nutrient-efficient ones such as tuber crops instead of cereals (e.g., northeastern India), and ill-conceived alternatives to shifting agriculture are factors that have an adverse impact on *in situ* conservation of crop biodiversity.

Realizing that biodiversity decreases as habitats change from forest to traditional agriculture and then on to modern agriculture, a variety of models for the loss of biodiversity under varied intensities of management regimes are proposed (Fig. 1). It seems obvious that the biodiversity decline is the sharpest somewhere in the area close to the middle intensity of management (Fig. 1, curve IV), therefore, concentrating at this level of management is crucial for sustainable agriculture, with emphasis on biodiversity.

B. Land Degradation and Desertification

The complex tropical rain forests are often extremely fragile systems. First, these forests, which have developed over centuries, often rest on highly infertile soil, with biomass as the chief storage compartment for nutrients. Second, in oligotrophic areas, stability is ensured by the presence of a thick surface root mat (15–20 tons/ha) that picks up nutrients released from decaying leaf litter on the min-

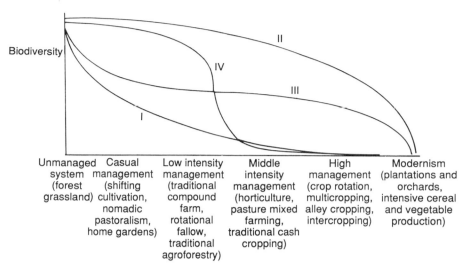

FIGURE I Biodiversity changes (four patterns) as related to agroecosystem types and intensity of management. Curves I and II represent two extreme, unlikely possibilities. Curve III is a softer version of the ecologist's expectation whereas curve IV seems to be more likely and is the most interesting from the point of view of biodiversity conservation and sustainable management of agroecosystems (from Swift *et al.*, 1994).

eral soil and pumps these nutrients back up into the living biomass before they have a chance to enter the mineral soil. Frequent and large-scale perturbations, as are occurring now, could upset the delicate balance in nutrient cycling.

The first step in site degradation is the replacement of forests by an arrested weed stage. Large tracts of forested land in the Asian tropics, for example, are taken over by the grass *Imperata cylindrica*. Exotic weed invasion is yet another major consequence of frequent perturbation under short agricultural cycles in many parts of the world. In extreme cases, the end result is a balded, totally desertified landscape (e.g., northeastern India). The consequence of this is less land area for shifting agriculture and a further shortening of the cycle. [*See* DESERTIFICATION, CAUSES AND PROCESSES; NUTRIENT CYCLING IN FORESTS.]

C. Agroecosystem Function

Traditionally, shifting agriculture has performed a variety of functions. Apart from the productive functions, these mixed cropping systems are based on strong internal controls instead of external energy subsidies, in the face of perturbation and risks as perceived by the farmer. This stability arises

largely out of biodiversity in space (plot to landscape levels) and time (within a given year to decades). This is reflected through (1) sustainable soil fertility through nutrient cycling, (2) reduction in pests and diseases through population interaction mechanisms, and (3) meeting the socio–cultural–economic functions associated with the complex social relationships of human groups. Shortening of the agricultural cycle at the plot level and site degradation at the landscape level have often damaged these internal controls, necessitating external subsidies that are difficult to come by and, consequently, leading to a sharp decline in productivity (e.g., many regions in Asia and Africa). This, in turn, has resulted in serious social disruptions among traditional societies in different parts of the world.

D. CO$_2$ Emission and Climate Change

Deforestation and CO$_2$ emission as a result of biomass burning contribute to climate change per se. If, according to a modest world estimate, it is assumed that about 200,000 km$_2$ is affected by shifting cultivation, involving 20 million families per year (with 20–45 kg ash released per 3 ha plot by an average family), the biomass that is burned and the CO_2 emitted could be substantial. Add on to this deforestation for other reasons such as permanent agriculture, cattle ranching, and timber harvest; carbon sequestration in the tropical forests is adversely affected, with implications for mitigating CO_2 increase on a global scale.

IV. ALTERNATES TO SHIFTING AGRICULTURE

Three different pathways for sustainable agriculture exist: (1) evolution by incremental change, (2) restoration through the contour management, and (3) development of energy intensive modern agriculture.

A. Incremental Change

Building on traditional technology is one of the options for shifting agriculture, at least in the short term. Maintaining a longer cycle length of 10 or more years where feasible and transferring better shifting cultivation technology from one area to another are quick solutions recognized for northeast India.

Strengthening the tree component of this weakened agroforestry system due to shortening of the cycle and using biologically valuable and socially acceptable tree species (e.g., the nitrogen-fixing, *Alnus nepalensis* tree species in northeast India recoups about 600 kg of nitrogen lost during one cropping season, which would normally require a minimum of 10 years of fallow phase and only one-half of this is recovered under a 5-year cycle) (Table II) are some of the short-term options. In parts of

South America, in Venezuela, natural secondary forest succession has been suggested as a model to replace traditional shifting agriculture: beans, corn, sugarcane, and pineapple in the first year, followed by woody yucca, cashew, or papaya, followed by larger trees such as Brazil nut and jackfruit. At another level, the home garden concept could be the model for developing a plantation economy (found in many Asian and Latin American countries).

B. Contour Management

As a possible medium strategy, this type of management acknowledges and works with the ecological forces that provide the base on which the system must be built, while acknowledging the social, economic, and cultural requirements of the farming communities. Working with nature instead of dominating it, this approach seeks active planning, keeping in mind the nature of the background ecosystem. Slope management has long been a major element of farming in the western Pacific region and in the uplands of the Asian tropics. The sloping agricultural land technology (SALT) developed by the Midanao Baptist Rural Life Center in early 1980s in the southern part of the Philippines is based on planting field and perennial crops in 3- to 5-meter bands between double rows of nitrogen-fixing trees and shrubs planted on contours for soil conservation. The crop species include rice, maize, tomatoes, and beans, whereas the perennials may be cocoa, coffee, banana, and citrus. The contour lines are planted with *Leucaena leucocephala*, *Flemingia macrophylla*, and *Desmodium rensonia*. As a soil

TABLE II

Net Change of Nitrogen ($\times 10^3 \text{ kg}^{-1}$) in a Year of Cropping in Soil under Three Cycles of Shifting Agriculture in Northeast India[a]

	15-year cycle	10-year cycle	5-year cycle	
			First year crop	Second year crop
Soil pool before burning	7.68	7.74	6.40	5.98
Soil pool at the end of cropping	7.04	7.15	5.98	5.60
Net difference	0.64	0.59	0.42	0.38

[a] From Ramakrishnan (1992).

conservation measure, SALT technology has been tested successfully in many countries in the Asian tropics but social problems need to be overcome for large-scale acceptance. These are related to (1) land tenure difficulties and (2) heavy initial investments followed by maintenance costs.

C. Energy Intensive Agriculture

Where feasible, on flat valley lands, more intensively managed monocropping systems could be developed, arising out of the existing traditional sedentary agriculture. This extreme pathway attempts to convert the natural ecosystem into one that contains only those biological and chemical elements that the planner desires, almost irrespective of the background ecological conditions, e.g., the "green revolution model."

D. Participatory Management of Forests

Involving local communities in managing forestry and nontimber forest product-related activities for cash economy is one of the alternates that has had limited success (as in the Chiang Mai region in Thailand), but which requires farmer's involvement in a much larger economic system. Unlike the typical "taungya," the benefits largely go directly to the farmer instead of to the forest industry. The concept of joint forest management committees, involving local community participation along with nongovernmental voluntary agencies and the forest department, now being promoted in India and elsewhere in the tropics is a step forward and is aimed at linking sustainable livelihood for local communities with forest management.

E. Landscape Management

In the ultimate analysis, as a long-term strategy, it may be desirable to have a mosaic of agroecosystem types using all of the pathways just mentioned, along with plantation systems, coexisting with natural ecosystem types, managed or unmanaged. Maintenance of the overall sustainability of the system requires the patchwork mosaic that would, albeit inadvertently, be the best plan for effectively managing natural resources in the shifting agriculture-affected areas. This approach for sustainable agriculture in a landscape has just started receiving attention.

Glossary

Biological diversity Refers not only to species and subspecific types at a regional level, but also to plot, ecosystem, and landscape level complexities.

Home garden One of the ways in which humans in the humid tropics have tried to imitate nature, through incorporation of trees and other perennials as an elaborately constructed traditional multipurpose system along with annuals, and often involving animals (number of species could be up to 80 or more).

Shifting agriculture A complex system with wide variations in cropping and cultivation practices. However, the basic features remain: slashing of the vegetation from the fallow phase and burning of the dried smaller slash, with the larger stems being removed for domestic uses. It is a complex set of systems, with wide variations in the cropping procedures and cropping patterns.

Shifting agricultural cycle The length of the fallow phase between two successive croppings on the same site is one cycle.

Taungya A land management system with origins in southeast Asia that combines agricultural crops with forest plantations. Peasants are allowed to grow crops for the first few years along with tree seedlings, but no wages are paid to the farmer for their labor.

Bibliography

Brookfield, H., and Padoch, C. (1994). Appreciating agroforestry: A look at the dynamism and diversity of indigenous farming practices. *Environment* **36,** 6–11 and 37–45.

Dale, V. H., Houghton, R. A., Grainger, A., Lugo, A. E., and Brown, S. (1993). Emissions of greenhouse gases from tropical deforestation and the subsequent use of the land. *In* "Sustainable Agriculture and the Environment in the Humid Tropics," pp. 215–262. Washington, D.C.: National Research Council, National Academy Press.

Gliessman, S. R., ed. (1990). "Agroecology: Researching the Ecological Basis for Sustainable Agriculture." New York: Springer-Verlag.

Kunstadter, P., Chapman, E. C., and Sabhasri, S., eds. (1978). "Farmers in the Forest." Honolulu: Univ. Press of Hawaii.

Pratap, T., and Watson, H. R. (1994). "Sloping Agricultural Land Technology (SALT): A Regenerative Option for Sustainable Mountain Farming." ICIMOD Occasional Paper No. 23, International Centre for Integrated Mountain Development, Kathmandu, Nepal.

National Research Council (USA) (1982). "Ecological Aspects of Development in the Humid Tropics." Washington D.C.: National Academy Press.

Nye, P. H., and Greenland, D. J. (1960). "The Soil under Shifting Cultivation." Technical Communication No. 51, Commonwealth Bureau of Soils, Harpenden, England.

Proctor, J., ed. (1989). "Mineral Nutrients in Tropical Forest and Savanna Ecosystems." Oxford: Blackwell.

Ramakrishnan, P. S. (1992a). Tropical forests: Exploitation, conservation and management. *Impact Sci. Soc.* **42**(166), 149–162.

Ramakrishnan, P. S. (1992b). "Shifting Agriculture and Sustainable Development: An Interdisciplinary Study from North-Eastern India." UNESCO-MAB Series, Paris, Parthenon Publ., Carnforth, Lancs. UK [Republished by Oxford Univ. Press, New Delhi, 1993]

Ramakrishnan, P. S., Campbell, J., Demierre, L., Gyi, A., Malhotra, K. C., Mehndiratta, S., Rai, S. N., and Sashidharan, E. M. (1994). "Ecosystem Rehabilitation of the Rural Landscape in South and Central Asia: An Analysis of Issues." New Delhi: UNESCO (ROSTCA).

Ruthenberg, H. (1971). "Farming Systems in the Tropics." Oxford: Clarendon Press.

Solbrig, O. T., van Emden, H. M., and van Oordt, P. G. W. J. (1992). "Biodiversity and Global Change," Monograph 8. Paris: International Union Biol. Sci. (IUBS).

Spencer, J. E. (1966). "Shifting Cultivation in Southeastern Asia," Publication in Geography, No. 199. Berkeley: University of California.

Swift, M. J., Vandermeer, J., Ramakrishnan, P. S., Anderson, J. M., Ong, C. K., and Hawkins, B. (1994). Biodiversity and agroecosystem function. *In* "Biodiversity and Ecosystem Function" (E. D. Schultz and H. A. Mooney, eds.) Chichester: John Wiley (in press).

Watters, G. A. (1971). "Shifting Cultivation in Latin America," F.A.O. Forestry Development Paper No. 17. Rome: F.A.O.

Woomer, P. L., and Swift, M. J., eds. (1994). "The Biological Management of Tropical Soil Fertility." Chichester, UK: Wiley-Sayce Publ.

Soil Ecosystems

H. A. Verhoef

Vrije Universiteit, Amsterdam

I. Introduction
II. Soil Ecosystem Structure
III. Soil Ecosystem Function
IV. Disturbances of Soil Ecosystems

An ecosystem can be characterized by the integrated and largely self-maintained functioning of a diverse community of organisms in terms of energy flow and nutrient cycling within a range of physical environments. In the soil ecosystem these major functional processes, energy flow and nutrient cycling, interact the most strongly. In extreme physical environments and at low nutrient availability (which may be a result of air pollution) the rate of soil processes determines the productivity of the whole terrestrial ecosystem. Soil animals and microorganisms are the important catalysts of the decomposition and mineralization of the organic input into the soil ecosystem, which consists mainly of plant and, to a lesser extent, animal residues. Disturbances of the functioning of the soil ecosystem by human impacts can only be understood and cured if the functioning of soil biota in interaction with the abiotic soil is known.

I. INTRODUCTION

Soil ecosystems, although part of larger systems, can be regarded as ecosystems in their own right. There is, however, a clear difference with other ecosystems. Since soil algae are the only organisms with the ability to capture solar energy, the ecosystem depends on other forms of energy input from the outside. These are either of organic origin (living plant roots or plant and animal debris) or of inorganic origin (e.g., NH_4^+ for nitrifying bacteria). The soil ecosystem can be divided into an abiotic part and a biotic part.

The *abiotic* components include the mineral fraction, the organic matter, the soil water, and the soil atmosphere. In *mineral fraction,* the composition of the various sized particles, the texture, is an important characteristic, whereas the density of the charges on the surfaces of clay minerals and organic particles determines the exchange capacity of a soil for the ionized soil solution. This in turn influences the bioavailability of the ions.

Soil organic matter consists of a cellular fraction and of humus. The former is constituted from particulate matter formed by the action of decomposer organisms (see later). The latter has an amorphous character, is a mixture of complex polymeric molecules, synthesized during decomposition, and forms colloidal particles which often associate with the mineral fraction. A well-known soil classification for temperate soils is based on the different humus types that can be distinguished: mor, moder, mull-like moder, and mull soils. These soils not only differ in organic matter content and composition but also in physical conditions and soil biota (see Table I). This classification is widely used

TABLE I

Characteristics of Different Humus Soil Types

Mor → Moder →	Mull-like moder → Mull
pH: acid	pH: nearly neutral or slightly alkaline
CEC[a]: low	CEC: high
C:N: >20	C:N: <15
Cellular fraction: high	Cellular fraction: low
Conifers, heathland	Deciduous trees, grasses
Fungal mycelium: abundant	Mycelium: less obvious
Bacterial counts: low	Bacterial counts: high
Earthworms: absent	Earthworms: present
Mesofauna: high numbers	Mesofauna: low numbers

[a] Cation exchange capacity.

by ecologists and foresters, but not by agricultural scientists, as most arable soils are of the mull type.

Soil water determines, in relation to the structural soil components, the survival of soil biota, their migration, the bioavailability of nutrients, and, in interaction with soil air, the *soil atmosphere*.

The *biotic* components are often classified based on size into the microbiota (algae, protozoa, fungi, and bacteria), the mesobiota (nematodes, springtails, small arthropods, and enchytraeid worms), and the macrobiota (larger arthropods and enchytraeids, earthworms, and molluscs). Burrowing rodents, reptiles, amphibia, and even plant roots can be added to the macrobiota.

This article discusses organisms in the context of the structure and function of the ecosystem. Structure and function are discussed separately in the following paragraphs in more detail.

II. SOIL ECOSYSTEM STRUCTURE

The characterization of an ecosystem by its structural components demands a thorough and detailed knowledge of the taxonomy of the inhabitants of the soil. The very fact that they live, often indiscernibly, in the soil is a reason for their obscurity. This certainly accounts for soil bacteria and protozoa and, to a lesser extent, for fungi, nematodes, enchytraeids, and microarthropods. There are different ways to express the numerical structure of an ecosystem, e.g., *species composition* provides in-

formation about the species richness of a specific taxonomic group. However, this parameter overlooks the fact that some species are rare whereas others are common. Therefore, diversity indices are developed, taking into account not only the species richness (the total number of species in the ecosystem) but also the number of individuals per species. In this way *species diversity* is expressed. Sometimes the term *relative abundance* is used, in which the abundance of a species is expressed as a percentage of the total abundance of the taxonomic group this species belongs to. *Genetic diversity* is another parameter that expresses the number and frequency of alleles within a species, but up until now has not been applied to the soil ecosystem. The *trophic structure,* which can be established for the soil ecosystem, indicates how many kinds of organisms occupy each trophic level from producers to top predators. The number of links in all food chains in an ecosystem gives the *food chain length,* which can be used as a structural soil ecosystem parameter. The construction of such a food chain is based on well-defined trophic levels: the species on a specific trophic level all feed only on species in the trophic level just below it and are fed on at most by species in the level just above it. However, in most ecosystems the trophic relationships among the species are seldom so clear. Therefore, this network of trophic interactions is more appropriately termed a *food web* (see Fig. 1).

Food web descriptions have been constructed that quantify the flow of energy and matter through

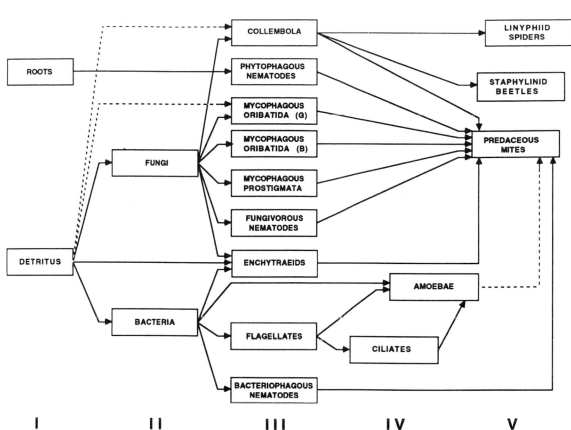

FIGURE 1 Food web of a forest soil ecosystem; I–V indicate the five trophic levels. G, grazers; B, browsers. [From Berg and Verhoef.]

soil ecosystems and the biological processes that regulate these flows. Such quantification reveals which species dominate biomass and energy and contributes to the understanding of the soil ecosystem function.

III. SOIL ECOSYSTEM FUNCTION

The diversity in forms and species is much greater than the diversity in functions. The following four soil ecosystem functions can be distinguished: *primary* production, which is mainly root production; *secondary* production; *decomposition;* and *humification.*

A. Primary Production

Primary production is the biomass produced by plants, the primary producers. Production depends strongly on abiotic factors such as humidity, temperature, light, energy supply (CO_2), nutrient availability, and soil structure and on biotic factors. So it is not surprising to find differences for the different climatic zones (see Table II). In tundras and temperate grasslands, a proportionally high root biomass has been found. The measuring of belowground biomass and primary production raises many technical questions. (The story goes that one attempt to extract tree roots for measurement from a woodland site in California, using dynamite, resulted in most of the root material landing in Nevada.) [*See* RHIZOSPHERE ECOPHYSIOLOGY.]

B. Secondary Production

This type of production is defined as the production of new biomass by heterotrophic organisms. In general, a positive relation between primary and

TABLE II
Production and Decomposition of Six Ecosystem Types

	Tundra	Boreal forest	Temperate deciduous forest	Temperate grassland	Savanna	Tropical forest
NPP[a] (tons ha^{-1} year^{-1})	1.5	7.5	11.5	7.5	9.5	30
Biomass (tons ha^{-1})	10	200	350	18	45	500
Photosynthetic (%)	13	7	1	17	12	8
Wood (%)	12	71	74	0	60	74
Root (%)	75	22	25	83	28	18
Decomposition rate (k year^{-1})	0.03	0.21	0.77	1.5	3.2	6.0

[a] Net primary production.

secondary production can be expected. In the soil ecosystem many bacteria, fungi, and soil animals are unable to manufacture the necessary complex, energy-rich compounds from simple organic molecules. Besides using living parts of the primary production, the roots, the organisms live mainly off of dead parts. Parts of the plant biomass die and enter the dead organic material (DOM) pool, just like parts that are incompletely digested by herbivores. The heterotrophic belowground community is called the *decomposer* community, as most of it is involved in the process of decomposition (see later). The first trophic level, the DOM, is consumed by the second trophic level, microorganisms and detritivorous animals, and those are then eaten by the third trophic level, microbivores and carnivores, and so on. The transfer efficiency of the energy flow through this decomposer community is very high, not because of the high energy use efficiency of the organisms, but because of the ongoing *recycling* of the organic matter through this community, until most material has been metabolized.

C. Decomposition

Decomposition involves the breakdown of DOM, during which process minerals and soluble organic substances are released by three constituent processes: *leaching, fragmentation,* and *biodegradation*. The release of nutrients is called *mineralization*.

I. Leaching

During the first phases of decomposition, leaching releases water-soluble compounds as the plant cells die off, releasing their contents. This involves amino acids, sugars, simple phenolics, and ions such as K, Na, Ca, and Mg. This process alone causes a weight loss of 10–20%.

2. Fragmentation

Soil animals play an important role in this process. By eating the DOM they can change the size, the morphological and chemical structure, the pH, and the water content of that material and enrich it with microorganisms, which greatly enhances the breakdown.

3. Biodegradation

This is a process in which both microorganisms and soil animals are involved: the enzymatic breakdown leads to end products as CO_2 and minerals and intermediate metabolites which are polymerized to more stable high molecular compounds, the humic substances.

The rate of decomposition is determined by the interaction of three process variables: climate, chemical composition of DOM, and the composition of the decomposer community. Decomposition can be studied by placing litter in nylon mesh litterbags, burying them in the soil, and following their weight loss. By following the decomposition of "standard" pine litter (originating from one location) over a climate gradient, it has been found that 80% of variation in mass loss was caused by soil temperature and 55% by evapotranspiration. The importance of chemical composition and that of the decomposer community for the rate of decomposition is illustrated in Fig. 2.

The decomposition rate in this study was also measured by means of litterbags. The differences between the breakdown curves of oak and beech are caused by their different chemical compositions. Bags with a large mesh (7-mm opening) admitted most soil invertebrates and all microorganisms, whereas bags with a smaller mesh (0.5-mm opening) excluded all but microorganisms, small nematodes, mites, and springtails. Larger animals have a clear effect on the breakdown.

The rate of decomposition is characterized by the decomposition constant k, which indicates the amount of decomposed organic material per total amount of litter present, per time. Constant k runs

from 0.2 per year for litter in a boreal forest to 6.0 per year for litter in a tropical forest (see Table II), indicating that the breakdown of boreal litter is much slower than that from tropical forests.

D. Humification

The humification process runs synchronously with the decomposition process and takes place in three ways: via direct humification, microbial humification, and indirect humification. During direct humification, fresh organic material, mainly lignins, is transformed into humines by so-called "brown-rot fungi." Microbial humification concerns the formation of humic substances (melanins) in biologically active, well-aerated soils. Indirect humification leads to the production of unsoluble fulvic and humic acids out of soluble precursors (polyphenols). These phenolic precursors can be leached out of the litter material, high molecular metabolites from the litter, or *de novo* biosynthesis by mi-

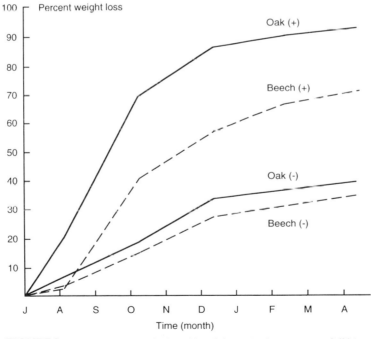

FIGURE 2 Decomposition of oak and beech leaves in the presence of all biota (+) and with the exclusion of larger (micro)arthropods (−). [After Edwards, C. A., and Heath, G. W. (1963). The role of soil animals in breakdown of leaf material. *In* "Soil Organisms," J. Doeksen and J. van der Drift (eds), pp. 76–84. North–Holland, Amsterdam.]

croorganisms out of amino acids, proteins, and sugars.

The residence times for humus in the soil are estimated at hundreds or even thousands of years, implying that these materials are almost inert.

IV. DISTURBANCES OF SOIL ECOSYSTEMS

Because modern man is perturbing natural and agricultural systems at an ever-increasing rate, it is essential to know how ecosystems respond to such perturbations and how they are likely to respond in the future.

The disturbances can be diverse, e.g., management practices related to forestry, agriculture, and recreation, and include harvesting and application of fertilizers and pesticides. These disturbances can affect both structure and function of the soil ecosystem. It is essential to mention here that any kind of disturbance due to a stressor, which is detectable at the soil ecosystem level, is the final feature of a whole series of effects that start at the individual level, through the population level, up to the community level, and then to the ecosystem level. This means that every adverse effect on the ecosystem level starts at the individual level. However, not all adverse effects on the individual level manifest itself at the ecosystem level, especially the function of the ecosystem. An example of this is when groups of organisms are killed, their functions are taken over by other organisms, resulting in the *functional redundancy* of an ecosystem. As a consequence the particular ecosystem function remains unchanged, but structural changes have taken place. In ecosystem management, sustaining the ecosystem function often has a higher priority than sustaining the ecosystem structure. It must be kept in mind that in an ecosystem in which essential functions are performed by a strongly reduced community, the sensibility for other adverse effects increases, leaving a very instabile ecosystem. The following stressors cause different forms of disturbance of the soil ecosystem structure.

High concentrations of heavy metals are able to significantly change *species composition* and *abun-*

dance in bacteria, fungi, and soil animals such as earthworms, nematodes, and springtails. Acidification and eutrophication of the soil also change species composition and relative abundance in fungi and microarthropods such as springtails and mites. Similar effects are found for cutting in deciduous and coniferous forests and for input of harvest residues in agricultural soils. Pesticide application in grassland soils affects *food chain length* and thereby the soil ecosystem stability.

Examples of disturbance of the soil ecosystem function concern stressors such as: atmospheric deposition [e.g. $(NH_4)_2SO_4$], atmospheric composition (e.g., increasing CO_2 level), and temperature increase, they all effect primary production, secondary production, and decomposition in a significant way. It should be noted that these stressors often operate synchronously, e.g., the synchronous occurrence of increased CO_2 levels and temperature will simultaneously increase the decomposition rate, whereas N deposition and CO_2 increase have, respectively, a negative and a positive effect on the decomposition rate. If these stressors operate synchronously the outcome depends on the dominating stressor.

In summary, soil ecosystems are characterized both by their structure and by their function, and the monitoring of specific ecological structural parameters can be helpful in indicating functional changes before they become irreversible.

Glossary

Decomposition Breakdown of complex, energy-rich organic molecules to simple inorganic constituents.
Food chain An abstract representation of the links between consumers and consumed populations, e.g., plant–herbivore–carnivore.
Food web Representation of feeding relationships in an ecosystem (component) that includes all the links revealed by dietary analysis.
Functional redundancy Phenomenon that specific functions of organisms can be taken over by other organisms.
Primary production Biomass produced by plants, the primary producers.
Secondary production Biomass produced by heterotrophic organisms.
Trophic structure Organization of an ecosystem described in terms of energy flow through its various trophic levels.

Bibliography

Ellenberg, H., Mayer, R., and Schauermann, J. (1986). "Ökosystemforschung: Ergebnisse des Solling-Projekts." Germany: Ulmer Verlag.

Fitter, A. H., Atkinson, D., Read, D. J., and Usher, M. B., eds. (1985). "Ecological Interactions in Soil; Plants, Microbes and Animals." Oxford: Blackwell.

Richards, B. N. (1987). "The Microbiology of Terrestrial Ecosystems." New York: Longman.

Teller, A., Mathy, P., and Jeffers, J. N. R., eds. (1992). "Responses of Forest Ecosystems to Environmental Changes." London: Elsevier.

Verhoef, H. A., and Best, E. P. H., eds. (1990). Decomposition processes in terrestrial and aquatic ecosystems. *Biogeochemistry* **11,** 3.

Soil Management in the Tropics

George D. Smith

Queensland Department of Primary Industries

I. Introduction
II. Factors in Managing Tropical Soils
III. Soil Properties and Processes in Tropical Soils
IV. Tropical Soil Management Options
V. Conclusion

Soil management is what agriculturalists do to soil in order to grow plants to meet society's needs ranging from food and fiber to drugs and complex industrial chemicals. It can also include practices to prevent the off-site movement of materials that are in some way harmful to society. Increasing population pressures and the need to maintain or improve the quality of life means that production systems in the tropics are coming under increasing pressure. The challenge facing soil managers is to assure adequate production, with acceptable levels of risk and without degradation of soil and water resources over the long term.

I. INTRODUCTION

Plant growth is influenced by soil physical, chemical, and biological properties. These are determined by the parent material from which the soil has been formed and the soil-forming processes. Tropical soils have formed from a wide range of parent materials under various conditions and their properties vary enormously. Plant growth is also influenced by climatic conditions, particularly temperature, and the quantity and distribution of water must be suitable. Climatic conditions in the tropics range from arid deserts to humid forests. The suc-

cessful growth of useful plants depends on providing suitable conditions for growth of the plant and also controlling or discouraging competition from weeds and attack by pests and diseases that are often intensified in the tropics. While there has been some form of agriculture in most tropical regions, modern intensive argiculture has generally been developed for temperate regions. The climate, soil types, and production systems (as well as socioeconomic and infrastructure support systems) in temperate regions are quite different than those in tropical regions. Soil management technologies suitable for temperate regions cannot be directly transferred to, and tropical soil management technologies must be adapted to or developed in, tropical regions. [*See* SOIL ECOSYSTEMS.]

Management means making inputs to change the soil in some way so as to improve the chances of a desired result. In soil management there are three basic goals: to increase the level of production (i.e., high yield per unit area), to maintain stable production with a low risk of yield loss (i.e., some yield from every crop), and to maintain the soil resource. Each can be considered over a short-term time scale (i.e., a season) or a long-term time scale (i.e., over many years). Management inputs can range from physical energy used in tilling the soil by hand, to complex strategies to change factors that limit or threaten crop yield or that help to preserve soil quality over the long term.

Tropical soil–climate–plant systems are complex and many factors interact. To have successful crop production, all factors that influence likely outcomes for the crop have to be considered and manipulated or controlled if possible. If any are neglected or if too much faith is placed in a particular input, yield may be lost and the investment in inputs may not be recovered. A balanced, whole-system, over-the-long-term strategy is essential. Management strategies or technologies that rely on heavy emphasis on some particular component (such as particular fertilizers or alternative methodologies) are often fashionable. For example, some innovative or alternative system may be promoted as better than a conventional system. While possibly offering short-term benefits, the unquestioning adoption of fashionable concepts may be unwise in the long term if they are unrealistically biased or lead to some unexpected form of soil degradation.

The key to selecting a sound management practice is knowledge of soil resources; planning so that they are used within the climate–soil–crop capabilities; monitoring of resource conditions to detect changes over time; research and development to improve technologies; and communication of the knowledge to farmers so that innovations can be adopted. Ideally, managers need to be able to predict the immediate and cumulative effect of management inputs on soil properties and soil and plant processes. Negative changes in soil and soil-related systems, because of management inputs, may be difficult to detect because they are slow or are masked by the effect of new, more productive technologies. Also, because agricultural systems are variable and complex (often extending over large space and time scales), it is difficult for the human mind to assimilate all available information and make unbiased best-practice recommendations for success in the long term. More and more decision support tools based on system simulation models are being developed and used. These tools usually have many submodels of component processes and can be used to simulate soil processes, crop yield, and system responses and to estimate probabilities by taking into account long-term (or generated) weather records.

II. FACTORS IN MANAGING TROPICAL SOILS

A. Soil Classification and Tropical Soil Types

There are various systems of soil classification. Classifications place soils in categories depending on how their properties and characteristics meet particular criteria for the class. Soil types in a particular category might be expected to behave in a similar way in response to management practices. Unfortunately, there are limits to the usefulness of soil classifications for more detailed management decisions. This is because criteria on which higher orders of the classification are based are not closely linked with factors affecting the soil processes which management needs to influence or because information on the classification of lower orders in the hierarchy is not available at the required small scale. Also, there is no truly international soil classification system as most systems have been developed to suit the soils of particular countries.

The U.S. Soil Taxonomy suffers from these shortcomings but is widely used for broadscale classifications. In this classification there are four main soil types in tropical regions: Alfisols, Oxisols, Ultisols, and Vertisols. Ultisols have a relatively low base status and a B horizon enriched in clay. Alfisols also have a clay-enriched B horizon but they have a higher base status. Both Ultisols and Alfisols are found in tropical regions in Africa, India, Australia, and South America. Oxisols are red, highly weathered soils with a high proportion of aluminium and iron oxides and a relatively high clay content (predominantly nonexpanding 1:1 lattice kaolinitic clay). There are large areas of oxisols in tropical regions of Africa and South America. Vertisols are dark, swelling clay soils that crack when dry and have a relatively high content of expanding 2:1 lattice smectitic clay. Large areas of Vertisols are found in the semiarid regions of Africa, India, and Australia.

B. Tropical Climates

Tropical climates range from arid to humid; some are seasonally arid and seasonally humid. Various climatic classifications exist through which climate at any particular point can be ascribed to a particular category in a hierarchy. Although it is simple to use the Tropics of Capricorn and Cancer as limits to tropical regions, the climate near these meridians, as well as being variable, changes only gradually with distance; areas in the subtropics may have a climate identical to areas in the tropics. The main climatic factors controlling crop production are the amount of photosynthetically active radiant energy, amount and seasonal distribution of rainfall, amount of heat, and extent of extremes (for example, frosts, heat waves, drought, floods, or strong winds). Photosynthetically active radiation may be limiting due to clouds in rainy seasons. The amount of heat is seldom limiting in the tropics but extremes in availability of water often have an impact on yield and risk in crop production. Soil management interacts with climate in so far as management practices can condition the effect of climatic factors, for example, by increasing water use efficiency or reducing waterlogging.

C. Soil Management Requirements for Tropical Crops

An extensive range of crops can be grown in tropical regions. The list includes rice, sorghum, millet, cassava, soy beans, pigeon pea, chick-pea, peanuts (groundnuts), cotton, maize, castor, annual and perennial pastures, and agroforestry species. Lowland rice has special requirements as it performs best in waterlogged, chemically reduced conditions. These conditions are usually provided by mechanical puddling to reduce internal drainage, combined with ponding to provide a submerged soil through most of the life of the crop. Ponding controls weeds and provides suitable soil physical and chemical conditions for rice.

Other crops vary in their tolerance of waterlogging; most prefer well-drained conditions. Key requirements of most crops are provision of suitable seedbed conditions to assure establishment of an adequate plant population, suitable conditions for root growth and development, control of weeds, sufficient nutrients, favorable soil biology, reserves of soil moisture, and adequate drainage. Soil disturbance with various tillage implements has been

commonly used to provide these conditions for annual crops. Research has shown that tillage leads to soil degradation through a decline in organic matter and an increased exposure to erosion; in recent times the trend has been toward reduced tillage systems. In some areas the lack of even animal power limits the ability of the farmer to provide reasonable seed bed preparation.

III. SOIL PROPERTIES AND PROCESSES IN TROPICAL SOILS

Because the goal of soil management is to influence soil processes, an understanding of basic soil properties and processes is useful. There are several disciplinary areas in soil science. Only a brief outline is given here.

A. Soil Physical Aspects

Soils consist of solid, liquid, and gas phases. The solid phase is made up of particles which are classed, according to an international system, in size ranges of clay (<0.002 mm), silt (0.002 to 0.02 mm), fine sand (0.02 to 0.2 mm), and coarse sand (0.2 to 2 mm). The relative proportion of each size determines the texture of the soil and has a strong influence on the physical behavior of the soil. Clay size particles behave as colloids, have extremely high surface area per unit weight, and respond to surface and electrical forces in the soil water system. Particle shape also varies—different clay minerals take the form of broad thin sheets, stacks of sheets, or rolled sheets—but is generally less important than particle size. The various particle sizes pack together in response to a range of forces. The resulting microscale packing and bonding together of particles determine the macroscale structure and the degree of aggregation of the soil. Most physical processes that soil management attempts to influence take place in the pore system created by the arrangement of the soil particles. The size distribution and continuity of the pore system, as well as its stability under external influences, are critical factors in soil management.

Soil physical aspects are important because they determine the physical environment for biological activities (heat, air, and water supply and physical support) and how easy it is for soil managers to manipulate soil processes to provide suitable conditions for plant growth and to control processes that would otherwise lead to physical degradation. The intent of soil management is to even out extremes due to climatic effects or to inherent soil constraints. Important physical factors include the following.

I. Heat Relations

These are largely determined by climate and elevation but management practices can play a part, for example, soils covered with mulch will warm and also dry more slowly than bare soil. In cool climates this can have an adverse effect on biological processes such as germination and seedling emergence or disease incidence, in warm climates it may be beneficial.

2. Soil Water Relations

Key aspects of soil water relations are infiltration, plant available water capacity (PAWC), drainage, and evaporation. Being porous, particulate materials with a large internal surface area and a microstructural pore system, soils can hold water and retain it against drainage by gravity. If water displaces all air the soil is saturated; drainage under gravity allows air to reenter parts of the pore system. As drainage proceeds, a stage [sometimes called "field capacity" (FC)] is reached when the surface tension forces holding water in the soil match the effects of gravity. Plant roots can continue to withdraw water to meet transpiration needs. As transpiration continues, a further stage [sometimes called "permanent wilting point" (PWP)] is reached where the plant cannot meet transpiration needs. The difference in water content (usually expressed in millimeters) between FC and PWP is a measure of PAWC. PAWC depends on soil properties and the species and stage of growth of the plant. Under rain-fed conditions, it determines the soil's ability to supply water to meet biological needs between rainfall events; in irrigated conditions, it determines irrigation frequency. Evaporation from saturated soil proceeds at the same rate as from a free water surface, but

as the water content falls the rate slows. However, water is lost by evaporation at levels well below PWP. The air dry water content depends on the relative humidity of the atmosphere and the temperature. FC, PWP, or air dry water content are not single value physical constants; they vary depending on the conditions. The soil water content is assumed to be zero after drying to constant weight at 100°C. The infiltration rate is a measure of the soil's ability to take in water under rain or irrigation. Water supplied to the surface at a rate above the infiltration rate will pond on the surface and run off. Surface drainage refers to water draining away laterally on the surface whereas internal drainage refers to water moving downward and through the soil into underlying layers.

3. Aeration

Larger pores drain above FC, allowing air to flow into the macropore network and allowing the interchange of gases with the atmosphere to occur. The pore network is determined by soil structure. Aeration is favored by management practices that create heterogeneous, rather than homogeneous, pore sizes and network of continuous pores. If drainage is restricted, soil is said to be "waterlogged." A chain of essentially anaerobic processes begins and imposes particular conditions on biological systems.

4. Soil Strength

All organisms have to fit within the physical structure of their environment. Plants need to be able to stand upright against wind and to be able to develop an adequate root system. This means that the soil pore system must be suitable and the soil strength adequate but not excessive. The formation of surface crust, development of plow pans, or other restrictive layers are examples of problems which may need to be managed.

B. Soil Chemical Aspects

Soil chemical aspects influence the physical behavior of soils as well as their suitability as a medium for soil biology and for plant growth. Soil chemistry is dictated by the chemical nature of the constituents present, including crystalline and amorphous mineral constituents present as various size particles with associated surface ions, organic materials, and soluble and relatively insoluble salts. These are in turn determined by the nature of the parent material, position in the landscape, the type of geomorphic and pedogenetic processes, temperature, rainfall, the ecosystem, and the overall soil age. Key aspects of soil chemistry in relation to soil management are the supply and availability of plant nutrients in the soil. The major elements needed by plants are nitrogen, phosphorus, and potassium; trace elements needed are calcium, magnesium, sulfur, sodium, copper, zinc, boron, iron, and molybdenum. Roots absorb these nutrients in various forms from the soil solution or from the surface of soil particles. The availability of nutrients depends on the total amount present, solubility of the form under the prevailing conditions, and ability of plant roots to access the nutrients in the prevailing physical conditions. The soil reaction or pH (acidity or alkalinity) influences the solubility and availability of nutrients. Required nutrients are generally readily available near neutral pH (6.5 to 7.5).

1. Soil Organic Matter

Organic matter plays a vital role in soil chemistry, physics, and biology. It serves as a store of the essential nutrients which may be taken up by the plant and later returned to the soil as plant, animal, or microbial remains which, through a sequence of many interacting soil processes, are broken down into a vast array of chemical compounds. The sequence of processes is referred to as a "cycle"; study of the nitrogen, phosphorus, and sulfur cycles is important in soil science. The nitrogen cycle in particular is complex and has important implications for soil management. Two processes that are critical for availability of nitrogen are mineralization and denitrification. In normal aerobic soil environments, mineralization processes convert organic forms of nitrogen into the soluble nitrate form in which it can enter other soil processes such as plant uptake, leaching deeper into the soil, or denitrification. Denitrification occurs when soil microorganisms run short of oxygen, but growth conditions (energy substrates, temperature) are

otherwise suitable. The organisms use soil nitrate as a source of oxygen and, by reducing it, release gaseous nitrogen which is lost to the atmosphere.

2. Soil Organic Matter and Soil Physical Properties

Organic matter also affects the physical behavior of soils. Most mineral soils have less than 5% by weight, but decomposed organic matter has a large surface area per unit weight and a range of complex surface electrical properties which interact with mineral particles. Organic matter, by increasing interparticle bonding, improves soil aggregation which improves soil structure. Aggregation depends on two processes: bonding to link particles together and fracturing to form a discontinuity which defines aggregates. Organic matter aids both of these processes. Organic matter found in virgin soils has built up over many years and the amount present is usually related to the amount of annual rainfall which strongly influences plant growth under natural conditions. When land is cleared and cultivated, the amount of organic material returned to the soil annually may change. In addition, many farming systems rely on tillage to manipulate soil structure, control weeds, and provide drainage. Tillage accelerates the rate of breakdown (mineralization) of organic matter and it is usual for soil organic matter levels to fall under tillage-based systems. It eventually reaches an equilibrium level much lower than in the natural soil. The type of organic compounds present will change also.

3. Clay Minerals and Soil Physical Properties

The type of clay minerals predominating in a soil depends on pedogenetic processes. Under the strong weathering conditions that lead to the formation of Oxisols, larger and less active kaolinitic minerals are formed, along with iron and aluminium oxides, which reduce the activity of clay minerals. Such soils are essentially "rigid" soils. In contrast, under more arid conditions that lead to the formation of Vertisols, smaller, more active smectite type clay minerals predominate. These have large surface areas and, although under some conditions organic matter or calcium carbonate can restrict activity, they are usually highly active. Wetting and drying causes swelling and shrinking in response to energy fields associated with forces holding water in the soil. The results of these very strong forces are most commonly seen as gross cracks in dry soils, but they can also be evident as serious damage to highways and major buildings. While relatively small amounts of organic matter can have a strong influence on the structure of Oxisols, the structure in Vertisols is largely determined by the behavior of the clay. In rigid soils some form of mechanical energy has to be applied if soil structure is to be changed quickly. In nonrigid soils structure can change very quickly due to wetting or drying and there is considerable scope for soil managers to work in harmony with these natural shrink–swell processes.

4. Colloidal Phenomena and Soil Physical Properties

The colloidal or electrokinetic behavior of clay particles in the wet state and as the soil dries has a strong influence on the structure of dry soil. If individual clay particles tend to repel one another (dispersion), the pore size distribution will be small; if, in contrast, particles attract one another and clump together (flocculation), more larger size pores will form. Since many beneficial processes occur in larger size pores, flocculation is highly desirable. For example, dispersive soils form surface seals and crusts and are excessively cloddy; flocculated soils have a continuous network of larger pores and are more granular and friable. Only a small proportion of the clay particles in a soil are ever free to truly disperse; these are usually close to the soil surface. The input of mechanical energy, such as from raindrop impact or a tillage machine, is usually needed, although wetting alone may be sufficient in some soils. Clay dispersion or dispersive tendencies cause many of the problems faced by soil managers. The type of exchangeable cations present on the surface of clay particles and the electrolyte concentration of the surrounding solution determines whether or not flocculation or dispersion prevails. Divalent cations such as calcium and magnesium reduce interparticle repulsion, as does increasing the electrolyte concentration. Increasing the proportion of monovalent cations (usually sodium) on the particle surface and reducing the electolyte concentration favors dispersion. The nature of exchangeable cations and soluble electrolytes in the soil can change as a result of management practices (such as use of a particular irrigation water or addition of a soil amendment such as lime or gypsum). The term "sodic soil" (earlier "alkali") is used to describe soils that show strong adverse effects due to dispersion. In the United States the criteria were taken as an exchangeable sodium percentage (ESP) of 15; in Australia this was revised downward to an ESP of 6. Unfortunately, criteria based on ESP are misleading because dispersion is not controlled by ESP alone; the term "dispersive" soil would be more appropriate.

5. Toxicities

Soil management also has to contend with possible toxicities that interfere with plant growth or biological activities, including nutrient imbalances associated with excessively alkaline or acid conditions, excess levels of particular elements (such as boron, aluminium, manganese, selenium) residual effects of herbicides or insecticides, or excess levels of soluble salts.

6. Salinity

Salinity can have adverse effects because of high concentrations of particular ions or because it has an osmotic effect which interferes with the normal physiological functioning of soil plants and animals. Soil salinity is a major management problem in many irrigated agricultural soils, particularly where it has not been anticipated and guarded against in management strategies. Disruption of hydrologic equilibria in landscapes due to the clearing of vegetation is also increasingly leading to rising water tables, seepage, and salinization of rain-fed agricultural lands.

C. Soil Biological Aspects

Tropical soils are home to a wide range of organisms ranging in size from micrometers to several centimeters. The types of organisms found in particular soils depend on the soil and environmental conditions which affect habitat suitability. Soil managers have tended to ignore soil biology and become concerned only when problems arise due to the harmful effects of disease pathogens or insect pests. In many cases, particular farming practices provide a habitat that favors the build up of a troublesome species or soil ecosystem and discourages useful species. Worse still, it has been common to seek a "magic bullet" chemical solution to biological problems instead of attempting to manage a balanced ecosystem which might provide ongoing benefits and more sustainable systems. In recent years, there has been a greater focus on balanced or integrated ecosystem management within agricultural systems. An example is the swing away from tillage-based systems toward management systems which use biological agents to provide the desired conditions since tillage can be harmful to some useful soil organisms such as earthworms and mycorrhiza. Ongoing research and development are essential to ensure that sustainable systems, acceptable to farmers, are developed. For example, farmers rely on tillage to prepare seedbeds on soils that are poorly aggregated and tend to crust and seal. Research has shown that management practices, such as mulching with straw to stimulate the activity of soil macrofauna, which create useful pore systems, can provide more productive alternatives. But these innovations are unlikely to be acceptable to farmers in some countries who prefer to sell or use straw for other purposes or to burn straw as a disease control measure. It has long been known that crop and ley pasture rotations can improve production from subsequent crops. Mechanisms underlying the response are complex and are not all due to soil biological effects. For example, the effect may be due to removal of the host which breaks a disease pathogen life cycle or to reduced levels of extraction of a nutrient such as nitrogen.

D. Soils as Systems in Time and Space

Soil management, to be effective, must consider the soil as a whole system. It is not adequate to simply consider physical, chemical, or biological effects alone. A large number of the processes occurring in soils interact. While it might be appro-

priate to consider individual components in order to resolve a particular problem, the remedial management practices have to be integrated into a workable and acceptable system for farmers. Scale, both in time and space, is also important in soil management. If management practices alter soil processes taking place on a scale of micrometers the eventual consequences may extend through landscapes and extend over several kilometers. This sequence of processes may take many years before harmful consequences develop. Similarly, the decline of soil organic matter or the slow build up of acidity may take many years; when a crisis arises due to the accumulated effect of these changes (a crisis of system sustainability), very expensive remedies may be required. For example, the buildup in organic matter due to changed practices can be so slow that it may be undetectable for several years. Also, the consequences of soil processes, although not harmful for crop production, may be harmful to society because of off-site effects, for example, nitrate leaching into groundwater that is used for human consumption.

IV. TROPICAL SOIL MANAGEMENT OPTIONS

In some quarters there is a preference for soil management "technology packages" or prescribed recipes that farmers should follow without question. Because seasons and markets change it is better for farmers to have flexible systems that can be quickly adjusted to meet changing circumstances. The normal changes would be to select different management options or practices. The decision should aim for favorable outcomes and be based on a scientific understanding of soil and crop processes.

A. Traditional Practices

Traditional practices are those that have been used, essentially unchanged, for very many years. They vary depending on the climate, soils, and the crops being grown. Where water supplies allowed rice to be grown, a rice monoculture may have persisted more or less permanently for many years. In other areas, a form of shifting cultivation was practiced. Natural vegetation was cleared and crops were grown for some years until fertility declined and yields fell. A new area was then cleared and the previous area was left to regenerate as a bush fallow. In other areas, the traditional practice has been nomadic grazing where people and their livestock move large distances, sometimes in a set pattern, to find pasture for grazing. An element common to most traditional practices is low risk of failure. They might not be highly productive, but they do meet community needs in all but exceptional years. With increasing populations in tropical regions, there is greater pressure to use land more intensively to produce more food. Traditional practices are being changed. Land formerly under natural bush or forests or used for extensive grazing has now been developed for permanent arable farming. [*See* Shifting Cultivation.]

B. Conventional Practices

Conventional practices are the practices followed by most farmers in recent times. Conventional practices are still evolving as adjustments are made to maintain yield, meet markets, and reduce risk in the face of declining soil organic matter and soil degradation through erosion and acidification. In the initial stages, newly cleared lands are usually relatively fertile, as this fertility declines it is usual for farmers to use fertilizers. Further changes will be needed to ensure development of systems that are sustainable. An example is cereal production on Vertisols in Queensland, Australia. Winter cereals have been grown in a monoculture, in a predominantly summer rainfall environment, by relying on the substantial water-holding capacity of Vertisols. For many years, the cropping system was based on a winter crop–summer bare fallow (possibly extending over 18 months). Crop residues were burnt after harvest and the soil was tilled after each rainfall event to break up any surface crust, to create surface roughness to trap water, and to kill weeds. This tilled fallow also built up mineralized nitrate in the soil which met the needs of the winter cereal when sown. This system was found to predispose the soil to water erosion and eventually fertility

declined as soil organic matter levels dropped. The conventional practice has now changed to one where tillage has been greatly reduced, crop residues are retained to protect the soil from erosion, chemicals are used to control weeds, and fertilizers are used to supply nutrients. There are signs that new conventional practices are evolving as farmers react to rising costs of inputs, disease carryover on crop residues, and greater awareness of better ways to manage soil ecosystems and to reduce climatic risk.

Another example is available from research and development studies at the International Crops Research Institute for the Semi-Arid Tropics (ICRISAT) in India. In the traditional farming system, the land was fallowed in the rainy season and a crop was sown near the end of the rains to mature on stored soil moisture. This system underused available resources. An improved technology package for Vertisols was developed at ICRISAT and has a number of innovative components: two crops per year are grown by sowing the first into dry soil before the rains come and the second at the end of the rains; a broadbed and furrow surface configuration (similar to a traditional practice in Ethiopia) designed to provide surface drainage and reduce erosion; an animal-powered wheeled tool carrier to carry out all tillage operations; improved crop varieties; scope for supplementary irrigation; and prescribed fertilizer inputs. This system greatly increased food yield potential on these soils. Farmer adoption has varied. Some components were taken up quickly while others are still being adapted and are finding their way into conventional systems. The real value of such research and development is that it makes farmers and scientists aware of the production potential of soil resources.

C. Tillage

Tillage is used to create a particular soil structural condition or to physically damage or destroy weeds. For many years tillage has been synonymous with farming and there is a tendency in some areas to till the soil without really analyzing whether the operation is necessary. Physical energy is applied through some form of implement; the energy source can be from a human, an animal, or some type of traction engine. Designs of ground-engaging tools are many and varied depending on the action of the tool and the purpose of the tillage, for example, moldboard and disc plows, rotary tillers, and tines. Moldboard plows are a common implement in temperate agriculture; they usually invert a slice of soil and are very useful for turning in crop residues. These plows have been used in tropical regions where enough traction power is available, but energy costs have risen and the benefits of systems with less intensive soil disturbance have been realized by soil managers. There is thus a tendency to reduce the number of tillage operations and the degree of soil disturbance.

In general, a tillage operation should be carried out only if it has a definite purpose and if the equipment available will give the desired result. To be effective, soil moisture conditions must be suitable; tillage of soils that are too wet leads to compaction and formation of plow pans (which of course may be desirable in rice paddies) whereas in soils that are too dry it leads to cloddiness. Also, the depth of working for tined implements is important as the soil should lift easily and break out. Even in so-called zero or no-till systems, the sowing method often disturbs a high proportion of the soil surface. In general, disturbance at sowing should be minimized so that continuity to the surface, of pore structure created by biological mechanisms, is maintained. Tillage on Vertisols is a particular case. Structure in these soils can change very quickly due to wetting and drying and is largely determined by aggregation mechanisms related to clay type and amount. It is often necessary to have a bed of loose aggregates for soil-leveling purposes or to form pits, ridges, or beds to improve water storage or surface drainage. For the best effect with the minimum input of energy, the implement should simply separate aggregates along lines of weakness produced by wetting and drying. A similar result can be obtained with "biological" tillage of rigid soils if plant root systems can be used to define aggregates.

Deep tillage is sometimes used to break up restrictive layers in the subsoil and to improve root penetration. Responses depend on subsequent seasonal conditions. It is usually beneficial if there is a clear target layer that restricts water storage or crop root growth. It is unlikely to be beneficial in wet seasons because there is no shortage of available water. Deep-rooted crops sown in sequence in the same row position can improve root penetration into subsoil gravel layers.

D. Mulching

A mulch is a cover of some material on the soil surface. The material can be organic, such as crop residues, farmyard manure, straw, sawdust, or factory wastes; it can be sand, gravel, or stones; or it can be an extruded film or a fabric woven from various fibers. Mulches protect the soil interface from the atmosphere, slowing and controlling processes. They provide a better habitat for many soil organisms because they stabilize temperature and moisture conditions and possibly provide food sources. Mulches serve a key role in protecting the soil surface from raindrop impact. Raindrop impact on bare soil has sufficient energy, at the high water contents prevailing in the surface layer under rain, to break down aggregates and disperse clay particles. Raindrops also compact the surface layer which, coupled with suction from within the soil below, can form an impermeable seal. If a surface seal does form, further rain will run off and water that might otherwise be stored for crop use and turned into yield is lost. Runoff can erode soil, and mulches reduce the erosive power of runoff. The effectiveness of mulches in preventing erosion is usually because infiltration rates are maintained through a continuous (protected) pore structure. It is not necessary to have a complete soil cover to maintain infiltration and reduce erosion. Under most circumstances a 20–30% surface cover is enough to have a substantial effect.

Mulches slow evaporation, but the benefits for crop production depend on the frequency of rain. In many subtropical areas, rain falls infrequently and in relatively small amounts. It does not penetrate far into the soil, particularly in Vertisols, and is readily lost by evaporation. In these situations mulches are unlikely to have much effect as the accumulation of soil water over time depends more on climate than on soil management. However, the soil surface must be managed so that when heavy rain does fall, it is not lost as runoff. Developing soil management practices to improve food production per unit of rainfall is a major challenge for scientists working on soil management in the more arid tropical regions.

Mulches also perform a valuable role in extending the period of good conditions for germination and in protecting emerging seedlings from heat blast and wind damage.

E. Crop Rotation

A crop rotation is a repeated sequence or cycle of crops within a cropping system. In contrast to a monoculture (e.g., sorghum–fallow–sorghum), crops placing different demands on the soil are rotated (e.g., sorghum–fallow–maize–groundnuts). Rotations are usually utilized to improve the nutrient supply or to break a disease cycle, but unexplained benefits often occur. Rotations should be designed so that each crop provides suitable conditions for the next crop, e.g., herbicides that have a potential residual effect should be avoided and crops that host similar insect pests or diseases should not be grown consecutively. Legumes are particularly useful in rotations because they fix nitrogen from the air into forms that are available to soil ecosystems. In some cases (e.g., grain legumes), they simply ease the demand for nitrogen from within the soil so that subsequent crops benefit whereas in other cases (e.g., forage or pasture legumes) they add nitrogen to the soil system and over time may change the carbon/nitrogen ratio in soil organic matter.

F. Drainage

Drainage in wet periods is critically important in maintaining adequate soil aeration. In irrigated agriculture, it is probably more important to have good drainage than it is to have a good water supply. In rain-fed agriculture, periods of waterlogging in the wet season can be disastrous on soils with poor internal drainage such as Vertisols. Internal drainage is the movement of water through the soil and out of the root zone into underlying layers whereas surface drainage is the movement of water off the field in a system of flow channels. In soils with high infiltration rates, drainage is usually only a problem if there is a restrictive layer or throttle somewhere in the profile or if a generalized water table rises into the root zone. If a water table rises in irrigated systems it may be because of leakage from supply channels. Installation of underground drainage systems may be necessary to prevent salinization. In some rain-fed areas, localized shallow water tables may occur in the wet season or clearing may disrupt hydrological equilibria. Localized groundwater interception drains are a possible solution. In well-drained soils, the leaching of dissolved nutrients beyond the root zone can be a problem. Vertisols usually have slow internal drainage and are often flat or have no well-defined drainage lines. On these soils surface drainage networks have to be provided by smoothing or grading the surface and by creating microrelief so that plants have a raised, aerated zone and water can flow off the field. This can be done by heaping the soil into mounds, forming ridge and furrow or bed and furrow systems. Because there are periods of intense rain in the tropics, loss of water as surface runoff is inevitable. The drainage system must allow the water to flow away without causing erosion. On sloping lands this is usually done by shortening the slope length with a series of graded drains (usually on a grade of <0.5%) that convey water to grassed or otherwise protected waterways that convey it down the slope.

G. Water Harvesting

Water harvesting is used in drier regions to concentrate the scarce supply of water and to make it available to plants that would not otherwise grow. A part of the land surface is managed so that it will shed water that flows to intake areas around the plants. The scale of harvesting systems varies. In some cases water is stored and used for irrigation in dry weather. A particular form of water harvesting uses surface configurations such as pits and tied ridges to trap water that would otherwise run off and allow it to infiltrate into the soil. The effectiveness depends on the nature of rainfall in the following season. In wet seasons there may be a crop yield loss due to waterlogging.

H. Water Erosion Control

Factors affecting soil erosion by water are summarized by the universal soil loss equation (USLE):

$$A = R \cdot K \cdot L \cdot S \cdot C \cdot P,$$

where A is the average annual soil loss per unit area, R is a rainfall factor depending on the intensity and duration of rainfall, K is a soil erodibility factor (reduced by aggregation), L is a slope length factor, S is a slope steepness factor, C is a crop or cover factor, and P is a factor that varies according to the erosion control practice used. On sloping lands, the conventional approach has been to design and construct a drainage network based on mechanically constructed terraces or bunds. These are formed across the slope, almost on the contour but with a slight grade. This series of cross-slope drains reduces slope length and ensures that large flow volumes will concentrate only at places that can be given special protection (waterways). The problem with this approach has been that it focuses on the engineered structures as the control measure rather than the management practices used between the graded channels or bunds. Soil erosion will continue between the bunds unless farming practices minimize tillage and maximize surface cover. Erosion, particularly in semiarid regions, is usually episodic. Studies have shown that most erosion is caused by infrequent catastrophic events. Unless the land between graded bunds is well managed, they will have little effect on erosion rates in catastrophic events. Because engineered structures are inappropriate for some land subdivision and tenure systems, and because some farmers do not maintain them, there has been increasing interest in vegetative bunds or barriers [using, for example, vetiver grass (*Vetiveria zizanioides*)] that filter water and are self-maintaining. On less sloping lands such as floodplains, strip-cropping is an effective measure. Strips of various crops and fallow land are aligned at right angles to the direction of water flow, slowing velocity and reducing soil loss.

I. Wind Erosion

Wind erosion is a serious problem on many arid soils. Control measures include providing windbreaks to slow wind speed and to provide cover on the soil surface to protect erodible particles. Techniques include leaving anchored crop residues and planting windbreaking strips of trees, shrubs, crops, or grass. Sometimes tillage is used to create a rough surface, particularly as a short-term or emergency measure.

J. Fertilizers

Many soils are naturally deficient in some plant nutrients or have levels which constrain crop yield potential. In other soils, nutritional constraints increase over time as soil organic matter reserves decline. Fertilizers are commonly applied to overcome these constraints. Because fertilizers are expensive, farmers apply only the optimum amount for the prevailing conditions. Recommended rates are usually based on soil analyses, the results of which have been calibrated by field experiments so that the yield response for a given crop can be reliably estimated. Another approach is to base the estimate on experience with a particular cropping system. In risky environments, farmers are reluctant to invest in fertilizers because other factors such as drought or flood might rob them of any return. Denitrification during a short period of waterlogging can cause all the nitrates in a soil to disappear as gaseous nitrogen. Also, there is increasing concern about the movement of nutrients away from farmlands into ground and surface water supplies. All of these factors are encouraging farmers to take a much more strategic approach to fertilizer use. The tendency is to apply lower rates but to apply them so that losses are low and nutrients are available at critical times.

K. Soil Ameliorants, Amendments, and Conditioners

These materials are applied to soils to change physical or chemical properties to make management easier or to alleviate a constraint. Examples are lime applied to reduce soil acidity, gypsum applied to increase electrolyte concentration so as to improve soil structure, and farmyard manure or straw applied to increase soil organic matter, stimulate biological activity, and supply nutrients. Most ameliorants are applied at a relatively high rate compared with fertilizers, i.e., several tonnes per hectare. A special group of soil conditioners are commercially promoted chemicals, usually synthetic, that are recommended to be applied at relatively small or even trace amounts. It is often difficult to assess whether there will be an economic return on the investment of an ameliorant because so much depends on the mode of action, soil properties, subsequent seasonal conditions, and residual effects. When deciding on substantial investments in "miracle chemicals" that are to be applied at trace amount rates, the crop-limiting factors and what responses can be reasonably expected from an equivalent investment in any other input should be taken into account. If they are used, one or more untreated trial strips, with identical management in other respects, should be carefully compared with the treated area.

L. Organic Farming

Organic farming uses systems that do not utilize commercial fertilizers or chemical pesticides and are supposed to be more in harmony with nature. While commonly regarded as more sustainable than modern intensive farming systems, organic systems must also meet the test of continued stability. For example, organic systems that rely only on inherent soil organic matter to supply nitrogen requirements will not be sustainable. In the long term, the systems successfully meeting society's needs in tropical regions are more likely to require a flexible and balanced combination of inputs.

M. Use of Soils for Waste Disposal

Agricultural soils are being increasingly used for waste disposal. Management of such soils in the tropics and elsewhere must take into account the cumulative effect of substances being added, on soil and product quality, and the amount and composition of runoff and deep drainage. Also, the impact of added water on the water relations of the soil and landscape needs to be assessed against long-term weather records.

N. Computer Simulation of Effects of Management Practices on Soil Processes

Because the climate in tropical regions, particularly in less humid regions, is highly variable, it is impossible to forecast the impact of particular management practices in the coming season. Managers are usually influenced by memories of the most recent seasons; even long-term averages are of little use. Field experiments that test innovations face similar difficulties and need to be maintained for many years in order to sample a representative range of seasonal conditions; in the meantime they may become obsolete. It is possible to develop computer simulations of the various processes involved in production systems, including crop growth models and hydrological and erosion process models. Submodels of the various processes must be validated in field experiments that show the effect of the management practices. Building the models into interacting systems allows the results of research and development studies in various fields to be integrated and to "value-add" to the investment in these studies. By running the system simulation using long-term weather records, probabilities for particular outcomes can be assessed. Although there are many gaps in available submodels, such simulations provide powerful decision support tools for considering and comparing management options in practical farming systems, assessing the sustainability of systems, and deciding priorities for research studies.

O. People Participation

The secret of improved soil management is acceptance and adoption by practicing farmers. If there is to be a change of practices, it will depend on farmers having ownership for, and commitment to, the need for change and what it involves. They will then be prepared to change attitudes and take the necessary actions. This can be achieved by working with groups of farmers and their families to analyze the present system, identify problems and research needs, constraints on adoption of changed practices, and courses of remedial action.

V. CONCLUSION

Soil management practices fundamentally underpin the health and well-being of a society. This is particularly so in the tropics. The future of burgeoning populations in the tropical parts of the earth's surface depends on the development of sustainable agricultural production systems. Tropical soils are a fragile resource and need to be managed carefully to maintain stability and fertility. A major international research effort is being made through the International Agricultural Research Centers, through many developed country universities, and through the national research programs found in most countries. Proven sustainable systems remain limited in number and the research should be maintained. Whole-system studies are needed to show the effects of interacting processes both in agricultural fields and in off-site land and water systems.

Glossary

Biological tillage Effect of soil biota or plant roots altering soil structure.

Bunds Term used in Asia for a small man-made earthen bank or ridge.

Decision support tool Any source of knowledge and information to help make a decision (e.g., book, guidelines, computer model, expert system).

Deep tillage A tillage operation designed to penetrate deeper than usual to disrupt a plow pan or restrictive layer.

Dispersion Process of soil structure breakdown in which individual clay size (<0.002 mm) particles are separated from each other and may show mutual repulsion in an aqueous system.

Exchangeable cations Cations held by electrical attraction on the surface of clay particles and which can readily exchange with cations in the soil solution.

Flocculation Process whereby the mutual repulsion between individual clay particles is reversed so that dispersed particles in suspension tend to agglomerate together and settle out.

Hydrologic equilibria A state where all the water-related processes in a landscape balance one another to maintain a particular range of moisture conditions.

Mechanical puddling Destruction of soil structure by breaking up soil aggregates into ultimate particles by applying mechanical energy when the soil is very wet.

Pedogenetic processes Soil-forming processes such as carbonation, hydrolysis, freezing, and thawing.

Photosynthetically active radiation Radiation which can be used by chlorophyll in plant cells to produce energy and fix carbon.

Plow pans Dense, compact layers formed under the tilled layer because of soil compaction and smearing by wheels or tillage implements.

Rigid soils Soils, usually sandy, that do not shrink on drying or swell on wetting.

Smectitic (or montmorillonitic) clay Crystalline silicate mineral particles <0.002 mm in size which have an expandable lattice arrangement, usually found in relatively less weathered soils.

Surface seal A thin (usually <3 mm) layer of disrupted aggregates and particles that restricts water infiltration.

Tool carrier A bar for mounting ground-working tools or sowing or fertilizing devices as part of an animal- or tractor-drawn machine.

Vegetative bunds Lines of plants that take the place of an earthen bund.

Water harvesting A system that promotes the drainage of runoff into a storage system.

Water use efficiency A measure of the amount of dry matter produced by plants per unit of water transpired; sometimes referred to as yield per unit of rainfall received.

Bibliography

Coughlan, K. J. (1984). The structure of vertisols. *In* "The Properties and Utilisation of Cracking Clay Soils" (J. W. McGarity, E. H. Hoult, and H. B. So eds.), pp. 87–96. Armidale: University of New England.

Edwards, C. A., Lal, R., Madden, P., Miller, R. H., and House, G., eds. (1990). "Sustainable Agricultural Systems." Ankeny, IA: Soil and Water Conservation Society.

El-Swaify, S. A., Pathak, P., Rego, T. J., and Singh, S. (1985). Soil management for optimised productivity under rainfed conditions in the semi-arid tropics. *Adv. Soil Sci.* **1**, 1–64.

Freebairn, D. M., Littleboy, M., Smith, G. D., and Coughlan, K. J. (1991). Optimising soil surface management in response to climatic risk. *In* "Climatic Risk in Crop Production: Models and Management for the Semiarid Tropics and Subtropics" (R. C. Muchow and J. A. Bellamy, eds.), pp. 283–306. CAB International.

Lal, R. (1977). The soil and water conservation problem in Africa: Ecological differences and management problems. *In* "Soil Conservation and Management in the Humid Tropics" (D. J. Greenland and R. Lal, eds.), pp. 93–97. New York: Wiley.

Littleboy, M., Silburn, D. M., Freebairn, D. M. Woodruff, D. R., and Hammer, G. L. (1992). Impact of soil erosion on production in cropping systems. I. Development and validation of a simulation model. *Aust. J. Soil Res.* **30**(5), 757–774.

Smith, G. D., Yule, D. F., and Coughlan, K. J. (1984). Research on soil physical factors in crop production on Vertisols in Queenland, Australia. *In* "Proceedings ACIAR-IBSRAM Workshop on Soils Research in the Tropics, 12–16 September, 1983," pp. 87–104. Townsville, Australia.

Swindale, L. D. (1982). Distribution and use of arable soils in the semi-arid tropics. *In* "Transactions 12th International Congress of Soil Science. New Delhi," Vol. 1, pp. 67–100.

Swindale, L. D., and Miranda, S. M. (1984). The distribution and management in dryland agriculture of Vertisols in the semi-arid tropics. *In* "The Properties and Utilisation of Cracking Clay Soils" (J. W. McGarity, E. H. Hoult, and H. B. So eds.), pp. 316–323. Armidale: University of New England.

Speciation

Douglas J. Futuyma
State University of New York

I. Diversity and Speciation
II. Species
III. Characteristics of Species
IV. Mechanisms of Speciation
V. Consequences of Speciation

Speciation is the evolutionary process by which two or more descendant species are formed from an immediately ancestral species. The definition of speciation hinges on the meaning of "species." Most evolutionary biologists use the "biological species concept," according to which a species is a population or group of populations that actually or potentially interbreed and is prevented by its biological properties from freely interbreeding with other such groups. The process of speciation therefore resides in the evolution of differences among populations that prevent interbreeding, i.e., the exchange of genes between them.

I. DIVERSITY AND SPECIATION

Perhaps 1.4 million living species have been named, but the total number of species alive today is certainly far greater—surely at least 10 to 15 million, and possibly as high as 30 million! Vast numbers of species of insects, mites, and nematodes await description, as well as many thousands that inhabit the deep sea and other almost unexplored habitats such as the interstices between sand grains, where a diverse fauna of tiny animals thrives. Living species, moreover, compose less than one percent of those that have existed in the past. All these species form a great "tree of life," or phylogeny, of organisms descended from a single common ancestor. This diversity has been generated by two related processes: anagenesis, the transformation of characteristics of each individual evolutionary lineage; and cladogenesis, the branching of ancestral lineages into two or more descendants. Each branch in phylogeny originated in the division of one species into two or more, which, it is generally agreed, were at first very similar, but which, embarking on different paths of change, anagenetically accumulated such differences that their descendants are classified as different genera, families, or still higher categories. Speciation is the fundamental cause of diversity. [*See* EVOLUTION AND EXTINCTION.]

Changes in the number of species in a clade (i.e., a group such as beetles that is descended from a single ancestor) depend on both the rate of speciation and the rate of extinction. Given an estimate of the age of a group from the first appearance of fossils, and an estimate of the number of species (N) at some time t million years later (such as the present), the average rate of increase (R) in species number per preexisting species can be estimated from the expression $N = N_0 \exp (Rt)$, assuming that the initial number of species, N_0, is 1. These rates of increase, R, vary greatly. According to S. M. Stanley, they are lower in bivalves, for example ($R = 0.06$), than in bovid ruminants ($R = 0.15$), murid rodents ($R = 0.35$), and colubrid snakes (R perhaps as high as 0.56 per million years). The net rate of increase, R, is the speciation rate (S) per species less the extinction rate (E), which can sometimes be estimated from the fossil record. Stanley finds that in rapidly radiating clades, rates of speciation and extinction are correlated, and that S varies among groups; for instance, S is in the range of 0.43 to 0.93 new species per preexisting species of mammal, per million years, but only 0.09 to 0.15 for bivalves. [*See* EVOLUTIONARY TAXONOMY VERSUS CLADISM; SPECIATION AND ADAPTIVE RADIATION.]

II. SPECIES

A. Concepts of Species and Speciation

The definition of "species" (Latin for "kind") has been and still is controversial for two centuries, partly because of real biological ambiguities and partly because it is used in two research traditions—taxonomy and evolutionary biology—with somewhat different histories, concepts, and aims. Pre-Darwinian taxonomists viewed species as "kinds" separately created. Holding an "essentialist" or "typological" philosophy inherited from Plato and Aristotle, according to which each "kind" has an immutable ideal "essence," they sought to ignore variation among individual organisms, and classified each individual based on its morphological conformity to the species' ideal "type."

Later taxonomists became aware of the considerable morphological variation within species, especially among specimens taken from different localities (geographic variation); a continuum of degree of difference from slight to great exists. In some instances, very different-looking forms, such as striped and ringed snakes, were found in litters from the same mother and so were acknowledged as conspecific (the phenomenon of polymorphism). Early in this century, it was further found that morphologically indistinguishable forms of organisms such as mosquitoes, fruit flies, flycatchers, and *Paramecium* differed in ecological and behav-

ioral features, chromosome structure, and often slightly in morphology if examined closely enough. Such "sibling species" indicated that morphological differences were not sufficient for distinguishing species. Confronted with evidence that forms do or do not interbreed, almost all taxonomists gave precedence to this criterion over the morphological degree of difference. [*See* SPECIES DIVERSITY.]

In the late 1930s and the 1940s, there emerged from such taxonomic and biological evidence and from the genetical theory of evolution being developed at that time, the concept that community versus discontinuity of reproduction is the defining concept of species. The evolutionary geneticists Theodosius Dobzansky and Hermann Muller, and especially the systematist Ernst Mayr, rather successfully promulgated the *biological species concept,* which Mayr defined in 1942 as: "species are groups of actually or potentially interbreeding populations, which are reproductively isolated from other such groups." "Reproductive isolation" need not mean hybrid sterility: many species can produce fertile offspring if forced to cross, yet do not interbreed in nature because of differences in behavior or other features (see later). The biological differences that confer reproductive isolation were termed "isolating mechanisms" by Dobzhansky.

The biological species concept (BSC) has been widely adopted by evolutionary biologists because it best describes patterns of variation in most organisms. Moreover, reproductive isolation plays a critical role in evolutionary theory. As long as populations are capable of free interbreeding, genetic differences between them either cannot develop to a substantial degree (if they actually interbreed freely) or once developed may be lost (if segregated populations meet and interbreed). Thus reproductive isolation confers on populations the ability to diverge genetically, following independent evolutionary trajectories, to any degree. Without speciation, there would be precious little diversity, and indeed little evolution.

Many biologists have objected to the BSC. Some objections arise from the two overlapping but distinct uses of "species" in contemporary biology. To most evolutionary biologists, the species is an *evolutionary concept,* as described earlier. But the species is also a *taxonomic category:* organisms with names such as *Escherichia coli, Homo sapiens,* and the red oak *Quercus rubra* are taxa in the category "species," just as *Homo* and *Quercus* are taxa in the category "genus." Many "taxonomic species" do not fit the BSC. For example, predominantly asexual organisms such as *Escherichia coli* are taxonomic but not biological species, simply because the domain of the BSC is restricted to sexually reproducing organisms. Different stages in a single evolving lineage may be named by paleontologists as different taxonomic species ("chronospecies"), such as *H. sapiens* and its antecedent *Homo erectus,* but the domain of the BSC is restricted to fairly narrow intervals of time within which populations might exchange genes. Some systematists have proposed a "phylogenetic species concept," according to which, if a phylogenetic tree can be constructed for a set of populations, any population characterized by a new ("derived") evolutionary feature will be designated a species, whether it interbreeds with other populations or not. Defenders of the BSC maintain that when the biology of organisms places them within the domain of the BSC, it is a useful definition that should be adopted. Organisms that do not fall within its domain, such as asexual forms, have different evolutionary dynamics, require different criteria for designation of species (or perhaps should be described in other terms entirely), and do not vitiate the BSC within its proper domain.

Some opponents of the BSC identify the concept with Dobzhansky's term "isolating mechanism." They hold that "mechanism" implies an adaptive function favored by natural selection (e.g., the heart is a mechanism for circulating blood) and that there is little evidence that biological barriers to gene exchange have evolved in order to accomplish this effect (see later). These authors (e.g., H. Paterson) argue that species should be defined not by reproductive isolation, but instead by their sharing in a common "specific mate recognition system." Adherents to the BSC view this as merely the opposite side of the same coin, and object only to these authors' exclusion of hybrid sterility and inviability as possible (though not universal) criteria of species difference.

The principal real difficulty of the BSC is that the extent of gene exchange (or its converse, repro-

ductive isolation) is not an all-or-none affair: populations with adjacent (*parapatric*) or coexisting (*sympatric*) geographic distributions may exhibit partial reproductive isolation in varying degree. Sympatric forms sometimes interbreed to some extent, yet retain their identity by and large, with most individuals readily classifiable. Sympatric hybridization appears to be more frequent among plants than animals (for example, the red oak *Quercus rubra* hybridizes with several closely related species). Parapatric forms often form a narrow "hybrid zone" where they meet, yet are fully distinctive shortly distant from the zone, implying that many genes from the one population (or "semispecies") are eliminated from the "gene pool" of the other. The fire-bellied and yellow-bellied toads (*Bombina bombina* and *B. variegata*), each distributed broadly in Europe, hybridize in a region only about 20 km wide. Such cases, in which it is rather arbitrary whether one species or two be recognized, are expected if, and indeed demonstrate that, the genetic divergence that causes speciation is a gradual process. They do not invalidate the BSC where it *does* apply (which is to say, for the majority of sympatric species). It simply must be acknowledged that not all populations fit into discrete definitional categories.

B. Definition versus Diagnosis of Species

A practical, although not a conceptual, drawback of the BSC is that *allopatric* populations, those that occupy separate geographic areas, can be shown to be different species only by placing samples together and finding that they fail to mate or that the hybrids are inviable or sterile. Such experiments are more feasible with some organisms (e.g., *Drosophila* or rapidly growing plants) than with others. To determine whether a sample of specimens from the same place consists of one or more than one species does not, however, require breeding experiments. Although reproductive discontinuity *defines* species, they may be recognized by the consequences of reproductive isolation, namely discontinuities in genetically based characteristics of any kind. A single discrete difference, such as stripes versus rings in the California king snake (*Lampro-*

peltis getulus), may be a single-gene polymorphism in a single species; but if several such characters, whether of morphology, behavior, or other features, segregate the sample into two or more discrete groups, they show reproductive isolation. This is a consequence of simple principles of population genetics. If the frequencies of two alleles at a locus (*A, A'*) are *p* and *q* (where *p* + *q* = 1), the proportion of heterozygotes should be 2*pq* under random mating. A strong deficiency of heterozygotes, or of intermediate phenotypes if neither allele is dominant, is evidence of some reproductive discontinuity. Moreover, recombination causes alleles at two or more variable loci (*A* and *B*) to become uncorrelated in a single randomly mating population; thus if a strong association is found between *A* and *B* and between *A'* and *B'*, these combinations (*AABB* and *A'A'B'B'*) are likely to mark distinct species. Thus if loci *A* and *B* affect different characters, a sample of two species will form two "character clusters"; if they affect the same (polygenic) character such as bristle number in flies, the character may show a bimodal distribution. Most sympatric species have been named by taxonomists on the basis of discrete sets of phenotypic differences that betoken genetic discontinuity and are indeed good species.

III. CHARACTERISTICS OF SPECIES

A. Barriers to Gene Flow

Species status depends on biological differences that potentially or actually impede gene flow between populations. These barriers ("isolating mechanisms") fall roughly into prezygotic and postzygotic categories (Table I). Prezygotic barriers, those that reduce the likelihood of the union of gametes, include ecological differences, such as mating on different host plants in the case of many host-specific insects; temporal differences, such as different flowering seasons in many groups of plants; gametic incompatibility, such as the failure of eggs and sperm, cast into water by many marine invertebrates, to "recognize" each other; and ethological ("sexual") isolation, especially important in

animals. Differences in mating signal and response are exemplified by the mating calls of frogs and crickets, the sexual pheromones (chemical signals) of many insects and mammals, and courtship behaviors and associated visual and/or vocal signals in many birds, fishes, and flies. These are the chief barriers to interbreeding in nature, and many such species (e.g., of ducks) can be hybridized, producing fertile offspring, in the zoo or laboratory. In flowering plants, the analogue of ethological isolation is attraction of different pollinating animals or differences in flower form that ensure deposition of pollen on conspecific stigmas instead of on those of other species.

Postzygotic barriers, arising from incompatible genetic programs for development, include the inviability or infertility (sterility) of F_1 or backcross hybrids; both inviability and infertility can vary in degree from slight to complete. A conspicuous feature of postzygotic barriers has been termed "Haldane's rule:" if there is a sexual difference in hybrids in the degree of inviability or infertility, this is more accentuated in the sex with only one X chromosome. Thus postzygotic isolation is typically more pronounced in male than in female hybrids (e.g., in flies and mammals), but the reverse holds in butterflies and birds, in which the female is XY and the male is XX.

B. Hybrid Zones

Hybrid zones, such as in *Bombina* toads, have been important in studies of speciation. The width of a hybrid zone is directly proportional to the average distance of dispersal of individuals of each species, or of hybrids, into the range of the other and is inversely proportional to the strength of selection against the hybrid gene combinations. Such selection may arise through differences in the ecological environment of the two populations, within which certain genes of the other species are disadvantageous, or it may arise because of the intrinsically lower fitness of hybrid gene combinations. For instance, heterozygosity for a certain locus or chromosome may lower viability or fertility, as when the populations are fixed (homozygous) for different inversions or other rearrangements of a chro-

TABLE I

Biological Barriers to Gene Exchange

I. Prezygotic barriers (preventing union of gamete nuclei)
   1. Potential mates do not meet (isolation by habitat or by season, time of day of mating).
   2. Potential mates meet but do not mate.
      a. Animals: ethological (= behavioral or "sexual" isolation).
      b. Plants: transfer of pollen reduced by different pollinators or flower form.
   3. Mating occurs (or is attempted) but gametes do not unite.
      a. Animals: insemination prevented by mechanical or structural barriers (when fertilization is internal), or chemical (e.g., cell surface) incompatibility between gametes.
      b. Plants: physiological (chemical) incompatibility between pollen and pistil, or between gamete cells.
II. Postzygotic barriers (reduction in fitness of hybrid zygotes)
   1. Egg is fertilized but zygote mortality is high (inviability, due to developmental incompatability, ecological selection, or both).
   2. Viable adult hybrids have reduced fertility (sterility, of one or both sexes).
      a. Incompatibility among genes, or between gene combinations and maternal cytoplasm, prevents development of viable gametes.
      b. Proportion of viable gametes reduced by aneuploidy (unbalanced chromosome complement) resulting from failure of chromosomes to pair or from structural rearrangements yielding unbalanced gene complements.
      c. Offspring of hybrids have reduced viability.

mosome, and the heterozygous F_1 hybrid suffers low fertility because of aneuploid segregation. This selection against the heterozygote prevents either chromosome from increasing in frequency when it is spread from one population to the other by interbreeding. Selection against hybrids may often arise from the inferior adaptive value of hybrid combinations of genes (an instance of *epistasis*); for example, AA and AA' may confer high fitness in combination with genotype BB at a second locus, and the combination $A'A'B'B'$ may likewise have high fitness, whereas other gene combinations may reduce viability or fertility. In this instance, it is likely that the prevalent genotypes in the two populations will be $AABB$ and $A'A'B'B'$, respectively.

The fitness of hybrid gene combinations in some instances depends on environmental factors. Thus it is not uncommon to find plant species (e.g., certain irises) that are reproductively isolated in natural environments, in which each species is adapted to a different habitat, but which form "hybrid swarms" where habitats have been greatly disturbed, providing intermediate microhabitats within which backcross hybrids thrive.

If F_1 hybrids are capable of some reproduction, and so by backcrossing conduct genes from each semispecies into the other, an allele that reduces the fitness of backcross hybrids will attain only low frequency in the semispecies into which it is introduced, and there will be a steep *cline* of decreasing allele frequency away from the hybrid zone (Fig. 1). This is also true at closely linked loci, but the cline will be less steep at loci that are more loosely linked to a selected locus because recombination in backcross progeny dissociates selected

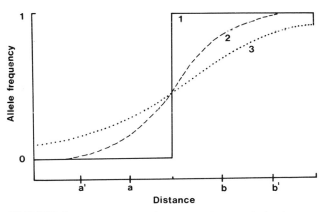

FIGURE 1 Idealized allele frequency clines in hybrid zones. Curve 1 (solid): No gene exchange between parapatric populations due to complete pre- or postzygotic reproductive isolation. Curve 2 (dashed): A steep cline across interval *a–b* due to strong selection relative to gene flow. Curve 3 (dotted): A shallow cline across interval *a'–b'* due to weaker selection relative to gene flow. Curves such as 2 and 3 may be displayed by different loci in a single hybrid zone.

from nonselected alleles, so the latter flow more freely between the semispecies. The spatial profile of neutral alleles at a marker locus therefore suggests whether or not it is closely linked to a selected locus that contributes to reproductive isolation. By extension, steep clines at few versus many neutral marker loci, distributed throughout the genome, can tell whether reproductive isolation is caused by only a few genes or many. In the case of the *Bombina* hybrid zone, J. Szymura and N. Barton found that most electrophoretic markers are characteristic of one or the other species and decline rapidly in frequency past the narrow hybrid zone; they concluded that almost every marker locus is therefore rather closely linked to a selected locus and that the number of such loci is at least 55. Barton likewise concluded that at least 150 loci each slightly diminish hybrid fitness and contribute to reproductive isolation between "races" of the grasshopper *Podisma pedestris*.

C. The Genetic Basis of Species Differences

The genetic differences between two closely related species include those that contribute to pre- or postzygotic reproductive isolation (i.e., those that cause the populations to be different species) and those that do not. The latter may include selectively neutral differences, such as perhaps many allozymes, and selectively important differences such as those that underlie ecological differentiation, but which are not necessarily causal in speciation. For any class of genes, the overall degree of genetic difference between two populations that do not exchange genes (either because of allopatry or reproductive isolation) should increase over time because of genetic drift and/or natural selection. Even if selection maintains the same mean state of a *character* in both species, its underlying genetic basis is likely to turn over. (For example, the same mean for an additively inherited polygenic character can be produced by any of many gene combinations that may be selectively equivalent, so allele frequencies at the constituent loci can change by genetic drift.) The genetic basis of speciation is better sought among recently differentiated populations or species than among long-differentiated taxa that have acquired further differentiation subsequent to the speciation event.

An enormous literature on electrophoretic (allozyme) differences among populations and closely related species has shown that (a) the degree of differentiation constitutes a continuum, from slight differences among conspecific populations, to greater differentiation as we pass through "subspecies," semispecies, and full species; and (b) the level of difference between reproductively isolated species differs greatly from one comparison to another. Molecular investigations show almost no differences between species in certain groups, such as some African cichlid fishes that are nonetheless strongly differentiated in ecological habits, behavior, and color pattern. In these and other such groups, molecular and other evidence indicate that speciation has been both rapid and recent, and that reproductive isolation is conferred primarily by sexual and/or ecological isolation, with postzygotic inviability or infertility playing a small role, if any.

Investigation of the genetic basis of morphological or other phenotypic differences among species has been limited to cases in which at least some fertile hybrids can be obtained and backcrossed. An extensive classical literature, treating plants especially, showed that the individual morphological characters that distinguish species typically are inherited as polygenic traits, as is usually true also of character variation within species. In many cases, the character is inherited more or less additively, but strong epistatic effects are sometimes evident. Among the most striking of these is the observation, first made by the great geneticist A. H. Sturtevant, that a feature which is identical in two species may nevertheless have a very different genetic foundation: hybrids between *Drosophila melanogaster* and *D. simulans* have aberrant, highly variable patterns of dorsal bristles, even though the two species have identical, almost invariant patterns. This observation implies incompatible interactions between different sets of genes governing the development of the bristles (and of the nervous system, of which these structures are part). In the same vein, a gene encoding a black spot in one species of platyfish (*Poecilia*) causes melanotic tumors when crossed into another species, implying that the interaction with "alien" genes disrupts its normal developmental function. These and many like observations led Mayr, Dobzhansky, and others to conclude that speciation entails the evolution of different, *coadapted,* systems of interacting genes. According to this view, the evolution of a given genetic difference imposes selection for "modifiers"—alleles at other loci—which interact favorably with the primary gene to yield harmonious function. Instances such as the platyfish pigment gene show, further, that genes detected by their morphological effects can have pleiotropic effects on development and thus on hybrids' viability or fertility.

Hybrid infertility is sometimes attributable solely to chromosome differences that prevent proper meiotic pairing and segregation. In crosses between some plant species, such as between certain primroses and between radish (*Raphanus sativus*) and cabbage (*Brassica oleracea*) that have the same chromosome number, the diploid F_1 hybrid is almost entirely sterile and the chromosomes do not pair. Doubling the chromosomes of the F_1, however, results in a tetraploid hybrid with the same genetic constitution as the diploid, but in which each chromosome can pair with its identical partner. These hybrids are fully fertile, showing that the diploid's sterility is caused by structural chromosome differences instead of differences at individual genes. Each of the chromosome differences that commonly distinguish related species probably lowers hybrid fertility only slightly. Chromosome differentiation does not necessarily accompany speciation; some species of Hawaiian *Drosophila,* for example, have identical chromosome structure.

Postzygotic barriers such as sterility are often, and prezygotic barriers almost always, based on genic differences instead of gross chromosomal rearrangements. The genetic basis of such characters can be analyzed indirectly, as Barton did from clinal patterns of gene frequency, or more directly by backcrossing hybrids. Some such studies have used the methods of quantitative genetics, which provide very inexact estimates of the "effective num-

ber" of gene differences from data on the means and variances of a character in F_1, F_2, and backcross generations. For example, Hawaiian species of *Drosophila* have diversified greatly in male secondary sexual characteristics on the legs, head, and mouthparts. *Drosophila heteroneura* differs from its close relative *D. silvestris* in that the male's head is greatly expanded to either side. *Drosophila silvestris* itself is divided into two allopatric "races," one of which has only 1 or 2, and the other about 30, long bristles on the male's foreleg, used to brush the female during courtship. An analysis of backcross hybrids of the races of *D. silvestris* provided evidence that about 30% of the difference in bristle number is due to one or more X-linked genes and the remainder to genes on at least two of the five autosomes. The three "effective genes" are certainly an underestimate because the method cannot detect additional linked genes or genes with small effects. A similar analysis of the head shape difference between *D. silvestris* and *D. heteroneura* concluded that at least six to eight gene differences are involved.

The other method of genetic analysis, almost limited so far to certain species of *Drosophila*, takes advantage of mutant markers on most or all of the chromosomes (or even chromosome arms) and of the lack of recombination in males. This enables an investigator to identify, among the backcross progeny, individuals with different mixtures of chromosomes from each species. Differences in morphology, fertility, or viability among these classes enable one to determine which marked chromosome(s) contributes to the difference between species. These studies, in several species groups of *Drosophila,* have generally shown that for characters such as male and female fertility and male genitalic morphology, the number of gene differences is as great as the number that the method can detect (i.e., at least one gene difference per chromosome arm, if each arm carries a mutant marker) (Fig. 2).

These polygenic differences suggest, as does variation in the degree of reproductive isolation among populations in many taxa, that reproductive isolation evolves gradually, by the successive substitution of many allelic differences between species. However, studies of populations that have diverged more recently, such as *Drosophila pseudo-*

FIGURE 2 Male sterility in backcross hybrids with different mixtures of *Drosophila pseudoobscura* (white) and *D. persimilis* (black) chromosomes. The X chromosome has the greatest effect (compare genotypes 1–8 with 9–16), but each autosome (chromosomes 2, 3, and 4) also has an effect (e.g., genotypes 1 vs 2 and 7 vs 8).

obscura from the United States versus Colombia, suggest that only a few gene differences may confer substantial intersterility. An interesting case is the European corn borer (*Ostrinia "nubilalis"*), which consists of two "pheromone races." A single gene difference determines whether the female moth releases one or another blend of sex pheromones, and another single gene difference determines the male's almost exclusive attraction to one or the other. Perhaps speciation is effected by fewer gene substitutions than previous evidence has suggested.

Most studies of the genetic basis of isolation indicate that the genes occupy consistent chromosomal sites, which together with other evidence indicates that (contrary to some speculations) transposable elements probably do not contribute to speciation. On the other hand, cytoplasmically inherited factors often cause inviability or infertility in insects, if combined with the nuclear genome of another species. These factors are usually endosymbiotic bacteria.

It should be stressed that the postzygotic isolation analyzed in many genetic studies may not be the effective barrier to gene flow among these species in nature, where prezygotic isolation is usual.

IV. MECHANISMS OF SPECIATION

A. Geography of Speciation

Speciation usually takes too long to observe directly. Hence our understanding of the mechanisms of speciation must rely less on experiments than on observations of patterns in nature, interpreted through theory. The applicability of one or another theoretical model, however, depends on data, which together with models help us to evaluate the likelihood of alternative explanations in any given case.

For example, population genetic models show that absent strong selection, a barrier to free interbreeding is necessary if strong interlocus associations, such as those among the several or many loci that contribute to a pre- or postzygotic isolating "mechanism," are to develop. Such a barrier could in principle be provided by a distributional gap between two initially genetically similar populations, by selection counteracting gene flow among neighboring populations, or by a genetic polymorphism that creates *assortative mating* or some other interruption to gene exchange within a population. These possibilities constitute the *allopatric, parapatric,* and *sympatric* models of speciation.

Allopatric speciation is unquestionably a major, if not the major, mode of speciation. From a theoretical point of view, the forces of genetic drift and natural selection will inevitably engender genetic differences among separated populations that confer reproductive isolation, given long enough time. The empirical evidence includes: (1) Conspecific populations vary in genetic composition, to a greater or lesser degree, in every kind of feature, including those that confer reproductive isolation. (2) The proliferation of closely related species is frequently correlated with the abundance of topographical or habitat barriers to gene flow. A species is often separated from its closest relative by such a barrier, as in fishes on either side of the Isthmus of Panama. Particularly strong evidence is afforded by island archipelagoes, in which each island is typically inhabited by more species than are single isolated islands; for example, Cocos Island, distant from the Galápagos archipelago, has only one species of Darwin's finch, whereas most of the Galápagos Islands harbor several. This pattern is attributed to speciation by isolation on different islands, followed by recolonization and sympatry. (3) Genetic differentiation among populations is most pronounced in organisms with a low capacity for dispersal and gene flow, such as land snails and wingless grasshoppers.

Hybrid zones between populations (such as *Bombina*) that are differentiated at many loci are thought often to represent *secondary contact* between formerly allopatric populations that, subsequent to differentiation, have expanded their range.

A continuum among two extreme patterns of allopatric speciation has been posited by Mayr and other authors. One extreme is the so-called "dumbbell" or vicariance model, in which a widely distributed species is sundered into two widely distributed entities by the emergence of a topographic barrier such as a mountain range or body of water. The two entities then diverge. At the other extreme, a population differentiates in a restricted locality after colonization from the main body of the species. The latter mode (peripatric speciation) has been considered by Mayr and others to be more frequent, based on observations that strongly differentiated populations and distinct species commonly have very limited distributions peripheral to the more widely distributed but more uniform "parent" species. Speciation by "vicariance" appears often to be slower than in localized populations, but few if any studies have demonstrated the effect on the speciation rate of population size per se by controlling for levels of gene flow and time since isolation (both difficult to determine).

The incidence of parapatric and sympatric speciation is highly controversial. Parapatric speciation might happen if sufficiently strong selection for different genes in different habitats were to counteract gene flow, engendering genetic differences sufficient to confer reproductive isolation. Among the few documented examples is divergence of grass populations adapted to soils impregnated with toxic metals from mine wastes; these have diverged not only in metal tolerance, but also in flowering time and the rate of self-fertilization, both of which confer partial reproductive isolation from populations on nearby normal soils. However, most hybrid zones between parapatric semispecies seem to have been formed by secondary contact rather than differentiation *in situ* and appear not to evolve the prezygotic reproductive barriers that characterize fully formed species.

Speciation by polyploidy is universally accepted as a mechanism of sympatric speciation because a polyploid plant is instantly postzygotically isolated from the parent diploid. Sympatric speciation by genic divergence is far more controversial, and population genetic theory implies that it is unlikely. A common model envisions that in, say, an insect that feeds on two plant species, each homozygote (AA, aa) is best adapted to a different plant, with the heterozygote (Aa) less well adapted. Selection might then favor associative mating (AA with AA, aa with aa), yielding adapted progeny. If mate preference depends on another locus (B, b), however, the rarer of these alleles can increase in frequency (forming two partially reproductively isolated subpopulations) only if an association (linkage disequilibrium) develops between alleles at the two loci (i.e., if the AB and ab combinations are more prevalent than Ab and aB). This can happen only if recombination is counteracted by strong selection on the A locus, and is facilitated if the basis for associative mating is attraction to and mating on the different plants.

One case which may illustrate the first step toward sympatric speciation is the true fruit fly *Rhagoletis pomonella*, in which samples from apple and hawthorn trees differ genetically in the time of emergence, which is correlated with the fruiting phenology of their hosts. In few other instances, however, is there evidence of strong selection of different genotypes on different plants, so the requirements for sympatric speciation may seldom be met.

B. Causes of Speciation

The causes of speciation are the area of greatest ignorance and controversy. The central questions are whether the genetic differences that confer reproductive isolation are caused by genetic drift (chance), natural selection, or some combination of the two; and if selection is primary, what its agents may be. The critical point is that speciation is the formation of gene complexes, differing typically at two or more loci, that are incompatible in heterozygous condition. If the heterozygote *Aa* has lower fitness than genotypes *AA* and *aa,* either allele is reduced in frequency by natural selection if it is rare. The problem then is to explain how two populations, both initially *AA* (or *aa*), can diverge so that the alternative allele is fixed in one of them. Divergence at many such loci (or chromosome rearrangements) poses the same problem. In an "adaptive landscape" that represents the mean fitness of genotypes in a population as a function of gene frequencies at one or more loci (Fig. 3), two species may be considered to occupy distinct peaks, with the low-fitness valley between them representing a population of hybrids. As long as the landscape (dictated by the fitnesses of the various genotypes under a certain constellation of environmental selective factors) is fixed, changes in gene frequency that move a population downslope are opposed by natural selection. An increase in the frequency of a rare allele, deleterious in a heterozygous condition, would be such a prohibited change. The problem then is how a population shifts from one adaptive peak to another.

There are two theoretical solutions (Fig. 4). One is for the allele frequencies to change rapidly in opposition to natural selection so that the genetic composition "leaps" across an adaptive valley to the slope of another peak. This might be accomplished by genetic drift, which in a sufficiently small population can counteract natural selection. Natural selection, which increases mean fitness, would then move the population up hill to a new, adaptive, genetic constitution (Fig. 4B). This is the "peak shift" model. An analogue to this was proposed by Mayr in 1954, who, seeking to explain the apparently rapid divergence of small peripheral populations, postulated that a population founded by a few colonists would by chance carry aberrant gene frequencies at certain loci. Because of epistatic interactions, gene frequency changes at other loci (modifiers) would occur under the guidance of natural selection, so that the initially random changes at some loci would set in motion the evolution of a new system of coadapted genes. Mayr's "founder effect speciation" or "peripatric speciation" is therefore a mixed model of random genetic drift and natural selection.

The other theoretical solution for getting populations onto different adaptive peaks is to change the

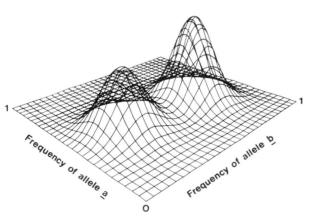

FIGURE 3 An "adaptive landscape" with two fitness peaks. The height of each point on the surface is the mean fitness of individuals in a hypothetical population with a specified gene frequency at each of two loci, specified by the horizontal axes. Gene frequencies change by natural selection so that a population evolves only up slope toward a peak.

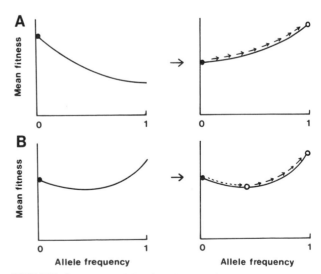

FIGURE 4 Two models of evolution of a population (circle) to a new genetic equilibrium on a one-locus adaptive landscape. (A) Adaptive shift. A changed environment alters relative fitnesses, and the population evolves solely by natural selection. (B) Peak shift. Genetic drift moves the population (broken arrow) to the slope of a second peak, and natural selection (solid arrows) then moves it up slope. The ecological environment has not changed.

landscape: in populations in different localities with different environments, different genotypes are favored by natural selection, i.e., they evolve on different adaptive landscapes (Fig. 4A). Because of its relative simplicity, this "adaptive divergence" model is favored by many population geneticists, who consider founder effect speciation unlikely.

Under either model, the impact of genetic changes on reproductive isolation must be viewed as an *incidental* (pleiotropic) effect instead of having evolved to serve the specific function of reproductive isolation. This is evident from the observation that allopatric populations, not faced with the "threat" of hybridizing, diverge in both pre- and postzygotic characters. The possibility exists that the reproductive isolation acquired in allopatry may be reinforced by natural selection if the populations expand and meet.

What is the evidence that the genetic changes responsible for speciation are caused by natural selection? Surprisingly, there is precious little. Genes conferring copper tolerance in a mine-associated population of monkeyflower (*Mimulus guttatus*) are lethal in hybrids with nontolerant populations; in the ground finches (*Geospiza*) of the Galápagos Islands, differences in beak size, an adaptation to feeding on different kinds of seeds, are also the signals for mate choice and sexual isolation. Although related species usually differ in many adaptive characteristics related to their different ecology, it is not known if, in general, these genetic differences are pleiotropically responsible for reproductive isolation. If sympatric speciation by adaptation to different resources were common, such instances of speciation would be driven by natural selection; but it is not known if this is the case. In several experiments with *Drosophila,* divergent selection among or within laboratory populations (for characteristics such as bristle number, food preference, and temperature tolerance) has brought about incipient sexual isolation; but in one such experiment, sexual isolation was at least as great among replicate populations subjected to the same environment as among those that adapted to different environments. The role of ecological selection in speciation is still not known.

In many groups of animals, *sexual selection,* based on variation in success at obtaining mates, may be a potent agent of speciation. Given genetic variation in a male display character and in female preference among variant male phenotypes, a "runaway" process may theoretically lead to an extraordinary elaboration of seemingly arbitrary features. Moreover, random differences among populations in the frequencies of alleles for various such features and female responses may initiate rapid runaway divergence. The consequence will be prezygotic, sexual isolation. The effects of sexual selection are conspicuous in many groups that have speciated rapidly and prolifically, such as cichlid fishes in the African Rift valley lakes, birds of paradise in New Guinea, hummingbirds in tropical America, and *Drosophila* in the Hawaiian islands. In all these cases, the variety of male display organs and behaviors is extraordinary; birds of paradise, for example, are variously adorned with elaborate modifications of the feathers of the tail, back, breast, and head. A comparable runaway process of coevolving signals and responses between plants and pollinators may be responsible for the high speciation rate in orchids, in which astonishing modifications of the flowers promote pollination by different insects.

Although at least some degree of reproductive isolation evolves while populations are allopatric, Dobzhansky postulated that it would become accentuated when incipient species meet, for selection should favor individuals which, by avoiding hybridization, produce fitter offspring. Dobzhansky therefore viewed barriers to gene exchange as "mechanisms" selected, at least in part, for isolation. (This hypothesis can apply only to prezygotic barriers since alleles that increase sterility cannot increase in frequency by natural selection for reproductive isolation.) Population genetic models show that this hypothesis is less plausible than it sounds; as in models of sympatric speciation, it requires tight linkage between genes that lower hybrid fitness and those responsible for assortative mating, which if loosely linked will become dissociated by backcrossing. Moreover, gene flow from neighboring populations can counteract selection within a narrow hybrid zone, and selection for modifiers that improve the fitness of hybrids could eliminate the raison d'être for discrimination in mating.

The hypothesis of reinforcement predicts that sexual isolation, or differences in associated display characters, should be more pronounced between sympatric populations of two species than between allopatric populations of these species, which are not subject to selection for assortative mating. Although numerous such comparisons have been made, in only a few cases has the expected pattern (termed reproductive *character displacement*) been found. One such instance is provided by a pair of species of Australian tree frogs, in which males from sympatric populations differ more in the mating call than do allopatric populations. Many investigators have experimentally measured sexual and postzygotic isolation among various populations and species of *Drosophila*. Using the level of allozyme differentiation as a measure of time since divergence, J. Coyne and H. A. Orr found, in a compilation of this literature, that among allopatric taxa, both sexual and postzygotic isolation increase gradually over time at similar rates; but among sympatric taxa, sexual isolation increases faster than the level of hybrid sterility or inviability (Fig. 5). This is, to date, the best evidence that selection

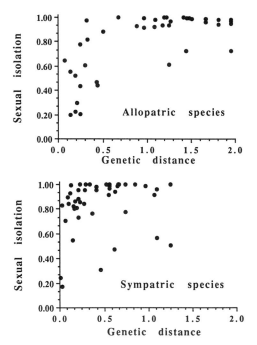

FIGURE 5 Genetic distance versus degree of sexual isolation between pairs of allopatric (top) and sympatric (bottom) *Drosophila* species. Genetic distance, a measure of divergence in allozyme frequencies, is correlated with time since isolation. At low genetic distance (e.g., 0.2), a higher level of sexual isolation is displayed by more sympatric than allopatric pairs, suggesting reinforcement in sympatry.

to avoid hybridization may generally enhance reproductive isolation. The paucity of such evidence, incidentally, argues against the prevalence of parapatric speciation.

The greatest contemporary controversy about causes of speciation concerns peak shift models, in which genetic drift in an initially small population initiates a shift to a new genetic composition. This concept arose from evidence of the prevalence of epistatically coadapted gene pools, implying that gene frequency changes at some loci might constitute selection for change at still other loci, and from the apparently rapid divergence of small, localized populations, e.g., of Hawaiian *Drosophila* and various Pacific island birds. These observations, however, do not constitute *evidence* for peak shifts because new coadapted genomes can evolve solely by natural selection (e.g., in a changed environment) and because small populations typically occupy small ranges, which, being ecologically more homogeneous than wider regions, should differ more in species composition and other environmental agents of selection.

Experiments will probably be necessary to elucidate the role of peak shifts in the evolution of reproductive isolation. For example, E. Bryant and L. Meffert subjected large experimental populations of houseflies (*Musca domestica*) to repeated "bottlenecks" of small size. The bottlenecked populations diverged in elements of male courtship behavior and exhibited some slight sexual isolation both among each other and with reference to a large control population. More research will be needed, however, before the controversy about peak shifts reaches resolution.

C. Ecology of Speciation

Ecology plays a threefold role in speciation: it affects the opportunity for spatial isolation, selection for genetic divergence, and the fate of newly formed species. The opportunity for spatial isolation of populations depends on dispersal and on how exacting the organism's ecological requirements are. Species with high rates of long-distance dispersal, such as the coconut, red mangrove, and many aquatic birds, are typically widely distributed and show little propensity for speciation compared to organisms that disperse little. Highly specialized species are often distributed as isolated populations in widely scattered suitable habitats; for example, this may be one explanation for the great numbers of closely related species in many groups of host-specialized herbivorous insects.

It has been noted that the role of ecological agents of natural selection in the evolution of reproductive isolation is little understood. Ecology, finally, is important in the fate of newly formed species. The principle that excessively similar competing species cannot indefinitely coexist (the competitive exclusion principle) undoubtedly governs many interactions. Allopatrically formed species that expand their range will not overlap, according to this principle, unless they differ sufficiently in resource use (or, in general, that the species' densities not be limited as if they constituted a single population). Surely many newly formed species must be too similar to coexist; this is doubtless reflected by the parapatric distributions of many close relatives (e.g., of birds in New Guinea) that do not interpenetrate each other's range. In at least some instances, such as among the ground finches of the Galápagos, species respond to competition by evolving differences in resource use (ecological character displacement) and can then stably coexist. Such ecological diversification is typical of many of the most prolifically speciating groups, such as African lake cichlids, among which closely related species occupy an extraordinary variety of specialized niches.

V. CONSEQUENCES OF SPECIATION

A. Rates of Speciation

The phrase "speciation rate" can mean either the rate at which species grow in number by speciation alone or the rate at which populations acquire reproductive isolation, i.e., become different species. The first is limited by the second. For instance, Stanley estimates the mean speciation rate (S) for several bivalve families to be 0.09 to 0.15 per million years, where S is the per capita rate of increase in species number, less the extinction rate estimated from the fossil record. The time required for the number of species to double by speciation alone is then $(\ln 2)/S$, or 7.7 to 4.6 million years. The average value for several groups of mammals is 1.6 to 0.75 million years. These are very approximate estimates of the mean time between successive speciation events in a clade; the actual time required for each such event (evolution of reproduction isolation) could be less. But if it required a minimum of, say 3 million years for each speciation event to transpire, the number of mammal species presumably could not have increased as rapidly as it seems, in fact, to have done.

The time required for speciation to transpire has been estimated, for recently originated species, from geological data on the age of barriers or of habitats, and from levels of genetic divergence (especially allozyme data) which are also calibrated by geological data. Instances of very rapid speciation include the African lake cichlids: 170 species, apparently from a single ancestor, have evolved in Lake Victoria, which is 0.50–0.75 million years old; and five species have arisen in a small satellite lake that is only 4000 years old. During the Pliocene, more than 30 endemic genera of limnocardiid snails diversified from a single ancestor in less than 3 million years in the present region of the Caspian sea. The island of Hawaii, 800,000 to 1 million years old, has endemic species of *Drosophila* derived from ancestors on the older islands. Among *Drosophila* species generally, reproductive isolation is reached, on average, at levels of genetic difference (calculated from allozyme data) thought to correspond to 1.5 to 3.5 million years of separation, although some pairs of species have arisen in less than 1 million years. Instantaneous speciation by polyploidy is of course the most rapid of all. A new species of cordgrass (*Spartina anglica*) has arisen by allopolyploidy within this century.

Factors conducive to rapid speciation are likely to include: (a) low vagility, fostering genetic divergence of spatially segregated populations; (b) strong sexual selection, perhaps especially in forms with complex precopulatory behavior; (c) ecological opportunity for divergence in different "niches," which may enhance selection for genetic change, ecological isolation, and persistence of newly formed species; and (d) perhaps features of the genetic system, such as strong epistasis, that facilitate peak shifts.

B. Macroevolutionary Consequences of Speciation

Among the important consequences of speciation for evolutionary history writ large, the most self-evident is that it is the source of diversity. An individual species can, by genetic polymorphism and alternative developmental pathways, fill an appreciable diversity of niches, but the range of intraspecific variation is far more limited than the variation among species.

Another effect of speciation is that while ancestral characters are transformed in some lineages, they may be retained in others. Thus living club mosses, silverfish, and lungfishes retain features that would have been erased among plants, insects, and vertebrates by anagenetic evolution had speciation not occurred. The persistence of "primitive" character states enables us to trace phylogenies and sometimes to infer the course of evolution of characteristics, even in the absence of a fossil record.

The most controversial issue is the extent to which anagenetic evolution may depend on or be facilitated by speciation. In 1972, N. Eldredge and S. J. Gould proposed *punctuated equilibrium* as both a description of a pattern they claimed is common in the fossil record and as a hypothesis about the evolutionary process. The pattern is one of *stasis*—a lineage often displays no substantial change for several millions of years—*punctuated* by rapid shifts in one or more morphological features (such as the number of rows of eye lenses in trilobites). The explanatory hypothesis (as distinct from observed pattern) of punctuated equilibrium is that a lineage is prevented from evolving, by internal genetic constraints, except when a localized population undergoes a peak shift in genetic constitution, during which both morphology and reproductive isolation evolve. This hypothesis applies to the fossil record Mayr's theory of founder effect (peripatric) speciation; indeed, Mayr had foreshadowed the punctuated equilibrium hypothesis in 1954.

Both Eldredge and Gould and also S. Stanley concluded that if departures from the ancestral morphology require speciation (splitting), then long-term evolutionary trends (anagenesis) may be attributable not to simple evolution within a single unbranching lineage, but to differences in rates of speciation and extinction among species in a clade. For example, several lineages of gastropods display an evolutionary trend away from the planktonic juvenile state, apparently because of a greater speciation rate in forms with more sedentary juveniles. Stanley termed this "species selection," a higher-level analogue of natural selection among individual organisms.

The hypothesis of punctuated equilibrium predicts that morphological shifts should be accompanied by speciation (splitting). Although several such instances have been described by paleontologists, other cases merely show the rapid (i.e., on the order of 10,000 to 100,000 years) evolution of certain features in nondividing lineages. Mere changes in the evolutionary rate are explicable as simple effects of natural selection in an altered environment. Paleontologists are strongly divided on the prevalence of the pattern that Eldredge and Gould claimed, and population geneticists are almost universally opposed to the hypothesis that genetic constraints prevent evolution except in association with speciation. They cite as counterevidence the abundance of genetic variation within populations, the many examples of rapid adaptation observed in natural and experimental populations, and the prevalence of geographic variation within species, much of which has demonstrably originated very recently.

An alternative explanation of the paleontological patterns has been suggested by D. Futuyma, who pointed out that although spatially segregated populations may diverge greatly in morphology and ecological roles, the divergent characters of a local population will ultimately be lost by interbreeding with more widely distributed phenotypes because populations move about the landscape in response to climatic and other environmental changes, and so ultimately will have the opportunity to interbreed (Fig. 6). If a divergent local population has achieved reproductive isolation, however, it can

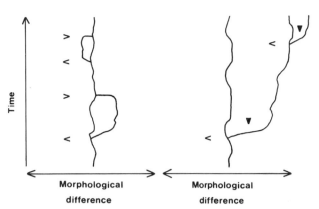

FIGURE 6 The possible role of speciation in anagenesis. (Left) A large, morphologically stable species buds off (<) allopatric isolates, which diverge but then lose their divergent character by subsequent interbreeding (>) with the parent species. (Right) An allopatrically formed (<) isolate retains its divergent features if it evolves reproductive isolation (▼) and can subsequently generate still more divergent allopatric (<) populations that repeat the process.

retain its distinctive ecology and morphology even when sympatric with the parent form; its divergence from the ancestral state, instead of being lost by recombination, can persist long enough both to be registered in the fossil record and to be modified still further in the same manner. Thus speciation may facilitate anagenesis not by freeing populations from the bonds of internal constraints, but by acting, like a mountaineer's pitons, to preserve successive steps in evolution away from the ancestral condition.

C. Species, Speciation, and Environmental Biology

The existence of reproductive isolation among populations, i.e., the existence of species, has many practical consequences for human affairs and for conservation. For example, the potential for gene flow among populations of a widespread species (i.e., the lack of reproductive isolation) has consequences in epidemiology. A gene for insecticide resistance in the mosquito *Culex pipiens* in Africa, Asia, and North America is identical at the molecular level, evidence that it originated as a single mutation that has been spread worldwide by gene flow.

Recognizing that an apparently single species actually consists of sibling species with different eco-

logical characteristics can be exceedingly important. Some of the mosquito species formerly included under "*Anopheles maculipennis*" carry malaria whereas others do not. In agricultural ecology, it is critical to recognize sibling species of insects with different host plant associations, and of parasitic insects that may be used for biological control of pests, but which differ in the hosts they attack.

The loss of biological diversity, because of habitat destruction, pollution, and other onslaughts originating in the growth of human populations, is one of the great contemporary tragedies. Human activity is probably responsible for the extinction of about 20% of bird species, and about 11% of the surviving species are endangered. Many of the extraordinary cichlid species of Lake Victoria have been extinguished by a predatory fish introduced for sport. Even by conservative estimates, the current rate of extinction is probably 1000 times the average throughout evolutionary history. Each such species is a unique gene pool, a unique book in the world library of genetic diversity. Some of these gene pools can be useful as agents of pest control, as sources of food or medicine, or as banks of useful genes that modern genetic technology could use for improving domesticated plants and animals. But utility is the less important reason for conservation. The astonishing diversity, the exquisite adaptations, and the profound intricacy and beauty of species are justification enough to save them, as we would a national gallery of art. Each extinction is an irreversible impoverishment, and the rate of speciation is far too slow to replenish it.

Glossary

Allozyme One of several forms of an enzyme, encoded by different alleles; usually distinguished by electrophoretic mobility.

Allopatric Allopatric, parapatric, and sympatric populations occupy separate, adjacent, and the same geographic areas, respectively. Allopatric populations need only be distant enough for gene flow between them to be slight.

Aneuploid A cell or organism with an unbalanced chromosome complement, i.e., an excess or deficiency of one or more chromosomes.

Assortative mating Nonrandom mating among like phenotypes.

Cline A gradual change in an allele frequency or in the mean of a character over a geographic transect.

Epistasis A synergistic effect of genes at two or more loci, whereby their joint effect differs from the sum of their individual effects.

Fitness Average contribution of one allele or genotype to the population in succeeding generations compared with that of other alleles or genotypes.

Linkage disequilibrium Association of alleles at two or more loci, at a different frequency than predicted from the individual allele frequencies: an excess or deficiency of certain gene combinations.

Pleiotropy Multiple phenotypic effects of a gene.

Polymorphism Existence within a population of two or more genotypes for a locus or trait, the rarest exceeding some arbitrarily low frequency (e.g., 1%).

Polyploid A cell or organism with more than two sets of chromosomes.

Sexual selection Variation in fitness, usually among males, due to differences in the number of matings achieved as a consequence of competition among males or of female "choice" among male phenotypes.

Sibling species Species that are difficult or impossible to distinguish by morphological features.

Bibliography

Atchley, W. R., and Woodruff, D. S., eds. (1981). "Evolution and Speciation: Essays in Honor of M. J. D. White." Cambridge: Cambridge Univ. Press.

Coyne, J. A. (1992). Genetics and speciation. *Nature* **355**, 511–515.

Futuyma, D. J. (1986). "Evolutionary Biology," 2nd Ed. Sunderland, MA: Sinauer.

Futuyma, D. J. (1987). On the role of species in anagenesis. *Am. Nat.* **130**, 465–473.

Harrison, R. G. (1990). Hybrid zones: Windows on evolutionary processes. *In* "Oxford Surveys in Evolutionary Biology" (D. J. Futuyma and J. Antonovics, eds.), Vol. 7, pp. 69–128. New York: Oxford Univ. Press.

Mayr, E. (1970). "Populations, Species, and Evolution." Cambridge, MA: Belknap Press.

Otte, D., and Endler, J. A., (eds.) (1989). "Speciation and Its Consequences." Sunderland, MA: Sinauer.

Stanley, S. M. (1979). "Macroevolution: Pattern and Process." San Francisco: Freeman.

Templeton, A. R. (1981). Mechanisms of speciation: A population genetic approach. *Annu. Rev. Ecol. Syst.* **12**, 23–48.

Speciation and Adaptive Radiation

Mark R. Macnair

University of Exeter, United Kingdom

Species are groups of organisms that are adapted to live in particular ecological niches, and are reproductively isolated from other species occupying the same or different ecological niches. Speciation is the process by which a group bifurcates into two groups, each with different attributes. As a lineage splits into two or more species with different ecological niches, the range of ecologies displayed by the lineage increases. This phenomenon is known as adaptive radiation.

I. INTRODUCTION

The process of speciation is still not fully understood, and there is no universal definition of what constitutes a species. This partly reflects the limits of our knowledge, but it also results from the fact that there are likely to be a number of different speciation mechanisms, which have differing consequences for adaptive and reproductive radiation. This article explores specifically the relationship between adaptive radiation into new ecological roles and the acquisition of reproductive isolation. [*See* SPECIATION.]

II. WHAT ARE SPECIES?

One of the problems of defining and explaining speciation is that there is no one, clear definition of a species. A number of species attributes can be recognized.

A. Morphological and Ecological Differentiation

Species are normally morphologically distinct: it is indeed how we recognize and describe them. These morphological differences are often associated with characters that adapt the species to particular ecological roles. Ecological theory suggests that two species that occupy exactly the same ecological niche (i.e., utilize the same limiting resources) cannot coexist in the same place, so that, in general, for two species to be able to coexist, some degree of ecological differentiation must have occurred.

However, sibling species are species in which very little or no morphological differentiation has occurred, and two species whose distribution does not overlap are able to have identical ecological niches. It is thus not possible to define species objec-

369

tively by their degree of morphological or ecological differentiation.

B. Reproductive Isolation

Species that coexist within a single geographical area (sympatric species) must be sufficiently reproductively isolated so that their two gene pools can remain separated and their genetic differentiation maintained. It is this characteristic that is emphasized above all others in the biological species concept, which sees the acquisition of reproductive isolation as the key step in the formation of species. Isolating barriers are traditionally classified into pre- and postzygotic (see Table I), depending on whether the barrier operates prior to the formation of a zygote. Postzygotic barriers involve the wastage of the female's gametes and, although they can prevent gene flow between populations, involve a large loss of fitness to individual females. [See EVOLUTION AND EXTINCTION.]

Thus coexisting sympatric species must be differentiated in two ways: ecologically and reproductively.

C. Genetic Differentiation

Species obviously differ genetically. It is ultimately differences in genes that cause all the morphological, ecological, and reproductive differences between species. It is impossible, however, to specify how much differentiation at the DNA level is required to induce speciation. Many studies have looked at this characteristic using differences in allozyme frequencies as a measure of genetic difference. (Allozymes are electrophoretically distinguishable alleles at a single protein locus.) While there is frequently a correlation between genetic distance and taxonomic rank *within* a particular group of organisms, there is a poor correlation when all organisms are considered. Thus in the *Drosophila willistoni* group, different populations of the same species have average genetic distances of 0.031, different sibling species of 0.581, and morphologically different species of 1.056. In contrast, different populations of the same species of cichlid fishes in Lake Malawi have genetic distances of 0.045 on average (i.e., similar to *Drosophila*), but species in different genera have genetic distances of only 0.076 (i.e., only slightly more than differences between populations and much less than differences between species in *Drosophila*). It appears that genetic differences as measured by these allozyme techniques are largely determined by time since separation instead of by differences in morphology or ecology. [See SPECIATION.]

TABLE I
Mechanisms of Reproductive Isolation

Prezygotic	
Ecological	Species do not meet because they occupy different microenvironments within an area
Ethological	Animals: courtship or other signal/receiver system differ; plants: pollinators differ
Temporal	Breeding seasons differ
Mechanical	Animals: Structural differences make successful mating difficult; plants: floral shape differences mean that pollinators do not transfer pollen efficiently
Breeding system	Selfing species are isolated from outbreeders
Gametic	Plants: pollen unable to grown down style; animals: sperm destroyed by female, or unable to locate egg because chemical attractants unrecognized
Postzygotic	
Hybrid inviability	Hybrid dies between fertilization and adulthood
Hybrid sterility	Hybrid is viable, but more or less infertile
Hybrid breakdown	Hybrid is viable and fertile, but progeny are inviable or infertile.

D. Chromosomal Differentiation

Most species are different at the chromosomal level, having differences in either chromosome number or organization. Many of these differences, although not all, could result in postzygotic reproductive isolation, because the chromosomes are unable to pair and separate regularly at meiosis to produce balanced gametes. Chromosomal differences could also lead directly to morphological or ecological differentiation, if the chromosomal rearrangements have resulted in a change in the expression or regulation of genes and their products.

III. THE PROBLEM OF SPECIATION

The process of speciation involves the splitting of a lineage, so that a clade that was originally one potentially cohesive gene pool, with all individuals able to mate and form fertile offspring with all others, becomes split into two cohesive gene pools. Within each of these, all individuals can mate and form fertile offspring, but they are not able to do so with members of the other. Both the starting and end points of this process are easy to visualize and comprehend: it is the gap in between that is conceptually difficult. There has to be a way to see how the population can move steadily from the one state to the other, invoking only the normal population genetic processes of natural selection and genetic drift. The problem is that for many reproductive-isolating barriers, there is an adaptive valley between the two states, so that while both beginning and end points are adaptive and stable, there is a valley of reduced fitness between the two. Standard models predict that natural selection will oppose and prevent the population traversing this valley (natural selection causes populations to increase average fitness, and so only moves population "uphill" on a landscape of fitness). Note that not all reproductive-isolating mechanisms have such an adaptive valley. In order to understand how the various models overcome the problem of the adaptive valley, it is helpful to analyze the various isolating mechanisms in more detail and to classify them according to how the valley arises.

A. Isolating Barriers Leading to Adaptive Valleys

1. Heterozygote Disadvantage

Many chromosomal rearrangements or changes in number lead to partial or total infertility in heterozygotes. Single translocations or inversions may lead to only small reductions in fitness (in the order of 5–10%) but even such a disadvantage will be strongly selected against. Multiple rearrangements lead to larger infertility, but presumably could have evolved one by one. These sorts of chromosomal changes cannot spread easily in large populations because when they first arise, they are heterozygous; homozygotes (with restored fitness) will not be common until the rearrangement has increased considerably in frequency, which it cannot do.

In plants especially, changes in the ploidy level are commonly involved in speciation. These chromosomal changes can, and do, happen in a single step, and the new plant is frequently fertile. Thus polyploidization provides a mechanism for instantaneously crossing the adaptive valley between the two stable states (say diploid and tetraploid). However, a cross between a diploid and a tetraploid will produce a triploid, which is normally more or less sterile. Thus, if an outcrossing tetraploid arises as a single individual in a population of outcrossing diploids, then all its progeny will be triploids, and the individual will be of low fitness and not survive. Single step speciation via polyploidy is more probable if the tetraploid is self-fertile.

A heterozygote disadvantage for single gene changes is unusual and so may not be important in the evolution of most of the physiological, behavioral, or morphological differences between species.

2. Epistasis

Many genes may interact to give a disadvantage to hybrids. Epistatic interactions of the sort that Ab and aB are fit combinations, but AB or ab are unfit, produce an adaptive valley. A population that is fixed for the A and b alleles cannot be invaded by either an a allele or a B allele. An example of this sort of interaction is given by signal/receiver type mating systems. Many animals may find their mates by the females emitting a signal (say a phero-mone) that the male receives and responds to. It is likely that different genes are responsible for the signal and the receiver. A gene that altered the signal in the female would be selected against because males would recognize it less well; a gene that altered the receiver would be selected against since males possessing it would fail to respond properly to the signal. There is thus stabilizing selection for both characters. Yet many species are primarily isolated by mate recognition systems that conform to this general signal/receiver categorization. For instance, most cases of ethological isolation (see Table I) in animals, mechanical isolation, or gametic isolation will conform to this model. Note that this does not mean that these species differ in single genes for the signal and receiver (indeed they almost certainly differ in many loci contributing to both parts of the system), but that these differences must have evolved by the spread of individual genes that had these properties.

B. Isolating Barriers Not Involving Adaptive Valleys

1. Epistasis

Many pre- or postzygotic barriers may involve epistatic interactions that do not involve an adaptive valley in order to evolve. For instance, many postzygotic barriers may involve the complementary action of genes so that *AB* is lethal, but *ab*, *Ab*, and *aB* are viable genotypes. It is then possible for an ancestral population that is of genotype *ab* to evolve into two populations, *Ab* and *aB*, without either going through an adaptive valley, but the hybrids between the two are inviable. Again, it is not being suggested that species are generally isolated by simple genetic systems of this type (though systems with these properties are known in plants), but isolation could well arise by the spread of many pairs of genes, each with this type of interaction, and none of which requires an adaptive valley.

Hybrid inviability or lack of fitness is often ascribed to the breakdown of "coadapted gene complexes." These are combinations of genes that work well together, but not when they are broken down

into their component genes. As a simple illustration, this might be if *ABC* and *abc* are fit combinations (coadapted complexes), but the combinations *AbC, aBc, abC,* and *aBC* are not. *ABC* can evolve from *abc* by the successive spread of first *A,* then *B,* and finally *C*. At no stage is an adaptive valley crossed, but the final product is isolated from the starting population, since the hybrid between the two populations will generate the unfit recombinant types.

2. Adaptive Characters

Some isolating barriers can be produced without the spread of any genes that cause inviability or a necessary loss of fitness in hybrids. These are isolating barriers that follow as a consequence of adaptation to an ecological niche. Thus many barriers involved in ecological isolation, temporal isolation, or a change in pollinators in plants can evolve in direct response to adaptation to prevailing conditions, producing isolation as a by-product of adaptation, and without any conceptual problem caused by crossing an adaptive valley.

C. Isolating Barriers as Characters

The biological species concept emphasises reproductive isolation as being the important process defining speciation: the evolution of isolation is a necessary and sufficient condition for the occurrence of a speciation event. It is very easy therefore to imagine that isolation itself is a character that is fashioned by evolution. This would be a mistake. The majority of speciation models see isolation as being an epiphenomenon, a by-product of other processes leading to differentiation and adaptation. All barriers involving postzygotic isolation and hybrid breakdown can only evolve as the by-product of genetic differentiation.

The only exception to this is the so-called "Wallace effect," or reinforcement. If two populations are sufficiently differentiated so that the hybrids between them are of reduced fitness, either because of postzygotic barriers or because the hybrids are not adapted to prevailing environmental conditions, then there could be selection for genes that caused individuals within each population to mate only with each other. In this case selection for genes whose immediate phenotype is "isolation" is being postulated: the initial isolating barrier is being reinforced. Reinforcement is a very controversial phenomenon, and there are few specific examples where it has been shown to be the most probable cause of prezygotic isolation. On the other hand, Coyne and Orr have conducted a large-scale survey of *Drosophila* in which they compared the strength of pre- and postzygotic isolation in pairs of allopatric and sympatric species. While the strength of postzygotic isolation was the same in both groups, prezygotic isolation was stronger in sympatric pairs. These results are consistent with the theory of reinforcement.

IV. MODELS OF SPECIATION

A. Allopatric Speciation

The classic allopatric model of speciation serves as sort of "null hypothesis"—a model that clearly has great explanatory power and is undoubtedly correct in many cases. In this model, a widespread species becomes separated into two large populations by some geographic barrier. During the period when the geographical barrier exists, the two populations evolve independently. Either because they adapt to different conditions in each refuge or because the genes that spread in each population, although functionally equivalent, are different, genetic differences arise between the two populations. Genetic coadaptation within each population is expected. If the geographical barrier breaks down, and the two populations reencounter each other, then isolation may prove to have evolved so that the two gene pools may remain distinct despite contact. Isolation by either pre- or postzygotic mechanisms, or both, could have occurred. If only postzygotic isolation has evolved, reinforcement may occur and prezygotic mechanisms evolve.

For instance, over the past $2^1/_2$ million years, a number of ice ages have caused the ice to advance across most of northern Europe, retreating northwards in the interglacials. During the ice ages, species were driven southeast and southwest, per-

sisting in refugia in southern Europe. Populations detached in Spain were isolated from those in refugia in Greece or Asia minor. Many taxa that are now widespread in Europe have Western and Eastern races that probably evolved during one or more of these periods, including crows, mice, and grasshoppers.

This model explains the evolution of isolating barriers that do not involve adaptive valleys, but it does not explain the evolution of those, such as chromosomal rearrangements or changes to signal/receiver mating systems, that do require an adaptive valley to be crossed. Separation of the two populations geographically does not enable either to evolve by natural selection into the adaptive valley. For instance, stabilizing selection will still act in each population on signal/receiver systems and so prevent them diverging. Only if the genes that produce the signal, say, also govern other characters that are subject to selection in one of the two populations but not the other, can divergence of this mate recognition system occur. The signal will evolve in the one population in response to the selection acting on the other characters that are genetically correlated with it and will cause the receiver to evolve to track the changes in the signal. Note, however, that isolation between the two populations has followed as a by-product of the adaptation of the signal with respect to its other properties, not its mate recognition character.

The speed of speciation under this model, and the degree of adaptive radiation produced, depends on the environmental differences encountered by the species on either side of the geographical barrier. If the two environments are essentially the same, then the two populations of the species should continue to occupy the same ecological niche in the two habitats, and divergence of the two populations, and ultimately speciation, is expected to be slow. Genetic differences will accumulate gradually by both genetic drift and selection, and eventually, after thousands of generations, the two populations may have evolved an intrinsic barrier to gene flow. Adaptive radiation, however, will not necessarily have taken place. On the other hand, if the geographical barrier has also resulted in the two populations occupying habitats that differ in physical or biological features, then adaptive differentiation can occur between the two populations. The speed of differentiation will depend partly on how different the habitats are, and thus on the strength of natural selection. Many empirical studies have shown that rapid differentiation can occur when selection is strong. In this situation, adaptive radiation is expected to occur. Note, however, that the adaptive radiation occurs not as a result of the allopatric distribution, but because of the environmental differences between the two habitats. Geographical separation may increase the likelihood that the environments differ and may allow differentiation to occur, particularly in mate recognition systems, that could not occur in the face of gene flow if selection is weak, but it is not a necessary condition for the differentiation of populations.

B. Founder Effect Speciation

There have been a number of models that have emphasized the role of extreme genetic drift associated with small numbers of individuals founding a new population. This could be caused by the migration of a small number of individuals (ultimately a single pregnant female or a single self-compatible hermaphrodite plant) into a novel habitat or the bottleneck caused by the dramatic reduction in population numbers caused by extreme environmental variation.

The important point of these models is that large random fluctuations in gene frequencies caused by genetic drift in small populations are possible. Surviving individuals will be close relatives, allowing considerable inbreeding. These conditions are ones in which the evolution of isolating barriers involving the fixation of genes through adaptive valleys may be possible. For instance, the fixation of chromosomal rearrangements with heterozygote disadvantage will be possible because inbreeding will lead to the rapid formation of homozygotes. Genetic drift could result in the disruption of a signal/receiver system so that it changed significantly.

These models have been particularly invoked to explain the relatively rapid speciation and adaptive radiation of various organisms that have occurred on isolated islands. For instance, the Hawaiian is-

lands have enormous radiations of many groups of organisms, e.g., *Drosophila* and crickets among insects and *Bidens* and tarweeds among plants. Very few colonizers were probably responsible for establishing the species on the islands in the first place (Hawaii is probably the most isolated archipelago in the world), and interisland migration is also likely to be low. So the conditions exist for founder effects, but it is not clear that this is the only or most important factor. Isolated islands have a very different floral and faunal composition than continental environments. The migrating population experiences a very different environment from that whence it came; this is likely to lead to strong natural selection for adaptation to the novel environment. Under these circumstances it is impossible to empirically disentangle the effects of selection and drift. Theoretical studies of the conditions under which drift and founder effects may be important have not made a convincing case for their importance.

C. Sympatric/Parapatric Speciation

Various models have been proposed in which an interruption to gene flow by geographical separation is not required. In parapatric models, the two ends of a species' distribution are subjected to selection in different directions, so that different coadapted gene complexes accumulate in the two areas. In between a hybrid zone is formed, and under certain circumstances genes giving isolation can be selected for in this region, by reinforcement. In sympatric models, a single species splits into two within a single geographical area in response to the availability of two ecological niches. In practice, since niches will almost always show some spatial distribution, as soon as any differentiation has occurred some geographical substructuring of the population will begin. This has led some to assert that this sort of speciation is simply microallopatric. This is disingenuous. The important distinction between the strict allopatric models and the sympatric/parapatric models is whether selection is able to produce significant differentiation within a single species in the presence of gene flow. If it is not, then gene flow must be eliminated entirely before speciation can be initiated (allopatric models). If it is, then the degree of geographical substructuring required depends on the relative strengths of selection and gene flow, and the nature of the adaptations to the prevailing environmental conditions.

Many sympatric speciation models have modeled the conditions for the spread of genes giving reproductive isolation when their sole phenotypic effect is isolation, as in reinforcement. The conditions generally turn out to be rather restrictive. However, in other models the reproductive isolation follows as a by-product of the adaptation to the novel niche. In this case, if selection is strong enough to give adaptation, speciation will follow automatically. Examples include changes in flowering time or breeding season; breeding system alteration and pollinator shifts in plants; female oviposition and mate location in animals.

V. EXAMPLES OF ADAPTIVE RADIATIONS

A. Cichlid Fishes of the Great Lakes of Africa

Over the past 6 million years a number of lakes have been formed in Central Africa. Lake Victoria is the youngest (0.5–0.75 million years), whereas the rift valley lakes, Lake Malawi and Lake Tanganika, are older (1–2 and 6–10 million years, respectively). All three lakes have produced an amazing radiation of cichlid fishes, with several hundred species recognized in each lake. The majority of these species are endemic to the lakes and are often endemic to very small areas of the lake. The species differ markedly in both coloration (which may indicate sexually selected characters important in ethological reproductive barriers) and morphological characters, particularly jaw and dentition characters, that enable many of the species to occupy very different ecological niches.

The processes producing this diversity and adaptive radiation are becoming increasingly understood. Early discussions emphasized the need for strict allopatry, and much ingenuity was exercised in producing models of how such allopatry could be obtained in a single lake. For instance, it was suggested if the level of the lake dropped, then this could cause fragmentation of the lake into a number of lagoons or basins, each of which would be isolated from the others. Species marooned in one lagoon could then speciate allopatrically from sister species separated in a different lagoon. That this model may be correct in some cases is shown by the case of Lake Nagugabo. This is a water body separated from Lake Victoria for about 3500–4000 years. Several species have been found in this lake that are distinct from sister species in the main body of Lake Victoria. It does not require many cycles of isolation when the lake level drops, and remixing when the lake level rises again, to produce an impressive degree of radiation.

However, this model does not work quite so convincingly in Lake Malawi or Lake Tanganika. Although the water levels in these lakes have undoubtedly varied considerably over time, it is not obvious that this variation would produce the fragmentation of the lake required for this strict allopatric model. Some model of intralacustrine speciation is more likely in these cases. In Lake Malawi, there are over 200 species of a group of cichlids called the Mbuna; these are rock-dwelling fishes that appear to be weakly dispersing. Frequently endemic species of Mbuna are restricted to a single isolated rock outcrop. The fishes show great ecological adaptation to particular depths of water and choice of food. Suitable habitat may be isolated from other similar areas by sand, mud, or deep water that does not provide a substrate for Mbuna. Rare migrations of a species from one outcrop to another would provide sufficient geographical isolation for the populations to evolve independently. The migrants will be few in number (in mouth-brooding species, perhaps a single brooding female). The small population size will produce genetic drift, while the faunal composition, nature of the outcrop, depth of water, etc. might place sufficient selective pressures on the new population that in adapting to them it diverges sufficiently from the ancestral population to become a new endemic species.

This model combines many of the features of allopatric and nonallopatric models. The sedentary nature of the Mbuna restricts gene flow between rock outcrops. However, the rapid evolution of novel species is believed to follow from adaptation to local conditions. If selection is strong enough, it is not necessary to postulate that migration between rocks is as infrequent as required by an allopatric model. Various observations make it likely that at least some migration takes place between outcrops. Artificial reefs established in areas of inhospitable habitat swiftly become colonized by Mbuna, even if by a restricted number of species. The water levels of the lake can vary quite considerably over a relatively short period of time: even a drop of 10 m could expose novel rocks that were previously too deep or render less hospitable shallow rock outcrops. The fishes will obviously have to migrate to track the change in water level. It is clear that changes in level of this order of magnitude are not uncommon—between 1915 and 1980 the lake has increased in depth by 7.2 m. All of these observations suggest that the near total elimination of gene flow postulated in allopatric models may be unlikely.

The species differ in feeding habits and other ecological specializations, but also dramatically in color. This factor makes them very attractive to aquarists. Since cichlids frequently have a complex courtship behavior, it has been suggested that sexual selection acting on male coloration has been an important factor in the rapid speciation in this group instead of adaptation to specific ecological factors. However, recent evidence has challenged this suggestion, and the reasons for the diversity in coloration remain unknown.

The rapidity with which this group has radiated is remarkable. The case of Lake Nagugabo has already been mentioned, where several species have evolved in the past 4000 years. But there is evidence that significant speciation has even occurred within the past 200 years. Between about 1500 and 1850 the level of Lake Malawi appears to have been between 120 and 150 m lower than it is now (see Fig. 1). Much of the southern end of the lake must have been dry, and Likoma Island would have been connected to the mainland. The lake must have filled up to approximately present levels during the 19th century. In doing so, various islands in the southern part of the lake became available for colonization and now house a considerable number of endemic species (Fig. 1).

The cichlid fishes of the great lakes of Africa may be a model for the evolution of many other famous radiations, particularly those associated with island groups, such as the Hawaiian *Drosophila,* Darwin's finches on the Galapagos, etc. Ecological space has been relatively unfilled because only a limited number of organisms have colonized the island or lake initially. Diversification of habitat, the availability of niches, and the lack of competition have promoted the radiation of groups that in other places have been constrained by the existence of other, perhaps better adapted, species. The colonization of allopatric microhabitats by small numbers of migrants may be contributory factors to the radiations, but should not be seen as the reason for them.

B. Host Races in Insects

There are many examples of insect groups where closely related and sympatric species are highly adapted to live on specific host plants. Adaptations include camouflage, emergence time to coincide with optimal resource value, and biochemical adaptation to cope with plant defenses. Mating is frequently on or associated with the host. The allopatric hypothesis for their evolution suggests that the species evolved on their respective hosts when the hosts had an allopatric distribution and that sympatry for both host and insect is secondary. This explanation is probably valid in many cases, but there are equally many instances where there is no evidence that the hosts were ever allopatrically distributed. The simple explanation is that the insect species speciated in response to the availability of a novel host.

One good example of the process in action has been provided by the evolution of host races of the true fruit fly, *Rhagoletis pomonella.* Originally it seems that this species only infested the fruits of the hawthorn in North America, but in 1864 it was discovered to also be attacking apples, which of course are an introduced species to North America. The apple race spread gradually across the United States from its point of origin, and because this is a pest of economic importance, its spread is well documented. In this century further host shifts to plums and cherries have been recorded.

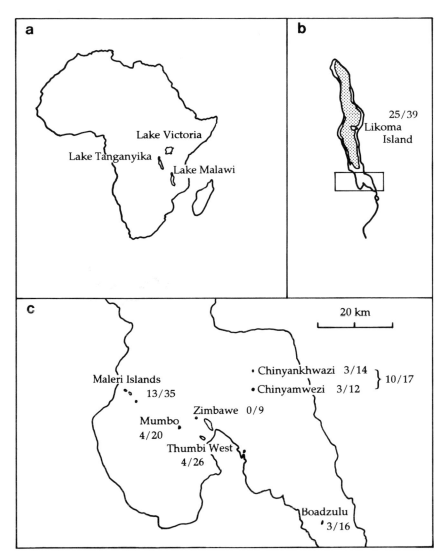

FIGURE I Adaptive radiation of cichlid fishes in the great lakes of Africa. (a) Lake Victoria has 172 species of cichlid, all endemic; Lake Tanganyika has 190 species, 99% endemic; and Lake Malawi has more than 250 species, more than 99% endemic. (b) Map of Lake Malawi, indicating the extent of the lake 150 years ago (stippled area). The location of Likoma Island is also indicated. (c) Detail of the southern area of Lake Malawi. The numbers of cichlid species endemic to each island are given, together with the total number of cichlid species found at the island (as endemic/total). [b and c are adapted from R. B. Owen, *et al.* (1990). *Proc. Roy. Soc. B* **240,** 519–553.]

R. pomonella still infests haws, and it is possible to compare the flies attacking haws with those attacking apples. It is clear that they have differentiated genetically and that many of the changes are adaptive. Females show host choice, in that they prefer to lay eggs in the fruit in which they themselves developed, and males are likewise attracted to their fruit of origin. Because mating takes place on the host fruit, this leads to assortative mating within host races. The development time of the maggots differs also, thus generating a differential emergence time for adults, which causes partial temporal reproductive isolation.

The criticism of sympatric speciation has always been that gene flow between the two emerging races will swamp any divergence, particularly in genes that cause reproductive isolation. This is probably true for genes whose only effects are, say, assortative mating. However, in *Rhagoletis,* the assortative mating occurs as a by-product of the flies showing a habitat preference. Now the habitat preference is going to be adaptive in itself, since it will help to protect the genes that adapt the flies to growing in the two different hosts. It therefore can spread for this reason, and incidentally give the assortative mating that will promote further divergence and speciation. Modeling has shown this process to be plausible.

C. Edaphic Endemics in Plants

The distribution of many plant species is profoundly affected by edaphic factors, that is the mineral, structural, and water-retaining characteristics of the soil. Some plants are excluded from soils with a particular characteristic, while others are only found on such soils. Plants that are restricted to soils of a particular type are called edaphic endemics.

One of the most dramatic examples of soils harboring edaphic endemics is the case of serpentine soils. Serpentine (ultramaphic) rocks are ultrabasic rocks that are rich in various ferromagnesian silicate-rich minerals. They weather to produce soils that are basic; contain high levels of various heavy metals, particularly nickel, chromium and cobalt; and have a high magnesium to calcium ratio. Calcium is a very important mineral to plants and is normally present at higher concentrations than magnesium. Serpentine soils also tend to be free draining and thus prone to drought.

The unusual geology of these soils has resulted in many cases of extreme endemism associated with them. For example, 10% of the endemic flora of California is found on serpentine soils, but these only represent less than 1% of the land area of the state. It is possible to recognize two sorts of plants growing on such soils. The first are endemics, species that within an area are only found on these soils (though in some cases they may be found elsewhere on normal soils). The second are normal species, that are found both on and off the serpentine. In the latter case, it is likely that in most cases

they are present on the serpentine as local adapted ecotypes; populations from normal soils would find it difficult to survive on the serpentine. Such species have been called *bodenwag* species.

Two types of endemic have been recognized. The first are paleoendemics. These are species that have been present on the serpentine for very long periods of time, and their sister species on normal soil have disappeared. It is impossible in these cases to determine whether, when they first arose, they were endemics or ecotypes of a *bodenwag* species. Their survival presumably reflects the fact that the factors causing their sister species to go globally or locally extinct do not operate so strongly on the serpentine. This could occur if, for instance, the sister species had been outcompeted by a species that was unable to evolve the ability to grow on the serpentine. The second type of endemic are neoendemics: species that have evolved comparatively recently, and the sister or progenitor species are still present, either as *bodenwag* species or as local nonserpentine species.

The interesting question in the evolution of neoendemics is, what is the difference between the endemic species and an ecotype of a *bodenwag* species? Presumably an endemic always has to go through the stage of being an ecotype before it evolves into the endemic. The allopatric hypothesis would suggest that an ecotype evolves, and then the progenitor nonadapted populations from which the ecotype evolved become locally extinct. The ecotype is now allopatrically isolated from the rest of the species and so it is able to evolve distinctive features that make it specifically different from its progenitor when conditions alter to allow the progenitor to recolonize the area. This hypothesis is essentially the same as the standard explanation for paleoendemics.

An alternative hypothesis emerges from the study of a model of the evolution of serpentine endemics: heavy-metal tolerance. Many areas of the world have become contaminated by high concentrations of heavy metals, particularly copper, zinc, lead, and arsenic. These metals are very phytotoxic, and the soils appear to pose similar problems for plants as do serpentine soils (Fig. 2). However, the soils are never completely uncolonized;

FIGURE 2 (a) The Idria serpentine barrens of California. (b) The tailings at Devon Great Consols Mine, near Tavistock, Devon, UK. The mine produced large quantities of copper and arsenic between about 1860 and 1902.

some plant species can evolve tolerance to the heavy metal contamination. However, not all plant species are able to do so; only a limited subset of the local flora will typically be found on an abandoned mine site. The absence of many species is caused by the fact that they do not have the genetic variability to evolve metal tolerance since they do not possess the necessary genes. The mine environment poses a different set of problems to species than the normal soils: not only is there less interspecific competition, but also the soil differs in many edaphic factors, such as water or essential mineral content. The mine populations therefore evolve adaptations to these factors also. Metal-tolerant ecotypes can vary in many adaptations apart from just metal tolerance. In some circumstances these adaptations may also give some reproductive isolation from the progenitor population. For instance, mines (and serpentine soils) are normally drier than normal soils. For many plants, an adaptation to drought is to flower early. Early flowering enables the ecotype to avoid the drought, but it also partially or totally prevents crossing between the ecotype and normal populations. The mine environment may also be lacking in pollinators or have a different guild of pollinators. In such a case, a change in the breeding system (e.g., by evolving self-fertilization) or a change in flower color or shape, so attracting a different pollinator, may be adaptive. Either adaptation will provide reproductive isolation from the normal population and begin the process of speciation.

All stages in this model have been empirically demonstrated in populations that have evolved tolerance to heavy metals in the recent past, and it is probable that many serpentine species have also evolved this way. Obviously, the absence of many species that would have been present on normal soils, but are excluded because they lack the genetic variability to evolve tolerance in the first place, means that many ecological niches are vacant on serpentine soils. In the longer term it is expected that species will diverge still further from their progenitors to occupy these niches, enabling the endemic species to coexist with an ecotype of the progenitor *bodenwag* species that has not changed the breeding system or flowering time.

VI. CONCLUSIONS

Speciation is a process that depends on a number of factors, particularly the strength of selection, the degree of geographical separation and extrinsic barriers to gene flow, and chance. Chance includes individual rare events, such as colonization, as well as more ongoing processes such as genetic drift. It would be wrong to assert that all factors are equally important in all cases; there is almost certainly not one mechanism of speciation. Continued study will be required to determine the relative importance of the classical model of slow allopatric speciation, or the other models producing rapid adaptive speci-

ation. It is not necessarily easy, however, to disentangle the various factors and determine what are the critical factors in any one example.

For instance, in 1916 two rock wallabies from temperate Australia were introduced into tropical Hawaii. They escaped from captivity and formed a small feral colony. Eighty years later they are very different from any known Australian population of rock wallabies. They have much shorter fur, which is paler than Australian species. The animals are smaller and the skull shape is different. Many of the changes that have occurred can be interpreted as adaptations to living in a tropical instead of a temperate environment. Various questions arise when one tries to interpret this observation. Have the wallabies formed a new species? What are the roles of geographical isolation (Hawaii is very isolated from Australia), adaptation (tropical habitat vs temperate habitat), and small population size in the differentiation of this population? Is the speed of differentiation observed here unusual or unexpected?

Whether these wallabies have speciated or not on Hawaii is not possible to determine without testing for reproductive isolation from their progenitor population, and this is unknown (and may be extinct). It would seem reasonable to suggest, however, that the population is speciating. The role of allopatry in this case is that the transport to a distant habitat has moved from one climate to another, which has imposed adaptive demands on the population; it is not clear that a small amount of gene flow, had it been possible, would have prevented the evolution of the adaptive characters seen here. The small size of the founding population would normally be thought to have inhibited adaptation, because of the restrictions on genetic diversity that must have occurred, instead of promoting it. The many other cases of rapid divergence and speciation seen in cichlid fishes, *Rhagoletis* fruit flies, and edaphic endemics show that this example may not be unusual.

Anthropogenic changes to the biosphere mean that the habitats of many species must be being disrupted by as great a factor as any of the examples considered here. Rapid adaptation in the face of this change in habitat must be expected by some, but not all, species. This may mean that some regeneration of biodiversity will take place to compensate for the loss of biodiversity seen as a result of human destruction of natural ecosystems, and that the speed of regeneration may be measured in hundreds rather than millions of years in some individual cases. Overall, however, the dominant mode of speciation must be slow divergence in allopatry, which means that the restoration of the full diversity of the biosphere will not occur until long after *Homo* itself has gone extinct.

Glossary

Adaptation A character evolved by natural selection to improve the ability of an organism to live in its current habitat.

Allopatric Populations are allopatric if they are separated by sufficient distance or inhospitable habitat that no migration takes place between them under normal circumstances.

Epistasis Interaction between two different genes, so that the effect of an allelic change at one locus depends on the allelic state at the other.

Founder effect Special case of genetic drift (qv) when a small number of founders gives rise to a population.

Genetic drift Change in gene frequency by random sampling variation from generation to generation.

Natural selection Change in gene frequency caused by one allele leading to greater average survival or reproductive productivity in carriers of that allele than of an alternative.

Parapatric Populations are parapatric when they do not overlap, but abut along a zone of contact.

Sibling species Two species that are reproductively and possibly ecologically distinct, but which are morphologically very similar or identical.

Sympatric Populations are sympatric if they coexist within one geographical locality.

Bibliography

Baker, A. J. M., Proctor, J., and Reeves, R. D., eds. (1992). "The Vegetation of Ultramaphic (Serpentine) Soils." Intercept, Andover, Hants.

Echelle, A. A., and Kornfield, I. (1984). "The Evolution of Fish Species Flocks." Orono: Univ. of Maine Press.

Futuyma, D. J. (1986). "Evolutionary Biology." Sunderland, MA: Sinauer.

Otte, D., and Endler, J. A., eds. (1989). "Speciation and Its Consequences." Sunderland, MA: Sinauer.

Ridley, M. (1993). "Evolution." Oxford: Blackwell.

Species Diversity

Rosie Trevelyan and Mark Pagel

University of Oxford

Ecologists characterize communities of species according to indices of species diversity. Historically, indices of species diversity have combined information on the number of different species that are present in a region or habitat with information on the commoness or rarity of those species; some species tend to be common, they are represented by many individuals, whereas other species tend to be rare. More recently, however, a much simpler measure of diversity known as species richness has gained favor. Species richness is just a count of the number of different species in a community. This measure is easily interpretable and avoids the problem inherent to other indices of diversity of how to assign weights to the varying degrees of commoness and rarity among species. Species diversity and species richness are used synonymously in this article to describe the worldwide trends in the distribution of species, their ecological causes, and the implications for conservation.

I. NUMBER OF SPECIES ON EARTH

Discussion of the diversity of species living on Earth and of how that diversity is distributed around the globe must be measured against our knowledge of the total numbers of species living today. Biologists have identified and named approximately 1.7 million different extant species. The list of named species is dominated by simple organisms such as algae and bacteria, viruses, slime molds, fungi, protozoa such as the malaria-causing plasmodium, and nematodes. Insects and higher plants are also well represented, accounting for perhaps half of the known species. Mammal and bird species receive attention wholly out of proportion to their numbers. Nearly all members of these two groups are probably known, yet they number only 4500 and 9000 of the known species, respectively. [*See* BIODIVERSITY.]

This is the diversity about which ecologists have some knowledge; much more may still await discovery. Estimates of the total number of different species that may inhabit the earth range over an order of magnitude. At the low end, estimates place the total number of species at approximately 12.5 million, whereas other estimates are as high as 100 million. Moreover, a sobering thought for those wishing to enumerate all species is that almost every species is a host to some form of parasite. The disparity in estimates of the total number of species on earth partly reflects differences among taxonomists in the rules they use to classify species, as well as different definitions of what constitutes a species. Disagreements also occur over how to extrapolate from known numbers of species to realistic estimates of actual numbers.

Biologists employ a range of methods to estimate total numbers of species from smaller numbers of sampled species. One method is based on the effort put into obtaining a sample of species. The number of species discovered in a sample increases as the amount of effort put into obtaining the sample increases. Typically, the number of new species obtained rises sharply at first, later slowing despite continuing efforts. The point at which an asymptote would be reached, given sufficient sampling, provides an estimate of the total number of species that exist in that area. A similar technique employs the fact that the number of different species increases in a known way as the body size of the species decreases; there are, for example, many more species of beetle (300,000 are known, prompting J. B. S. Haldane to remark that God showed "an inordinate fondness for beetles" when creating the Earth) than there are species of mammal. So numerous are the smaller species, it has been estimated that to a first rough approximation all species are insects! This technique, then, makes guesses about the numbers of very small species not yet identified by extrapolating the known relationship between body size and numbers of species. A recent method makes use of ideas gained from the fractal geometry of living systems. The number of habitat niches in an "environment" such as the edge of a leaf—roughly the number of different ways of making a living along that edge—may increase rapidly as the environment is studied at finer and finer resolutions.

By current estimates, taxonomists and other researchers have identified perhaps as few as 2% of species living on earth. This poses a challenge to biologists because understanding the factors that influence both the number of species on earth and their regional patterns of diversity is central to principles of ecological study as well as to environmental management strategies.

II. WORLDWIDE PATTERNS OF SPECIES DIVERSITY

A. Number of Species Found in a Given Area

The number of different species found in a given area varies greatly across the globe and even between different kinds of habitat. Current estimates suggest that over 90% of all species may live in tropical moist forests, even though such forests comprise only 7% of the world's surface area. The best known pattern in the global distribution of species is that the number of species per unit area increases from the poles to the equator. This latitudinal gradient in species diversity was first identified by Wallace in 1878. Since then it has become accepted as an almost universal law. Mammals, birds, frogs, and large numbers of other vertebrates, plants (trees and orchids), arthropods, tunicates, molluscs, planktonic foraminiferans, and even deep-sea dwelling benthic organisms of the North and South Atlantic all exhibit latitudinal gradients in species diversity (Table I).

Highlighting latitudinal gradients is another way of saying that most species exist in or near the tropics. The number of species of ant declines from approximately 222 to 3 along a line drawn from Brazil to Alaska. Benthic organisms of the North Atlantic make their living in the ooze some hundreds to thousands of meters beneath the surface. The number of different species of these organisms is estimated to decline from around 25 to 40 in samples near the equator to near zero in the far North. Approximately 150 to 225 different North and Central American mammal species reside in the area between 10 and 20° N latitude (Southern Mexico and Central America), as many as half of which are bats. However, this number declines to somewhere between 50 and 75 at latitudes near the border between Canada and the United States, and drops to fewer than 5 species at around 70° N latitude. This far north, the class Mammalia is represented in North America principally by polar bears, moose, caribou, and wolves. Mammals of the palearctic and landbirds of North and South America

TABLE I

List of Major Taxonomic Groups Showing Latitudinal Gradients in Species Richness or
Geographic Range Size

Taxonomic group	Region of study	Latitudinal gradient in	
		Species diversity	Geographic range size
Landbirds	New World	+	+
	Australia	−	
Mammals	North America	+	+
	Palearctic	+	+
	Australia	−	−
Lizards	Nearctic	+	
	Australia	−	
Anurans	Global	+	
Lizards and amphibians	New World	+	+
Snakes	Global	+	
Fish	North America	+	+
Ichneumonid parasitoids	Global	−	
Papilionid butterflies	Global	+	
Dragonflies	Global	+	
Freshwater crustacea	New World	+	+
Molluscs	Nearctic	+	+
Permian brachiopods	Nearctic	+	
Tunicates	Global	+	
Helminth parasites	Global	+	
Corals	Global	+	
Planktonic foraminifera	Nearctic	+	
Trees	Palearctic	+	+
Orchids	New World	+	+

Note. +, species diversity decreases at higher latitudes, geographic range size increases at higher latitudes; −, no significant relationship; blank, no study reported.

exhibit broadly similar trends. Some exceptions to the latitudinal rule have been discovered. For example, Australian birds, Australian mammals, and Australian lizards do not demonstrate clear latitudinal gradients in species diversity. This may be because vast tracts of Australia, spanning wide bands of latitude, are desert.

Paralleling the latitudinal gradients in species diversity are elevational gradients. Fewer species of birds, mammals, reptiles, amphibians, insects, and trees exist at high altitudes than at lower altitudes. Longitudinal trends in species diversity, first investigated by George Gaylord Simpson in the 1960s, have recently received renewed attention. The number of mammal species in a given area of the western parts of North America is about double that found in the eastern parts. This phenomenon

is observed independently of the latitude of the species, and remains even after taking into account increases in diversity at the west coast of North America. Although the longitudinal effects are not as great as those for latitude, this result has important implications for discussions of the causes of species diversity. A range of environmental features change with latitude, so isolating which of these factors is responsible for variation in diversity is difficult. Ecological factors within any one latitudinal zone may vary much less, however. This means that studies of longitudinal gradients in diversity may be more successful in identifying the forces responsible for that diversity.

Darwin noted that islands are relatively bereft of species compared with equivalent areas of a mainland. Robert MacArthur and E. O. Wilson advanced their theory of island biogeography to explain this observation. According to this theory, the numer of species on an island is partly determined by the balance between immigration of new species and extinction of resident species. Islands are less likely to be successfully colonized than are areas of the mainland, particularly with decreasing size and distance from the mainland, which provides a continuous pool of new individuals. These processes may explain why the numbers of species of land birds are lower on islands and why fewer species are found on the smaller islands. Although Darwin's observations were of oceanic islands, the "island effect" also seems to apply to any isolated habitat such as a mountain top, tree, or individual animal (as a host to parasites).

B. Size of a Species' Geographic Range

A species' geographic range is the area over which the species is found. The size of species' geographic ranges turns out to be a good inverse measure of species diversity. Individual species tend to have small geographic ranges in areas of high species diversity, whereas the reverse is true in areas of low species diversity (Table I). The geographical range sizes of trees, marine molluscs, freshwater crustaceans, fish, reptiles, amphibians, birds, and mammals all increase the further north from the equator these species are located. So pronounced is this effect that, along with the latitudinal gradient in diversity, approximately 50% of all North American mammal species are found in Central American latitudes, and usually in areas much smaller than the size of Nicaragua or Honduras. This has strong implications for conservation: comparatively small environmental changes, such as clearing areas of tropical forest, imperil the existence of many species.

The phenomenon of a latitudinal gradient in geographic range size has come to be known as Rapoport's rule, after the Argentinian ecologist who first popularized the idea. Rapoport's rule also applies to elevational gradients in species diversity. Trees, insects, reptiles, birds, and mammals all increase their geographic extent with increasing elevation. Similarly, the longitudinal gradients in diversity

for North American mammals are accompanied by gradients in range sizes: geographic range sizes are smaller in the West.

III. CAUSES OF VARIATION IN SPECIES DIVERSITY

The repeated patterns in the numbers and geographic ranges of species suggest that general causal processes underly species diversity. However, the diversity of explanations for these patterns rivals that of the diversity of species in lower latitudes; at least 21 causal explanations for species diversity have been proposed (see List 1). As yet, ecologists have not reached a consensus about which explanations are best, and even the better ideas often provide only a partial explanation of the patterns.

List 1
Explanations for Trends in Species Diversity

Climatic/environmental explanations
  Seasonality
  Environmental stability
  Environmental predictability
  Harshness
  Aridity
  Area
Habitat-based explanations
  Habitat diversity
  Patchiness
Energy
  Productivity
  Temperature
Species' geographic range sizes
  Species range sizes
  Spillover effects
Evolutionary time
  Length of evolutionary time
  Differential origination and/or extinction rates
Biotic factors
  Competition
  Predation
  Niche width
  Host diversity
  Epiphyte load
  Mutualism
  Biotic spatial heterogeneity

Note. Many of these explanations are not mutually exclusive and are often used in combination within the same hypothesis.

A. Climatic Variability

An organism must be able to survive the range of environmental conditions it normally experiences. How can this self-evidently true statement explain patterns of species diversity? Species living at higher latitudes experience a wider range of climatic conditions, including wider seasonal as well as diurnal fluctuations, than species living at lower latitudes. Consequently, proponents of the climatic variability idea argue that high-latitude organisms, unlike their tropical counterparts, cannot specialize on a narrow set of environmental conditions. Natural selection will therefore favor broad climatic tolerances among organisms that live at higher latitudes. Because of this, these species will tend to be generalists, they will be less restricted in their habitat use, and as a result will have broad latitudinal ranges. This in turn will lead to low species diversity. Conversely, low-latitude organisms, faced with much narrower swings of climate, will tend to specialize on a particular habitat and will therefore have smaller geographic ranges. Specialization thus means that more species can coexist in a given area at low latitudes than at higher latitudes.

More northerly dwelling North American and Palearctic mammals do appear to be generalists compared to species that live in the southern latitudes. Species at higher latitudes occupy a greater number of habitats in relation to the number of habitats available to them compared to species at lower latitudes, which tend to specialize. More direct evidence regarding the climate variability explanation comes from research in which climate variables are measured directly. Lizard species diversity in the United States correlates strongly with "sunfall," although climatic variability does not explain bird species diversity as well. The geographic range sizes of North American mammals are positively correlated with annual temperature range.

Despite its successes the climatic variation idea is not without its critics. Coral reefs in tropical shallow seas typically have great diversity but are proving to be not as environmentally stable as previously thought. However, such arguments have not been tested by comparing these sites with an equivalent area of temperate shallow seas. If it is found that conditions fluctuate more widely in temperate seas, which contain far fewer species of coral, then the climatic variability hypotheses may still apply. Another problem with the climatic variation hypothesis is its unstated assumption that large numbers of different generalist species cannot occupy the same region. Proponents must demonstrate that some sort of process of competitive niche exclusion operates among species. [*See* CORAL REEF ECOSYSTEMS.]

B. Diversity of Habitats

The diversity of terrestrial habitats in North America is greater in the south and west than in the north and east. This northeast to southwest habitat gradient coincides remarkably with the gradients in the numbers of species. Despite there being fewer different habitats at more northern latitudes, the species that reside there occupy a greater proportion of those that are available. The opposite is true in southern latitudes. Species there tend to occupy fewer of the relatively larger number of habitats that are available. These trends support the view that species in higher latitudes are generalists, whereas those in the south tend to specialize.

Daniel Janzen, an ecologist, pointed out that a consequence of greater specialization by southern species is that "mountain passes may be higher in the tropics." If tropical species are less accustomed to variation in climate, then topographical relief poses a greater barrier to these species' movements. The result is an even greater isolation of species and increased specialization. Parts of the southwestern regions of the United States are purported to contain more subspecies of mammals than any other comparably sized continental area in the world.

The steep latitudinal gradients in species richness among deep-sea benthic organisms may seem incompatible with the climatic variation and habitat diversity hypotheses. However, the received view of the deep sea as a homogenous environment relatively unaffected by the environmental gradients measured in surface waters may prove wrong. Environmental gradients which parallel surface gradients exist, even at depths of up to several thousand meters.

Despite the successes of habitat diversity in predicting trends in species diversity, studies of the idea must answer a crucial objection. Ecologists often base their definitions of habitats at least partly on the organisms that inhabit them. Therefore, arguments invoking habitat variation to explain variation in species diversity may be circular. For example, studies of animal diversity often define habitats by vegetation types, yet the selective forces that shape the diversity of the plants comprising these habitats also shape the diversity of the animals that occupy those habitats. As a result, the level of habitat diversity will automatically be paralleled by measures of species diversity because both are a consequence of the same forces.

Habitats can be defined by characteristics such as geology (including drainage systems) and other abiotic measurements that are independent of the organisms under study. Correlations between these measures of habitat diversity and species diversity have been recorded for freshwater organisms. Lacking this, attempts must be made to show that the behavior of organisms differs independently of the definition of the habitat. As mentioned earlier, mammals living in the northern regions of North America make use of a greater proportion of the habitats available to them than do more southern species.

C. Microhabitats and the Edge Effect

Species richness at low latitudes and elevations could be enriched by a spillover process from adjacent populations. If the microhabitats of low-latitude species are ecologically narrow, then some part of the population at the edge of the habitat is likely to spill over into an adjoining and presumably less favorable habitat. Individuals that disperse into less favorable areas will probably not be locally self-sustaining because they will be outcompeted by the local specialists. However, the proximity to each other of many small and specialized habitats will produce a continuous stream of immigrants. These "edge effects" will artificially inflate the species richness of a given area. The tendency for spe-

cies to have larger ranges in higher latitudes will reduce the edge effects and thereby reduce species richness.

Few empirical tests of edge effects have been conducted. North American "twitchers" (enthusiastic bird watchers) report greater numbers of accidental and visitor bird species further south. Higher proportions of decapod crustaceans in the tropics are vagrants, persisting through constant recolonization. However, these are indirect approaches to the hypothesis, and direct comparisons between communities of high and low species richness, taking into account environmental differences, would be fruitful. Ultimately, however, edge effects can never hope to explain more than an exaggeration of existing latitudinal gradients.

D. Gradients in Energy

Does energy supply in the environment limit the numbers of different species (as distinct from the number of individuals) that inhabit a given area? The most popular measure of the energy that is available to a habitat is the primary productivity of that habitat. Primary productivity is the amount of new biomass produced over some unit of time. Primary productivity is difficult to measure directly and so researchers frequently employ indirect measures. The diversity of North American tree species, for example, is strongly linked to productivity as measured by levels of evapotranspiration. North American vertebrate species richness, including birds, mammals, and reptiles, is also correlated with measures of productivity. Variations in energy levels with altitude or depth may also explain the gradients associated with those two dimensions.

Associations between measures of productivity on the one hand and geographic gradients in diversity on the other hand leave out an important intermediate assumption. The energy supply hypothesis fails to give a clear explanation for why merely pouring more energy into a system will produce greater diversity of species rather than promote a single or a few species at the expense of others. Furthermore, not only the total amount of energy but also the variation in daily or seasonal energy supply varies with latitude. Until such effects as these can be untangled, the role of the total amount of energy in producing species diversity remains uncertain.

E. Evolutionary Time

Some ecologists point out that increased habitat diversity in the tropics cannot alone explain the greater richness of tropical species. Even comparable habitat types seem to support more species in tropical than in extratropical latitudes. An intriguing explanation for this is that the number of species in a particular sample may be directly related to the length of uninterrupted (i.e., evolutionary) time through which that biota has evolved. Tropical habitats have been free of the disturbances wrought by recent ice ages longer than temperate or more northerly habitats. The diversity of freshwater lake species is greater in older lakes than in newer lakes. [*See* SPECIATION.]

Stacked against the evolutionary time point of view is the idea that the length of time per se is not as important for determining species diversity as is the dynamical interaction between origination and extinction of species. If these processes happen over sufficiently short periods of time, differences among habitats in the length of time since they were last disturbed become unimportant. The tropics probably do provide a more favorable milieu for the rapid origination and evolutionary diversification of species. Post-Paleozoic fossil marine invertebrates show higher origination rates in tropical regions than in temperate regions. Such observations may be partly probabilistic—the greater species richness in tropical regions may itself promote greater origination rates.

A simpler phenomenological objection to the evolutionary time idea is that it cannot explain longitudinal gradients in species diversity or the lack of latitudinal gradients in some regions. Species diversity gradients do exist across areas of similar evolutionary age.

F. Biotic Factors

Biologists have served up a host of additional ideas to explain relationships between biotic components of an ecological community and its species richness (see List 1). Greater tropical species richness is variously attributed to increased competition or predation, or conversely to mutualism. Reef-building corals create habitats with high degrees of biotic and spatial heterogeneity. This could explain the high levels of species diversity in these communities. However, many of these biotic factors are circular as explanations for diversity: their predictions depend strongly on having greater number of species in such areas in the first place. For example, the number and breadth of niches in an environment may determine the number of species that environment can support. However, the definition of a niche is traditionally based on the life style of the organism, and therefore a higher number of niches in an area will result merely from the greater number of organisms measured. Similarly, the number of epiphyte species (species that live on plants, including trees) in lowland tropical forests is thought to increase species diversity by increasing the numbers of diseased and weakened trees. Such trees are more likely to fall and thereby create disturbances in the habitat. This, in turn, creates opportunities for new species. Again, however, this effect fails to provide an explanation for the high diversity of epiphytes, and in any case is not applicable to large-scale trends in species diversity in other habitats.

IV. TAXONOMIC AND GEOGRAPHIC BIASES IN OUR KNOWLEDGE

Most of our ideas on species diversity and its causes are based on knowledge of a very small minority of taxonomic groups. Some of the best known patterns in species diversity come from studies of birds and mammals. These groups, along with the rest of the vertebrates, comprise only about 4% of all recorded animal species. Bearing in mind that recorded species may represent as few as 2% of the total number of species, our knowledge is based on a small sample indeed.

Conservative estimates suggest that fungi may be the most species-rich group after the insects. Would studies of species diversity gradients in fungi or nematodes support or challenge our established views on large-scale diversity patterns and their causal processes? Evidence from well-studied animal and plant taxa in Britain shows that areas rich in species for one group, such as birds, may be depauperate for a different group, such as the butterflies or liverworts.

Large biases in the geographical location and the kinds of environments in which species diversity has been studied also exist. Most of the well-known gradients in species richness come from studies conducted in northern temperate latitudes.

V. IMPLICATIONS FOR CONSERVATION

The current rate of extinction of the world's species lends an urgency to the task of understanding species diversity. Conservation programs always represent a compromise either between the interests of conflicting human groups or between the conflicting demands of a single group. Knowledge of the global patterns in the distribution and diversity of species can help to inform the debate that arises out of such conflict. If general trends in diversity can be identified, conservationists will not have to study every species on earth before making rational decisions about their management.

Glossary

Benthic organisms Bottom-dwelling organisms of the seas or lakes.

Biomass Total amount of living material in a sample or community of a specified size.

Competitive niche exclusion Idea that a species may exclude other species from a niche by outcompeting them.

Evapotranspiration Total amount of water liberated into the atmosphere by the living materials in a community.

Niche A way of making a living in an environment. A grazing animal fills a grass-eating niche.

Primary productivity Net amount of biomass produced in a community of specified size over some specified period of time.

Species abundance Number of individuals in each of the species found in a sample. Some species may be represented by many individuals while others may not.

Species diversity Index that combines information on the number of species in a sample with the relative abundances of each species; used synonymously with species richness (see Species Abundance and Species Richness).

Species richness Number of different species found in a sample.

Twitcher Individual who is obsessed with seeing every bird species. More common in Britain than in the United States.

Bibliography

Begon, M., Harper, J. L., and Townsend, C. R. (1986). "Ecology, Individuals Populations and Communities." Oxford: Blackwell.

Brown, J. H., and Maurer, B. A. (1989). Macroecology: The division of food and space among species on continents. *Science* **243**, 1145–1150.

Jablonski, D. (1993). The tropics as a source of evolutionary novelty through geological time. *Nature* **364**, 142–144.

Janzen, D. H. (1967). Why mountain passes are higher in the tropics. *Am. Nat.* **113**, 81–101.

May, R. M. (1990). How many species? *Phil. Trans. Roy. Soc. B* **330**, 293–304.

Pagel, M., May, R. M., and Collie, A. R. (1991). Ecological aspects of the geographical distribution and diversity of mammalian species. *Am. Nat.* **137**, 791–815.

Pianka, E. R. (1989). Latitudinal gradients in species diversity. *Trends Ecol. Evol.* **4**, 223.

Prendergast, J. R., Quinn, R. M., Lawton, J. H., Eversham, B. C., and Gibbons, D. W. (1993). Rare species, the coincidence of diversity hotspots and conservation strategies. *Nature* **365**, 335–337.

Rapoport, E. H. (1982). "Areography: Geographical Strategies of Species." Oxford: Pergamon Press.

Rex, M. A., Stuart, C. T., Hessler, R. R. Allen, J. A., Sanders, H. L., and Wilson, G. D. F. (1993). Global-scale latitudinal patterns of species diversity in the deep-sea benthos. *Nature* **365**, 636–639.

Stevens, G. C. (1989). The latitudinal gradient in geographical range: How so many species coexist in the tropics. *Am. Nat.* **133**, 240–256.

Stevens, G. C. (1992). The elevational gradient in altitudinal range: An extension of Rapoport's rule to altitude. *Am. Nat.* **140**, 893–911.

World Conservation Monitoring Centre (1992) "Global Biodiversity; Status of the Earth's Living Resources." London: Chapman and Hall.

Systematics

Sandra J. Carlson

University of California

Systematics is the scientific study of the diversity and genealogy of life. Studies of diversity distinguish differences among organisms in size, shape, genetic structure, behavior, physiology, or any of a wide range of features that characterize living things. Organisms are given unique names to acknowledge consistent, discernible differences from other organisms. Rather than emphasizing differences, studies of genealogy seek to discover evolutionary relationships among organisms. Groups of individuals, and groups of groups of individuals, are recognized according to their relationship through the process of common descent. Our perception of the variety and complexity of life is structured fundamentally by how we recognize distinctions between and connectedness among organisms. Understanding the theory and practice of systematics is therefore of primary importance in the study of biodiversity and environmental and evolutionary biology.

I. WHAT IS SYSTEMATICS?

Environmental biology can be thought of as the study and interpretation of a continuous and ever-changing conversation between organisms and environments, occurring today and over many millions of years of geologic time. We tend to think of "the environment" as a more or less static backdrop for the activities and interactions among organisms over time. Not so. Organisms may actively create and change the environment in which they live; environmental change in turn permits organismic evolution to occur. Consider some of the "major events" in the history of life: the evolution of life, the origin of multicellular organisms and skeletonization, the origin of land plants, the diversification of marine plankton, and the evolution of human beings and culture. In each instance, the interaction between organism and environment resulted in a fundamental change in both organism *and* environment: abrupt increases in atmospheric oxygen levels, the "locking up" of certain ions in skeletons that are released upon the death of those skeletonized organisms, and the direct manipulation of the environment by humans. [*See* EVOLUTION AND EXTINCTION.]

How does systematics fit into this conversation? Evolution, the biological process of descent with modification, generates the remarkable diversity of life and structures genealogical relationships among living things. In a very real sense, evolution can be thought of as the fundamental *biological* process that regulates the Earth. Because all organisms are related to one another through the process of evolutionary descent and because the evolutionary process itself is driven by the interactions among organisms and environment, the study of systematics is central to the study of environmental biology in its broadest sense.

To understand the language in which the conversation between organisms and environments takes place, it is necessary to develop an informed appreciation for processes and patterns of biological *and* geological evolution. Both require a strong sense of "deep time" and the relationship among past, present, and future. This article focuses on the biological side of the conversation, but it must be kept in mind that the Earth is a complex and dynamic system with a long history. The present Earth ecosystem is, to a great extent, a product of its past history. It is essential to study Earth history and consider fossil evidence in order to adopt a truly historical perspective on systematics. Over 99% of the Earth's biota is extinct. The extant (living) flora and fauna provides such a small and nonrepresentative sample of "all life" that restricting systematic investigations to the living biota produces a very incomplete picture.

Systematics is a discipline that allows us to make distinctions between and connections among organisms, and groups of organisms, according to evolutionary principles. Because evolutionary processes structure the diversity of life, the primary goal of systematics is to infer phylogenetic (genealogical) relationships among species and reconstruct explicit hypotheses of phylogeny (Fig. 1A). A phylogeny is a hypothesis about a pattern of common ancestry that reflects the evolutionary process of common descent. Groups of species that represent complete systems of common ancestry are called *clades;* they include an ancestral species and *all* of its descendants. By the nature of their formation, clades nest within other clades in a hierarchical pattern that can be represented in a branching diagram. Phylogenetic inference is a process of discovery by which clades are identified and their internested structure elucidated. [*See* EVOLUTIONARY TAXONOMY VERSUS CLADISM.]

Taxonomy is a branch of systematics that concerns the theory and practice of naming and classifying organisms. Classifications place organisms

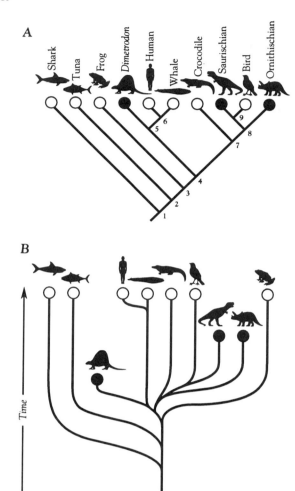

FIGURE 1 (A) Cladogram illustrating phylogenetic relationships among 10 familiar kinds of animals, internested in 9 (numbered) clades. Open circles denote living taxa; filled circles denote extinct taxa. The common names and Linnean binomials of the terminal taxa are shark *(Carcharodon carcharias);* tuna *(Thunnus albacares);* frog *(Rana pipiens);* fin-backed pelycosaur *(Dimetrodon grandis);* human *(Homo sapiens);* whale *(Balaenoptera musculus);* crocodile *(Crocodylus acutus);* saurischian dinosaur *(Tyrannosaurus rex);* bird *(Melospiza melodea);* and ornithischian dinosaur *(Triceratops horridus).* One or two of the shared derived characters (synapomorphies) that distinguish each numbered node on the cladogram: (1) vertebral column; (2) bony internal skeleton; (3) four limbs, five fingers and toes; (4) amniotic egg (with protective membranes); (5) lower temporal fenestra (opening) in skull; (6) mammary glands; (7) antorbital fenestra in skull; (8) reduced fourth and fifth digits, regionalized vertebrae; (9) "lizard-hipped" pelvis, long neck and hands. (B) Phylogenetic tree congruent with cladogram in A.

and groups of organisms into some kind of order. They can be structured to serve any of several different purposes: as systems of organization, information retrieval systems, indices of observed morphological diversity, or as hierarchical patterns of common ancestry. Classifications may organize entities according to arbitrary criteria, as names in a telephone book are arranged alphabetically, or by their origin, as books in a library are arranged by author rather than subject or title. Most evolutionary biologists today recognize the advantages, both practical and philosophical, of classifications that reflect the phylogenetic hierarchy of life instead of some more arbitrary grouping criteria. For this reason, reconstructing phylogenetic patterns and using those patterns to construct classifications are among the most significant goals of evolutionary biologists and paleontologists interested in biodiversity and environmental biology.

II. ROLE OF SYSTEMATICS IN BIOLOGY

A. Naming Organisms

Human curiosity and survival require that we communicate with one another about the vast and remarkable diversity of life that surrounds us. For this reason, we give names to organisms that differ in ways that we decide merit recognition. Organisms vary in countless dimensions, including, among many others, spatial (biogeographic), temporal (over ontogenetic and evolutionary scales), behavioral, and morphological (in size and shape). Organisms that provide food, medicine, or shelter, or those posing a threat to health or safety are usually among the first to be named by native people in a particular region. Different local "environments," however defined, require different vocabularies of names. People living far from the sea have little need to learn names for marine organisms but may find it necessary to recognize and give names to very subtle differences among the plants and animals of the mountains, deserts, or plains. Similarly, Eskimos are said to have over 18 words for snow, while English-speaking people have only one word and pygmies have none.

Many different folk taxonomies have been developed over the centuries to name and organize organic diversity in different local regions around the world. They have had great value in helping to inventory the world around us, prevent us from making life-threatening mistakes, and increase our practical and aesthetic appreciation of the unique differences and similarities among organisms. These classifications tend to focus largely on morphology (size, shape, color, etc.), but they often reflect habitat and behavior as well. Interestingly, the kinds of organisms recognized and given common names by nonscientists frequently correspond to species named by scientists who subscribe to the biological species concept. This species concept, formulated by Ernst Mayr in 1942, is the one most widely adopted by biologists today. It states that a species is a group of interbreeding natural populations of organisms that are reproductively isolated from other such groups. Note that morphology is not explicitly included in the definition, only reproductive behavior. Some years later, the definition was extended to include reference to the specific ecological niche occupied by each species, acknowledging the ongoing conversation between organism and environment. It has been suggested that the correspondence between commonsense "kinds" and biological species indicates the "naturalness" of species as the basic morphologic/ecologic unit in evolution. [See SPECIATION.]

Giving common or scientific names to different kinds of organisms we encounter is only one step in the process of comprehending the biological environment in which we live. A minimum of from 10 to 100 million different species are estimated to be living today, only 1.5 million of which have been named formally. We estimate roughly that nearly 100 times this number of fossil species existed. The magnitude of biodiversity is truly staggering and requires some kind of conceptual organization. Given the infinite ways in which organisms can vary, what rules should we use to classify them?

B. Organizing the Names

Aristotle (384–322 B.C.) was one of the first naturalists to write extensively about his observations of the diversity and organization of life, particularly marine invertebrates. Although not an evolutionist himself, his notion of an "unbroken sequence" in nature—from inanimate objects through plants to animals, with humans as the crowning achievement of divine creation—encouraged evolutionary thinking and was later converted by 18th century biologists into the concept of the *scala naturae*. The concept emphasizes both continuity and progress within the organic world (Fig. 2). In this world view, "lower" organisms (e.g., fishes, among the vertebrates) are considered to be inferior, both literally and figuratively, to the "higher" organisms in the Great Chain of Being.

With the rise of Christianity, many 18th and 19th century natural historians recognized species as distinct and well-defined units of creation. Species were homogeneous "natural kinds" of organisms that remained unchanged over time and were connected to one another in the organic world like links in a chain. According to Carl Linné (a Swedish botanist; 1707–1778) and his contemporaries, the purpose of naming species and classifying them into groups was to better understand God and the laws of nature by discerning the orderly plan of creation.

By now it should be clear that taxonomic principles and methods were not derived historically from evolutionary principles of common descent. In fact, the Linnean system of classification that biologists and paleontologists continue to use today was established well before evolution was recognized as the process structuring patterns of phylogeny and the diversity of life. Jean Baptiste Lamarck

(a French zoologist; 1744–1829) was instrumental in demonstrating that evolution occurs—species could and did change over time. Later still, Charles Darwin (a British naturalist; 1809–1882) articulated a mechanism by which evolutionary change can occur, namely the process of natural selection.

Despite the fundamental problem with its pre-evolutionary *structure,* the Linnean system of classification provides a standard, global reference of names of organisms. A particular organism can be identified uniquely by its Linnean binomial, even if it has been given different common names in different cultures. For example, *Strix aluco* is called the cat owl in Sweden, the wood owl in Germany, and the tawny owl in Britain. Under the Linnean system, organisms are referred to by a species name and a genus name, listing the genus name (capitalized) first. By convention, genus and species names are always italicized or underlined. Consider the earliest fossil bird, *Archaeopteryx lithographica*. *Archaeopteryx* is the genus name meaning "ancient wing;" *lithographica* is the species name that refers to the discovery of the fossil in lithographic limestone near the town of Solnhofen in Germany. Linnean names often refer to some feature or special characteristic of the organism or serve as a tribute to a person or place. A type specimen, deposited in a public museum or repository, is designated for each named species. This specimen serves, very imperfectly (considering the range of intraspecific variability, as we now understand it), as a tangible reference with which unidentified specimens can be compared.

In the Linnean system, hierarchical ranks above the species level are called higher taxa, and these ranks are given names. Species are grouped into a genus, genera into a family, families into an order, orders into a class, classes into a phylum, and phyla into one of several kingdoms (Fig. 3C). The grouping criteria have been based historically on the possession of shared characters. In other words, taxa have been defined by characters possessed by the organisms considered to be members of the taxon. Higher taxa are thus commonly defined as "classes" (groups or types) of organisms rather than as systems of common ancestry. The higher the taxon rank in the Linnean hierarchy, the greater

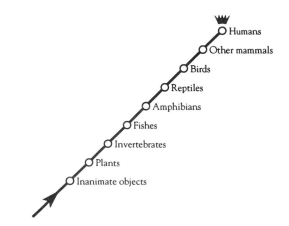

FIGURE 2 One interpretation of the *scala naturae,* also referred to as the Great Chain of Being.

A

Vertebrata (node 1)
  Chondrichthyes (shark)
  Teleostomi (node 2)
    Osteichthyes (tuna)
    Tetrapoda (node 3)
      Amphibia (frog)
      Amniota (node 4)
        Synapsida (node 5)
          Dimetrodon
          Mammalia (node 6)
            Primates (humans)
            Cetacea (whales)
        Archosauria (node 7)
          Crocodilia (crocodile)
          Dinosauria (node 8)
            Saurischia (node 9)
              Ceratosauria (*T. rex*)
              Aves (bird)
            Ornithischia (*T. horridus*)

B

Subphylum Vertebrata
  Class Chondrichthyes
    Subclass Elasmobranchii
      Order Selachii (shark)
  Class Osteichthyes
    Subclass Actinopterygii
      Order Perciformes (tuna)
  Class Amphibia
    Subclass Lissamphibia
      Order Anura (frog)
  Class Reptilia
    Subclass Archosauria
      Order Crocodilia (crocodile)
      Order Saurischia (*T. rex*)
      Order Ornithischia (*T. horridus*)
    Subclass Synapsida
      Order Pelycosauria (*Dimetrodon*)
  Class Aves
    Subclass Neornithes
      Order Passeriformes (bird)
  Class Mammalia
    Subclass Eutheria
      Order Primates (human)
      Order Cetacea (whale)

C

Kingdom Animalia
  Phylum Chordata
    Subphylum Vertebrata
      Class Aves
        Subclass Archaeornithes
          Order Archaeopterygiformes
            Family Archaeopterygidae
              Genus *Archaeopteryx*
                species *lithographica*

FIGURE 3 (A) Phylogenetic ("cladistic") classification. No ranks are named; the hierarchy of relationships is expressed as an indented list. Each indentation refers to a node in the cladogram in Fig. 1A; this classification of higher taxa is completely congruent with the internested pattern of clades illustrated there. In this example, the taxon Amniota is defined as the most recent common ancestor of humans and birds and all of its descendants, whether known or not yet discovered. This is a node-based definition. Amniota can also be defined as humans and birds and all tetrapods that share more recent common ancestry with them than with frogs (stem-based definition). This definition can accommodate fossils, discovered subsequent to the definition of the taxon, that are located between nodes 3 and 4. An apomorphy-based definition of Amniota is the clade that includes all tetrapods with an amniotic egg. Apomorphy-based definitions must be sensitive to evolutionary character transformations (e.g., whales are tetrapods, but lack four limbs). (B) Traditional classification. Note that the class Reptilia, as characterized here, is a paraphyletic group that includes some, but not all (i.e., birds), of the descendants from a common ancestor. (C) Complete Linnean hierarchy of *Archaeopteryx lithographica*.

the difference (in morphology, genotype, etc.) or disparity among ranks. Two phyla are considered to be more different from one another than are two classes in the same phylum; two classes more different than two orders in the same class, and so on (Fig. 3B). Even single species, if very different from other species in a named higher taxon, have been placed in a higher taxon all their own. The aardvark (*Orycteropus afer*), for example, is the only living species in the order Tubulidentata (although numerous extinct species have been discovered and named). These monospecific families or orders are generally named to acknowledge their great mor-

phological distinctiveness from other families or orders.

Because evolutionary principles do not structure the Linnean system of classification, not all named higher taxa are true clades (for example, Reptilia in Figs. 1A and 3B). This can present a problem when higher taxa are treated uncritically as if they *were* clades, because paraphyletic or polyphyletic higher taxa (see Glossary) are not complete systems of common ancestry as clades are. The roles that "nonclades" play in macroevolutionary phenomena can be ambiguous and confusing (see later). The Linnean hierarchy can only serve as a rough approximation of the evolutionary hierarchy of life. In a truly phylogenetic taxonomy, taxon names have explicit phylogenetic significance, and the hierarchy of classification reflects the nested pattern of common ancestry.

Although many systematists acknowledge that biological classification should reflect patterns of phylogeny, the principles and methods of classification have not changed since Linnaeus' time. Proposals have recently been made, however, to shift the perspective and the practice of the Linnean system fundamentally to reflect phylogeny (see references). Until (and if) these proposals are adopted, systematists interested in understanding the phylogenetic significance of named higher taxa face at least two major tasks. The first is to investigate the clade status of higher taxa named under the Linnean system—e.g., is the class Reptilia a clade or not? If not, what do nonmonophyletic higher taxa signify? The second is to evaluate the comparability of equivalent ranks in the taxonomic hierarchy—e.g., how does an order of mammals compare in diversity and genealogy to an order of angiosperms or an order of molluscs? *Can* they be compared, and if not, of what value are the ranks themselves?

III. SYSTEMATIC METHODOLOGY

Phylogenetic systematics is a discipline that orders entities (species and clades) according to their patterns of common ancestry. Character homology and polarity (see Sections III,A,2, and III,A,3) are two basic elements of phylogenetic methodology because the evolutionary transformation of characters can reveal patterns of common ancestry only if those characters are both homologous and apomorphic (also referred to as "derived"). The proximal goal of phylogenetic inference is to identify monophyletic groups (clades). Interpreting relationships among clades in terms of ecology, functional morphology, and developmental biology, we can reach a more biologically meaningful understanding of the evolution of character complexes, the phylogenetic status of named higher taxa, and the strength and nature of character support for individual clades.

A. Phylogenetic Inference

The methodology of phylogenetic systematics is explicit, which makes the process of reconstructing phylogeny comprehensible and the results more easily testable. [Although some systematists continue to equate phylogenetic systematics with "cladistics," significant philosophical and operational differences between the two methods are now widely recognized.] In a phylogenetic analysis, the taxa whose phylogenetic relationships are being investigated (called the "ingroup") are chosen, a list of characters and character states that vary among the taxa is drawn up, each taxon is coded for each character, a method is chosen for determining the direction of character change in evolution, the analysis is performed (often with the aid of computer algorithms), and the results are evaluated and interpreted.

I. Taxa

The entities used as terminal taxa in an analysis are themselves clades (monophyletic groups), or are assumed to be clades (Fig. 1). Clades are recognized by synapomorphies—characters that are both shared (homologous) and derived (relative to a primitive state). Paraphyletic or polyphyletic groups are not acceptable terminal taxa because they represent incomplete systems of common ancestry. Fossils and extant taxa may be combined in an analysis, as long as they represent monophyletic groups.

2. Character Homology

Characters are attributes of organisms. They may be morphological, genetic, ecological, behavioral, physiological, etc. Examples of characters include leg length, shell mineralogy, type of mating dance, or position in a gene sequence. Any characters that can be inherited from a common ancestor and are observed to vary among the ingroup can be analyzed. Character states distinguish different variations of a given character. The variation can be characterized by counting or measuring or describing it qualitatively. Smooth, lobed, or toothed are different states of the character "leaf margin type." Adenine (A), thymine (T), cytosine (C), and guanine (G) are the four nucleotides that are possible character states at each position (locus) in a genetic sequence (DNA, or deoxyribonucleic acid); the position in the sequence is itself the character. A data set composed exclusively of one type of character (molecular, for example) can serve as a semi-independent test of phylogenetic relationships suggested by a different data set (for the same ingroup) composed of a different type of character (morphological, for example).

There are two types of characters in phylogenetic inference: homologous and homoplastic. Homologous characters share common ancestry and are useful in reconstructing phylogeny. They are often, but not always, similar to one another in appearance. Homology of features can be tested before doing an analysis by comparing their development and location on organisms. Hair, for example, develops in the same manner, from the same kind of cells located in the skin in all mammals; hair is homologous in kangaroos, humans, platypus, etc. Homologues may be difficult to recognize, however. The process of evolution itself, descent with modification, can obscure similarity among homologues, most dramatically through the loss of features. Vertebrate wings are highly modified forearms. Snakes are tetrapods that have entirely lost their limbs in most living species.

Homoplastic characters are quite similar in appearance, but have evolved from different ancestors by processes of convergence or parallelism. They are also called analogous characters because their similarity is based on analogy (similar function) instead of homology (common ancestry). Convergent characters evolve in taxa that share only rather distant common ancestry. Human eyes and octopus eyes are convergent; they develop very differently and have evolved independently in two distantly related groups of animals, vertebrates and molluscs. Similar characters that evolve independently in two (or more) closely related clades are said to have evolved in parallel. Radial ribs as an exterior ornament on the shell have evolved in parallel in many different clades of bivalved molluscs. Homoplastic characters are not helpful in phylogenetic inference and can confuse patterns of relationship indicated by homologous characters because

they are often in conflict with them. Unfortunately, it is not always easy to distinguish homology from homoplasy. Hypotheses of character homology can be tested by examining their distribution on a cladogram relative to other, known homologues.

3. Character Polarity

Characters change from one state to another through the process of evolution. The direction of evolutionary character transformation is referred to as character polarity. The initial (old) state is said to be primitive, or plesiomorphic; the new or novel state is derived, or apomorphic, relative to the primitive state. Derived states are more useful than primitive states in determining phylogenetic relationships because they are more specific and more informative. To locate a particular car in a large parking lot, for example, knowing the model, year, and color is much more helpful than simply knowing the company that made the car. In a biological example, birds have (among other characters) four limbs, two of which have been modified as feathered wings. Having four limbs is a primitive character (symplesiomorphy) that birds share with all tetrapods and does not uniquely identify birds as a clade within tetrapods. Feathered wings, however, are modified forelimbs; this is a synapomorphy that distinguishes birds from all other tetrapods that lack feathered wings. Because primitive (unfeathered forelimb) and derived (feathered forelimb = wing) states are determined relative to some reference frame, their status changes if the level of analysis changes. Having four limbs is primitive for birds, a clade within the tetrapods, but it is a derived feature of the tetrapods, relative to fishes that lack four limbs (Fig. 1).

Several methods of polarity determination exist; none is infallible. Outgroup methods have proved most useful and are used most commonly today. The outgroup includes taxa (preferably more than one) presumed to be most closely related to the entire ingroup on the basis of a previous phylogenetic analysis. The character state present in the outgroups(s) is usually the primitive state, whereas the state present in the ingroup is usually derived. It may be difficult to choose informative outgroups if the ingroup taxa share a large number of highly derived features and as a group are quite different from their closest relatives. Or the character states in the outgroup may be highly derived relative to those of the ingroup, as the platypus is relative to living marsupial and placental mammals. The traditional paleontological (or stratigraphical) method of determining character polarity infers the order of character change from primitive to derived relative to the order of appearance of successive character states in the fossil record. However, geological biases in the fossil record can confuse the pattern; more primitive taxa might appear later in the fossil record than more derived taxa simply because they were not preserved in the earlier sediments. According to ontogenetic polarity, a feature that appears earlier in ontogeny is more general, and thus more primitive, than a feature appearing later in ontogeny. Insertions and deletions in the developmental pattern complicate the use of ontogeny alone to polarize character transformation. In other words, ontogeny does not always recapitulate phylogeny. Other polarity methods use biogeographical distributions or structural or functional analysis to polarize character change in evolution.

4. Phylogenetic Analysis

Several computer algorithms have been developed over the past decade or so (PAUP, MacClade, PHYLIP, Henning86) that enable large numbers of taxa and characters to be analyzed much more quickly and accurately than by hand. Each program searches for the "optimal" pattern, according to some optimality criteria, of relationships among the terminal taxa, given the distribution of character states in the data matrix. Parsimony and maximum likelihood are two criteria commonly used in constructing patterns of relationship among taxa. According to the principle of parsimony, the simplest explanation (shortest branching pattern or cladogram) to account for a given set of data is the preferred explanation, all other things being equal. More complex explanations (longer cladograms) require a greater number of ad hoc hypotheses and are thus less preferable. Parsimony is generally used as a methodological tool, but does not necessarily imply that the process of evolution giving rise to

the most parsimonious pattern of relationships was itself parsimonious. Maximum likelihood methods are statistical estimation methods. The most *probable* branching pattern, instead of the shortest and simplest, is chosen from among all possible patterns, given certain assumptions about the process of evolutionary change.

5. Other Methods

Although many systematists today use phylogenetic systematic (or "cladistic") methods, other methods are still in use, including numerical taxonomy (phenetics) and evolutionary systematics. Both the philosophy and methodology of phylogenetics differ from these other methods. Whereas phylogeneticists consciously avoid similarities that are primitive or homoplastic, and thus misleading from an evolutionary perspective, pheneticists utilize overall similarity, irrespective of its cause, to cluster organisms. Also, phylogeneticists determine polarity and typically "root" cladograms according to some reference point (using outgroups, stratigraphically oldest fossils, developmental sequence, etc.). Shark and tuna root the tetrapod clade in Fig. 1A. Phenograms (phenetic cluster diagrams) are not rooted. Instead, they diagrammatically represent distances, as the amount of morphological or genetic difference between taxa; taxa that cluster together closely in a phenogram are more similar to one another than taxa that are far apart.

The fundamental goals of evolutionary systematics and phylogenetic systematics are the same: identifying evolutionary patterns that result from evolutionary processes. However, the two methods differ in their philosophies of classification. Phylogenetic systematists do not accept paraphyletic taxa as valid taxonomic groups because they represent incomplete (partial) systems of common ancestry. Evolutionary systematists claim that paraphyletic taxa are acceptable because they have interpretable evolutionary histories, representing grades of evolution that share characteristic morphological and/or paleontological (temporal) similarities. For this reason, they are reluctant to embrace strictly genealogical classifications. For example, they recognize birds as a taxon distinct from dinosaurs (see Fig. 3B), despite their shared ancestry, because of their morphological differences and largely different fossil/stratigraphical records. Evolutionary systematists typically represent genealogical relationships in phylogenetic trees, which explicitly include a time dimension, rather than cladograms, which illustrate time only in a relative sense (Fig. 1). Their methods of phylogenetic analysis often combine those of phylogenetics and phenetics, so that their results are often not as rigorously genealogical as those of phylogenetics.

B. Phylogenetic Taxonomy

Because the process of evolution generates "natural" groups of organisms, species, and higher taxa united by genealogy, the evolutionary principle of common descent is the central tenet from which systematic principles and methods should be derived. For this reason, it is most informative to name groups of species on the basis of their genealogy. Recall that phylogenies are hierarchical and that clades are internested. Giving names to the hierarchy of clades would produce a taxonomy that reflects the process of descent with modification (Fig. 3A).

Traditional definitions of the taxon Mammalia, for example, are character-based—all vertebrates that have hair, various novel skeletal features, mammary glands, and give birth to live young are classified as mammals. But what about whales, which lack hair, or platypuses, which lay eggs? Are they not mammals? Most systematists do not adhere strictly to character-based definitions, in which organisms that do not possess the defining characters are not members of the group. We could instead define Mammalia, with reference to Fig. 1A alone, as the most recent common ancestor of whales and humans and all of its descendants. This is called a node-based definition because it defines the taxon at the basal node of a clade (Fig. 4). We

could also choose to define Mammalia as whales and humans and all synapsids that share more recent common ancestry with them than with *Dimetrodon*. This is referred to as a stem-based definition. Apomorphy-based definitions are the most similar to character-based definitions, but focus on a key feature that is both shared and derived. We could choose to define Mammalia, for example, by the possession of hair. Some fossil synapsids (cynodonts) have been discovered with impressions of hair surrounding the skeleton. They are more primitive than the taxon Mammalia just defined according to the node-based definition but by the apomorphy-based definition, they too would be members of Mammalia. Whales, lacking hair, would be acknowledged as mammals that have secondarily lost this apomorphy through the process of evolution itself. Whatever phylogenetic definition is agreed upon, it is clear that the definition refers explicitly to a hypothesis of phylogenetic relationships expressed in a cladogram.

C. Why Phylogenetic Taxonomy?

What are the advantages of a phylogenetic taxonomy with respect to evolutionary and environmental biology? All evolutionary aspects of comparative biology can be structured and interpreted with respect to genealogy. A taxonomy that reflects phylogenetic relationships can be used to address macroevolutionary questions about long-term changes in biodiversity, ecology, physiology, function and behavior, biogeography, ontogeny and development, and many other phenomena. For example, using patterns of taxonomic diversity through time to infer pattern and process in macroevolution (selectivity of extinctions, for example) is valid only if the taxa are clades (complete systems of common ancestry). Strong inferences about functional and morphological evolution can be made only when character homology and analogy are distinguished. It makes an enormously significant difference in our understanding of the evolution of locomotion or physiology or reproductive behavior to know if flight, homeothermy (warm-bloodedness), or parental care has evolved more than once and, if so, from what primitive states. Heterochrony, the

FIGURE 4 Simple cladograms. (A) Node-based higher taxon, (B) stem-based higher taxon, and (C) apomorphy-based higher taxon. [Redrawn from de Queiroz and Gauthier (1990).]

study of the evolutionary consequences of changes in developmental rates and timing, requires information on both developmental and phylogenetic patterns. Whenever functional, developmental, or phylogenetic "constraints" are implicated in shaping patterns of evolutionary change, some knowledge of genealogical relationships is necessary to provide a basis for comparison. Knowledge about the ancestry of organisms provides an essential foundation for understanding the environment (physical or biological) in which evolutionary changes have occurred.

To expand on just one example from functional morphology, we can examine the evolution of flight. Many different kinds of organisms, both animals and plants (seeds), possess wing-like structures that facilitate either powered or passive movement through the air. Most people realize that all organisms with wings do not share a close common ancestry. We recognize that birds have evolved flight from terrestrial dinosaurs, pterosaurs from a different group of terrestrial reptiles, bats from terrestrial mammals, bees and other winged insects from wingless insects; the examples are numerous and familiar. Although flight is a complex adaptation to life in the air, involving major morphological and behavioral changes, we know that it has evolved many times independently because of our knowledge of the systematic relationships among aerial organisms and their terrestrial ancestors. The wonder of flight is enhanced by our understanding of the many different ways in which it has evolved over geologic time, throughout the organic world. Similar examples can be recalled for the evolution of life on land (in both plants and animals), return to an aquatic mode of life (whales, penguins), the evolution of flightlessness, and so on. We often think in terms of systematics without being fully aware of its subtle yet powerful ability to structure the way we observe and interpret life around us. Understanding the phenomenon of convergence and adaptation in evolution requires a perspective from systematic biology. No other perspective can provide the pattern of common ancestry that reveals patterns in adaptation and functional morphology.

IV. RELEVANCE OF SYSTEMATICS TO ENVIRONMENTAL BIOLOGY

Systematics provides an evolutionary foundation to environmental biology. It gives us a vocabulary of names to communicate about the diversity of life and reveals the historical pattern of genealogical relationships among organisms in conversation with their environment. Organisms have played critically important roles in the evolution of Earth's hydrosphere, atmosphere, and geosphere at least for the past 3.5 billion years, and organismal evolution has been affected, in turn, by those environmental changes.

Organisms function as chemical factories, processing, manipulating, and storing ions that are present in their habitats. Physiological processes (respiration, digestion, etc.) cycle ions in the short term. Biomineralization, the process of making skeletons, locks up calcium, carbonate, phosphate, silica, and many other ions during the life of an organism and beyond. Some of these ions are released when the organisms die and reequilibrate with their environment, but many are stored in the fossilized skeletons and are not released for many millions of years. Clues about the structure and composition of ancient environments are preserved in fossil skeletons. Major, minor, and trace element concentrations and the composition of stable isotopes, particularly carbon and oxygen, reflect important characteristics of the environments in which the skeletons were formed.

Organisms are primary engines in the global carbon and oxygen cycles. Photosynthetic organisms (plants) extract carbon dioxide from and release oxygen to the atmosphere on a daily basis. Non-photosynthetic organisms (like humans) breathe in oxygen and breathe out carbon dioxide. Longer term carbon reservoirs exist in the biomass of living organisms, in fossil fuels, and in limestone formations in the geologic record worldwide. Plate tectonic activity cycles these biologically mediated geological deposits by consuming them at subduction zones, and releasing carbon dioxide and other compounds in volcanic gases and at spreading ridges, to be used by living organisms once again.

Variations in ocean chemistry and the composition of the atmosphere and lithosphere affect and are affected by the origin and evolution of particular clades of plants and animals at particular points of time throughout geological history. A stable oxygenated environment did not exist until approximately 2 billion years ago, more than 1 billion years after the first photosynthetic organisms appear in the fossil record. Diversification in the Mesozoic era of the marine plankton that form deep sea oozes (carbonates and silicates) coating downgoing slabs in subduction complexes fundamentally changed the kinematics of subduction and the geochemistry of subduction-related volcanism. The vast majority of our energy resources derive from the fossilized remains of once-living organisms: coal from Paleozoic coal swamp floras and oil and gas from Mesozoic and Cenozoic marine microorganisms. Fluctuations over geological time in the relative diversity and abundance of organisms that contribute to hydrocarbon formation can only be studied with help from a taxonomic vocabulary of names and with some sense of the genealogical relationships among those organisms (*See* EVOLUTIONARY HISTORY OF BIODIVERSITY).

Living and fossil organisms are the most important generators, recorders, and barometers of environmental changes. Different organisms with different developmental, ecological, and phylogenetic histories play identifiably different roles in environmental biology. We can better appreciate the ecological, physiological, and biomineralogical variety of life, and the effect that life has on the Earth, with an understanding of systematic biology.

V. SYSTEMATICS AND HUMAN LIFE

Unobtrusive but ubiquitous, systematics plays a critical role in our lives every day. Our primary means of understanding organisms, both today and in the geological past, is through systematics—having access to a vocabulary of names organized according to evolutionary principles. We grow and hunt organisms to get nourishment and to create shelter and clothing from them. Organisms carry diseases such as malaria (transmitted by *Anopheles* mosquitos) and pneumonia (often from bacteria) and toxins like the deadly nightshade plant (*Atropa belladona*), to name just a few. Organisms also help us fight disease and ease pain. The antibiotic penicillin was developed from the *Penicillium* mold. Aspirin contains salicylic acid, a pain reliever that exists naturally in the willow *Salix*. Having a more complete understanding of the systematics of organisms important in our lives means that we can develop new and better vaccines, crop plants, pest management plans, and natural resource management programs more efficiently and effectively.

The study of systematics puts in clear perspective the fact that humans are only a single species in the grand diversity and genealogy of life. We are a species that has obviously had an enormous impact on the Earth during our brief several hundred thousand years here. And we will undoubtedly continue to affect other species by our activities, causing the extinction of many and perhaps encouraging the origination of others. Whatever our role in environmental biology, systematics educates us that human beings do not represent the pinnacle of evolutionary success, the ultimate result of progress in evolution. Instead, we are just one species among millions of others making a living on the Earth today. Because we are part of a vast web of interdependent relationships, it is wise to remember that the fate of the rest of the biosphere determines our own fate.

Glossary

Ancestry Organisms reproduce and make more organisms. The "parents" are referred to as ancestors, whereas the offspring are referred to as descendants. Organisms (or clades) that evolved from the same ancestor are said to share common ancestry accomplished by the process of common descent.

Apomorphy Apo means "away from;" morph refers to shape. Apomorphies are homologous character states that are derived relative to a primitive (former) ancestral state. Synapomorphy is a derived character shared (through common ancestry) among organisms or taxa.

Clade A monophyletic (single "tribe") group, including all the descendants from a single common ancestor; they are complete systems of common ancestry. Usually refers to a group of species, although some systematists claim that individual species are also clades. A cladogram is a branching diagram that illustrates the internested pattern of clades (Fig. 1A).

Evolution Process of descent with modification (change or transformation).

Homology Features of organisms that correspond in structure, location, and/or development because they have evolved from a common ancestor. As forearms, bird wings, human arms, and seal front flippers are homologues.

Homoplasy Features of organisms that are similar in form because of the evolutionary processes of convergence, parallelism, or reversal, *not* the process of common ancestry. Wings of birds, bats, and insects are homoplastic features.

Most recent common ancestor Refers to the most recent in a sequence of common ancestors. Birds and humans share common ancestry, but it is a more remote common ancestry than that shared by birds and other saurischian dinosaurs (Fig. 1A).

Natural selection Differences in mortality and reproductive success from one generation to the next due to the interaction of organisms with their habitats.

Ontogeny Life cycle of an individual.

Paraphyly Paraphyletic taxa include some, but not all, of the descendants from a common ancestor. They are defined by shared primitive characters instead of shared derived characters (which are more informative) and commonly represent only the ancestral portions of clades. The traditional class Reptilia (Fig. 3B) is a paraphyletic taxon.

Phylogeny Genealogy of a group of organisms; the lines of descent from a common ancestor. The study of, or the pattern of, genealogy.

Plesiomorphy Plesio means "near." Plesiomorphies are homologous character states that are primitive, or closer to the ancestral state. Symplesiomorphy is a primitive character shared among organisms or taxa.

Polarity Direction of character state transformation in evolution, from primitive to derived states.

Polyphyly Polyphyletic taxa includes species that do not share most recent common ancestry. A higher taxon defined by the possession of wings, that would include birds, bats, and insects, is a polyphyletic taxon. They are not evolutionarily informative groups.

Taxon A group of organisms (or species or clades) that is given a name. The plural of taxon is taxa.

Bibliography

American Society of Plant Taxonomists, Society of Systematic Biologists, and the Willi Hennig Society. (1993). "Systematics Agenda 2000: Integrating Biological Diversity and Societal Needs." New York.

Association of Systematics Collections. (1988). "Systematics: Relevance, Resources, Services, and Management: A Bibliography." Washington, D.C.

Ax, P. (1987). "The Phylogenetic System: The Systematization of Organisms on the Basis of Their Phylogenesis." Chichester: Wiley.

Donoghue, M. J., and Cantino, P. D. (1988). Paraphyly, ancestors, and the goals of taxonomy: A botanical defense of cladism. *Bot. Rev.* **54,** 107–128.

Mayr, E. (1982). "The Growth of Biological Thought: Diversity, Evolution, and Inheritance." Cambridge, MA: Harvard Univ. Press.

Mayr, E, and Ashlock, P. D. (1991). "Principles of Systematic Zoology," 2nd Ed. New York: MacGraw-Hill.

O'Hara, R. J. (1992). Telling the tree: Narrative representation and the study of evolutionary history. *Biol. Philos.* **7,** 135–160.

Panchen, A. L. (1992). "Classification, Evolution, and the Nature of Biology." Cambridge: Cambridge Univ. Press.

de Queiroz, K. (1988). Systematics and the Darwinian revolution. *Philos. Sci.* **55,** 238–259.

de Queiroz, K., and Gauthier, J. (1990). Phylogeny as a central principle in taxonomy: Phylogenetic definitions of taxon names. *Syst. Zool.* **39,** 307–322.

de Queiroz, K., and Gauthier, J. (1992). Phylogenetic taxonomy. *Annu. Rev. Ecol. Syst.* **23,** 449–480.

Swofford, D. L., and Olsen, G. J. (1990). Phylogeny reconstruction. *In* "Molecular Systematics" (D. M. Hillis and C. Moritz, eds.), pp. 411–501.

Wilson, E. O. (1992). "The Diversity of Life." Cambridge, MA: Harvard Univ. Press.

Terrestrial Halophytes

Edward P. Glenn
University of Arizona

I. Measurement and Expression of Salinity in Soil and Water
II. Diversity of Halophytes
III. Examples of Important Halophyte Habitats
IV. Salt Tolerance and Ecophysiology of Halophytes
V. Halophyte Agronomics and Utilization

Halophytes are salt-tolerant higher plants, including land plants and submerged forms. Terrestrial halophytes, the subject of this article, grow in many different, naturally saline ecosystems: coastal salt marshes, mangrove swamps, coastal salt pans, inland salt marshes, saline seeps, and salt deserts. They also grow on land that has become salinized by human activity. Halophytes are of environmental importance from several perspectives. Their high rates of primary production form the base of the food chain in coastal salt marshes and mangrove swamps, many of which have been altered or destroyed by human activity. They are important browse plants in desert and range ecosystems. They are used in rehabilitation projects to return salinized soils to productive use and are components of constructed wetlands. Halophytes have been developed as new forage and food crops for saltwater irrigation. They are also studied as models of salt tolerance to understand what traits need to be introduced into crop plants to improve their performance on saline soils. The article treats the biology of halophytes, but the first section reviews the most common units used to express the salt content of soil and water, as the different units of salinity expression can be confusing even to workers in the field.

I. MEASUREMENT AND EXPRESSION OF SALINITY IN SOIL AND WATER

Salinity is not a precise chemical term. Most saline water or soil contains a mixture of salts. Salts in solution are dissociated into cations and anions; for example, NaCl in solution is dissociated into Na^+ and Cl^-. Depending on the salt and its concentration, complete dissociation does not always occur. [*See* ASPECTS OF THE ENVIRONMENTAL CHEMISTRY OF ELEMENTS.]

The predominant salt in seawater is NaCl, but there are significant quantities of $MgSO_4$, KCl, and other salts as well (Table I). In soils and waters that have been salinized by contact or dilution with seawater, the proportion of individual salts is usually close to the ratios in Table I even though the absolute concentrations may differ over a wide range. Even in inland saline environments, the ratio of salts may be essentially similar to that of seawater, as most terrestrial salt deposits were of marine origin. For example, ancient seas evaporated to deposit the salts in the Great Salt Lake basin of Utah in the United States, whereas salts in the Australian soils originated from marine aerosols blown inland from the Antarctic Ocean. Less commonly, the predominant salts may differ from these

TABLE I

TABLE I

Concentration of Major Ions in Seawater (g/kg seawater) Normalized to 35‰ Salinity

Ion	g/kg	Molar ratio
Chloride	19.35	1.00
Sodium	10.76	0.85
Magnesium	1.29	0.027
Sulfate	2.71	0.028
Calcium	0.41	0.010
Potassium	0.40	0.010
Bicarbonate	0.15	0.002

TABLE II

Approximate Conversion Factors for Commonly Used Salinity Measurements[a]

Method:	Osmotic pressure		Weight basis (ppt)	Molal basis[b] (moles/liter)	Electrical conductivity[c] (dS/m)
Unit:	−MPa	−bars			
−MPa	1.00	10	13.9	0.24	23.1
−bar	0.10	1.00	1.39	0.024	2.31
ppt	0.072	0.72	1.00	0.017	1.66
moles/liter	4.17	41.7	58	1.00	96.3
dS/m	0.043	0.43	0.602	0.0104	1.00

[a] All conversions are based on a completely dissociated solution of NaCl at standard temperature and pressure. The conversion factors are not valid for salts other than NaCl and are intended to be used as an aid in reading and comparing literature reports rather than in experimental work.
[b] To convert moles/kg to osmoles/kg multiply by 2.
[c] Electrical conductivity is not linear with salt concentration so the conversion value is not accurate at salinity above 10 ppt.

in seawater composition. The salt lakes of the Great Rift Valley in Africa, for example, are high in $NaHCO_3$ rather than NaCl, and consequently are extremely alkaline in addition to being saline. [*See* SOIL ECOSYSTEMS.]

The salt content of water and soil is often expressed on a weight basis (e.g., g/kg or ppt—parts per thousand) without regard to the type of salts present. Salinity may also be expressed on a molar basis if the specific salt composition is known, or in terms of molar equivalents of the dissociated ions. Salts enhance the conductivity of electricity in water and salinity is often reported in terms of the electrical conductivity (EC) of a solution or soil extract, in units of dS m^{-1}, where S is the siemens, a unit of conductance. Sometimes salinity is expressed as the osmotic potential of a solution in units of negative pressure (in bars or pascals). The most common salinity measurement units and approximate conversion factors are given in Table II.

Salt content in natural water sources varies from near zero (rainwater) to greater than 220 ppt at saturation with NaCl (e.g., in evaporating tidal pools). Seawater in open oceans is 32–35 ppt whereas along desert coasts it may be as high as 40 ppt owing to evaporation and uneven mixing near shore. On the other hand, inland seas such as the Black Sea and the Caspian Sea have lower-than-oceanic salinities from dilution by river waters flowing into the seas.

Salts can concentrate in soils to many times the level present in the water supply. For example, a saline water table lying close to the soil surface can cause a thick salt crust to form on the soil surface due to the capillary rise of the saline solution followed by evaporation of water and depositon of the salts at the surface. This is the mechanism by which salt flats form, either naturally or through improper irrigation practices. Salts also concentrate in the soil as water dries up or is lost to evapotranspiration. In irrigated soils, mean salt content in the soil solution is often two to three times higher than in the irrigation supply because of evaportranspiration.

As a soil dries out, the residual water becomes increasingly more difficult for plant roots to absorb because it is held more tightly within the soil matrix (the water molecules are bound to the soil particles). Soil salinity and moisture content together determine the total water potential in the soil, usually expressed as negative bars of pressure similar to the expression of osmotic potential of solutions. This is the force the plant roots have to act against to extract water. Dry, saline, desert soils may have extremely low water potential in the range −50 to −95 bars, and present a more serious challenge to plant growth than salinity alone.

II. DIVERSITY OF HALOPHYTES

There are 2–3 million hectares of coastal wetlands and mangrove swamps in the world and over 400 million hectares of inland saline soils. In addition to naturally saline habitats, approximately 46 million hectares of land have been salinized by human activity, principally through irrigation projects that add salts to the soil over time and raise saline water tables under fields because of inadequate drainage. Halophytes are the main or only form of vegetation on many of these soils.

Halophytes are not a single taxonomic grouping and they come in many different growth forms. Their ranks include annuals, perennials, grasses, trees, shrubs, forbs, and succulents. Representative species of halophytes are shown in Figs. 1–5. It has been roughly estimated that there are 2000–4000 species of halophytes. However, no complete list of halophyte species has been compiled and botanists still differ on the basic question of "what is a halophyte?", as many gradations of salt tolerance exist among higher plants.

FIGURE I *Distichlis palmeri,* a salt marsh grass endemic to the Colorado River delta, the seeds of which were harvested by the Cocopah Indians as a seawater grain.

Halophytism is thought to have evolved several times separately among higher plant families. Except for two families of ferns that have halophyte species, almost all halophytes are angiosperms. Over 100 of the 350 flowering plant families contain halophytes, but over half the genera with halophytes belong to just 20 families. The most prominent halophytic family is the Chenopodiaceae, which has 44 halophytic genera containing some 300 species (although more than half the genera in this family are not halophytic). Other important dicot families containing halophytes are Asteraceae, Aizoaceae, Apiaceae, Papilionaceae, Euphorbiaceae, Scrophulariaceae, and Caryophyllaceae. Halophytes are perhaps less well represented among monocot families, with most genera occurring in the Poaceae (the salt grasses), the Cyperaceae (sedges, with triangular, solid stems), and the Juncaceae (rushes, with round, hollow stems).

Several different classification schemes for halophytes have been proposed based on edaphic or climatic factors (where they grow), plant associations (what plants they grow with), and their degree of salt tolerance (which is usually inferred by where they grow). Different "physiotypes" of halophytes have also been recognized, based on a species' tendency to accumulate or exclude sodium, develop succulence, or manifest other physiological responses to salt stress. However, there is no comprehensive classification scheme for halophytes in common use.

Halophytes adapted to the most saline environments are called euhalophytes. These species, typified by members of the Chenopodiaceae but occurring in other families as well, grow in seawater marshes, salt deserts, and other hypersaline environments. They actually require at least some salt in the water or soil for optimal growth. Some, such as *Salicornia* spp., have been considered to be obligate halophytes because their growth on fresh water is severely impaired. However, most if not all halophytes can complete their life cycle in fresh water so the term "obligate halophyte" should probably not be used.

Plants that normally grow in less saline environments are called miohalophytes. They show their best growth in fresh water but can grow and repro-

FIGURE 2 *Atriplex paludosa* (a saltbush), a xerohalophyte exhibiting salt, drought, and heat tolerance.

FIGURE 3 *Phragmites australis* (common reed), a wetland miohalophyte that can tolerate some salt in the water but that grows best in fresh water.

FIGURE 4 *Avicennia germinans* (black mangrove), a hygrohalophyte that tolerates anerobic soil and standing water.

FIGURE 5 *Salicornia bigelovii,* an annual succulent that has been explored as a seawater oilseed crop.

duce in brackish water, and many can tolerate some exposure to pure seawater. These species, scattered in dozens of families, typically grow in dilutions of seawater such as are found in tidal rivers and brackish ponds at the backs of coastal marsh systems, or in basically freshwater systems that receive occasional tidal inundation.

Xerohalophytes are desert shrubs that can tolerate both drought and salinity stress. They are typified by the genus *Atriplex,* which dominates large areas of arid shrubland around the world. At the opposite extreme, hygrohalophytes require permanently moist soil or standing water. They are found in the saline wetlands of the world. *Scirpus, Juncus,* mangroves, and *Spartina* grass are examples of hygrohalophytes.

At the low end of the salinity scale, halophytes grade into salt-tolerant glycophytes. *Typha* (cattail), for example, is basically a freshwater wetland genus but it can grow in water with up to 4–5 ppt salts, and its tubers can withstand immersion in full seawater for several months and still produce new shoots when freshwater conditons return to the marsh.

III. EXAMPLES OF IMPORTANT HALOPHYTE HABITATS

A. Temperate Zone Coastal Marshes

Grass-dominated tidal salt marshes indent the coastline of all continents outside of the tropical zone, which is dominated by mangrove estuaries. A salt marsh usually forms at the point where a river enters the sea. Sediments carried by the river deposit on the sea bottom at the point of entry, creating, if the river is of significant size, vast areas of shifting mudflats dissected by tidal channels. There is a net accretion of sediments into an active marsh system leading to the building of a delta. The shifting terrain of a temperate salt marsh is colonized by a rather restricted number of halophyte genera, many of worldwide distribution. [*See* WETLANDS ECOLOGY.]

Various species of *Spartina* (cordgrasses) are the dominant vegetation of many western hemisphere temperate marshes (Fig. 6). The rhizomatous *Spartina* plants grow so thickly that few other species can grow within the stands. Along the Atlantic Coast of North America from Newfoundland to northern Florida and in the Gulf of Mexico, two *Spartina* species dominate the marshes: *S. alterniflora* (smooth cordgrass) and *S. patens* (salt meadow cordgrass). The former, tall species grows in the low and middle regions of the intertidal zone and along the creek channels, and the latter, shorter species grows mainly in the high zones that are inundated by tides less frequently and to a shallower depth than the lower zones.

In a series of important studies conducted in Georgia salt marshes, Eugene Odum established that net productivity of *Spartina* marshes was directly related to the degree of tidal irrigation that the plants received. *Spartina* in the low marsh received frequent and vigorous tidal flushing, grew tall (over 2 m), and produced up to 4000 g m^{-2} yr^{-1} dry matter, equal to the most productive conventional crops such as sugarcane. The middle zone of the marsh received gentle tidal irrigation and supported medium-height *Spartina* producing 2300 g m^{-2} yr^{-1} dry matter, and the high zone received infrequent irrigation and supported short

FIGURE 6 A coastal *Spartina* marsh on a tributary of the York River, Virginia, showing *Spartina alterniflora*. (Courtesy of Carole McIvor.)

cordgrass producing only 750 g m^{-2} yr^{-1} dry matter.

These findings are significant, for the salt marsh is often viewed as a stressed environment owing to the saline, anaerobic substrate, yet the halophyte community is highly productive, taking advantage of the "energy subsidy" provided by the tides. Besides providing water, tidal flushing distributes nutrients to the plants. Odum viewed *Spartina* marshes as pulse-stabilized climax communities rather than successional stages leading to some other vegetation form. Adaptations that allow *Spartina* to be so productive in this environment include aerenchyma tissue to aerate the root zone and the C$_4$ pathway for photosynthesis, which limits the amount of water that must be transpired (and salts that must be handled) per unit dry matter production.

The North American *Spartina* marshes on the Pacific Coast are dominated by *S. foliosa*, a medium-height cordgrass, from San Francisco Bay southward to Guererro Negro, Baja California. Even though San Francisco Bay was the northern limit for the native *Spartina*, in the 1970s the Atlantic species, *Spartina alterniflora*, was introduced into South San Francisco Bay. It proved to be more aggressive and cold tolerant than the native species and is now invading marshes north into the state

of Washington, where it is regarded as a pest species that disrupts the natural marsh ecosystems of those areas.

Similarly, European marshes did not originally contain *S. alterniflora*, but is was accidentally introduced into a marsh at Poole, England, in the nineteenth century. It hybridized with its European cogener, *S. maritima*, to produce a new species, *S. townsendii*, a male-sterile hybrid that spread vegetatively throughout the British Isles. Eventually another new species, *S. anglica*, developed from *S. townsendii* by doubling of the chromosomes. *Spartina anglica* is fertile and can spread by seeds. The two new species continue to spread throughout the English and western European marsh systems. Elsewhere in European marshes, grasses such as *Agrostis* occupy the niche held by *Spartina* in the New World marshes. In the southern hemisphere, *Spartina brasiliensis* and *S. montevidensis* dominate the South American temperate marshes and are not found elsewhere.

Puccinellia and other shortgrass species occur with *Spartina* in north temperate marshes and extend beyond the *Spartina* zone into the polar regions, where the growing season is short. At the southern end of the *Spartina* range, short salt grasses such as *Distichlis spicata* and *Sporobolus virginicus* occur with *Spartina* in the more southern marshes and extend into the mangrove zone. A particularly tough salt grass is the low, wiry *Monanthocloe littoralis,* found at the back of many southern, New World marshes, which is almost impossible to walk over with bare feet.

Although they may be dominated by grasses, temperate marshes are the home of other halophyte species as well. Succulent species, such as *Salicornia, Suaeda, Allenrolfia, Sesuvium,* and *Batis* are well represented in nearly all temperate marshes. Perennial *Salicornia* grows as shrubs in the high zones whereas annual species such as *S. europaea* and *S. bigelovii* colonize the low zones, especially areas of shifting sediment that have not yet been colonized by *Spartina* or other grasses. Nonsucculent or slightly succulent flowering plants from genera such as *Frankenia, Cressa,* and *Aster* grow on the higher, more stable portions of the intertidal plain. Rushes, sedges, and reeds are found mainly at the backs of the marshes in the supralittoral zone (above the normal highest tide level), where there is some freshwater influence, as these species are less salt tolerant than the intertidal zone grasses and succulents.

Overall, salt marshes do not support a great deal of floristic species diversity compared to freshwater ecosystems: a mature marsh system may contain 15–25 species of halophytes in the intertidal zones, with diversity increasing in the supralittoral. On the other hand, their high rates of productivity support a great deal of faunistic diversity.

Temperate coastal marshes were historically used for grazing or cutting hay, but in the last 100 years they have been cleared, diked, and filled for "higher uses" at an alarming rate around the world. The reclaimed land is often used for agriculture, housing, or industrial development. Another reason for loss of coastal wetlands is diversion of river flows for upstream uses. The wetlands of the great desert river deltas such as the Nile, Indus, and Colorado have been greatly reduced from lack of fresh water influx; the water is now used for upstream irrigation and hardly any flow reaches the sea in most years.

It is now recognized that coastal wetlands play many important ecological roles: they are hatcheries and nurseries for marine fish and crustaceans; provide breeding grounds for migratory water-fowl; act as biological filters for removing excess nutrients from river waters flowing to the sea; and form the base of a detrital food chain that enriches the marine environment outside the marsh. Contrary to the trend over the last 100 years, there is now perhaps more interest in finding ways to restore coastal wetlands than in filling them in for development, at least in the industrialized countries. Large plantings of dominant species such as *Spartina* have been established in disturbed marshes and constructed wetlands. *Spartina* plants are easily established by transplanting rhizome sections from established stands into new areas to be vegetated. However, these artificial wetlands at first lack species diversity and cannot be expected to replace natural wetlands in all their ecological roles. [*See* MARINE PRODUCTIVITY.]

B. Mangrove Estuaries

Mangroves, like halophytes in general, are not a natural taxomonic grouping. The term refers to trees or woody shrubs that live in muddy, wet soils in tropical tidewaters. The term mangrove refers to the plants, whereas the term mangal refers to the entire community of organisms living in the mangrove-dominated ecosystem. Mangroves once grew along 60–75% of the tropical coastlines, but many stands have been cleared, similar to the temperate marshes. Approximately 1,650,000 hectares of mangrove forests remain in the world.

The geographic distribution of mangroves is controlled by five basic requirements: (1) tropical temperatures where the mean temperature in the coldest month is 20°C or above; (2) fine-grained alluvium such as soft mud composed of silt, clay, and organic matter, typical of deltaic deposits at the mouths of rivers; (3) protection from strong wave and tide action; (4) some degree of salt in the environment (mangroves can live without salt but do not compete well against inland forest species under freshwater conditions); and (5) a wide horizontal tidal range to allow formation of a wide belt of alluvium on which mangroves can establish.

The best-developed mangrove forests lie in the equatorial band between the Tropic of Cancer and the Tropic of Capricorn, but because of regional climate differences, the mangrove lines in the north and south hemispheres are wavy rather than straight. For example, mangroves grow in Texas and Louisiana in the Gulf of Mexico, and along the southeast coast of Australia, well outside the tropical zones. Mangroves cannot survive freezing and in locations such as the Gulf of California, which has a continental climate, the occasional killing frost rather than the mean monthly winter temperature limits the northern distribution of mangroves.

There are approximately 60 mangrove species in 14 plant families, although as with the term halophyte, the question "what is a mangrove?" has been disputed. There are two main world groupings of species: the Indo-Pacific or eastern mangroves and the Atlantic or western mangroves. The Cape of Good Hope in South Africa appears to have been an uncrossable barrier separating eastern and western populations during the period of maximum speciation (the vast expanse of the Pacific Ocean formed the barrier on the other side of the world). Perhaps the three most important mangrove genera worldwide are *Avicennia* (black or honey mangrove in the Americas, white mangrove in West Africa), *Rhizophora* (red mangrove in the Americas and Africa) (Fig. 7), and *Laguncularis* (white mangrove in the Americas). In the family Avicenniaceae, *Avicennia marina* is the most widely distributed species in the

east, and its counterpart in the west is *Avicennia germinans*. In the Rhizophoraceae, *Rhizophora mangle* and *R. racemosa* are the common western species whereas *R. apiculata* and *R. mucronata* are the main eastern *Rhizophora*. In the Combretaceae, *Laguncularis racemosa* is found in the west, whereas species of the related genus *Lumnitzera* are found in the east.

Mangrove species often grow in zones within a mangal, with *Rhizophora* species being the most seaward, *Avicennia* growing behind the *Rhizophora* zone, and *Laguncularis* occurring at the backs of the mangal. The zonation pattern roughly follows the salt tolerance of the species. In moist climates, mangrove forests often grade into freshwater swamps at their backs. Species of *Conocarpus* (e.g., *C. erectus*, button mangrove), another genus in the Combretaceae, may grow behind the *Laguncularis* zone in such mangals, in low-salinity water or fresh water.

In desert climates, sometimes only a single species such as *Rhizophora mangle* or *Avicennia marina* or at most a few species are found in a mangal. In moist climates, species diversity is higher. Maximum species diversity occurs in the humid, Southeast Asian mangals. The genera *Bruguiera* and *Ceriops* (family Rhizophoraceae), *Xylocarpus* (family Meliaceae), and *Sonneratia* (family Sonneratiaceae) are important Asian genera that are absent or nearly so from the west.

FIGURE 7   A mangrove swamp in Everglades National Park, Florida, showing *Rhizophora mangle* (dwarf red mangroves). (Courtesy of Carole McIvor.)

Other halophyte species besides mangroves occur in mangals, and the species diversity among the nonmangrove flora is again higher in moist than in dry climates. In the desert mangals of the Red Sea and Persian Gulf, *A. marina* is often the only mangrove present with a meager understory of *Suaeda, Salicornia,* and *Limonium* plants filling out the flora. Grazing pressure by camels in addition to climatic factors has reduced species diversity in those marshes. In fact, camels not only eat the understory species but the mangroves as well, stripping the trees clean of their leaves.

Mangroves have many special adaptations to their environment. Mangrove leaves are often specialized to store water and many species are high in tannins, which may help the plants resist fungal attack in their continually moist environment. Mangal soils are often anaerobic and mangroves are usually not deeply rooted. Some mangroves have aerenchymotous prop roots or pneumatophores, woody projections rising out of the mud containing aerenchyma tissue, which act as snorkels to bring oxygen below the soil line. Many species, especially the pioneering species that occupy the seaward fringe, also exhibit vivapary, or seed germination on the parent plant. Such pregermination facilitates plant establishment, for germination of seeds in seawater is difficult, even for halophytes. The floating, dagger-shaped propagules of *Avicennia marina* are a common feature of the tropical beachwash wherever this species occurs. When the propagules float into sheltered, shallow water, the woody root, which points down in the water, penetrates the substrate and the plants begin to grow. The propagules can be collected and pushed into the soil as a simple method of artificially establishing new stands.

Mangrove swamps usually have lower rates of net primary production than temperate marshes but rates as high as 2000 g m^{-2} yr^{-1} have been reported, similar to the middle-zone of a Georgia *Spartina* marsh. Mangals are as important to the ecology of the tropical coastal zone as *Spartina* marshes are to the temperate coastal zone. They serve as an ecotone between land and sea and trap sediments entering the sea, which helps build new shoreline. At the seaward edge they are home and breeding grounds for crabs, shrimp, mollusks, and fish, including many of commercial importance. At their landward ends they are home to amphibians, crocodiles, and air-breathing fish. They are nesting and foraging areas for numerous bird species. They capture excess nutrients in runoff from the land and recycle them into the biosphere via the detrital food chain. In moist climates the canopy of the mangrove forest is essentially a freshwater environment, in which rainwater collects in puddles on the leaves and stems and supports a variety of algae, insects, and protozoa.

Traditionally, the greatest use of mangroves was for lumber or firewood. This activity has progressed from a subsistence practice to an industrial-scale activity, in which entire mangrove forests are clear-cut and the trees milled to wood chips, to be used in power generation in remote locations. Besides clear-cutting, the most serious threats to mangrove forests are diversion of fresh water, land reclamation for agriculture and aquaculture, filling for industrial development, siltation from land runoff, mining, conversion for solar salt ponds, and die-off from chemical, domestic, and petroleum pollutants (Fig. 8).

Mangrove swamps are receiving protection from further development in some areas. In Dade County, Florida, the total fines for destroying a single mangrove tree can exceed $40,000. However, penalties are not only monetary, for the offender must replant mangroves on an area larger than was destroyed. As a consequence of such strict enforcement, there are now specialists in mangrove propagation and good knowledge has been gained on how to establish or reestablish mangrove stands. Efforts to replant mangroves are also under way in the Middle East, especially in the Persian Gulf, which once supported large stands of *Avicennia marina*. Mangroves can be established by directly planting propagules or by establishing them in a nursery and then transplanting them into the field.

C. Desert Halophytes

Inland salinized land occurs on all the continents but is concentrated in the arid zones, where evaporation exceeds precipitation, thereby creating a potential

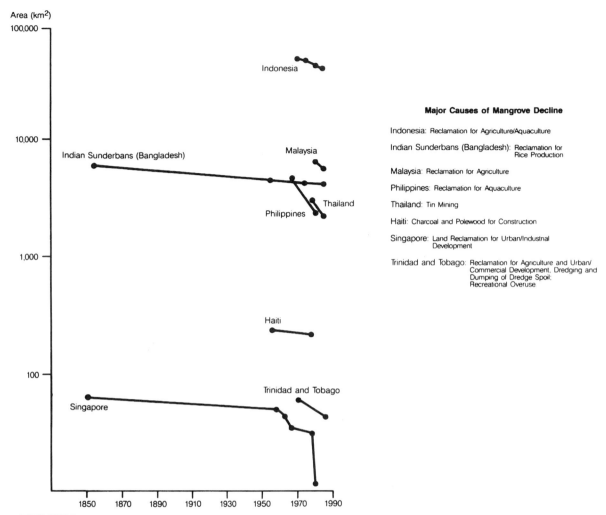

Major Causes of Mangrove Decline

Indonesia: Reclamation for Agriculture/Aquaculture

Indian Sunderbans (Bangladesh): Reclamation for Rice Production

Malaysia: Reclamation for Agriculture

Philippines: Reclamation for Aquaculture

Thailand: Tin Mining

Haiti: Charcoal and Polewood for Construction

Singapore: Land Reclamation for Urban/Industrial Development

Trinidad and Tobago: Reclamation for Agriculture and Urban/Commercial Development, Dredging and Dumping of Dredge Spoil; Recreational Overuse

FIGURE 8 Decline in mangrove stands in selected countries, 1850–1990. [From "World Resources" (1986) by permission.]

for the accumulation of salts in the topsoil, aquifers, and surface waters. Xerohalophyte communities dominate millions of hectares of desert shrubland, salt flats, and saline seeps around the world. In some locations, halophytes are the most important source of rangeland forage for animal production. [*See* DESERTS.]

Examples of important desert halophyte types around the world are the leafy saltbushes (genus *Atriplex*); succulents, in the genera *Suaeda, Spergularia, Sesuvium, Salicornia, Arthrocnonemum,* and *Allenrolfia;* grasses and reeds such as *Sporobolus, Puccinellia, Agropyron, Distichlis, Scirpus,* and *Juncus;* and herbaceous or shrubby plants in genera such as *Frankenia, Aster,* and *Limonium.* These genera all have coastal counterparts, and it is believed

that salt tolerance evolved first among coastal species and then spread inland. A salt-tolerant tree that dominates many riparian and other inland saline areas but is not considered to be a marine species is *Tamarix pentandra* (salt ceder); but even this species can be found in the backs of estuaries, where there is some freshwater influence to dilute the seawater.

Desert halophyte habitats lack the energy subsidy of the tides, hence rates of productivity tend to be lower than coastal systems and are often related directly to the amount of rainfall at a site. The Rainfall Use Efficiency (RUE) has been quantified for perennial shrub drylands as 0.25–0.40 g dry matter m^{-2} for each millimeter of annual rainfall. *Atriplex* stands often exceed this value owing to the presence of saline groundwater, which they exploit for growth beyond what is supported by rainfall alone. They may achieve standing biomass values of 20 tons ha^{-1} and have dry matter productivity rates of 300–500 g m^{-2} yr^{-1} even in hyperarid zones, which are lower than coastal systems but higher than surrounding nonhalophyte desert communities (Fig. 9).

Aside from their salinity tolerance, desert halophytes often have adaptations to heat, drought, and grazing pressure that are absent from coastal species. Many species have the C$_4$ photosynthetic pathway, which not only leads to greater water use efficiency but enables photosynthesis to proceed at higher temperatures than does the C$_3$ pathway. Some *Atriplex* species can carry out photosynthesis above 40°C. Salt secretions via bladders, trichomes, or glands on the leaves serve a dual function of eliminating excess salts from the apoplast (mostly cell walls) of leaves and creating a reflective, silvery coating to control heat absorption by the epidermis. Desert halophyte seeds may have inhibitors that must be leached away before they can germinate, an adaptation that ensures they will not germinate until they have been exposed to a sufficiently heavy rain to support growth. The "seed bank" of desert saline soil is often surprisingly high. When samples of these soils are moistened, they may be found to contain several thousand viable seeds per square meter of ground surface.

Many desert species have concentrations of saponins, tannins, or salts in their leaf and stem tissues that discourage grazing animals. Halophytes are generally regarded as a "reserve browse" on rangelands, which the animals will eat only after more palatable grass species are gone. *Atriplex* shrubs are valuable in maintaining grazing animals through the dry season in the rangelands of North Africa, Australia, and North America. Even the succulent halophytes may be eaten when no other browse is available, though their high salt and water content limits their energy value.

Like the coastal systems, dryland plant communities including halophytes are undergoing degradation from human activity. Forty-three percent of the Earth's land surface is arid or semiarid. The United Nations Environment Program Panel on Desertification estimates that 70% of the drylands have already undergone moderate to severe desertification and further desertification is proceeding at a rate of 3.5% yr^{-1}. Salinization, soil erosion, and loss of soil fertility are the effects of destructive land use practices such as irrigating without proper drainage, overgrazing, clear-cutting shrublands for firewood, and burning crop residues rather than returning them to the soil. Population pressure is perhaps the chief underlying cause of exploitative land use practices in the drylands. Population growth in the dryland areas (except Australia) has occurred at a rate of 2.5–4.0% yr^{-1} over the past 30 years, approximately the same as the rate of desertification.

Though some natural halophyte communities in the drylands have been disappearing, new halophyte communities have arisen on land salinized by human activity. The development of saline seeps in Western Australian wheat land illustrates how secondary salinization occurs. The first contributing factor was the clearing of the native shrubs and grasses from the low lands to plant wheat. A wheat crop transpires less water over a year than the native vegetation, hence the water table under the fields began to rise, bringing up salt-laden groundwater into the root zone and creating saline, wet spots in the fields where crops would not grow. The second contributing factor was clear-cutting of the *Eucalyptus* forests on the slopes surrounding the wheat lands (the so-called intake regions). As the uplands lost the trees that transpire water arriving as rainfall,

FIGURE 9 A stand of desert halophytes growing in a saline seep in the coastal Sonoran Desert, showing the saltbushes *Atriplex lentiformis* (large plants) and *Atriplex barclayana* (smaller plants).

excess water infiltrated into deeper soil layers and ultimately resurfaced in low spots in the farmers' wheat fields, aggravating the problems of waterlogging and salinity.

The saline seeps support diverse halophyte communities of *Atriplex, Mareana,* and other Chenopodiaceae shrubs. After concluding that it was too expensive to attempt to reclaim the seeps for wheat production, under the leadership of Clive Malcolm of the Western Australia Department of Agriculture, the farmers learned to utilize the seeps for sheep browse in the dry season. Today, the halophyte communities in Western Australian wheat fields are regarded as valuable resources even as the underlying problems are attacked through programs to reforest the highlands. Seeding and transplanting machines have been built to facilitate the establishment of halophytes in the seeps, and commercial subcontractors specialize in planting halophytes on farmland. This knowledge is being spread to other countries, such as Pakistan, which have similar problems.

Halophytes also grow within salinized irrigation districts, where river salts have accumulated over years of excess irrigation and inadequate drainage. They are often seen in fallow fields or in low spots and at the tailwater end of irrigated fields where salts accumulate. Formerly regarded as pest species in such places as the San Joaquin Valley of California, farmers are now deliberately planting halophytes on their salinized soils and irrigating them with saline drainage water from their crop fields. The halophytes reduce the amount of drain water that must be disposed, help absorb toxic elements from the drain water, and can be used as a source of animal feed. Desert saltbushes have also become important landscape plants in many arid regions because of their low water requirements and the attractive, silvery appearance. Many nurseries in arid regions of the United States now carry *Atriplex* and other xerohalophytes.

IV. SALT TOLERANCE AND ECOPHYSIOLOGY OF HALOPHYTES

Understanding how halophytes tolerate salts is important to understanding how they are distributed and the limitations to their growth. Furthermore, because 30% of the world's irrigated area has become too saline for normal crop growth, considerable effort has been invested in trying to improve the salt tolerance of cultivated plants through selective breeding. Unfortunately, progress has been slow, in part because the genetic factors that impart salt tolerance to a plant are not well understood. Finding out how halophytes handle salts and applying the findings to crop plants is an area of active research in plant physiology.

One of the earliest observations about salt stress on plants was that it is a form of "physiological drought." Salty water tends to draw water out of a less-saline solution across a semipermeable membrane (such as a cell membrane). Hence, if the external solution is saltier than the root, stem, or leaf cells exposed to that solution, water will flow out of the cells, leading to their dehydration and death, similar to the effects of drought on a plant. The essential similarity between drought and salinity stress is seen by the fact that xerohalophytes such as *Atriplex* handle the two stress factors by the same set of adaptations.

The basic method by which plants or any other organism adjusts to osmotic stress is to increase the osmotic potential of the cell sap through the accumulation of salts and/or organic compounds, so that water continues to flow inward. Halophilic bacteria accumulate large amounts of salts in their cytoplasm to counteract the external salt level; their enzyme systems are specially adapted to operate under high salt levels. Single-celled algae such as *Dunaliella,* on the other hand, accumulate large amounts of organic compounds such as glycerol as the internal osmoticum.

Halophytes use both methods of osmotic adjustment at the cellular level. They accumulate large amounts of salts, chiefly NaCl, in their cell vacuoles, which creates positive turgor pressure to keep the cell expanded. However, they accumulate normal metabolites that are overproduced, such as glycine-betaine, in their cytoplasm to balance the vacuolar salt level. The organic compounds are called "compatible osmotica." Some of them appear to function in nitrogen storage as well as osmotic adjustment. The enzyme systems of halophytes do not appear to be specially adapted to operate under high salinities, and high salt levels are confined to the vacuoles, which are considered a "nonliving" portion of the cell.

Because they are emergent plants, halophytes face a special problem compared to organisms that are surrounded by saline solution. Land plants must sustain a constant flow of water through the plant in order to grow. They typically transpire 200–500 g of water from their leaf surfaces for each gram of dry matter they produce in photosynthesis (this loss of water is an unavoidable consequence of opening leaf stomata to admit CO_2 for photosynthesis). If the water available in the soil contains salts and if these were taken up passively in the transpiration stream, the aboveground tissues would quickly accumulate far more salt than the plant could handle. For example, if the external solution contained just 1% NaCl and 500 g of water was transpired per gram of new growth, then 5 g of NaCl would be accumulated for each gram of organic dry matter production—the plant would be mostly salt.

To prevent this occurrence, all halophytes are, foremost, efficient excluders of salts, leaving almost all the salt behind in the soil while admitting water, and taking into the plant only the amount of salt that is needed for osmotic adjustment. Even when the external solution is highly saline (up to 1 M NaCl), the salt content in the xylem sap is low (10–20 mM NaCl). This small amount of salt is accumulated in the leaf cell vacuoles to maintain a positive flow of water into the leaf cells. Any excess salts in the xylem stream that enter the apoplast can be excreted from the leaf by special salt glands or bladders. Alternatively, succulent halophytes can dilute or concentrate salts depending on osmotic needs by gaining or losing water from the leaf cells. A simplified, schematic view of the flow of water and salt through a halophyte is shown in Fig. 10.

Although the basic mechanisms of salt tolerance appear to be similar among halophytes, they differ markedly in their secondary adaptations to salinity. For example, halophytes differ in their ability to germinate on saline solution. Seeds of most species undergo induced dormancy when exposed to highly saline solutions, but they retain the ability to germinate when exposed to freshwater conditions.

On the other hand, a few species, notably the annual succulents that must rapidly colonize exposed areas of salt marsh each spring, germinate readily on full-strength seawater. Irwin Ungar has compiled a comprehensive review of ecophysiological studies on the different adaptive strategies of halophytes (see Bibliography).

Even though many of the hypotheses about salt tolerance remain to be tested, it is generally believed that high salt tolerance does not depend on just one or a few genes but involves many adaptations at all levels of the plant, and that functional salt tolerance is an attribute of whole plants rather than discrete cells, tissues, or organs. Hence, one approach to increasing the range of salinity under which crops can be grown involves domesticating halophytes rather than transferring salt tolerance to conventional crop plants.

V. HALOPHYTE AGRONOMICS AND UTILIZATION

Although not normally considered food plants, some halophytes have actually been utilized by indigenous people for many centuries. The Seri Indians of the coastal Sonoran Desert harvested the seeds of eelgrass (*Zostera marina*) to make bread, and the Cocopahs harvested the grain of the salt grass *Distichlis palmeri* each summer in the Colorado Delta as a main source of summer subsistence (see article by Richard Felger in Bibliography).

Since the 1960s, efforts have been under way to domesticate halophytes as new crops for food, fiber, and animal feed production on saline soils. Several hundred thousand hectares of saline desert soil have been planted with *Atriplex* in North Africa, for example, to provide additional browse for sheep. Experiments first undertaken in Israel and now in the United States, Mexico, and Saudi Arabia have attempted to use seawater for direct crop production. Several dozen halophyte species have been screened for productivity under seawater irrigation, and the most productive yield as much dry matter as conventional crops under freshwater irrigation (Table III).

Given the high productivity of halophytes even on seawater, efforts have been expanded to de-

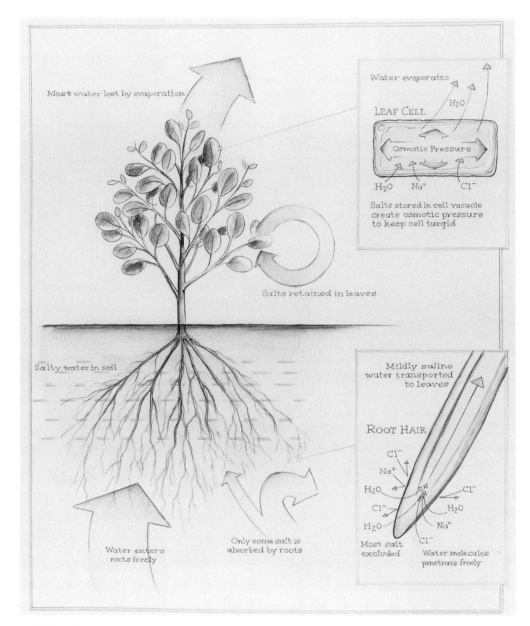

FIGURE 10   Schematic view of how halophytes handle salt and water flow through the plant. [From *EPRI Journal*, by permission.]

velop useful products from halophytes. Despite their salt content, halophytes can be used to replace conventional forage materials in ruminant diets at up to 50% inclusion level. *Atriplex* plants have received the most attention as potential new crops, but the succulent, annual salt marsh halophyte *Salicornia bigelovii* may be the first commercial halophyte species to be grown on seawater (Fig. 11). It is an oilseed crop, yielding 1.5–2.5 tons ha^{-1} of seed and 15–20 tons ha^{-1} of straw on seawater, similar to the seed and biomass yields of soybeans on fresh water. The seed is 30% oil and 35% protein and, unlike the straw, is low in salt. The oil is rich in the polyunsaturated fat linoleic acid, similar to safflower oil; the meal left after pressing out the oil is 42% protein and can be used in animal diets as a protein supplement. The straw can be fed to sheep, goats, or

TABLE III
Productivity of Halophytes Grown on Seawater in a Coastal Desert Environment[a]

Species	Plant type	Usable product	Yield
Atriplex lentiformis	Saltbush	Forage	1794
Batis maritima	Succulent	Forage	1738
Atriplex linearis	Saltbush	Forage	1723
Distichlis palmeri	Salt grass	Forage	1364
Salicornia bigelovii	Succulent		
Straw		Forage	2021
Seed		Oilseed	203

[a] Data were collected on an experimental seawater farm at Puerto Penasco, Sonora, Mexico, from 1981 to 1993. Yield values are in g dry weight m^{-2} yr^{-1}.

cattle at 30% of their diet or, after partially desalting by soaking and pressing, it can be fed as 50% of their diet. After 10 years of experimental plantings, *S. bigelovii* is now being grown on seawater in Mexico and Saudi Arabia on several hundred ha prepatory to commercial production.

Recently, an International Halophyte Network has been proposed to coordinate the many projects utilizing halophyte germplasm for food, fuel, feed, and restoration work. The conclusion reached by several research groups around the world is that, when properly prepared and utilized, halophytes can be as useful as conventional feed ingredients in animal diets, despite the fact that animals may not prefer halophytes as browse plants in a range setting. Halophytes have also been contemplated as

biomass crops to sequester carbon from the atmosphere, similar to the proposal to use trees to sequester carbon. The advantage of halophytes is that they can be grown on the vast areas of saline soil and along coastal deserts that are not presently used for forestry or food production. Hence, the importance of halophytes in human affairs is apt to increase in the coming decades.

Glossary

Euhalophyte *(eu=*true, Greek) Most salt-tolerant halophytes that exhibit growth stimulation by NaCl and inhabit hypersaline environments such as seawater marshes and salt flats.

Glycophyte *(glyco=*sweet, Greek) Nonhalophytes that grow preferentially in fresh water, although some can tolerate mildly saline soils and water supplies.

Halophyte *(halo=*salt, *phyte=*plant, Greek) Salt-tolerant higher plants.

Hygrohalophyte *(hugros=*moist, Greek) Halophytes that are only found in permanently moist soil or standing water; they often are especially adapted to anaerobic soil conditions and have aerenchyma tissue to transport oxygen to the root zone.

Mangal Entire ecosystem associated with a mangrove forest, including the mangroves and their associated flora and fauna.

Miohalophyte *(meion=*less, Greek) Halophytes that grow in brackish rather than full seawater habitats; they show their best growth in fresh water but can tolerate highly saline water to various degrees.

Xerohalophyte *(xero=*dry, Greek) Desert halophytes, typified by the genus *Atriplex,* adapted for drought as well as salt tolerance.

Bibliography

Felger, R. S. (1979). Ancient crops for the 21st century. *In* "New Agricultural Crops" (G. Ritchie, ed.). Boulder, Colo.: Westview Press.

Glenn, E. P., O'Leary, J. W., Watson, M. C., Thompson, T. L., and Kuehl, R. O. (1991). *Salicornia bigelovii* Torr.: An oilseed halophyte for seawater irrigation. *Science* **251,** 1065–1067.

Le Houerou, H. N. (1986). Salt tolerant plants of economic value in the Mediterranean Basin. *Reclamation and Revegetation Research* **5,** 319–341.

Lieth, H., and Al Masoom, A., eds. (1993). "Towards the Rational Use of High Salinity Tolerant Plants. Volume 1: Deliberations About High Salinity Tolerant Plants and Ecosystems. Volume 2: Agriculture and Forestry under

FIGURE 11 Harvest of the oilseed halophyte *Salicornia Bigelovii* at Kino Bay, Mexico.

Marginal Soil Water Conditions," Dordrecht, The Netherlands: Kluwer Academic.

Reimold, R. J., and Queen, W. H. (1974). "Ecology of Halophytes." New York: Academic Press.

Squires, V. R., and Ayoub, A., eds. (1994). "Halophytes as a Resource for Livestock and for Rehabilitation of Degraded Land," Dordrecht, The Netherlands: Kluwer Academic.

Thompson, J. D. (1991). The biology of an invasive plant. *BioScience* **41,** 393–401.

Ungar, I. (1991). "Ecophysiology of Vascular Halophytes." Boca Raton, Fla.: CRC Press.

Territoriality

Judy Stamps

University of California

I. INTRODUCTION

Over 50 years ago, Noble defined a territory as "any defended area." Since then, this definition has been refined and elaborated, but not by much. These days, most workers focus on space that is defended for a long period relative to an animal's life span, in order to differentiate territory defense from the defense of ephemeral resources such as food items or basking sites. In addition, the initial assumption that a territory is defended against any member of the same species has been relaxed to include species in which defense is directed at particular age, sex, or size class of conspecifics. For instance, male house mice defend exclusive territories against other dominant males, but permit adult females, subordinate males, and juveniles to share their territory. Territory defense may even be directed at selected members of other species, as in algae-eating reef fish that defend their territories against other species with similar feeding habits.

Similarly, although territorial defense was originally defined in terms of aggressive acts such as chases or fights, it is now acknowledged that other types of behavior patterns may also reduce the tendency of intruders to enter a territory. Conspicuous advertisement signals such as cricket songs, lizard push-ups, or rabbit scent marks fall into this category. More subtle forms of territory defense include changes in space use patterns that increase an owner's ability to detect intruders, e.g., the use of elevated song posts or perches, or regular patrols along the boundaries of the territory.

II. TERRITORIAL DIVERSITY

Early workers were impressed by the variety and diversity of territorial behavior within and among taxonomic groups. To date, territorial behavior has been described in a bewildering array of animals, including cnidarians, limpets, insects, crustaceans, spiders, fish, frogs, lizards, mammals, and birds. Considerable variation often exists even within a taxon in territory size and the type of activities that occur within the confines of the defended area. For example, many birds defend "all purpose" territories, in which an adult male and female raise offspring, forage, and survive the onslaughts of inclement weather, predators, and disease. However, other types of territories are also common in birds. Herring gulls raise their young within small breeding territories but forage elsewhere; rufous hummingbirds defend areas containing flowers during a stopover on the autumn migration; and male sage grouse defend small territories in clusters called leks, which females visit to choose a mating partner before leaving to raise the offspring on their own. [*See* BIRD COMMUNITIES.]

There is also considerable variation within and among taxa in the age, sex, and number of owner–defenders per territory. In species with extensive parental care, dependent offspring often share their parent's territory, as in ants, squirrels, and songbirds. However, juvenile territoriality has been reported in a variety of insects, fish, lizards, and other groups in which individuals are self-sufficient from the time of hatching. Territory defense by breeding males has received the most attention from field workers, but female territoriality is also widespread, as is territory defense by non-breeding adults of either sex. Finally, there are a surprising number of species in which two or more owners share and defend the same territory. For instance, shared defense by a male–female pair is common not only in birds, but also in other socially monogamous territorial species such as gibbons, butterfly fish, and prairie voles. Group territories defended by multiple owners have been observed in a variety of species, including wolves, parrotfish, and ants.

III. ECONOMIC APPROACHES TO TERRITORIALITY

Early descriptive studies of territoriality emphasized variation in territorial behavior among animals, but this emphasis on diversity understandably discouraged workers from trying to formulate global models that might apply to a wide range of species. A swing back toward generality was made in the late 1960s and early 1970s with the development of economic models of territorial behavior and habitat selection. Although originally designed with birds in mind, the economic approach has been applied with considerable success to territorial animals from many different taxa.

Economic models of territory defense assume that the defense of space is in fact a surrogate for the defense of resources located within that space, where "resource" can be loosely defined as any environmental factor that enhances growth, sur-

vival, or reproduction and that is in short supply relative to the number of potential users. At the simplest level, animals may defend resources required for their own use, for example, a food supply or a refuge from potential predators. Many animals also defend resources required for the growth and survival of their offspring, and in red squirrels, offspring receive a portion of their mother's territory upon independence. More circuitous relationships between resources and reproductive success have been observed in other species, for example, when male dragonflies defend territories containing resources required by females, and hence gain access to mates that they would not otherwise enjoy.

The economic approach to territoriality assumes that individuals defend territories when the benefits of territory defense exceed the costs, where benefits and costs are both measured in terms of lifetime reproductive success. However, since it is usually impossible to measure lifetime reproductive success under field conditions, most workers rely on indirect indices of the benefits and costs of territorial behavior. For instance, the benefits of a feeding territory might be estimated by food density or quality, or the benefits of a breeding territory by the number of independent young produced per territory. Along the same lines, estimates of the cost of territory defense typically rely on indirect indices such as the number or intensity of attacks directed at intruders or neighbors, or the rate at which owners produce territorial advertisement signals.

The economic approach has been most successful in predicting territorial behavior in simple systems, in which important resources can be easily quantified and in which owners alter their behavior in response to short-term changes in resource levels or defense costs. Many now classic studies of territorial behavior focused on nectivorous birds during the nonbreeding season, in which flowers and their energy contents could be readily measured and in which territory owners were prepared to change their territory sizes on a daily basis. However, simple economic models have been less successful in predicting the behavior of animals that defend "all purpose" territories over longer periods of time.

Such animals frequently do not alter territory sizes in response to short-term changes in resource levels, leading to suggestions that the defense of long-term territories with multiple resources might involve different types of costs and benefits than apply to the short-term defense of a single resource.

IV. PROXIMATE MECHANISMS

At the same time as the economic approach was applied to "ultimate" questions about the function and evolution of territorial behavior, other workers have been studying the proximate mechanisms that are involved in territory acquisition and maintenance. This literature also reveals remarkable convergence in patterns of territorial behavior across a wide range of taxonomic groups. One of the most striking of these general phenomena is the "prior residency advantage," which refers to the fact that the owner virtually always wins when competing with intruders for possession of a territory. Although superior size, strength, or weaponry usually determine success in contests if both opponents are new to the area in question, owners typically win disputes even with newcomers larger, older, or stronger than themselves. Similarly, if two owners both become familiar with the same area prior to encountering one another, contests are often exceptionally prolonged and vicious. The parallels between the prior residency advantage in animals and the "home court advantage" in human sports and warfare are intriguing, but the proximate and ultimate reasons for both of these phenomena are still poorly understood.

Another widely observed pattern in territorial species is the "dear enemy" phenomenon. Although owners are usually quite aggressive to unfamiliar intruders, they are often less so to familiar neighbors. Interactions between long-term neighbors sometimes have almost amiable overtones, involving the exchange of territorial advertisement signals over a mutual territorial boundary instead of the vigorous chases or fights typical of interactions between owners and intruders, or of the period of territorial establishment. Some avian neighbors even cooperate in expelling intruders or warning one another about the presence of intruders or predators in the neighborhood. These observations and recent theoretical studies imply that territorial behavior is not exclusively competitive and aggressive, and suggest that territory owners might benefit by living near other owners within territorial neighborhoods.

V. NEW DIRECTIONS

Currently, researchers are attempting to combine proximate and ultimate approaches to the study of territorial behavior. For instance, laboratory and field studies show that the prior residency advantage and the dear enemy effect both require time to develop. Biologists working at the proximate level are examining neural and endocrinological processes that might contribute to the development of these characteristic behavior patterns. At the ultimate level, the prior residency advantage and the dear enemy effect imply that familiarity with an area and its inhabitants might increase the value of space to prospective territory owners and help explain some previously puzzling aspects of territorial behavior. For example, if animals require time to become familiar with an area and establish social relationships with neighbors, then one would not expect owners to lightly abandon or switch territories in response to short-term fluctuations in resource levels. Other workers are examining widely held assumptions about the factors that determine success in territory acquisition and maintenance. While modelers have assumed that prowess, size, or strength determine success in territorial contests, recent field studies suggest that persistence and luck also play an important role. Indeed, some lizards win space without ever winning a fight, by persisting in an area until the previous owner gives it up.

Another topic of current interest is the relationship between territoriality and conservation biology. Many endangered vertebrates are territorial, and the success of many conservation programs hinges on the ability of humans to induce these species to live in areas designated as refuges or reserves.

However, unless these attempts are based on an understanding of how animals normally choose habitats and establish territories, they will be doomed to failure. One recent suggestion is that potential settlers in territorial species may be attracted instead of repelled by the advertisement signals produced by established territory owners, implying that behavioral "decoys" might be used to coax settlers to establish territories in vacant but suitable habitats. This is but one example in which information about the processes and consequences of territorial behavior may help shield some of the earth's fauna from the consequences of human overpopulation and exploitation.

Glossary

Prior residency advantage A phenomenon in which residents usually win aggressive contests with intruders.

Resident An animal that lives in a territory and defends it against neighbors or intruders.

Resource An environmental factor that increases individual growth, survival, or reproduction and that is in short supply relative to the number of potential users.

Territory An area defended against members of the same or other species.

Bibliography

Alcock, J. (1993). "Animal Behavior," 5th Ed. Sunderland, MA: Sinauer.

Carter, C. S., and Getz, L. L. (1993). Monogamy and the prairie vole. *Sci. Am.* June, 100–106.

Davies, N. B., and Houston, A. I. (1984). Territory economics. *In* "Behavioral Ecology: An Evolutionary Approach" (J. R. Krebs and N. B. Davies, eds.), pp. 148–169. Sunderland, MA: Sinauer.

Reed, J. M., and Dobson, A. P. (1993). Behavioural constraints and conservation biology: conspecific attraction and recruitment. *Trends Ecol. Evol.* **8**, 253–256.

Stamps, J. A. (1994). Territorial behavior: Testing the assumptions. *Adv. Study Behav.* **23**, 173–232.

Traditional Conservation Practices

Madhav Gadgil
Indian Institute of Science

I. Optimal Foraging
II. Deliberate Restraints
III. Refugia
IV. Keystone Species
V. Other Practices
VI. Prospects

The history of biodiversity on earth over the last 50,000 years or so, dominated by modern *Homo sapiens* with symbolic languages, is one of gradual encroachment of humans and their artifacts on the biosphere, leading to a steady depletion of natural biological diversity. But history is not an entirely monotonous process of destruction of nature at the hands of humans. Although humans have more often than not liquidated the capital stock of biological resources, reducing *biodiversity*, they have exhibited deliberate restraints on resource *harvests* that have promoted the *sustainable use* of biological resources and the *conservation* of biodiversity in many different places and times. Such restraints have often become part of the cultural traditions of prescientific societies.

I. OPTIMAL FORAGING

Humans harvesting natural populations would be expected to concentrate on species that yield the most returns per unit of time or the amount of physical effort or that involve minimal risk. Only when they cannot fulfill their nutritional or other resource needs with the most desirable of such species do they turn to others less desirable. Indeed, *optimal foraging,* the branch of ecological theory that addresses how animals garner resources, suggests that hunter-gatherers should arrange potential resource species in a sequence, grading them on the basis of returns per unit effort (or time or risk), utilizing only as many at the top of the sequence as necessary to fulfill their total requirements. Large flightless birds such as moas are apt to have headed such a list, perhaps along with large fleshy fruit like durian or jackfruit, whereas rats or grass seeds might be further down on the list. As a result, humans would harvest the most desirable items first, shifting to less desirable ones only when the more desirable options were exhausted.

Such a process would cause the *extinctions* of several plant and animal species when humans first colonized any land mass. The most favored prey species, such as large herbivores, would be especially liable for *overexploitation*. Indeed there is a strong suggestion that the *first colonizers* of Americas drove a whole set of such species, for example, mammoths and horses, to extinction. Similar processes are believed to have taken place on islands such as Madagascar where larger species of lemurs were wiped out and Pacific islands with the extermination of giant flightless birds. Subsequent colonizations by technologically more advanced people were also accompanied by further waves of overexploitation. Thus the European colonization of the Americas led to waves of overhunting or overfishing proceeding from east to west. [*See* EVOLUTION AND EXTINCTION.]

II. DELIBERATE RESTRAINTS

Humans are not always so profligate. Once an area is settled, humans begin to realize over generations how their harvests and their manipulations of the environment have an impact on the resource populations. They are able to perceive resource depletions that follow nonsustainable harvesting pressures and recoveries when, for some reason, harvesting pressures are reduced. Furthermore, when living in a given locality generation after generation, they might perceive self-interest in moderating harvesting efforts so that harvests are improved in the long run. This would be particularly so if they are largely dependent on harvests from a circumscribed area and are using technologies that do not lead to higher and higher levels of harvesting efficiencies. Many scall-scale, autonomous *hunter-gatherer* or *horticulturalist societies* settled in a region where they could control well-defined terrtories from which most resources were garnered fulfill these conditions. Such people may then have had the knowledge base, the motivation, and the capability of moderating resource harvests so as to serve long-term group interests in sustaining and conserving natural diversity.

III. REFUGIA

Examples of a variety of such practices of moderating harvesting efforts are known. These include establishing *refugia* immune from any harvests. These refugia may take the form of *sacred groves,* ponds, or pools along river courses or lagoons of coral islands where entire living communities are totally protected from any human interference. In 1880, Dietrich Brandis, the first Inspector General of Forests of India noted that the entire subcontinent was covered by a network of sacred groves

representing every vegetation type. Remnants of this network persist to this day and serve to conserve, for instance, rain forest species that have disappeared from the human-impacted vegetation on the Western Ghats and the West Coast of India. Thus, the Botanical Survey of India discovered a new species in the first Indian record of a genus of a leguminous climber, *Kunstleria keralensis,* in a sacred grove on the thickly populated coastal areas of the state of Kerala. In addition, the only surviving population of a freshwater turtle, *Trionyx nigricans,* survives in a sacred pond dedicated to a moslem saint in Bangladesh. [*See* COMMUNITY-BASED MANAGEMENT OF COMMON PROPERTY RESOURCES.]

IV. KEYSTONE SPECIES

The use of restraints may protect *keystone species* that support the existence of a range of other species. Thus, fig trees belonging to the genus *Ficus* are recognized as important sources of fleshy fruits that are available in seasons when no other species produce fleshy fruits in the tropical forest communities. *Ficus* thereby promotes the existence of a number of insect, bird, bat, and primate species for which they serve as a critical resource in a pinch period. All species of *Ficus* are now widely protected as sacred trees through much of tropical Asia and Africa, where the local communities seem perfectly aware of its ecological role. [*See* KEYSTONE SPECIES.]

V. OTHER PRACTICES

The restraints also include protecting *critical life history stages* that are especially vulnerable to overharvest. Thus, phasepardhis, a nomadic hunting tribe of Western India, have the tradition of letting loose any pregnant does or fawns of antelope or deer caught in their snares, whereas egrets, stocks, herons, pelicans, ibises, and cormorants at their colonial nesting grounds are given immunity from hunting over most of India, although these birds are considered fair game in the nonbreeding season. In the New Guinea highlands the hunting of birds of paradise may be suspended when there are signs of population depletion. Indeed it has been remarked that Pacific Islanders have traditionally employed every conservation practice invented by modern fishery managers.

VI. PROSPECTS

Such traditional conservation practices have tended to be rejected as superstitious by modern organized religions like Christianity and as primitive and unscientific by modern resource managers. There is, however, a resurgence of interest in and a move toward respecting these conservation practices. This is, for instance, reflected in specific provisions of the International Convention on Biological Diversity. [See CONSERVATION AGREEMENTS, INTERNATIONAL.]

Glossary

Conservation practice Deliberate restraint on harvesting levels promoting long-term persistence of a resource population.

Keystone species Species whose persistence in a community promotes the occurrence of a large number of other species.

Optimal foraging Patterns of harvest that permit the animal to maximize its short-term gains from resource consumption.

Refugium A locality in which the population of a species is completely immune from being harvested.

Bibliography

Berkes, F., ed. (1989). "Common Property Resources: Ecology and Community Based Sustainable Development." London: Belhaven Press.

Gadgil, M., Berkes, F., and Folke, C. (1993). Indigenous knowledge for biodiversity conservation. *Ambio* **22**(2–3), 151–156.

Johannes, R. E. (1978). Traditional conservation methods in Oceania and their demise. *Annu. Rev. Ecol. Syst.* **9**, 349–364.

Kellert, S. R., and Wilson, E. O., eds. (1993). "The Biophilia Hypothesis." Washington, D. C: Island Press.

Oldfield, M. L., and Alcorn, J. B., eds. (1991). "Biodiversity: Culture, Conservation, and Ecodevelopment." Colorado/Oxford: Westview Press.

Warren, D. M., Slikkerveer, L. J., and Brokensha, D., eds. (1994). "Indigenous Knowledge Systems: The Cultural Dimension of Development." Zed London: Zed Books, (in press).

Tragedy of the Commons

Garrett Hardin

University of California

When an unmanaged natural resource is used in common, each user is rewarded for seeking maximum, personal gain, while commonizing and postponing costs, resulting in behavior that appears individually "rational," but which will eventually bring ruin to all who depend on that resource. It is important to recognize the difference between a managed and an unmanaged common, and that a common can be managed in many ways—formally, informally, effectively, or not. A global common will eventually bring ruin worldwide.

I. INTRODUCTION

Philosophers of science, whether professional or amateur, frequently sound the call for more interdisciplinary investigations. They have a good point: in this day of excessive specialization, many fascinating questions are likely to fall between the cracks. But the difficulties of publishing interdisciplinary findings must not be underestimated. Faced with more meritorious solicitations than their journal has room for, referees and editors may unconsciously seize upon any excuse for rejecting a candidate paper. On the other side, the author, foreseeing difficulty in placing his material, may choose a nontraditional form of publication at the outset, thus shielding his ego from a possible rejection. Something of this sort may have happened when W. F. Lloyd was considering how to publicize his theory of the commons and its application to population problems. In any event he published his theory in 1833 as part of a series of lectures given at Oxford. The volume seems to have excited no notice whatever. For more than a century it was cited only rarely. In 1953—120 years after the original publication—a massive review of the population literature dismissed Lloyd's theory in only 43 words, and then it got the thrust of the argument wrong by 180 degrees! Not until 1968 was the theory developed at length, and published in *Science,* a favorable platform. Within a very few years specialists were including the 1968 paper in anthologies in the fields of economics, sociology, philosophy, ethics, public health, political science, and the law, not to mention ecology and the environmental sciences. It seems unlikely that the idea will ever again be lost sight of.

It must be admitted that the intellectual climate of the 19th century was not favorable for the reception of the basic idea of the unmanaged commons. After Adam Smith published "The Wealth of Nations" in 1776, laissez faire, "do your own thing," as we might say now, became the prevailing spirit in influential scholarly quarters. Optimism was the order of the day and was made the reason for a policy of noninterventionism in public matters. Let each man pursue his own interest, economists said, and the interests of all will be best served in the long run.

II. THE COMMONS

Lloyd refuted the optimism of laissez faire. Whenever a number of herdsmen turn their cattle loose in a pasture that is jointly owned, the common will soon be ruined. Why? Because the pasture has a limited carrying capacity (to use a modern term), and each herdsman gets the full benefit of adding to his herd, while the disbenefits arising from overexploitation of the resource (e.g., soil erosion) are shared by all the herdsmen. Fractional losses are not enough to deter aggressive cattle owners, so all the exploiters suffer in a common, or what today is called an *unmanaged* common. [*See* COMMUNITY-BASED MANAGEMENT OF COMMON PROPERTY RESOURCES.]

III. ALTERNATIVES TO COMMONS

Alternatives to the unmanaged commons can be classified under two headings. In privatism, the resource is subdivided into many private properties. Each owner is responsible for the management of his plot: those who manage well, prosper; those who manage poorly, suffer. In socialism, the resource is "common property," but the property owners ("the people") appoint a manager to control its exploitation. Theoretically, an incompetent manager can be fired. In practice, when "the people" is a nation of many millions, it is all too easy for empowered managers (bureaucrats) to survive by hiding their mistakes.

Both privatism and socialism can either succeed or fail. But, except in the smallest of communities, *commonism* cannot succeed. An unmanaged common fails because it rewards individual exploiters for making the wrong decisions—wrong for the group as a whole, and wrong for themselves, in

the long run. Freedom in the commons does *not* produce a stable prosperity. This is Lloyd's revolutionary point. Popular prophets, intoxicated by laissez faire, simply could not hear Lloyd.

IV. APPARENT EXCEPTIONS

Apparent exceptions to the theory need to be accounted for. When a resource is present in abundance, an unmanaged common may actually be the most efficient. The general rule of "freedom of the seas" has, for centuries, led to the economical exploitation of oceanic fisheries. Today, of course, overfishing has depleted everything from whales to sardines. Even the "wide blue sky," something that once seemed limitless, having been treated as a common by ever larger numbers of people, has now been degraded to the point of "greenhouse" warming and ozone thinning.

A second "scale effect" must be kept in mind. People of the Hutterite faith in northwestern United States and adjacent Canada live by the Christian ideal that (ironically) Karl Marx expressed best: "From each according to his abilities, to each according to his needs." Farms are owned in common; everybody is supposed to pitch in and do his share of the work, while taking out no more than a fair share of the products. Conscientious Hutterites make a nominally unmanaged common work, but only so long as the operational unit is less than 150 people. As the number approaches this, more and more commune members shirk their tasks. (Perhaps we should say a community below 150 really *is* managed—managed by conscience.)

If such devout and hard-working people cannot make an unmanaged common work, there is no reason to think that anyone can. It is a long way from 150 to the millions that make up a modern nation. Scale effect rules out the unmanaged commons as an important political possibility in the modern world. Modern nations are a changeable hodgepodge of socialism and privatism.

V. RECOGNIZING A MANAGED COMMON

Some ecologists have failed to see subtle signs of management in traditional societies. For instance, the survival of the Turkana people in Africa under a system of common ownership of grazing has been cited as an instance of the success of commonism. Yet the same observers have conceded that access to resources was effectively controlled by the elders of the tribe. Such a managed common presents no problem to the theory of unmanaged commons.

VI. GLOBAL COMMONS

The survival of today's industrialized nations is now threatened by a commonization under another name. "A world without borders" is really a poetically named unmanaged commons. Marx's political recipe, "To each according to his needs," implies that *needs create rights*. Such rights can be fulfilled on a global scale only if national borders are effectively liquidated. The resulting poverty will accelerate the destruction of environmental wealth. Gresham's law of economics "bad money drives out good" has, under a global system of laissez faire, its equivalent in the environmental sphere: "Low environmental standards drive out high." If there are no borders, poverty will displace wealth all over the globe.

"To each according to his needs" is an immensely seductive phrase to religious people, but in a world without national population controls it is a sure recipe for disaster. Those who are really concerned with the environment—concerned with the well-being of posterity—must give the carrying capacity of the environment precedence over discontinuous human needs, however much these needs may tug at our heartstrings. Of every impulse to globalize wealth the ecologist must ask his ultimate question,

"And then what?" What happens after globalized wealth degenerates into globalized poverty? What happens then to the environment for which posterity will hold us responsible?

Glossary

Carrying capacity Maximum population of a species of organisms that can be sustainably maintained over a long period of time without degradation of the territory.

Freedom of the seas A poetic assertion that the oceans are an unmanaged common.

Managed common A territory, the ownership of which is vested in the entire community, which appoints managers (bureaucrats) to manage it; socialism.

Open access system Synonym for an unmanaged common.

Scale effects What follows whenever the ratio of two (or more) variables changes with a change in size or number.

Unmanaged common Property vested in the community as a whole, which exercises no control over access to it.

Bibliography

Abernethy, V. D. (1993). "Population Politics: The Choices that Shape Our Future." New York: Plenum Press.

Berkes, F., ed. (1989). "Common Property Resources: Ecology and Community-Based Sustainable Development." London: Belhaven Press.

Bromley, D. W. (1992). "Making the Commons Work: Theory, Practice and Policy." San Francisco: Institute for Contemporary Studies.

Hardin, G. (1993). "Living within Limits: Ecology, Economics, and Population Taboos." Oxford: Oxford Univ. Press.

Ostrom, E. (1990). "Governing the Commons: The Evolution of Institutions for Collective Actions." Cambridge: Cambridge Univ. Press.

Stevenson, G. G. (1991). "Common Property Economics: A General Theory and Land Use Applications." Cambridge: Cambridge Univ. Press.

Tundra, Arctic and Sub-Arctic

R. L. Jefferies

University of Toronto

"Tundra" which is derived from the Finnish work "tunturi" (treeless heights) includes vegetation types that occur in the Arctic (and alpine regions) beyond the climatic limit of trees. The vegetation types range from tall shrub to dwarf shrub, heath to forb, graminoid, moss and lichen communities. These communities grade into polar semidesert in the extreme north where the plant cover dominated by cushion plants, mosses, and lichens is low (<50%). The term "tundra" has two meanings. It refers to the tundra zone, the biome north of the taiga forests of Eurasia and North America as just indicated. This definition, which is used in this article, includes all vegetation types within the zone. The second definition, which is more specific, includes only those types of plant communities that are not found in other biomes. This more restricted definition excludes azonal plant communities, such as mires, and patches of forest distributed in the tundra. The generally accepted definition of the Arctic among terrestrial biologists is the land beyond the climatic limit of trees. This definition is broadly equivalent to that of the tundra zone, but climatologists, oceanographers, and geographers do use other definitions.

I. THE NORTHERN TUNDRA ZONE

The circumpolar Arctic occupies approximately 2,560,000 km² in Russia and the Scandinavian countries, 2,167,000 km² in Greenland and Iceland, 360,000 km² in Alaska, and 2,480,000 km² in Canada. These areas include both low and high arctic tundra and polar semideserts and deserts (see below). The geomorphologies of the Eurasian and North American arctic regions are dissimilar. The unbroken land mass as far north as 78°N in Eurasia is characterized by vegetation zones based on a strong northward gradient in temperature, floristic composition, and plant life form. In contrast, no unbroken land mass extends to 78°N in North America. This, coupled with the effects of glaciation, the presence of mountain ranges, and the islands of the Canadian Arctic Archipelago, have produced a vegetational mosaic reflecting patterns of geomorphology, temperature, and precipitation across the region. [*See* NORTHERN POLAR ECOSYSTEMS.]

Despite these differences between the two continents in geomorphology and climatic patterns, there are similarities in the vegetation, and an integrated classification scheme for arctic vegetation in North America and Eurasia has been proposed in which the Arctic is divided into the Low and High Arctic with subdivisions within each major division (Tables I and II).

II. VEGETATION TYPES OF LOW AND HIGH ARCTIC

A. Forest Tundra

The increasingly shallower and colder active layer of the soil as one moves northward in the region of continuous permafrost (see later) provides a rooting zone that is less favorable for tree growth. Forests become open and trees grow shorter and are restricted to sheltered river valleys. The loss of *Picea* (spruce), *Larix* (larch), and *Betula* (birch) marks the transition from forest tundra to tundra.

B. Low Arctic: Shrub Tundra

Beyond the forest tundra are shrub tundra communities dominated by *Betula nana, B. glandulosa, Salix glauca, S. pulchra, S. lanata,* and *Alnus fruticosa* which grow from 2 to 5 m. The ground vegetation includes *Eriophorum vaginatum,* dwarf heath shrubs (*Ledum palustre, Vaccinium vitis-idaea, V. uliginosum, Empetrum hermophroditum,* and *Cassiope tetragona* that are up to 25 cm high), and an abundance of lichens and mosses. The tundra

in central and eastern Canada contains little shrub tundra, probably because of low snow cover and abrasive winter winds. A large number of boreal species are present in these southern low arctic tundra communities.

C. Low Arctic: Sedge–Dwarf Shrub Tundras

These northern, low arctic tundras are marked by the loss of many boreal species. Most shrubs mentioned in Section II,B are restricted to favorable localities. With the exception of *C. tetragona* and *V. vitis-idaea,* many of the dwarf heath shrubs become rare in the northern sections of the Low Arctic. The plant communities contain numerous grass and forb species and abundant lichens including *Cladina arbuscula, C. rangiferina,* and *C. gracilis.* The main mosses are *Tomenthypnum nitens, Hylocomium splendens,* and *Aulacomnium turgidum.*

D. Low Arctic: Tussock–Dwarf Shrub Tundras

Tundras dominated by *E. vaginatum* tussocks together with the sedges *Carex lugens* and *C. consimilis,* dwarf shrub and heath species, mosses, and lichens occupy land north of the sedge–dwarf shrub tundra in Alaska, the Yukon Territory, and in western Siberia. Tussock tundra is more limited in the central and eastern Canadian Arctic, probably because of the presence of thin soils of low nutrient status.

TABLE I
Ecological Comparisons of the High Arctic and the Low Arctic[a]

(1)	A shorter growing season (2 to 2.5 months vs 3 to 4 months).
(2)	July mean temperature 4 to 8°C vs 4 to 11°C and degree days above 0°C (150 to 600°C vs 300 to 900°C).
(3)	Eight species of land mammals vs 10 to 15 species.
(4)	Peary's caribou vs barren-ground caribou, and a greater importance of musk-ox.
(5)	Ten to 20 nesting species of birds vs 30 to 60 species.
(6)	A flora of 50 to 115 species of vascular plants compared to 600 species in the Low Arctic.
(7)	A greater number of bryophytes with only a few species of *Sphagnum* present in the High Arctic.
(8)	A major shift in plant communities in the High Arctic to those dominated by cushion, rosette, and graminoid plants. Plant cover of vascular plants is 30% or less.

[a] Modified from L. C. Bliss (1988).

TABLE II

North American Classification		Eurasian Classification[a]	
High Arctic	Polar desert (herb–cryptogam) Polar semidesert (cryptogam–herb, cushion plant–cryptogam, mire)	Polar desert zone	Polar desert (cryptogam–herb) Arctic tundra (dwarf shrub–herb)
		Tundra zone	Typical tundra (sedge–dwarf shrub, polygonal mires)
Low Arctic	Tundra (low shrub–sedge, tussock–dwarf shrub, mire)		Southern tundra (low shrub–sedge, tussock–dwarf shrub, mire)
Taiga	Forest–tundra Taiga	Taiga	Forest–tundra Taiga

[a] After Bliss and Matveyeva (1992; see Chapin *et al.*, 1992).

E. Low Arctic: Mires

Wetland plant communities are common on the flat coastal plains of the arctic mainland coast. The dominant sedges are *Carex aquatilis, C. rariflora, C. chordorrhiza, Eriophorum angustifolium,* and *E. scheuchzeri.* Grasses include *Dupontia fisheri, Arctophila fulva,* and *Actagrostis latifolia.* In shallow ponds, *Menyanthes trifoliata, Equisetum limnosum,* and *Potentilla palustris* are common. These sedge-dominated mires also occur to a more limited extent in the central and eastern Canadian Arctic and in Greenland. Across the Russian Federation mires are common, dominated by similar species as those just listed and by various mosses, including species of *Drepanocladus, Meesia, Calliergon,* and *Polytrichum.* On large peat hillocks, dense thickets of *Betula nana* or *B. excelis* are common. [*See* WETLANDS ECOLOGY.]

F. High Arctic

In North America a major shift in the structure of arctic vegetation as well as floristics occurs northward from mainland Canada to the Islands of the Canadian Arctic Archipelago. The low shrub, dwarf heath shrubs, and cotton grass tussock tundra communities are replaced by cushion plants (*Dryas integrifolia, Saxifraga oppositifolia*), prostrate shrubs of *Salix arctica,* and rosette species of *Saxifraga, Draba,* and *Minuartia* as well as by cryptogams, although these low arctic communities may persist in protected areas. In the Russian Arctic the changes are more gradual and involve subtle shifts in the species composition of communities northward.

G. High Arctic: Mires

High arctic mire communities are structurally and floristically related to those in the Low Arctic. These wetlands are a major feeding habitat for musk-oxen, caribou, and bird populations. *Carex stans* is abundant together with *Eriophorium scheuchzeri, E. triste,* and *Dupontia fisheri.* Comparable mire communities occur in the Russian Federation and are similar to those of the Low Arctic, except that *B. nana, S. pulchra,* and *V. vitis-idaea* are absent.

H. High Arctic: Polar Semideserts

Mats of *Dryas, Salix arctica,* and clumps of *Carex rupestris* and associated species of *Draba* and *Saxifraga* are common in the southern Canadian arctic islands and in Greenland, particularly on well-drained neutral to alkaline soils. Plant cover is discontinuous (50% or less). In ice-wedge depressions, *C. tetragonia* is often present. In northern Russia, graminoids (*Alopecuris alpinus, Deschampsia borealis,* and *Luzula confusa*) account for 5% cover, willows (*Salix polaris, S. arctica*) for 15%, and cryptogams (*Hylocomium, Aulacomnium,* and *Tomenthypnum*) species for 30–40%.

I. High Arctic: Cryptogam–Herb Communities

On sandy to clay loam soils in the central and western Queen Elizabeth Islands, a species-rich cryptogam–herb community is present. Bryophytes and lichens are abundant together with species of *Draba, Minuartia, Saxifraga, Luzula,* and *Alo-*

pecurus alpinus. Similar diverse plant communities occur at Cape Chelyuskin, Franz Josef-Land Archipelago, and Severnaya Zemlya Archipelago in the Russian Federation where plant cover can reach 60%. The diversity of species and the predominance of mosses and lichens result in much more vegetation than the barren polar deserts of the islands of the Canadian Arctic Archipelago.

J. High Arctic: Polar Deserts

Large areas of the Queen Elizabeth Islands are devoid or nearly devoid of plants. Low summer temperatures, a short growing season, and dry soils all act to restrict plant growth. At most sites vascular plants contribute more to plant cover and species richness than do cryptogams. Vascular plant cover is about 2% and the corresponding value for lichens and mosses is about 1%. The most common vascular plants include species of *Draba*, *Saxifraga*, and *Minuartia* as well as *Papaver radicatum* and *Puccinellia angustata*. Cryptogams include species of *Lecanora*, *Dermatocarpon*, and *Thammolia*. Within these barren landscapes, small areas of richer vegetation exist in snowflush zones. Russian polar deserts are restricted to sections of northern islands in the Arctic Ocean where the mean temperature of the warmest month is about 2°C. Russian biologists usually use a more general definition of polar deserts which includes polar semideserts than their North American counterparts.

III. ORIGINS OF THE TUNDRA FLORA AND FAUNA

Although the arctic vascular flora can be considered a single unit (892 species in 230 genera and 66 families), approximately 45% of all vascular species also occur in alpine regions and the distribution of species within the Arctic is uneven. At some localities the flora consists overwhelmingly of nonvascular plant species (ca. 70%). Based on paleoecological evidence (where available), a number of tundra habitats, such as the coastal tundra of the Alaskan North Slope and tussock tundra in Alaska, apparently have changed little in the last 5000–6000 years.

In relation to other biomes the tundra is young, it developed in the early Pleistocene, although tundra faunas and floras apparently appeared in late Miocene or early Pliocene (~18–15 million yr B.P.). In the late Tertiary period, deciduous and mixed deciduous and coniferous forests (*Juglans*, *Tilia*, *Carya*, *Corylus*, *Larix*, *Picea*, *Pinus*, *Tsuga*) occurred to 71°N, and on Axel Heiberg Island, NWT, Canada, coniferous forests of *Picea*, *Pinus*, *Tsuga*, *Larix*, and *Thuya* were present. Tundra floras and faunas probably evolved in the highlands of central Asia and the Rocky Mountains in early Pliocene but circumpolar arctic and alpine tundras did not develop until the Pleistocene. Much of the Arctic and most alpine landscapes were glaciated in the Wisconsin (Weichselian) glaciation. However, recent studies suggest that the contemporary Arctic flora may have been able to derive at least some of its genetic inheritance from populations of plants that survived in small isolated communities in ice-free refugia at least during the latter stages of the Weichselian glaciation. The region of Beringia (the North Slope of Alaska, the northern Yukon Territory, western Banks Island, and eastern Siberia) escaped glaciation. The phytogeographical evidence indicates that the flora spread out from these centers as well as north from the Rocky Mountains with the retreat of the ice at the end of the Wisconsin glaciation. In Beringia, around 25,000 to 15,000 B.P., herbaceous tundra vegetation with a low percentage of plant cover existed. The flora was dominated by grasses, sedges, and forbs, but large areas of cotton grass and dwarf shrub heath were of minor extent or absent at that time. In the northern Yukon at about 15,000 to 12,000 B.P., major changes occurred in the vegetation from herb tundra to closed/open woodland at low elevations and to herb tundra in the uplands. Associated with these changes in the vegetation were changes in the late Pleistocene fauna which consisted of woolly mammoth, horse, Dall sheep, caribou, bison, camel, musk-ox, lion, and elk. The species were replaced or partially replaced by caribou and moose. The extinctions and reductions in species of large mammals probably reflected climatic changes that brought about changes in vegetation and an increase in hunting pressure from human populations. These topics are an active area of research.

IV. ANIMAL POPULATIONS

Despite the limited primary production (see later) of tundra ecosystems, they can sustain large populations of mammalian and avian grazers, such as musk-oxen, reindeer/caribou, lemmings, voles, and greater and lesser snow geese. The impact of herbivory on these tundra ecosystems may be considerable. Although herbivores in many ecosystems consume only about 10% of aboveground primary production, lemmings and lesser snow geese can consume between 50 and 90%. Thus in some tundra grazing systems, such as coastal salt marshes and the wet sedge meadows of the Alaskan coastal plain, vertebrate herbivores, assisted by their gut decomposers, replace free-living decomposers as the primary consumers and reducers of primary production.

A. Musk-ox (*Ovibos moschatus*)

Largely as a result of hunting, musk-oxen (*O. moschatus*) in historical times were restricted to high arctic oases and coastal plains in arctic North America where plant production was high. Because animals have been translocated and the range has expanded, the animals are present in other areas as well today. At the southern extremity of their range their diet is primarily willows but graminoid species become increasingly important further north. In summer, musk-oxen feed extensively in sedge meadows where plant production exceeds the forage consumed. In winter the animals feed on dormant plant tissues ("hayed-off" grasses/sedges) and leafless willow twigs. The nutritional quality of these "hayed-off" graminoids is higher than that of comparable plants from temperate regions. The unavailability of this low quality forage ("hayed-off" graminoids) in years of deep snow results in increased mortality and females failing to reproduce. The animals cannot travel far in deep snow. Thus density-dependent and independent processes interact to determine population size.

The reproductive cycle is strongly linked to forage availability. In most areas young are born several weeks before the period of plant growth. Milk production by cows during early lactation is sustained by body reserves. Ovulation and conception depend heavily on body condition before the rut at the end of summer. From January to March, the metabolic rate and growth are at seasonal lows.

B. Caribou and Reindeer (*Rangifer tarandus*)

The various subspecies of *R. tarandus* (reindeer/caribou) occupy habitats from the taiga to the high arctic islands in North America and Eurasia. Latitudinal differences in winter and summer diets indicate that the animals forage selectively. Populations of *Rangifer* from the High Arctic and northern Greenland feed on graminoids, woody plants, forbs, and lichens. Seasonal migration routes are short and the animals can survive without lichens as a dominant part of their diet, in contrast to caribou and reindeer in the Low Arctic, which may migrate up to 1000 km from tundra to taiga where they forage for terrestrial lichens. The animals are able to dig through snow (<60 cm) in search of forage, unless it is hard. Caribou have narrow mouths and are selective feeders, unlike musk-oxen which are bulk feeders. Rates of forage intake in summer depend on the growth habit of plants and their abundance. Willows and sedges are important components of the diet and the animals are less able to process food high in fiber, such as dried sedges and grasses, compared to musk-oxen. In addition, food items such as mushrooms are important components of diet. Animal performance (daily weight gain) is strongly influenced by temporal and spatial grazing patterns, selection of plants high in nutrients, and avoidance of plants rich in secondary compounds (e.g., dwarfshrub plants). The metabolic rate of the animals falls in winter, but the lichens on which the animals feed are rich in energy sources but low in protein. For *Rangifer*, as for *Ovibos*, the carrying capacity is probably largely determined by the availability of winter forage and winter range conditions. The animals change their migration routes to exploit seasonal differences in forage availability. The recovery of tundra vegetation after heavy exploitation by reindeer can take 15 to 20 years (e.g., lichen pasture). Precisely when animals return to a particular site is not known, the return time may be on the order of years, as for willow ptarmigan (*Lagopus lagopus*), which would allow the vegetation to recover.

C. Arctic Rodents

The three groups of rodents in the Arctic are lemmings (*Dicrostonyx, Lemmus*), voles (*Microtus*), and ground squirrels (*Spermophilus*). Some of these microtine rodents show cyclic fluctuations in population density, particularly in more northern latitudes which are probably a consequence of the nutritional quality of available forage. Lemmings and voles show strong dietary preferences. In northern Alaska, the brown lemming (*L. sibiricus*) feeds on monocotyledons and mosses, the collared lemming (*D. groenlandicus*) eats the leaves of dicotyledonous shrubs, and the tundra vole (*M. oeconomus*) feeds on both monocotyledons and dicotyledons. Arctic rodents in other areas show similar preferences. Ground squirrels prefer herbaceous dicotyledons, particularly legumes and plants with a high water content; in winter the animals hibernate.

Satisfactory evidence that densities of arctic rodents are related to the quality of available forage is difficult to obtain. General levels of microtine herbivore populations may be set by food availability. Sudden declines in numbers of microtines may be primarily caused by resource limitation and the buildup of secondary plant compounds, but declines can be exacerbated by increased predation when numbers of rodents are high. An adequate depth of winter snow is also needed for the survival of these animals. Small herbivores with a high reproductive potential are able to take advantage of a single favorable season so that their numbers increase temporarily beyond the system's capacity to support sudden population surges with catastrophic consequences.

D. Lesser Snow Geese (*Anser caerulescens caerulescens*) and Other Tundra-Breeding Birds

Large numbers of different species of duck and geese breed in the coastal areas of the Arctic in early summer and migrate southward in early fall. The growth of goslings is rapid. For example, goslings of lesser snow geese grow from 80 g at hatch to 1500 g in less than 7 weeks, with the adults regaining the loss of body weight (≤40%) associated with egg laying and incubation. In some coastal salt marsh habitats, the rapid gains in body weight are achieved because the geese modulate nitrogen flow in a system that is nitrogen-limited. The geese are colonial and they produce droppings approximately once every 4 min while feeding, effectively accelerating the nitrogen cycle by providing available nitrogen for plant growth. The intense grazing by birds in the colony results in the consumption of up to 90% of the aboveground primary production. Cyanobacteria colonize the surface of bare sediments among graminoid shoots and fix atmospheric nitrogen which replaces most of the nitrogen incorporated into the geese.

Increases in goose populations largely as a result of changing agricultural and conservation practices in both winter and spring staging areas are leading to the destruction of coastal marshes. Grubbing for roots and rhizomes in early spring before the onset of aboveground growth destroys the summer grazing pasture. In addition, increased evaporation from damaged swards leads to the buildup of hypersaline conditions which further checks the growth of plants and the availability of forage. This destruction of coastal vegetation not only has a detrimental effect on snow goose populations but also has adverse effects on populations of shore birds and ducks, such as widgeon (*Anas americana*) which are grazers.

Arctic birds may be classified into two groups, the first group is more or less confined to this particular environment while the second comprises species that breed in tundra habitats during the summer but spend the winter elsewhere. Winter residents include willow and rock ptarmigan, snowy owl, raven, gyrfalcon, and arctic redpoll. The predominant feature of tundra bird communities is the large number of wader species, all of which are entirely or primarily insect predators. These include golden plover, knot, sanderling, dunlin, and many other species of sandpipers. In addition, there are a significant number of species of waterfowl, particularly in coastal lowlands. Passerines as a group are relatively unimportant. Most of the species in the latter group that nest in the tundra are boreal forest and forest–tundra species. However, there are a number of characteristic arctic passerines such as the Lapland bunting and the snow bunting. The former species may reach high nesting densities (100 km^{-2}). There are a number of avian predators that prey on rodents, shore birds, and eggs of shore birds and waterfowl, including the different jaeger (skua) species, short-eared owl, gyrfalcon, and species of gulls.

E. Invertebrates

Insects are the dominant group of invertebrate herbivores at tundra sites, although in general the density, standing crop, and number of invertebrate species are low.

Lepidopterans are widely distributed throughout the Arctic. Among the mayflies (Ephemeroptera), dragonflies (Odonata), and stoneflies (Piecoptera) there are few arctic endemics and no species reaches the High Arctic. The aphids (Aphididae) are the best represented family with about 20 species. In the Arctic the number of generations of different forms of these insects is reduced and the alternation of host plants is eliminated. Beetles (Coleoptera) are underrepresented in the fauna. The Diptera (true flies) are well represented: 270 species in 39 families of which 20% are found in the High Arctic. The crane flies (Tipulidae) have larvae that feed on organic detritus. Of the mosquitoes (Culicidae), few are true tundra species. Adult females can mature a small batch of eggs autogenously (no protein meal). Most blackflies (Simuliidae) do not extend their ranges far onto the tundra, and nearly all species of blackflies are also autogenous in these northern habitats. Nonbiting insect midges make up over half of the insect diversity in these regions. The life spans of individuals are very long (almost 2–3 years, a few as along as 6 years), although the adults may live only a few weeks. The tundra invertebrates are highly selective in their choice of food plants; deciduous shrubs appear to be the preferred food source (especially willows) whereas graminoids and evergreens rank the lowest. The low preference for some species probably reflects the large amounts of fiber, silica, or secondary com-

pounds in their leaves as well as the low nutrient content of the tissues. Within soils nematodes are abundant (ca. 10^6 m^{-2}), although their biomass rarely is greater than 1 g dry wt m^{-2}. In the High Arctic, however, the biomass of soil invertebrates per unit area far exceeds that of other terrestrial animals.

F. Carnivores

The gray wolf probably has the greatest natural range of any living mammal except humans. The size of the animal combined with its habit of traveling in packs enables it to feed on larger species of prey. When wolves and ungulates inhabit the same range, ungulates are usually the main prey but predation is generally selective. Very young, old, or sick animals are taken. Wolf populations have suffered periodic persecution. For example, the total wolf population in Siberia in 1973 was estimated at only 3000 animals.

Wolf packs in North America tend to remain in a home range; the maximum densities of wolves occur where there are high densities of caribou.

The arctic fox is virtually restricted to the arctic biome. During winter the animals disperse widely with many moving out on the pack ice and following polar bears to feed on the remains of their kills. The arctic fox is dimorphic in winter; some individuals are white, whereas others are dark, bluish gray. In parts of Greenland, 90% of animals may be of the blue phase, whereas in Siberia the corresponding percentage is only 3 to 4%. The density of dens is about one den per 30–50 km^{-2} on average, although much higher densities have been recorded. Pups are born in late spring after a gestation period of 50–57 days. In any 1 year only about one-third of young females (1–2 years) breed, whereas about 85% of the older females breed. In inland habitats, lemmings and small mammals are the main source of diet, but coastal animals eat fish, birds and their eggs (particularly geese), and sea mammals.

The grizzly bear is rare in the Russian Federation and it occurs mainly on the Chukotka peninsula. Its range extends across Alaska and Canada almost to the Hudson Bay. The northern tundra regions constitute the northern limit of its range.

During winter dormancy the drop in body temperature of bears is about 5°C and the animals do not feed, urinate, or defecate but utilize fat reserves laid down before entering the den, usually in October. Breeding and production of young are related to the nutritional status and weight of females instead of age. The average age when young are first produced is about 8 years and litter sizes range from one to three. Mortality of cubs is high, about 46%. Grizzly bears require large expanses of tundra with considerable habitat diversity. They are omnivorous and eat a wide selection of plants, rodents, fish, and carrion. The home range of male bears is about 700 km² and of females with cubs about 300 km² in the Brooks Range in Alaska.

Polar bears, which are a marine mammal, are found up to 300 km offshore from land masses and they have been recorded as far north as 88°N. The worldwide population of bears is about 25,000, which is divided into a number of more or less discrete populations. Pregnant female bears come ashore in summer, and in fall construct maternity dens in peat and snow banks in arctic coastal areas which include northwest and northeast Greenland, Wrangel Island, eastern Svalbard, Franz Josef Land, and Cape Churchill, Manitoba. After the mother and cubs leave the den in early spring, they move onto sea ice where their diet consists of ringed and bearded seals. In summer when coastal pack ice melts in many areas, all bears spend these months ashore. At this time they are omnivorous and their diet includes seaweed, berries, twigs and leaves, eggs, moulting birds, and carcasses washed up on the shore. When the seas freeze the bears return to the sea ice.

V. ECOLOGICAL PROCESSES IN TUNDRA ECOSYSTEMS

A. The Physical Environment

The climate of the Arctic is characterized by a near continuous cover of snow and ice for 8 to 10 months of the year and a short, largely snow-free growing season of continuous light of 2 to 4 months. Both the vegetation and the arctic air masses respond to net radiation which varies between 670 and 800 mJ m⁻² year⁻¹ in the Low Arctic and 200 to 400 mJ m⁻² year⁻¹ in the High Arctic. In the far north of Greenland or Ellesmere Island it falls close to zero. Much of the growth of vegetation occurs after summer solstice when daily totals of photosynthetically-active radiation are declining.

Winter snow cover averages 20 to 50 cm over upland areas, whereas in low-lying areas it may reach 5 m in depth. Although most snow disappears by mid- to late June, snow banks may persist until August. There is also a strong correlation between shrub height and snow depth; tall shrubs are confined to river bottoms and similar sites where snow depth is up to 5 m or more.

The term "permafrost" refers to soil, water, and rock that remains below 0°C for more than 2 years. The depth of the permafrost in the High Arctic is 400 to 700 m, whereas it is about 500–600 m deep in the region of the Mackenzie River Delta and 400 m along the Alaskan coastal plain. Arctic soils thaw to a depth of 3 m in summer depending on the soil texture (the so-called active layer is 10–50 cm in peats and fine-textured soils; up to 3 m in depth in gravels and sands). In August, September, and October refreezing occurs from the top down and the bottom up in the active soil layer. Water trapped under pressure is forced to the surface, forming soil frost boils.

Permafrost and the many freeze–thaw cycles of the upper few centimeters of soil ($\approx$20) that occur annually produce a number of surface features, including sorted and nonsorted stone and gravel circles on flat terrain and stone and gravel stripes where there is a slope (>3 to 5°). Polygons, earth hummocks, and solifluction steps are also characteristic landscape features. Each feature, which may be from 10 cm to 10 m or more in diameter, affects soil development and plant distribution. Soil hummocks (30 cm high) probably result from intense soil churning during the spring and fall freeze–thaw cycles. Needle ice also may form in fine textured soils in spring and fall, causing the lifting of surface soils which may inhibit seedling establishment.

B. Soils

Soils are less developed than in temperate regions. Reduced plant cover, a short growing season, the presence of permafrost, and the churning of some soils all restrict soil development. Chemical weathering processes are very slow and soil-forming processes proceed at greatly reduced rates compared to those in temperate latitudes. The process of podzolization is limited in the Arctic to well-drained soils with a deep active layer. Where dwarf heath shrubs grow, a weak podsol may develop that is associated with the slow decomposition of plant litter and the buildup of humic acids. Soils that develop under poorly drained lowlands remain saturated throughout the summer and peat accumulates where rates of decomposition are low and the soils are constantly wet. The surface layers are often acidic. Where some drainage occurs, the B mineral horizon is often gray colored or mottled with iron and aluminium (gleying). Within the High Arctic, rates of soil-forming processes are further reduced. The limited sedge–moss or grass–moss communities have thin peats that overlie these gleyed soils in poorly drained sites. Slopes containing cushion plant–*Dryas*–cryptogam communities frequently have well-drained, carbonate soils (pH 7.5–8.5) with only a thin layer of organic matter; these rendzina soils are deficient in nitrogen and phosphorus. Arctic brown soils occur where lichen–*Dryas* communities grow. They produce a mull-type humus in which the parent mineral matter, rich in iron and clay, is mixed with organic matter. They are characteristic of regions where igneous rocks are present. [*See* SOIL ECOSYSTEMS.]

C. Plant Community Dynamics

Plant successional processes are not a conspicuous feature of tundra landscapes. Species, such as different graminoids that are characteristic of early successional habitats created as a result of disturbance, are also present in late successional plant communities. The time scales for the establishment of plant communities vary from the order of 100 years in the High Arctic to 25 years in the Low Arctic. In riverine communities, pioneer herb communities are initially established but are replaced by willows and alder (*Alnus crispa*). Away from river banks where the soil active layer is shallower and the soils colder, cotton grass–dwarf shrub heath tundra and wet sedge tundra dominate. On coastal sand dunes, *Elymus arenarius* and *Salix* species colonize the dunes and are followed by *D. integrifolia* and *Arctostaphylos rubra*. In time, a lichen–heath tundra develops on dry sites and in wet areas *Dupontia fisheri, Eriophorum angustifolium, E. scheuchzeri,* and *Carex aquatilis* are common. Thaw lakes in the Alaskan coastal plain may drain as a result of physical processes which initiate a sequence of vegetational changes involving the replacement of *Arctophila fulva* by *C. aquatilis* and then vice versa as the thaw ponds reestablish over a period of 1000 years.

Other disturbances include fire burns. Recovery of cotton grass–dwarf shrub heath and low shrub vegetation is rapid following a fire. The heavy flowering of cotton grass is associated with the rapid sprouting of shrubs. In some sites seedlings of graminoid plants also appear where there is a persistent seed bank. In contrast, heath species, lichens, and mosses recover more slowly. Much of this flowering and regrowth is associated with nutrient release (N and P) from the burn. At the tree line, fire may eliminate tree seedlings, unless subsequent years are warm, thereby causing the tundra to advance south. Russian scientists call this pyrogenic tundra. [*See* FIRE ECOLOGY.]

The effects of abrasion by windblown snow on woody plant growth at the tree line can be dramatic. The present position of the tree line is, in part, determined by soil conditions as discussed earlier and, in part, by these adverse effects on tree growth. The snow consists of small hard, sharp crystals that are extremely abrasive when driven by winter gales. Stunted trees frequently possess many branches within 30–50 cm off the ground. Above this height, up to 1 to 2 m, the trees are devoid of branches as a result of snow and ice abrasion. If the tree or shrub grows beyond this height branches can develop. The lower branches are protected by the soft snow of early winter. In addition there may be a layer of these branches in the soft snow and peaty substratum from which daughter vertical shoots develop to produce a stand of plants that act as a windbreak and snow trap. This stunted tree growth with a basal layering of branches and the main vertical axis devoid of branches up to 2 m is known as Krummholtz growth.

D. Standing Crop and Plant Production

Standing crop and net annual primary production are greatest in tall shrub communities. Total net annual production is between 250 and 400 g m^{-2} if allowance is made for root production. The aboveground standing crop of low shrub tundra communities, cotton grass–dwarf shrub heath, and wet sedge–moss tundras, is similar; net annual primary production is in the range of 125 to 250 g m^{-2}. Cryptogams in these communities make a significant contribution to standing crop and primary production (<25%).

In High Arctic plant communities there are marked decreases in both standing crop and net annual primary production compared to corresponding values for the Low Arctic. The shorter growing season, reduced degree days, colder soils, and less available nitrogen, phosphorus, and water all limit plant growth. Cushion plants have a large amount of dead material within the plant cushion. In cryptogam–herb communities, vascular plants contribute little to the aboveground biomass. Plant production is significantly higher in wet meadow communities where water is more readily available. In addition, cyanobacteria actively fix atmospheric nitrogen in these communities, thereby contributing to the nitrogen economy of these meadows.

Within the Low Arctic the same constraints limit net primary production but the limitations of the different resources are less severe.

E. Life Histories of Plants: Reproduction, Germination, and Plant Establishment

Generally, arctic plants are long-lived and the annual habit is very rare (e.g., *Koenigia islandica*) or is not represented in most communities. Genetic variation based on morphological or ecophysio-logical differentiation of populations is widespread among arctic plant species, many of which are widely distributed over large geographical areas. In the Low Arctic many perennials of widespread genera (*Carex, Puccinellia, Dupontia, Eriophorum, Petasites, Betula, Empetrum,* and *Salix*) undergo extensive clonal growth, although individual plants flower and set seed on occasion. Many species in the genera *Arnica, Hieracium, Poa, Potentilla,* and *Taraxacum* are apomitic (viable seeds are formed by asexual reproduction). Some grasses such as *Festuca vivpara* are viviparous, forming small vegetative bulbils within floral bracts in place of flowers. The bulbils root and produce new clones of parent plants. In most tundra habitats the establishment of plants by seed is a rare event. However, where disturbances occur the seedling establishment of graminoids is common. The warmer-exposed surfaces may facilitate germination, as the optimum temperature for germination of many arctic plants is comparatively high (15–30°C). Within the High Arctic rates of seed germination are particularly low. Germination is often restricted to bryophyte mats or to cracks in the soil surface, where water, nutrients, and the absence of needle ice provide suitable conditions for germination. Estimates of the size of the seed bank for various localities range between 400 and 3500 seeds m^{-2}, although in some communities the value may be much lower.

F. Plant Phenology

Seasonal patterns of plant growth and development are similar to those of temperate latitudes, except that the growing season is appreciably shorter. Leaf initiation and leaf expansion in deciduous forbs and shrubs are nearly synchronous because of buds that formed the previous year. In contrast, graminoid growth may be continuous throughout the growing season, especially in plants with an indeterminate growth system. The leaves of some species overwinter or remain green at the leaf bases. Studies indicate that the leaves of *Saxifraga oppositifolia* encased in ice maintain turgid green leaves not only throughout this period of anoxia but also on reexposure to air. Hence, such plants are tolerant of frequent freeze–thaw cycles. Growth is initiated as soon as the snow melts. Root growth often begins when the soils warm above 0°C. The temporal partitioning of root and shoot growth may reflect the partitioning of limiting resources (carbon and nutrients) at different stages during the growing season. The cushion and mat form of plant growth, including both live and attached dead leaves, creates a boundary layer so that temperatures of leaves within the boundary layer may be 10°C above ambient. In a similar manner the parabolic-shaped flowers of *Ranunculus* and *Papaver* result in the temperature of the center of the flower being 5–10°C above air temperature so that insect pollination is favored and the higher temperatures aid seed and fruit development. Daily growth rates of individual graminoid plants may be as high as those of temperate graminoids but the short season limits overall net primary production.

G. Physiological Characteristics of Plant Growth

Deciduous shrubs require large carbon gains each summer to produce a new leaf surface. This group of plants has relatively high photosynthetic rates (10–30 mg CO_2 dm^{-2} hr^{-1}) and high leaf conductances. In addition, rates of root growth, root respiration, and uptake of nitrogen and phosphorus are relatively high in species such as *S. pulchra* and *B. nana* compared to those of evergreen species. The latter have low photosynthesis and growth rates, water conductance rates are low, and the leaves remain alive for up to 5 years. Plants of many of these evergreen species produce large quantities of secondary metabolites (tannins, alkaloids) that deter herbivores.

The graminoids are the dominant growth form in large areas of the Arctic. They have relatively high photosynthetic rates (15–20 mg CO_2 dm^{-2} hr^{-1}), high rates of leaf conductance, and they can grow in soils that are often water saturated. As a group, the root biomass exceeds shoot biomass by two to three times. In *E. vaginatum*, net photosynthesis is low (2–3 mg CO_2 dm^{-2}hr^{-1}) and the entire root system is replaced annually. This represents a large energy drain, but individual tussocks may live up to 180 years. Most of the root development occurs in the organic layer of the soil. Cushion plants are also long-lived (100–150 years). Although photosynthetic rates are low, the leaves may function for 2 years or more but remain on the plant for many years before decomposing. There is a limited nutrient pool at the base of the shoots where most active roots are located. Root : shoot ratios are low in both cushion and rosette plants (ca. 0.6) in contrast to the graminoids mentioned earlier where the ratio may be 10 to 20.

In these arctic soils temperatures are only just above 0°C, except at the immediate soil surface. Rates of mineralization of soil organic matter as a consequence are very low. Rates of net nitrogen mineralization range between 0.02 and 0.6 g m^{-2} year^{-1}. Corresponding rates in boreal forest soils are one to two orders of magnitude higher. In addition to the low temperatures, dry soils may also inhibit bacterial decomposition processes. As a consequence the form of combined inorganic nitrogen most frequently found in soils is ammonium (ions). Nitrate does occur, especially late in the summer when soils are warmer and better drained. Plants appear able to utilize both inorganic forms, although the rates of uptake may be low. In addition, the mycorrhizal associations of Ericaceous plants secrete exopeptidases that break up soil proteinaceous material into peptidases and amino acids that are absorbed by the fungal hyphae, so that the nitrogen cycle is bypassed. It has been shown that sedges, such as *E. angustifolium,* which are devoid of mycorrhizal associations and which grow in cold water and saturated soils where decomposition processes are slow, also are capable of absorbing amino acids and peptides directly from the environment. Other mechanisms that enable these plants to conserve elements such as nitrogen and phosphorus include an efficient recycling of these elements (<80%) from senescing leaves. In addition, there is evidence that some of these plants (*Astragalus alpinus, Oxytropis* spp., *Dryas octopetala*) have nitrogen-fixing associations with microorganisms and that bryophyte communities contain free-living cyanobacteria (*Nostoc*) growing within the mat that fix nitrogen. Hence, there are a number of mechanisms that conserve the supply of essential elements to sustain the growth of these arctic plants during the limited snow-free season.

H. Plant–Animal Interactions: Herbivory

Both tundra and boreal forest herbivores are generalists in selecting forage species. They utilize plants from a limited flora and supplement their food supplies with animal foods (insects, eggs). There is no common pattern of diet selection across tundra regions for a given herbivore. As a result of their foraging activities, there is relatively little evidence that these herbivores do influence the plant species assemblages in the different communities in the long term or alter the trajectory of plant succession. However, they may have considerable influence on the vegetational composition in the short term at the local scale. Successional processes may be slowed, as in the case of lesser snow geese foraging on salt marsh swards; in the High Arctic, greater snow geese eat sedge rhizomes, slowing community development. The exceptions to these generalizations invariably involve the influence of human agencies that have directly or indirectly altered herbivore numbers in creating instability in plant communities, soil conditions, and the herbivore populations themselves. The relatively simple trophic ladders, as distinct from trophic webs, make these northern ecosystems particularly susceptible to such perturbations. At larger scales of space and time other types of disturbances, such as fire, wind, snow, and ice action, have a marked effect on both the rate and trajectory of plant succession, largely as a result of changes in nutrient availability. These disturbances are necessary triggers for subsequent changes in vegetation which, in turn, attract intense foraging from herbivores seeking forage of high nutritional quality. Abiotic disturbances at the landscape level are therefore often essential for intense foraging by herbivores at the community level.

I. Human Populations and Tundra Landscapes

The Inuit, Nenet, and Chukchi or their ancestors have occupied the region from Siberia to Greenland for at least 5000 years. They are highly evolved hunters who in the past have been highly dependent on the availability of arctic wildlife, unlike other hunter–gatherers elsewhere who rely on plants more than animals. Many of the hunting implements are unique to their cultures. In addition, the nomadic way of life, elaborate methods of food storage, versatile diets and sharing of meat at the community level, and methods of population control enabled these peoples to carry out a successful subsistence economy, primarily based on arctic coastal environments (harvesting of land and sea animal resources).

Since the 18th and 19th centuries when Europeans and North Americans made extensive links with the Inuit there have been considerable changes, as they have become increasingly tied to a cash economy. The subsistence economy has all but disappeared. The initial extensive contacts were made by the whaling industry and soon after trapping became an important component of the cash-orientated economy. The latter has since diminished, except in a few communities. Today, part-time or full-time wage employment and the sale of furs, skins, fish, and handicrafts provide the cash income for these people who now live in large settlements. Heavy corporate and individual income subsidies are often necessary to meet the requirements of an expanding population. Despite these changes the Inuit still are heavily dependent on the land and inshore areas for resources. The use of firearms has allowed them to increase local food supplies for villagers and dogs. Most of the food imports into the settlements from the south are carbohydrate-based foods. The extent to which local economies are based on returns from wildlife in one way or another is highly dependent on the availability of animal resources in the vicinity of villages. For example, such resources are few on Ellesmere Island compared to Banks Island, NWT, Canada. In addition, the use of snowmobiles to hunt has led to a decrease in dog teams, and it often requires a monetary subsidy to maintain and operate such machines in contrast to the animal packs. The expanding human population and the availability of modern equipment to harvest wildlife have put considerable strain on the ability of these northern terrestrial and aquatic animal populations to sustain numbers.

In short, the role of the top carnivore in these tundra ecosystems has changed dramatically since the 19th century with an ever increasing energy subsidy from southern sources. The full effects of these changes on these northern societies and on wildlife populations are not well understood. More comprehensive land-use regulations and satisfactory management policies are needed. It is hoped that the settlement of land claims and the establishment of Nations of the First Peoples (e.g., Nunavut in the Eastern Canadian Arctic) will lead to rapid changes in conservation and management.

VI. CONSERVATION AND RESOURCE MANAGEMENT OF TUNDRA ECOSYSTEMS

A. Northern Circumpolar Protected Areas

Protected areas range considerably in size from the Greenland National Park (70 million ha) and two Alaskan areas of over 7 million ha to 15 small bird sanctuaries on Svalbard that are only 19,000 ha in total area. The categories and objectives of protected areas differ from country to country. Of sites established for nature conservation there are 63 known protected areas within the Arctic Circle, each over a 1000 ha. There is also a wide variability in the size and number of reserves in each country. Canada and the Russian Federation have fewer areas protected than smaller Nordic countries. The total area of protected tundra communities is just above 1 million km^2, located in 24 reserves. Northern lands are a neglected domain and the establishment of protected areas has come rather late to these areas. In Canada, for example, where 40% of the land surface lies north of latitude 60, only 1.7% of this area is national park but other areas are afforded some protection that are not national parks. In Alaska, with the passing of the Alaska National Interest Lands Conservation Act in 1980, 28% of the land surface in the state is permanently protected. As in Canada, the number and size of protected areas in the Russian Federation are small compared to the size of the country. Currently, the parks of Laplandski, Kandalakshskii, and Taymyr and the Wrangel Island refuge are established in the North.

Analyses of protected areas indicate that in the Nearctic regions there are gaps in the reserve coverage in Kamchatka, Greenland, and the Canadian Archipelago based on biogeographical considerations. Additional areas of high and low arctic tundra and polar desert and ice caps require some form of protection. In all countries there are proposals to extend the system of national parks and protected areas as finances allow.

Systems of land classification indicating priorities in different countries are not easily comparable. For example, in addition to national parks in Canada, there are managed wildlife areas, bird sanctuaries, and protected cultural heritage sites, all of which are given some degree of protection. There are clearly many legitimate claims on land for resource exploitation of various kinds and more importantly as the homeland for the different peoples of the First Nations. Hence, any system of protected areas must be part of the integrated planning of the region as a whole.

B. Industrial Development

During the last three decades the fast expanding mineral, oil, and gas operations in tundra regions have increased concern about the disturbance to and destruction of these ecosystems. The construction of the Prudhoe oil field and the trans-Alaskan pipeline was a major event in arctic development and the ecological impact studies associated with its development have wide applicability for other developments within the tundra zone. The pipeline is 1250 km in length and the associated haul road stretches 577 km from the Yukon River to Prudhoe Bay. Much of the route is underlain by permafrost. Because of the need for an environmental impact statement, much research was done in an area that had previously been poorly studied. The two major ecological problems associated with an industrial development of this type are the effects of disturbance and the problem of the revegetation of disturbed sites.

Early exploration of the Arctic has resulted in many scars produced by tracked vehicles. Research on the effect of the haul road development showed that plant productivity increased in the vicinity of the road associated with greater nutrient availability. A change in species composition occurred, leading to a greater proportion of graminoids in the vegetation. The greater thaw depth and higher soil temperature do not appear to be primary factors increasing productivity. A large amount of dust was and is blown from the road by passing vehicles. Mosses and, to a lesser extent, lichens are most susceptible to this change in edaphic conditions. Other plants that are susceptible include *Cassiope tetragona* and *Lycopodium annotinum*.

Oil spills were common along the route. Over 16000 spills have been recorded, many of which were only a few gallons of spilled oil. Diesel spills have been the most damaging to the vegetation. The general conclusion reached in a number of studies was that although vegetation was damaged by oil, recovery of the native vegetation over time ensured that revegetation occurred. Much of the toxic low molecular weight fraction was highly volatile and was lost from sites in 48 hr. Again, mosses and lichens are susceptible to this toxicity, particularly in dry sites.

Revegetation of disturbed areas has been based on the use of fertilizers, mulch, and grass seed mixtures. The presence of an organic mat is essential for insulating the permafrost and for nutrient cycling. Terrain stabilization has been achieved but rehabitation with native plants has been slow, even after a number of years. Careful applications of fertilizer may allow the different species to grow, depending on their nutritional requirements.

The effect of disturbance on mammals was of considerable concern, particularly as the pipeline lay along the migration path of caribou (*R. tarandus*). Adult bulls were seen in the vicinity of the pipeline much more frequently than cows, and cows with calves were generally absent from the corridor. However, grazing animals were attracted to the fertilized revegetated areas. Moose appeared to be unaffected by the presence of the pipeline.

The early melt along the haul road has attracted migrating waterfowl before other areas become free of snow.

All of these results are preliminary and the full ecological impact can only be assessed after a long-term monitoring of the environment.

C. Other Examples of Industrial and Natural Pollution

Since the mid-1950s, there has been a marked increase in levels of air pollution in arctic regions. One example is arctic haze which is mainly composed of fine droplets of organic and inorganic sulfur compounds, including ammonium sulfate and bisulfate. It is most pronounced during January to April each year. Studies have indicated that the sulfur is derived from anthropogenic sources, most likely from Europe, rather than from local anthropogenic or biogenic sources.

There are a number of mines in tundra regions where metals are extracted from mineral deposits. However, there are sites where high concentrations of heavy metals occur naturally in soils and where tundra vegetation is fumigated with sulfur dioxide. These sites are on the arctic coast at the Smoking Hills, east of Tuktoyaktuk (NWT, Canada), and on the west coast of Greenland where lignite deposits in cliff faces have burnt continuously since Europeans first recorded the phenomenon. The fumigations deposit large amounts of heavy metals and sulfur dioxide on the tundra in the vicinity of the burn. The pH of the soil is well below 4, yet populations of *Artemisia* and *Arctophila* can grow under these extreme conditions. Water in nearby ponds may have a pH of less than 3, yet species of the water flea, *Daphnia,* can grow and reproduce under these conditions. This natural experiment has provided information on suitable plants to revegetate metal spoil tips.

D. Global Warming and Tundra Landscapes

Based on predictions of computer models, global warming due to the accumulation of greenhouse gases could lead to the polar amplification of annual mean air temperature by as much as 5°C as a result of a reduced energy requirement to melt snow and ice and the altered ground albedo. These increases in temperature could increase the depth of the soil active layer, prolong the growing season, increase rates of heterotrophic soil respiration and soil mineralization, and increase primary production. Which of these processes will increase most rapidly is uncertain. The tundra may become a sink for carbon dioxide if production increases more rapidly. Alternatively, if soil respiration increases more rapidly because of high rates of decomposition as the active layer deepens, the tundra may become a source of carbon dioxide, further enhancing the accumulation of atmospheric carbon dioxide. A new equilibrium in the carbon balance between air and land may be established after centuries. Since the carbon cycle interacts strongly with cycles of other elements, such as nitrogen and phosphorus, the outcome of an increase in temperature on the carbon budget of tundra ecosystems may be modified by changes in other elemental cycles. Most of the carbon in arctic ecosystems is present in the wet tundra of the Low Arctic, particularly the wet sedge mires and the tussock-dwarf shrub tundras. These areas contain about 2% of the total terrestrial carbon stocks of the world, or approximately 43 Pg carbon (1 Pg = 10^{15} g). [*See* GREENHOUSE GASES IN THE EARTH'S ATMOSPHERE.]

In contrast, the drier shrub tundras and the polar semideserts contain only 18 Pg of carbon. More than 94% of all carbon, nitrogen, and phosphorus in these tundra ecosystems resides in soil organic matter. The combined effects of low temperature and high soil moisture limit decomposition and lead to very slow turnover rates of soil organic matter. The application of fertilizer readily increases the primary production of tundra plants. Because of severe nitrogen limitation for plant growth, the overall carbon balance of tundra ecosystems is largely determined by the carbon losses associated with soil nitrogen mineralization (soil heterotrophic respiration) versus carbon gains associated with nitrogen uptake (photosynthesis and plant growth).

With increased global warming, tundra ecosystems and polar deserts are likely to become more restricted than at present. For example, boreal forest trees and tall shrubs are likely to extend northward along river valleys. Changes in the species composition of tundra plant communities or in the relative abundance of the different species are also likely to occur. The multiplicity of closely related forms of different plant genera within the Arctic provides the potential for the formation of new hybrid genotypes from the existing gene pool in the event of global warming. There is no evidence to suggest that tundra landscapes will disappear or that the essential biological characteristics of this biome will be lost.

Glossary

High Arctic oases Areas in the High Arctic where local conditions permit high levels of biological productivity.
Plant cover Proportion of the ground occupied by perpendicular projection onto it of the aerial parts of individuals of the species under consideration. It is usually expressed as a percentage of the total area of the sampling frame.
Ungulates Any large group of mammals that have hooves.
Microtine rodents Tundra voles and Arctic and brown lemmings.

Bibliography

Bliss, L. C., ed. (1977). "Truelove Lowland, Devon Island, Canada: A High Arctic Ecosystem," pp. 1–703. Edmonton: University of Alberta Press.
Bliss, L. C. (1988). Arctic tundra and polar desert biome. *In* "North American Terrestrial Vegetation" (M. G. Barbour and W. D. Billings, eds.), pp. 1–32. Cambridge: Cambridge University Press.
Bliss, L. C., Heal, O. W., and Moore, J. J., eds. (1981). "Tundra Ecosystems: A Comparative Analysis," pp. 1–813. Cambridge: Cambridge University Press.
Brown, J., Miller, P. C., Tieszen, L. L., and Burnell, F. L., eds. (1980). "An Arctic Ecosystem: The Coastal Tundra at Barrow, Alaska," pp. 1–571. Stroudsburg, PA: Dowden, Hutchinson and Ross.
Chapin, F. S., III., Jefferies, R. L., Reynolds, J. F., Shaver, G. R., and Svoboda, J., eds. (1992). "Arctic Ecosystems in a Changing Climate: An Ecophysiological Perspective," pp. 1–469. New York: Academic Press.

Chernov, Y. I. (1985). "The Living Tundra" (translated by D. Löve), pp. 1–213. Cambridge: Cambridge University Press.

Nelson, J. G., Needham, R., and Norton, L., eds. (1987). "Arctic Heritage," pp. 1–653. Ottawa, Canada: Association of Canadian Universities for Northern Studies.

Sage, B. (1986). "The Arctic and its Wildlife," p. 190. New York: Facts on File Publications.

Shaver, G. R., Billings, W. D., Chapin, F. S., III., Giblin, A. E., Nadelhoffer, K. J., Oechel, W. C., and Rastetter, E. B. (1992). Global change and the carbon balance of Arctic ecosystems. *BioScience* **42,** 433–441.

Virgin Forests and Endangered Species, Northern Spotted Owl and Mt. Graham Red Squirrel

M. P. North

University of Washington

The dwindling supply of virgin forests in the United States has endangered many species dependent on old-forest conditions. Most of the remaining virgin forests are in western states and are highly fragmented by timber harvesting. The controversies surrounding the Mt. Graham red squirrel and the northern spotted owl illustrate the synergetic development of our scientific, social, and legal responses to species conservation. A cluster of telescope observatories has been built within the red squirrel's mountaintop habitat after a congressional exemption abbreviated regulatory review. In the Pacific Northwest, an involved legal and political controversy over the spotted owl has halted most logging and has prompted the development of a regional ecosystem management plan. The two cases exemplify how difficult it is to achieve effective conservation or public consensus if the debate is narrowed by the legal procedures of the Endangered Species Act to a single species issue. The move from species to ecosystem management in the spotted owl controversy provides a model for more effective conservation of our forests and their endangered wildlife.

I. INTRODUCTION

"[Man] fells the forests and drains the marshes. . . The wilds become villages, and the villages towns. The American, the daily witness of such wonders, does not see anything astonishing in all this. This incredible destruction, this even more surprising growth, seem to him the usual progress of things in this world. He gets accustomed to it as to the unalterable order of nature."

Alexis de Tocqueville (1831)

Virgin forests, which covered about 385 million ha, or 50% of the contiguous United States before the arrival of Europeans, were viewed by early settlers as both an inexhaustible bounty of timber and an obstacle to civilized settlement and agriculture. The cutting and clearing of the American landscape has since occurred with every increasing energy. Most of the remaining virgin forests in the lower 48 states are found in three regions that were among the last areas to be settled as Europeans migrated west: northern California and western

Oregon and Washington. By one estimate made in 1991, these remnant virgin forests total just 1.53 million ha, or less than 0.4% of the original forest cover.

A. Virgin Forests

Virgin forests, by definition, are forest stands that have not been disturbed by human activity, in particular timber harvesting. In areas with good growing conditions, several centuries of undisturbed growth produce the large trees, snags, and down logs called old growth. Virgin forests, however, also include forests that never develop old-growth characteristics because they grow at high elevation, in nutrient-poor soil, or in an area frequently disturbed by fire, wind, insects, or pathogens. [See FOREST PATHOLOGY.]

To a historical observer the cutting of virgin forests may not appear so dramatic because many areas of virgin forest have been reforested. While urban and suburban growth have diminished forests in some areas, other regions, such as rural New England and the Southeast, have increased forest cover on abandoned agricultural lands. What has changed is the structure, composition, and function of these forests, the attributes that distinguish virgin from second-growth forests. [See FOREST STAND REGENERATION, NATURAL AND ARTIFICIAL.]

Regardless of tree size, the stand structure of virgin forests differs from second growth by having more foliage layers, more dead wood, and gaps in the canopy cover where trees have died (Fig. 1). The composition of most virgin forests has a greater diversity of tree and understory species, much more variety in the size of woody structures, and higher spatial heterogeneity than second growth. A key functional difference between the two forest types is that virgin forests provide dead wood that is important for nutrient cycling, soil and fungal processes, and provide habitat for many birds, small mammals, amphibians, and insects. [See NUTRIENT CYCLING IN FORESTS.]

In contrast, the structure, composition, and function of second-growth forests are managed more for economical than ecological products. To this end, the ecosystem is simplified to increase wood production. In its extreme form, second-growth forests become even-aged, single-species plantations. Forest succession is truncated into repeat rotations of young, fast-growing trees. Most timber plantations are cut when the trees are between 40 and 90 years old, as they approach the peak of their mean annual increment in wood volume. These forests never develop the large, tall trees, complex canopies, or accumulation of dead woody material common in old forests. Land owners cannot afford to manage forests for the long-growing period needed to produce large-diameter trees, snags, and down logs. For these reasons, the cutting of virgin forests is not the harvest of a renewable resource, it is the replacement of one kind of forest ecosystem with another of very different character.

Although it is not considered profitable to create old-forest conditions, existing virgin forests are highly prized by the timber industry. Such forests typically contain a higher volume of wood per hectare than second growth and much of it is of very high quality. Futhermore, there are no management costs and stands are ready to harvest. America's large logging companies have migrated from New England to the West Coast in step with a retreating line of virgin forests.

As the forests were cut in the eastern and central states, animals that threatened livestock and those dependent on virgin forest conditions were gradually extirpated. By 1855, Henry David Thoreau described New England as an "emasculated country" where "the nobler animals have been exterminated." While Thoreau's lament was written more than a century ago, forest species extinction has become a general public concern only recently when most of the remnant virgin forest is pressed into a sliver of land against the Pacific Ocean.

B. Species Endangerment

The cutting of virgin forests is not detrimental to all forest wildlife. In fact, species that prefer open or brushy areas, such as some songbirds, may increase as old forests are cut. Other "edge" species, such as deer (*Odocoileus* spp.) and elk (*Cervus* spp.), thrive in forests that have a combination of young and old stands for browsing and bedding down.

FIGURE 1 Sunlight from a gap in a virgin forest illuminates a stand with large trees, high understory species diversity, and a complex canopy structure.

The species threatened by the cutting of virgin forests are those dependent on interior, old-forest conditions. These are often rarely seen, nongame animals that have developed specialized niches associated with virgin forests. Other species are dependent on virgin forests simply because they require forest habitat free of human disturbance. For example, in the Pacific Northwest, Roosevelt elk (*C. elaphus*) populations have increased on the Olympic Peninsula as clearcuts adjacent to the Olympic National Park have created more edge habitat. In the same area, populations of the Olympic Torrent salamander (*Rhyacotriton olympicus*) are believed to have fallen sharply because of stream siltation and warmer water temperatures related to logging of headwater streams.

Two laws constitute the principal legal protection provided threatened species. The Endangered Species Act (ESA) of 1973 charges the U.S. Fish and Wildlife Service (FWS) with evaluating species for listing using only biological considerations in its review. If a species is listed, all actions that might jeopardize the species or its habitat must be reviewed. The National Forest Management Act (NFMA), passed in 1976, charges the U.S. Forest Service to manage habitat "to maintain viable populations of existing native and desired non-native vertebrate species." The ESA applies to all species, including plants and invertebrates. Language in sections of the NFMA implies that plant diversity and invertebrate communities should be considered, but there is no consensus on this interpretation. [*See* WILDLIFE MANAGEMENT.]

The ecology and politics of species endangered by the loss of virgin forests are as different as the individual animals involved (Table I). While the biology and management of each species are unique, a common pattern in all of the controversies is the strain of conflicting public demands and wildlife needs on an increasingly rare ecosystem, virgin forests.

The most common causes of virgin forest clearing are development and timber harvesting. Two different cases illustrate each of these pressures on virgin forests: The Mt. Graham red squirrel (*Tamiasciurus hudsonicus grahamensis*) in the Pinaleno Mountains of Arizona and the northern spotted owl (*Strix occidentalis caurina*) in the Pacific Northwest.

The red squirrel conflict is an interesting example of development in the singular, isolated habitat of

TABLE I

A Sample of Species Associated with Virgin Forests That Have Been Listed as Threatened or Endangered by the U.S. Fish and Wildlife Service[a]

Scientific name	Common name	Habitat
Campephilus principalis	Ivory-billed woodpecker	Bottomland hardwoods and old-growth long-leaf pine
Dendroica chrysopaira	Golden-cheeked warbler	Mature mixed forests of oak and Ashe juniper in Texas
Felis concolor coryi	Florida panther	Undisturbed dense, subtropical forests of Florida
Felis concolor couguar	Eastern cougar (possibly extinct)	Undisturbed forests of North Carolina and the Virginias
Glaucomys sabrinus coloratus and *fuscus*	Carolina and Virginia northern flying squirrel	Old-growth transition zone between conifer and hardwood forests
Phaeognathus hubrichti	Red hills salamander	Ravine slopes in mature hardwood forests of Alabama
Picoides borealis	Red-cockaded woodpecker	Old-growth pine stands in southeastern states
Plethodon nettingi	Cheat Mountain salamander	Spruce/birch forests of West Virginia
Rangifer tarandus caribou	Woodland caribou	In winter, old-growth cedar/hemlock forests of northern Washington–Idaho border into Canada
Sciurus niger cinereus	Delmarva Peninsula fox squirrel	Undisturbed forest/shrub ecotone in Maryland and Delaware
Strix occidentalis lucida	Mexican spotted owl	Old-growth forests of Southern Utah and Colorado, Arizona, New Mexico, and northern Mexico

[a] Many more species on the candidate list await the agency's review.

a species with a small population. If the scale and economics involved are smaller than those of the spotted owl controversy, the issue highlights the limits of current regulations for protecting species when the impacts of development are unknown. In contrast, few have questioned the effects of logging on spotted owls. Instead the issue shows how conflict over the scale, cost, and methods of conserving a species can escalate to a debate over the fundamental management of a region's forest ecosystems. The stories of these two species point out the limits of our ability to conserve species and possible remedies to improve conservation efforts in the future.

II. THE MT. GRAHAM RED SQUIRREL AND ITS ISLAND FOREST HABITAT

At the end of the last ice age, about 12,000 years ago, the Sonoran Desert advanced into the Arizona lowlands and mountaintops became a refugium for the spruce/fir forests that once covered the surrounding valleys. In these mountaintop islands of Pleistocene forest, a number of plants and animals evolved into distinct subspecies. The Mt. Graham red squirrel, a

subspecies of the red squirrel common throughout North America, is endemic to a small subalpine zone in the Pinaleno Mountains of southeastern Arizona. Mt. Graham also contains the southern extreme of the Engelman spruce/corkbark fir (*Picea engelmanii/ Abies lasiocarpa* var. *arizonica*) association, as well as the northern extreme of several rare plants and reptiles from Mexico's Sierra Madre Occidental.

A. Habitat and Population

The Mt. Graham red squirrel is a small grayish-brown arboreal rodent that weighs about 230 g (Fig. 2). The squirrel's main diet is conifer cone seeds and fungal fruiting bodies (both "mushrooms" and "truffles"). Squirrels show a strong preference for the stand structure, microclimate, and species diversity found in the old-growth spruce/fir and Douglas-fir (*Pseudotsuga menziesii*) forests above 2800 m elevation. The high elevation forests have closed canopies that keep the understory cool, damp, and dark. These conditions may increase fungal productivity and prevent cones stored in squirrel middens from drying out and losing their seeds. The closed canopy may also

FIGURE 2 A Mt. Graham red squirrel. [Photograph courtesy of Bob Miles, Arizona Game & Fish Department. Used with permission.]

afford overhead protection from raptors and a network of connected branches for safe movement through the forest. Large snags common in the spruce/fir forest provide nesting spots and additional cone storage areas. Logs are used by the squirrel as safe runways and subnevean cone storage sites. The high conifer diversity in the combined Douglas-fir and spruce/fir zone may be particularly important because a poor cone crop is unlikely to occur in all species in the same year. These conditions are only found in the isolated, mountaintop virgin forests. Few red squirrels have been found in lower-elevation Ponderosa pine (*Pinus ponderosa*) forests where canopy cover is low and conditions are drier and warmer.

The Mt. Graham red squirrel is susceptible to extreme population swings and extinction because it occupies a single, isolated habitat. Before the arrival of Europeans, the population may have numbered about 1000 individuals. In the spring of 1990 the population dipped to 132 squirrels, but by the fall of 1993, a census of middens estimated 375 squirrels. Scientists believe that most of the year-to-year fluctuation is due to the abundance of the annual cone crop. The concentration of the squirrel in a small area also makes it susceptible to any habitat changes. For example, a fire in the mid-

1970s on Mt. Graham's west peak may have caused the extirpation of the local squirrel population by isolating the subpopulation and destroying available habitat. Given these conditions, some biologists suggest that management should not focus on achieving a stable squirrel population but instead try to actively keep the squirrel's population as high as possible. Estimates suggest that the current habitat, under optimal conditions, could support up to 650 squirrels.

B. Telescopes and Politics

Mt. Graham is in the Coronado National Forest, managed by the U.S. Forest Service. The four peaks that make up Mt. Graham have approximately 4750 ha of potential squirrel habitat, of which only 830 ha is given a good to excellent habitat rating. The threat to the red squirrel's habitat is an astrophysics project sponsored by the University of Arizona. The university wants to build seven telescopes on the mountain top because it has little light pollution and ideal dry air conditions for nonvisible wavelength telescopes. The initial proposal called for buildings to be constructed on about 3 ha of land, improving an existing road, and restricting public access to an area around the observatories.

The university first proposed the telescope project in 1980. Although the university had expected to begin building by 1982, construction was delayed while the U.S. Forest Service prepared a draft environmental impact statement and solicited public input. In June of 1987 the U.S. Fish and Wildlife Service (FWS) listed the Mt. Graham red squirrel as an endangered species. In the following year the FWS issued a "statement of jeopardy" for the squirrel and suggested three "reasonable and prudent" alternatives. The Forest Service selected a recommended alternative which allowed three telescopes to be built on Emerald Peak about 2 km west of High Peak, the site originally proposed. The three observatories would occupy 3 ha within a 49-ha preserve. This alternative also provided an option for building four more telescopes in the future if red squirrel populations are observed to be unaffected by the first three telescopes.

The Forest Service, noting that the selected alternative would force closure of some areas to the public, informed the university that a final decision could not be made until more public comment was considered. The university, eager to begin construction, persuaded the Arizona congressional delegation to attach an amendment to the Arizona–Idaho Conservation Act of 1988 exempting the observatories from continued evaluation. Construction began shortly after passage of the act and in September of 1993 two of the first three telescopes were dedicated. In August of 1994, a court injunction halted construction of the third telescope after the university moved the building site approximately 300 m to another peak on the mountaintop. Eighteen environmental organizations filed the lawsuit, claiming the relocation site is outside the area exempted in the 1988 Congressional Act.

To date, any construction impacts on the squirrel have not been observed. The squirrel population has increased, in large part, biologists believe, because cone crops have been high in the last few years. Construction of the four additional telescopes will require several years of high squirrel populations and then a full NEPA evaluation.

In the early stages of the red squirrel conflict public reaction was limited to a few highly interested groups such as hunters, hikers, environmental organizations, and the university. The FWS final recovery plan received only 21 written comments. The muted response may reflect the small size of the affected area and the uncertainty about how observatory construction would impact the squirrel population. These factors made it possible for the university to politically shorten a regulatory process it found cumbersome and slow. The recent court injunction, however, indicates political and public support is building for red squirrel protection.

When the FWS lists a species as "endangered" or "threatened," only biological considerations are supposed to influence a species' management plan. Investigations by Congress and the U.S. General Accounting Office, however, found that the selection of Emerald Peak was based on nonscientific considerations. It is clear that without a broad base of public and political support, species conservation can be strongly influenced by economic and political forces as well as by biological concerns. The conflict of all these forces has been dramatically played out in the largest and most costly endangered species conflict since passage of the Endangered Species Act.

III. THE NORTHERN SPOTTED OWL AND OLD-GROWTH FORESTS

The controversy in the Pacific Northwest over old-growth forests and the northern spotted owl has involved science, economics, politics, and public response on a regional, if not national, scale. Much of the debate illustrates how the management and perception of endangered species have changed in the last 20 years since the ESA was passed. Although there are many factors that have molded the spotted owl/old-growth controversy, four influences are particularly important: research on the spotted owl, ecological studies of old-growth forests, changes in the timber industry, and a change in society's perception of forests.

A. Spotted Owl Research

Three subspecies of the spotted owl are recognized. The Mexican spotted owl (*Strix occidentalis lucida*), genetically distinct from the other subspecies, resides in northern Mexico, Arizona, New Mexico, and southern Utah and Colorado. Similar in its habitat requirements to the northern subspecies, it was listed as a threatened species by the FWS in 1993. The California spotted owl (*Strix occidentalis occidentalis*), found in coastal forests south of San Francisco and in the Sierra Nevada south of the Pit River, is considered distinct from the northern subspecies because of morphological, not genetic, differences. The northern spotted owl (*Strix occidentalis caurina*) ranges from northern California into southern British Columbia in the Cascade Mountains and westside forests. It is the best studied of the three subspecies, and plans developed for its management may provide a framework for managing the other two subspecies.

Although the first written report of a spotted owl was made in 1860, the rarely observed bird (Fig. 3) was not scientifically studied until the 1970s by Eric Forsman. In those preliminary studies, Forsman, an Oregon State University graduate student, immediately noted two characteristics of the owl that would eventually generate most of the debate surrounding the species: spotted owls are found predominantly in old-growth forests, and they have large home ranges.

Scientists have since refined the first observation, emphasizing that the spotted owl is closely associated with specific forest stand structures rather than an age class of forest. Two important forest structures for the owl, multilayer canopies and large snags, however, are often found only in old growth. These features may facilitate prey capture and provide cavities for one of the owl's staple prey species, the northern flying squirrel (*Glaucomys sabrinus*) (Fig. 4). Some old-growth forests also have a greater abundance and diversity of "truffles" than second-growth forests. Truffles are an important

FIGURE 4 A northern flying squirrel with a truffle, which comprises 90% of the squirrel's diet. The abundance of truffles in old growth may attract flying squirrels, which in turn may improve owl foraging success.

food source for many small mammals, and may influence the local abundance of several of the owl's prey species.

In the course of the spotted owl controversy, reports have circulated that owls have been found in young forests. In general, these reports result from three situations: (1) Spotted owls have been found using younger forests that have old-growth remnant patches in the core activity area, (2) juvenile owls have been located in young forests as they search for unoccupied habitat territory (see the following discussion), and (3) in northern California, spotted owls will use redwood (*Sequoia sempervirens*) forests as young as 60 years old. These forests, however, because of favorable growing conditions, may produce large trees with some old-growth attributes within 50 to 60 years of being logged.

Changes in forest type and the owl's diet are believed to influence home range size. One contro-

FIGURE 3 An adult northern spotted owl.

versial theory is that as the number, biomass, and availability of prey decreases, owls need larger home ranges. For instance, in the mixed conifer forests of northern California and southern Oregon, the owl feeds mostly on bushy-tailed and dusky-footed woodrats (*Neotoma cinera* and *N. fuscipes*) and on northern flying squirrels. The combined weight and density of the three species give an average available biomass of 388 g per ha. The owl's median home range size in this region varies from 400 to 3000 ha. In the western hemlock (*Tsuga heterophylla*), Douglas-fir forests of Washington's Olympic Peninsula and North Cascades, the owl's prey is mainly flying squirrels with an available average biomass of 61 g per ha. The home range size in Washington varies from 2500 to 12,000 ha. Other scientists who disagree with the preybase theory note that some studies have found high densities of woodrats and flying squirrels in young forests. They argue that forest structure and foraging success are also an important influence on home range size and the owl's preference for old-growth forests.

Spotted owls, which may live for 12 years or more, reach sexual maturity by their second year. Mating usually occurs in March and the two eggs hatch in May. Owlets remain in the nest for about 35 days after hatching. By September the adult owls stop feeding their young and the juveniles disperse to search for their own territory. Before dispersal, about one-third of the owlets die; starvation is frequently the cause. Survival is further jeopardized as juveniles search for old-growth stands unoccupied by other owls. In the Pacific Northwest most timber has been harvested on public land in dispersed 8–16 ha (20–40 acre) clearcuts (Fig. 5). The resulting landscape pattern forces juveniles to cross open areas where they are especially susceptible to predation by the great horned owl (*Bubo virginianus*).

The importance of juvenile dispersal and territory establishment prompted scientists to think of owl conservation using metapopulation models. The whole nothern spotted owl population was analyzed as a dynamic set of subpopulations where local extinction and recolonization depend on successful juvenile dispersal and adult migration. Owl conservation studies applied the principles of island biogeography to landscape ecology. Owl activity centers were viewed as islands surrounded by a forest matrix. The matrix could help or hinder juvenile dispersal depending on how many clearcuts perforated the forest and the stand structure of the intact forest cover.

A diminishing supply of old-growth forests makes it more difficult for juveniles to find suitable unoccupied territory. Biologists have speculated that this lack of old growth may explain why owls sometimes occupy young forests or crowd at high densities in fragmented old-growth stands. This crowding of birds into remaining habitat, referred to as "packing," may reduce local prey availability and decrease the reproductive success of both "floater" birds and the resident owl pair. Scientists realized that measures of owl density may not be a good indicator of population health or habitat quality.

As timber sales were halted, scientists were pressured to provide information on the spotted owl. Thousands of individuals have been located by "calling surveys" and more than 100 owls were fitted with radio telemetry packages so their habitat preference, home range size, and dispersal patterns could be mapped. As data accumulated, biologists noted that few juveniles survived their search for new habitat. Although scientists were gaining an unprecedented understanding of a species' complex biology, it was clear they were also witnessing a declining spotted owl population.

B. Old-Growth Ecology

Owl biologists, however, were not the only scientists exploring the old-growth forests. In fact, much of the debate surrounding the owl is really an expression of concern for the old-growth ecosystem. Certainly much of the growing public concern over old growth developed in response to scientific studies of how old-growth forests differ from young-managed plantations. As recently as the early 1970s old growth was viewed as overmature, decadent timber, a "biological desert" shunned by most animals. Beginning in the late 1960s, however, funding from the International

FIGURE 5 A pattern of small, scattered clear-cuts in the Pacific Northwest exposes dispersing owls to predation and fragments the remaining old-growth habitat.

Biological Program and other sources enabled more extensive studies of old growth. The resulting research changed some of the most fundamental thinking about forest ecosystems.

Typically, managed forests eliminate the development of large woody structures by confining forest succession to early scral stages. Large, old trees provide a more complex canopy structure for modifying the microclimate, reduce the flood potential of rain-on-snow runoff, and increase ephiphyte and invertebrate diversity. In particular, studies started to focus on the functional role of snags, dead wood, and decomposing processes (Fig. 6). Although forest management had concentrated on young tree growth and fiber production, researchers now realized that perhaps only half of a tree's ecological functions occur while it is alive. Snags provide insects for wood-boring birds and cavities for many birds and arboreal mammals. Down logs are important habitat for woodland salamanders and most detritivores, and function as nutrient and moisture reservoirs. Old growth also provides high understory plant diversity, an important source of logs for stabilizing stream beds, a conservative nutrient cycle, and unique above-and below-ground conditions for canopy invertebrates and soil microfauna.

C. The Timber Industry

A growing awareness of old-growth ecosystems and the spotted owl's imperiled status were not the only factors that provoked the owl/old-growth debate. Fundamental changes in the region's timber industry were underway, squeezing timber workers, rural communities, and the markets on which both depend.

The western half of Oregon and Washington have some of the world's most productive timber land. Early settlers marveled at Douglas-fir trees more than 3 m in diameter and up to 90 m tall. The supply of timber seemed inexhaustible. In 1865, on the shores of Washington State's Puget Sound, the newly arrived A.S. Mercer claimed, "The supply of logs for lumber will only be exhausted when the mountains and valleys surrounding the Sound are destroyed by some great calamity of nature."

Early settlers began harvesting the timber that could be easily floated to market. In the 1890s, when the last of the virgin pine stands were ex-

FIGURE 6   An illustration of an old-growth Douglas-fir and western hemlock forest near Stephenson, Washington. This stand displays many of the characteristics of old-growth forests such as (A) a complex, multilayered canopy; (B) a gap with regenerating trees created when an overstory tree died; (C) a large snag with a cavity; (D) an understory of shade-tolerant trees including western yew (*Taxus brevifolia*); (E) a tipup mound and large, windthrown log; and (F) trees growing on a "nurse" log. [Illustration by Robert Van Pelt.]

hausted in the upper Midwest, many large timber companies moved to the Pacific Northwest. Logging changed from a frontier economy that anyone with an axe and pair of oxen could pursue, to an organized industry with company-owned towns and overseas exports. In the 1950s, privately owned, old-growth timber began to run out and federal land managers increased their annual harvest to sustain the local lumber economies. Forest Service managers were directed to set annual harvest at a sustainable yield, the level at which the forest would regenerate timber at the rate it was being cut. Harvest rates on many forests, however, were not sustainable because optimistic growth and yield projections were used, and the congressionally dictated sale quantity increased each year. In the 1960s and 1970s, rural logging economies

boomed in the Pacific Northwest, as fully half of the total National Forest Service timber harvest came from western Oregon and Washington alone. Timber supply experts began to point out that old-growth timber would run out before most maturing second-growth stands would reach a merchantable size.

While the timber supply condition was becoming ominous, changes in technology and world markets were concurrently impacting logging economies. Improvements in milling mechanization reduced the number of workers needed in lumber manufacturing. Only mills that modernized their equipment could afford to stay competitive and many small milling operations closed. For the mills that did mechanize, timber was still scarce because Asian markets were paying higher prices for raw logs than were U.S. mills. All of these changes pushed to shrink the logging economy at the time when the issue of the spotted owl was taking hold.

In the Pacific Northwest, logging is not only a means of making a living, it is also a part of the rural community's cultural heritage. Many loggers felt the economic slowdown was an assault on both their livelihood and on their way of life. As timber sales slowed and the controversy became increasingly portrayed in the media as "jobs versus owls," the timber communities vented their frustration on the spotted owl and environmentalists. One reason the controversy has become so polarized and difficult to resolve is that forest mismanagement, changes in timber markets, and threats to a way of life have all been simplified to a single visceral symbol, the spotted owl.

D. Society's Perception of Forests

If the spotted owl became a simplified symbol for change in many timber communities, the same could be said of many urbanite's response to the issue. As America has changed from a rural, resource-dependent economy to a service and information-based society, forests are less a source of money and food than an escape from urban work and stress. This demographic shift from rural to urban has been accompanied by a perceptual shift in the utility of forests. Forests, once viewed as stands of timber for working and hunting, are now understood as complex ecosystems with some attributes and wildlife species that may be renewable only after many centuries. Much of this shift can be explained by the percolation of spotted owl and old-growth scientific research into the public's understanding of forest ecosystems. In this context, forests were best left as undisturbed environments for wildlife and nonconsumptive recreation. The new aesthetic might best be summed by the popular Sierra Club adage, "Take only pictures, leave only footprints."

Forests also embody more than the sum of their biological parts. Vast woodlands are part of the unique American landscape that was woven into the cultural heritage distinguishing early Ameri-

cans from their European roots. The French writer Chateaubriand noted this heritage when he remarked, "There is nothing old in America excepting the woods . . . they are certainly the equivalent for monuments and ancestors." This sense of forests as an historical legacy fueled some of the passionate debate in the spotted owl controversy. Cutting old growth was described as the destruction of nature's cathedral and the loss of a future generation's rightful heritage. In this scenario, loggers were cast as a selfish lot, bent on profiting from the final decimation of an American treasure.

E. Political and Legal Actions

For environmentalists concerned with preserving old growth, the best political tool was neither a spiritual appeal nor lobbying for the recreational value of the forest. Since its passage in 1973, the Endangered Species Act has proven most effective for reserving land because economic and social impacts of protecting an endangered species are not supposed to be weighed in determining whether or not a species is listed. As one environmentalist put it, "If the spotted owl hadn't come along, it would have been necessary to invent it."

Spotted owl and old-growth research, changes in the timber industry, and society's perception of forests all contributed to push the spotted owl/old-growth debate into the political and legal arena. As early as 1981 the Portland Regional Office of the FWS described the owl as "vulnerable," but concluded the species did not qualify for listing at the time. In the mid 1980s the Bureau of Land Management (BLM) and the Forest Service developed and revised spotted owl management plans which in turn were challenged by conservation groups. In 1987 the FWS was again petitioned to list the spotted owl as endangered. Their decision not to list the owl was appealed by conservation groups in 1988. The U.S. Federal Court in Seattle ruled that the decision against listing was not based on biological studies and it ordered the FWS to reexamine the owl for listing.

By 1988, Washington and Oregon state agencies had listed the spotted owl as endangered and threatened, respectively. In 1989 an agreement among the Bureau of Land Management, the Forest Service, the Fish and Wildlife Service, and the National Park Service established an Interagency Spotted Owl Scientific Committee. The committee produced a spotted owl management plan in May 1990, commonly referred to as the Thomas report after the Committee's chairman Jack Ward Thomas. The report proved to be a landmark in species conservation plans by taking a landscape-level approach to managing northern spotted owl populations. The committee developed a region-wide plan calling for 5.8 million acres to be set aside for owl habitat and outlined a management guideline for the connecting areas between reserves to provide for owl dispersal. The so-called "50–11–40" rule called for 50% of the land base between owl reserves to be maintained in stands with an average tree diameter of 11 inches and at least 40% canopy cover.

In the summer of 1990 the U.S. Fish and Wildlife Service listed the northern spotted owl as a threatened species. With this listing, federal land management agencies were required to develop new spotted owl environmental impact statements. In the interim most timber sales on federal lands were halted. In the summer of 1992 the marbled murrelet (*Brachyramphus marmoratus*) was listed by the FWS as a threatened species, further restricting all timber sales in a 50-mile-wide band along the ocean coast where the sea bird nests. Environmentalists were also pressing the Forest Service under the 1976 National Forest Management Act to include viability assessments of other old-growth associated species in their plans. There was also mounting public concern about logging impacts on the headwater streams of several salmon runs that were in critical condition.

The resulting legal challenges polarized the environmentalists and timber industry until issue discussion and compromise became impossible. The newly elected Clinton administration convened a timber summit in April of 1993 in an attempt to break the deadlock and hear all sides of the issue. Following the summit, a team of scientists was directed to develop an array of alternatives for managing the region's forests that would provide for

the viability of all species associated with old forests. The resulting set of 10 alternatives, including the selected option 9, was open to public comment and more than 100,000 responses were received. The timber industry objected to the plan because the region's annual federal timber sale was reduced from 4.5 to 1.2 billion board feet (a board foot is equal to a 1-inch thick board measuring 1 foot by 1 foot). Environmental groups oppose the plan because it allows some continued logging of old growth and some salvage logging in owl habitat. Federal agencies jointly developed plans to implement the option 9 recommendations on the lands under their jurisdiction. In the fall of 1994 a federal court reviewed the joint plan and decided it satisfied the requirements of NFMA and ESA. In the absence of congressional action the federal court decision will be the final arbiter in the spotted owl controversy.

What started with an effort to slow old-growth harvests and give the spotted owl federal protection has developed into a legal tangle about the region's forests and NFMA's requirement to provide for the viability of all vertebrate species threatened by the loss of old growth. Certainly the regulations, legal reviews, and political maneuverings are costly, slow, and often shrill. Yet for all the inefficiency and possible inequities, the controversy has focused research effort and public scrutiny on the old-growth ecosystem. Initially, scientists and environmentalists concentrated on the spotted owl. However, with closer scrutiny, the diversity and complexity of old forests became clear. It is not just the spotted owl that is threatened by the disappearance of old growth. Other species include the fischer (*Martes pennanti*), Vaux's swift (*Chaetura vauxi*), Northwestern salamander (*Ambystoma gracile*), and hooded lancetooth snail (*Ancotrema voyanum*) to name but a few of the estimated 1098 terrestrial species (excluding arthropods) at risk (Table II). Cutting old-growth forests does not simply eliminate big trees, but also complex canopies, large snags with cavities, big down logs, and diverse understory plants. The dwindling supply of old growth also means the loss of a refuge from urban stress, an environment where the human imprint has not yet been stamped. Clearcuts are

TABLE II

Estimates of the Number of Species Closely Associated with Old-Growth Forests in the Pacific Northwest (PNW)[a]

Group	Number of species associated with old growth	Comments
Fungi	527	One hundred and nine species are endemic to the PNW.
Lichens	157	Many of these species have been extirpated or are in decline in Europe and eastern North America.
Bryophytes	106	Thirty-two species are endemic to the PNW.
Vascular plants	124	Many of these species are poor dispersers, needing large rotting nurse logs or specific fungi.
Mollusks	102	Many rare land and freshwater snails with high endemism. Eight species proposed for federal listing.
Amphibians	18	Mostly salamanders that require large logs or undisturbed riparian areas.
Birds	38	Many cavity nesters that use large snags.
Mammals (excluding bats)	15	Most species are in the Rodentia order. The marten, fisher, and red tree vole are the most threatened.
Bats	11	Seven are *Myotis* spp. Roost sites in crevices, old tree hollows, and under bark are important habitats.
Arthropods	7000+	An estimate since little is known about most species and only approximately 75% of regional species have been described.

[a] These estimates were made by the Forest Ecosystem Management Assessment Team (FEMAT), a scientific report developed following the timber summit.

replanted and will regenerate, but second-growth forests are not managed to provide old-growth attributes.

IV. CONSERVING VIRGIN FORESTS AND ENDANGERED SPECIES

Although the Endangered Species Act has been a powerful legal means of protecting wildlife, it has directed conservation efforts into a species-by-species approach. When the ESA was passed in 1973, immediate legal protection was needed for such species as the peregrine falcon (*Falco peregrinus annatum*) endangered by exposure to the pesticide DDT. This approach was essential triage for the hemorrhaging of a long-neglected environment. However, scientific studies of species have always emphasized the connectivity between all species and the structure, function, and composition of their habitat. The spotted owl controversy is a good example of how ecological research can help expand the debate beyond a single species approach.

The plan that evolved from the timber summit's team of scientists was appropriately called forest ecosystem management, emphasizing that species viability is best ensured by managing for ecosystem integrity within a connected landscape. What is threatened is the old-growth ecosystem of which the spotted owl is but one component. The plan makes another important distinction by stressing that low-elevation, old-growth ecosystems are the most threatened. Ecosystem management at a regional scale can evaluate whether existing reserve areas provide a representative sample of the natural range of ecosystems. For example, most U.S. Park and Wilderness areas protect scenic, high-elevation ecosystems, while species-rich lowlands are underrepresented. By changing the currency of conservation from the species to the ecosystem, management will preserve more diverse habitats and their associated wildlife.

Another important contribution of the spotted owl debate is the development of a species conservation plan at a regional scale. The large habitat size and dispersal needs of the spotted owl required management planning over a landscape. The Thomas and FEMAT reports are the first large-scale application of theories from the new fields of conservation biology and landscape ecology. The effective conservation of many species requires planning for species dispersal and movement, and the influence of a landscape's matrix on habitat reserves.

The spotted owl debate has also accelerated changes in forest management. All of the agency plans submitted for court consideration adopted "new forestry" practices. These practices emphasize that forests should be managed to minimize the logging impacts on ecosystem processes and wildlife habitat. On harvest sites, large live trees, snags, and logs will be left to increase the structural diversity and habitat potential of the regenerating forest. This practice more closely mimics the effects of wildfire and windstorm disturbances on a forest than clearcutting. Scientists are working to identify the density, arrangement, and kinds of woody structures essential to threatened species. When a forest is harvested, managers focus on what is to be left on site instead of simply calculating the timber volume that can be clearcut and extracted.

At the landscape level, "new forestry" may aggregate harvest areas to minimize forest fragmentation and retain habitat connectivity. Management plans examine logging impacts on watersheds, the effects of road construction on erosion and fragmentation, and cumulative changes on the landscape matrix.

The politics and public debate surrounding the Mt. Graham red squirrel and the northern spotted owl are a telling narrative of our fledgling struggle to conserve endangered species when there is little time or room for error. In both instances, a reactive approach to an endangered species framed each conflict as one of immediate economic loss against the uncertain future loss of a species. In the absence of broad-based public support, the politics and economics of the moment strongly influence how an endangered species debate is resolved. In the spotted owl controversy, however, the scope of the debate kept expanding as regulations and legal challenges prevented an expedient resolution. The evolving controversy eventually spawned the nation's first attempt at regional ecosystem management. To stay ahead of the species extinction curve, this example of ecosystem management may be a promising pro-

active model for conserving what remains of our virgin forests and their endangered species.

Glossary

Ecosystem management A strategy to manage ecosystems for all associated species instead of a species-by-species approach to conservation.

Forest fragmentation Splitting a continuous old-forest landscape into a mosaic with younger-age stands.

Landscape matrix Dominant cover type in the landscape that binds smaller, different types together.

Metapopulation A species population made up of subset populations linked through migration. A habitat unoccupied following a local extinction can be recolonized by immigration from another subpopulation.

New forestry Forest management that minimizes the impact of timber harvesting on ecosystem integrity and landscape connectivity.

Old growth Virgin forest that has been free of disturbance long enough to develop large, woody structures and senescent mortality.

Packing Crowding of species into a dwindling supply of habitat that can depress prey abundance and successful reproduction.

Sustainable yield A timber harvest rate at which a region's forest regenerates the same volume of wood that is cut each year.

Succession Progressive change in forest composition and structure with age; broadly defined by four seral or successive stages: stand initiation (the forest regenerates), stem exclusion (self-thinning), understory reinitiation (dying trees produce canopy gaps and light for establishing trees), and old growth (tree growth balances mortality losses).

Virgin forest Forest of any age that has not been altered by human activity.

Bibliography

Carey, A. B., Horton, S. P., and Biswell, B. L. (1992). Northern spotted owls: Influence of prey base and landscape character. *Ecol. Monogr.* **62,** 223–250.

Flather, C. H., Joyce, L. A., and Bloomgarden, C. A. (1994). "Species Endangerment Patterns in the United States." USDA Forest Service, RM-GTR-241.

Forsman, E. D., Meslow, E. C., and Wight, H. M. (1984). "Distribution and Biology of the Spotted Owl in Oregon." *Wildl. Monogr.* **87,** 1–64.

Harris, L. D. (1984). "The Fragmented Forest." Chicago, IL: The University of Chicago Press.

Ruggiero, L. F., Aubry, K. B., Carey, A. B., and Huff, M. H., eds. (1991). "Wildlife and Vegetation of Unmanaged Douglas-Fir Forests." USDA Forest Service, PNW-GTR-285.

Thomas, J. W., Forsman, E. D., Lint, J. B., Meslow, E. C., Noon, B. R., and Verner, J. (1990). "A Conservation Strategy for the Northern Spotted Owl." Report of the interagency committee to address the conservation strategy of the northern spotted owl. Portland, OR: United States Forest Service.

Thomas, J. W. (leader). (1993). "Forest Ecosystem Management: An Ecological, Economic, and Social Assess-ment." Report of the Forest Ecosystem Management Assessment Team. Government Publications.

U.S. Fish and Wildlife Service. (1992). "Mount Graham Red Squirrel Recovery Plan." Albuquerque, NM: U.S. Fish and Wildlife Service.

Warshall, P. (1994). The biopolitics of the Mt. Graham red squirrel (*Tamiasciuris hudsonicus grahamensis*). *Cons. Bio.* **8,** 977–988.

Williams, M. (1989). "Americans and Their Forests: A Historical Geography." New York: Cambridge University Press.

Wastewater Treatment Biology

N. F. Gray
Trinity College, University of Dublin

I. Introduction
II. Secondary Treatment
III. Fixed Film Reactors
IV. Activated Sludge Processes
V. Other Biological Systems

After preliminary and primary treatment, wastewaters still contain significant amounts of colloidal and dissolved material that needs to be removed before discharge to a watercourse. This is achieved by secondary treatment, sometimes called aerobic biological treatment.

I. INTRODUCTION

Secondary treatment is a biological process where the settled wastewater enters a specially designed reactor and under aerobic conditions organic matter is utilized by microorganisms. The reactor provides a suitable environment for a microbial population to develop, and as long as oxygen and food, in the form of settled sewage, are supplied, then the biological oxidation process will continue. Biological treatment is primarily due to bacteria, which form the basic trophic level in the reactor food chain. The biological conversion of soluble material into dense microbial biomass has essentially purified the wastewater and all that is subsequently required is to separate the microorganisms from the treated effluent by settlement. Secondary sedimentation is essentially the same as primary sedimentation

except that the sludge is composed of biological cells instead of gross fecal solids.

II. SECONDARY TREATMENT

Methods of purification in secondary treatment units are similar to the *self-purification process* that occurs naturally in rivers and streams, and involves many of the same organisms. Removal of organic matter from settled wastewaters is carried out by heterotrophic microorganisms, predominately bacteria but also fungi. The microorganisms break down the organic matter by two distinct processes, biological oxidation and biosynthesis, both of which result in the removal of organic matter from solution. Oxidation or respiration results in the formation of mineralized end products which remain in solution and are discharged in the final effluent, while biosynthesis converts the colloidal and soluble organic matter into a particulate biomass (new cells) which can then be subsequently removed by settlement. If the food supply, in the form of organic matter, becomes limiting, then the microbial cell tissue will be endogenously respired (autooxidation) by the microorganisms to obtain energy for maintenance. All three processes occur simultaneously in the reactor and can be expressed stoichiometrically as:

Oxidation:
$$COHNS + O_2 + bacteria \rightarrow CO_2 + NH_3 +$$
(organic matter)

$$\text{other end products} + \text{energy}$$

Biosynthesis:
$$COHNS + O_2 + bacteria \rightarrow C_5H_7NO_2$$
(organic matter) (new cells)

Autooxidation:
$$C_5H_7NO_2 + 5O_2 \rightarrow 5CO_2 + NH_3 + 2H_2O + \text{energy}$$
(bacteria)

In natural waters soluble organic matter is principally removed by oxidation and biosynthesis, but in the intensified microbial ecosystem of the biological treatment plant, adsorption is perhaps the major removal mechanism, with material adsorbed and agglomerated onto the dense microbial mass. The adsorptive property of the microbial biomass is particularly useful as it is also able to remove from solution nonbiodegradable pollutants present in the wastewater such as synthetic organics, metallic salts, and even radioactive substances. The degree to which each removal mechanism contributes to overall purification depends on the treatment system used, its mode of operation, and the materials present in the wastewater. [See WASTEWATER TREATMENT FOR INORGANICS.]

In nature, heterotrophic microorganisms occur either as thin films (periphyton) growing over rocks and plants, or in fact over any stable surface, or as individual or groups of organisms suspended in the water. These natural habitats of aquatic heterotrophs are utilized in wastewater treatment to produce two very different types of biological units: one uses attached growths while the other uses suspended microbial growths. The design criteria for secondary treatment units are selected to create ideal habitats to support the appropriate community of organisms responsible for the purification of wastewater, so attached and suspended microbial growth systems require fundamentally different types of reactors. Both treatment systems depend on a mixed culture of microorganisms, but grazing organisms are also involved so that a complete ecosystem is formed within the reactor, each with distinct trophic levels. In its simplest form the reactor food chain comprises:

Heterotrophic bacteria and fungi → holozoic protozoa → rotifers and nematodes → insects and worms → birds.

Due to the nature of the reactor, suspended growth processes have fewer trophic levels than attached growth systems (Fig. 1). These man-made ecosystems are completely controlled by operational practice, and are limited by food (organic loading) and oxygen (ventilation/aeration) availability.

Chemical engineers have been able to manipulate the natural process of self-purification and, by supplying ideal conditions and unlimited opportunities for metabolism and growth, have intensified and accelerated this biological process to provide a range of secondary wastewater treatment systems. However, a number of basic criteria must be satisfied by the design. In order to achieve a rate of oxidation well above that found in nature, a much denser biomass in terms of cells per unit volume must be maintained within the reactor. This results in increased oxygen demands which must be met in full, so as not to limit the rate of microbial oxidation. Essentially this is done by increasing the air–water interface. The wastewater containing the polluting matter must be brought into contact with a dense population of suitable microorganisms for a sufficient time under aerobic conditions, to allow oxidation and removal of unwanted material to the desired degree. Finally, inhibitory and toxic substances must not be allowed to reach harmful concentrations in the reactor.

The main methods of biological treatment rely on aerobic oxidation. To ensure that oxidation proceeds quickly it is important that as much oxygen as possible comes into contact with the wastewater so that the aerobic microorganisms can break down the organic matter at maximum efficiency. Secondary treatment units of wastewater treatment plants are designed to bring this about. Oxidation is achieved

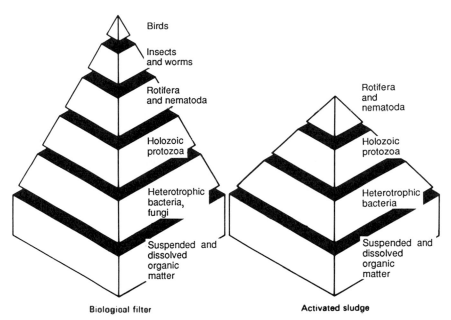

FIGURE I Comparison of the food pyramids for fixed film and completely mixed systems. Reproduced with permission from Academic Press.

by three main methods: (a) by spreading the sewage into a thin film of liquid with a large surface area so that all the required oxygen can be supplied by gaseous diffusion; (b) by aerating the sewage by pumping in bubbles of air or stirring vigorously; and (c) by relying on algae present to produce oxygen by photosynthesis.

In systems where the microorganisms are attached, a stable surface must be available. Suitable surfaces are provided by a range of media such as graded aggregate or molded plastic and even wooden slats, retained in a special reactor, on which a dense microbial biomass layer or film develops. These reactors are generally categorized as fixed-film reactors; the most widely used type is the percolating or trickling filter. Organic matter is removed by the wastewater as it flows in a thin layer over the biological film covering the static medium. Oxygen is provided by natural ventilation which moves through the bed of medium via the interstices supplying oxygen to all parts of the bed. The oxygen diffuses into the thin layer of wastewater to the aerobic microorganisms below. The final effluent not only contains the waste products of this biological activity, mainly mineralized compounds such as nitrates and phosphates, but also particles of displaced

film and grazing organisms flushed from the medium. These are separated from the clarified effluent by settlement, and the separated biomass is disposed as secondary sludge (Fig. 2).

In suspended growth or completely mixed system processes the microorganisms are either free-living or flocculated to form small active particles or flocs which contain a variety of microorganisms, including bacteria, fungi, and protozoa. These flocs are mixed with wastewater in a simple tank reactor, called an aeration basin or tank, by aerators that not only supply oxygen but also keep the microbial biomass in suspension to ensure maximum contact between the microorganisms and the nutrients in the wastewater. The organic matter in the wastewater is taken out of solution by contact with the active suspended biomass. The purified wastewater is displaced from the reactor by the incoming flow of settled wastewater and contains a large quantity of microorganisms and flocs. The active biomass is separated from the clarified effluent by secondary settlement, but if all the biomass is disposed of as waste sludge the concentration of active biomass in the aeration basin rapidly falls to such a low density that little purification occurs. Therefore the biomass, called activated sludge, is returned to the reac-

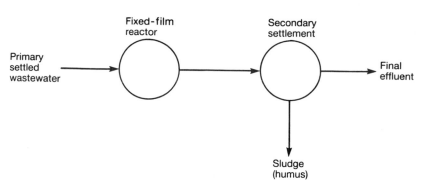

FIGURE 2 Layout of a fixed-film reactor.

tor to maintain a high density of active biological solids, ensuring a maximum rate of biological oxidation (Fig. 3). Excess biomass, not required to maintain the optimum microbial density in the reactor, is disposed of as surplus sludge.

Biological wastewater treatment differs from more traditional fermentation processes, such as the production of bakers' yeast, in a number of important ways. Principally, wastewater treatment is aimed at removing unwanted material while the commerical fermentation processes are all production systems. These production fermentations use highly developed, specialized strains of particular microorganisms to synthesize the required end product, whereas in wastewater treatment a mixture of microorganisms are used. These are largely self-selecting and nearly all the organisms that can contribute to substrate removal are welcome. Unlike commerical fermenters, wastewater reactors are not aseptic and, because production fermenters require highly controlled conditions, they are more com-

plex and comparatively more expensive than those used in wastewater treatment.

The secondary treatment phase may be composed of other biological systems, both aerobic and anaerobic, or may incorporate a mixture of several systems. Among the more unusual biological systems used for municipal treatment that fall outside the fixed film or completely mixed categories are aerobic and anaerobic lagoons, reed beds, wetlands, and land treatment.

III. FIXED FILM REACTORS

A. Basis of the Process

In fixed film reactors, the microbial biomass is present as a film that grows over the surface of an inert and solid medium. Purification is achieved when the wastewater is brought into contact with this microbial film. Because the active biomass is

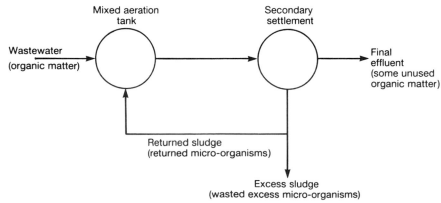

FIGURE 3 Layout of a mixed reactor (activated sludge) with sludge return.

largely retained within this reactor, there is no need to recirculate any displaced biomass back to the reactor in order to maintain a sufficient density of microorganisms, as is the case with completely mixed systems. The required contact between the film and wastewater is achieved in most fixed-film reactors by allowing the wastewater to pass over the stationary medium on which the film has developed. However, it is not essential for the medium to be stationary, and in more recently developed reactors, such a rotating biological contactors (RBCs) and fluidized beds, the medium itself moves through the wastewater. Fixed-film reactors can be designed as secondary treatment processes to partially treat (high-rate filtration) or fully treat (low-rate filtration) either screened or more generally settled wastewater. While they are normally aerobic, for carbonaceous removal and nitrification, anoxic and anaerobic filters are used for denitrification and for treating moderately strong organic wastewaters, respectively. While anoxic and anaerobic filters are submerged reactors, i.e., the medium is permanently submerged under the wastewater, submerged aerobic filters are less common as they require a diffused aeration system to maintain a sufficient oxygen concentration within the reactor. However, submerged aerobic filters are

particularly useful where loadings are intermittent and where difficulty occurs in maintaining the minimum "wetting" loading in order to prevent the filter from drying up and the microbial film dying.

The most widely used fixed-film reactor is the trickling or percolating filter. In its simplest form it is composed of a bed of graded hard material known as the medium, about 2.0 m in depth (Fig. 4). The settled wastewater is spread evenly over the surface of the bed by a distribution system that can be used to regulate the volume and frequency of application of the wastewater. The filter has a ventilating system to ensure free access of air to the bed, which passes through the interstices or voids of the medium ensuring that all parts of the filter have sufficient oxygen. The movement of air is by natural draft and, depending on the relative temperature difference between the air and inside the filter, can be in an upward or downward direction. The treated effluent passes through a layer of drainage tiles which supports the medium and flows away to the secondary settlement tank.

B. Film Development

The microbial film takes 3–4 weeks to become established during the summer and up to 2 months

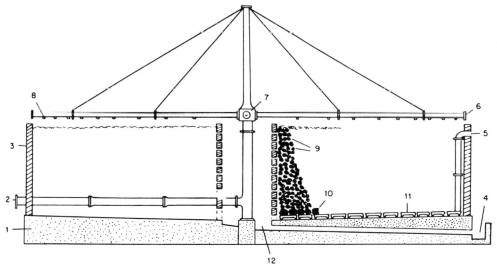

FIGURE 4 Basic constructional features of a conventional percolating (trickling) filter. (1) Foundation floor, (2) feed pipe, (3) retaining wall, (4) effluent channel, (5) ventilation pipe, (6) distributor arm, (7) rotary seal, (8) jets, (9) main bed of medium, (10) base layer of larger medium, (11) drainage tiles, and (12) central well for effluent collection. Reproduced with permission from Academic Press.

in the winter. Unlike many industrial wastewaters, it is not necessary to seed domestic wastewaters treated by percolating filters as all the necessary microorganisms are present in the sewage itself, with the dipteran grazers flying onto the filter and colonizing it. The film only develops on the surfaces that receive a constant supply of nutrients, so the effectiveness of media to redistribute the wastewater within the filter, to prevent channeling, and to promote maximum wetting of the medium is an important factor affecting performance. The film is a complex community of bacteria, fungi, protozoa, and other mesofauna, plus a wide diversity of macroinvertebrates such as enchytraeids and lumbricid worms, dipteran fly larvae, and a host of other groups which all actively graze the film.

The organic matter in the wastewater is degraded aerobically by heterotrophic microorganisms which dominate the film. The film has a spongy structure that is made more porous by the feeding activities of the grazing fauna that are continually burrowing through it. The wastewater passes over the surface of the film and to some extent through it, although this depends on film thickness and the hydraulic loading (Fig. 5). In low-rate filters a large proportion of the wastewater may be flowing

through the film matrix at any one time, and it is the physical straining action of this matrix that allows such systems to produce extremely clear effluents. Another advantage is that the greater the proportion of wastewater that flows through the film the greater the microorganism:wastewater contact time, which is known as the retention time or hydraulic retention time (HRT). The higher the hydraulic loading the greater the proportion of the wastewater passing over the surface of the film which results in a lower HRT and a slightly inferior final effluent.

Oxygen diffuses from the air in the interstices, first into the liquid and then into the film. Conversely, carbon dioxide and the end products of aerobic metabolism diffuse in the other direction. The thickness of the film is critical, as the oxygen can ony diffuse to a certain depth before being utilized, leaving the deeper areas of the film either anoxic or anaerobic. Only the surface layer of the film is efficient in terms of oxidation, so only a thin layer of film is required for efficient purification, in fact the optimum thickness in terms of performance efficiency is only 0.15 mm. This means that it is the total surface area of active film that is important and not the total biomass of the film. As the thickness continues to increase, most of the soluble nutrients are utilized before they can reach the lower microorganisms forcing them into an endogenous phase of growth. This has the effect of destabilizing the film as the lower microorganisms lose their ability to hold onto the surface of the medium resulting in portions of the film within the filter becoming detached and washed away in the wastewater flow, a process known as sloughing. Although thick film growths do not reduce the efficiency of the filter, excessive growth can reduce the volume of the interstices, reducing ventilation and even blocking them completely, preventing the movement of wastewater. Severe clogging of the interstices is known as ponding and is normally associated with the surface of the filter when whole areas of the surface may become flooded.

The accumulation of film within a filter bed follows a seasonal pattern: thin in summer because of high metabolic rates and grazing rates but thick in the winter when the growth rate of the microorganisms is reduced, as is the activity of the grazing fauna.

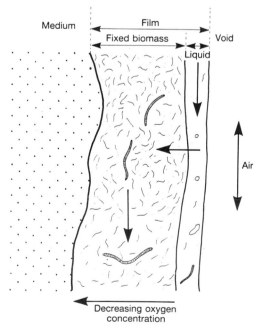

FIGURE 5   Structure of the film in a fixed-film reactor.

As the temperature increases in the spring there is a discernible sloughing of the film that has accumulated over the winter months (Fig. 6). Temperature is also an important factor in film accumulation. Below 10°C the rate of film accumulation increases rapidly. At higher temperatures a greater proportion of the BOD removed by adsorption is oxidized, and therefore fewer solids accumulate. The rate of oxidation decreases as the temperature falls, although the rate of adsorption remains unaltered. Therefore, at lower temperatures there is a gradual increase in solids accumulation which eventually results in clogged filters. Although grazers suppress maximum film accumulation and maintain minimum film accumulation for a large period of time after sloughing, temperature primarily controls film accumulation. While hydraulic loading in low-rate filters is of little significance compared to the action of macroinvertebrate grazers in controlling film, as the hydraulic loading increases, as is the case after modifications to the process such as recirculation or double filtration, then physical scouring of the film by the wastewater becomes increasingly important.

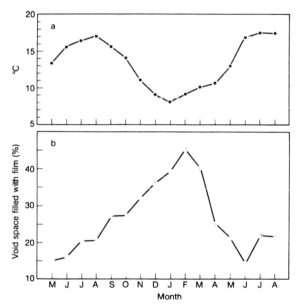

FIGURE 6 Seasonal variation in the temperature of sewage applied to a percolating filter and the proportion of void space in the filter filled with film and water. (a) Monthly average temperature of applied sewage and (b) monthly average proportion of voids filled with film. The quantity of film and water is measured *in situ* by the neutron scattering technique. Reproduced with permission of Academic Press.

In high-rate filters, especially those employing modular plastic media, the high hydraulic loading controls the film development by scouring the film from the smooth-surfaced media as it reaches critical thickness. In such filters sloughing tends to occur on a more regular basis rather than seasonally.

Solids discharged from filters are of three types: flocculated solids, detached fragments of the accumulated film, and the grazing fauna, their feces, and fragments of their bodies. These solids are collectively known as humus and, like all secondary solids, they require further treatment after settlement. The mode of operation influences the nature of the humus sludge. A high-rate operation produces a humus sludge mainly composed of flocculated solids and detached fragments of film, whereas a low-rate filter sludge contains a large proportion of grazing fauna and animal fragments. Sludges containing animal fragments and grazing fauna are more stable than sludges from high-rate systems where the grazing fauna is absent or reduced. The production of humus varies seasonally with mean production rates for low-rate filters between 0.20 and 0.25 kg humus per kg BOD removed, varying from 0.1 kg kg^{-1} in the summer to a maximum of 0.5 kg kg^{-1} during the spring sloughing period. In high-rate systems a greater volume of sludge is produced because of shorter HRT resulting in less mineralization. The humus production does not vary seasonally to the same extent with a mean production rate of 0.35 kg kg^{-1}, although this is dependent on the nature of the influent wastewater. Sludge production is much less compared to activated sludge, being more stabilized and containing less water (Table I), although as loading increases the mode of purification in percolating filters approaches that of the activated sludge process and the sludge alters accordingly both in quality and quantity.

Nitrification is a two-stage process with ammonia oxidized to nitrite by bacteria of the genus *Nitrosomonas* and nitrite to nitrate by *Nitrobacter* spp. In low-rate single pass filters containing a 50-mm stone medium, virtually full nitrification can be obtained throughout the year with a specific ammonia removal rate of between 120 and 180 mg m^{-2}day^{-1} when loaded at 0.1 kg BOD

TABLE I

Typical Sludge Volumes and Characteristics from Secondary Treatment
Processes Compared to Primary Sludge

Source	Volume (liter per head day^{-1})	Dry solids (kg per head day^{-1})	Moisture content (%)
Primary sedimentation	1.1	0.05	95.5
Low-rate percolating filtration	0.23	0.014	93.9
High-rate percolating filtration	0.30	0.018	94.0
Activated sludge (wasted)	2.4	0.036	98.5

$m^{-3}day^{-1}$, resulting in a final effluent low in ammonia but rich in nitrate. In a percolating filter, nitrifying bacteria tend to become established later than heterotrophs, with *Nitrosomonas* established before *Nitrobacter* as ammonia is abundant. So the first sign of nitrification in a filter is the production of nitrite rather than nitrate. The number of nitrifying bacteria and the level of nitrifying activity increase with depth, resulting in the upper level of single pass filters being dominated by heterotrophs and the lower section containing a proportionately higher number of nitrifying bacteria. The reason for this apparent stratification is due to a number of factors. The autotrophic bacteria responsible for nitrification are slow growing compared to heterotrophs and have an even more reduced growth rate in competitive situations. So in the upper layers of the bed where there is abundant organic matter, the heterotrophs dominate. Nitrifiers are extremely sensitive to toxic compounds in the wastewater, especially heavy metals, and so the presence of such compounds limits the growth of the bacteria until the compounds have been removed from the wastewater by adsorption by the heterotrophic film as it passes through the filter bed. Nitrifying bacteria are strict aerobes and so are inhibited by reduced aerobic conditions caused by high heterotrophic activity and since nitrification is a high oxygen-comsuming process, adequate supplies of air are required. When loadings are increased, extending the depth of heterotrophic activity, nitrifying bacteria are overgrown and inactivated by the quicker growing heterotrophic bacteria. Nitrifica-

tion is virtually eliminated by hydraulic loadings of domestic wastewater of >2.5 $m^3m^{-3}day^{-1}$ due to enhanced heterotrophic growth extending throughout the depth of the bed and effectively pushing the nitrifying organisms out of the filter. Temperature also has a marked influence on nitrification.

IV. ACTIVATED SLUDGE PROCESSES

A. Basis of Process

Activated sludge is currently the most widely used biological wastewater treatment process in the developed world, treating both domestic and industrial wastewaters. It consists of two phases: aeration and sludge settlement. Originally it was developed as a batch process with aeration and settlement taking place within the same tank. Now a continuous system is used with no settlement allowed within the aeration tank and two separate units are used, one for aeration and another for sedimentation (Fig. 3). In the first phase, wastewater is added to the aeration tank containing the mixed microbial population and air is added either by surface agitation or via diffusers using compressed air. The aeration has a dual function, to supply oxygen to the aerobic microorganisms in the reactor for respiration and to keep the microbial biomass in a continuous state of agitated suspension, ensuring maximum contact between the microorganisms and the wastewater. This continuous mixing action is important not only to ensure adequate food for the microorganisms, but also to achieve a maximum air–water oxygen concentration gradient to enhance mass transfer, and to help disperse metabolic end products from within the biomass. As the settled wastewater enters the aeration tank it displaces the mixed liquor (the mixture of wastewater and microbial biomass) into a sedimentation tank. This is the second stage, where the flocculated biomass settles rapidly out of suspension to form a sludge with the clarified effluent, which is virtually free from solids, subsequently discharged as the final effluent. In the conventional activated sludge process, between 0.5 and 0.8 kg dry weight of sludge is produced for every kilogram of BOD removed. The sludge is like a weak slurry containing between 0.5 and 2.0% dry solids, and so can be easily pumped. As the solids content increases the viscosity rapidly becomes greater, although under normal operating conditions activated sludge is difficult to consolidate to >4% dry solids by gravity alone. Most of the activated sludge is returned to the aeration tank to act as an inoculum of microorganisms, ensuring that there is an adequate microbial population to fully oxidize the wastewater during its retention within the aeration tank. The excess sludge requires further treatment prior to disposal. The concentration of biomass in the aeration tank is known as mixed liquor suspended solids (MLSS) and in conventional activated sludge systems is maintained at 3500 mg liter^{-1}.

Ideally the activated sludge process should be operated as close to a food-limited condition as possible to encourage endogenous respiration when the microorganisms untilize their own cellular contents, reducing the quantity of biomass produced. During the endogenous respiration phase the respiratory rate falls to a minimum which is sufficient for cell maintenance only. However, under normal operating conditions the growth of the microbial population and the accumulation of non-biodegradable solids results in an increase in the amount of activated sludge produced. The removal mechanism, assimilation or mineralization, can be selected by using specific operating conditions with certain advantages and disadvantages. For example, the most rapid removal of nutrients is achieved by removing organic matter by assimilation only

(i.e. adsorption) in which it is precipitated in the form of biomass. Such processes produce considerable surplus sludge which requires a higher proportion of the operating costs to be spent on sludge separation and disposal. Complete oxidation (mineralization) of wastewater is much slower and requires long aeration periods. So although much less sludge is produced, which reduces the sludge handling costs, aeration costs are much higher.

B. Process Microbiology

The basic operational unit of activated sludge is the floc. Under the microscope, activated sludge is composed of discrete clumps of microorganisms known as flocs, which vary both in shape and size. Good flocculant growth is important for the successful operation of the process, so that suspended, colloidal, and ionic matter in the wastewater can be removed by adsorption and agglomeration in the aeration tank, and subsequently in the sedimentation tank for the rapid and efficient separation of sludge from the treated effluent. There is a rapid agglomeration of suspended and colloidal matter onto the flocs as soon as the sludge and wastewater are mixed (enmeshment) which results in a sharp fall in the residual BOD of the wastewater. The volatile matter content of flocs is generally high, between 60 and 90%, although this depends on the nature of the wastewater and the amount of fine suspended and colloidal inert matter present. The adsorption capacity of the floc depends on the availability of suitable cell surfaces. Once all the adsorption sites are occupied the floc has a very reduced capacity for adsorbing further material until it has metabolized that already absorbed. Breakdown and assimilation of the agglomerated material proceeds more slowly (stabilization). So if the hydraulic retention time of the wastewater (HRT) in the aeration tank is too short, there will be a progressive reduction in BOD removal as there will have been insufficient time for adsorbed material to be stabilized. Although all flocs have a specific gravity of >1.0, only well-formed flocs in the sedimentation tank settle out of suspension with smaller flocs and dispersed microorganisms being carried out of the tank with the final effluent. Removal efficiencies in sedimentation tanks are significantly lower for smaller particles, especially those <20 μm in diameter. The settling velocity of a floc is a linear function of the cross-sectional diameter of the largest dimension of the floc. Also, the porosity of flocs increases as a function of the largest dimension, with the rate of increase falling dramatically after the floc exceeds 200 μm. The process depends on continuous reinoculation with recycled settled sludge, so the system will only select floc-forming organisms that rapidly settle in the sedimentation tank. So the process is microbially self-regulating with the best flocs recirculated.

Individual flocs are complex biochemical units. Each floc is a cluster of several million heterotrophic bacteria bound together with some inert organic and inorganic material. There is a wide range of particle sizes in the activated sludge process ranging from individual bacteria of between 0.5 and 5.0 μm up to large flocs greater than 1 mm (= 1000 μm) in diameter. The maximum size of flocs is dependent on their physical strength and the degree of shear exerted by the turbulence caused by the aeration system in the aeration tank, although flocs are generally between 50 and 350 μm in diameter.

The process of floc formation is far from being understood. Originally it was attributed to the slime-forming bacterium *Zoogloea ramigera;* however, many other bacteria and protozoa are now known to be associated with floc formation. Bacteria, protozoa, and detritus are either attached to the surface of the floc or embedded in some form of material forming a matrix. The exact nature of this flocculating material is still not known, although it appears largely bacterial in origin. The material can be readily extracted from activated sludge and constitutes a significant portion of the dry weight of the sludge, up to 10%. This material is a polymer that can be composed of a number of organic compounds such as polysaccharides, amino polysaccharides, and protein. Lipids may also be present, but the exact nature of these flocculating polymers depends on the species of bacteria or protozoa producing it. Each polymer has varying surface properties and charges that influence not only the settling characteristics but also the water-

binding properties of the floc. The polymer not only gives the floc components cohesion, it also allows suspended particles in the waste to bind to the floc by adsorption. As a result, the polymer has a critical role in the operation of the activated sludge process. These extracellular polymers (ECPs) are not food reserves like poly-β-hydroxybutrate and so are not easily decomposed. The surface charges on the microbial cells and bridge formation by polyvalent cations also contribute to flocculation. The examination of flocs by electron microscopy has shown that granular and amorphous materials are present and that fine cellulose fibrils are formed. These cellulose fibrils can only be seen by taking ultrathin sections of sludge flocs and examining them with a transmission electron microscopy at magnifications in excess of × 20,000. Young flocs contain actively growing and dividing heterotrophic bacteria with a high rate of metabolism. In contrast, older flocs have a lower proportion of viable cells, being composed mainly of dead cells surrounded by a viable bacterial layer. Although the majority of these cells are no longer viable, they retain active enzyme systems. Older flocs have a reduced rate of metabolism but as they are physically larger they settle far more readily than younger flocs which are often associated with poor settleability. In fact the viability of microorganisms making up flocs may be as low as 5–20%. As the floc ages the slower growing autotrophs become established, especially the nitrifying bacteria, so the concept of sludge age is important in terms of overall efficiency.

Flocs undergo a secondary colonization by other microorganisms such as protozoa, nematodes, and rotifers. The ciliate protozoa are considered particularly important as they feed on dispersed bacteria reducing the turbidity of the final effluent, while the higher trophic levels present graze on the floc itself reducing the overall biomass. A well-flocculated sludge is in a state of dynamic equilibrium between the flocs aggregating into larger flocs and being broken up into smaller flocs by the shear stress imposed by the aeration system. Flocs have strong binding forces, based on calcium, magnesium, or other multivalent cations, and can tolerate quite high shear stresses. The surface area of flocs has been calculated by dye adsorption and ranges between 43 and 155 $m^2 g^{-1}$ of floc. The radius of flocs can then be calculated from the specific surface area, assuming that it is a homogeneous mass of microorganisms. So the lowest surface area of 43 $m^2 g^{-1}$ has a floc radius of 0.58 μm. However, in practice, flocs are considerably larger than this, indicating that the high specific surface area is due to flocs being porous. This is confirmed by electron microscopy which shows that flocs have a spongy appearance. The porous nature explains why flocs are so good at adsorbing particulate matter, also why the diffusion rate of nutrients and oxygen into the center of the floc is greater than if flocs were homogeneous masses of bacteria. There are, however, areas of the floc that are either food or oxygen limited. For example, gradients of oxygen concentration have been measured, decreasing toward the center of flocs, with the availability of oxygen limited by the rate of diffusion in larger flocs. The oxygen uptake rate of mixed liquor is significantly increased by breaking up the flocs by homogenization, thus effectively removing any diffusion limitation.

Floc morphology of lightly and heavily loaded activated sludge plants varies in a characteristic way. For example, lightly loaded plants have compact flocs with a darker central core or inclusions that are made up primarily of inorganic material such as iron hydroxide, calcium phosphate, and aluminium hydroxide, along with nonbiodegradable organic material. The lighter, less dense outer regions of such flocs are composed of active microorganisms. This is because these flocs are older and the floc has undergone repeated periods of active growth and subsequent starvation resulting in a compact floc with the older nondegraded material in the center. In contrast, heavily loaded plants often form finger-like growths (zoogloeal) in which individual bacteria are embedded in a transparent matrix. It is assumed that growth is rapid since the age of flocs is relatively short and comprised almost entirely of active bacteria. In practice, however, variability is so great that both inert inclusions and finger-like bacterial growths radiating from flocs are often seen together, so the diagnostic value of such features remains uncertain.

The most important function in the activated sludge process is the flocculant nature of the microbial biomass. Not only do the flocs have to be efficient in the adsorption and subsequent absorption of the organic fraction of the wastewater, but they also have to rapidly and effectively separate from the treated effluent within the sedimentation tank. Any change in the operation of the reactor leads to changes in the nature of the flocs which can adversely affect the overall process in a number of ways, most notably in poor settlement resulting in turbid effluents and a loss of microbial biomass.

In general, good flocculation is associated with low-rate activated sludge processes and poor flocculation with high-rate processes. However, considerable variation is seen between flocs from different sludges and it is this variation that causes operational difficulties. For example, toxic discharges, nutritional imbalances, or changes in the microbial ecology of the process can all alter the surface chemistry of the flocs which in turn can influence the settlement characteristics of the activated sludge.

C. Operational Problems

The most common operational difficulties associated with activated sludge are concerned with the separation of sludge from the clarified wastewater in the sedimentation tank. The ability of activated sludge to separate is normally measured by an index of settleability such as the sludge volume index (SVI) or the sitrred specific volume index (SSVI). Problems in sludge settlement can be caused by bulking, deflocculation, pin-point flocs, foaming, or denitrification. These terms describe the effects, although their definitions are rather imprecise and there is some overlap.

With the exception of denitrification, all settleability problems can be traced back to the structure of the activated sludge floc. Floc structure has been subdivided into two distinct categories: micro- and macrostructure (Fig. 7). Microstructure is where the flocs are small (<75 μm in diameter), spherical, compact, and relatively weak. They are composed of floc-forming bacteria and are formed by aggregation and bioflocculation where individual microorganisms adhere to one another to form large aggregates. The structure of such flocs is termed weak because in the turbulent conditions of the aeration tank they can easily be sheared into smaller particles. While such flocs rapidly settle the smaller aggregates that have been sheared from the larger flocs, which take longer to settle, may well be carried out of the sedimentation tank in the final effluent, increasing the BOD and giving the clarified effluent a high turbidity. When filamentous microorganisms are present the flocs take on a macrostructure, with the microorganisms aggregating around the filaments, making larger flocs of irregular shape and able to withstand high shear forces within the aeration tank due to this extra support.

In structural terms, bulking is due to flocs having an excessive macrostructure, so much so that filamentous organisms are present in large numbers. Bulking is a phenomenon where filamentous organisms extend from the flocs into the bulk solution interfering with settlement and the subsequent compaction of the activated sludge, with a SVI > 150 ml g^{-1}. The poor settleability extends the sludge blanket so that large flocs are carried out from the sedimentation tank with an increase in both the suspended solids and BOD concentration of the final effluent. This results in much thinner sludges being returned to the aeration tank with a low MLSS, so that maintaining the required concentration of biomass within the tank becomes progressively more difficult, leading to a fall off in performance. In the ideal floc, where the SVI is between 80 and 120 ml g^{-1} and the final effluent is largely free from suspended solids and turbidity, the filamentous and floc-forming organisms are balanced. The filamentous organisms are retained largely within the floc giving it strength and a definite structure. Although a few filaments may protrude from the floc, they are sufficiently scarce and of reduced length not to interfere with settlement. In contrast the flocs comprising a bulking sludge have large numbers of filaments protruding from the floc, with two types of bulking flocs discernible (Fig. 7). Fairly compact flocs with long filaments growing out of the floc and linking individual flocs together (bridging) to form a meshwork

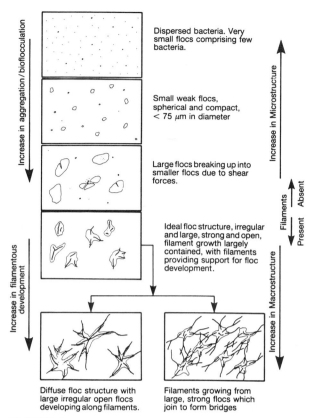

FIGURE 7 The floc structure of micro- and macrostructures.

of filaments and flocs. Alternatively, flocs with a more open (diffuse) structure that is formed by bacteria agglomerating along the length of the filament to form rather thin, spindly flocs of a large size. The type of floc formed, the type of compaction, and the settling interference caused depend on the type of filamentous organisms present. About 25 different filamentous bacteria cause activated sludge bulking. A number of fungi and algae also cause bulking, although they are not normally found as dominant organisms in activated sludge.

V. OTHER BIOLOGICAL SYSTEMS

A. Stabilization Ponds

A stabilization pond is any natural or man-made lentic (enclosed) body of water where organic waste is oxidized by natural activity. These ponds are classified into three groups: anaerobic ponds and lagoons, oxidation ponds, and aeration lagoons. Oxidation ponds are aerobic systems where the oxygen required by the heterotrophic bacteria is provided not only by transfer from the atmosphere but also by photosynthetic algae. In facultative ponds the algae use the inorganic compounds released by aerobic and facultative bacteria for growth using sunlight for energy. They release oxygen into solution which in turn is utilized by the bacteria, completing the symbiotic cycle (Fig. 8). The algae in facultative ponds are restricted to the euphotic zone, which is often only a few centimeters deep, depending on the organic loading and whether it is day or night. Such ponds have a complex ecology with many predator–prey associations, with phyto- and zooplantonic forms predominating. Ponds can be used to produce fish, which feed either directly on the algal biomass or on the intermediate grazers. High-rate aerobic stabilization ponds are not designed for optimum purification of wastewater but for algal production. The algae are harvested for biomass or as single cell protein. Green algae such as *Chlorella* and *Scenedesmus* have a protein content of 50% (dry weight) compared to 60–70% for the blue-green alga *Spiru-*

lina. Light availability is the most critical factor controlling algal growth and so such ponds are restricted to those areas where there is plenty of sunshine.

B. Plants

The controlled culture of aquatic plants is becoming widely used in wastewater treatment. The plants employed can be classified into submerged algae and plants, floating macrophytes, and emergent vegetation.

The role of algae has already been discussed in relation to stabilization ponds. Floating macrophytes (duckweeds, water ferns, and water hyacinth) use atmospheric oxygen and carbon dioxide, but obtain the remaining nutrients they require from the water. The water hyacinth *(Eichhornia crassipes)*, a rhizomatous plant with large glossy green leaves and a feathery unbranched root system, is most widely used. Hyacinths are grown in special lagoons and need to be harvested periodically. They are used mainly for tertiary treatment, although they are also employed for secondary treatment. The roots provide a substrate on which heterotrophic micro-

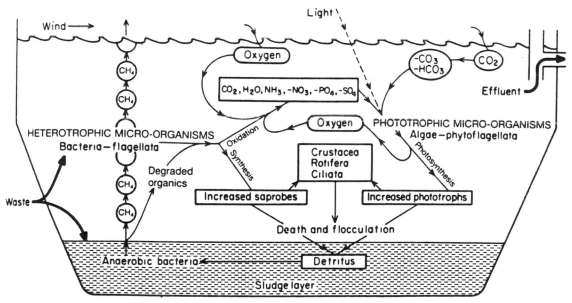

FIGURE 8 Summary of heterotrophic and phototrophic activities in a facultative pond that results in the complete stabilization of organic waste. Reproduced with permission from Academic Press.

organisms thrive. The most improvement in BOD and suspended solids is due to settlement rather than biological activity. The plants exhibit luxury uptake of nutrients, far in excess of normal requirements, and also accumulate other components in the wastewater such as heavy metals and synthetic organic compounds. The uptake and accumulation of such compounds can be disadvantageous when the concentration in the plant tissue prohibits the subsequent use of the biomass for conversion into energy by digestion or for use as a feed supplement.

The most widely used application of plants in wastewater treatment employs emergent vegetation in artificial wetlands or reed beds. Constructed wetlands are a low-cost alternative to conventional treatment processes in both capital and operational terms, often operating without any mechanical or electrical equipment. The process uses the root zone method of treatment. Emergent plants are grown in beds of soil or gravel retained by an impermeable subsurface barrier (clay or synthetic liner). The base of the bed has a slope of 2–8% to encourage settled sewage to pass horizontally through the soil (Fig. 9). The depth of the bed depends on the root penetration of the plants used, e.g., cattails, 300 mm; reeds, 600 mm; and bulrushes, 750 mm. The soil and the roots provide a large potential surface area for the growth of heterotrophic microorganisms. The long roots of the vegetation penetrate deep into the substrate.

The internal gas spaces of the plant, an adaption to its partially submerged existence, transport oxygen to the saturated rhizomes and roots. Thus an aerobic zone is created within an otherwise waterlogged sediment. Within the rhizosphere there are also anoxic and reduced areas. This combination allows both aerobic and anaerobic bacteria to survive, allowing carbonaceous oxidation, nitrification, denitrification, and anaerobic degradation to occur. The plants take up nutrients and store them in their tissues which are removed from the system when harvested. Finally, the substrate has a sorption potential to remove metals from solution. The removal potential exceeds 90% for most metals and these become immobilized in the anaerobic mud layer in the base of the filter as metal sulfides. Excessive accumulation eventually inhibits the process. During the winter the emergent part of the vegetation dies back and so treatment is restricted to the root zone only. The root zone method is becoming popular for the treatment of a range of nondomestic wastewaters, including landfill leachate, acid mine drainage, pulp mill effluents, and pharmaceutical wastewaters. Because of the relatively low loading rates that can be applied, constructed wetlands require large areas of land compared to conventional processes, and so are normally restricted to treating domestic wastewater from small communities. However, because of their ability to withstand a wide range of operating conditions and their high potential for wildlife con-

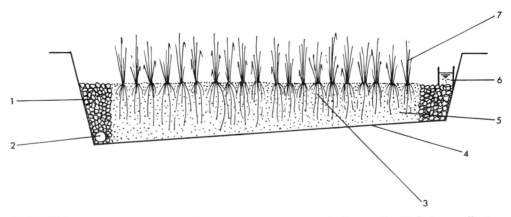

FIGURE 9 Basic features of a reed bed. (1) Drainage zone composed of large rocks, (2) drainage collection for treated effluent, (3) roots and rhizomes, (4) impervious liner, (5) gravel or soil, (6) influent sewage distribution system, and (7) reeds.

servation, this system has become popular with the industrial sector as part of an overall environmental management strategy.

C. Anaerobic Treatment

Anaerobic degradation occurs in the absence of oxygen. The basic difference between aerobic and anaerobic oxidation is that in the aerobic system oxygen is the ultimate hydrogen acceptor with a large release of energy, but in anaerobic systems the ultimate hydrogen acceptor may be nitrate, sulfate, or an organic compound with a much lower release of energy. Briefly, the process of anaerobic decomposition involves four discrete stages (Fig. 10). The first stage is the hydrolysis of high-molecular-weight carbohydrates, fats, and proteins, which are often insoluble, by enzymatic action into soluble polymers. The second stage involves acid-forming bacteria which convert the soluble polymers into a range of organic acids (acetic, butyric, and propionic acids), alcohols, hydrogen, and carbon dioxide. Acetic acid, hydrogen, and carbon dioxide are the only end products of the acid production that can be converted directly into methane by methanogenic bacteria. A third stage is present when the organic acids and alcohols are converted to acetic acid by acetogenic bacteria. It is in the final phase, which is

perhaps the most sensitive to inhibition, that methanogenic bacteria convert the acetic acid to methane. Although methane is also produced from hydrogen and carbon dioxide, in practice about 70% of the methane produced is from acetic acid. Obviously the methanogenic stage is totally dependent on the production of acetic acid and so it is the third stage, the acetogenic phase, that is the rate-limiting step in any anaerobic process. A large number of anaerobic processes are available, including anaerobic lagoons, digesters, and filters. Anaerobic lagoons and filters are normally used to treat strong organic wastes, although further treatment using an aerobic process is normally required to reduce the BOD and ammonia concentrations in the final effluent prior to discharge.

Glossary

Activated sludge Flocculant microbial biomass produced when wastewater is aerated continuously.

Aerobic oxidation Breakdown of organic matter under aerobic conditions by bacteria to form stabilized end products such as carbon dioxide, nitrate, sulfate, and water.

Anaerobic degradation Breakdown of organic matter in the absence of oxygen by anaerobic bacteria.

Carbonaceous oxidation Oxidation of organic matter to carbon dioxide by aerobic bacteria.

Denitrification Microbial reduction of nitrate to nitrogen gas.

Nitrification Oxidation of ammonia to nitrite and nitrate by autotrophic bacteria.

Root zone method System of treatment using reeds or other emergent plants in a special reactor similar to a natural wetland.

Sludge loading rate Weight of BOD (wastewater) applied to a reactor in relation to the weight of biomass (microorganisms) present to oxidize it.

Stabilization pond Any natural or artificial enclosed water body used for the stabilization of organic matter in wastewater.

Trickling or percolating filter A bed of inert medium over which wastewater is distributed. The organic matter in the wastewater is oxidized by contact with a microbial layer that develops over the surface of the medium.

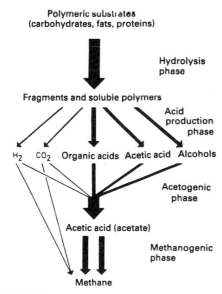

FIGURE 10 Major steps in anaerobic decomposition.

Bibliography

Cooper, P. F., and Findlater, B. C., eds. (1990). "Constructed Wetlands in Water Pollution Control." Oxford: Pergamon Press.

Curds, C. R., and Hawkes, H. A., eds. (1975). "Ecological Aspects of Used Water Treatment. I. The Organisms and Their Ecology." London: Academic Press.

Fox, C., Fitzgerald, P. R., and Lue-Hing, C. (1981). "Sewage Organisms: A Color Atlas." Chicago: Metropolitan Sanitary District of Chicago.

Gray, N. F. (1992). "Biology of Wastewater Treatment." Oxford: Oxford University Press.

Gray, N. F. (1990). "Activated Sludge: Theory and Practice." Oxford: Oxford University Press.

Mudrack, K., and Kunst, S. L. (1986). "Biology of Sewage Treatment and Water Pollution Control." Chichester: Ellis Horwood.

Wastewater Treatment for Inorganics

Li-Yin Lin

City of Niagara Falls, New York

Inorganics are very important to environmental biology. Some inorganics (e.g., oxygen, nitrogen, phosphorus, trace metals) are the essential elements in cell metabolism. For instance, oxygen is vital for respiratory processes in aerobic organisms, and nitrogen and phosphorus are nutrients for biomass production. Trace metals are required as constituents of enzymes or as cofactors. Some inorganics also exist in high amounts in organisms, such as sodium (Na), potassium (K), and calcium (Ca). Nevertheless, inorganics can also be toxic to human health and aquatic environments depending on their concentration. Research has found that metals cause some degree of anemia in fish, and the New York State Department of Environmental Conservation reported that human health is adversely affected by the consumption of mercury-contaminated fish. Therefore, wastewater treatment for inorganics, which originate mostly from industrial, agricultural, and domestic pollution sources, has been required to protect human health and water resources.

I. INTRODUCTION

The level of wastewater treatment is dependent on the water quality requirements of receiving streams. In turn water quality is based on the use of water, such as public water supply, industrial/agriculture water supply, recreation, water power, and navigation. Table I lists the surface water inorganic criteria for public water supplies. The treatment processes for inorganic substances, including metals, cyanide, nitrogen compounds, phosphorus compounds, and sulfur compounds, are discussed in this article. [See WASTEWATER TREATMENT BIOLOGY.]

II. METALS AND CYANIDE

Trace metals are required in cell metabolism. However, excess metals as well as cyanide can affect enzymes and then impede the fundamental biochemical reactions. Control of metals and cyanide, therefore, is needed to protect human health and the environment. Most metals and cyanide are primarily from direct industrial discharges, such as plating, steeling, and pigment industries. Section 307 of the 1987 Clean Water Act lists antimony, arsenic, beryllium, cadmium, chromium, copper, lead, mercury, nickel, selenium, silver, thallium, zinc, and their compounds plus cyanide as priority pollutants. Treatment of these priority pollutants is required prior to discharge.

TABLE I
EPA Interim Primary Drinking Water
Standards

Parameter	Maximum level (mg/liter)
Arsenic	0.05
Barium	1.0
Cadmium	0.01
Chromium	0.05
Flouride	1.4–2.4
Lead	0.05
Mercury	0.002
Nitrate (as N)	10
Selenium	0.01
Silver	0.05

A. Metals Treatment

Both biological and physical/chemical process can remove metals in wastewater. The commonly used biological processes for metals are activated sludge and trickling filter. The U.S. Environmental Protection Agency (EPA) studied the fate of priority pollutants in the late 1970s in publicly owned treatment works that mostly have biological treatment process. Results showed that both activated sludge and trickling filter have a good removal rate on metals. However, it should be noted that metals are not degraded in the biological process, but only transferred to sludge streams.

The physical/chemical processes that are commonly adopted by industries to remove metals in wastewater include chemical precipitation and coprecipitation, carbon adsorption, ion exchange, filtration, resin adsorption, sedimentation, and steam stripping.

1. Chemical Precipitation and Coprecipitation

The process of adding chemicals to enhance the settling of pollutants has been used for years. Chemical precipitation includes hydroxide precipitation, sulfide precipitation, and carbonate precipitation. Hydroxide precipitation acts through the addition of lime or other caustic to attain the pH of minimum metal solubility:

$$M^{2+}\begin{Bmatrix} CO_3 \\ SO_4 \\ Cl_2 \end{Bmatrix} + Ca(OH)_2 \rightarrow M(OH)_2 \downarrow + Ca^{2+}\begin{Bmatrix} CO_3 \downarrow \\ SO_4 \downarrow \\ Cl_2 \end{Bmatrix}$$

$$M^{2+}\begin{Bmatrix} CO_3 \\ SO_4 \\ Cl_2 \end{Bmatrix} + NaOH \rightarrow M(OH)_2 \downarrow + Na_2\begin{Bmatrix} CO_3 \\ SO_4 \\ Cl_2 \end{Bmatrix}$$

Sulfide precipitation is accomplished by adding sulfide compounds for metals removal:

$$M^{2+}\begin{Bmatrix} CO_3 \\ SO_4 \\ Cl_2 \end{Bmatrix} + \begin{Bmatrix} Na_2 \\ H_2 \\ Fe \end{Bmatrix}S \rightarrow MS \downarrow + \begin{Bmatrix} Na_2 \\ H_2 \\ Fe \end{Bmatrix}\begin{Bmatrix} CO_3 \\ SO_4 \\ Cl_2 \end{Bmatrix}$$

Carbonate precipitation is especially effective for the removal of lead and nickel:

$$M^{2+}\begin{Bmatrix} SO_4 \\ Cl_2 \end{Bmatrix} + Na_2CO_3 \rightarrow MCO_3 \downarrow + Na_2\begin{Bmatrix} SO_4 \\ Cl_2 \end{Bmatrix}$$

Pretreatment is needed for the wastewater containing cyanide and ammonia, which will interfere with carbonate precipitation. Coprecipitants act as coagulating material and to adsorb metals in wastewater. The chemicals most commonly used as coagulants are alum [$Al_2(SO_4)_3 \cdot 14H_2O$], ferric chloride [$FeCl_3$], Ferrous sulfate [$FeSO_4$], ferric sulfate [$Fe_2(SO_4)_3$], and sodium aluminate [$Na_2Al_2O_4 \cdot 3H_2O$]. Polymer or bentonite clay are often used as coagulant aids to assist in achieving a better settling.

2. Ion Exchange

An ion-exchange resin is another alternative for treating metals in wastewater. The cation ion-exchange resin is normally applied for metal removal. The stoichiometric reaction of cation exchange is

$$\underset{\text{solution}}{A^{n+}} + \underset{\text{resin}}{n(R^-)B^+} \rightleftharpoons \underset{\text{solution}}{nB^+} + \underset{\text{resin}}{(R^-)_nA^{n+}}$$

The exchangeable counterion of an acidic cation resin may be either hydrogen or some monovalent cation such as sodium.

B. Cyanide Treatment

Both physical/chemical and biological methods can be used for cyanide removal. The treatment methods include alkaline chlorination, ozone oxidation, electrolysis, hexavalent chromium oxidation, and the activated sludge process. The generally accepted technique for industrial waste treatment of cyanide compounds is the physicochemical method of alkaline chlorination.

1. Alkaline Chlorination

Cyanide is destroyed by oxidation under alkaline conditions with chlorine or hydrochloride. The detailed reactions are

$$Cl_2 + H_2O <=> HOCl + HCl \tag{1}$$
$$CN^- + HOCl <=> CNCl + OH^- \tag{2}$$
$$CNCl + OH^- <=> Cl^-HCNO \tag{3}$$

III. NITROGEN

Nitrogen is abundant on the earth. For instance, close to 80% of air is nitrogen, and approximately 12.5% of microbial cells is nitrogen. It can exist in many forms, such as nitrogen gas (N_2), ammonia (NH_3) or ammonium (NH_4^+), nitrate (NO_3^-), nitrite (NO_2^-), or organic nitrogen. Biological reactions can change it from one form to another. The nitrogen cycle is illustrated in Fig. 1.

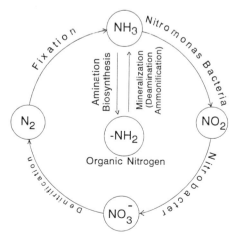

FIGURE I The nitrogen cycle.

Nitrogen compounds can also be a nuisance to human health and the environment. The following are the reasons why treatment of nitrogen compounds is required.

1. Nitrogen ion is a public health hazard in water consumed by infants. Nitrate reacts with blood and interferes in the blood's capacity to carry oxygen, this may lead to the "blue baby" syndrome.
2. Free ammonia is toxic to fish and other aquatic organisms.
3. One germ of ammonia (or ammonium ion) consumes 4.33 germ of oxygen. The dissolved oxygen in receiving water will be depleted by the presence of ammonia.
4. Nitrogen is one of the essential constituents to biological growth. The presence of nitrogen promotes growth in water and causes eutrophication.

The main sources of nitrogen in wastewater are from human and animal wastes. Nitrogen in sewage is composed of about 60% organic nitrogen and 40% ammonium nitrogen with very little nitrite or nitrate. Biological treatment systems are commonly used for nitrogen treatment. Physical/chemical processes, including breakpoint chlorination, selective ion exchange, and air stripping, are also employed for nitrogen removal.

A. Biological Treatment

Biological treatment is the most attractive nitrogen control technology. In many cases, biological treatment for nitrogen is also a cost-effective alternative. Two biological reactions are involved in nitrogen treatment: nitrification and denitrification. Organic nitrogen is first biodegraded to ammonia nitrogen, which is then transformed either by biosynthesis or by nitrification and denitrification.

1. Biosynthesis

$$NH_4^+ + 1.83O_2 + 1.98HCO_3^- \rightarrow 0.98NO_3^-$$
$$+ 0.021C_5H_7NO_2 + 1.88H_2CO_3 + 1.04H_2O$$

2. Nitrification and Denitrification

a. Nitrification

$$2NH_4^+ + 3O_2 \xrightarrow[\text{bacteria}]{\text{Nitrosomonas}} 2NO_2^- + 2H_2O$$
$$+ 4H^+ + \text{new cells}$$

$$2NO_2^- + O_2 \xrightarrow{\text{Nitrobacter}} 2NO_3^- + \text{new cells}$$

b. Denitrification

$$NO_3^- + \text{carbon source} \xrightarrow{\text{Endogenous}} N_2 + H_2O +$$
$$CO_2 + OH^-$$

The biological nitrogen removal systems include single-sludge and two-sludge systems, oxidation ditch, trickling filters, and rotating biological contactors.

B. Physical/Chemical Treatment

I. Breakpoint Chlorination

When wastewater contains ammonia nitrogen, ammonia will react with chlorine and eliminate the disinfection efficiency in chlorination. Breakpoint chlorination is the reverse of the theory to add enough chlorine in the waste stream to oxidize ammonia-nitrogen to nitrogen gas. The graphical representation of the chemical reaction between ammonium and chlorine is known as the breakpoint curve, which is illustrated in Fig. 2. The overall reaction is

$$NH_4^+ + 1.5Cl_2 \rightarrow 0.5N_2 + 4H^+ + 3Cl^-$$

Dechlorination may be needed to remove excess chlorine in treated water.

2. Selected Ion Exchange

Ion exchange is another alternative for ammonium removal. The ion-exchange bed should have a high selectivity for the ammonium over other cations in the wastewater. The natural zeolite clinophlolite has been found suitable for this application. Regen-

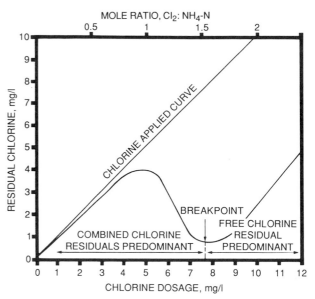

FIGURE 2   The chlorine–ammonium breakpoint curve.

eration is also required when the ion-exchange column is saturated.

3. Air Stripping

Ammonium ions become ammonia gas in high pH:

$$NH_4^+ + OH^- \rightleftharpoons NH_3 + H_2O$$

The process of air stripping works by raising the pH of wastewater to pH 10.5 to 11.5 and providing sufficient air flow to strip the ammonia gas from wastewater.

IV. PHOSPHORUS

Phosphorus is an essential nutrient to the photosynthesis of aquatic plants. Excess phosphorus added to a waterway, however, produces a nuisance quantity of aquatic plants and causes eutrophication. Therefore, control of phosphorus is required to prevent eutrophication. The sources of phosphorus in wastewater include (1) residential wastes (body wastes and kitchen wastes), (2) agricultural runoff, (3) detergent industries, and (4) commercial cleaning/laundry services. Phosphorus can occur in various forms: orthophosphate, condensed phosphate (pyro-, meta-, poly-), and organic phosphate. Currently, the standard of phosphorus discharge is 2 ppm or less depending on the receiving water. On the Great Lakes Basin, the total effluent phosphorus concentration cannot exceed 1 mg/liter.

A. Chemical Treatment

Phosphorus removal can be accomplished by the addition of lime [$Ca(OH)_2$] or the salts of aluminum or iron [alum, $Al_2(SO_4)_3 \cdot 18H_2O$; sodium aluminate, $NaAlO_2$; ferric chloride, $FeCl_3$; ferric sulfate, $Fe_2(SO_4)_3$; ferrous sulfate, $FeSO_4$; and ferrous chloride, $FeCl_2$]. These salts react with soluble phosphorus and form insoluble compounds. The insoluble compounds can be flocculated with or without coagulant aid (such as a polymer) to facilitate separation by sedimentation. Filtration may follow to achieve even better removal.

B. Biological Treatment

In the conventional biological treatment process, microbial solids have about 1.5 to 2% phosphorus. About 10–30% of all phosphorus can be removed as a result of sludge wasting in the process. The biological phosphorus treatment process is a modification of the conventional biological treatment system. This system grows biomass that has a much higher cellular phosphorus content, 3–6%. Therefore, more phosphorus can be removed from the liquid stream to the waste solids, which yields a lower effluent phosphorus concentration. The biological phosphorus removal systems A/O and Phosstrip are commonly used in the United States.

V. SULFUR

Sulfur compounds are widely distributed in nature and can exist either as the pure element or as the mineral. The sulfur cycle, shown in Fig. 3, may help one understand the transformations that occur. Elemental sulfur is mostly deposited deep underground. It is odorless and has been extensively used in industries, such as the chemical industry (e.g., for making sulfuric acid) and the rubber industry. However, pollution may occur as a result

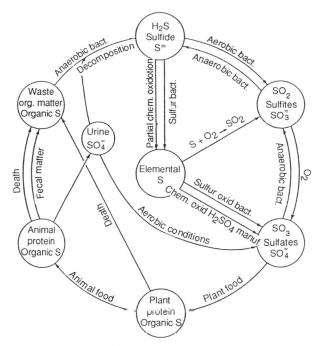

FIGURE 3 The sulfur cycle.

of using sulfur compounds or materials containing sulfur. For instance, burning high-sulfur coal and oil produces the air pollutant SO_2, which causes acid rain. Some sulfur compounds (e.g., hydrogen sulfide and dimethyl sulfide) cause nuisance odor problems.

A. Sulfates

The sulfate ion is one of the major anions occurring in natural waters. Sulfates are important in both water and wastewater fields. Because of its cathartic effect upon humans and the tendency to form hard scales in boilers and heat exchangers, the recommended upper limit of sulfates is 250 mg/liter in the water supply. Sulfates are also of considerable concern in wastewater, because they are indirectly responsible for two serious problems often associated with its handling and treatment: odor problems and sewer corrosion problems.

This occurs because sulfates can serve as a source of oxygen (the electron acceptor) for biological oxidations produced by anaerobic bacteria in the absence of dissolved oxygen and nitrates. Under anaerobic conditions, the sulfate ion is reduced to sulfide ion. The sulfide ion becomes the basis for un-ionized H_2S (hydrogen sulfide), an odorous compound, at pH levels below 8.

B. Sulfides

Sulfides are derived from two sources:

a. Sulfate reduction

$$SO_4^{-2} + \text{organic matter} \xrightarrow[\text{bacteria}]{\text{Anaerobic}} S^{2-} + H_2O + CO_2$$

b. Organic compound reductions: H_2S can be produced by the anaerobic decomposition of organic sulfur compounds, for example, amino acids.

The most commonly known sulfide compound is hydrogen sulfide. It has the following important characteristics: low odor thresholds of 0.5 ppb; obnoxious order, like rotten egg odor; corrosiveness; and toxicity. Control of this toxic and nuisance gas is required. The known methods of H_2S control involve reduction of H_2S production and the collection and treatment H_2S.

1. Reduction of H_2S Production

As mentioned previously, H_2S is often formed as a result of anaerobic bioactivities. Therefore, killing the bacteria or limiting their number is an effective means of reducing H_2S production. The commonly used disinfectants include chlorine, ozone, chlorine dioxide, hypochlorites, and hydrogen peroxide.

2. Collection and Treatment of H_2S

The method of collecting and treating H_2S is considered the most positive form of H_2S control and is gaining wider acceptance. The procedure consists of localizing H_2S, covering unit process facilities to prevent the gases from reaching the atmosphere, and treating H_2S by wet scrubbing, incineration or catalytic combustion, carbon adsorption, and chemical oxidation.

Glossary

Biological treatment Process that relies on mixed microorganics to break down organics in wastewater; this process has been commonly used in publicly owned treatment works.

Ion exchange Process used to remove either cations or anions in wastewater; both natural and artificial ion-exchange resins are available.

Physical/chemical treatment Process that includes chemical precipitation, carbon adsorption, ion exchange, chlorination, and other procedures as a way to change the physical properties of wastewater to separate pollutants or change pollutant characteristics.

Bibliography

DEC (1980). "Trend in Levels of Several Known Chemical Contaminants in Fish from New York State Waters," DEC Publication Technical Report 80-2. Albany, New York, June 1980.

EPA (1974). "Physical–Chemical Nitrogen Removal," EPA Technology Transfer Seminar Publication. Washington, D.C.: EPA.

EPA (1980). "Fate of Priority Pollutants in Publicly Owned Treatment Works," EPA-44-/1-80-301. Washington, D.C.: EPA.

EPA (1985). "Odor and Corrosion Control in Sanitary Sewerage Systems and Treatment Plant," EPA/625/1-85/018. Washington, D.C.: EPA.

Groronszy, M. C., Eckenfelder, W. W., and Froelich, E. (1992). "Wastewater—A Guide to Industrial Pretreatment," Vol. 6, pp. 78–83, Chemical Engineering, June, 1992. New York: McGraw-Hill, Inc.

International Joint Commission, Great Lakes Regional Office "1985 Report on Great Lakes Water Quality Report" Great Lakes Water Quality Board.

Lin, L. Y. (1989). "Odor Control Study Report." Niagara Falls, N.Y.: City of Niagara Falls, Department of Wastewater Facilities.

Water Quality Improvement Functions of Wetlands

Donald A. Hammer

Hammer Resources, Inc.

Natural wetlands receive inputs of nutrients and energy from waters flowing off surrounding upland regions. The complex of plants and animals in depressional wetlands have adapted to and, in fact, thrive on these imported substances. Over time, we have discovered how natural wetlands process, recycle, and trap these nutrients and energy. This knowledge has enabled us to deliberately design constructed wetlands that emulate the structure and function of natural wetlands to provide low-cost, low-maintenance, but highly effective treatment for a variety of wastewaters. In turn, our information on constructed wetlands readily transfers to improved understanding of water quality improvement in natural wetlands.

I. INTRODUCTION

Natural wetland systems provide many functional benefits to our society but virtually all can be grouped into three broad categories—life support,

hydrologic buffering, and water quality improvement. Most of us are familiar with the many types and large numbers of animals, especially birds, that are dependent on wetlands, but how many realize that the crayfish industry in Louisiana and the shellfish and much of the finfish industry along our coasts rely on wetlands habitat? Wetlands also reduce flooding along rivers and streams by reducing and desynchronizing peak runoff through slowing floodwater velocities. At the other extreme, delayed flows emanating from wetlands augment base flows in streams and rivers and so maintain levels essential for aquatic life. Finally, contaminated waters flowing through natural wetlands are cleansed by a combination of physical, chemical, and biological activities and emerge as clean water. [*See* WETLANDS ECOLOGY.]

Wetland ecosystems have intrinsic abilities to modify or trap a wide spectrum of waterborne substances commonly considered pollutants or contaminants. Doubtless our ancestors perceived and exploited these abilities, but in more recent times, casual observations fostered renewed interest leading to investigations that documented changes in concentrations of various materials after processing by natural wetland systems. Much of the early work on constructed wetlands for wastewater treatment was stimulated by observing this purification phenomenon in wetland systems. [*See* WASTEWATER TREATMENT BIOLOGY.]

Many observers have noticed accelerated soil erosion after heavy rains wash across unvegetated soils and some were fortunate to encounter situations where silt-laden waters transiting natural wetland systems were readily compared with unprocessed waters. The striking visual differences were easily verified by sampling and analysis and the information became an important component in a communal body of knowledge on natural wetland ecology. Most ecologists believed that this phenomenon was widespread and a few even suggested that it might occur on a large scale, although little documentation was available. A few years ago, I observed an example of water quality improvements in river waters by a natural wetland system on a very large scale.

The Pantanal of western Brazil and adjacent portions of Paraguay and Bolivia is a large basin bordered by high plateaus on the east and north and a moderate mountain range on the west. Runoff from these regions causes much of the 11,000,000-hectare area to be flooded from December to June and a significant portion is permanently wet. Although Pantanal means marsh or swampland in Portuguese, the region is more correctly considered a large plain with a considerable amount of permanently inundated area in old river channels, meanders, lakes, smaller depressions, and potholes. Many rivers enter the Pantanal from the eastern and northern highlands, gradually disappearing and then reforming on the western and southern boundaries and draining off to the south. Over geological time, alluvial deposits of highlands silt have gradually transformed a flat or concave basin floor into a convex domelike surface with higher elevations in the center and lower on the margins.

Doubtless this region provided important water improvement functions since tectonic forces created the original basin. The accumulated deposits that formed the present cross-sectional profile are dramatic evidence of previous beneficial modification of inflowing river waters. But accelerated erosion and pollution from clearing and agricultural activities and other anthropogenic sources have tremendously increased the contaminant loading of rivers draining the adjoining plateaus. The Rio Taquiri alone carriers over 30,000 metric tonnes of silt per day plus a variety of agrochemicals from soybean fields on the eastern plateau. Other rivers transport lower silt loads but most receive untreated sewage and industrial and mining pollution before reaching the Pantanal. For example, one iron mill uses 4.8 kg of detergent per day for washing ore stacks along the Rio Correntes, gold miners lose 36,000 kg/year of mercury along the Rio Couros and the Rio Aqua Branca, and eight alcohol distilleries (fuel) discharge 3600 m^3/day of organic waste ("vinhoto") into rivers draining the northern plateau. The combined impact of increased pollutant loadings has caused recent hydrological and biological changes in the upper reaches of the Pantanal that are of concern to Brazilians and conservationists around the world. [*See* RIVER ECOLOGY.]

But the amazing fact is that alarmingly high concentrations of silt and pollutants in inflowing river waters are reduced to innocuous levels in waters of rivers draining the region. The Rio Taquiri is a fairly wide, deep, and fast-flowing river that drops off the plateau to the east, coursing out into the Pantanal, and rather quickly its width, depth, and velocity are reduced. A third of the way into the Pantanal, numerous small braided streams arise flowing perpendicularly out of the Rio Taquiri into the adjacent regions. Progressively increasing water loss with penetration into the Pantanal drastically reduces the Rio Taquiri until it almost disappears. A similar pattern is evident in the course of the Rio Aquidauana. Many of the rivers flowing into the Pantanal virtually disappear because of sheet flow and dissemination through very small waterways in this vast wetland region. In fact the Pantanal functions as an 11,000,000-hectare sponge that absorbs inflowing waters, cleanses them of impurities, and slowly releases clean water through minor streams that aggregate into larger rivers along the southern and western boundaries. This large natural wetland complex transforms heavily polluted influent waters into clean waters collected by the Rio Paraguay and used throughout much of southern South America. Not only does it provide clean water, but the slow release of waters collected during the rainy season augments base flow in the Rio Paraguay during the dry portion of the year, supporting adequate year-round supplies for communities, navigation, and other human and natural uses. On a large scale as well as in local areas, natural wetlands can perform substantial improvements in water quality and quantity despite abnormal conditions. Even though this system is grossly overloaded and significant changes have occurred in wetlands of the upper Pantanal, the natural wetland complex still provides valuable water improvements for downstream rivers.

Natural wetlands provide effective, free treatment for many types of water pollution. Wetlands can effectively remove or convert large quantities of pollutants from point sources (municipal and certain industrial wastewater effluents) and nonpoint sources (mine, agricultural, and urban runoff), including organic matter, suspended solids, metals, and excess nutrients. Natural filtration, sedimentation, adsorption, microbial decomposition, and other processes help clear the water of many pollutants. Some are physically or chemically immobilized and remain there permanently unless disturbed. Chemical reactions and biological decomposition break down complex compounds into simpler substances. Through absorption and assimilation, wetland plants remove some nutrients for biomass production. One important by-product of plant physiology is the release of oxygen, which increases the dissolved oxygen content of the water and also of the soil in the immediate vicinity of plant roots. This increases the capacity of the system for aerobic bacterial decomposition of pollutants as well as its capacity for supporting a wide range of oxygen-using aquatic organisms, some of which directly or indirectly utilize additional pollutants.

Some nutrients are held in the wetlands system and recycled through successive seasons of plant growth, death, and decay. If water leaves the system through seepage to groundwater, filtration through soils, peat, or other substrates removes remaining nutrients and other pollutants. If water leaves over the surface in winter, excess nutrients released from decaying plant tissues have less effect on downstream waters during the nongrowing season. Nutrients trapped in substrate and plant tissues during the growing season do not contribute to noxious algae blooms and excessive aquatic weed growths in downstream rivers and lakes.

It is well known that natural wetlands can remove iron, manganese, and other metals from acid drainage—they have been doing it over geological time periods. In fact, accumulations of limonite, or bog iron, were mined as the source of ore for this country's first ironworks. Limonite deposits are most common in the bog regions of Connecticut, Massachusetts, Pennsylvania, New York, and elsewhere along the Appalachians. Although now of limited economic importance in the United States, bog iron is still a significant source of iron ore in northern Europe.

Because wetlands treatment systems transform or remove pollutants from inflowing waters, the ultimate fate of certain substances within the wetlands ecosystem is of more than academic interest. Depending on the source, influents may contain various natural and anthropogenic organic compounds, heavy metals, pathogenic organisms, and salts. A few materials (i.e., selenium, arsenic) are selectively taken up by plants but most are precipitated or complexed within the substrate. Generally, only 4–5% of the nutrient loading on a wetlands system is incorporated into plant or animal tissue. However, some metals may occur in plants at relatively high concentrations. For example, iron levels as high as 5000 mg/kg and manganese levels up to 4100 mg/kg were present in cattail leaves and stems grown in experimental cells that were heavily loaded with acid mine drainage. However, similar concentrations occurred in cattail parts from a natural wetlands not receiving acid drainage. In either case, only traces of other metals were present. Copper was below detection limits in cattail from a natural marsh but averaged 6.1 mg/kg in cattail from two municipal wastewater treatment systems. But higher concentrations of lead were found in a natural cattail stand (1.7 mg/kg) than in cattail from the muncipal systems (0.3 mg/kg).

Though only low levels of potential toxic metals occurred in these samples, long-term effects of relatively high levels of iron and manganese are not known. In the short term, iron and manganese did not appear to have detrimental effects on cattail growth and vitality in the experimental cells. In fact, plants in the upstream portion of each cell were more robust than plants in the lower sections. Upper portions of each cell received raw inflowing acid mine drainage that probably contained small concentrations of micronutrients in addition to substantially higher concentrations of iron and manganese. Differential robustness within each cell was likely due to micronutrient uptake in the upstream portion and limited micronutrient availability to plants in the lower sections.

Constructed wetlands have recently received considerable attention as low-cost, efficient means to cleanup many types of wastewater. Though the concept of deliberately using wetlands for water purification has developed only within the last 20 years, in reality human societies have indirectly used natural wetlands for waste management for thousands of years. Humans have dumped wastes into nearby streams or wetland areas since prehistoric times. And as they do for natural ecosystems, wetlands processed these wastes and discharged relatively clean water. However, as human populations increased and concentrated in towns and later cities, the increased quantity of wastes discharged into these areas soon overloaded and damaged the wetlands, destroying their ability to remove waterborne pollutants. Excessive human wastes unmodified by wetlands damaged aquatic life in rivers, bays, and oceans and threatened drinking water supplies. Consequently, various methods for treating wastes were developed, starting with simple holding ponds and progressing through lagoons to more complex mechanical treatment plants to the conventional wastewater treatment systems in use today.

A. Definitions

Constructed wetlands, in contrast to natural wetlands, are human-made systems that are designed, built, and operated to emulate natural wetlands or functions of natural wetlands for human desires and needs. Constructed wetlands as used for wastewater treatment may include swamps (wet regions dominated by trees, shrubs, and other woody vegetation) or bogs (low-nutrient, acidic waters dominated by *Sphagnum* or other mosses), but most commonly they are designed to emulate marshes. Marshes are shallow-water regions dominated by emergent herbaceous vegetation—cattails, bulrushes, rushes, and reeds. Many marshes also have deeper regions with submergent and/or floating leaved macrophytes. Marshes are adapted to a tremendous variety of soil and climatic conditions and some marsh plants occur on every continent except Antarctica. Marshes are also adapted to a wide range of water quality conditions as well as substantial fluctuations in water flows and depths. Although bogs and swamps have been used for wastewater treatment, both are difficult to establish or manage and both have fairly specific water quality and quantity requirements. Alterations in either are likely to cause undesirable changes in the structure and function of bogs and swamps. [*See* Wetland Restoration and Creation.]

A few of the wetlands descriptors have unfortunately been used synonymously and need precise definition to ensure common understanding.

Natural wetlands are those areas wherein at least periodically the land supports predominantly hydrophytes and the substrate is predominantly undrained hydric soil or the substrate is nonsoil and is saturated with water or covered by shallow water at some time during the growing season of each year. Natural wetlands have and continue to support hydric soils and wetland flora and fauna.

Restored wetlands are areas that previously supported a natural wetland ecosystem but were modified or changed, eliminating typical flora and fauna, and used for other purposes, but then subsequently altered to return poorly drained soils and wetland flora and fauna to enhance life support, flood control, recreational, educational, or other functional values.

Created wetlands formerly had well-drained soils supporting terrestrial flora and fauna but have been deliberately modified to establish the requisite hydrological conditions, producing poorly drained soils and wetland flora and fauna to enhance life support, flood control, recreational, educational, or other functional values.

Constructed wetlands consist of former terrestrial environments that have been modified to create poorly drained soils and wetland flora and fauna for the *primary purpose of contaminant or pollutant removal from wastewater*. Constructed wetlands are essentially wastewater treatment systems and are designed and operated as such, though many systems do support other functional values. Constructed wetlands are designed to transform many pollutants into gaseous forms for release to the long-term biogeochemical reservoir in the atmosphere (i.e., nitrogen; Fig. 1) or to trap others in the substrate (i.e., phosphorus, metals).

Floating aquatics systems are a related type of natural treatment system that consist of specialized applications of floating plants, that is, the water hyacinth (*Eichhornia*) or duckweed (*Lemna*) systems. These are *not* constructed wetlands because they use a different conceptual design. In floating aquatics systems, the vegetation is used to take up the remaining nutrients after primary or secondary conventional treatment, and the vegetation is harvested to remove the nutrients from the system. In these systems, maintenance costs—primarily harvesting and disposal of plant biomass—may be very high.

The vast majority of wetlands constructed for wastewater treatment are classified as surface-flow systems, that is, influent waters flow across and largely above the surface of the substrate materials. Substrates are generally native soils (Fig. 2). In the other class—subsurface-flow systems—waters flowing through the system theoretically pass entirely within the substrate and free water is not visible. Substrates in subsurface-flow systems are typically various sizes of gravel, crushed rock, or soil. Relatively few subsurface-flow systems treating municipal wastes are operating in North America but most of the European municipal systems are subsurface-flow, soil-based systems. With a few exceptions only, surface-flow systems have been used for mine drainage, agricultural waste, urban stormwater, industrial wastewaters, or other applications to date. Because many of the operating subsurface-flow systems have experienced serious clogging problems, only surface-flow systems can be recommended for anything more than tertiary polishing of effluents with low concentrations of nutrients in small-scale applications such as individual home septic tank systems. In addition, subsurface-flow systems have not been shown to reliably accomplish nutrient (nitrogen and phosphorus) removal.

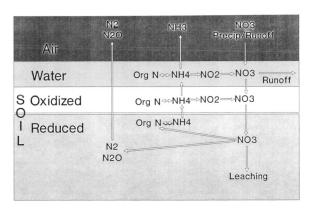

FIGURE 1 Nitrogen transformation processes in wetlands soils. [Adapted from Faulkner and Richardson, (1989). *In* "Constructed Wetlands for Wastewater Treatment," (D. A. Hammer, Ed.), p. 831. Chelsea, Michigan: Lewis Pub., Inc.]

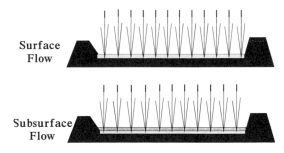

(Root Zone, Reed Bed, Rock Reed, etc.)

FIGURE 2 Surface-flow and subsurface-flow constructed wetlands differ largely on the basis of substrate media.

II. DEVELOPMENTAL HISTORY

Early ecological studies of the processes, reservoirs, cycling, and substance dynamics in wetlands revealed that natural wetlands receive and process many forms of nutrients and energy transported from adjacent upland regions by inflowing waters. Ironically, many of these investigators set out to evaluate detrimental impacts to wetlands from receiving various forms of wastewater. Most were surprised to discover that little impact was evident, but instead the wetlands removed many of the contaminants from the inflowing waters and discharged relatively clean waters. Both saltwater and freshwater wetlands were shown to have very important roles in the natural cycling of organic and inorganic materials. Concurrently, these studies showed that the immense productivity of wetlands was related to the inputs of nutrients and energy from terrestrial environments. In addition, the important role of wetlands as short-term and long-term sinks (reservoirs) became apparent because wetlands are often the major reducing compartment in the environment. The transformation of nutrient and energy inputs to increased biomass and innocuous products within the wetland system, while concurrently discharging cleansed waters, even led to some workers labeling wetlands the "kidneys" of the environment.

G. E. Hutchinson, E. P. Odum, and H. T. Odum and their students conducted extensive investigations of element cycles and summarized the studies of others in landmark publications from 1944 through 1971 that became the foundation for understanding material transformations in wetland ecosystems and the important role of wetlands in the landscape. Productivity evaluations of Wisconsin marshes by John Kadlec and of Danish *Phragmites* marshes by Hans-Henrik Schierup, and symposia at the Savannah River Ecology Laboratory in 1976 and in 1965 at Pallanza, Italy, also contributed to the information base on material processing and transformations in wetland ecosystems.

Over many years our knowledge of water purification in natural wetlands has slowly increased, leading to fairly rapid developments in constructed wetlands technology. Ironically, current and future research on wastewater treatment in constructed wetlands systems is likely to substantially improve our understanding of this important functional value in natural wetlands.

A. Municipal Wastewaters

The first work deliberately investigating wastewater treatment by wetlands plants was conducted by Kathe Seidel at the Max Planck Institute in Plon, Germany. In 1952, Dr. Seidel explored the removal of phenols from wastewater by *Scirpus lacustris* and she began testing dairy wastewater treatment with *S. lacustris* in 1956. From 1955 through the late 1970s she published numerous studies on water and wastewater treatment with wetlands plants. A student, R. Kickuth, continued her experimental work but also initiated monitoring at the Othfresen system in 1975, where municipal wastewater was discharged into a natural *Phragmites* stand. Although the Othfresen results have been questioned, Dr. Kickuth and his team popularized the concept in Europe, resulting in nearly 300 municipal and industrial waste treatment systems.

In the early 1970s, John and Robert Kadlec at the University of Michigan and Howard Odum and Kathy Ewel at the University of Florida began exploring the use of natural wetlands for wastewater treatment. Their efforts, along with those of Sterns, Spangler, Fetter, Sloey, and Whigham in Wisconsin, Valiela and Teal in New England, and others in addressing the impacts of wastewater effluents on natural wetlands, led to deliberate attempts to construct wetlands for wastewater treatment by Max Small at the Brookhaven National Laboratory in New York, and to develop various aquatic plant systems by Ray Dinges in Texas and Billy Wolverton at NASA in Mississippi. Subsequently, intensive pilot studies were conducted by Richard Gersberg at Santee, California, and Robert Gearheart at Arcata, California, along with early operational systems at Listowel, Ontario; Iselin, Pennsylvania; and Arcata, California.

B. Acid Drainage Modification

Similarly, wetlands modification of acid drainage was first documented during studies to evaluate detrimental impacts to wetlands caused by mine drainage by B. Huntsman in Ohio and R. Wieder and G. Lang in West Virginia. Subsequently, the teams of Wieder, Lang, *et al.* and Huntsman *et al.* initiated natural bog monitoring studies along with pilot experiments and small-scale constructed bogs designed to treat acid drainage waters. In the early 1980s, Ben Pesavento constructed a number of operating systems with a "let's learn as we go" approach that significantly contributed to widespread implementation of wetlands for acid drainage control in Pennsylvania.

Since natural wetlands treatment of mine drainage was found in *Sphagnum*-dominated bogs, most experimental work and early operating systems were attempts to construct bogs, but they had highly variable results because basic information on creation and management of bogs was severely limited. In 1984, the initial Tennessee Valley Authority (TVA) team of D. Hammer, W. Pearse, and D. Tomljanovich designed *Typha* marshes for acid drainage treatment because considerable information was available for marshes and marshes were likely precursors to bogs in the Ohio and Pennsylvania natural wetlands. Subsequently, most acid drainage treatment wetlands have been *Typha* or *Scirpus* surface-flow marshes and numerous reports on experiments and operating systems appeared during the middle to late 1980s and over 800 systems are now operating in Appalachia alone. [*See* Bog Ecology.]

C. Other Major Events

Other notable firsts include the use of constructed wetlands for treating:

1956–livestock wastewaters–experimental (K. Seidel)

1975–petroleum refinery wastewaters–operational (D. Litchfield)

1978–textile mill wastewaters–operational (R. Kickuth)

1978–acid mine drainage–experimental (B. Huntsman)

1979–fish rearing pond discharge–operational (D. Hammer and P. Rogers)

1982–acid mine drainage–operational (B. Pesavento)

1982–reduction of lake eutrophication–experimental (R. Reddy)

1982–urban stormwater runoff–operational (M. Silverman)

1983–pulp/paper mill wastewaters–experimental (R. Thut)

1985–photochemical laboratory wastewaters–experimental (B. Wolverton)

1985–seafood processing wastewater–experimental (V. Guida and I. Kugelman)

1988–compost leachate–operational (U. Pauly)

1988–landfill leachate–experimental (N. Trautman and K. Porter)

1988–livestock wastewaters–operational (D. Hammer and B. Pullin)

1988–river cleanup–operational (D. Hay)

1989–sugar beet processing plant wastewaters–operational (P. Anderson and D. Hammer)

1989–reduction of lake eutrophication–operational (F. Szilagyi)

1990–harbor dredged materials–experimental (U. Pauly)

1991–pulp/paper mill wastewaters–operational (R. Thut and D. Hammer)

Relevant conferences and symposia featuring wetlands and other aquatic plant systems have been held every few years, starting in 1972 with J. Tourbier and R. Pierson's conference on "Biological Wastewater Treatment" in Philadelphia, followed by the University of California-Davis wastewater aquaculture conference in 1979. The AWWA Water Reuse Symposium held in San Diego in 1984 included several sessions featuring wetlands treatment projects. Beginning in 1976, a series of conferences focusing on wetlands systems have been held, including the University of Michigan, Ann Arbor, conference in 1976, the workshop on ecological impacts at the University of Massachusetts in 1984, the University of Florida conference in Orlando in 1986, the TVA international conference in Chattanooga, the Humboldt State University conference in Arcata, California, in 1988, the second international conference in Cambridge, U.K., in 1990, a meeting in Pensacola, Florida, in 1991, the third international conference in Sydney, Australia, in 1992, and the fourth international conference in Guangzhou, People's Republic of China, in 1994.

III. WETLANDS PURIFICATION PROCESSES

Wetlands accomplish water quality improvement through a variety of physical, chemical, and biological processes operating independently in some circumstances and interacting in others (Table I). Vegetation obstructing the flow and reducing the velocity enhances sedimentation, and many substances of concern are associated with sediments because of clay particle adsorption phenomena. Increased water surface area for gas exchange and algal production improves dissolved oxygen content for decomposition of organic compounds and oxidation of metallic ions. But the most important

TABLE I
Contaminant Removal Processes in Wetlands[a]

Process	Contaminant
Biological	
Microbial metabolism	Colloidal and settleable solids, refractory organics, nutrients, metals, biodegradable organics (BOD$_5$, COD)
Predation	Pathogens
Natural die-off	Pathogens
Plant metabolism	Refractory organics, pathogens
Plant adsorption	Nutrients, metals, refractory organics
Chemical	
Precipitation	Metals, salts
Adsorption	Metals, salts
Oxidation/reduction	Refractory organics, metals
Volatilization	Refractory organics, nutrients
Physical	
Sedimentation	Colloidal and settleable solids (includes metals, biodegradable organics, pathogens, refractory organics, nutrients)
Filtration	Colloidal and settleable solids
Adsorption	Coloidal solids
Photolysis	Refractory organics, pathogens

[a] Adapted from Stowell, R., Ludwig, R., Colt, J., and Tchobanoglous, G. (1981). Concepts in aquatic treatment system design. *J. Environ. Eng.* **107**, 919–940.

process is similar to decomposition occurring in most conventional treatment plants—only the scale of the treatment area and composition of the microbial populations are likely to be different. In both cases an optimal environment is created and maintained for microorganisms that conduct desirable transformations of water pollutants. Maintaining that environment in the small treatment area of a conventional package plant requires substantial inputs of energy and labor. Wetland systems use larger land areas and natural energy inputs to establish self-maintaining treatment systems providing environments for many more types of microorganisms because of the diversity of microenvironments in a wetland. The latter, along with a larger treatment area, frequently provides more complete reduction and lower discharge concentrations of waterborne contaminants. Most removal or transformation of organic substances in municipal wastewaters or metallic ions in acid mine drainage is accomplished by microbes—algae, fungi, protozoa, and bacteria. Wetlands, like conventional treatment systems, simply provide suitable environments for abundant populations of these microbial populations.

A. Components

Water purification functions of wetlands are dependent on four major components—vegetation, water column, substrates, and microbial populations. Plant stems and leaves in the water column reduce flow velocities, increasing time for chemical reactions and facilitating sedimentation—the physical and chemical processes. However, the principal function of vegetation in wetlands systems is to create additional environments for microbial populations, that is, substantial quantities of area—reactive surface—for attachment of microbes. Plants also increase the amount of aerobic microbial environment in the substrate incidental to the unique adaptation that allows wetlands plants to thrive in saturated soils. Most plants are unable to survive in waterlogged soils because their roots cannot obtain oxygen in the anaerobic conditions rapidly created after inundation. However, hydrophytic or wetgrowing plants have specialized structures in their leaves, stems, and roots somewhat analogous to a mass of breathing tubes that conduct atmospheric gases, including oxygen, down into the roots. Because the outer covering on the root hairs is not a perfect seal, oxygen leaks out creating a thin-film aerobic region—the rhizosphere—around each and every root hair. The larger region outside the rhizosphere remains anaerobic but the juxtaposition of a large, in aggregate, thin-film aerobic region surrounded by an anaerobic region may be important to transformations of nitrogenous compounds and other substances.

We have known for some time that plant architecture using temperature differentials between various portions of the plant increases gas exchange beyond levels expected from passive (air tube) transport. But attempts to compute oxygen mass introduced into the substrates by radial oxygen loss have been confounded by a number of variables that lack precise definition. Although earlier experiments suggested that plants with deep root structures had higher removal efficiencies, recent results suggest that species with dense, fibrous though shallow roots have lower radial oxygen loss per unit of plant biomass but may input larger quantities of oxygen into the substrate because they tend to grow in denser stands. However, *Scirpus cyperinus* (dense stands, fibrous, shallow roots) had lower removal efficiencies than did *Scirpus validus* or *Typha latifolia/angustifolia* (deep rhizomatus roots) in two systems treating municipal waste in western Kentucky. But widely fluctuating loading rates and clogged gravel substrates in these systems confused comparative evaluations and only root depth and root dry weight were related to discharge water chemistry. German scientists did not observe an increase in dissolved oxygen in water flowing through the root horizon of experimental substrates. More recent work has suggested that in wetlands receiving municipal wastewater, much of the radial oxygen loss is used by microbes within the rhizosphere and little is available for reducing the biochemical oxygen demand inherent in concentrated organic wastes.

More importantly, plants create and maintain the litter/humus layer that functions as a thin-film bioreactor. As plants grow and die, leaves and stems falling to the surface of the substrate create multiple layers of organic debris—the litter/detritus/humus/peat component of wetlands (Fig. 3). This accumulation of differentially decomposed biomass creates highly porous substrate layers that provide a substantial amount of attachment surface for microbial organisms. Decomposing biomass also provides a durable, readily available carbon source for microbial populations. The water quality improvement function in constructed and natural wetlands is principally dependent on the high conductivity of this litter/humus layer and the large surface area for microbial attachment. Wetlands vegetation substantially increases the amount of environment (aerobic and anaerobic) available for microbial populations both above and below the surface. Consequently, the most important role of the plants is to simply grow and die, which explains why removal performance efficiencies of different plant species tend to be similar within broad ecological categories, that is, emergents, submergents, and wet meadow plant species. This also explains why vegetation harvesting is not needed and could, in fact, be detrimental to pollutant removal performance of a wetland.

In constructed wetlands, plants generally take up only very small quantities (<5%) of the nutrients or other substances removed from the influent waters. Organic pollutants are often reduced to elemental forms with some (nitrogen, carbon, hydrogen) be-

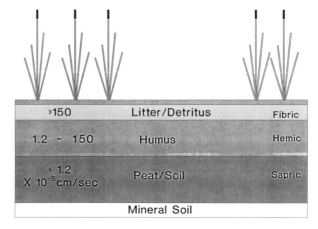

FIGURE 3 After death, marsh vegetation falls to the surface of the substrate, creating the litter/detritus layers that provide enormous quantities of reactive surface area and attachment sites for microbial organisms.

ing released to the long-term bio-geo-chemical reservoir in the atmosphere. Inorganic materials are transformed to innocuous substances and/or precipitated and trapped in the substrates. By contrast, floating aquatic systems (water hyacinth and *Lemna*) must incorporate periodic plant harvesting to slightly increase nutrient removal at considerable operating expense. Unfortunately, floating aquatics systems enjoyed wide use before constructed wetlands were developed and the concept that the plants extract nutrients in floating aquatic systems has erroneously carried over into common understandings of processes in natural and constructed wetlands.

Microbes—bacteria, fungi, algae, and protozoa—alter contaminant substances to obtain nutrients or energy to carry out their life cycles. In addition, many naturally occurring microbial groups are predatory and will forage on pathogenic organisms. The effectiveness of wetlands in water purification is dependent on developing and maintaining suitable environments for desirable microbial populations. Fortunately, these microbes are ubiquitous, naturally occurring in most waters, and likely to have larger populations in wetlands and waters contaminated with nutrient or energy sources. Only rarely, with very unusual pollutants, is inoculation of specific type or strain of microbes needed.

Substrates—various soils, clays, peat, sand, or gravel—provide physical support for plants, reactive surface area for complexing cations, anions, and some compounds, and attachment surfaces for microbial populations. Surface water (the water column) and subsurface water provide media for chemical reactions, transport substances and gases to microbial populations, carry off by-products, and provide the environment for biochemical processes of plants and microbes.

Invertebrate and vertebrate animals harvest nutrients and energy by feeding on microbes and macrophytic vegetation, recycling and in some cases transporting substances outside the wetlands system. Functionally, these components have limited roles in pollutant transformations but they often provide substantial ancillary benefits (recreation/education) in successful systems. In addition vertebrate and invertebrate animals serve as highly visible indicators of the health and well-being of a marsh ecosystem, providing the first signs of system malfunction to a trained observer. Some invertebrates and many vertebrates occupy upper trophic levels within the system that are dependent on robust, healthy populations of microscopic and macroscopic organisms in the critical lower levels. Declines in lower-level populations (including those involved in pollutant transformations) are reflected in changes in more visible animals in the higher levels. However, observations on types and numbers of indicator species must be carefully interpreted by an experienced wetlands ecologist because certain species thrive in overloaded, poorly operating systems.

IV. ADVANTAGES AND DISADVANTAGES

In comparison with conventional wastewater treatment systems, constructed wetlands have many advantages and a few disadvantages:

Advantages	Disadvantages
Low construction costs	Larger land area
Low operating costs	Optimal design factors lacking
Consistent, reliable	Engineers and regulators are unfamiliar with the technology
Simple operation	
Energy efficient	Project costs are not attractive to engineering firms
Advanced level treatment	
No sludge or chemical handling	Potential mosquito production
Accept load variations	
Attractive to wildlife	
Aesthetically pleasing	

A. Advantages

Advantages of constructed wetlands include relatively low construction costs—essentially grading, dike construction, and vegetation planting with little steel or concrete—and low operating cost—maintenance consists of monitoring water levels and plant vitality, collecting NPDES samples, and grounds maintenance (mowing dikes and roadways). Properly designed and constructed systems do not require chemical additions, internal pumping, sludge handling, or other procedures of conventional treatment systems. Neither do they require plant harvesting as do the floating aquatics systems.

In addition, construction does not require specialized expertise or expensive materials because construction primarily consists of grading the cells level, building low dikes or berms, and placing piping—PVC or concrete—depending on the volume of flow. Construction skills and required materials are available locally in most parts of the world. Operation is also simple and can be taught to local peoples with rudimentary educational skills. Basically, constructed wetlands are designed to be self-maintaining, requiring only periodic inspections to ensure integrity of pipes, water controls, and dikes.

Typically construction costs for constructed wetlands range from one-tenth to one-half as costly as comparable conventional treatment systems (Fig. 4). Wastewater treatment efficiencies are very good, especially for BOD_5, TSS, and fecal coliform bacteria with common discharge values of 10–20 mg/liter of BOD_5 and TSS and 50–100 colonies per 100 ml of fecal coliforms. With proper design and adequate treatment area, removal of nitrogen compounds and phosphorus is readily accomplished. Metallic ion removal even from acid waters is excellent and slight increases in pH are common when influent water is moderately acidic. But little decrease in acidity has been demonstrated for strongly acidic source waters.

As can be expected, performance varies with different designs, wastewater sources, and amount and type of pretreatment and treatment area/retention times, with most variation related to the type of system and treatment area/retention times. Though early subsurface-flow, gravel systems showed considerable promise for BOD_5 and TSS removals, severe clogging problems within 2–3 years of initial operation has caused failure of many systems. In addition, the anaerobic conditions in subsurface-flow systems do not support nitrification (Fig. 5). Consequently, subsurface-flow systems have not been successful for nutrient removals. Constructed wetlands are also amenable to substantial fluctuations in loading rates, capable of adapting to weekly and annual fluctuations in flows, for example, from high schools and urban stormwater runoff.

Constructed wetlands can provide ancillary benefits in the form of wildlife habitat, recreational and environmental space, or simply urban greenspace. Recreational activities derive from vertebrates, larger invertebrates, and to some extent vegetational components. Amateur naturalists and environmental educators quickly identify and exploit the educational and recreational benefits of a nearby simulated marsh. Systems located near urban areas may also provide greenspace benefits or simply open, natural areas that attract a variety of low-intensity recreational users-walking, jogging, picnicking, relaxing, and so on. Treatment system

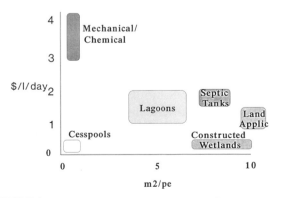

FIGURE 4 A comparison of construction costs for various types of wastewater treatment systems.

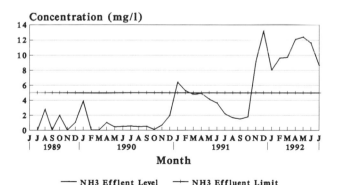

FIGURE 5 Subsurface-flow gravel wetlands remove ammonia-nitrogen during initial operation through cation exchange, but after the substrate cation exchange capacity is saturated, nitrogen removal stops. This wetland was constructed in Mandeville, Louisiana.

operators, pleased with the attention and support received from local citizens, usually welcome ancillary uses. More importantly, many realize that recreational benefits and pleasing aesthetics of wetland systems may reduce opposition to siting new or expanded systems.

However, wetlands constructed for wastewater treatment, at least initially, are comparatively simple, often single-plant-species systems. A properly designed and constructed cell with adequate treatment area covered in a dense stand of *Typha* or *Scirpus* will efficiently remove target contaminants from influent waters while providing habitat for a few muskrats (*Ondatra*), blackbirds, and some songbirds but little else. If operated at maximum efficiency for wastewater treatment, it will not have adequate capacity to store floodwaters nor can it release substantial quantities to amplify low stream flows in dry conditions. Wastewater treatment has been maximized through optimized design and operating criteria and all other functional values have been subordinated. But the water improvement function is still efficient and enduring even though other wetland functional values are substantially reduced or nonexistent.

B. Disadvantages

Constructed wetlands are land intensive—they require much more land area than do package treatment plants—and they require relatively level surfaces. Where land costs are high, such as in larger cities or in very rugged terrain that would need considerable grading, constructed wetlands may be more expensive to construct than conventional systems, although lower operating costs over a 20- to 30-year plant life time must be factored into the decision process. For example, current design recommendations specify 1.5–5 hectares of treatment area per 1000 m³/day of flow depending on the level of pretreatment and the desired discharge limits.

In addition, present design, construction, and operating criteria are imprecise—the reason for the range of treatment area requirements often recommended. And wetland systems, either natural or constructed, are complex, dynamic systems about which we have only limited understanding. However, a number of experimental and operating systems throughout the world are beginning to accumulate the data base from which precise design and operating criteria are being developed.

Another disadvantage is delayed operational status. Because peak removal efficiencies of constructed wetlands are dependent on vegetation and microbial population establishment and growth, design removal efficiencies are not likely to be attained until after two or perhaps three growing seasons. Completing construction and planting simply does not translate into full operational status and potential users must plan to gradually phase in wetlands operation concurrent with phaseout of the existing conventional system. Proper operation of the wastewater flow and/or existing conventional plant during this phase can be critical to avoid overloading the new wetland.

Treatment system longevity is poorly documented as few operating-scale systems have been in operation for more than 15 years. Because these systems simulate natural wetlands ecosystems that have functioned to purify water for thousands of years, I expect that system efficiency is not likely to be detrimentally impacted by age, but artificial constraints may require modifications or restarting after some period of time. Litter/detritus accumulation rates have been measured at 2–3 cm/year in municipal systems with no loss of treatment efficiencies. Therefore designs should incorporate this accumulation factor in dike height specifications and dikes should have one meter of freeboard for a 30-year operating lifetime or greater for longer operational status. At that point, the system may need to be cleaned out and restarted, and after testing to identify possible toxic substances, accumulated litter may be composted or land-applied similar to conventional sludge.

With other applications, for example, acid mine drainage, accumulated deposits may be useful in recycling or mining. A small wetland in northeast Alabama removes over 5 metric tonnes of iron each year from influent seep waters. Between 800 and 900 constructed wetlands currently treat acid mine drainage in Appalachia alone and mining companies are building 50–60 a year. In 40 or 50 years these small wetlands will have duplicated the "bog iron" sources that provided the foundation for the original iron-processing industry in Europe and North America. Our grandchildren may find these concentrated iron deposits much less expensive to process than taconite or other low-grade iron ore sources.

Finally, improperly designed or operated constructed wetlands could create pest problems—mosquitoes or rodents—that may cause adverse impacts on local residents. Both are easily avoided with appropriate designs and operating procedures.

V. CONSTRUCTED WETLANDS WASTEWATER TREATMENT SYSTEMS

Design, types, and composition of constructed wetlands vary considerably depending on when the project was designed (later designs include improvements) project objectives, wastewater applications, geographic/climatic location, and designer experience. The original designs attempted to emulate natural wetlands with subsequent "improvements" by adding artificial media, "best" plant species, and mechanical devices (recirculating pumps, greenhouses, etc.) in attempts to improve efficiency. However, many of the "improvements" resulted in more costly systems to build, operate, and maintain and failed to fulfill some of the original objectives.

Our objectives in designing constructed wetlands for wastewater treatment include developing a system that is:

1. capable of providing high-level treatment and discharging relatively clean water;
2. inexpensive to build;
3. largely self-maintaining, requiring little or no operation/maintenance time or expense;
4. manageable by operators with very limited training; and
5. capable of providing aesthetic/recreational/educational benefits.

Consequently, some attempts to reduce the size (required treatment area) to the absolute minimum,

to install artificial media (gravel), to use vertical flow and/or batch loading, to add recirculation and/or aeration, or to cover the wetlands with a structure (greenhouse) fail to meet our objectives. Virtually all of these may reduce the initial construction cost but all will increase operating costs. More importantly, all these modifications will increase operational complexity and costs, thereby eliminating the original attributes that made constructed wetlands for wastewater treatment highly attractive. A simple, slightly larger system may be more expensive to construct but it will be much easier and less costly to operate. Projecting operational costs over a 20- to 30-year plant lifetime often results in lower total costs for the larger system that was more expensive to initially construct. Furthermore, probability for failure increases almost directly with increasing complexity. Adding pumps, piping, artificial media, greenhouses, fans, and so on lays the basis for failure of a critical component at an inopportune time (midnight Saturday night!) and dramatically increases the likelihood that operator inattention will increase the probability of failure.

For municipal and livestock wastewaters, preliminary treatment is provided by a single-stage or multistage lagoon system designed to achieve a 40–50% reduction of the BOD_5 and TSS loading in the wastewater stream. If a lagoon is not present or practical, at least a settling basin designed to remove solids, grit, and debris must be located upstream of the wetlands. For urban stormwater treatment, a sediment/retention pond provides the requisite preliminary treatment. Strongly acid waters often require pretreatment in a limestone anoxic buffer (LAB) for successful performance of the wetland.

Wetland site selection is generally controlled by the desire to provide gravity flow for wastewater to the system, between system components, and within each component of the system to eliminate costs and maintenance of pumping wastewaters. Similarly, water control structures or devices consist of simple "T" pipe structures along the length of inlet distribution piping and swiveling "elbow" piping or flashboard/stoplog constructs for discharge control structures. Neither clogs as easily or requires adjustment or other complex maintenance typical of ball or gate valve flow control devices.

A. System Components

Some of the most successful designs for municipal wastewater treatment consist of a number of distinct components sequentially located in each cell as follows:

Emergent Marsh: The first compartment is a shallow basin with densely growing marsh vegetation—typically cattail (*Typha*), bulrush (*Scirpus validus* or *cyperinus*), reed (*Phragmites*), or rushes (*Juncus, Eleocharis*)—in 10–20 cm of water (Fig. 6). The marsh functions in BOD_5, suspended solids (TSS), metals, pathogens, and complex organics removal as well as in ammonification. Initial operating water depth is 8–10 cm above the soil surface, gradually increasing during the next 15–20 years to 15–20 cm as peat accumulates in the marsh.

Pond: The second compartment is a constructed pond with 0.75–1.0 m water depths similar to an aerobic lagoon/oxidation pond. Duckweed (*Lemna*) grows on the surface of the pond and various algae within the water column. Submerged pondweeds with linear, filiform leaves (*Potamogeton, Ceratophyllum, Elodea, Vallisneria*) are planted in shallow portions of the pond to increase microbial attachment surface area. The pond functions in further reduction of BOD_5 and most significantly for nitrification and denitrification. Operating depth is typically 0.6–0.9 m throughout the years of operation.

Emergent Marsh: The third compartment is a shallow basin with densely growing marsh vegetation—typically cattail (*Typha*), bulrush (*Scirpus validus* or *cyperinus*), reed (*Phragmites*), or rushes (*Juncus, Eleocharis, Vallisneria*)—in 10–20 cm of water. The marsh functions in BOD_5, suspended solids (primarily algae and *Lemna*), metals, pathogens, and complex organics removal as well as in denitrification. Initial operating water depth is 8–10 cm above the soil surface, gradually increasing during the next 15–20 years to 15–20 cm as peat accumulates in the marsh. Comparatively the marsh functions most efficiently for BOD_5, TSS, and pathogen removal, but the pond, because of the greater amount of oxidized environment, is more efficient at transforming ammonia to nitrogen gases.

B. Size and Configurations

Because most wastewater treatment wetlands have been designed for minimum size and cost to provide the required level of pollutant removal, maximizing effective treatment area and reducing short-circuiting or unused treatment area results in rectangular shapes (Fig. 7). Generally inlet distribution and outlet collection piping is located completely across the upper or lower end of the cell and a rectangular shape theoretically enhances

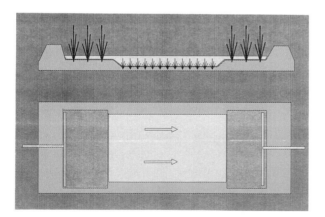

FIGURE 6 The marsh–pond–marsh design concept to enhance ammonia removal in constructed wetlands treatment systems. The alternating zones of shallow-water/emergent plants and deeper-water/submergent plants provide the combination of environments required for nitrification/denitrification.

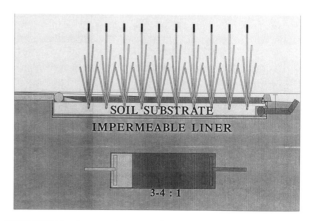

FIGURE 7 A typical rectangular wetlands cell with inlet and outlet distribution piping and an impermeable liner.

broad sheet flow across the width of the cell. Circular or elliptical shapes would likely have unused portions in each cell, but a "V" shape pointing downstream has been successfully used. A few projects have used linear systems contoured into the hillside because little level land was available and these tend to blend into the surroundings better than standard rectangular shapes. Generally wastewater treatment wetlands should have a 3–4:1 length to width ratio and rectangular shape if minimal treatment area is available. However, more aesthetic configurations would be practical if the wetland area is increased and ancillary benefits will increase with larger size.

Larger size has added benefits in that contaminant loading is less, loading fluctuations are more readily accommodated, and greater diversity of plant and animal species is possible, hence improving the water purification function as well as increasing other beneficial functions. Although planners tend to design the smallest wetland with the fewest components, that is, reduce the treatment system to the bare essentials, for most wastewater treatment projects, treatment efficiency and system resiliency increase with added size and biological complexity. Small simple systems are vulnerable to upset from fluctuating loading rates or pest outbreaks in important vegetation components. Too small and too simple often causes higher operating and maintenance costs and hampers system performance.

Numerous studies have shown that BOD_5 and TSS removal occurs in the upper portion of most constructed wetlands cells and is to a large extent related to removal of suspended materials through filtration, settling, and precipitation (Fig. 8). However, concentrations of nutrients, metals, and pathogens often remain above desired levels beyond the cell midpoint. Consequently, if only BOD_5, TSS, and moderate pathogen removal is required, square cells or even aspect ratios (length:width) of <1 are used. But if high removal efficiency is needed for BOD_5, TSS, pathogens, and/or nutrients and metals, cell aspect ratios should approximate 4:1, that is, cell length is four times greater than cell width. Wide, short cells perform fairly well for BOD_5 and TSS removal but poorly for ammonia removal. Higher aspect

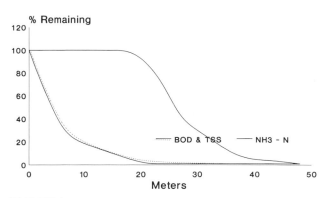

FIGURE 8 Concentrations of BOD_5 and TSS have been shown to rapidly decline in the upper section of many wetlands cells, but NH_3 concentrations remain high until most of the BOD_5 and TSS have been removed and dissolved oxygen levels increase.

ratios (>5:1) substantially increase construction costs without commensurate increase in removal efficiency. In addition, cells with high aspect ratios tend to be grossly overloaded in the upper portions, with adverse impacts on the vegetation and cell aesthetics, appropriately loaded through the middle portion, and under-loaded in the lower section.

Each component of the wetland system is basically a shallow pond or lagoon. Design and construction techniques used for farm ponds or treatment lagoons are appropriate for general features such as dikes, berms, and typical flashboard/ stoplog water control devices with two important exceptions. Dike freeboard must accommodate an organic matter (peat) accumulation rate of 2–3 cm/ year in the marshes and extend 30 cm above normal water level elevation for each 10 years of projected operation. Adequate freeboard and water level control is also necessary to provide capacity for flow beneath expected thickness of ice cover in colder climates.

Bottom slopes for treatment marshes and ponds are essentially flat. Width slope for the marshes must be flat to ensure equal flow distribution of wastewaters. Length slope may not be >0.02% in the marshes. Some earlier designs incorporated slopes of 0.5% and rarely as high as 4% primarily in subsurface flow systems, in the belief that increasing the elevation difference between the upper and lower ends of the cell would provide the "head" to force waters through persistently clogging gravel substrates. Most systems with >0.05%

slope and an aspect ratio <1:1 have experienced severe problems in maintaining wetlands vegetation. This is not surprising, because as little as 0.1% slope on a 300-m-long cell creates a water level difference of 30 cm between the upper and lower ends and operators find it impossible to maintain the desired 4- to 10-cm water depths needed by most emergent wetland plants throughout the length of the cell. Attempts to establish 6–10 cm at the midpoint mean that the upper end dries out and the lower end is too deep, with wetland plant mortality occurring above and below the midpoint.

C. Application Rates

Design recommendations should be based on mass loading rather than hydraulic loading rates. Because constructed wetlands are basically thin-film reactors, with an immense quantity of bioreactive surface, it is not surprising that comparisons of hydraulic loading rates or retention times show much poorer correlations with treatment efficiencies than do comparisons of contaminant mass loading rates with effective surface (treatment) area in these systems. Consequently, designs are based on mass loading rather than hydraulic loadings or retention times.

Application rates for treating municipal wastewaters in temperate climates, after preliminary treatment (at least settling or comminution), are as follows:

1. To meet secondary discharge standards of BOD_5 and TSS <30 mg/liter; fecal coliforms <200 CFU/100 ml; pH 6–9; and dissolved oxygen >4 mg/liter.

BOD_5 and total suspended solids (TSS) <100 kg/ha/day
Hydraulic loading <1000 m^3/ha/day
Retention time >5 days

2. To meet advanced discharge standards of BOD_5 and TSS <20 mg/liter; fecal coliforms <100 CFU/100 ml; pH 6–9; dissolved oxygen >4 mg/liter; NH_3 <2 mg/liter; and PO_4 <1 mg/liter.

BOD_5 and TSS <70 kg/ha/day
TKN or NH_3 <3 kg/ha/day
PO_4 <0.1 kg/ha/day
Hydraulic loading <500 m^3/ha/day
Retention time >10 days

In tropical regions, higher average annual temperatures will generally allow for higher loading rates, but our experience in the tropics is inadequate to develop guidelines at present. In arctic regions, larger treatment area is needed to compensate for reduced performance in cold periods of the year.

This is the required effective treatment area measured from dike toe to dike toe, not dike crown to dike crown. To avoid discharging untreated wastewaters if a single component fails, this treatment area is divided into parallel cells, that is, at least two marsh/pond/marsh cells with approximately equal treatment areas. Flow control devices are designed to divide normal flows between the parallel units but should have the capability to route all flow to one cell in case one unit must be removed from service for maintenance or repair.

Constructed wetlands treating livestock wastewaters, following primary treatment in lagoons, generally use the same loading rates. However, owing to the large quantity of solids common to livestock wastes, a lagoon or settling pond is essential for primary treatment prior to the constructed wetlands. Designers of a few previous livestock systems failed to include primary treatment and the wetlands rapidly filled with solids. In addition, the livestock lagoon discharge may have such high ammonia concentrations that dilution with clean water may be required to prevent damage to wetlands vegetation.

Industrial wastewaters with "normal" organic pollutants, suspended solids, and nutrients are loaded at similar rates. However, various salts, metals, complex organics, dyes, and "color" need substantially lower loading rates. For example, significant "color" removals often require retention times in excess of 30 days (and commensurate loading rates) and most heavy metals are loaded at <1 kg/ha/day and some as low as 0.1 kg/ha/day. On-site, individual home systems may be designed with the BOD_5 and TSS for advanced discharge standards or simply scaled to provide 3–4 m²/pe (person equivalent) in the household.

For acid mine drainage waters, recommended effective treatment areas vary with influent pH levels. Loading rates should not exceed 40 kg/ha/day for iron to produce an effluent level of <4 mg/liter total iron even though influent pH may be as low as 4.0. Manganese loadings are less clear and appear to be confounded by reduction of total iron within the system and influent pH levels. If adequate treatment area is available to reduce iron levels to <4 mg/liter and influent pH is >5.0, then a maximum of 4 kg/ha/day appears to be appropriate to produce discharge levels of <2 mg/liter manganese. If the influent pH is <5.0 and the buffering capacity low, a LAB trench will likely be needed. In that case, the manganese loading rate should not exceed 2 kg/ha/day to achieve effluent concentrations of <2 mg/liter.

D. Plant Materials

Wetlands vegetation substantially increases the amount of habitat available for microbial populations in the water column, in the litter/humus layer, and in the rhizosphere. Attributes of preferred plant species include:

1. adaptation to local climate and soils (native species);
2. tolerance to pollutants in the wastewater;
3. high biomass production;
4. perennial species;
5. rapid growth and colonization;
6. nonweedy, aesthetic habit; and
7. values for wildlife habitat.

Marshes are planted with *Typha* or *Scirpus* on 1-m centers during the first half of the growing season. *Potamogeton* in weighted cotton mesh bags and *Lemna* may be simply placed in the pond. Water lilies are planted similar to the emergent marsh species, consequently, it may be necessary to drain down the pond prior to planting of these species.

Cattail	*Typha angustifolia, T. domingensis,* or *T. latifolia*
Bulrush	*Scirpus* spp.
Rush	*Juncus, Cyperus, Fimbristylis, Eleocharis* spp.
Arrowhead	*Sagittaria* spp.
Iris	*Iris versicolor, I. pseudacorus*
Plantain	*Alisma* spp.
Giant reed	*Phragmites australis*
Submergents (pondweeds and lilies)	*Potamogeton, Ceratophyllum, Najas, Nelumbo, Nymphaea, Nymphoides,* etc.

Planting *Scirpus* or *Typha* is similar to planting any garden plant, but careful supervision is important because plants with damaged roots or plants placed incorrectly may not survive. Planting materials are often obtained locally, for example, cattail, reed, and in some cases bulrush are common in many depressions and roadside ditches. However, depending on labor costs, digging local materials may be more costly than purchasing plant stocks from a nursery or supplier. After planting is finished, water levels are gradually raised to normal operating elevations as the plantings grow, but water levels must not overtop new growth during the first growing season. Emergent plants are not as susceptible to drowning after the first growing season or in waters with relatively high dissolved oxygen content.

Alternatively, cattail and to some extent the other species can be established by simply manipulating the water levels at the appropriate time of the year. Or cattail and others may be seeded, but annual germination rates are very low, that is, on the order of 3–5% per year. Unfortunately, this method is dependent on natural means of seed dispersal and germination and may require more than one growing season to develop a dense stand.

Mosquito fish (*Gambusia*) and minnows (*Pimephales*) or other bait and/or top-feeding fish are often introduced into the marsh and pond after operating water depths have been stabilized. Bottom-feeding fish (carp, *Cyprinus;* catfish, *Ictalerus*) are not used as low-turbidity water is important to survival of submergent vegetation and various treatment processes.

E. Water Control Structures

Some wetlands have been designed with only an overflow spillway to prevent excessive water levels that may damage dikes. Many constructed wetlands treating acid drainage in remote areas rely solely on overflow spillways because water control structures are expensive and susceptible to vandalism. A few small constructed wetlands treating livestock waste or row-crop runoff also lack control structures for similar reasons. However, management of these systems is only possible with earth-moving equipment and little can be done to provide the critical disturbance element that would retard the inevitable succession to a terrestrial system.

Depth and duration of flooding control much of the plant community and directly and indirectly the animal community in wetland ecosystems. Consequently, wetland management primarily consists of water level manipulation and water control structures in the dikes are essential tools for managing constructed wetland systems. The principal use is for setting a specific elevation to maintain a desired water depth in the upstream pool. In some types of established wetlands, that elevation may not change during the year or over the course of many years. In others it may be necessary to raise or lower the level during the growing or nongrowing seasons to simulate natural hydrologic cycles. Infrequently, the water control will be used to drastically lower and perhaps gradually raise the pool level to foster germination and establishment of wetlands plants. It may also be used to completely drain the pool for dike repair or other needed maintenance or for deep flooding to retard or reverse successional changes. Generally a certain water depth will be maintained for weeks, months, or even years and the type of water control structure design reflects expected operation and the principal function.

Various types of valves or penstocks are often used for very large systems with high flow capacities and flows are regulated by partially obstructing the opening (reducing the functional diameter of the pipe). With a given pressure and above very small volumes, valves provide accurate regulation of flow *volumes*. At minimal flows, valves are susceptible to clogging, requiring frequent flushing of the blockage.

Valves are designed to regulate *volume* of flow not water level *elevations*. Two types of water controls designed to regulate water elevations have been in use in smaller systems for many years. Perhaps the oldest and most widely used is the "flashboard" or "stoplog" type. The other, the swiveling pipe, is the less expensive and perhaps the simplest structure providing excellent capacity for water level regulation in small systems.

F. Operation

Wetlands systems are relatively simple to operate if gravity flow is incorporated in all aspects of the design and the discharge water control structure is a flashboard/stoplog type or an elbow pipe structure. Screw-gate discharge controls require constant adjustment to maintain the desired water level elevation in the cell because of fluctuating inflows and precipitation. Fairly precise water level control is important in enhancing growing conditions for desirable plant species and controlling weeds if necessary. Constructed wetlands are often operated with clean water or very low strength wastewater for the first month or two after planting, followed by gradually introducing stronger concentrations or higher flows during the next 6 months.

Normal operating water depth is 10–20 cm in each marsh component and 0.7–1.0 m in the pond. Proper water depth and careful regulation is the most critical factor for plant survival during the first year after planting. Many plantings have failed because of mistaken concepts that wetlands plants need or can survive in deep water. Small, new plants lack extensive root, stem, and leaf systems with aerenchyma channels to transport oxygen to the roots. Consequently, flooding often causes more problems for wetland plants during the first growing season than too little water, especially if the water has low dissolved oxygen content.

After plantings have become well established, operation and maintenance typically consists of driving around the dikes at least once each week to check for any dike erosion, seepage, or animal damage, mowing the dikes to improve visitor access and aesthetics, and collecting water quality samples as needed. Routine inspections ensure appropriate flows through the piping, check vegetation health and vigor, and water levels in each component and dikes and flow control structures.

G. Monitoring

An assessment of the performance of a constructed wetlands is dependent on accurate determinations of the effluent volume of wastewater, the concentrations of various pollutants, and comparison of those values with influent parameters and/or regulatory discharge permit limits. Therefore, operators must have accurate information on the volume of flow and the pollutant concentrations in that volume of water. Simple "V" notched weir plates for flow volume measurement and automated samplers or grab samples are adequate for water quality analysis and performance evaluations. System performance is evaluated on the basis of removal/transformation of pollutants, and influent and effluent monitoring provides the basic data for comparison.

In addition to water quality parameters, monitoring biological components can provide valuable information on the health and vigor and, consequently, successful operation of the constructed wetlands. The monitoring of major biological components requires only weekly walking or driving around system dikes inspecting for stress signs or damage to the wetland vegetation—chlorosis, poor growth, stunted, off-color, insect damage, and so on. Concurrently, operators can casually inventory bird, frog, and other large animal populations to detect any significant changes that could suggest problems in the system.

VI. WASTEWATER APPLICATIONS OF CONSTRUCTED WETLANDS

Constructed wetlands currently treat wastewaters from towns and cities, campgrounds, visitor centers and individual homes, acid drainage from strip mines, deep mines, and ash storage ponds, urban stormwater runoff, livestock production facilities, row-crop runoff, landfill leachate, tanneries, textile, food processing, and pulp paper plants, petroleum refineries, harbor dredge material spoil sites,

and many other industrial sources on every continent except Antarctica. Operating systems are located from sea level to over 1500 m and from the tropics to subarctic regions in Ontario and the Scandinavian countries. Because operation is dependent on chemical and biological processes, pollutant removal efficiencies decline somewhat during low temperatures, but lower loading rates in cold climates produce discharge levels that are well below permit limits (Fig. 9).

A small scale is not limiting as constructed wetlands operate at individual home sites treating discharges from failed septic tank systems, small or large housing developments, towns, and small cities. However, land costs may be prohibitive in larger cities and conventional treatment alternatives may be more cost-efficient. Mining, industrial, and agricultural applications may vary from very small waste streams to millions of cubic meters of wastewater per day.

A. Municipal

Many of the public-owned treatment systems serving communities of under 10,000 population in the U.S. violate discharge standards and similar conditions exist across the country. Public concern with clean water supplies results in increasing pressure to enforce existing standards and further restrict permissible discharges to receiving waters.

Because small communities are often located along small or even intermittent streams, permitted

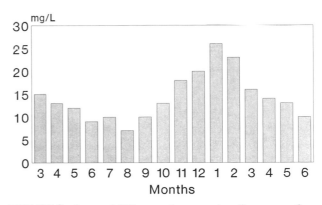

FIGURE 9 Seasonal differences in removal performance reflects temperature impacts on biochemical reaction rates. Systems operating in cold climates must have lower loading rates to ensure adequate treatment during the coldest portion of the year.

discharges are often more stringent than for cities discharging into major rivers, and so discharge limits for small treatment systems are likely to be lowered even more.

In the past, small towns benefited less from financial assistance programs because the concern, and rightly so, was to address the major problems first. But now and in the near future, the availability of construction grants is uncertain and many communities face the prospect of largely self-financing new, expensive wastewater treatment facilities. Some communities have the financial resources, but the majority of small towns have very limited ability to finance a multimillion dollar upgrade or new construction project.

Consequently, many towns are searching for some means out of their dilemma and an inexpensive, efficient process—constructed wetlands—is attractive. But constructed wetlands are not a panacea for all wastewater treatment problems. Although experimental work has been under way for over 40 years, the technology is still in its infancy and much remains to be learned about optimal design, construction, and operation. Most previous work has been directed toward municipal wastewater treatment and our information is adequate to *conservatively* design and operate systems for that use. Within the last 10 years a number of experimental and operating systems treating acid mine drainage have provided similar information. But the substantial potential for treating nonpoint sources, especially urban stormwater runoff and agricultural wastewaters, industrial wastes, leachates, and even failed septic tank drain fields at individual home sites, remains to be fully developed.

I. A Recent Design for Minot, North Dakota

The Minot constructed wetlands system provides polishing for wastewaters from some 43,000 residents. Minot previously used a five-cell facultative lagoon system (278 ha), but odor problems and the fact that Minot's discharge represents most of the flow in the Souris River during dry periods led to incorporation of very low (1 mg/liter) NH_3 parameters in the discharge permit. System upgrading included sludge removal and installing aerators in the first lagoons and construction of a four-cell marsh–pond–marsh wetlands with 51.2 ha of effective treatment area (wetted surface). The design was patterned after an earlier livestock wastewater treatment wetland built in northern Mississippi. Total project cost was $1.6 million compared to an estimated $7 million for a conventional treatment system. The wetlands was built in late fall 1990 and began operation in the summer of 1991.

Minot's northerly location (48°15′N) causes subfreezing air temperatures from November through March and ice cover on water bodies often approaches thicknesses of 1 m. Consequently, system design included 180 days of storage in existing lagoons with flows to the wetlands of 22,000 $m^3/$ day during the 180-day growing season. Design considerations included anticipated influent levels of 20 mg/liter for BOD_5, 20 mg/liter for TSS, 7 mg/liter for NH_3, 30 mg/liter for TKN, minimum influent levels of dissolved oxygen >3 mg/ liter, and desired wetlands discharge levels for NH_3 of <1.0 mg/liter. NH_3 was the critical design parameter and loading rates were established at 3 kg/ ha/day NH_3, although occasional peak loadings of BOD_5 and TSS needed to be accommodated.

Highest removal efficiencies for NH_3, BOD_5, and TSS were anticipated with alternating zones of dense emergent vegetation (i.e., *Typha* and *Scirpus*) in shallow water and submergent vegetation (plants growing under and up to the surface with long linear leaves and large amounts of surface area) in the deep sections. Creating wildlife habitat to provide opportunities for outdoor recreation and environmental education and to improve public acceptance of the wastewater treatment system was also desirable.

The design included four cells 177 m wide and 747 m long (length to width ratios of 4.2 : 1) with marsh–pond–marsh–pond–marsh zones sequentially within each cell (Fig. 10). Marsh zone A was designed to function for BOD_5 and TSS removal and ammonification. Design operating depth was 15 cm in the 100-m-long zone and it was planted in *Typha latifolia*. Pond zone B functioned for nitrification, had operating depth of 60 cm in the 221-m-long zone, and was planted with *Potamogeton pectinatus* and *Vallisneria americana*. Marsh

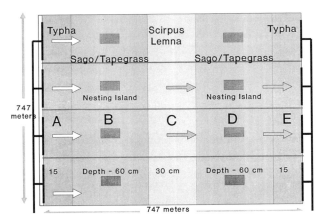

FIGURE 10 The Minot constructed wetlands system includes multiple shallow-water/emergent plants and deeper-water/submergent plants to provide high removal rates for NH₃ and fecal coliform bacteria.

zone C functioned for nitrification/denitrification and nutrient removal with operating depths of 30 cm in the 122-m-long segment and was planted with *Scirpus validus* and *Lemna*. Pond zone D functioned for nitrification/denitrification and nutrient removal, and to increase dissolved oxygen with operating depths of 60 cm in the 221-m-long zone planted with *Potamogeton pectinatus* and *Vallisneria americana*. The final segment, marsh zone E, functioned for denitrification, TSS (algae, *Lemna*), and fecal coliform removal in the 100-m-long segment with operating depths of 15 cm planted in *Typha latifolia*. Small islands were included in pond zones B and D for spoil disposal and wildlife nesting/loafing sites. Transition zones between zones A and B and zones D and E have 6:1 slopes with planted *Sagittaria latifolia*.

Monthly flows to the system have averaged 15,835 m³/day but ranged from 1896 to 28,761, with the flow equally distributed to all four cells in 1991, all flow applied to only cell number 4 to foster vegetation growth in the other cells from April to June 1992, and then applied to cells 1 and 4 through 1993.

Average influent levels for BOD_5 have been close to expected (12.8 mg/liter), but TSS levels have been higher (34.8 mg/liter) and NH₃ levels have been lower (2.4 mg/liter). Discharge values have been excellent: BOD_5 = 8.7 mg/liter, TSS = 16.2 mg/liter, NH₃ = 1.0 mg/liter (but 0.48 mg/

liter excluding November and December values when water temperatures were 0.5–1.0°C), and coliforms = 17.3 CFU/100 ml *without chlorination* (Fig. 11). Highest discharge NH₃ levels (4.62 mg/liter) occurred during cold weather when water temperatures approached freezing and when all flow was applied to only one cell (0.77 mg/liter) and the loading rate increased to 8.2 kg/ha/day NH₃. Average loading of 1.6 kg/ha/day NH₃ has been well within design limits. Effluent NH₃ concentrations were not related to hydraulic, BOD_5, or TSS loading rates but closely followed NH₃ loadings.

The Minot wetlands system has performed very well for normal wastewater discharge parameters. It has also provided excellent wetlands wildlife habitat and through cooperative efforts from the U.S. Fish and Wildlife Service, nesting platforms and other habitat amenities have been installed. Alternating depths and different vegetation types that optimize nitrogen removal also provides excellent combinations of shelter and feeding areas for many different kinds of wetlands and upland wildlife. The design has also accommodated additional wildlife inputs of fecal coliforms without chlorination.

2. Orlando, Florida

The Iron Bridge Regional Wastewater Treatment Facility consists of about 490 ha in 17 cells arranged in three components designed to treat 90,840 m³/day of municipal wastewater. The wet prairie (deep

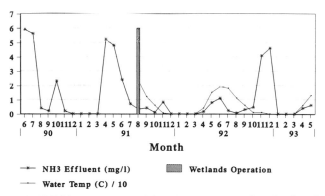

FIGURE 11 Operation of the Minot constructed wetlands polishing system has substantially improved the system's discharge quality, but NH₃ effluent concentrations were high during very cold weather when water temperatures were < 1°C.

marsh) was designed primarily for nutrient (nitrogen/phosphorus) removal and had a low initial plant diversity. The mixed marsh was designed to provide for nutrient polishing and a diverse wildlife habitat. The hardwood swamp serves primarily to provide wildlife habitat. A total of 20 different plant species was planted in the latter two stratum, and about 120 additional species have since developed from the seed bank. The facility has successfully reduced nitrogen discharge levels to <1 mg/liter and phosphorus levels to <0.1 mg/liter. Bird surveys indicate that 126 species utilize portions of the site at some point during the year, including 3 federally listed species. In addition, the numbers and types of small mammal, reptile, amphibian, and benthic macroinvertebrate species are similar to patterns seen for adjacent natural wetlands.

3. South Dakota Municipalities

Communities and engineers face severe climatic, physical, economic, and regulatory constraints in solving wastewater problems in South Dakota. Air temperatures range from $-30°C$ in winter to over $38°C$ in summer, and annual average precipitation varies from 63 cm in the southeast to 33 cm in the northwest. In 1987 a constructed wetlands was built to upgrade wastewater stabilization pond effluent at Kadoka, South Dakota. Since that time, over 40 systems have begun operation in South Dakota with the larger wetland systems at Huron (pop. 12,448), Belle Fourche (4335) in the northern Black Hills, Sisseton (2181), and Volga (1263). Many operate as total retention systems during the winter (150-day storage). The largest system at Huron treats 3785 m^3 of industrial meat-processing wastes in addition to domestic flows. Huron's wastewater treatment facilities include anaerobic lagoons, 110 ha of stabilization/storage ponds, and 134 ha of constructed wetlands to meet stringent ammonia limits for occasional zero flows in the receiving stream.

4. Others

The town of Iselin, Pennsylvania, has used a marsh–pond–meadow surface-flow constructed wetlands for its wastewater treatment since 1983. Construction cost was $174,000 for a system treat-

ing average daily flows of 26 m^3/day. Effluent concentrations have averaged 7.4 mg/liter for BOD_5, 19 mg/liter for suspended solids, 3.3 mg/liter for ammonia, 2.6 mg/liter for total phosphorus, and 150/100 ml for fecal coliforms.

The City of Arcata, California, has used a constructed wetlands system for treating its domestic wastewater since 1979. Average monthly BOD_5 and suspended solids values from the Arcata system (three marshes in series with total area of 10.5 ha have been loaded with oxidation pond effluent at 9452–11,355 m^3/day) by summer 1987 were consistently below Arcata STP discharge standards. These values have continued to improve, approaching a value of 10 mg/liter each for BOD_5 and suspended solids. Recent studies have also shown that economic benefits from tourism use of the treatment system have strongly influenced the economy of the city.

Since 1979, secondarily treated effluent has been used to create and maintain 113 ha of combined wastewater treatment/wildlife habitat wetlands in northeastern Arizona. These constructed wetlands have provided benefits of low-cost, efficient treatment of municipal wastewater for three local communities, highly productive marsh ecosystems, and nesting habitat for sensitive species of wildlife. This is an excellent example of wastewater reuse in arid climates.

B. Acid Drainage Remediation

Acid drainage is most commonly associated with strip or deep mining operations, but in fact it may follow any land disturbance in regions with moderate or high iron, aluminum, and/or manganese contents in soil materials. Acid drainage is generated by percolation of precipitation through soils wherein oxidation of iron sulfides has occurred, producing water that is acidic with variable concentrations of iron, sulfate, manganese, aluminum, suspended and dissolved solids, and occasionally color. Acid drainage frequently emerges as a seep or small spring, but in some cases where large areas and/or previous streams or waterways were disturbed, flows may be substantial. Conventional control techniques consist of land re-forming, including waterway modifications, and chemical treatment, basically addition of caustic soda (sodium hydroxide solution). Both are costly and chemical treatment requires long-term commitments by the operators. The ability of constructed wetlands to provide treatment for acid drainage was demonstrated in the early 1980s, leading to widespread application in coal and hard-rock mining areas and to a lesser extent in association with large construction projects. Currently over 800 constructed wetlands treat acid mine drainage in Appalachia alone.

Acid drainage wetlands are often located in remote, mountainous areas, and design, construction, and operation tend to differ from other wetlands treatment systems. Rough terrain dramatically increases earth-moving requirements and vandal-proof, self-maintaining features become critically important. High relief limits opportunities to create parallel cells and these systems commonly have cells or basins in series. Forming level or flat floors is also often impractical. Generally, a series of marshes/ponds is built by merely constructing dikes or berms at appropriate sites downstream from the acid seep/spring, resulting in shallow- and deep-water portions within each cell (Fig. 12). Initially, most treatment occurs in the upper, shallow areas, but over time, accumulation of precipitated deposits gradually shifts the shallow reaches farther and farther downstream, trans-

forming the deeper sections. Proper sizing of the cells to provide adequate treatment area as well as storage for deposited materials ensures operation over tens of years. In many cases, acid drainage becomes a normal spring over time as the reservoir of iron sulfide in the disturbed material is depleted. Simple overflow spillways, sized to accommodate 10-year 24-hour storm events, provide adequate water level management, resist unauthorized manipulation, and protect against high-flow damage.

Strongly acid waters (pH < 4) or waters with very high iron concentrations (>300 mg/liter) often require direct chemical modification either before or after removal of metallic ions with a constructed wetlands. The LAB (limestone anoxic buffer) system is designed to add buffering capacity to seep waters, raising the pH if initially very low or if the pH declines as a result of substantial iron precipitation. It consists of a trench dug into the seep, filled with high-calcium limestone rock, and capped with synthetic and clay materials. Size of the trench and the amount of limestone rock are dependent on flow volume, initial pH, iron concentration, and desired operating life, with typical LAB installations sized to provide buffering capability for 20–30 years.

Acid drainage wetlands range in size from a few square meters to over 100 ha depending on average flows, water quality parameters, storm flows, and available area. Shapes also vary because of topography, hydrology, and available lands, with many systems being simply a series of cells down through a small valley, although some systems have several cells terraced down a slope in very steep terrain.

Water depth and bottom slope vary considerably between systems and often within a single system, though most have depths ranging from 0.1 to 3 m because few wetland plants thrive in deeper waters. Shallower waters decrease retention times and long-term storage capacity. Deeper water inhibits many wetland plant species and increases the amount of anoxic, reducing conditions in the water column, which can reduce treatment effectiveness of the system. Preferred plants include *Typha latifolia* and *Scirpus validus,* though various species of *Scirpus, Juncus, Carex,* and *Eleocharis* have been used. Not surprisingly, after 2–5 species are planted

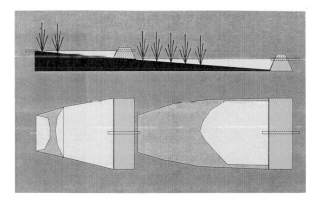

FIGURE 12 Mine drainage wetlands often have slight slope owing to the nature of the terrain in which seeps are located. Initially, treatment occurs in the upper shallow portion of each cell, but as masses of iron are deposited, the shallow zone moves farther and farther down the cell.

in a new system, considerable natural invasion occurs and a 4- to 5-year-old wetlands may have 60–90 wetland plant species.

Construction costs vary widely as some have been carefully designed and constructed whereas others were built and planted when workers and equipment were not scheduled for other tasks. One survey suggested that average cost ranged from $2.50–$15.00/m^2 of effective treatment area with variation due to intensity of effort and system size, because larger systems achieve economies of scale. Generally, construction costs approximated chemical and maintenance costs using conventional treatment for 1–2 years, that is, construction costs are often recovered within 2 years. For carefully designed systems, design and project management represented 20%, equipment and supplies 35%, and labor 45% of total costs. Operating costs normally consist of a monthly casual inspection and sample collection for NPDES reporting purposes.

Performance of acid drainage wetlands varies widely owing to types of designs, design and construction techniques, age of the system (general improvement with age), and drainage water quality parameters. Many early designs (and some recent) have not performed well and were subsequently modified to incorporate more recent information. Strongly acid or waters with high concentrations of metallic ions caused "failures" of some systems until a LAB was added. Generally, most systems successfully meet NPDES discharge standards and many kilometers of streams and rivers are gradually recovering with aquatic life returning after an absence of 50–100 years. Constructed wetlands have provided inexpensive treatment for acid waters that formerly impacted over 17,000 km of streams in Appalachia.

C. Industrial Wastewaters

I. The Mandan Oil Refinery

A small existing wetlands complex was expanded in 1972 after a careful evaluation revealed that it would cost $250,000 versus costs of alternate systems ranging from $1 million to $3 million at Amoco's Mandan, North Dakota, oil refinery to provide adequate treatment to meet stringent discharge standards for all of the process water, wastewater, and runoff for the refinery. A variety of contaminant concentrations are reduced by 90–100% between the API separator effluent and the final discharge point. Reductions in parameter concentrations of 36–99.9% are obtained in the primary lagoon. Concentrations are further reduced by 70–100% in the constructed wetlands and discharges to the Missouri River are well below NPDES limits. Superior wastewater quality and cost-effectiveness are only two advantages of the refinery's artificial wetlands. One hundred eighty-four species of plant life have been identified in this area, including numerous wetland plants and grasses such as reed canary grass, wild rice, musk grass, cattails, and bulrushes. Ponds were stocked with rainbow trout, bass, and bluegill. Results from annual necropsy analysis on rainbow trout up to 2.3 kg have been normal. Over 190 species of birds have been observed, and 51 species nest in the area. In addition to geese, numerous other wildlife, such as pheasants, grouse, partridge, deer, fox, badger, skunk, and raccoons, also inhabit this area. This system visibly demonstrates that industrial and environmental concerns can exist harmoniously.

2. Paper Mills

A 35-ha constructed wetlands began operation in 1991 at Weyerhaeuser's mill in Mississippi that reduces BOD_5 and TSS levels by 60 and 90%, respectively. Georgia-Pacific Corporation's Pulp Operations in Mississippi and Pope and Talbott's mill in southern Oregon have experimental wetlands systems consisting of $1/8$- to $1/4$-ha ponds constructed with natural substrate and planted with native vegetation on the mill sites. These inexpensive systems have demonstrated significant polishing of mill wastewaters, with the following reductions of contaminant parameters: BOD_5, 45%; TSS, 97%; nitrate, 88%; phosphorus, 85%; TKN, 83%; and NH_3, 81%.

3. Sugar Beet Processing

American Crystal Sugar Company (ACSC) built a 63.6-ha constructed wetlands in 1991 to polish wastewater from a mechanical plant and primary/secondary treatment lagoons at their factory near Hillsboro, North Dakota (Fig. 13). The system was operated at one-half design capacity in 1992 (2839 m^3/day) and loadings were increased to design limits in 1993. Discharge concentrations have been well below the permit limits of 25 and 25 for BOD_5 and TSS, even though factory operation is heaviest during the winter after the fall harvest season. More than 40 species of waterfowl have been identified in the system. ACSC has a second wetlands under construction at their Drayton, North Dakota, factory.

D. Nonpoint Source Pollution

Public concern over the last 20 years resulted in state and federal legislation regulating discharges and providing financial assistance for municipal treatment facilities. Substantial progress in treating point sources has and is continuing to occur, especially in larger cities and with major industries. And widespread implementation of wetlands treatment technology may accomplish similar objectives with small community and small industry sources. However, anticipated improvements in the nation's waters have not been realized, and recent evaluations reflect a growing concern over nonpoint sources (NPS), especially agricultural waste and urban stormwater runoff. These principal contributors to nonpoint source problems have been

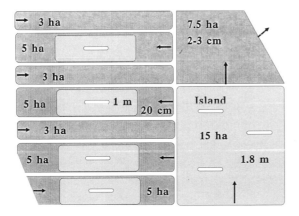

FIGURE 13 The American Crystal Sugar Company-constructed wetlands near Hillsboro, North Dakota, successfully polishes sugar beet processing wastewaters through the cold winter months as only BOD_5 and TSS reductions are needed.

difficult to remedy with conventional wastewater treatment methods.

Nonpoint source pollution emanating from agricultural operations, urban areas, failed home septic tank drain fields, mining, and a host of other land-disturbing activities continues to detrimentally impact 30–50% of our nation's waterways. Increasing focus on agricultural activities resulting in NPS pollution probably results not only from reduction or elimination of other sources but also from real changes in waste loading in receiving streams because of changing animal husbandry practices. Free-ranging livestock (including poultry) at relatively low population densities caused little aquatic pollution because wastes were widely dispersed and natural soil systems recycled nutrients on site. However, the historical and continuing tendency to confine livestock in ever-smaller areas to improve production efficiency also concentrates animal waste loading with subsequent runoff to nearby streams. Concurrent removal of woody and nonwoody riparian vegetation to increase efficiency by using all available acreage or incidental to livestock grazing and loafing has eliminated the buffer strip that formerly protected streams from direct pollutant impacts. Unfortunately, reversing either trend is unlikely under the economic pressures impacting today's farmers.

In cities, addition and expansion of impervious surface area from new housing, shopping, and office/industrial complexes continue to increase total and peak flows of urban runoff to receiving streams. Not only does this runoff contain petrochemicals and heavy metals, but overuse of fertilizers and pesticides in urban and suburban settings commonly results in higher pollutant loading in urban stormwater runoff than from comparable areas of agricultural fields.

Farmers, miners, homeowners, and developers are unlikely to purchase and operate package treatment plants nor can cities construct and operate conventional systems to treat stormwaters running off streets, parking lots, and other impervious surfaces. Requiring a hog producer to purchase a multimillion dollar treatment plant to deal with the waste from 1000 hogs (similar organic loading but much more concentrated than waste from a city of

1000 residents) is unrealistic, as many farmers are heavily in debt with marginally profitable operations. Nor can a city afford, justify, or even operate conventional treatment facilities to adequately treat very large but sporadically occurring flows after heavy rainfall events.

Constructed wetlands waste treatment systems seem amendable to the substantial range of hydraulic and pollutant loading, temporal fluctuations, dispersed nature, and the need for low-cost, low-technology systems acceptable to farmers, developers, and communities. On the farm, planting and maintenance requirements differ little from skills needed in growing other crops and land costs are lower. For cities with higher land costs, combined greenspace/recreational areas and stormwater treatment systems with the ability to accommodate widely fluctuating influents may be acceptable because of the multiple, instead of single, uses of properly designed treatment areas. In addition, constructed wetlands have demonstrated abilities to remove hydrocarbons and metals without the disruptive impacts these substances cause in conventional treatment facilities. And a small wetland with pleasant flowering plants—irises, orchids, and ferns—may treat the black water emerging in a homeowner's lawn from a failed septic tank drain field as well as enhance the aesthetics of the yard.

I. Livestock Waste

The Sand Mountain swine waste treatment system was built in 1988 to evaluate treatment performance and to develop design/operating criteria at Auburn's Sand Mountain Agricultural Experiment Station in northeast Alabama. The system receives effluent from a secondary lagoon treating waste from approximately 500 hogs. The design included a small farm pond for flood protection and dilution water, a mixing pond, and two replicates of five individual cells containing different vegetation (Fig. 14). Different loading rates are applied to different cells to test plant survival and removal efficiencies under fairly extreme conditions (>100 mg/liter NH_3). Wetlands construction and planting costs were:

Removal of old lagoon	$3,500

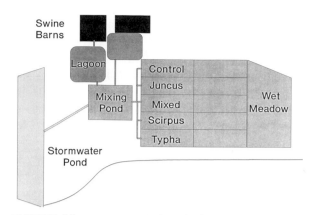

FIGURE 14 An experimental wetlands wastewater treatment system for a 500-hog operation at Auburn University's Sand Mountain Experiment Station that was originally designed to examine removal performance efficiencies of different species of vegetation.

Grading and dikes	1,400
Pipe and materials	400
Labor for pipe installation	500
Plants and planting labor	7,000
Total	$12,800

Obviously, a farmer could construct a comparable wastewater treatment system (excluding the old lagoon removal) and reduce the direct planting costs by limited planting and appropriate water level manipulation to establish dense stands in two or three growing seasons. Total cost might approximate $3000.

Influent BOD_5 levels ranged from 19.2 to 99.0 mg/liter, averaging 63.7 mg/liter, and a average effluent values of BOD_5 varied from 4.9 to 17.6 mg/liter. BOD_5 removal rate by the entire wetlands system averaged 90.4%. Suspended solids influent values varied by an order of magnitude (i.e., 21 to 210 mg/liter). Nonetheless, average effluent levels from the wetland cells did not vary and the average removal rate of suspended solids by the treatment wetlands system was 91.4%. Average density of fecal coliforms entering the upper wetland cells was 1.75×10^5 CFU/100 ml. Discharge values ranged from 0 to 3.1×10^3 CFU/100 ml with an average 2.7×10^3 CFU/100 ml, and overall the system reduced fecal coliforms by 99.4%. Influent ammonia nitrogen levels varied from 14.4 to 106.3 mg/liter and averaged 54.7 mg/

liter. Similarly, influent TKN concentrations ranged from 25.3 to 121.3 mg/liter with a mean of 69.8 mg/liter. Total ammonia nitrogen and phosphorus removal rates were 93.6 and 75.9%, respectively.

Treatment effectiveness is dependent on system design criteria. Swine waste at the Sand Mountain Agricultural Experiment Station receives primary treatment in three lagoons that are expected to reduce BOD_5 by 60%. The 500-animal swine operation is estimated to produce 90 kg BOD_5/day reduced to 36 kg BOD_5/day in the final lagoon discharge. Minimum treatment area for 36 kg BOD_5/day at 150 m^2/kg BOD_5/day is 5400 m^2. The wetlands treatment system has 3600 m^2 within the wetland cells and 2100 m^2 in the wet meadow, for a total treatment area of 5700 m^2 or 158 m^2/kg BOD_5/day.

The upper farm pond was a less expensive alternative to excavating a floodwater channel alongside the wetlands cell complex that serendipitously provided an opportunity to demonstrate simple treatment for truly NPS runoff. Contoured terraces in the winter pasture above and alongside the farm pond direct virtually all the runoff from the pasture to the upper end of the pond. High nutrient concentrations in pasture runoff supported an excessive algae bloom the first spring and summer after construction, and pond overflow discharging into a ditch flowing alongside the wetlands cell complex was poor quality. However, by merely installing a three-strand, barbwire fence across the upper end and west side of the pond and planting wetlands vegetation on the pond margin and in the discharge channel, algae blooms and poor-quality water in the ditch were eliminated in subsequent years. The shallow upper reaches of the pond are densely covered with *Fimbristylis, Carex, Juncus,* and *Scirpus,* which remove nutrients from the pasture runoff, thus eliminating impacts to the farm pond and downstream waters.

In 1990, a newer design was used in building a waste treatment system for a 500-hog operation at Mississippi State's Pontotoc Experiment Station in northeast Mississippi. The Pontotoc constructed wetlands system is a two-cell marsh–pond–marsh design (total area of 820 m^2) polishing effluents from primary and secondary lagoons (Fig. 15). Standard overland flow strips were added to compare removal efficiencies with the wetlands cells and to provide a means for land application of the discharge rather than a piped, surface discharge.

Average flows of 6 m^3/day included 35 mg/liter BOD_5, 108 mg/liter NH_3, and 43 mg/liter PO_4. The system was constructed in 1990 and allowed to stabilize before data collection was initiated in spring 1992. Cell 1 has *Typha latifolia* in the shallow regions and *Potamogeton pectinatus* in the deep portion; cell 2 has *Eleocharis dulcis* in the shallow portions. The critical design parameter was NH_3 and loading rates were established at 10 kg/ha/day for NH_3. However cell 1 has received flows of 5 m^3/day and cell 2 flows of 9.7 m^3/day, producing NH_3 loading rates of 14.6 and 26.4 kg/ha/day for cells 1 and 2, respectively. Consequently, effluent NH_3 levels have averaged 27.5 and 48 mg/liter for cells 1 and 2, yielding 75 and 56% removal rates, but an effluent NH_3 level that is much too high for most receiving waters.

2. Row-crop Runoff

In northern Maine, runoff from potato fields jeopardizes cold, deep lakes with a lake trout and land-locked salmon sport fishery that is economically significant to Aroostook County. In 1990 two demonstration nutrient/sediment control systems were designed and built in watersheds with other best management practices (BMPs) for erosion control

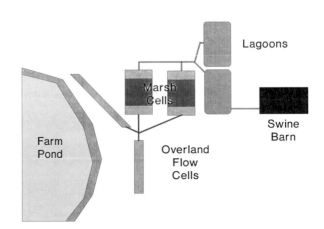

FIGURE 15 An operational wetlands wastewater treatment system for a 500-hog operation at Mississippi State University's Pontotoc Experiment Station.

already in place. The nutrient/sediment system consists of a sedimentation ditch bermed on the lower side leading to an overland flow meadow, followed by a cattail marsh, a pond, and a final polishing meadow (Fig. 16). Initially, five of these systems were constructed in northern Maine on drainage areas of 36–160 acres cultivated in potatoes. Some 30 additional systems were constructed in 1991 and 1992.

In practice, nutrient/sediment control systems have removed 90–100% of suspended solids, 85–100% of total phosphorus, 90–100% of BOD_5, and 80–90% of total nitrogen, and the treatment systems provide black duck breeding habitat as well as bait-fish rearing sites. Construction costs ranged from \$13,500 to \$23,000 for 8- and 67-ha watersheds, respectively. These costs, amortized over the expected life of the system, plus annual maintenance, average only \$50–\$60/ha/year for the contributing watershed.

3. Urban Stormwater

In contrast to relatively constant flow rates and pollutant concentrations in municipal or mine wastewaters, urban stormwater runoff has widely fluctuating volumes with high sediment loads and generally lower although temporally variable organic and metallic ion concentrations. Variable concentrations of petroleum-based hydrocarbons are also common in urban stormwater runoff. Typically organic loading is 14–15% of the levels in municipal waste and stormwater may have up to

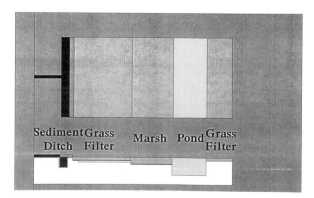

FIGURE 16 The USDA, Soil Conservation Service-designed nutrient/sediment removal systems provide high levels of control of row-crop runoff in northern Maine.

60 mg/liter iron, 30 mg/liter lead 15 mg/liter copper, and less than 5 mg/liter of zinc, cadmium, manganese, chromium, and nickel. These concentrations are generally lower than the 200–400 mg/liter iron and 30–40 mg/liter manganese and other metallic ions successfully removed from acid mine drainage in constructed wetlands system. Almost 90% of the sediment and contaminant load in stormwater occurs in the "first flush," typically the runoff following the first 2–4 cm of rainfall. Water quality of subsequent runoff is much improved.

Sediment is an important component of stormwater not only because treatment systems must be able to remove and retain waterborne sediment but also because many other contaminants in stormwater are closely associated with sediment due to clay particle adsorption phenomena. In simplest terms, removing the sediment load will provide a substantial amount of treatment for stormwater runoff and a dense stand of vegetation is very effective at trapping silt. In fact, a 10-m band of trees, shrubs, and other vegetation between a bare field and a stream will remove 70% of the sediment before the water reaches the stream. Of course, complete removal will require a larger vegetated area and dissolved substances must be treated in other ways.

Constructed wetlands systems may be designed to provide stormwater runoff treatment to very high standards depending on contaminant loading and treatment area available. A system designed to meet NPDES discharge standards for contaminants in typical urban stormwater runoff included a small permanent pond, a temporary runoff storage area, an emergent marsh, a wet meadow and appropriate water control structures and spillways. Only 4% of a 20-ha watershed with 50% impervious surface was dedicated to single-use treatment area and 8% of the total watershed included in the treatment system was available for a variety of other multipurpose uses. The core component for removal of dissolved substances was a small cattail marsh and a meadow occupying 4% of the watershed. In conjunction with riverside greenbelts, small, strategically located wetlands will provide treatment of stormwater runoff to high discharge standards.

Florida regulations require the treatment of stormwater from all new development or redevelopment. Natural and constructed wetlands frequently are incorporated into comprehensive stormwater management systems, which provide not only flood control and pollution treatment, but frequently ancillary benefits such as open space, recreation, wildlife habitat, and aesthetics. Over 200 wetlands stormwater management systems have been constructed in Florida in the last few years. These systems range in size from 1 to 900 ha, serving agriculture and small residential subdivisions, office parks and shopping centers, and regional stormwater systems for entire drainages.

Natural wetlands may provide less expensive treatment than constructing or restoring wetlands within cities. Protecting existing riparian wetlands vegetation and riverside greenbelts or parkways will enhance corridor aesthetics, improve recreational opportunities, modify urban microclimates, and provide maintenance-free treatment of stormwater runoff. Old channels, oxbows, or sloughs can be a major component of the treatment process and also provide educational and recreational opportunities if allowed to remain in their natural state. Planning, zoning, or acquisition to protect a buffer of existing wetlands vegetation (trees, shrubs, and herbaceous plants) along all waterways is the least-cost method for treating stormwater runoff and other common pollutants (Fig. 17). In conjunction with riverside greenbelts, small, strategically located wetlands will provide treatment of stormwater runoff to high discharge standards.

Where previous development has eliminated the riparian band, restoration and subsequent protection have been necessary. In many cases, protecting this zone for a few years to allow native vegetation to become reestablished may be all that is required. If more immediate results were desirable, suitable trees and shrubs can be planted.

VII. THE FUTURE OF WETLANDS WATER QUALITY IMPROVEMENT

Much can be learned from both natural wetlands receiving wastewaters and from constructed systems with a few years of operating history. Despite some attempts to reduce wetlands treatment systems to minimal components and treatment areas with seemingly most efficient combinations of substrate, vegetation, and loading rates, most *successful* systems are indistinguishable on casual examination from natural marshes. In fact, poorly performing systems that I have visited did not appear to be viable marsh ecosystems. Generally the absence of an important component, attribute, or characteristic was obvious to anyone with experience in natural marsh ecosystems. Conversely, successful systems are often quite similar to a natural marsh and it is beginning to appear that the basis for design of wetlands constructed for wastewater treatment should be to simulate the structure and functions of a natural marsh ecosystem.

Because constructed wetlands are open, outdoor systems, they receive inputs of animal and plant life from adjacent areas and from distant sites and over time are likely to become more and more similar to the natural wetlands in a region. Though we may design and build a system with a specific substrate and only one or two plant species currently thought to be highly efficient, eventually many types of plants and animals will take up residence. Consequently, a constructed wetland is likely to become more similar to a natural wetland as the system matures and ages. To prevent these invasions and attempt to maintain a monoculture would be difficult and costly and may be self-defeating. Living organisms that become established in an operating system are not likely to detrimentally impact treatment efficiency and may very well improve system operation. In addition, maintaining a monoculture is difficult, as any farmer knows, because the single-species stand or monoculture is often susceptible to disease, insects, or grazing animals. For example, cattail in a marsh treating acid water seepage from coal ash storage ponds was devastated by an outbreak of armyworms during the first year of operation.

This is *not* to suggest that natural wetlands are casually usable for wastewater treatment. We know too little of the complex interactions among a myriad of components in natural systems and too little of optimal treatment area requirements or applica-

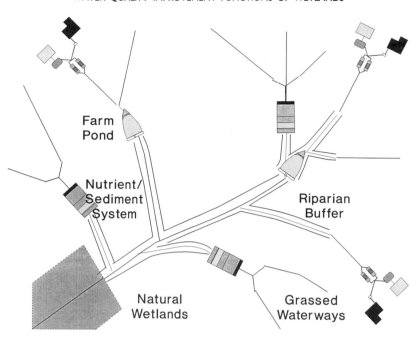

FIGURE 17 A combination of constructed and natural (restored) wetlands can provide high levels of treatment for farm wastes, row-crop runoff, and urban stormwater runoff.

tion rates to risk damaging natural wetlands. These systems are too valuable to lose as research on natural wetlands systems will continue to increase our understanding and ability to design and build constructed wetlands for specific purposes. Intact natural wetlands also provide a host of other benefits to society.

If we examine the history of the development of constructed wetlands for wastewater treatment, it is obvious that astute observations and careful monitoring of natural wetlands receiving polluted waters were the basis for early proposals for the use of wetlands in wastewater treatment. Subsequently, Max Small (marsh–pond–meadow) and others designed constructed wetlands patterned closely after natural systems. However, as always, we thought we could improve on Mother Nature, and some designs began to appear that were dependent on a highly conductive substrate, the engineered soil or gravel bed types. But as we discovered clogging problems with horizontal flow, gravel-based systems, modifications included applying the wastewater to the upper surface in a vertical flow design. Later, recirculating was added and some have even included forced aeration with or without a greenhouse enclosing the entire system. Unfortunately,

most "advances" since the early marsh–pond–meadow design have failed to achieve the original objective of developing a simple, low-cost, low-maintenance, and effective wastewater treatment system. The latest designs (marsh–pond–marsh) are quite simply an attempt to return to the first developments that patterned constructed wetlands after natural wetlands. In fact, one might observe that we have come full circle in developing this technology (Fig. 18). We started out copying natu-

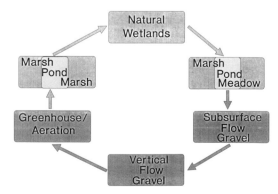

FIGURE 18 The circular history of development in the constructed wetlands wastewater treatment technology progress from systems that emulate natural wetlands through complex, artificial designs and eventually returning to designs that simulate natural wetlands.

ral wetlands, increased complexity in construction and operation but failed to attain better efficiencies, and recently returned to the original concept that attempted to emulate natural wetlands. The "newest" but in fact "old" concepts have proven to be the least costly to build, have higher removal efficiencies for a wider variety of pollutants, are less costly and complex to operate, and also provide substantial ancillary benefits.

In retrospect, "improvements" from the earliest designs that emulated natural wetlands deviated from the original objectives, which were to provide a wastewater treatment system with:

1. low construction cost (capital costs);
2. self-maintaining attributes (little time, expense, or operating skill); and
3. high pollutant transformation capabilities.

With increasing design complexity to improve treatment efficiency (and perhaps reduce required treatment areas), our designs added substantial construction and operating costs and complexity such that some recent types with recirculation, vertical flow in a batch loading mode, and, in a few cases, complete enclosures have become as costly to build and operate as conventional treatment systems. Operational complexity has increased concurrently and many of these complex systems are conceptually, physically, and financially similar to trickling filters or other conventional wastewater treatment systems. The original objectives of low-cost, efficient, self-maintaining systems have been lost in attempts to improve removal efficiencies.

Contaminant removal processes in wastewater treatment wetlands are similar to microbial transformations present in conventional package treatment plants, lagoons, or other conventional wastewater treatment systems. Conventional systems require large inputs of energy, complex operating procedures, and subsequent costs to maintain optimal environmental conditions for microbial populations in a small treatment area. The low capital and operating costs, efficiency, and self-maintaining attributes of wetland treatment systems result from a complex of plants, water, and microbial populations in a large enough land area to be self-sustaining. It may be less costly to construct a minimally sized, least-component wetlands treatment system, but operational costs to maintain that system could easily negate initial cost savings. For small communities, farms, mines, and some industries, a conservatively designed and biologically complex system may provide more efficient treatment, greater longevity, and reduced operating requirements and costs.

Much of the previous work on constructed wetlands has occurred in North America and Europe, where applications to various types of wastewater have provided substantial cost savings. However, recent applications in the developing nations have yielded operating demonstrations of a low-cost, efficient technology that can easily be constructed and operated with local skills and materials. These systems do not require costly imports of high-technology equipment or skills that further drain the capital reserves of developing nations. Consequently, constructed wetlands are likely to provide the means whereby many developing nations can reasonably address their water pollution problems.

Glossary

Ammonification Decomposition of proteinaceous organic matter yielding ammonia nitrogen (NH_3) along with other products.

BOD$_5$ Five-day biochemical oxygen demand, that is, a measurement of the amount of oxygen consumed by microbial organisms decomposing/oxidizing the included organic material during a 5-day incubation period.

CFU Colony-forming units, that is, a measurement (inventory) of the number of bacterial colonies formed on standard growth media.

Comminution Reducing particle sizes in wastewater through grinding and screening.

Denitrification Biochemical process of reducing nitrogen (N) from nitrate (NO_3), producing nitrogen gases (N_2 and N_2O).

Effluent/outflowing Liquids flowing out of a system, lake, wetlands, etc.

Eutrophication Process whereby a body of water (lake or stream) contains abnormally high nutrient (nitrogen and phosphate) levels that permit excessive populations of microorganisms (especially algae) to become established. Typically water flowing in carry the high nutrient loads from upland regions into the lake or river.

Hydrophyte Wet-growing plants.

Influent/inflowing Liquids flowing into a system, lake, wetlands, etc.

Leachate Liquid emerging from solid material such as disturbed soils or landfills. It originates with precipitation percolating down through the solid material, typically dissolving certain substances (contaminants) and discharging near the base or margin of the solid materials.

Macrophyte Larger plants, that is, of a size greater than microscopic levels.

Microbe Microscopic organisms including very small invertebrates, algae, bacteria, protozoa, and fungi.

Monoculture Deliberate cultivation of an area to favor a single species of plant, for example, agricultural fields such as wheat or corn.

Monotypic A single type, for example, a plant community composed of only one species of plant.

NH$_3$ Nitrogen in the form of ammonia.

Nitrification Oxidation of ammonia nitrogen (NH$_3$) to nitrite (NO$_2$) and nitrate (NO$_3$) by bacteria.

NPDES The National Pollution Discharge Elimination System; this legislation providing for regulation of point source discharges to U.S. waters.

POTW Publically owned treatment works.

Pretreatment Initial treatment of wastewaters often consisting of settling and perhaps comminution.

Primary treatment Typically settling, comminution, and initial decomposition in septic tanks, lagoons, or oxidation ponds.

STP Sewage treatment plant.

TDS Total dissolved solids.

Trophic layer Ecological groupings of organisms such as primary producer, primary consumer, secondary consumer, decomposer, and so on.

TSS Total suspended solids.

Bibliography

Anon. (1992). "Wetlands Systems in Water Pollution Control." Sydney: Australian Water and Wastewater Association.

Cooper, P. C., and Findlater, B. C. (1990). "Constructed Wetlands in Water Pollution Control." Oxford, England: Pergamon Press.

Godfrey, P. J., Kaynor, E. R., and Pelczarski, S., eds. (1985). "Ecological Considerations in Wetlands Treatment of Municipal Wastewaters." New York: Van Nostrand Reinhold.

Good, R. E., Whigham, D. F., and Simpson, R. L., eds. (1978). "Freshwater Wetlands (Ecological Processes and Management Potential)." New York: Academic Press.

Hammer, D. A., ed. (1989). "Constructed Wetlands for Wastewater Treatment." Chelsea, Mich.: Lewis Publishers.

Hammer, D. A. (1991). "Creating Freshwater Wetlands." Chelsea, Mich.: Lewis Publishers.

Mitsch, W. J., and Gosselink, J. G. (1993). "Wetlands," 2nd ed., New York: Van Nostrand Reinhold.

Moshiri, G. A., ed. (1993). "Constructed Wetlands for Water Quality Improvement." Chelsea, Mich.: Lewis Publishers.

Reddy, K. R., and Smith, W. H., eds. (1987). "Aquatic Plants for Water Treatment and Resource Recovery." Orlando, Fla.: Magnolia Publishing.

Wetland Restoration and Creation

R. W. Tiner

University of Massachusetts

I. Introduction
II. Why Restore and Create Wetlands?
III. Wetland Restoration and Creation vs Protection
IV. Defining Project Goals and Objectives
V. Project Design
VI. Wetland Restoration and Creation Techniques
VII. Monitoring and Evaluation

Wetland restoration has been variously defined, but it is most often considered the process, or the result, of returning either a former wetland (now a nonwetland) to a functioning wetland of some type or a degraded, contaminated, or functionally impaired wetland to its prealtered condition. A restored wetland in the former case may or may not be similar to the "original" wetland type, but the net result is an increase or gain in wetland acreage. The other type of restoration seeks to reconstruct the wetland type that existed prior to disturbance, with no net increase in wetland acreage. This latter process may be better termed wetland rehabilitation, where a damaged system is restored to normal. Rehabilitation could include eradication of exotic and/or pest species. Wetland restoration does not include changing the condition of an existing unaltered wetland to improve or strengthen one or more functions. This action is called wetland enhancement and it usually changes the wetland type for the benefit of a particular function, e.g., converting a wet meadow to a marsh–pond complex to improve waterfowl habitat. With more than half of the wetlands in the coterminous United States destroyed, largely through agricultural conversion and hydrologic alteration, there are many

opportunities for wetland restoration in the lower 48 states. Wetland creation is the process of constructing a wetland in an upland (nonwetland) area. To build a wetland where one never existed requires creating hydrologic conditions that promote the establishment and successful reproduction of hydrophytes, the development of hydric soils, and the performance of desired wetland functions (e.g., flood storage, shoreline stabilization, pollution abatement, and wetland wildlife habitat). In some cases, wetlands have been unintentionally created by seepage from impoundments, elevated local water tables from irrigation projects, restricted drainage from undersized culverts or lack of sufficient culverts along roads, or other altered drainage patterns that make a site wetter than it was prior to the activity.

I. INTRODUCTION

Wetlands are permanently flooded shallow water areas or periodically flooded or saturated lands. They range from margins of lakes and ponds and lands flooded by the tides to areas that are seasonally saturated for extended periods near the soil surface in most years. Common types include salt marshes, mangrove swamps, tidal flats, inland marshes, wet meadows, prairie potholes, playas, fens, shrub swamps, bogs, pocosins, wooded swamps, certain bottomland hardwood forests, muskegs, and shallow ponds. Wetlands are now considered among the world's most valuable natural resources, providing a wealth of benefits to society, essentially free of charge. These values include fish and shellfish production, wildlife habitat for valued species, temporary flood water storage, shoreline stabilization, water quality renovation, and water-based recreational opportunities. Despite these and other valuable functions, wetlands have been viewed by many cultures as wastelands whose highest purpose could only be attained through conversion to other uses, such as filling for commercial and residential real estate and drainage for cropland or silviculture. As a result of this attitude, many wetlands have been destroyed or significantly degraded in the United States and elsewhere. In the coterminous United States alone, over half of the wetlands that existed prior to European colonization have disappeared. Most of these lost wetlands were either converted to agricultural land (Fig. 1) or destroyed through hydrologic alteration (e.g., channeliza-

tion/drainage projects and regulated river flows), while filling and dredging were major causes of coastal wetland destruction. [*See* WETLANDS ECOLOGY.]

Public opinion in the United States toward wetlands began changing in the 1960s and 1970s, due largely to scientific reports of wetland functions and values and of increasing losses and threats to natural wetlands. This led some states to enact laws to protect coastal wetlands and later, inland wetlands to varying degrees. The federal government also strengthened its role in wetland protection through the Clean Water Act. These laws and corresponding regulations usually require that persons seeking to alter wetlands first obtain permission or a permit from the applicable regulatory agency (e.g., Corps of Engineers or state wetland agency) prior to commencing work. This gives the government an opportunity to consider and evaluate the potential environmental impact of the proposed work, with the intent of avoiding or minimizing wetland loss or degradation. In obtaining a permit, the permittee may be required to restore or create wetlands to lessen the environmental impact of the project. This compensatory requirement is usually the last step in a sequential impact minimization

FIGURE 1 Millions of acres of former wetlands have been converted to cropland like this former bottomland hardwood swamp in the lower Mississippi river valley. In many cases, government-sponsored flood control/channelization projects have accelerated such conversions.

process called "mitigation." Ideally, an alternatives analysis should initially be performed to establish that the purpose of a project can only be met by altering the wetland at the proposed site. Then, the impact of the project on wetlands at the site should be minimized to the extent necessary to reasonably satisfy the project's intended goal. Finally, unavoidable wetland losses resulting from project construction should be compensated for by wetland restoration, creation, or other measures. Thus, regulations developed in accordance with various environmental laws serve as a major catalyst for wetland restoration and creation.

II. WHY RESTORE AND CREATE WETLANDS?

Wetland restoration and creation projects offer opportunities to maintain and improve the status of wetlands and their functions. These efforts may, in some way, help offset the cumulative functional losses resulting from past wetland alterations. The major reasons for wetland restoration and creation are (1) to compensate for the impact of proposed wetland alteration permitted through the regulatory process, (2) to repair damaged systems, and (3) to increase wetland acreage above current levels for the purpose of gaining wetland functions that benefit society. The U.S. federal government and some states have adopted an environmental policy of "no net loss" of wetlands. Wetland restoration and creation are vital to achieving this goal, since it is clear that certain wetland alterations will continue to be permitted through government wetland regulatory programs. These programs are designed, in part, to minimize adverse impacts on wetland resources from new development. Consequently, government agencies, industry, landowners, and developers seeking permits to modify wetlands for various projects are often required to mitigate for unavoidable wetland losses as part of the conditions for receiving a government permit. Many of these permitted activities involve both wetland restoration and creation as mitigation for altered wetlands. Other government programs dedicated to encouraging wetland restoration on private property through positive incentives include the U.S. Fish and Wildlife Service's Partners for Wildlife Program, U.S. Department of Agriculture's Conservation and Wetland Reserve Programs, and state-run programs. State and federal wildlife management agencies have been active in wetland enhancement and perhaps, to a lesser degree, in wetland restoration and creation for at least 50 years. They have diked and impounded wetlands and created ponds on public lands for the benefit of certain wildlife, mainly waterfowl. Enhancement projects typically promote one wetland function, while diminishing some other functions.

Nongovernment organizations and private organizations like Ducks Unlimited have been involved in similar projects to improve waterfowl habitat, for example. Wetlands have been touted as natural water quality filters, so artificial wetlands are now being built for municipal, industrial, and agricultural wastewater treatment in North America and Europe. These wetlands are used mainly to polish secondary effluent and for tertiary treatment and stormwater management. Some wetlands are being constructed to control nonpoint source pollution from agricultural lands. In mining regions of the Appalachians in the eastern United States, wetlands have been created to neutralize acidic water runoff from coal mines. In Asia, wetlands have been created to produce the staple, rice, that feeds millions of the world's population. To a lesser degree, private individuals have created wetlands largely to protect their property from shoreline erosion or have restored wetlands for wildlife habitat due to their personal interest in wetland wildlife conservation. Ponds are perhaps the most common and widespread wetland type that private landowners construct without government incentives, although the government has programs to provide partial funding or technical assistance for farm ponds and catfish ponds. This happens chiefly because the value of a pond directly benefits the landowner. One can fish, hunt, obtain water, engage in water-based recreation, or simple enjoy the sight and sounds of pond life, whereas the benefits of many vegetated wetlands are often less tangible. Coastal wetlands and streamside wetlands have been created by private landowners because they derive a direct benefit—protection of property through shoreline stabilization and erosion control—from these wetlands.

III. WETLAND RESTORATION AND CREATION VS PROTECTION

Wetlands form on the landscape where there is frequently an excess of water for prolonged periods. Many wetlands are associated with floodplains of major rivers, low-lying plains along coastal waters, and depressions surrounded by upland where surface water collects. Yet all wetlands do not fit this pattern. Some wetlands have been established on terrains with varying slopes in areas of groundwater discharge. These sites include springs, seeps, and drainage ways. Since wetlands naturally occur in landscape positions where water accumulates or periodically overflows, protection of these naturally occurring wetlands should be favored over wetland restoration and creation. Existing wetlands are performing wetland functions and it is far easier to preserve wetland functions and perhaps to enhance these functions than to create wetlands where they never existed or, to a lesser extent, to restore lost wetlands. This, in essence, is the foundation for the government's sequencing steps for mitigating wetland impacts of proposed projects. If wetland losses can be avoided, they should be. If not, then wetland losses should be minimized and compensated for through wetland restoration, creation, and enhancement. These efforts, however, should not be used as a substitute for protecting naturally functioning wetlands.

If establishment of new wetlands or additional wetland acreage is the objective, then the choice is between wetland restoration and creation. Wetland restoration is usually more likely to successfully establish wetlands than wetland creation because (1) the affected area is a former wetland, (2) it is naturally in the right landscape position in the watershed to accumulate water, (3) it has a seedbank for hydrophyte reestablishment, (4) it possesses drained hydric soils, and (5) it has a modified hydrology that may be restored to its original wetland hydrology. In contrast, wetland creation requires establishing wetland hydrology in an area that did not support wetland. Creating wetland hydrology sufficient to make a wetland is usually a much more difficult task than restoring wetland hydrology. Evaluations of created wetlands built to replace natural wetlands suggest a low success rate, while the potential for success by restoration efforts is much higher. There also remains much question whether created wetlands are functionally equivalent or similar to the natural wetlands they are intended to replace.

IV. DEFINING PROJECT GOALS AND OBJECTIVES

The primary goal of any restoration or creation project should be to establish an area that provides wetland functions—a wetland. Wetlands restored in response to government regulatory requirements are usually intended to provide multiple functions like natural wetlands. Other restored wetlands often have a single purpose, with increasing waterfowl habitat being perhaps the most common and widespread goal in North America. Ideally, the restored wetland should have a self-sustaining hydrology that requires little or no maintenance. However, in practice, there may be considerable operation and maintenance associated with restored wetlands, especially if water levels must be managed to maximize waterfowl use. Created wetlands may be built to replace lost functions of natural wetlands destroyed by various developments or may be designed to perform a special function that society wants, such as wastewater treatment, stormwater management, or erosion control. In the former case, the objective should also be to establish a self-sustaining wetland like a restored wetland with multiple purposes, while the management objectives in the latter typically involve the installation of water control structures, requiring considerable operational and maintenance costs to achieve the desired function.

Without clearly stated objectives, it is virtually impossible to evaluate the success of wetland restoration and creation projects. The lack of specified objectives of many restoration projects has often led critics to claim that these projects are no more than pond creation or wetland enhancement. Ideally, every restoration or creation project should have a written plan detailing the specific goals and objectives and a set of measurable parameters to evaluate project success. This is especially needed for mitigation projects and government-sponsored projects that are designed to replace lost wetland functions. Information from such plans is vital to better understanding why a particular project succeeded or failed and to allow others to reproduce successful results. Documentation for most past projects is either poor or nonexistent, providing little specific guidance on successful methods and problems to be avoided for future projects.

The first and perhaps most important question in any restoration or creation project is what type of wetland is desired? The answer may be easy to determine if the restoration/creation project is being performed to replace functions of lost wetlands; in-kind replacement is the norm. If the desired wetland type differs from the one destroyed (out-of-kind replacement), there must be a good reason for this. Perhaps the intent is to restore wetlands that have been subject to heavy historical losses or to establish wetlands with a particularly high value for one or more functions (e.g., waterfowl habitat, endangered species habitat, or flood storage). The former requires knowledge of wetland status and trends. The U.S. Fish and Wildlife Service has published national statistics on wetland trends and also has similar data available for some specific geographic areas. This information is vital to knowing what wetlands are in greatest need of restoration. Government agencies may choose to restore entire ecosystems, such as the Kissimmee River and the Everglades of Florida or the bottomland hardwood swamps of the lower Mississippi alluvial plain, or may opt to restore individual wetlands in high priority watersheds or regions, such as prairie potholes in the upper Midwest. In some cases, wetlands may simply be restored or created where there is a willing participant, but this is probably the least desirable option from an ecosystem management standpoint. Restoring a wetland requires knowledge of its condition prior to alteration. For recent disturbances, this may be accomplished through conventional photointerpretation techniques. Aerial photos predating the perturbation may be examined to determine the previous wetland type. Such photography is available for many areas back into the late 1930s and early 1940s.

After deciding what type of wetland to restore or create, a number of questions arise. The 10 questions listed here are examples of questions that will aid in site selection and project design.

1. What government regulations may apply to construction activities required by the restoration/creation project? (Federal Clean Water Act, state wetland laws, local zoning bylaws, etc.)

2. Where are the lands suitable for restoration or creation and are they available for the project? (Former wetland sites, uplands with high water tables, existing land use, land ownership, water rights, etc.)

3. Given the above, should the project be a wetland restoration or a wetland creation?

4. What should the project size be?

5. If the project is being initiated in response to a government permit for wetland alteration, should the restoration/creation be located on-site, off-site in watershed, or off-site out of watershed?

6. What hydrologic conditions are to be established? (Water depth, flooding frequency and duration, seasonal water tables, tidal flow regime, etc.)

7. What plant communities are desired? (Dominant species, diversity, ratio of vegetation to open water, etc.)

8. How much time should be allowed for wetland vegetation to establish?

9. What faunal species and kind of animal use are desired? (Feeding, nesting, brood rearing, wildlife travel corridors, etc.)

10. What is an acceptable risk of structural failure (such as washed out culverts or eroded berms) that would require repair at some frequency (e.g., 5, 10, 25, 50, or 100 years)?

The answers to these questions will likely vary depending on whether the objective is to establish a naturally functioning wetland or to restore or create a wetland for a specific purpose.

Several sources of existing information are available to help answer some of the preceding questions. These sources include National Wetlands Inventory (NWI) maps (U.S. Fish and Wildlife Service), soil survey reports (U.S.D.A. Soil Conservation Service), state water summaries (U.S. Geological Survey), climate data (U.S. Weather Bureau), and state or local wetland maps. By comparing NWI maps with a soil survey report, for example, potential sites for restoration may be detected (i.e., hydric soil map units without NWI wetlands). Soil survey reports may also be used to find upland sites with high potential for wetland creation. Field inspections are required for evaluating the actual site potential for wetland restoration or creation. These sources provide useful background information to aid in identifying potential sites.

V. PROJECT DESIGN

Successful designs for restoration and creation projects require assessment of both on-site and off-site environmental conditions and application of current knowledge of wetland formation processes and wetland functions. In restoring or creating wetlands, site selection is the first step toward project success. If the project is being initiated in response to a regulatory requirement, the regulatory agency will usually provide guidance on the on-site/off-site and the wetland restoration vs creation questions. Sites best suited for restoration are hydrologically modified former wetlands where wetland hydrology is easy to reestablish (Fig. 2). Such sites possess drained hydric soils and contain a natural seedbank of hydrophytic species that should greatly improve the chances for successful revegetation. For constructed wetlands, site selection is most critical, especially if attempting to create a wetland that is somewhat functionally equivalent to the one destroyed. The location should be one where wetland hydrologic conditions can be efficiently and effectively replicated. The best sites probably are adjacent to existing wetlands or water bodies where lowering the ground surface through excavation will expose the affected area to wetland hydrologic conditions. The hydrology of these created wetlands may be surface water- and/or groundwater-driven or artificially controlled. Similarly, wetland basins may be created by excavating an isolated depression to a point at or below the local groundwater table. This is how many ponds are built. These artificial wetlands are essentially groundwater-driven systems, although contributions of surface water through runoff are variable, depending on the size of the upstream watershed. Other ponds and wetlands may be created by impounding a natural valley to collect surface water. For any surface water-driven wetland, the quantity and quality of the inflowing water are of utmost importance. Too much suspended sediment can accelerate basin filling and affect plant composition, wetland hydrology, and associated wetland functions. Contaminated waters could produce disastrous consequences for certain wetland functions, especially fish and wildlife habitat. Wetlands may also be created within water bodies by depositing fill material and stabilizing this material with wetland plants and/or man-made erosion control fabrics. This is frequently done in creating coastal marshes on dredged material disposal sites and to stabilize eroding shorelines along freshwater rivers, lakes, and streams. In addition to hydrology (water quantity and seasonal fluctuations in water levels) and other water-related properties (e.g., water chemistry), other features to consider in site selection include local topography, soil properties (e.g., texture, permeability, fertility, erodibility, and underlying substrates), degree of site exposure to wave action (for sites in large water bodies), the ratio of the acreage of wetlands/water bodies in the watershed to the total watershed acreage, predominant land use in the watershed, and land use adjacent to the site.

Once a site is selected, a number of other environmental parameters should be evaluated, including channel slope or gradient, ground elevations at the site (through topographic surveys), the proximity to other wetlands, and the presence of populations of exotic and/or potential pest plant species. These and other factors (e.g., culvert sizes above and below the project site) will provide valuable information for (1) establishing the scope and effect of the project (e.g., size and shape of wetland and planned hydrology), (2) determining externalities

FIGURE 2 Former wetlands drained by open ditches are among the easiest wetlands to restore. Many areas similar to the one shown in North Dakota are suitable for marsh and wet meadow (prairie pothole) restoration. The federal government is actively engaged in wetland restoration in this area.

that may affect project success (e.g., storm flows), (3) developing contingency plans to get rid of excess water and for drawdown, and (4) evaluating potential impacts to adjacent properties and downstream areas.

Knowledge of wetland formation processes and wetland functions provides the basis for determining the critical elements to restore or construct wetlands. All wetlands are not functionally equivalent. Some wetlands have higher capacities to perform certain functions than others. Consequently, the functional analysis of wetlands proposed for alteration provides the foundation for determining what mitigation should be required in regulatory cases. Functional evaluations of neighboring wetlands can also provide useful information for project design, especially when the project is not the result of a regulatory action. Specific designs can be drafted to accentuate specific functions, if desirable. Some points to consider in designing wetland restoration and creation projects include: (1) ratio of open water habitat to vegetated wetland, (2) wetland type and desired plant community composition, (3) method of revegetation (e.g., natural recruitment of plants, dressing topsoil with hydric soil containing natural seedbank, or seeding/planting),

(4) sources of planting stock (e.g., transplanted local stocks, nursery-grown native plants, or horticultural varieties), (6) planting time, (7) soil fertility and organic matter composition, and (8) desired hydrology (e.g., frequency and duration, sheet flow, channelized flow, and amount of water management).

Documentation of vital site characteristics and functional design specifications will greatly help in evaluating project success and in replicating successful results for other areas. It must be remembered, however, that excellent project design also requires proper implementation to achieve success. Designed elevations may be perfect, but if site preparation fails to attain these levels, then the project may be doomed. In wetlands, for example, small changes in elevation can make an enormous difference in the environmental conditions that greatly affect plant establishment, survival, and reproduction.

VI. WETLAND RESTORATION AND CREATION TECHNIQUES

Establishing the appropriate hydrologic regime is critical for all wetland restoration/creation projects. Hydrology is the driving force that creates, maintains, and largely determines functions for wetlands in nature. Replicating wetland hydrology is vital to the success of any project. This is mainly accomplished by three methods: (1) controlling ground surface elevations (chiefly for creation and restoration of filled wetlands; Figure 3), (2) regulating the water depths and duration through a combination of earthen dikes and water control devices (for restoration, enhancement, and creation), and (3) destroying existing drainage structures (mainly for restoration). The first is done by excavating soil to a level where permanent or periodic flooding or prolonged soil saturation will occur. The second action involves installation of water control devices (e.g., gates, valves, riserboards, or stoplogs) to attain desired water levels in the diked area (impoundment). The final method requires plugging drainage ditches or breaking tiles to effectively demolish the current drainage system and restore wetland hydrology. Achieving and maintaining the desired hydrology are probably the greatest obstacles facing restoration and creation efforts. Unpredicted low water tables, extremes in climatic conditions (e.g., droughts and floods), improper site grading and slopes, coarse-textured soils, and erosion contribute to this problem.

After planning the desired hydrology, attention usually focuses on establishing a wetland plant community of a particular type. Species composition, maintenance of genetic diversity of local wetland ecotype stocks, seed/seedling sources (including salvage plants from wetlands planned for alteration and hydric soils with natural seedbanks from donor wetlands), plant material handling, planting techniques, spacing requirements, planting/seeding times, fertilization, substrate/soil type, plant survival and reproduction, and control of exotic and pest species are among the major issues facing wetland restoration and creation projects. Herbivory by insects, geese, muskrats, and rabbits, for example, is also a potentially significant issue that must be dealt with by some projects.

Techniques for restoring and creating wetlands often vary with the desired wetland type due to plant species requirements and different environmental conditions. Aquatic beds, estuarine wetlands (salt/brackish marshes and mangrove swamps), and palustrine wetlands (inland marshes, swamps, and bogs) all have somewhat unique circumstances to deal with. The following paragraphs of this section address some of these differences. Pond creation will not be discussed, but it should be recognized that ponds have been successfully created by many cultures throughout the course of human history.

Efforts to restore aquatic bed vegetation require establishing plant communities in permanent shallow water. Critical environmental factors for site selection, besides water depth and substrate, may include turbidity, sedimentation, thermal pollution, and oil/chemical pollutants. Many projects in large water bodies, especially estuaries, major rivers, and large lakes, must address the effects of water currents and wave action. Low-energy sites are best suited for the establishment of aquatic beds. Much restoration has been performed in estuaries. Typical species involved in these projects are eel grass (*Zostera marina*) and widgeon-grass (*Ruppia maritima*) in northern U.S. estuarine waters and turtle-grass (*Thalassia testudinum*), manatee-grass (*Cymodocea filiformis*), shoal-grass (*Halodule wrightii*), and sea-grasses (*Halophila* spp.) in southern waters. Although seeds may be planted directly into the substrate, the planting of individual specimens or plugs is more typical. Most plants are collected from local populations to maintain genetic diversity and fitness for local environmental conditions. Plantings may be anchored in some fashion (e.g., steel staples or biodegradable meshes) to prevent washouts. In general, the best sites for restoration are former sites where water quality has improved or elevations are now suitable for reestablishment. In freshwater systems, target species for restored or created aquatic beds include white water lily (*Nymphaea odorata*), spatterdock (*Nuphar* spp.), and pondweeds (*Potamogeton* spp.). At created sites, flooding is required after seed germination or transplanting. If planting needs to be stabilized to preventing uprooting, biodegradable meshes may be used.

Restoration of estuarine marshes is limited since most of the historic losses were due to dredging and/or filling which eliminated these habitats. It is

FIGURE 3 Wetland restoration and creation may require soil removal to the seasonal high water table to restore or create wetland hydrology. (A) Grading of a project site nears completion in June. (B) Site just 3 months later, after seeding and planting over 1500 shrubs. While the results look impressive, this project will require monitoring over several years to determine whether the project has achieved its objectives. (Photos courtesy of Fugro-East, Northborough, MA.)

not likely that dredged material will be returned to the created canals or channels or fill material removed from affected areas since most are occupied by buildings of various kinds. If, however, fill is recent and the activity was unauthorized, government regulators usually require the responsible individual to remove the fill and restore the affected wetland. Perhaps the greatest opportunities for restoration of estuarine wetlands are on dredged material disposal sites where former marshes were filled. Yet the costs for removing this material may be prohibitive given current technologies. Many coastal marshes have been impounded, yet most are managed as estuarine systems and could be returned to more naturally functioning wetlands, if desired. Other wetlands have restricted tidal flows that have reduced salinities creating brackish conditions, thereby allowing common reed (*Phragmites australis*), narrow-leaved cattail (*Typha angustifolia*), and other plants to invade and dominate many former salt marshes. This condition is easily remedied by enlarging culverts, breaching dikes, or through other means. Smooth cordgrass (*Spartina alterniflora*) returns quickly to these sites, provided that the subsidence of marsh soils has not been significant. These impounded and degraded marshes are suitable for rehabilitation which is often considered a form of restoration. Significant opportunities also exist for estuarine wetland creation to stabilize dredged material deposited in shallow water and tidal flats or to protect eroding shorelines from wave action. Some coastal states, including Maryland and Delaware, have actively encouraged private property owners to build estuarine wetlands to stabilize shorelines (Fig. 4) instead of constructing bulkheads and rip-rap structures for erosion control. Wetlands may be created through the excavation of uplands adjacent to estuarine marshes, with material removed to a level that promotes frequent tidal flooding. All these projects have little problem accessing hydrology since they are established along tidal embayments, rivers, or existing estuarine marshes. The main obstacles are attaining proper elevations, planting the various species at the right levels, and reducing the effects of wave action. Other important factors in project success are salinity, soil properties, site drainage characteristics, proper acclimation of nursery-grown stock prior to transplanting, nutrient availability (especially nitrogen), and controlling herbivory and human actions (e.g., foot traffic and ATVs). Snow geese and muskrats have caused major problems for some creation projects. Suitable species for estuarine marsh restoration and creation are halophytes (salt-tolerant plants), including smooth cordgrass for regularly flooded (low marsh) sites and salt-hay grass (*S. patens*), big cordgrass (*S. cynosuroides*), salt grass (*Distichlis spicata*), and black needlerush (*Juncus roemerianus*) for irregularly flooded (high marsh) sites on the U.S. Atlantic and Gulf coasts. On the U.S. West Coast, Pacific cordgrass (*S. foliosa*) is the major species planted. Direct seeding has been done. Alternatively, sprigs, seedlings, and plugs have been hand planted or mechanically planted.

Mangroves dominate the coastlines of the world's tropics. Mangroves have been planted for silviculture in the Philipines for about 200 years. Red mangrove (*Rhizophora mangle*) is the most widely planted species in Florida, whereas black mangrove (*Avicennia germinans*) and white mangrove (*Laguncularia racemosa*) have also been planted. Mature seedlings (propagules) are collected from local swamps. These propagules may be directly planted or planted aerially. In some cases, individual shrubs or trees may be planted, but this increases project costs. The best sites for restoration/creation are low-energy shorelines, sheltered from strong wave and current action. Attempts to establish mangroves in high-energy environments have a low potential for success. High salinities and elevated soil surface temperatures pose serious problems at some sites.

Inland marshes are among the easiest wetlands to restore and create and are probably the most widely established wetland type in the United States. Marshes represent early stages in hydrarch succession and are generally resilient and tolerant of disturbance. Prairie pothole marshes, for example, are well-adapted to drastic annual fluctuations in water levels and their high productivity is directly related to these dynamics. Since many of the former marshes were drained and converted to cropland, it may be relatively simple to restore wetland hydrology and, thereby, reestablish these wetlands. The soils contain a natural seedbank or reservoir of hydrophytic plants species, so once wetland hydrologic conditions return, these plants quickly recolonize the site. The buried seeds of hydrophytes may remain viable for centuries. Perhaps the easiest marshes to restore are small, isolated former wetland basins that have been drained by open ditches

FIGURE 4 Estuarine wetlands are being created along eroding shorelines in coastal waters. These projects involve planting halophytic species like smooth cordgrass (*Spartina alterniflora*) which usually form dense stands within 1 year. (A) A created marsh soon after planting, (B) the same area 1 year later.

(Fig. 2). Cleaning out the organic matter in the ditch and then placing an earthen plug in the ditch can bring back wetland hydrology. Basins with larger watersheds require culverts and/or spillways, plus erosion control measures (e.g., biodegradable meshes, rip-rap, and anti-seepage diaphragms) to prevent washouts. Former wetlands that are tile drained are more difficult and costly to restore. First, one has to locate the tiles, then destroy a portion of the tile system, and install a ditch plug with a spillway. It is probably advisable to till or disk restoration sites prior to restoring wetland hydrology, so that existing turf is broken down and thereby facilitating colonization by seedbank hydrophytes. Failure to do this may significantly slow revegetation. Many marsh restoration projects are actually wetland enhancements where the hydrology of an existing wetland, usually a wet meadow, is changed to that of a marsh by increasing the hydroperiod through a combination of dikes and water control structures. Marsh creation requires additional considerations. One must create a wetland basin in an upland site by excavation (Fig. 3) and/or impoundment of a natural valley. Ideally, if hydric soil (with its natural seedbank) can be brought from the altered site to the creation site, it may be easier to establish the desired plant community than by plantings or seeding. For all projects, first-year water levels are critical. With few exceptions, marsh plants tend to germinate best in moist to saturated soils. The seedlings of most species are very susceptible to early season flooding, so it is usually recommended that the site's initial hydrology be one of saturated soils (until seedlings attain some height) followed by shallow flooding (less than 1 inch of water). It is important not to inundate the entire plant during the first year. In future years, flooding depths can be gradually increased to reach the desired level. All projects should have contingency watering plans, especially for projects in arid and semiarid regions, to ensure favorable conditions for plant growth during the critical first year. Shallow wells may be installed if necessary. Marsh restoration and creation projects should consider producing a diversity of habitats, including islands and other wetland types. The hydrology required for the establishment of wet meadows is one of alternating prolonged periods of saturated soils with brief shallow flooding events. Because grazing may pose a problem for some meadows, exclusion fences may be required. Common plant species in freshwater marshes and meadows that have been used in restoration and creation include cattails (*Typha* spp.), bulrushes (*Scirpus* spp.), pickerelweed (*Pontederia cordata*), arrowheads (*Sagittaria* spp.), arrow arum (*Peltandra virginica*), sedges (*Carex* spp., *Eleocharis* spp., *Cyperus* spp.), reed canary grass (*Phalaris arundinacea*), and panic-grasses (*Panicum* spp.). Wetlands constructed for wastewater treatment in the United States and Europe have used the following

species which have recognized values for nutrient uptake (nitrogen removal) and assimilation suitable for this treatment: broad-leaved cattail (*T. latifolia*), soft-stemmed bulrush (*Scirpus validus*), tule or hard-stemmed bulrush (*S. acutus*), bulrush (*S. lacustris*), woolgrass (*S. cyperinus*), common reed (*Phragmites australis*, the principal species used in Europe), reed canary grass, water hyacinth (*Eichhornia crassipes*), and rushes (*Juncus* spp.). Building wetlands for this purpose requires much more elaborate design, operational, and maintenance considerations than creating wetlands for wildlife habitat. This added effort is needed to maximize the plant–soil interaction with wastewater for removing pollutants and microbial pathogens.

Many types of shrub swamps may be as easy to establish as marshes and wet meadows because of similar hydrologies. Site preparation is also similar (Fig. 3). The planting of seedlings is probably the most typical revegetation technique for shrubs, with some exceptions. Willow twig cuttings may be directly planted at restoration sites. Common species that have been used or may be suitable for restoration/creation projects include willows (*Salix* spp.), buttonbush (*Cephalanthus occidentalis*), dogwoods (*Cornus* spp.), alders (*Alnus* spp.), arrowwoods (*Viburnum* spp.), winterberries (*Ilex* spp.), swamp azalea (*Rhododendron viscosum*), sweet pepperbush (*Clethra alnifolia*), and highbush blueberry (*Vaccinium corymbosum*). Varieties and relatives of the latter species are widely cultivated on former wetland sites for berry production in New Jersey. Bogs, however, are a notable exception. Bogs characterized by ericaceous shrubs (e.g., leatherleaf, *Chamaedaphne calyculata*) are perhaps the most difficult wetland type to establish because of their unique soil chemistry and deep organic soils, although there is at least one report of an attempt to relocate a bog. Perhaps the dense shallow root system binds the organic soil and makes it possible to carefully remove a living carpet of bog vegetation, much like sod or turf mats used for establishing residential lawns. If possible, this probably would be a very labor-intensive and costly project. [*See* BOG ECOLOGY.]

Restoration of forested wetlands takes longer to successfully accomplish than for emergent wet-

FIGURE 5 Bottomland hardwood forest restoration in eastern Louisiana. It will take 10 years or more for the forest to develop, so these types of restoration projects require longer monitoring than restored marshes and wet meadows.

lands and most shrub swamps simply because it takes more time for trees to mature and a forest to reestablish. The most extensive forested wetland restoration projects involve bottomland hardwood forests (Fig. 5). Millions of acres of these wetlands have been converted to agriculture (e.g., soybean fields) in the southeastern United States, especially in the lower Mississippi alluvial plain. Consequently, the potential for restoring these forests is enormous. The best sites for restoration are poorly drained or frequently flooded cropland that is considered low value farmland because of excessive wetness and frequent crop failure. Species used in restoration depend on wildlife habitat/forestry objectives. Zonation patterns of bottomland plant communities correspond to elevational gradients and differences in the frequency and duration of flooding. Observing plant distribution in neighboring bottomland swamps provides valuable insight for species selection for proposed restoration sites. Typical southern bottomland species include bald cypress (*Taxodium distichum*), oaks (*Quercus* spp.), pecans (*Carya* spp.), ashes (*Fraxinus* spp.), elms (*Ulmus* spp.), sweet gum (*Liquidambar styraciflua*), black gum (*Nyssa sylvatica*), water hickory (*Carya aquatica*), silver maple (*Acer saccharinum*),

and sycamore (*Platanus occidentalis*). Bottomland reforestation involves site preparation (e.g., disking the soil to a foot or more to remove existing vegetation and control rodents, and soil fertilization, if necessary) prior to seeding or planting. Planting seedlings (over 1.5 feet tall) is the typical method, while direct seeding by hand or machines has also been done. Oaks, pecans, and other tree species with large seeds may be planted directly into the soil. Some tree species are suitable for direct planting of fresh twig cuttings, although most are prerooted following standard horticultural techniques prior to planting. These species include poplars (*Populus* spp.), sycamore, ashes, willows, and sweet gum. There are reports of relocating entire swamps where full-sized trees and associated shrubs were replanted at the new site. This practice is extremely limited, probably because of high costs. It is, however, the fastest way to create a forested wetland. Attempts to restore forested wetlands are limited elsewhere, perhaps, in part, because these types have not experienced the tremendous historical losses that their southern counterparts have and because government-sponsored restoration is focusing on other wetland types. Saplings of red maple (*Acer rubrum*) and other

FIGURE 6 This forested wetland project involves both restoration and creation to mitigate for wetland alteration during roadway expansion. The area has been planted with saplings of trees and shrubs typical of palustrine forests in the vicinity. It will take many years for this wetland to function as a forested wetland.

northern species have been planted with shrubs and other plants in red maple swamp restoration/creation projects in the Northeast. These projects are usually required as mitigation for permitted work in natural wetlands (Fig. 6).

VII. MONITORING AND EVALUATION

Due chiefly to the difficulty of establishing the desired hydrology, it is vital that project sites be monitored and evaluated for success. The hydrology of the newly established wetland should be monitored frequently during peak flows to ensure that design is working as planned. During such times, it is advisable to make a few on-site inspections each week. If the design has taken all significant environmental factors into account and the project is constructed exactly as drawn, the project will probably succeed. Practical experiences suggest, however, that it is easier to draft a good plan on paper than it is to build it on the ground. For this reason, it is advisable to record the as-built dimensions of

the project after construction. The final dimensions will largely determine the fate of the project.

Most projects that fail, do so mainly because they did not establish the desired hydrology. Besides poor project design, there are numerous other significant problems leading to project failure, including planting at inappropriate elevations, invasion by undesirable species, lack of organic matter in the soil, overcompaction of substrates, grazing by herbivores, vandalism, human traffic, and climatic extremes such as droughts, floods, and hurricanes.

Most problems arise in the first couple of years when the plants are establishing themselves, so it is imperative that all restoration and creation sites be monitored for at least 2 years. Detection of problems during this time will allow necessary adjustments to be made with minimal loss of desired wetland functions. Quarterly observations may be advisable during the first year for all projects. Documentation of observations and remedial actions taken is imperative. Marshes, wet meadows, and shrub swamps (excluding bogs) should probably be evaluated in years 1, 2, and 5 following project completion. Forested wetland restoration projects

FIGURE 7 Photos are often taken to show the evolution of a wetland restoration or creation project, but they provide little data on project success. To improve their value, photos should be taken from permanent locations. (A) Project during site preparation. (B) Site planting. (C) Restored wetland after 2 years. This project involved planting about 33,000 tubers of marsh herbs and more than 100 shrubs. Eighty percent vegetative cover was attained within two growing seasons. (Photos courtesy of Fugro-East, Northborough, MA.)

FIGURE 7 *Continued*

should be monitored for longer periods, perhaps 10 years at a minimum, to assess revegetation success. For these wetlands, a monitoring schedule might require site evaluation in years 1, 2, 5, and 10. Fifty years of monitoring is unreasonable as a requirement for government permits, but this length of time may actually be required to fully evaluate the success of restoring bogs and some forested wetlands.

What measures are used to evaluate project success? First, the project's goals and objectives should provide a means for determining appropriate criteria. What are the intended functions to be performed by the restored or created wetland? The answer to this question should be included in the project design plan. Comparisons between restored/created wetlands and natural wetlands of similar form and function are often useful for assessing project success, but do not expect them to be exactly alike. Since it is easier to evaluate form rather than function, most criteria used to evaluate restored or created wetlands are form related. Some commonly used parameters are size and shape of the wetland, type of wetland, interspersion of vegetation and open water, amount of shoreline or edge,

water depth and seasonal fluctuations in the water table (surface water and groundwater well monitoring), plant species composition (diversity) at different elevations, plant cover, weighted average of "wetland" species vs "non-wetland" species, stem density, plant height, aboveground and belowground biomass, basal area, seedling survival rates, number of volunteer plant species, reproductive success of plants, wildlife species, wildlife use, wildlife abundance, aquatic invertebrate diversity and biomass (for marshes), accumulation of organic matter in the soil, and water quality (e.g., nitrogen, phosphorus, and suspended solids). Wetlands created or restored for one particular function are evaluated relative to that goal. Panoramic photos are often taken to show the before and after condition of the project site, but they usually provide insufficient information to judge project success (Fig. 7). To be most beneficial, these photos should be taken from permanent locations so that periodic comparisons can be made. Low-altitude aerial photos acquired during the peak of the growing season over a series of years would show the extent and annual changes in vegetative cover and the open water to vegetation ratio at the site.

This could be supplemented with ground surveys to verify species composition and other parameters.

Careful documentation throughout the entire project is critical not only for evaluating project success, but also for being able to reproduce successful results in the future. This provides indispensable information on the do's and don'ts of wetland restoration and creation. Far too many projects have paid little or no attention to this aspect. Given the tremendous acceleration of wetland restoration/creation projects, it is imperative that projects be well-documented from the planning stage through the completion of monitoring. Only by meticulous recording and reporting will we be able to better understand the factors affecting both the successes and failures of wetland restoration and creation and to better design and construct future projects for improving the status of wetlands and increasing the valued functions they perform.

Glossary

Estuarine wetlands Wetlands periodically inundated by salt or brackish tidal waters (salinity above 0.5 parts per thousand), including salt and brackish marshes, mangrove swamps, salt barrens (salinas), and tidal flats.

Halophyte A plant adapted for life in saline soils and characteristic of salt and brackish tidal marshes and mangrove swamps, inland saline marshes and meadows in arid and semiarid regions, and salt flats.

Hydric soil Soil that is saturated, flooded, or ponded long enough during the growing season for anaerobic and reducing conditions to develop in the upper part and that typically supports the growth of hydrophytes; soil characteristic of marshes, swamps, bogs, and other wetlands.

Hydrophyte An individual plant adapted for life in water or in periodically flooded and/or saturated soils (hydric soils) that exhibit prolonged anaerobic conditions; plants growing in deepwater habitats and wetlands; may represent the entire population of a given species (obligate hydrophytes) or only a subset of individuals (e.g., wetland ecotypes) so adapted (facultative-type hydrophytes).

Palustrine wetlands Nontidal wetlands and freshwater tidal wetlands that are typically dominated by persistent vegetation, including marshes, wet meadows, prairie potholes, playas, fens, pocosins, shrub swamps, wooded swamps, certain bottomland hardwood forests, wet flatwoods, Carolina bays, hydric hammocks, muskegs, and wet tundra.

Wetland A vegetated or nonvegetated area that is permanently covered by shallow water (less than 6.6 ft or 2 m) or is periodically inundated and/or saturated near the soil surface by surface or groundwater at a frequency and duration usually sufficient to create prolonged anaerobiosis that favors the growth and reproduction of hydrophytes and the development of hydric soils; includes a diverse assemblage of wet habitats ranging from shallow aquatic habitats to seasonally saturated lands such as marshes, bogs, swamps, fens, prairie potholes, Carolina bays, pocosins, playas, vernal pools, ponds, tidal flats, wet flatwoods, hydric hammocks, and certain floodplain and bottomland forests.

Wetland creation Process or result of constructing a wetland where one did not exist; the process may be either intentional (e.g., to create a wetland for wastewater treatment) or accidental (e.g., seepage from an earthen impoundment), but the net result is a gain in wetland acreage.

Wetland enhancement Process or result of changing the existing condition of a wetland to improve one or more of its functions, with little or no change in wetland acreage; usually changes the wetland type (e.g., wet meadow to marsh).

Wetland rehabilitation Process or result of restoring a degraded, contaminated, functionally impaired, or otherwise damaged wetland to its original (prealtered) condition.

Wetland restoration Process or result of returning a former wetland (now nonwetland) to a functioning wetland of some other type which produces a net gain or increase in wetland acreage; also defined by some authors to include wetland rehabilitation and wetland enhancement activities (see preceding definitions).

Bibliography

Dennison, M., and Berry, J., eds. (1993). "Wetlands: Guide to Science, Law, and Technology." Park Ridge, NJ: Noyes Publications.

Hammer, D. A., ed. (1989). "Constructed Wetlands for Wastewater Treatment: Municipal, Industrial, and Agricultural." Chelsea, MI: Lewis Publishers, Inc.

Hammer, D. A. (1991). "Creating Freshwater Wetlands." Chelsea, MI: Lewis Publishers, Inc.

Kentula, M. E., Brooks, R. P., Gwin, S. E., Holland, C. C., Sherman, A. D., and Sifneos, J. C. (1992). "Wetlands: An Approach to Improving Decisionmaking in Wetland Restoration and Creation." Washington, D.C: Island Press.

Kusler, J. A., and Kentula, M. E., eds. (1989). "Wetland Creation and Restoration: The Status of the Science." Corvallis, OR: U.S. Environmental Protection Agency,

Environmental Research Lab. EPA 600/3-89/038A. (Also published by Island Press, Washington, D.C.)

Marble, A. D. (1990). "A Guide to Wetland Functional Design." Federal Highway Administration, McLean, VA. Rept. No. FHWA-IP-90-010. (Also published by Lewis Publishers, Chelsea, MI.)

Mitsch, W. J. (1992). Landscape design and the role of created, restored, and natural riparian wetlands in controlling nonpoint source pollution. *Ecol. Engineer.* **1,** 27–47.

National Research Council (1992). "Restoration of Aquatic Eco-systems." Washington, D.C., National Academy Press.

Schneller-McDonald, K., Ischinger, L. S., and Auble, G. T. (1990). "Wetland Creation and Restoration: Descriptions and Summary of the Literature." Washington, D.C: U.S. Department of Interior, Fish and Wildlife Service, Biol. Rept. 90(3).

Thayer, G. W., ed. (1992). "Restoring the Nation's Marine Environment." College Park, MD: Maryland Sea Grant Book.

USDA Soil Conservation Service (1992). "Chapter 13: Wetland Restoration Enhancement or Creation." Engineering Field Handbook, Washington, D.C., USDA publ. #210.

Wetlands Ecology

Curtis J. Richardson
Duke University

I. Introduction
II. What Are Wetlands?
III. Why Do So Many Types of Wetlands Exist?
IV. How Are Wetlands Created on the Landscape?
V. What Are the Ecological Values and Functions
of Wetlands?

Wetlands make up approximately 6% of the world's land surface and are found in every climate from the tropics to the frozen tundra. These ecosystems are transitional between terrestrial and aquatic systems and have a global similarity of excessive water supply such that these lands have water at or near the surface of the ground for much of the year. The native plants and animals living in wetlands are uniquely adapted to live under conditions of intermittent flooding, lack of oxygen (anoxia), and harsh (often toxic) conditions of reduced chemical species (e.g., H_2S instead of SO_4). Values ascribed to many wetlands include providing habitats for fishing, hunting, waterfowl, timber harvesting, wastewater treatment, and flood control, to name a few. The intensive conversion of wetlands to agriculture, forestry, and urban areas has resulted in the loss of 53% (89 to 42 million ha) of the wetland habitats in the conterminous United States during the period of 1780 to 1980. Europe has developed and exploited nearly all of its native wetlands. Third world countries in Africa, South America, and Asia are currently developing vast wetland areas for food and fiber. The extraction of these human values and the development of native wetlands result in the direct loss of most of the ecological functions found within these ecosystems. Key ecosystem functions on the landscape include water and nutrient storage; chemical transformations of nitrogen, phosphorus, sulfur, and carbon; high primary productivity; and low decomposition rates. At the population level, wetlands function as wildlife habitat, maintaining unique species and increased biodiversity. Finally, there is a lack of understanding that wetland values can only be sustained if ecosystem functions are maintained. These functions are lost once the wetland is drained.

I. INTRODUCTION

For thousands of years wetlands were viewed as wastelands responsible for disease and pestilence. In the 17th century the Englishman Colonel William Byrd surveyed and named the swamp between the Virginia and North Carolina border as the "Great Dismal Swamp." He described it as a "horrible desert, the foul damps ascend without ceasing, corrupt the air and render it unfit for respiration." This negative attitude toward wetlands was prevalent with Europeans and their descendants as they colonized most lands of the world and pushed native populations aside. With westernization came massive wetland drainage programs over the past few centuries, uncontrolled timber harvesting, wildlife removal, and the near extinction of many species. In the rush to exploit and extract resources from these wetlands and convert them to agriculture, there was a failure to recognize that the human values taken from wetlands (e.g., crops, timber, peat, waterfowl, hunting, fishing) directly diminished the wetlands' ability to function as a wetland on the landscape. The importance of ecological functions that wetlands provide on the landscape was not known until the concepts of modern ecosystem ecology were articulated by Tansley in England in the 1930s and by Eugene Odum in the United States in the 1940s. These concepts were not accepted by society until the 1960s and the 1970s. [*See* Bog Ecology.]

Ironically, many cultures, including ours, had survived and flourished only because of the great productivity of plant and animal life they utilized from wetlands. For example, more than half the world population's main grain staple is rice (*Oryza sativa*), a domesticated tropical reed swamp grass now grown on managed agricultural wetlands. The Spanish conquistadors found Aztec *chinampa* or a "floating-garden agriculture" of maize, beans, and peppers growing on floating beds of decomposing aquatic plants and peat in swampy wetlands of Mexico. In North America, lowland Britain, and northern Europe the wet meadows and salt marshes have for centuries provided grazing for cattle, hay production, and roof thatching. The Russians, Fins, and Irish have mined peatlands for energy for centuries. The great expanses of freshwater and tidal marshes of the world, like the Camargue in southern France, the Everglades in Florida, coastal salt marshes along the eastern and southern coast of the United States, and Europe's largest wetland, the Wadden Sea bordering the northern Netherlands, Germany, and Denmark, are critical to the waterfowl and fisheries populations of their respective regions. Estuaries throughout the world and the mangrove regions found in the equatorial belt (30°N to 30°S) are the planktonic algae and carbon food-chain base for the world's great shrimp, shellfish, and fisheries populations. Thus, it is evident that the wetlands of the world are extremely pro-

ductive and have contributed greatly to the survival of our species.

Although the term "wetland" encompasses an enormous diversity of plant and animal communities, these systems have one thing in common: their formation and characteristics are controlled by water. To understand what characterizes a wetland and the ecology of wetland ecosystems, we need to first define wetlands and then review what hydrologic, soil, and community properties separate a wetland from an aquatic or terrestrial ecosystem. To categorize the types of wetlands that exist on the landscape it is necessary to analyze the relative proportion of three water sources—precipitation, groundwater discharge, and lateral surface flow—that affect wetland hydrology, as well as determine their topographic position on the landscape. Another issue that must be assessed is the relationship between ecological functions found in wetlands and human values taken from wetlands. The terms *functions* and *values* are not synonymous, and the values of an ecosystem are derived directly from the existing and operating functions within the wetland ecosystem.

II. WHAT ARE WETLANDS?

Wetlands are not easily defined. It is a collective term used to describe a great diversity of ecosystems worldwide whose formation and existence are dominated by water. Permanently flooded deep water areas (generally greater than 2 m) are not considered wetlands. Wetlands form part of a continuous water gradient from the drier upland to the continuously wet open water ecosystems on the landscape. Thus, the upper and lower limits of a wetland are somewhat arbitrary; a fact causing considerable problems in the legal determination of what is or is not defined as the wetland area regulated under the U.S. Army Corps of Engineers (COE) Section 404 of the 1977 Clean Water Act. Historically, wetlands were defined by specialists like botanists and foresters who focused on plants adapted to flooding and/or saturated soil conditions. A hydrologist's definition emphasized the position of the water table relative to the ground surface over time. Wildlife biologists focused on waterfowl, wading birds, and game and fish species habitats. No formal or universally recognized definition of wetlands existed until the U.S. Fish and Wildlife Service (FWS) in 1979, after years of review, proposed a comprehensive definition in a report entitled "Classification of Wetlands and Deepwater Habitats of the United States." The wetland definition in this report was used as the criteria for a new National Wetlands Inventory and stated:

> Wetlands are lands transitional between terrestrial and aquatic systems where the water table is usually at or near the surface or the land is covered by shallow water. Wetlands must have one or more of the following three attributes: (1) at least periodically, the land supports predominantly hydrophytes (water-loving plants), (2) the substrate is predominately undrained hydric soil (wet soils), and (3) the substrate is nonsoil and is saturated with water or covered by shallow water at some time during the growing season of each year.

This definition is widely accepted today by scientists, land use planners, and state managers alike, and has been adopted internationally in many countries. This definition, although comprehensive and useful for the purposes of classification and mapping of wetlands, does not lend itself to the legal definition of wetlands that must be used to determine whether or not a wetland is a jurisdictional wetland for purposes of a dredge and fill permit issued by the COE under Section 404 of the Clean Water Act of 1977. In 1989, four federal agencies [COE, Environmental Protection Agency (EPA), Soil Conservation Service, and the FWS], collaborated and developed an "accepted" jurisdictional wetland definition which, unlike the FWS definition, is more restrictive since it must meet all three attributes. In general terms a jurisdictional wetland must have hydric or waterlogged soils, 50% or more of the dominant plants must be wetland plants, and the water table must be found within 30 cm of the ground surface for at least 5% of the growing season. The interagency legal definition of wetlands is as follows:

> The term "wetlands" means those areas that are inundated or saturated by surface or groundwater at a frequency and duration sufficient to support, and that under normal circumstances do support, a prevalence of vegetation typically adapted for life in saturated soil conditions. Wetlands generally include swamps, marshes, bogs, and similar areas (EPA, 40 CFR 230.3 and COE, 33 CFR 328.3).

With a scientifically sound definition of wetlands they can be classified into types. Many classification systems based on vegetation have been used worldwide but the most successful are based on hydrology and ecologically similar characteristics.

III. WHY DO SO MANY TYPES OF WETLANDS EXIST?

Part of the confusion is due to the fact that ecologically similar wetland types are simply referred to by many different names throughout the world. The generic term wetland is now used worldwide and includes ecosystems known regionally as bogs, bottomlands, carr, fens, floodplains, mangroves, marshes, mires, moors, muskegs, playas, peatlands, pocosins, potholes, reed swamps, sloughs, swamps, wet meadows, and wet prairies. Moreover, a swamp in North America refers to a wetland with trees or shrubs but in Europe they are called "carrs." The pocosins (Algonquin Indian word for swamp-on-a-hill) of the southeastern coastal plains of the United States are really nothing more than shrub bogs. Likewise, a bottomland refers to a floodplain wetlands generally along a stream, and both are often called riparian wetlands, even though this term represents only the area of streamside zone influence. The billabongs of Australia are known as lagoons or backswamps elsewhere. Mire comes from the Old Norse "myrr" and is used to describe peatlands in Europe. In Canada, peatlands are often called muskegs, another Algonquin Indian term, and in Germany and England they are called moors. Thus, the use of local and regional names greatly increases the confusion about the true number of wetland types.

A classification of distinct wetland types characterized according to their sources of water, nutrients, ecological similarities, and their topographic position on the landscape clarifies and separates wetland types. The key to ordering these systems is based on their dominant sources of water. For classification purposes the water inflows can be simplified to inputs from (1) precipitation, (2) groundwater discharge, and (3) surface and near-surface inflow (e.g., tides, overbank flow from stream channels, or overland flow) (Fig. 1a). A particular wetland may have only one dominant source of water input or a combination of sources. The relative positions of major wetland types, when scaled according to the relative importance of water sources, show that bogs and pocosins receive their water almost exclusively from rainfall (ombrotrophic) and as such are nutrient poor (oligotrophic). In contrast, fens, seeps, and some marsh types are controlled by groundwater inputs. These wetlands are nourished by minerals from the ground (minerotrophic) and are often more nutrient rich (eutrophic). The surface flow-dominated wetland types are very diverse and include swamps, salt marshes, and wetlands found along the fringe of lakes and streams. The relative position of these wetland types to each other on the landscape is best displayed when scaled out along axes of both water regime and nutrients (Fig. 1b). The nutrient content of the vegetation and soils is lowest in the bogs, and increases along the fen, marsh, and swamp gradient. In contrast, the amount of peat formation decreases along this gradient because of increased decay (decomposition) of organic matter in enriched sites with higher pH and longer dry periods. The biomass of trees in proportion to shrubs in bogs is generally low, but forest species increase to become the dominant in swamps. The duration of the hydroperiod (seasonal pattern of water level at or near the surface) ranges from permanently flooded (12 months) in the lacustrine (shallow lake) wetland to periods ranging from 6 to 9 months in the bogs and as little as 2 weeks in some swamp

systems. The seasonal magnitude of water level change ranges from as little as 0.5 m in bogs to over 7 m in tropical swamp forest systems. Thus, the amount of rainfall versus surface or groundwater can be used to separate the general types of wetlands, bog, fens, marshes, and swamps. The amount of peat formation and tree productivity are related to nutrient content, acidity, and length of time the soil is covered by water.

The modern FWS classification of wetlands divides wetland and deepwater habitats into five ecological systems: (1) marine, (2) estaurine, (3) riverine, (4) lacustrine, and (5) palustrine, with a number of subsystems and classes (Fig. 2). Freshwater wetlands comprise the last three systems and make up over 90% of the world's wetlands. The *marine* system consists of open ocean and its associated coastline. Mostly deepwater habitat, marine wetlands are limited to intertidal areas, rocky shores, beaches, and some coral reefs with salinities exceeding 30 parts per thousand. The *estuarine* system is more closely associated with land and is composed of salt and brackish tidal marshes, mangroves, swamps, and intertidal mud flats, as well as bays, sounds, and coastal rivers to where ocean-derived salts measure more than 30 parts per thousand and inland to where salts are less than 0.5 parts per thousand. The *riverine* system is limited to lotic (flowing) freshwater rivers and stream channels and is mainly a deepwater habitat. The *lacustrine* (lake) system includes wetlands situated in lentic (non-flowing) water bodies such as lakes, reservoirs, and deep pond habitats where trees, shrubs, and emergent plants do not make up more than 30% aerial coverage in areas less than 2 m in depth. The *palustrine* system comprises the vast majority (>90%) of the world's inland marshes, bogs, mires, and swamps and does not include any deepwater habitats.

IV. HOW ARE WETLANDS CREATED ON THE LANDSCAPE?

The climatic processes and water conditions that formed the swampy environments of the Carboniferous period are evident through fossil analysis

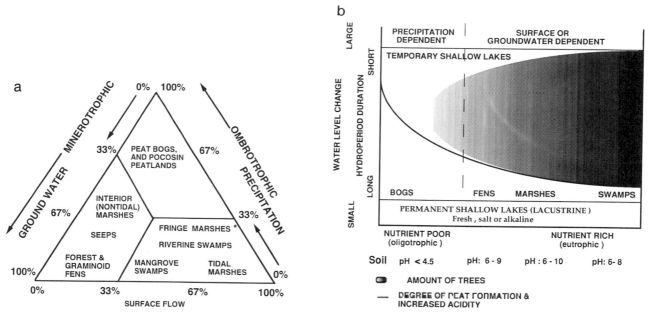

FIGURE I (a) Relative importance of the water source to major wetland types on the landscape. The percentage of water contributions to a wetland from precipitation, groundwater, or surface flow creates different types of wetlands. Ombrotrophic bogs lack groundwater contributions and upstream inflow. Minerotrophic wetlands, such as fens, receive water and minerals from groundwater. (★ = Includes both lacustrine and estuarine marshes.) [Modified from Brinson (1993).] (b) A model of the different types of wetlands found along a gradient of nutrients (X axis) and water regimes (Y axis). The swamps are dominated by trees under both long and short hydroperiods. The water level changes within wetland types vary from permanently flooded or long hydroperiods and small water level changes (lacustrine marshes) to those with short hydroperiods and large water level changes (tropical swamp forests). [Modified from Gopal et al. (1990).]

and the fuel (e.g., coal) that is burned in today's industrial society. Compared with mountains and most rivers, our present wetlands are young and dynamic ecosystems on the landscape. Most of the current peat-based wetlands date from the past 12,000 years (postglacial), although wetland habitats surely exist from the Pleistocene ice ages (2 million to 10,000 years ago) near lakes and former glaciers. Any process that produces a hollow or depression in the landscape and holds sufficient water may result in wetland formation. Wetlands are found in deserts near springs and in high rainfall or runoff areas of the mountains. Many of the northern bogs and prairie potholes (marshes) of the midwestern United States and Canada were formed in depressions left by buried ice blocks (kettle holes) from retreating glaciers approximately 10,000 years ago. The periodic flooding of rivers and streams lays down layers of silt and mud (alluvial deposits) along the banks and floodplains, creating bottomlands or swamp forests. Beavers also

play a vital role in creating thousands of acres of new wetlands each year by damming up streams.

Wetland formation can occur suddenly as when a major flood along the Mississippi in 1973 created new wetlands along the Atchafalaya delta or when debris dam up local rivers or streams. Much slower formation processes are at work in the Arctic where only the upper frozen layer of ice melts in the summer. Because of permafrost (permanently frozen ground), cool conditions, and low evaporation, the small amount of annual rainfall and ice melt creates waterlogged soils and in turn the world's most extensive peatlands, primarily in northern Russia, Canada, and Alaska. Man has also contributed greatly to the formation of wetlands in Great Britain and Europe by cutting the forests which caused the water table to rise and form many of the present day peatlands. Peat extraction for fuel in the Netherlands and building of fish ponds in Bohemia during the Middle Ages created man-made wetlands. The secret to the successful creation of functioning

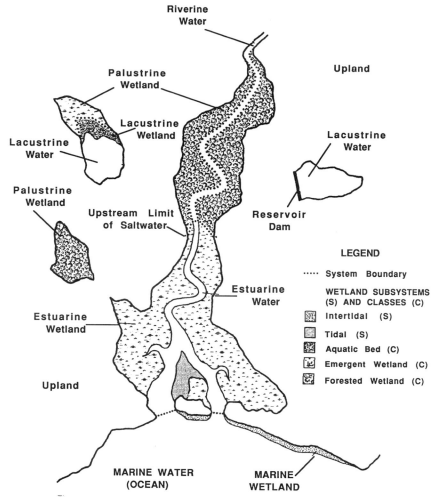

FIGURE 2 The five major wetland systems of the U.S. Fish and Wildlife Service classification system are displayed along with several subsystems and classes. (From Tiner 1985.) The general boundary and relationships among systems are shown.

wetlands on the landscape, whether man-made or natural, is primarily dependent on maintaining the hydrological integrity (i.e., maintaining the volume of water and the seasonal water level patterns within the wetland).

Wetlands usually are found in depressions or along rivers, lakes, and coastal waters where they are subjected to periodic flooding (Figs. 3a and 3b). They also can occur on slopes adjacent to groundwater seeps or can cover an entire landscape in areas like large portions of Ireland, northern Minnesota, or coastal North Carolina, where precipitation levels exceed *evapotranspiration* (ET) and blocked drainage creates peat soils, which can be over 90% water by volume. Surface water depressions re-

ceive precipitation and overland flow (Fig. 4a). Losses occur through ET, but with little downward seepage into the water table. Because groundwater depressions are in contact with the water table, they receive groundwater plus rainfall and overland flow (Fig. 4b). Seepage wetlands occur on slopes where groundwater flows near the surface, differing from groundwater depressions in that they have an outlet. (Fig. 4c) The size of these wetlands depends directly on the amount of groundwater and the overland discharge into these wetlands. Overflow wetlands receive water from river flooding, lake water, or daily tidal influences (Figs. 3a, 3b, and 4d). The water level in these wetlands closely follows the water levels or flooding frequency of the

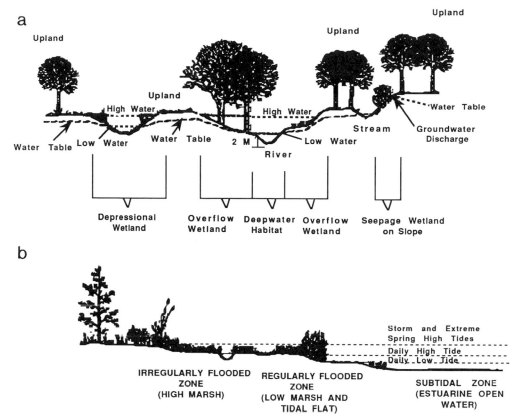

FIGURE 3 (a) Wetlands, deepwater habitats, and uplands on a landscape. Depressional, overflow, and seepage wetland types are due to distinct types of hydrologic input and topographic position on the landscape. [Modified from Tiner (1984).] (b) Cross-sectional diagram of a salt marsh showing daily low and high tide zones of influence as well as the area of irregularly flooded high marsh influenced by seasonal storm or wind tides.

water source. Clearly, the most important factors controlling the type of wetland found on the landscape are the sources of water inputs: precipitation, surface flow, or groundwater, and their topographic position on the landscape. Moreover, the sources of water control the amount of nutrients and salts the wetland receive. [*See* BOG ECOLOGY.]

V. WHAT ARE THE ECOLOGICAL VALUES AND FUNCTIONS OF WETLANDS?

The *value* of a wetland is an estimate, usually subjective, of the worth, merit, quality, or importance of a particular ecosystem or portion thereof. The term imposes an anthropocentric (human) focus, which connotes something of use or desirable to *Homo sapiens*. Values ascribed to many wetlands include providing habitats for fishing, hunting, waterfowl, timber harvesting, wastewater assimilation, water quality, and flood control, to name a few. Moreover, these perceived values directly arise from the ecological *functions* found within the wetlands. For example, wetlands under anaerobic conditions (no oxygen) process nitrate to N_2O and release the nitrogen as a gas to the atmosphere. This is an ecological function for wetlands. This function has value on the landscape in that wetlands can be successfully used to remove nitrogen from agricultural or municipal wastewater runoff. This is a value to society that is based directly on the ecological function of nitrogen cycling within the wetland ecosystem and is lost if the wetland is drained.

A general listing of the functions and values often attributed to natural wetlands is given in Table I.

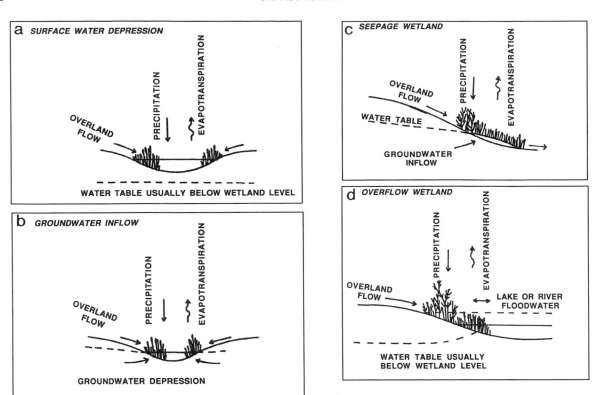

FIGURE 4   (a–d) Four major hydrologic input regimes that characterize wetlands with regard to water source, water table, and land forms. [Adapted from Novitzki (1979).]

The wetland functions are placed in five ecosystem-level categories with specific examples of these functions listed in the following paragraphs. The functions and values are placed in separate categories to try to clarify the differences between processes that wetland systems perform and the values that society extracts from these functions. Wetland values are derived directly from functions, and the specific relationships are noted by the numbers following the values given in Table I. The most important or direct relationship is listed first. Wetland functions that are potentially altered or destroyed when a certain value is obtained from the wetland can be directly assessed from this relationship. For example, timber production is a function of biological productivity (function 2 in Table I) that is directly affected by any impact or harvesting technique that reduces the net primary productivity of the site. Harvesting could also seriously affect biogeochemical cycling (3), decomposition (4), or other functions. The hunting value of a site is primarily related to the secondary productivity and status of the community/wildlife habitat (2,5), but

also may be seriously impacted by a reduction in the biological diversity of the site. Moreover, any activity that significantly impacts the hydrologic function (1) of the wetland in question may well result in loss of the wetland itself or a loss in the ability of the wetland to perform flood control, flood storage, and sediment and erosion control, as well as water quality, water supply, or even timber production values on the landscape. The importance of each function and value depends on the wetland type since not all functions and values are found in each wetland. A brief comparison and analysis of some key functions of native wetlands on the landscape are given next and give an index of their ecological importance in the biosphere.

A. Biological Productivity

Wetlands have some of the highest reported plant growth rates of any ecosystem in the world due to high nutrient and water inputs. Their rates greatly exceed grasslands, cultivated lands, and most forests (Fig. 5). Cattail, salt marsh, and reed grass

TABLE I

Attributes Generally Reported as Functions and Values of Wetland Ecosystems[a]

Wetland functions
1. Hydrologic flux and storage
   a. Aquifer (groundwater) recharge to wetland and/ or discharge from the ecosystem
   b. Water storage reservoir and regulator
   c. Regional stream hydrology (discharge and recharge)
   d. Regional climate control (evapotranspiration export, large scale atmospheric losses of H_2O)
2. Biological productivity
   a. Net primary productivity
   b. Carbon storage
   c. Carbon fixation
   d. Secondary productivity
3. Biogeochemical cycling and storage
   a. Nutrient source or sink on the landscape
   b. C, N, S, P, etc. transformations (oxidation/ reduction reactions)
   c. Denitrification
   d. Sediment and organic matter reservoir
4. Decomposition
   a. Carbon release (global climate impacts)
   b. Detritus output for aquatic organisms (downstream energy source)
   c. Mineralization and release of N, S, C, etc.
5. Community/wildlife habitat
   a. Habitat for species (unique and endangered)
   b. Habitat for algae, bacteria, fungi, fish, shellfish, wildlife, and wetland plants
   c. Biodiversity

Wetland values
1. Flood control (conveyance), flood storage (1,2)[b]
2. Sediment control (filter for waste) (3,2)
3. Wastewater treatment system (3,2)
4. Nutrient removal from agricultural runoff and wastewater systems (3,2)
5. Recreation (5,1)
6. Open space (1,2,5)
7. Visual–cultural (1,5)
8. Hunting (fur-bearers, beavers, muskrats) (5,2)
9. Preservation of flora and fauna (endemic, refuge) (5)
10. Timber production (2,1)
11. Shrub crops (cranberry and blueberry) (2,1)
12. Medical (streptomycin) (5,4)
13. Education and research (1–5)
14. Erosion control (1,2,3)
15. Food production (shrimp, fish, ducks) (2,5)
16. Historical, cultural, and archaeological resources (2)
17. Threatened, rare, endangered species habitat (5)
18. Water quality (3,1,4)
19. Water supply (1)
20. Global carbon storage (4,2)

[a] From Richardson, 1994.
[b] Wetland values that are directly related to wetland functions (1–5) or those functions that can be adversely affected by the overutilization of values. The order of the numbers suggests which primary function is most directly or first affected.

marshes are the most productive and nearly double the above- and belowground annual plant biomass production (annual increase in plant organic matter fixed from sunlight by photosynthesis) stored in sedge marshes or bogs, fens, and muskegs. Grasslands and croplands of the United States produce only one-fourth of the fiber of the most productive wetlands whereas upland forests produce about one-half. Only the tropical rain forest comes close to matching the primary productivity of the mangrove swamps or cattail and salt marshes. This biomass serves as the food for a great multitude of both aquatic and terrestrial animals as well as for many species that are endemic to wetlands (only exist in wetlands). Mammals such as moose, caribou, and muskrat graze on marsh plants. Waterfowl depend heavily on marsh plant seeds. When the plants die they decay and form plant fragments called detritus. This shredded organic material forms the base of the aquatic food chain that supports zooplankton (single-celled floating animals) and, in turn, shrimp, clams, and fish (Fig. 6). Detritus from wetlands is the food source for many aquatic insects important to commercial species like salmon, shrimp, and crabs and for the majority of nonmarine aquatic animals. In 1991, 3 million tons of fish were landed in the United States; 2 million tons were from species that spend part of their life cycle in estuarine ecosystems (estuarine dependent). Wetlands thus provide important habitat and food for many juvenile species of shrimp, crayfish, crabs, and fish, i.e., secondary productivity.

B. Biogeochemical Cycling

One of the key ecological functions of wetlands on the landscape is their ability to store, transform, and cycle nutrients. Wetlands are able to filter and hold 60 to 90% of the suspended solids and sediments from wastewater additions and agricultural water runoff prior to water discharge into streams and lakes. Wetlands maintain among the widest range of *oxidized* and *reduced* chemical states of all ecosystems due to their periodic flooding and drying cycles. The unique shift from aerobic (free oxygen present) to anaerobic (no free oxygen) soil conditions gives these systems the ability to process PO_4, NO_3, SO_4, and C and to release gases (N_2O,

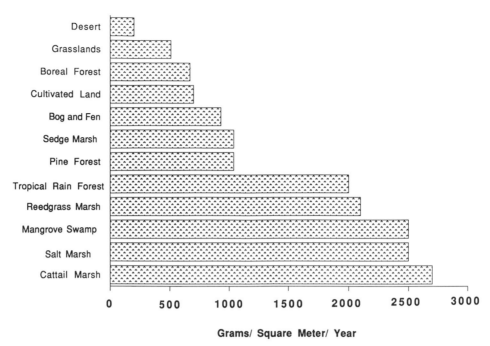

Grams/ Square Meter/ Year

FIGURE 5 A comparison of net annual primary plant productivity values for wetlands and terrestrial ecosystems. All values are in grams per square meter per year dry weight ($g/m^2/year$) for above- and belowground growth. [Values from Richardson (1979) and Maltby (1986).]

H_2S, CH_4, CO_2) into the atmosphere, thus affecting biosphere scale problems such as acid rain, global warming, and increasing greenhouse gas concentrations. A classic example of these transformations is shown by the role of wetlands on the nitrogen cycle (Fig. 7). Nitrogen transformations are a complex assortment of microbial-mediated processes and chemical reactions strongly influenced by the redox status (degree of reduced or anaerobic conditions) of the wetland soils. Most flooded wetland soils have a thin, oxidized layer at the surface caused by the proximity to the air or

higher dissolved oxygen in overlying floodwater. The reduced layer below controls the soil and deeper water column processes and prevents upland plants, bacteria, and animals from invading into these ecosystems due to a lack of oxygen for normal metabolism. Multiple biotic and abiotic transformations take place in these layers (involving seven valence states $+5$ to -3) and the formation of such oxidized nitrogen species as nitrate (NO_3^-), nitrite (NO_2^-), and nitrogen dioxide gas (N_2O) as well as the reduced ammonium ion (NH_4^+). Organic N is mineralized in both the re-

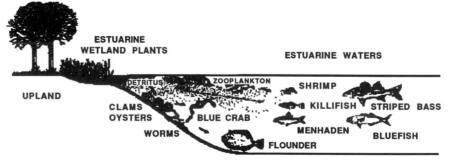

FIGURE 6 Simplified food chain from estuarine wetland plants through zooplankton, shrimp, and crabs to commercial and game fish species. [Adapted from Tiner (1984).]

duced and oxidized layers to NH_4^+. The thin oxidized zone is where NH_4^+ is oxidized to NO_3^- by bacteria (nitrification). In the reduced layer, NH_4^+ is stable and may be adsorbed to sediment or utilized by plants and microbes for growth. Any NO_3^- that diffuses downward in this zone is denitrified (microbially converted) to N_2 or N_2O and is released to the atmosphere as a gas. Thus, wetlands provide key ion transformations processes that keep nitrogen in global circulation. The most unique function of wetlands may be related to their ability to transform most essential life elements since these systems maintain the widest range of oxidation/reduction reactions on the landscape (Fig. 7). [*See* BIOGEOCHEMICAL CYCLES.]

C. Decomposition

Another unique feature of wetlands compared to terrestrial ecosystems is that dead plants and animals do not decompose as fast, especially in acid peatlands. The decay rate in peatlands is only one-half to one-quarter the rates of the more aerobic uplands. When leaves, small stems, and roots die they normally are shredded and decomposed by macroinvertebrates, earthworms, nematodes, and microbes. The complex organic matter, which is 45% carbon by weight, usually decays within 1 year in terrestrial ecosystems to CO_2 and H_2O and the mineral elements are released for new growth. This is not the case in wetlands. The slow decay rates in wetlands are due not only to a lack of oxygen, but low pH, calcium availability, and low soil temperatures often contribute. These anaerobic conditions eliminate most of the decomposer organisms, such as filamentous fungi, and the existing anaerobic bacteria obtain energy by inefficient processes, such as fermentation. Very reduced soils (i.e., no NO_3 or SO_4 present) release important greenhouse gases such as methane and hydrogen as "marsh gas." Decomposition is so greatly reduced in wetlands that many of these ecosystems slowly build up peat (undecomposed organic matter and minerals) at 1 to 2 mm per

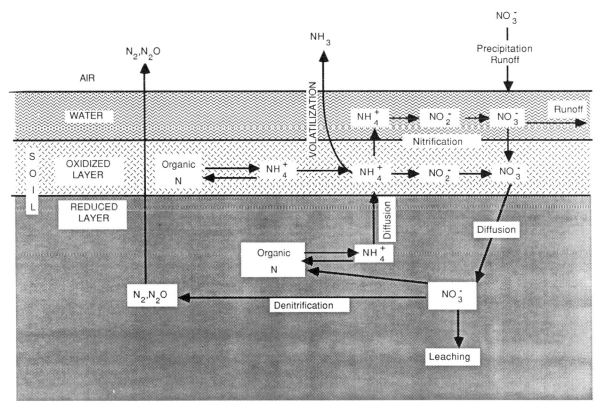

FIGURE 7 Nitrogen transformations in wetlands. [Adapted from Faulkner and Richardson (1989).]

year on average. As a result, peatlands store vast quantities of carbon and help balance the global carbon cycle. Not only have the peat-forming organisms and local biota been preserved in this peat over thousands of years, but pollen and archaeological artifacts including human remains have been found buried almost intact. One preserved human body over 2000 years old, known as the Tollund Man, was discovered in a peat bog in Denmark in 1950 with skin, whiskers, and fingernails intact. The analysis of pollen and macrofossils from peat soil from all over the world has enabled scientists to determine historical vegetation changes and climatic variations as well as human impacts on the landscape as far back as 275,000 years ago. Undisturbed wetlands are thus a time machine to past conditions.

D. Community/habitat

Wetlands are important year-around habitats for hundreds of bird species, amphibians, reptiles, and mammals, especially in the warmer climates. It has been estimated that 150 types of birds and 200 fish species are wetland dependent. The fur bearers such as the North American muskrat (*Ondatra zibethicus*) depend on wetland habitats and food like cattail (*Typha* spp.) roots, while other species like otter (*Lutra lutra*) and mink (*Lutreola lutreola*) heavily utilize these communities. In the tropical wetlands, alligators, crocodiles, and hippopotamus (*Alligator, Crocodilus, Hippopotamus,* and other genera) feed voraciously on animals and plants. The American crocodile, an endangered species, is now only found in mangroves and coastal water of Florida. The rare bog turtle (*Clemmys muhlenbergi*) and Pine Barrens tree frog (*Hyla andersoni*) only exist in a few acid bog and swamp habitats, and are threatened by the draining of these wetlands. Rare orchids (*Habenaria* spp.) and unusual insectivorous plants such as sundews (*Drosera* spp.), Venus flytraps (*Dionea muscipula*), and pitcher plants (*Sarracenia* spp.) only exist in wetland habitats.

Wetlands also provide important breeding grounds, overwintering areas, and feeding grounds for millions of migratory waterfowl and other birds. Sixty to 70% of the 10 to 12 million ducks that breed annually in North America utilize the prairie pothole region of the United States and Canada alone. The bottomland hardwood forests and marshes of the southeast are the winter home of millions of ducks that wing their way south along the meandering rivers and adjacent forested riparian wetlands. These bottomland hardwood forests also provide habitat for neotropical migrant species. A diverse number of habitats exist within each wetland based on water depth, amount of open water, and type of vegetation adapted to each zone. The species of birds and mammals have evolved so as to not compete for the same niche (food sources and nesting sites) (Fig. 8). Some ducks (dabblers) nest in upland sites or trees (wood ducks) and feed along the marsh edge and in shallow areas. Diving ducks (ruddy (*Oxyura jamaicensis*) and redheaded (*Aythya americana*)) nest over water and fish by diving. The muskrat (*Ondatra Zibethicus*) lives on cattail roots and shoots and lodges in deeper protected areas of the marsh. Wading birds (Heron and Egret etc.) usually nest in the wetlands and feed along the shore. Rails (*Rallus* spp.) live throughout the wetland. Loons (*Gavia* spp.) use the deeper waters, and grebes (*Podilymbus* spp.) prefer the marshy areas, especially during nesting season. The diversity of wetland types, as well as the variation within each wetland, thus provides essential habitats for animals and plants. Additionally, many upland animal species depend on wetlands for water and food, especially during drought periods.

E. Hydrology

The hydrologic function of wetlands on the landscape is determined by the type (e.g., bog, fen, or swamp), topographical position (i.e., proximity to large or small streams or lakes), its overall size, and its connection, if any, to groundwater. Many wetlands, including bogs, pocosins, peatlands, and marshes, function as "water pumps" on the landscape, sometimes losing two-thirds of their annual water by ET and leaving only 30% or less for annual runoff or groundwater recharge (Fig. 9). The proportion of water inflow compared to rainfall varies greatly; for example, swamps receive

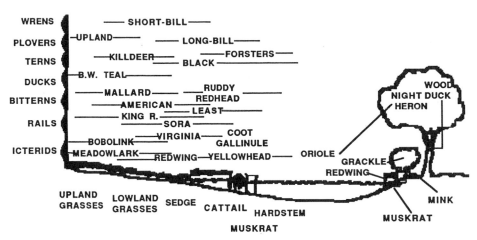

FIGURE 8 The specialized use of wetland habitats by different families of birds and mammals in a prairie pothole in the central United States. [Adapted from Weller (1994).]

most of their inputs by stream flow. They also gain and lose some water to the shallow aquifer beneath them. The stream flow out of swamps comprises the biggest loss, although the water can be held for months and released more slowly than in areas where the vegetation is removed. Bogs receive most of their water by rain and lose almost none to groundwater. The large portion of annual runoff from perched bogs (i.e., water table elevated above the regional water table) is in spring after winter snow melt or after the bog peat has filled with rain because of low winter ET rates. The peat soils of these systems act much like a "sponge" on the landscape by holding and slowly releasing vast quantities of rainfall when the peat is drier, and rapidly giving up water when the sponge is full and has little storage capacity. Many small bogs have no flowing outlets and lose almost all their water by ET. Because of their size, pocosins (often 10,000 ha or more) slowly lose water by surface and subsurface runoff that is distributed over broad surface areas to surrounding lakes, rivers, and estuaries. Fen systems are flow-through systems with runoff often exceeding ET, especially during ice thaw periods in the spring. The annual flow from fens connected to groundwater is very uniform through the year.

Wetlands can be an important source or sink of regional surface and groundwater supplies kilometers from the actual wetland itself. Studies have shown that the relationship between the regional water table and wetlands is more complex and dynamic than once thought and that seasonal reversals of groundwater flow among wetlands and between wetlands and lakes often occur. This suggests that the drainage of a wetland in one area may have serious consequences for water conditions in lakes, groundwater, other wetlands, or downstream areas that are not readily evident without detailed hydrologic studies. It has been demonstrated that peak stream flows are only 20% as large in basins with 40% lake and wetland areas as they are in similar basins with no lake or wetland area. A relationship between the amount of wetland area on the landscape and storm water flow peaks has been shown. These data suggest that when less than 10% of the watershed is in wetlands, significant peak flows occur.

It is clear that ecologists need to do a better job of quantifying wetland functions and relating them to human values so that it can be determined what will be lost when wetlands are altered or developed. The key ecological question that needs to be addressed under any development scenario is whether or not wetland functions have been significantly altered. For example, if water storage from a wetland is the important process on the landscape, then the question becomes: what will it "cost" to replace that function? Moreover, can the wetland function be replaced at all? Replacement costs are often far higher than the cost of selecting another area or avoiding the wetland functional loss. The billions

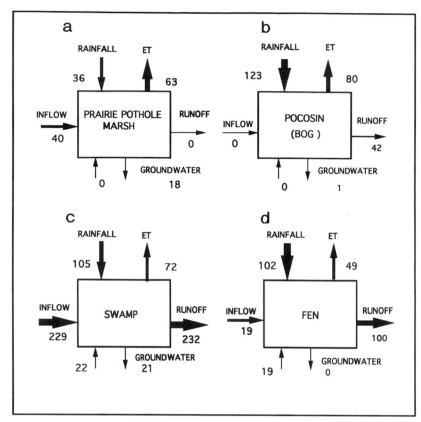

FIGURE 9 Annual water budgets for (a) a prairie pothole marsh in North Dakota, (b) a pocosin bog in North Carolina, (c) an alluvial cypress swamp in southern Illinois, and (d) a rich fen in North Wales, UK. All water flow values are reported in centimeters per year. [Data from Mitsch and Gosselink (1993) and Richardson (1994).]

of dollars of property and crop loss along the Mississippi and Missouri rivers in the Midwest during the record summer floods of 1993 is a case in point. States like Iowa and Missouri, which have developed nearly 90% of their wetlands, both suffered record flooding levels because of a lack of water storage capacity in natural wetlands as well as uncontrolled development in the floodplains themselves. Once all their ecological functions are understood, these former wastelands will become valuable wetlands. [See WETLAND RESTORATION AND CREATION.]

Glossary

Aerobic Occurring in the presence of free molecular oxygen. Obligate aerobic bacteria and fungi cannot live in the absence of free oxygen (anaerobic).

Alluvial Pertaining to alluvium or material transported by flowing water.

Ammonium A reduced inorganic univalent ion of nitrogen in the cation form (NH_4^+). Ammonia (NH_3) is a gas.

Biogeochemical cycling Interaction and integration of biological and geochemical cycles. Chemical elements (N, C, P, K, etc.) tend to circulate in the biosphere through characteristic paths from environment to organisms and back to the environment. These movements (cycling) of elements and inorganic compounds are essential to life, and wetlands play a key role in returning N, C, and S to the atmosphere.

Bog An acid peatland (often pH <4.5) that is nutrient poor because it is isolated from mineral-rich groundwater and receives most of its nutrients from rainfall (ombrotrophic). Peat is usually formed *in situ* under closed drainage and very low oxygen concentrations. The term is often used to mean peatlands in general. These wetlands can be dominated by Sphagnum moss or shrub species.

Decomposition Process of undergoing chemical breakdown. The accumulation of peat in a wetland is the result of incomplete destruction of the organic matter produced from primary production.

Evapotranspiration (ET) Combined amount of water that is lost from evaporation from any surface plus the loss of water by plant transpiration; usually expressed in the same units (centimeters or millimeters) as precipitation (rainfall).

Fen A circumneutral to basic peatland (pH 5.5–9) fed by groundwater (minerotrophic) and richer in nutrients than bogs. A meadow-like appearance due to dominant sedge and grass species, but often low shrubs or a sparse layer of trees exist.

Hydric soil Soil that is wet long enough to periodically produce anaerobic conditions, thereby influencing the growth of wetland plants and restricting the invasion of terrestrial plants. Hydric soils have been identified and mapped for most U.S. counties by the Soil Conservation Service. This is often used as one key piece of evidence to delineate wetland boundaries.

Hydrophyte Any plant growing in water or on a substrate that is at least periodically deficient in oxygen as a result of waterlogging. Plants typically found in wet habitats.

Hydroperiod Period of time (usually measured in days) water is at or near the surface of a wetland such that the ecosystem functions as a wetland. Maintaining the normal patterns of this hydrologic component is critical to the health of each wetland type. Water depth, duration, seasonality, and frequency of flooding characterize each wetland type.

Mangrove A general term for several halophytic (salt-loving) woody species found along tropical and subtropical coastlines of areas where climates have little or no frost; found primarily in the United States along the Atlantic and Gulf coasts of south Florida, Louisiana, and Texas. Two of the dominant genera, *Rhizophora* and *Avicennia,* like most mangrove species, are highly adapted to live in areas of high salinity, anaerobic soils, and intensive coastal storms.

Marsh Circumneutral to alkaline (pH 5.5–9) wetlands with grassy-like wet areas periodically inundated up to a depth of several meters. The soils are usually mineral and organic soils are often found, but peat does not usually accumulate. Dominant vegetation consists of a variety of nonwoody plants such as rushes, reeds, reedgrasses, and sedges. Where open water exists, floating aquatic plants flourish. Coastal marshes are inundated by fresh, brackish, and salt water.

Minerotrophic Nourished by mineral water. Generally refers to wetlands, mostly peatlands, that receive minerals from flowing or percolating subsurface or groundwater.

Nitrification Microbial transformation from ammonium to nitrite and from nitrite to nitrate; an energy-yielding aerobic process.

Ombrotrophic Peatlands (bog and mires) that depend on nutrients only from rainfall; they are usually extremely nutrient-poor systems (oligotrophic) with very low pH (<4), very slow plant growth, but vast stores of peat due to low decomposition rates.

Organic soils Term used to classify soils that have developed primarily from plant remains and contain >30%

organic matter. These soils are called histosols by the Soil Conservation Service.

Oxidation/reduction (redox) A quantitative measure of the tendency of substrates to oxidize or reduce susceptible substances such as iron, nitrate, or manganese. Oxidation is the loss of electrons and reduction is the addition of electrons. For example, ferric iron (FE^{3+}) is reduced to ferrous iron (Fe^{2+}) under flooded soil conditions once free O_2 and NO_3 are reduced. The scale expressed in millivolts (+400 MV represents aerated soils to −300 MV greatly reduced soils) is especially useful for describing the reducing conditions of wetland soils and the form of ions found within these environments.

Peat Organic material constituting peatlands, exclusive of live plant material and consisting largely of organic residues accumulated as a result of incomplete decomposition of dead plant roots, stems, and leaves; also contains a variable portion of mineral material left from the decaying processes as well as material transported to the peat by runoff or dust.

pH Negative log of the hydrogen (hydronium) ion concentration.

Pocosin Nutrient-poor evergreen shrub bogs (pH <4) of the southeastern coastal plain of the United States that frequently burn. These peat-based areas cover thousands of acres, are slightly domed (raised elevation) in the center, and sometimes have lakes present because of the extensive deep fires that have burned a hole in the peat to the mineral substrate. The dominant vegetation consists of evergreen shrubs (*Cyrilla* spp.; *Magnolia* spp., and pond pine (*Pinus serotina*).

Playas Shallow depressions similar to prairie potholes, but abundant on the southern high plains in Texas and New Mexico. They undergo annual and multiyear cycles of dry-out and filling.

Prairie pothole Depressional wetlands found in the upper midwestern United States and the plains provinces of Canada, which were formed primarily by glacial activity.

Primary production Weight of new organic matter created by photosynthesis or energy it represents. Observed increase in biomass (weight) of green plants over a period of time is net productivity.

Riparian Pertaining to the boundary between land and water. Normally represents the streamside zone or the zone of stream water influence. A term more often used in the western U.S. to refer to streamside wetlands.

Seepage A site where groundwater of a shallow aquifer discharges to the surface, often at the toe of a slope. Wetlands are often formed at these water release points.

Swamp Wet, forested minerotrophic wetlands where standing-to-gently flowing waters occur seasonally or persist for long periods on the surface. The waters are circumneutral to moderately acid and show little mineral or oxygen deficiency. Mineral or organic sediments or peat are deposited *in situ*. The dominant vegetation cover consists of coniferous or deciduous trees and tall shrubs.

Upland Land upslope or raised above a wetland. These terrestrial sites lack wetland characteristics or plant species.

Water table Upper surface of a zone of saturation. No water table exists where the surface is formed by an impermeable layer.

Bibliography

Brinson, M. M. (1993). "A Hydrogeomorphic Classification for Wetlands." Technical Report WRP-DE-4, U.S. Army Corps of Engineers, Waterways Experiment Station, Vicksburg, MS.

Cowardin, L. M., Carter, V., Golet, F. C., and LaRoe, E. T. (1979). "Classification of Wetlands and Deepwater Habitats of the United States." U.S. Fish and Wildlife Service, Department of Interior, Washington, D.C.

Faulkner, S. P., and Richardson, C. J. (1989). Physical and chemical characteristic of freshwater wetland soils. In "Constructed Wetlands for Wastewater Treatment: Municipal, Industrial and Agricultural" (D. A. Hamer, ed.), pp. 41–72. Chelsea, MI: Lewis Publishers Inc.

Good, R., Whigham, D., and Simpson, R. (1978). "Freshwater Wetlands: Ecological Processes and Management Potential." New York: Academic Press.

Gopal, B., J. Kvet, H. Loffler, V. Masing and B. C. Patten. 1990. "Chapter 2, Definition and Classification," in Wetlands and Shallow Continental Water Bodies, volume 1, pp. 9–15, edited by B. C. Patten et al. 1990. S. P. B. Academic Publishing, The Hague, The Netherlands.

Gorham, E. (1991). Northern peatlands: Role in the carbon cycle and probable responses to climatic warming. Ecol. Appl. **1**, 182–195.

Gore, A. J. P., ed. (1983). "Ecosystems of the World, Mires, Swamp, Bog, Fen, and Moor," Vols. 4A and 4B. Amsterdam: Elsevier Science Publishers.

Greeson, P. E., Clark, J. R., and Clark, J. E., eds. (1978). "Wetland Functions and Values: The State of our Understanding." Minneapolis, MN: American Water Resources Association.

Lyon, J. G. (1993). "Practical Handbook for Wetland Identification and Delineation." Boca Raton, FL: Lewis Publishers.

Maltby, E. (1986). "Waterlogged Wealth." London: Earthscan.

Mitsch, W. J., and Gosselink, J. G. (1993). "Wetlands," 2nd Ed. New York: Van Nostrand Reinhold.

Niering, W. A. (1985). Wetlands. In "The Audubon Society Nature Guide." New York: Alfred A. Knopf.

Richardson, C. J. (1983). Pocosins: Vanishing wastelands or valuable wetlands? Bioscience **33**, 626–633.

Richardson, C. J. (1985). Mechanisms controlling phosphorus retention capacity in freshwater wetlands. Science **228**, 1424–1427.

Richardson, C. J. (1989). Freshwater wetlands: Transformers, filters, or sinks? In "Freshwater Wetlands and Wildlife" (R. R. Sharitz and J. W. Gibbons, eds.), pp. 25–46. CONF-8603101, DOE Symposium Series No. 61, USDOE.

Richardson, C. J. (1994). Ecological functions and human values in wetlands: A framework for assessing impacts. Wetlands 1–9.

Teal, J. M., and Teal, M. (1969). "Life and Death of the Salt Marsh." Boston, MA: Little, Brown, and Co.

Tiner, R. W., Jr. (1984). "Wetlands of the United States: Current Status and Recent Trends." Washington, D.C. U.S. Fish and Wildlife Service.

U.S. Army Corps of Engineers. (1987). "Corps of Engineers Wetlands Delineation Manual," Technical Report Y-87-1, U.S. Army Engineer Waterways Experiment Station, Vicksburg, MS.

Weller, M. W. (1994). "Freshwater Marshes: Ecology and Wildlife Management," 3rd Ed. Minneapolis, MN: Univ. of Minnesota Press.

Wild Crops

Alan F. Raybould

Institute of Terrestrial Ecology, Dorset

I. Introduction
II. Collected Crops
III. Crop Relatives
IV. Feral Crops

Wild crops are plant species harvested from their natural environments, wild relatives of domesticated crops, and populations of domesticated crops growing in nonagricultural habitats. These categories are connected by the process of crop evolution. Initially the crop is collected from wild populations. Then to improve the yield, quality and/or harvesting of the product wild plants are taken and cultivated in a man-made (agricultural) environment. Plants in the new environment become adapted to the changed conditions and this process of domestication leads to a divergence between the plants in the man-made habitat (the domesticated crop) and those remaining in the wild (the crop relatives). In some cases the domesticated crop and the crop relatives remain sexually compatible and gene flow can occur between them. This exchange of genes can be valuable to crop breeders wishing to introduce traits from the relative into the crop, but can also lead to the evolution of weeds. Domesticated crops can become so specialized to growth in agriculture that they lose the ability to survive in the wild. Many crops, however, can escape to form feral populations in nonagricultural habitats, and in some cases the process of domestication is reversed and the crop becomes a naturalized component of the wild flora.

I. INTRODUCTION

A crop is a plant that is cultivated to yield a useful product. Cultivation encompasses a multitude of practices designed to improve the yield or quality of that product and is typically, although not exclusively, associated with man-made (agricultural) habitats. As well as being cultivated, crops are often domesticated. Domestication involves adaptation of the crop to man-made habitats and implies genetic change in the species such that it is more suited to cultivation in these habitats.

Wild crops can be defined as crops other than domesticated species growing in an agricultural habitat, and it is convenient to divide them into three categories. First are those plants that are undomesticated and grow in natural habitats. If any cultivation is practiced it is simple and tends to augment natural processes, and harvesting is simply the gathering of plant parts. These plants are sometimes referred to as "collected crops." [*See* AGRO-ECOLOGY.]

Another group of wild crops is species that are the direct wild descendants of progenitors of domesticated crops or are sexually compatible close relatives of domesticated species. The progenitor species would once have been a collected crop but their present day wild descendants may no longer be used as such (although they may be extremely valuable in crop breeding). These plants are described as "crop relatives."

A final group of wild crops is "feral crops." These plants have often been domesticated and cultivated in man-made habitats for long periods but have populations that can grow without cultivation. In many cases they are weeds of agriculture and habitats disturbed by man, and populations are maintained by spillage of seed or vegetative propagules from agriculture. In some instances, however, crops have established self-sustaining populations in natural habitats that are difficult to distinguish from undomesticated forms.

Each type of wild crop is described using examples drawn almost exclusively from Western north temperate zone (mainly European) agriculture and concentrates on plants used for food. Examples of wild crops from tropical areas (rain forests, for example) and plants that yield nonfood products (for instance, medicinal substances) can be found in other entries in this encyclopedia.

In addition to descriptions of the categories of wild crop, the interactions between wild and domesticated crops are discussed. For example, the evolution of crops from crop relatives, gene flow between crops and crop relatives, and the origins of weeds are considered.

II. COLLECTED CROPS

The ingenuity of man has found many and varied uses for plant material. It is difficult to discount any species as never having been used for some purpose. However, the very occasional use of a wild plant should not qualify it for categorization as a crop; it should be harvested in a relatively regular and systematic manner. Collected crops are not grown outside their natural habitats and have not been selected for improved performance.

The criteria used to distinguish collected crops from noncrops and domesticated crops are imprecise. Although plants that are grown commercially are easy to exclude, a difficulty arises with native species that were once agricultural crops but have become unfashionable and been abandoned as crops

on a commercial scale. These species may persist after cultivation and be harvested locally, and it is often not obvious whether they are truly wild or relicts of past agriculture. In general these species have been excluded unless the yield from wild populations always exceeded that from agricultural ones. Also the distinction between wild and domesticated forms of forage species is unclear because even though many animals are still raised on semi-natural pasture, forages have been improved by plant breeding. Natural pastures may have been seeded with these domesticated varieties at some time, but it is difficult to prove.

Table I shows several European species of wild plant that never entered agriculture widely but which are periodically harvested for food. The plants divide into three types: those producing edible fruits, those used as salad vegetables, and those used as culinary herbs. No grains or pulses could be identified as being collected crops in Europe.

Probably all progenitors of domesticated crops would once have been collected crops, and it is interesting to speculate on why species in Table I have not been domesticated, and on why certain types of food plant are not represented. In many cases the reason for nondomestication may be simply that the use of the crop was insufficient to warrant cultivation. This could be because the species is common and has only local and limited use or because the plant is unsuited to agricultural habitats. Many of the vegetables, such as dandelions and nettles and some of the soft fruits, are such species.

Another reason for nondomestication is that the quality of product from the species in agriculture is inferior to that collected from the wild. This is especially so in culinary herbs. For example, marjoram plants from the north of Europe and the wetter parts of the Mediterranean have a very insipid flavor. Only plants growing wild in hot, dry areas have the desired pungency and this has prevented the widespread domestication of the species.

The final reason for some useful wild plants not being brought into commercial cultivation is that they have been superseded by introduced plants of similar type that have better quality or higher yields. The wild strawberry of Europe (*Fragaria*

TABLE I

Examples of "Collected Crops" in the European Flora

Species	Description and uses
Rubus fruticosus Bramble (Rosaceae)	A climbing shrub with prickly biennial flowering stems. Fruit eaten raw or made into jam.
Rubus chamaemorus Cloudberry (Rosaceae)	A herb with nonspiny annual flowering stems and a creeping rhizome. Fruit used to make jam.
Fragaria vesca Wild strawberry (Rosaceae)	A low-growing perennial herb with long runners that root at nodes to form separate plants. Fruit eaten raw.
Sorbus domestica Service tree (Rosaceae)	Can grow as a tree up to 25 m in height, although in Europe it occurs as a shrub. Fruit used for making jam or alcoholic drinks.
Malus sylvestris Crab apple (Rosaceae)	A thorny tree or shrub from 2–10 m high. Fruit used for making jelly.
Prunus spinosa Sloe (Rosaceae)	A shrub from 1–4 m high, often forming dense thickets. Fruit used to flavor wine and gin.
Sambucus nigra Elder (Caprifoliaceae)	A shrub or small tree up to 10 m tall. Vigorous stems often extend from the base. Fruit and flowers used in making wine, fruit for making jam.
Vaccinium oxycoccus Cranberry (Ericaceae)	A deciduous shrub with creeping, prostrate stems. Fruit used to mak jellies and sauces.
V. vitis-idaea Cowberry (Ericaceae)	An evergreen shrub with creeping rhizomes and erect stems. Fruit used in making jam and jellies.
V. myrtillus Bilberry (Ericaceae)	A deciduous shrub with creeping rhizomes and erect stems. Fruits used to make jam and also used in confectionary.
Arbutus uned Strawberry tree (Ericaceae)	A tree or shrub up to 10 m tall. Fruit made into jam and distilled into alcoholic drinks.
Juniperus communis Juniper (Cupressaceae)	An evergreen shrub, often with a prostrate habit, although can be small trees up to 8 m in height. Berries used to flavor gin and other spirits. Also used to flavor meat dishes in many parts of Europe.
Uritca dioica Nettle (Urticaceae)	A perennial herb with both erect and prostrate stems, the latter of which can root at nodes to form separate plants. Leaves and stems blanched for use as a salad vegetable or added to soups and stews.
Crambe maritima Sea kale (Cruciferae)	A perennial herb with fleshy roots and erect stems. Blanched stems are used as a vegetable.
Hirschfeldia incana Hoary mustard (Cruciferae)	An annual, or occasionally short-lived perennial herb. Young inflorescences used like broccoli.

continues

Continued

Salicornia europaea Glasswort (Chenopodiaceae)	A small annual herb. Steamed for use as a vegetable.
Crithmum maritimum Rock samphire (Umbelliferae)	A branched perennial herb. Sometimes cooked as a vegetable but more often pickled.
Taraxacum officinale Dandelion (Compositae)	A perennial herb with a fleshy tap root and a basal rosette of leaves. Leaves soaked or blanched for use as a salad vegetable.
Origanum vulgare Wild marjoram (Labiatae)	A woody, rhizomatous perennial herb. Leaves dried for culinary use.
Rosmarinus officinalis Rosemary (Labiatae)	A small evergreen shrub. Fresh or dried leaves used as a culinary herb.
Salvia officinalis Sage (Labiatae)	A small shrub with erect stems. Fresh or dried leaves used as a culinary herb.
Thymus vulgaris Garden thyme (Labiatae)	A small, woody perennial herb. Dried leaves used as a culinary herb.
Allium scorodoprasum Sand leek (Liliaceae)	A small, perennial bulbous herb. Its bulbs are used as a flavoring in a similar fashion to garlic.
Tilia species Limes (Tiliaceae)	Large deciduous trees up to 40 m tall. Flowers used to make lime tea.

vesca) is hardly ever cultivated because the garden strawberry, *F.* × *ananassa,* a hybrid between two North American species, dominates commercial production because of its heavy cropping. This is despite it having a high susceptibility to bird damage, unlike *F. vesca,* and, to many people, an inferior flavor. Similarly, none of the native *Vaccinium* species is cultivated in Europe, whereas the American cranberry, *V. macrocarpon,* has been introduced extensively on account of its much larger fruits.

The introduction of new species also explains the lack of some categories of crop from collected crops in Europe. Introduced domesticated grains and pulses are so superior in performance to the native wild species that collecting from these species is simply not worthwhile, especially when the crop may have to be processed before use (milling grain, for example), using skills that have been lost because of the easy availability of ready-prepared products. The continued use of wild crops in the face of similar commercially produced products may depend on the retention of the ability to process them properly.

To summarize, it seems that the factors that keep collected crop plants from being domesticated are their limited use (especially if the species is common), their inferior quality when grown in agricultural habitats, and the prior import of species with superior performance. The next section examines species that have cleared these hurdles and have given rise to domesticated crop species.

III. CROP RELATIVES

Agricultural crops are derived from wild crops by domestication. Crop relatives are the present day wild descendants of those wild crops. They may or may not be regarded as the same species as the crop, depending on the extent of divergence during domestication, but they retain a degree of sexual compatibility with the crop.

A. Origins of Domestication

Dating the beginnings of domestication is difficult and has been done for relatively few crops. The earliest confirmed date for domestication is 7000 B.C. for wheat and barley in the Near East and maize, squashes, and *Phaseolus* beans in Mexico and Central America. The places of origin have been studied in more detail. Originally it was thought that the center of origin of a crop was where it had its maximum amount of genetic diversity. Further research has shown this criterion to be unreliable, as many crops were transferred over long distances in their early domestication and evolved into many varieties in new habitats. For example, Ethiopia is a center of diversity for several wheat species, although there is not a single wild relative of wheat in Ethiopia. Thus domestication is unlikely to have occurred there.

The more modern view is that the distribution of crop relatives indicates the center of origin of crops. However, several crops no longer have an extant wild progenitor and in some cases the wild relative may have been transferred to new regions with the early crop domesticates. Therefore, it is probably more realistic to describe centers where crops were first domesticated instead of centers of origin. The centers of domestication of some of the world's commonest crops are shown in Table II. It should be noted that these areas are both floristically rich and the centers of early civilizations.

B. Process of Domestication

The maintenance of natural populations, for example, collecting seed from a wild crop and sowing it in the same natural habitat, does not lead to domestication. For domestication to begin, the life cycle of the plant must be completed in the man-made environment. Seed from a population cultivated in such a habitat should be collected and sown into that habitat. Gradually the population begins to diverge genetically from the ancestral wild crop

TABLE II

Centers of Domestication of Some of the World's Most Important Crops

Center of domestication	Crops
China	Soybean, adzuki bean, apricot, peach, orange, tea
India/Malaysia	Rice, chick pea, cucumber, jute, pepper, yam, banana, coconut
Central Asia	Bread wheat, pea, chick pea, flax, carrot, radish, pear, apple, walnut
Near East	Bread and durum wheats, two-rowed barley, rye, lentil, pea, alfalfa, flax, melon, grape, apricot
Mediterranean	Durum wheat, oats, broad bean, cabbage, olive, lettuce
Ethiopia/Horn of Africa	Durum wheat, barley, chick pea, lentil, flax, castor bean, coffee
South Mexico/ Central America	Maize, common (*Phaseolus*) bean, pepper, cotton, squashes
Peru/Ecuador/Bolivia	Potato, sweet potato, lima bean, tomato, papaya, tobacco
Brazil/Paraguay	Peanut, rubber, pineapple

as traits advantageous in the new environment are selected (disruptive selection, see later).

The changes associated with domestication have been most thoroughly documented in the cereals (or at least those grown for grain which is harvested by threshing) and are summarized in Table III. Many of these changes, for example, loss of seed dispersal mechanisms, also apply to other crops harvested for seed. For example, domesticated legumes retain their seeds in the pod rather than broadcasting them by the twisting and splitting of the pod as occurs in wild relatives. It should be remembered, however, that the use of crop plants changes with time. It has been suggested that the first domestication of wheat and barley was actually as winter fodder for animals. Thus selection for characters listed in Table III may be a relatively modern phenomenon in these crops.

The amount of variation in crops also changes with domestication. When a plant is first domesticated the amount of variation may actually increase because the selection pressures present in the natural environment are relaxed. As domestication pro-

TABLE III
Changes in Cereals Produced by Domestication

Domestication process	Selective force	Adaptation	Phenotypic change
Maximization of seed-sowing density	Increased competition between seedlings	Loss of dormancy	Uniform population maturity
		Increased seedling vigor	Larger seed
	Seed maturity at harvest	Uniform seed ripening	Reduced branching of culms and inflorescences
Ease of harvesting seed	Reduction of seed loss due to dispersal	Retention of spikelets at maturity	Absence of abscission layer
	Increased yield	Increased seed production	Larger inflorescences and restoration of fertility in sterile spikelets

ceeds, however, uniformity is usually selected as a desirable trait. Also as the crop is transported from its center of domestication, it will be selected for characters that increase performance in the new conditions. Thus the general trend in crop domestication is for increasing variability between different regions but uniformity within regions. The uniformity, however, may only be apparent in the area where selection occurred. Many apparently uniform crops can exhibit a high degree of variation in new environments. For example, perennial ryegrass bred in Wales flowers for a very short time span when grown in the British Isles, but in California, under different day lengths and temperatures, flowering is over a much longer period and some plants never flower.

The divergence of a crop and its wild ancestor during domestication is an example of a process called disruptive selection, in which a species becomes adapted to different habitats and intermediate forms are selected against. In the early stages of domestication the crop and crop relative may have grown close together and be sexually compatible with each other. Thus hybridization could occur freely. Hybrids possessing characteristics of both wild and domesticated plants may be less fit than the parental species in both natural and agricultural environments. In this situation, mechanisms that prevent the recombination of crop and wild characters are selected.

One way in which this is achieved is by the tight genetic linkage of genes coding for characters that are under intense selection in either environment.

That is to say these characters tend to be inherited as a single unit. An example of this is the clustering of genes for female inflorescence morphology on a single chromosome in maize and its wild relatives (teosintes). The alternative is that mechanisms preventing hybridization are selected. This can be achieved in many ways. For example, the evolution of self-compatibility and self-pollination, the divergence of flowering times of the crop and crop relative, the evolution of attractants of different insect species in insect-pollinated plants, and the evolution of different chromosome numbers in the crop and the crop relative making hybrids inviable. Despite all of these mechanisms, however, many crops and their wild relatives retain some degree of sexual compatibility and aspects of this are discussed next with respect to crop breeding and the evolution of weeds.

C. Uses of Crop Relatives in Breeding

Crop relatives often possess characters not present in the domesticated crop. While the crop has been selected for growth in agricultural habitats, the crop relatives have been exposed to selection in natural environments. Selection under cultivation produces plants adapted to improved growing conditions and so traits for survival in the wild may be lost. These traits can include resistance to disease and tolerance of adverse conditions such as drought or soils with low nutrient content. The crop relatives often possess genetic variation for these characters, as they can increase fitness in many natural habitats, and, despite disruptive selection, they frequently retain some sexual compatibility with the domesticated crop.

The genetic variation present in a crop and its wild relatives is described as the gene pool of that crop. The primary gene pool is the domesticated species and wild forms that are fully interfertile with it. The secondary gene pool is wild species that can be crossed with the domesticate in breeding experiments to produce at least some fertile offspring. Fertility of the offspring is crucial because when a crop is hybridized with a wild relative, half of the genetic material in the hybrid comes from each parent, but often only a single character from the relative is required in the finished crop variety. To remove the unwanted wild relative traits, the hybrid is crossed with the crop parent, and offspring from this cross still possessing the desired character are selected and again crossed with the crop parent. By repeating this many times, a process called backcrossing, a plant virtually identical to the original crop, but with the extra trait from the relative, is produced.

Table IV illustrates the use of wild species in the primary and secondary gene pools of several major crops. It should be noted that many of the traits introduced into crops from wild species are disease resistances, reflecting the continual exposure of the species to pathogens. Also many of the traits listed are controlled by single genes bred into cultivars by backcrossing. Other characters controlled by several genes are more difficult to introduce without altering the original crop characters. In this instance crossing within the hybrid progeny population may be more effective than backcrossing.

Many crops can be crossed with distantly related species to form nonviable seeds. Nonviability is often due to failure of endosperm development. Modern tissue culture techniques now allow embryos from these seeds to be isolated and grown on special nutrient media that replicate the function of the endosperm and allow the production of whole plants. Providing the hybrid has some fertility and can be crossed with the crop parent, genes can be transferred to the crop from these distant relatives (the tertiary gene pool). Cereals and *Brassica* species are especially amenable to embryo cul-

ture techniques and the transfer of disease resistance from wild crucifer species to oilseed rape is one potential use of this technique.

Molecular biology is expanding the range of genetic variation accessible to plant breeders far beyond the tertiary gene pool. It is now feasible to isolate genes from any organism and transfer them to a crop plant, a process called genetic engineering or genetic modification. In effect, the potential gene pool of crops is limitless. For example, by isolation and transfer of the appropriate genes, plants have been modified to produce human antibodies, fish antifreeze proteins, and a type of plastic made in certain bacteria. It is unlikely, however, that crop relatives and breeding through hybridization will become redundant. Most of the genes that code for the traits illustrated in Table IV are completely uncharacterized and would take many years to isolate. It is far quicker and cheaper to transfer these genes to the crop by backcrossing, especially with the recent introduction of marker-assisted selection, where selection of hybrid progeny is by the presence of a "molecular marker" (see later) linked to the gene instead of the gene itself. Also the introduction of traits controlled by many genes is very inefficient if one isolates and transfers the genes individually. Thus crop relatives will probably always be a valuable source of genetic variation for crop breeding and should be carefully conserved (see chapter on genetic resources). The message from molecular biology is, perhaps, that all species should be regarded as crop relatives and be afforded the same treatment. [*See* MOLECULAR EVOLUTION.]

D. Gene Flow between Crops and Crop Relatives

If crops and their wild relatives are sexually compatible and if the crop is cultivated in areas where the relative grows, then gene flow can occur between them. The process of gene flow is more than the formation of occasional hybrids; it involves crossing of the crop and relative over many generations so that genes from one species become incorporated into the gene pool of the other. This process is termed introgression and is frequently invoked as a mechanism for the formation of

TABLE IV
Use of Wild Crop Relatives in Crop Breeding

Crop	Wild relative	Introduced character
Triticum aestivum (bread wheat)	*Agropyron* species	Frost resistance
	Ageilops ventricosa	Eye-spot[a] resistance
	Agropyron elongata and *Aegilops umbellata*	Leaf rust[a] resistance
	Triticum timopheevi	Mildew[a] resistance
Oryza sativa (rice)	*O. nivara*	Rice blast[a] and grassy stunt virus resistances
Zea mays (maize)	*Tripsacum dactyloides*	Top-firing (death of upper leaves in hot weather) resistance
Hordeum vulgare (barley)	*H. spontaneum*	Mildew[a] and leaf rust[a] resistances
Avena sativa (oat)	*A. sterilis*	Leaf rust[a] resistance
Solanum tuberosum (potato)	*S. demissum*	Late blight[a] resistance
	S. acaule	Potato leaf roll virus and virus X resistances
	S. vernei	Nematode resistance
Ipomoea batatas (sweet potato)	*I. trifida*	Nematode resistance
Glycine max (soybean)	*G. soja*	Cold tolerance
Helianthus annuus (sunflower)	*H. petiolaris*	Male sterility
Lycopersicon esculentum (tomato)	*L. pimpinellifolium*	*Fusarium* wilt[a] resistance
	L. hirsutum and *L. chmielewskii*	Increased fruit color
	L. chmielewski	Increased fruit-soluble solids
	L. peruvianum	Increased vitamin C
	L. cheesmanii	Thicker fruit skin (allows mechanized harvesting and transportation)
	All of the above species	Ten or more disease and nematode resistances
Brassica napus (oilseed rape)	*B. atlantica*	Leaf spot[a] resistance
Saccharum officinarum (sugarcane)	*S. spontaneum*	Red rot, smut,[a] and sugarcane mosaic virus resistances
Beta vulgaris (sugar beet)	*B. martima*	Leaf spot[a] resistance
Lactuca sativa (lettuce)	*L. saligna* and *L. serriola*	Corky root[b] resistance

[a] Fungal diseases.
[b] Bacterial diseases.

weeds. The inference is that crop genes transferred to the wild relative may cause the relative to mimic the crop so that eradication of the relative without killing the crop is difficult. Similarly, crop relative genes transferred to the crop may allow it to persist after cultivation to become a so-called volunteer weed or grow in disturbed habitats, such as roadsides, that have similarities with agricultural environments.

The process of introgression between wild and cultivated plants is notoriously difficult to prove for three reasons. First, the identification of a true wild relative can be a problem. Wild plants similar to the crop may be escapes from agriculture rather than undomesticated relatives. These "feral" crops

are described in more detail later. In many species "domesticate–weed–wild complexes" have been described where the distinction among the crop, its wild ancestor, feral forms of the crop, and the possible products of introgression among these forms is almost impossible, especially on morphology only.

Second, if a crop and its relative in a particular region share traits, it may indicate gene flow between them, but it could also be a sign of convergent evolution (two species evolving similar characters in response to similar selection pressures). In particular, agricultural practices may favor as weeds those wild plants most similar to the crop. Many presumed examples of introgression occur in areas where weeding of crops is by hand. Characters that distinguish the weed from the crop will be selected against and similar characters will be at an advantage. Without any gene flow occurring, wild plants can become almost indistinguishable from cultivated plants at the points in the life cycle when weeding is carried out. This is known as "crop mimicry."

The final problem is that the crop and its relative inherit similar traits from their common ancestor. This is particularly associated with crops and wild relatives, as the time since divergence from a common ancestor is relatively short compared with many other pairs of species.

Convergent evolution is often associated with morphological characters as they are frequently under intense selection. "Molecular markers," such as variants of enzymes (isozymes) or DNA fragments (restriction fragment polymorphisms; RFLPs), are often assumed to be neutral with respect to selection and hence are more reliable indicators of introgression. Neutrality overcomes the problem of convergent evolution but not inheritance from a common ancestor. Although neutrality allows the more rapid divergence of characters, finding suitable markers for the study of introgression between crops and crop relatives is difficult. Nevertheless some long-standing debates about the occurrence of introgression are being resolved, especially wih the advent of RFLP markers.

Several often-quoted examples of introgression are given in Table V. The preferred type of evidence to demonstrate introgression is the possession of unique markers in the crop and its relative, except in areas where they are in close contact, and in some cases it is possible to infer which way gene flow is occurring if only the crop or the relative has the marker of the other species. This type of evidence is available in *Beta, Chenopodium, Oryza,* and *Zea,* where molecular markers have been used, and in *Daucus* with simply inherited morphological markers. Even in *Beta,* however, there are problems. Although there is excellent evidence that cultivated beets do introgress with wild "weed" beets, it is still not clear whether the weed beets are genuinely undomesticated sea beet or long-established feral populations (see earlier discussion). The wild relatives of rice and radish in the United States are also either feral crop populations or weed species introduced with the crop. In most other assumed cases of introgression, however, the evidence is merely that the crop and its relative have similar morphology when grown in the same area, and in many cases the wild plants are weeds of the cultivated species. Crop mimicry should be suspected in such situations, and similar morphology should be treated only as circumstantial evidence for introgression.

It is interesting that in most cases where molecular markers have been used they support the circumstantial evidence, strengthening the case for introgression in the other examples. Nevertheless caution is necessary, as illustrated by the case of sunflowers in California. On morphological evidence it was proposed that a weedy race of *Helianthus bolanderi* had arisen through introgression of genes from the cultivated sunflower, *H. annuus,* into a race of *H. bolanderi* which grows on serpentine rocks. This is commonly cited as the classic example of introgression. However, it has been shown that serpentine *H. bolanderi* and *H. annuus* share more rare enzyme variants than do weedy *H. bolanderi* and *H. annuus* and that the divergence in the sequences of chloroplast DNA is similar among all three forms. This shows that weedy *H. bolanderi* is not the product of recent introgression but has an earlier origin in which the two forms of *H. bolanderi* split shortly after the appearance of *H. annuus* and *H. bolanderi* from a common ancestor.

There is a resurgence of interest in introgression because of the arrival of genetically modified crops. The transfer of modified genes, especially those conferring completely novel phenotypes, from crops to wild relatives is perceived as increasing the likelihood of the evolution of pernicious weeds. Many of the crops in Table V have been genetically modified so it is to be expected that high priority will be given to determining the extent and frequency of introgression with their wild relatives.

IV. FERAL CROPS

Highly domesticated crops lack many characters that are beneficial in natural environments because

TABLE V
Evidence for Gene Flow between Crops and Wild Relatives

Crop	Crop relative	Direction of gene flow	Location	Evidence[a]
Beta vulgaris (sugar beet)	Beta maritima(?)[b]	Both directions	South west France	Bolters[c] in crop. Nuclear, chloroplast, and mitochondrial RFLPs[d]
Cannabis sativa (hemp)	Wild C. sativa	Both directions?	North and South America, South Asia	General morphology only
Chenopodium quinoa	C. berlandieri	Crop → Wild	Washington state	Flowering time; leaf morphology; isozymes
Curcurbita pepo (gourds etc.)	C. texana	Both directions	Texas	Isozymes
Daucus carota ssp. sativus (carrot)	D. carota ssp. carota (wild carrot)	Wild → Crop (Crop → Wild?)	Netherlands	Bolters in crop; white roots in crop; inedible roots in crop (general morphology of wild carrots near carrot fields)
Helianthus annuus (sunflower)	H. bolanderi	Crop → Wild	California	General morphology only. Isozymes and chloroplast RFLPs do not indicate gene flow
Hordeum vulgare (barley)	"Weed barley"	Crop → Wild?	Middle East, Turkey, Iran	General morphology only
Humulus lupulus var. lupulus (European hop)	H. lupulus var. pubescens and H. lupulus var. cordifolius	Wild → Crop?	Midwest United States and Japan	General morphology only
Medicago sativa (alfalfa)	M. falcata	Crop → Wild	Turkey	Flower color; seed pod shape (≡ "general morphology only?")

Oryza glaberrima	O. breviligulata	Crop → Wild	Africa	Isozymes
Oryza sativa (rice)	"Red rice" (O. rufipogon?)[b]	Crop → Wild	Louisiana	Isozymes
	O. perennis	Crop → Wild	India	General morphology only
Pennisetum typhoides (pearl millet)	P. violaceum/P. mollissimum	Wild → Crop	Africa	Good evidence for hybridization from laboratory experiments, but intermediate forms in the wild may be crop mimics
Raphanus sativus (radish)	R. raphanastrum	Crop → Wild	California	General morphology only, Cytology but in experimental populations only
Secale cereale (rye)	S. montanum	Crop → Wild	California	Morphology, but genetically well-characterized traits such as seed color and spikelet shape and color
Sorghum bicolor (sorghum)	S. halpense	Crop → Wild	Oklahoma and other states	Cytology
Triticum sp. (a tetraploid wheat)	T. dicoccoides	Crop → Wild	Israel	General morphology only
Zea mays (maize)	Teosintes Annual teosintes are ssp. of Z. mays, perennials are different spp.	Both directions	Mexico/Central America	General morphology suggests extensive introgression. Isozymes and chlorplast RFLPs indicate limited introgression

[a] General morphology only indicates that gene flow is inferred from similarities of morphology. This should be treated as circumstantial evidence only; see text.

[b] The status of the wild species is not clear. They may be "true" wild plants or "feral" populations derived from the crop itself.

[c] Bolters are plants that flower in the first year of growth in a species that is usually biennial (flowers in the second year).

[d] Restriction fragment length polymorphisms; see text.

they have been eliminated by selective breeding. This effectively confines them to agricultural habitats. Nevertheless some domesticated crops can escape from cultivation and form ephemeral populations in disturbed (man-made) habitats or as agricultural weeds while others can maintain large more or less permanent populations in natural or seminatural habitats. The term feral is used to describe populations of plants that have reverted to the "wild" state following domestication to distinguish them from plants that have never been domesticated.

The extent to which major crops in the British Isles form feral populations is given in Table VI. Many crops have similar kinds of feral populations and they can be divided into four groups. First are crops such as the cereals and grain legumes. They all have a long history of domestication, are introduced species, and most have no close wild relative in the British flora. Prolonged domestication will have removed many traits which adapted them to natural habitats and a lack of close relatives may indicate that they are generally unsuited to habitats found in the British Isles. It is not surprising, therefore, that these crops form only ephemeral feral populations in areas where seeds are likely to be dispersed by agriculture or other human activities (Fig. 1). Plants such as potato may form more or less permanent populations by persisting through vegetative organs such as tubers, but it is doubtful whether these feral populations ever spread by seed set in the wild.

In the second group are introduced plants that have close wild relatives. Here feral populations seem to be persistent but it is not clear whether they are self-sustaining (by seed set in the population) or are maintained by seed spillage from agriculture. The classic case in the British Isles is oilseed rape, (*Brassica napus*) which is spreading rapidly, especially on roadsides. Alfalfa (*Medicago sativa*) seems to be a similar type of crop, but is more common in continental Europe, as does radish (*Raphanus sativus*), but here the evidence for self-sustaining populations is stronger as radish is cultivated on a smaller scale.

The third group of feral crops comprises those native plants that have been domesticated, but not

to the extent where they are readily distinguishable from undomesticated forms. The forage grasses and legumes are typical of this group. They tend to be cultivated under conditions similar to natural habitats so characters adapting them to the wild have been retained. This means that they can escape cultivation readily and appear very similar to the wild forms when they do escape. Also, in many parts of the country that are farmed relatively unintensively, the distinction between cultivated and natural habitats is not easy to draw. All of these factors make the true extent of feral populations of these crops uncertain.

The final group are plants that are not natives but occur throughout the country in natural habitats. *Brassica oleracea* (wild cabbage) grows on sea cliffs scattered around the British Isles (Fig. 2). Many populations, particularly those in the south, are quite unlike any cabbage cultivars and are often assumed to be natives. Nevertheless, it is known that cabbage can quickly revert to "wild type" morphology out of cultivation, and *B. oleracea* has been cultivated in the British Isles since Roman times. Also all the major populations are known to be in areas where cabbages have been cultivated. It seems, therefore, that "wild" cabbage may really be an ancient "feral" cabbage. Similarly, "wild" chicory (*Cichorium intybus*) in the British Isles may not be native but an escape from cultivation. This species grows on waste ground and roadsides and, like cabbage, could by no means be said to be a dominant plant where it occurs. Indeed in the British flora it seems that nonnative domesticated crops have not been particularly invasive. Nonnative plants that have produced large unmanageable populations in natural and seminatural habitats have tended to be relatively undomesticated species introduced as ornamental garden plants. Examples include *Rhododendron ponticum*, *Fallopia japonica* (Japanese knotweed), *Impatiens grandulifera* (Indian balsam), and *Heracleum mantegazzianum* (giant hogweed).

The British Isles in some respects may be unusual with regard to feral crops. In other parts of the world there is a greater variety of crops and invasible habitats, giving more opportunities for establishment of feral crop populations and the transfor-

TABLE VI

Examples of Crops Grown in the British Isles That Can Form Feral Populations

Crop	Status	Habitat when feral	Persistence
Cereals: *Triticum aestivum*, *Hordeum vulgare*, *Secale cereale*, *Avena sativa*, *Zea mays* (wheat, barley, rye, oat, maize)	Introduced	Tips, roadsides, waste ground, etc.	Ephemeral
Grain legumes: *Vicia faba*, *Phaseolus vulgaris*, *Pisum sativum* (broad and runner beans, pea)	Introduced	Field margins, tips, etc.	Ephemeral
Lycopersicon esculentum (tomato)	Introduced	Common around sewerage works	Ephemeral
Solanum tuberosum (potato)	Introduced	Common volunteer weed, tips, etc.	Persistent only through vegetative reproduction
Lactuca sativa (lettuce)	Introduced	Waste ground and abandoned fields	Ephemeral
Helianthus annuus (sunflower)	Introduced	Common on tips due to spillage of birdseed	Ephemeral
Linum usitatissimum (flax)	Introduced	As sunflower	Probably ephemeral
Brassica napus (oilseed rape)	Introduced	Increasing on roadsides, waste ground, etc.	Populations appear persistent. Not clear whether self-sustaining or maintained by seed spillage
Medicago sativa ssp. *sativa* (alfalfa)	Introduced	Roadsides	Similar to oilseed rape, although not as common
Raphanus sativus (radish)	Introduced	Field margins, gardens, tips, etc.	Possibly persistent in some areas
Daucus carota ssp. *sativus* (carrot)	Introduced	Abandoned fields, waste ground, etc.	Probably persistent in abandoned fields
Lolium species (ryegrass)	Native	Grassland, rough and waste ground, roadsides	Persistent, although the relative extent of native wild forms and domesticated forms is unknown
Trifolium species (clover)	Native	Similar to ryegrass	Similar to ryegrass
Beta vulgaris ssp. *vulgaris* (sugar beet)	Introduced	Common volunteer weed	May persist, although not certain as occurs mainly as a weed of agriculture and tends to be removed annually
Cichorium intybus (chicory)	Introduced	Roadsides and rough grassland	Persistent
Brassica rapa (turnip)	Probably introduced. Several sspp., e.g., *campestris* ("wild" turnip), *rapa* (cultivated turnip)	River banks and relic of cultivation	The "wild" ssp. is persistent, cultivated sspp. may be ephemeral
Brassica oleracea (cabbage)	Probably introduced. Several sspp., e.g., *oleracea* ("wild" cabbage), *capitata* (cabbage), and *botrytis* (cauliflower)	"Wild" forms on sea cliffs, cultivated forms on roadsides and relics of cultivation	"Wild" forms certainly persistent

FIGURE I　A feral population of wheat (*Triticum aestivum*) growing in a disturbed roadside habitat. A possible origin of this population is a straw mulch used to stabilize the bank after construction of the road.

FIGURE 2　Wild cabbage (*Brassica oleracea*) growing on sea cliifs in Dorset (UK).

mation from useful crop to troublesome weed. In Central Europe and North America one could probably find many more examples of crops reverting to the wild. Nevertheless the principles that highly domesticated crops only form transitory populations in disturbed habitats; that domesticated varieties of native species have extensive, but overlooked, feral populations; and that some plants with a long history of cultivation have naturalized populations that appear to be native are probably generally applicable.

As with introgression, the advent of genetic modification of plants has increased interest in feral crops, as some modifications may promote the formation of invasive feral populations. Study of the characters that allow crops to persist in nonagricultural habitats will become important for assessing how modified crops should be used.

Glossary

Backcrossing　Crossing of a hybrid with one of its parental species.

Disruptive selection　Selection of the extremes in a population with a continuous distribution of phenotypes until two discontinuous types are produced.

Domestication　Process whereby a wild plant becomes adapted to growth in man-made environments.

Feral crop　A domesticated crop growing in a nonagricultural environment.

Introgression　Spread of genes from one species into another; occurs by the repeated backcrossing (q.v.) of a hybrid with one of its parental species producing plants similar to one parent but possessing some characters of the other.

Isozymes Multiple forms of enzymes that catalyze the same chemical reaction. They can often be distinguished by separation in a electric field and can be used to detect genetic variation within or between species.

Molecular markers The generic term for markers such as isozymes (q.v.) and RFLPs (q.v.).

Restriction fragment length polymorphisms (RFLPs) Restriction enzymes cut DNA into fragments at specific sites. RFLPs are variations in the lengths of these fragments and as with isozymes are used to measure genetic variation within or between species.

Bibliography

Barrett, S. C. H. (1983). Crop mimicry in weeds. *Econ. Bot.* **37**, 255–282.

de Rougement, G. (1989). "A Field Guide to the Crops of Britain and Europe." London: Collins.

de Wet, J. M. J., and Harlan, J. R. (1975). Weeds and domesticates: Evolution in the man-made habitat. *Econ. Bot.* **29**, 99–107.

Grant, W. F., ed. (1984). "Plant Biosystematics." Toronto: Academic Press.

Harlan, J. R. (1971). Agricultural Origins: Centers and Noncenters.

Prescott-Allen, R., and Prescott-Allen, C. (1983). "Genes from the Wild." London: International Institute for Environment and Development.

Raybould, A. F., and Gray, A. J. (1993). Genetically modified crops and hybridization with wild relatives: A UK perspective. *J. Appl. Ecol.* **30**, 199–219.

Sukopp, H., and Sukopp, U. (1993). Ecological long-term effects of cultigens becoming feral and of naturalization of non-native species. *Experientia* **43**, 210–218.

Wildlife Management

S. D. Schemnitz
New Mexico State University

Wildlife management is the art and science of making the land produce diverse and abundant wildlife.

I. INTRODUCTION

Wildlife conservation involves protecting, preserving, managing, and studying wildlife. Wildlife management is an essential part of wildlife conservation. There are many definitions of wildlife management, but in simple terms it is applied ecology. Many other complex definitions have been written. One widely accepted definition is the application of ecological knowledge to populations of vertebrate animals and their plant associates in order to strike a balance between the populations and the needs of people. Reconciling the needs of wildlife and people is a difficult and rigorous problem.

In comparison to forestry and range management, wildlife management is a new profession, originating in the 1930s in North America. A key date from a historical perspective is 1933 with the publishing of the book "Game Management," authored by Aldo Leopold, the "father" of wildlife management and the first wildlife professor at the University of Wisconsin. Another noteworthy date was 1935 when the first Cooperative Wildlife Research Unit was established at Iowa State University. This program has blossomed, with wildlife research units at land grant universities in 40 states from Alaska to Florida. In 1937 the Pittman–Robertson Act provided the first major and continuing source of funds for wildlife management and research. Since inception, this excise tax, along with a similar fishing equipment tax, has provided $4 billion for fish and wildlife conservation and management projects.

Wildlife biologists, beginning with Herbert L. Stoddard's classic book on the bobwhite quail in the 1930s, were concerned with species habitat relationships. Early researchers focused on game species but have now broadened to a diverse variety of terrestrial wildlife. (*Wildlife* biologists strive to determine the food, cover, and water needs and environmental conditions such as temperature, precipitation, and the influence of predators and competitors that determine the occupancy of an area by a given species or *population*.) Good quality habitat is related to high rates of reproduction and survival of the occupants.

Wildlife biologists interested in wildlife management need to be familiar with the habits, life history, and behavior (ethology) of wild animals. A useful categorization of animals into k-selected species (low birth rates, low death rate, and long life span) versus r-selected species (high reproductive and death rate and short life span) is helpful and requires different management strategies.

Another key to the future of wildlife is an informed public. Because the public has many diverse attitudes about wildlife, a major goal of wildlife managers is to educate the public. Ideally, wildlife should be managed on the basis of biological facts, but all too often biopolitics become involved.

II. WILDLIFE VALUES

Wildlife provides many diverse and important values that many segments of the general public do not fully appreciate. Many people derive much enjoyment from the recreational pursuit of wildlife, including hunting, fishing, bird watching, and photography. A survey done in 1990 by the U.S. Fish and Wildlife showed that one of two Americans actively participate in recreation activities involving wildlife. These activities contribute substantially to the nation's economy. Economists quantify these expenditures via three basic methodologies: willingness to pay, the travel cost, and replacement value.

Another major wildlife value is biological. Examples include soil tillage by gophers, pollination by bats and hummingbirds, sanitation by scavenging vultures, regulation of water resources by beavers, culling of diseased and unfit animals by predators, and transport and caching of seeds by squirrels.

Wildlife also has scientific value. For example, Darwin developed his concept of organic evolution from studies of wild animals. The physiological stress responses of wild animals to crowding has provided insight into human psychoses. The behavior of wild primates has relevance to human behavior.

A widely recognized value of wildlife is commercially derived from the selling or trading of animals or their products. Meat, fur, and antlers are examples and are readily measurable in dollars. Wildlife tourism and viewing offer a source of economic return to both the public and private sectors. Game ranching for food and sport has become an important source of economic and recreation value of wildlife on private lands. It has become a major

incentive for private landowners to improve habitat conditions, thus wildlife populations maximize income.

In contrast, and more important but less easily quantified in monetary terms, are aesthetic values. The values people derive from the pleasures of observing and hearing wildlife are immense but difficult to quantify. The aesthetic values of wildlife contribute greatly to our quality of life and to our enjoyment of nature.

III. MANAGEMENT PRACTICES

A. Wildlife Planning

The comprehensive planning of wildlife programs has emerged as a major, essential management effort by wildlife agencies at the state and federal levels. Development of broad but clear mission statements, general goals, and specific objectives are part of the strategic plan. The accompanying operational plan spells out strategies and tentative budget estimates. Orderly planning with inputs and review by all agency personnel helps establish priorities for present and future agency programs. The plan necessitates the collection of basic quantitative inventory information on habitats, species population, and public use demands.

Plans are updated and revised at 5- to 10-year intervals and are reviewed and commented upon by various interested publics. Planning by wildlife agencies increases efficiency and effectiveness. Wildlife planning is not a quick solution to crisis problems, but rather an orderly thought and decision-making process to determine the past, present, and desired future.

B. Protection of Wildlife and Wildlife Law Enforcement

Legislation to protect endangered species from illegal commercial trade was approved with the passage of the Endangered Species Act of 1973 and the approval in 1975 of the CITES treaty (Convention on International Trade in Endangered Species). The Lacey Act of 1900 with recent strengthening amendments has provided legal protection for most other wildlife species. Severe penalties for violation of these wildlife law acts have served as deterrents. Uncontrolled illegal hunting is widespread in many third world countries in Africa and Latin America.

Protection of wildlife by law enforcement practices is a vital and essential management tool. New and innovative law enforcement methods include the use of stuffed animal decoys that have movable parts, reward systems for anonymous informers, and use of forensic science methods to distinguish blood, tissue, and hair. Undercover, covert sting operations to collect evidence, use of improved surveillance equipment (e.g., night vision scopes), and more intensive in-service agency training with upgraded minimum hiring requirements such as college degrees are now required. All of these improvements have made wildlife law enforcement personnel more professional and effective. An increased focus on wildlife law enforcement research has also been helpful.

Other deterrents to wildlife law violations are more stringent fines, confiscation of equipment and vehicles, and jail sentences for flagrant and repeated violators.

Wildlife law officers (game wardens) are getting more involved with enforcing general environmental regulations concerned with air and water pollution, littering, and solid waste violations. Numerically, these state conservation officers outnumber other wildlife personnel and thus serve an essential function in contacting the public. In addition, they serve as key public relations personnel for the various wildlife resource management agencies. At the state level, they are responsible for revenue received by enforcing licensing requirements, the main source of income support for various fish and wildlife management programs.

As commercialization and wildlife values increase, the rate of illegal harvest has accelerated. The need for more effective and efficient wildlife law enforcement has become essential. To accomplish this, widespread public support is needed.

C. Protection of Wildlife Habitats

The first national wildlife refuge (NWR) was established on Pelican Island, near Cape Canaveral, Florida in 1903 with the system expanding to 80 million acres at present. Intense management is practiced on NWRs using fire, water impoundment, and population control by hunting as primary procedures. [See NATURE PRESERVES.]

The management of a habitat includes the prescribed (controlled) use of fire to encourage the growth of nutritious vegetation of value to various wildlife. [See FIRE ECOLOGY.]

A major goal of wildlife management is to increase natality and decrease mortality. Methods of accomplishing this are numerous. A key approach is the provision of adequate food for females during the reproductive period. An adequate provision of suitable cover helps minimize loss from predation. Good interspersion of food and cover during winter ameliorates severe winter weather.

Grasslands cover nearly 50% of the world's arid and semiarid landscape and if properly managed are capable of wildlife production. Grasslands can be perpetuated as wildlife habitat if they are not over- and undergrazed and are periodically subjected to fire. Substitution of wild ungulates for domestic livestock holds promise as a means of economically utilizing grassland forage on a sustained yield basis without environmental degradation.

Wildlife habitats are becoming increasingly fragmented, especially for species with large home ranges, such as black bears.

Over geologic time, many species (e.g., dinosaurs) have naturally gone extinct; but the actions of man have drastically accelerated the rate of extinction. Maintaining biodiversity has been an increasingly important goal of wildlife biologists. [See BIODIVERSITY, PROCESSES OF LOSS.]

At an international level, it is essential to consider the World Conservation Strategy (WCS). This important policy document spells out a plan for the survival of man and nature (wildlife) on planet earth. The three main objectives of WCS are to maintain essential ecological processes and life support systems, preserve genetic (biotic) diversity, and ensure the sustainable utilization of species and ecosystems. [See CONSERVATION AGREEMENTS, INTERNATIONAL.]

Certain wildlife populations can be effectively managed by controlled hunting. Hunting is a substitute for natural predation, but animal rights groups are vigorously opposed to hunting as a management practice. These groups are a threat to wildlife management because they thrive on anti-use of wildlife emotion. Wildlife managers believe that some wildlife can be used on a sustainable basis (the third WCS objective).

The future of wildlife in the United States and worldwide will ultimately depend on a drastic change in the current J-shaped curve of the rapidly expanding human population growth. Stable populations should become an immediate worldwide goal. Quality wildlife habitats cannot be maintained with the ever-increasing needs of people for food and fuel. The implementation of sustainable agriculture and development is essential if we are to maintain diverse wildlife populations for future generations.

The start of this essential program was achieved with the RAMSAR Convention on Wetlands of International Importance signed by countries which help protect wetland sites throughout the world.

Participation of the private sector is essential. Organizations like the Nature Conservancy have been able to preserve millions of acres of unique ecosystems of ecological and aesthetic significance in the United States.

Emphasis needs to be placed on ecosystem management instead of species management. Currently, there are about 3800 major protected areas worldwide, totaling 1.75 million square miles. This is a good start but represents only about 3.3% of the earth's land area and 7.4% of the United States. Ideally, at least 10% of the planet's land area should be in a worldwide system of reserves, parks, and other protected areas.

Ideally, these areas should include a protected core area surrounded by buffer zones open to regulated hunting, grazing, logging, recreation, and other uses on a sustainable basis. Allowing limited and controlled human use will alleviate poaching, farming, urbanization, and other destructive uses. Local people need to be included in the planning stages of development of protected area management regulations. Revenues derived from tourism and other uses, such as sport hunting, need to be shared with local people in order to maintain their support and cooperation.

D. Fire and Wildlife Management

In recent years, wildlife managers have been striving, with some success, to convince the public that fire is not totally destructive to natural ecosystems and that it has a beneficial role in ecosystem management. This has been a difficult task since many people have been brainwashed by the "Smokey Bear" propaganda that fires are destructive. In reality, fire is a natural occurrence and excessive fire control is a man-induced intrusion.

Gradually, wildlife and other land managers are using fire as a means of rejuvenating and perpetuating desirable wildlife food and cover plants. The use of prescribed fires has increased the nutritive quality, germination, and growth rate of many valuable wildlife plants. Oftentimes, invading brush of low palatability is replaced by more valuable forbs and grasses. Wildlife biologists do deplore the destruction of large and vast wild fires.

E. Predator Management

Predators are complex and controversial. Policies toward predators have evolved from persecution and elimination toward scientific management and control. Initially, the European tradition that predators are enemies was perpetuated by the American pioneers. Grizzly bears, wolves, and mountain lions were the main targets of control. Animals were judged as good (game animals) and bad (predators). Fortunately, with the aid of Leopold, Errington, the Craigheads, and Durward Allen and their research findings, public sentiment and support have changed and shifted from the 1940s to the 1990s toward predators and predation as tolerable and acceptable components of wildlife communities.

Currently, the emphasis is on predator management of localized individual predators where human health and safety are involved. Forests, ranges, agricultural crops, and livestock may be protected from predators when economic and social benefits exceed the costs of control. The old days of predator extermination are fortunately a thing of the past.

F. Wildlife Disease Management

Wildlife disease specialists continue to expand their research efforts into the causes and control of various wildlife diseases. Disease problems are becoming more serious with waterfowl and various endangered species. As more birds are crowded into diminishing wetland habitats, such diseases as botulism, avian cholera, and tuberculosis pose an increasing threat to wintering populations where the birds are concentrated. Prompt surveillance and cleanup of carcasses will help prevent the spread and diminish the virulence of lethal disease outbreaks. Drainage or flushing of disease-contaminated waters will help curtail mortality. Avoiding the buildup of large populations in restricted areas will also help. Providing adequate food supplies to ensure adequate nutrition will help the birds resist stress and help minimize disease losses.

Minimizing contact between wild and domestic animals will help alleviate disease problems such as occurred with an outbreak of canine distemper that decimated the main wild population of the black-footed ferret, an endangered mammal. Careful quarantining and serologic surveys of imported exotic wild animals is another means of suppressing diseases.

Wild animals also pose a disease threat to humans by serving as reservoirs or carriers. For example, sylvatic plague in the Southwest is communicated by fleas from infected rodents. Tularemia infects humans via ticks from infected rabbits or muskrats. Lyme disease is a danger to humans by way of infected ticks from white-tailed deer. Rabies is another human–wildlife disease of major consequence that is transmitted to humans by the bite of infected mammals. Vaccination of domestic dogs and cats is a major method to suppress rabies.

G. Urban Wildlife Management

With nearly 75% of the United States public residing in urban areas, it is imperative that more emphasis be placed on maintaining and improving habitat conditions for urban wildlife. Sociological studies have indicated that most urban residents want open space and trees and desire contact with nature.

Until recently, urban wildlife management focused on nuisance wildlife such as rats, domestic pigeons, and house sparrows. Sound management must be based on research, but unfortunately little research effort has been devoted to urban wildlife research. With the establishment of the National Institute for Urban Wildlife in Maryland in 1973, increased emphasis has been placed on this previously ignored habitat.

Unique habitats that serve as valuable habitats for urban wildlife include cemeteries, golf courses, and university campuses. Urban parks provide small islands of habitat. Population densities of some urban wildlife species such as gray squirrels and raccoons often exceed population densities in rural areas. An essential need of wildlife in cities is habitat diversity to dilute the urban monoculture of buildings and streets. Homeowners can enhance their properties for wildlife by planting and fostering a variety of shrubs and trees, and providing nest boxes, bird feeders, and watering stations. Ponds, lakes, and riparian corridors along waterways are habitat meccas for an array of urban wildlife.

Wildlife problems do exist in urban areas. Roaming cats and dogs cause severe predatory losses on songbirds and small wild mammals. Raccoons raid garbage cans. Squirrels cause structural problems and chew wiring. Exploding populations of Canada geese soil golf course greens with their feces. Rabbits and white-tailed deer devour gardens and flowers. These examples of man–wildlife conflicts illustrate the need to manage wildlife populations in urban areas to minimize problems. One way to accomplish this is to promote educational programs so that city residents will understand ecological concepts influencing wildlife.

H. Threatened and Endangered Species (T and E) Management

With the passage of the Endangered Species Act in 1973, increased emphasis has been placed on minimizing the potential for extinction of various wildlife species. Endangered status has been caused by a variety of factors. Foremost has been habitat degradation and loss. Other causes include mortality from pesticides, disease, commercial and illegal hunting, and competition from exotic predators.

Increased funding has been made available for the development of recovery plans for T and E species by recovery teams. Other management efforts include captive breeding, habitat preservation, and control of competitors and predators. T and E species have developed an international scope with the Convention on International Trade in Endangered Species of Wild Fauna and Flora (CITES). Over 100 nations are members of CITES. Each member country must appoint a management authority to issue import and export permits for listed T and E species.

With the increasing rate of T and E species and lack of funds for research and management, increased support for focusing on endangered habitats rather than endangered species has arisen. Perpetuating habitats would safeguard and encompass numerous species and would be more efficient than single, T and E species efforts.

I. Animal Behavior and Wildlife Management

A knowledge of animal behavior assists wildlife managers in a number of basic activities such as conducting census, improving habitat, and selecting suitable biologically sound harvest regulations. For example, imprinting, a type of permanent learning acquired during a critical period in the early behavioral development of an animal, is a key factor in the habitat selection in many species of birds. Circadian rhythm, the regular fluctuations in bodily function and behavior during a 24-hr cycle, is a major consideration in conducting a census. Dispersal, the movement of young animals away from their natal home range, is a major basic behavior contributing to genetic variability, repopulation of deleted areas, and occupation of new areas of habitat when a suitable habitat become available as a result of habitat improvement efforts.

Territorial behavior is a key factor in distributing animals over the landscape, setting a limit on the sizes of breeding populations, and influencing auditory census results. Migratory behavior often requires international cooperation and is exemplified by treaties between nations for the protection of migratory birds.

J. Wildlife Management in Parks and Wilderness Areas

The management philosophy of parks presents a dilemma to wildlife biologists. Maintaining and protecting wildlife in parks and providing for the enjoyment of the parks by people often creates problems. Maintaining natural dynamic processes without direct human influence is difficult.

Policies on national parks dictate against the use of public hunting to control wildlife. Often, ungulates multiply due to a lack of natural predators to such high levels that they destroy and eliminate preferred foods. This often causes problems in the eastern United States where super abundant white-tailed deer deplete the native flora.

National parks are managed as "vignettes of primitive America." To do this is nearly impossible, especially with the extinction of many wildlife and plant species and the curtailment of wild fires. Often, research is lacking on what is natural and what is not.

Another difficult problem is to meet the demand of the public for wildlife-viewing opportunities without disrupting the behavior and environment of the animals. A perplexing problem is how to minimize close confrontations between people and dangerous wildlife such as grizzly bears and buffalo.

Wilderness areas in the United States are roadless areas, 5000 acres or more of natural vegetation undisturbed to any extent by man. Management guidelines for wildlife in wilderness should include the exclusion of motorized vehicles, emphasis on the perpetuation of threatened and endangered species, avoidance of man-made structures, and restoration of viable populations of native predators such as grizzly bears and wolves where appropriate. A minimum of man-induced habitat alterations will be allowed, and exotics will be extirpated whenever feasible.

K. Exotic Successes and Failures

Rearranging wildlife distribution has been widespread. Some of these introductions have been purposeful and others accidental. Some classic examples of destructive mammals include rats and mongooses. Other exotic mammals that have caused problems by competing with native United States species but have some redeeming qualities include nutria, sika deer, barbary sheep, reindeer, and Iranian ibex. Domestic mammals, burros, horses, and pigs have escaped becoming feral and causing serious problems.

Destructive avian exotics include pigeons, English sparrows, muscovy ducks, and starlings. Starlings have adversely impacted native cavity nesting species such as bluebirds and woodpeckers by successfully competing for nest sites. Beneficial exotic birds include cattle egrets, chukar partridges, ring-necked pheasants, and gray partridges. Many attempts at exotic introductions have been unsuccessful.

Exotic species are considered to be a threat on the basis of potential disease introduction, hybridization, habitat destruction, and competition. Once released, exotics often are difficult or impossible to control. Exotic introductions have roles in wildlife management, provided adequate safeguards, and constraints are taken.

IV. PROGRESS IN MANAGEMENT

Although endangered species lists are growing, many wildlife species, especially game animals, are increasing. Mammals such as bison, elk, white-tailed deer, pronghorn, and beaver have been particularly successful with increased numbers and geographic range distribution. Similarly, many avian species such as wood ducks, Canada and snow geese, wild turkeys, and mourning doves have been success stories, responding to their own adaptability, protection, and wildlife management practices. Successful management has included the acquisition of basic knowledge of the species and its habitat via research, determination of the key limiting factors, and application of appropriate action initiatives with strong public support.

V. CONSERVATION ETHIC AND WILDLIFE MANAGEMENT

Aldo Leopold, the father of wildlife management, advocated a conservation ethic in which man, land, and wildlife lived in a state of harmony. Leopold believed that humans should consider themselves as a part of the earth's ecosystem rather than the dominant species. This ecocentric view is essential to developing and maintaining a healthy land ethic of land stewardship and sustainable resource utilization. A wildlife manager's first concern should be for the welfare, perpetuation, and enhancement of the wildlife resource.

Glossary

Biological diversity Abundance and distribution of different animal and plant species in a given area.
Census Complete count of a species in a given area.
Density Number of plants or animals per unit area.
Ecosystem Biotic community plus the abiotic system (soil, air, water, and sunlight).
Edge effect Boundary between two distinct vegetation life forms where there often is an increased abundance and diversity of species.
Habitat Environment where an animal lives. The main components include food, cover, water, and space.
Predator An animal that kills and feeds on other animals.
Succession Replacement of one plant community with another over time.

Bibliography

Allen, D. L. (1974). "Our Wildlife Legacy." New York: Funk and Wagnall's.
Bailey, J. A. (1984). "Principles of Wildlife Management." New York: John Wiley.
Dasmann, R. E. (1981). "Wildlife Biology," 2nd Ed. New York: John Wiley.
Gilbert, F. F., and Dodds, D. C. (1987). "The Philosophy and Practice of Wildlife Management." Malabar, FL: Robert E. Krieger.
Giles, R. H., Jr. (1978). "Wildlife Management." San Francisco: W. H. Freeman.
Miller, G. T., Jr. (1990). "Resource Conservation and Management." Belmont, CA: Wadsworth.
Peek, J. M. (1986). "A Review of Wildlife Management." Englewood Cliffs, NJ: Prentice-Hall.
Robbins, C. T. (1983). "Wildlife Feeding and Nutrition." New York: Academic Press.
Robinson, W. L., and Bolen, E. G. (1989). "Wildlife Ecology and Management," 2nd Ed. New York: Macmillian.
Shaw, J. H. (1985). "Introduction to Wildlife Management." New York: McGraw-Hill.

Wildlife Radiotelemetry

Robert A. Garrott

University of Wisconsin—Madison

I. Introduction
II. History
III. Applications
IV. Transmitting Systems
V. Receiving Systems
VI. Tracking Techniques
VII. Satellite Telemetry
VIII. Data Analysis
IX. The Future

I. INTRODUCTION

Advances in technology, particularly miniaturization of electronic components, have allowed wildlife biologists to remotely monitor free-ranging or captive animals while they pursue their normal movements and activities. Transmitters, each with a unique identifying frequency, are attached to the animals, and signals from these transmitters are received by the biologists. Such use of radio-tracking equipment to obtain biological information about animals and their environments is known as wildlife radiotelemetry. [*See* WILDLIFE MANAGEMENT.]

II. HISTORY

Wildlife radio-tracking began in the 1950s, with the monitoring of physiological parameters of captive animals. By the end of the decade, telemetry for free-ranging animals was being developed. Initial devices cost thousands of dollars and were bulky, heavy, and unreliable. Although many biologist began experimenting with telemetry during the 1960s, teams of scientists from two institutions played key roles in the early development of this technology. Researchers at the University of Montana were instrumental in developing and testing telemetry techniques and applications for large animals such as bears, cervids, and raptors. During this same period, a research team at the University of Minnesota's Cedar Creek Natural History Area constructed and tested electronics for smaller animals, designed and constructed the first automated radio-tracking system, and developed computer programs for the analysis of telemetry data. Thirty years after these pioneering efforts, telemetry is a standard tool in wildlife research.

III. APPLICATIONS

Today telemetry is used throughout the world with a wide variety of applications on a diversity of species. It can be used on most vertebrates and large invertebrates, including crustaceans (crayfish, crabs), fishes (perch, pike, catfish, androgenous salmon, ocean fishes), amphibians (frogs), reptiles (snakes, turtles, alligators), birds (warblers, ducks, owls, eagles), and mammals ranging in size from mice to whales. Common applications include spatial studies such as home range, migration, dispersal movements, and fidelity of animals to specific areas. Studies of habitat preference or habitat use also employ telemetry techniques to gain insights into how animals utilize various habitats and what resources are important for survival and reproduction. Behavioral studies use telemetry by instrumenting animals so that they can found easily and observed. This technique is particularly useful when animals are approachable and live in semiopen areas (savannas of Africa, parks where animals are acclimated to people). In conjunction with behavioral and habitat studies, activity patterns can be investigated by either decoding the radio signals with automated equipment or by listening to the signals and interpreting the modulation and pulse frequency patterns to distinguish various activities (moving, stationary, feeding). Telemetry was initially developed for physiological applications and there continue to be many varied uses in this area. Information on body temperature, heart rate, and other physiological parameters are monitored by specific sensors. Special circuits in transmitters have also facilitated use of telemetry to address questions of survival and mortality of animals, allowing investigators to study population processes. These demographic questions generally require a large sample of instrumented animals to produce reliable estimates. Miniaturization and mass production of transmitters has reduced costs and made such studies feasible. Telemetry is also being employed to assist in refining population estimation procedures. In essence, to estimate population size an investigator instruments animals and determines what proportion are detected during censuses, correcting the total count accordingly.

IV. TRANSMITTING SYSTEMS

A telemetry system has two main components, a transmitter and a receiving unit. Transmitters are instrument packages that consist of electronic components that produce a radio signal, an antenna that radiates the signal, and a power source. Many problems plagued telemetry equipment in the early years. Electronics available were quite primitive and much of the equipment was built by untrained investigators, hence, equipment malfunctions and failures were common. Today there are at least two dozen commercial manufacturers of telemetry equipment throughout the world. These companies employ electronic engineers and design specialists that build transmitters with printed circuit boards and custom-built chips resulting in the production of very reliable telemetry equipment for wildlife investigations.

Transmitter electronic components are designed to emit a radio signal on a predetermined frequency. Use of radio frequencies is regulated by the government, thus only specific bands are available for wildlife use. Bands assigned to wildlife use are on the VHF (very high frequency) band, generally between 30 and 300 MHz. The most common frequencies used in wildlife work are 148–152 MHz, 159–161 MHz, 164–166 MHz, and 172–174 MHz. Occasionally, other frequencies are used. The transmitter's frequency affects a systems physical and operating characteristics. Signal penetration through matter such as dense vegetation is better if a low frequency is utilized. Thus in studies where the animals being investigated live in dense vegetation, in burrows, or in water it is often best to utilize frequencies in the 30 to 40 MHz range. Receiving antennas must be built on proportions ($1/2$, $1/4$, $1/8$) of the wavelength they are to receive. The lower the frequency the longer the wavelength, hence, the longer the antenna.

Transmitters can be constructed to emit either a pulsed or continuous signal; however, the industry standard is a pulsed signal. There are two important characteristics of a pulsed signal: pulse width (duration), which is usually 30–80 msec, and pulse rate, which generally is from 30 to 75 pulses per min. The longer the duration of the signal and the higher the signal pulse rate the greater the drain on the power source. Continuous signal transmission, therefore, has a high rate of power usage and is employed only for specialized applications such as physiological monitoring where continuous data are required. The operating life of standard pulsed transmitters ranges, from 20–30 days for the smallest transmitters weighing only a few grams to 6–10 years for transmitters weighing 200–500 g. Often, the power source is the limiting factor in transmitter construction. The most common power source is batteries. This limits how small and light a transmitter can be built and still have an effective transmitting range and longevity. Lithium cells are the industry standard; however, silver oxide and mercury cells are also used for the smallest transmitters. Photovoltaic or solar cells are used in specific applications where the species being studied are consistently exposed to the minimum amount of sunlight necessary to power the unit.

Transmitters can be designed to include many special circuit options. One commonly utilized option is the mercury tip-switch. This is a small glass tube containing a bead of liquid mercury, wire leads, and a timer. The bead of mercury moves back and forth within the tube, making contact with the wire leads, thus resetting the timer. This feature is useful in studies designed to monitor animal activity or mortality. When investigating mortality the timers are set for a specific duration (3–12 hr) and if the mercury tip-switch does not move enough to reset the timer within the designated period the pulse rate of the signal automatically changes, generally doubling the normal pulse rate. This alerts the investigator to the possibility that the animal is dead and allows the animal to be quickly found to determine cause of death and record other pertinent data. Activity studies utilize mercury tip-switches in another manner. They are positioned in predetermined planes and change the pulse rate continuously, depending on the body positioning and motion of the animal. The investigator interprets the different pulse rates as indicative of certain forms of behavior, posture, or activity. Specifically, the mercury tip-switch may be positioned in a certain plane that indicates whether the animal's head is up or down. This could address questions such as herbivore or raptor feeding activity.

Another useful transmitter circuit option is a temperature sensor (thermistor). Transmitters containing these sensors and attached externally to an animal monitor the air temperature near the animal while implanted transmitters monitor the animal's body temperature. The thermistors change the pulse rate of the radio signal in relation to the change in temperature. Studies directed at how animals adapt and respond to thermal extremes in their environment are applications of external thermistors. Physiological studies on thermoregulation, reproduction, and mortality are common applications that often require body temperature data. Heart-rate sensors can also be used in physiological studies. These are normally incorporated into the electronic units of implantable transmitters and require the placement of wire leads on either side of the heart. Again the signal pulse rate changes in relation to changes in heart rate. This option can be used to address questions such as animal response to stimuli and energetics. Other sensors to monitor various physiological parameters can be incorporated into transmitters. The sensors can either be incorporated within the transmitters and designed to be surgically implanted or can be attached to the organ through leads that are also attached to an external transmitter. Data can then be transmitted through coded pulses to remote receiving stations. Many of these studies work with animals that are in captive facilities. These transmitters are specialized and often have a limited signal transmitting range.

Atmospheric pressure sensors are commonly incorporated into transmitters designed for aquatic species. These sensors sample atmospheric pressure, at preprogrammed intervals, documenting the depths the animal is occupying. These sensors are often built into circuits equipped with internal memory chips that record the time associated with each pressure reading. The combination of a pressure sensor to record depth, timer, and a memory chip is known as a time depth recorder (TDR). TDRs do not generally transmit continuously, instead they store the data on an internal memory chip. This necessitates recovery of the transmitter to access the data. This can be done easily with some species such as seals that have traditional haulout beaches. TDR units have been used successfully on seals, sea otters, whales, dolphins, and penguins to gain information on time and depth of dives and how the animals feed. Technology advances have also led to satellite retrieval of information from the TDR's memory chip. Successful satellite retrieval is dependent, however, on adequate transmission links. Technology that will allow the investigator to remotely interrogate the TDR and obtain information from the memory chip on demand is currently being developed. When these methods are widely available, recapture to retrieve data will not be necessary.

Transmitters can incorporate photocells as circuit options for some applications. Photocells are light detectors that can monitor light conditions around an animal by varying the pulse rate of the signal to reflect presence or absence or intensity of light. This can aid investigators studying species that utilize burrows, caves, etc. Photocells can also turn a transmitter off at night and on during the day, effectively extending the transmitter battery life. Timers can be built into transmitters to serve a similar function by presetting to regulate the operating time of a transmitter (operate every third day, 8 hr/day, 1 week/month). There are many other less utilized special circuit options available to the telemetry user such as sensors to measure pH, muscle contractions, urination, auditory monitors, and sensors to detect salt water.

Advances in transmitter electronics have coincided with improvements in the materials available to house these crucial components. Early transmitter housings were often susceptible to moisture infiltration from the external environment or from body fluids in the case of implantable transmitters, causing corrosion of the electronic components and destroying the integrity of the circuits, which would result in transmitter failure. Today, however, three methods are used to reliably isolate the electrical circuits from the external environment. One method utilizes a polymer coating of electrical resins or dental acrylic. The electrical circuits, power source, and antenna connection are sealed by dipping them into the polymers. This isolation process is limited to very small packages that must be light in weight such as those used on small bird and mammal species. A second method is to seal the electrical components in a metal canister, often filled with an inert gas. This method is commonly used for medium to large transmitter packages that

can be carried by the larger species. The final technique is to pot the entire transmitter into a mold filled with a polymer. This is the most common commercial method employed to build implantable transmitters and most large transmitters. These transmitters are heavy and thus limited to use on medium to large species. All three isolation techniques are very reliable, with documented instances of transmitters remaining functional while submerged in water for months.

Almost all early transmitters were mounted on a collar that was attached to the neck of the species being investigated. Collar mounting continues to be a successful and appropriate attachment method for many species; however, a wide variety of new attachment methods have been developed for species that cannot be collared. These techniques have greatly added to the versatility of transmitter use and include harness systems, eartags, epoxy attachment, and surgical implants. A marine epoxy and specialized grid system are utilized on many marine mammals to attach the transmitter to the fur. This has been very successful; however, it necessitates transmitter replacement each time the animal molts as the transmitter is also molted. Implants require subcutaneous or intraperitoneal placement and require minor surgical procedures to insert. New innovations in transmitter collars have also been developed. Collar-mounted transmitters incorporate a tranquilizer delivery system that can be remotely controlled. Other collars incorporate systems that can activate devices that allow the collars to drop off. As a result of these attachment systems, transmitters on the market today are more biologically sound and applications have become more varied.

Transmitters need an antenna to radiate the signal. Various types of transmitting antennas can be built into units. The most common type is a short whip antenna. These are single lengths of wire or cable material that extend out of the transmitter unit and are generally used on collar-mounted transmitters intended for use on larger animal and bird species. Care must be exercised with this type antenna as animal grooming can cause damage. Another transmitting antenna configuration is the loop. The antenna is bent into a loop shape and is totally embedded into the collar that the transmitter is attached to. Coil antennas are essentially whip antennas that have been coiled tightly and are used primarily in implantable transmitters where other configurations cannot be used and are less efficient at signal radiation.

An underlying assumption of all telemetry studies is that animals are not affected by the instrumentation procedure or the instrument package. Although this is the ideal situation that all investigators strive toward, the reality is that it is inevitable that the capture, handling, and attachment of the instrument package have an impact on the animals. Impacts, may vary considerably, ranging from subtle changes in behavior, which are short lived, to long-term changes that may affect an animal's ability to survive and reproduce. Short-term behavioral effects include increased preening or grooming, increased activity, decreased feeding, and avoidance of water. Documented long-term impacts include hair and feather wear, skin irritation, weight loss, decreased mobility or agility, disruption of reproductive behaviors (incubation), and decreased survival. Many of the impacts of instrumentation can be attributed to the weight of the instrument packages. Weight is particularly significant for animals that are small or birds that depend on flight. Therefore, investigators tend to utilize as small a transmitter package as possible and follow the general rule of having the transmitter package not exceed 3% of the animal's body weight. Impacts should be considered prior to instrumentation to be sure that they are minimized and short term. Many telemetry vendors manufacture a wide variety of complete transmitter and attachment packages that have been extensively field tested. Therefore, for most applications, packages are available that can be relied on to not adversely affect animals in a manner that would compromise their well-being, the integrity of the study, or the quality of data. However, for those studies that require specialized equipment or for species that have not previously been instrumented, it is important to test transmitters and attachment techniques on captive animals to evaluate the feasibility and impacts of instrumentation before placing them on free-ranging animals.

V. RECEIVING SYSTEMS

The receiving unit consists of a radio receiver, a receiving antenna with attaching cable, earphones, and optional auxiliary devices that plug into the receiver to decode specific information being transmitted through changes in radio signal pulse and/or width. There are three main types of radio receivers: channel receivers, dial-in frequency receivers, and programmable scanning receivers. A channel receiver has a number of channels, each covering a frequency range of 10 to 200 kHz. The investigator selects the appropriate channel and uses a fine-tune knob to scan across that frequency band to search for the specific frequency of an instrumented animal. A dial-in frequency receiver has dials or toggle switches that allow the user to directly select each transmitter frequency. These two types of receivers are the most commonly used in wildlife telemetry, primarily because of ease of use and cost of the units (channel receivers $500 to $700, dial-in receivers $800 to $1500). The third type is the programmable scanning receiver. Hundreds of specific frequencies can be programmed into the memory of these radio receivers. The receiver can then automatically or manually scan through all the frequencies. These receivers are very useful when monitoring large numbers of animals and when aerial tracking is incorporated into the study. Costs range from $2000 to $2800.

Signal processors and recorders are auxiliary devices that can be connected to the receiver to automatically record data and/or interpret signals. Processors provide information on interpulse periods or pulse frequency and signal strength. These processors are commonly used for studies of body temperature, activity, or head–body position in which the pulse frequency changes in some relationship to the parameter being measured (pulse increases as body temperature increases). Output from the processors can be a liquid crystal display (LCD) or may be directly entered into a data recorder. Data recorders are used to continuously record signal information. Recorders are often connected to a programmable scanning receiver. Thus, when the receiver scans the preprogrammed frequencies at regular intervals, the data gathered from each scan are automatically recorded. There are two types of commonly used data recorders. A strip chart records data on a paper chart that must be manually processed and converted into a digital format. Also, it must be manually calibrated periodically to accurately record the time of day. Digital recorders directly convert signal attributes to a digital format. They have internal clocks and generally include software programs that facilitate data entry into an appropriate computer format. Costs vary greatly. Signal processors are approximately $600 to $1000, recorders cost $800 to $1500, and processor/recorder units with built-in computer systems and software cost from $2700 to $25,000.

Antennas are essential to wildlife telemetry as they are needed both in the transmitter to radiate the signal, as previously discussed, and as part of the receiving unit to receive the signal. Receiving antennas are separate pieces of equipment that are connected to the receiver of choice by a cable. Antenna element size is dependent on the radio frequency of the signal being received. There are four basic types of receiving antennas: loop, dipole, H, and Yagi. Loop antennas were commonly used in the early years of telemetry when transmitter frequencies were commonly between 20 and 40 MHz. The loop alleviated the large, cumbersome antennas that would have been required for receiving such low frequency signals. Today transmitter frequencies are generally higher and the loop is not commonly utilized. The dipole or whip antenna is essentially one element oriented in a vertical plane. It is omnidirectional, receiving radio signals in a 360° radius, thus, not useful when directional information is needed. The H and Yagi antennas are directional and consist of a horizontal boom and cross elements. There can be from 2 (for the H antenna) to 16 (largest Yagi) elements, with increasing numbers of elements providing increased directionality. Yagi antennas with more elements also have a higher gain, which is a measure of the ability of the antenna to receive the radio signal. Antennas for ground tracking can be hand held, mounted on a vehicle, or mounted on fixed towers. Antennas can also be mounted on the struts of an airplane or helicopter for aerial tracking. The

receiving antenna configuration and method of use are dictated by the quality and type of data useful to the investigator. Applications that require precise directional bearings and accurate animal locations utilize Yagi antennas with many elements. The most precise receiving systems utilize two Yagi antennas spaced 1 wavelength apart and mounted on a horizontal pipe or boom. Both antennas are pointed in the same direction and are connected through an auxiliary switchbox that can add the signals received from each antenna in a manner that makes the system very directional.

VI. TRACKING TECHNIQUES

One of the most common applications of telemetry is determining the location of an animal. The simplest method of determining an animal's location is called homing. The investigator homes in on the animal by using an antenna to determine the direction of the strongest signal and traveling toward the signal until the animal is observed. It is a useful technique for collecting data on animal location, animal behavior, etc. when radiotracking from the ground or from the air. Homing, however, is very time intensive and limits the number of animals that can be tracked or studied at one time. Care must be taken to avoid approaching animals too closely and causing behavioral modifications that can reduce the quality of data or bias the information one is obtaining.

Aerial tracking is useful when an investigator must locate large numbers of animals quickly and efficiently and in studies in remote or rugged terrain where access is poor and ground tracking animals long distances is difficult. To aerial track a single antenna may be attached to a wing strut or boom. The antenna points forward and the aircraft (fixed-wing or helicopter) simply flies in the direction of the strongest signal. An alternative method utilizes two antennas. One antenna is attached on either side of the aircraft, pointing perpendicular to the flight path. The two antennas are connected to a switchbox that allows the investigator to switch from the antenna on one side of the aircraft to the other while listening to the signal. In this way, the investigator can determine which side of the aircraft the animal is on and the pilot can alter the course until the animal is seen or until the pilot can fly a tight circle, keeping the strongest signal to the inside of the circle. If the investigator does not observe the animal, it is assumed that the animal is within the circle. Data suggest that errors associated with locations determined using a fixed-wing aircraft can range from <0.5 to >2 km. Factors affecting accuracy of aerial tracking include speed, type of aircraft, altitude, prevalence of navigational landmarks, location procedure used, and investigator fatigue and experience. It is wise to test the tracking system by placing reference transmitters at known points on the ground and locating them from the air to assess the accuracy of your locations. Substantial error also may be introduced by an inability to determine one's location when terrain is not distinguishable such as in studies of marine mammals and pelagic birds offshore or polar bears on the pack ice. In these cases navigational aids such as the Loran-C system or a Navstar global positioning system must be utilized. The Loran-C system is widely used, relatively inexpensive, and common equipment on most aircraft. Loran-C navigation relies on a ground-based transmitting system and is limited to specific geographic areas that are covered by the system. The Navstar global positioning system is an alternative navigational aid that uses a constellation of satellites to determine the location of a receiving unit. This system has gained popularity and can provide a quick and efficient means of determining an aircraft's position anywhere in the world.

Triangulation is another, more complicated technique for locating animals. In triangulation the investigator estimates the location of an animal by using two or more compass bearings obtained from receiving locations remote from the transmitter. To obtain a bearing the investigator uses a directional antenna and determines the direction of the strongest signal and then takes a compass reading to correspond to that direction. The location of the intersection of these compass bearings is the estimate of the actual location of the instrumented animal. Triangulation systems vary in complexity and sophistication. The simplest systems consist of a single person using a hand-held antenna and traveling to two different receiving locations to obtain bearings. Antennas can also be mounted on a vehicle to allow the investigator to travel quickly between receiving points. More sophisticated systems may involve a number of fixed telemetry towers (three to six) constructed using stacked Yagi antennas where bearings to an instrumented animal are obtained simultaneously. These systems produce much more accurate and precise positional data than simpler ones. Automated triangulation systems are also available. These systems are similar to the earlier example except instead of each tower being manned, the bearings are obtained using computer-driven automated equipment. These systems are expensive, have limited availability, and are rarely employed. All triangulation procedures provide only an estimate of an instrumented animal's location because the direction of the signal cannot be measured exactly. Sources of error include the geometric relationship between receiving and transmitting locations, terrain, equipment, sophistication and directionality of antennas, weather, power lines, and vegetation. The most problematic aspect of using triangulation to estimate the location of instrumented animals is the potential for the transmitted signal to be reflected by some feature of the environment, causing directional distortion of the signal. This is known as signal bounce and is common in irregular terrain where buildings, rock outcrops, cliffs, and other features provide a surface that may reflect radio signals. This can be confusing to the investigator and can cause substantial problems with data collection.

These limitations are inherent in all triangulation techniques and, thus, require the investigator to obtain a measure of the ability of the system to determine reliable locations. To do this, a probability area around each estimated animal location is calculated. The analytical procedure used to calculate the probability area requires testing the telemetry system to determine the approximate standard deviation of the compass bearings obtained from different receiving locations. The complexity and intensity of the testing required are dependent on the desired precision of the animal locations that the investigator is trying to obtain. One or 2 days of field testing on a study area could be adequate for a system that studies the migratory pattern of a species where errors of 1 to 1.5 km are acceptable. However, much more intensive field testing would be required to test a triangulation system in a study area where the investigator is trying to determine the vegetation communities utilized by individual animals as this question requires more exact animal locations. Triangulation systems are tested by distributing transmitters throughout the study area at known points and obtaining replicate compass bearings for each transmitter from all receiving locations within the study area. The radio-tracking protocol should be the same as that employed when tracking instrumented animals. Information obtained from testing can be used in a number of computer programs available to perform the complicated mathematics required to calculate the animal locations and estimate their probability area. In general, the precision of the animal location or the confidence one can have in triangulated locations is a function of the standard deviation of the directional bearings, the distance between the transmitter and receiving points, the angle of the bearing intersections, the number of bearings used, and information on signal alteration (reflection or bounce). Animal locations obtained from two compass bearings are generally not as precise as those obtained from three or more. Where signal bounce is a problem it is imperative to have at least three compass bearings for each animal location in order to detect reflected signals. However, compass bearings for each instrumented animal should be obtained within a short time frame to be sure the animal does not move while the investigator is taking the bearings. To summarize, all animal locations obtained utilizing triangulation procedures have some error. The magnitude of error can be substantial and is dependent on the sophistication of the receiving system, tracking techniques, and the characteristics of the environment. Investigators must recognize the strengths and weaknesses of their system and not extend interpretation of their data beyond what is warranted.

VII. SATELLITE TELEMETRY

An alternative technique for tracking animals involves the use of satellites. Within 10 years of the first successful telemetry study of free-ranging wildlife, investigators recognized that one of its limitations was the immense manpower investment required to go into the field, locate animals, and obtain and process data. Alternatives that could expedite this process were sought. One potentially promising solution is the use of satellite-receiving systems. This technology has been pursued since 1970 when an 11-kg satellite transmitter was attached to a free-ranging elk. This initial attempt was not practical and was in essence an adaptation of the existing hardware designed for the monitoring of weather variables. Today more sophisticated satellite transmitters called platform transmitter terminals (PTT) are designed specifically for wildlife applications. These units are much lighter, with some weighing as little as 50 g. They transmit signals to the Tiros-N satellite system. Tiros-N is primarily a weather satellite with access available for biological investigations. Tiros-N is designed for coverage at high latitudes (>60 latitude). Thus, the most practical telemetry applications are in the extreme northern and southern hemispheres. Animalborne satellite transmitters have been used extensively in the arctic, especially in northern North America, to facilitate scientific investigations in areas where the logistics of placing people in the field for data collection is extremely difficult and expensive. Satellite transmitters have been used successfully on a number of species found in these areas, including polar bears, musk-oxen, swans, and eagles. As a consequence, information has been obtained on these species that could not be practically collected in any other manner. Accurate elevation information is necessary for the algorithms used to calculate reliable locations. For example, an elevation error of 500 m may result in transmitter location errors of from 125 to 1000 m. This limitation is not too critical for species such as polar bears and sea turtles that are only found at sea level; however, it is a major consideration for species inhabiting mountainous terrain. Satellite telemetry applications are not widespread because of the system's limitation to high latitudes, the need for accurate elevation information, and the high costs which can exceed $4000 annually to monitor one animal. Much work is still needed in the areas of transmitter size and weight, power sources, satellite access, and costs before satellite telemetry can become a widely used tool for wildlife applications. Adaptation of the Navstar global positioning system is currently being explored by wildlife telemetry manufacturers as an alternative to Tiros-N transmitters. This emerging technology may ultimately prove to be more flexible, accurate, and less costly.

VIII. DATA ANALYSIS

Telemetry studies can generate large volumes of data quickly that are best handled with the use of computers. Initially, data were laboriously punched onto cards and then read into mainframe computers housed in centralized facilities such as universities. However, much telemetry work is done in remote locations, far from major supporting facilities. Hence, investigators had to wait until the end of the field season to analyze their data. The advent of inexpensive, powerful, personal computers has dramatically changed the ability of the field investigator to analyze data. Today, powerful battery-operated computers, no larger than a notebook, can be taken directly into the field to be used for data logging and analysis as the data are being collected. In addition to a wide variety of analytical outputs, these field computers can also provide visual displays of data such as triangulation bearings, base maps of the study area, and plots of previous locations of animals being radio tracked. This immediate feedback can enhance the quality of the study by aiding in decisions about how data should be collected and what data should be collected next. For example, an investigator can identify data that result in inaccurate triangulation locations and can immediately take corrective action, thus avoiding the loss of valuable data and field time.

Many analytical tools have been developed specifically for telemetry data and new applications are constantly being tested. It is also becoming increasingly common for authors of new analytical procedures to provide a computer program code for executing the analysis, greatly facilitating applications of these procedures and providing for widespread use and testing. No comprehensive programs for the analysis of telemetry data currently exist. However, a variety of programs for specialized applications or procedures are available and are circulated widely among telemetry users. Programs that deal with computation of various home range estimators are most common. There are also PC-based programs that deal with many other aspects of telemetry, including survival analysis and habitat utilization. Many of the applications of telemetry in wildlife studies address the biological questions of how animals use their environments and attempt to identify the attributes of areas intensively used or avoided. Such applications are dependent on analysis of animal locations with respect to maps depicting the various characteristics of an area. Geographical Information Systems (GIS) are sophisticated software packages that have been developed over the past decade for these types of spatial problems. Primary users of GIS programs are agencies and organizations dealing with natural resource inventory and stewardship; the same agencies that are likely to be conducting wildlife investigations involving telemetry. Many telemetry users recognize the potential of GIS to provide a better understanding of their data. The applications of GIS technology to the analytical problems inherent in spatial data routinely collected during telemetry studies will continue to expand as the software becomes more standardized and user friendly. Biologists and programmers are currently developing codes for specific GIS that can be built upon and transferred from one system to another. As this process continues and investigators become more familiar with the potential of GIS to tackle the complex questions of how species utilize space, it will become thoroughly integrated into telemetry studies and provide a powerful tool for interpreting data.

IX. THE FUTURE

The advent of wildlife telemetry in many ways revolutionized wildlife research and monitoring as this technology has provided a means of remotely studying many aspects of an animal's natural history with little or no anthropogenic influence. Prior to the development of this technology, it was extremely difficult, if not impossible, to study many species, including those that are shy and elusive, nocturnal, arboreal or fossorial, large, dangerous species, far-ranging pelagic or migratory species, and those that occupy remote, inaccessible, inhospitable, or densely vegetated areas. Today there are few areas where telemetry cannot be applied to address questions of biology, management, or conservation of a species. Continued improvements in electronics, power sources, circuit design and manufacture, microprocessors, and attachment techniques are resulting in smaller, smarter, and more sophisticated instrument packages that will be even more useful. The future promises to provide telemetry devices that can collect and archive large quantities of data. Continuing advancements in the integration of satellite technology into wildlife telemetry could eventually eliminate most field-oriented data collection. For many studies of the future, perhaps the only investigator-induced animal disturbance will be at the time of capture for instrumentation. Thereafter, the data may be acquired remotely via satellite and downloaded through standard communication links into modems that connect directly to the investigator's personal computer.

Equally important are the advances in the statistical and analytical tools available to handle the data generated from telemetry. Biometricians are developing advanced statistical models for the study of many aspects of a species biology and demographic characteristics. More comprehensive and user friendly software is being developed for the display and analysis of data. Finally, integration of new analytical utilities specifically designed for telemetry data with powerful GIS environments are beginning to provide unprecedented opportunities to integrate many aspects of the environment, providing a better understanding of how animals behave, survive, and reproduce. Wildlife telemetry has been an important tool for studying free-ranging animals for nearly 30 years and promises to become

even more eminent in wildlife investigations in the future.

Glossary

Geographical information systems Sophisticated computer software packages developed for the storage, analysis, and manipulation of spatial data.

Global positioning system A navigational aid that uses a constellation of satellites to determine the location of a receiving unit.

Homing The simplest method of determining an animal's location accomplished by using an antenna to determine the direction of the strongest signal and traveling toward the signal until the animal is observed.

Loran-C A navigation aid that relies on a ground-based transmitting system and is limited to specific geographic areas that are covered by the system.

Platform transmitter terminals Sophisticated satellite transmitters designed for collecting information on free-ranging animals.

Receiving units Systems that consists of a radio receiver, a receiving antenna with attaching cable, earphones, and optional auxiliary devices that plug into the receiver to decode specific information being transmitted through changes in radio signal pulse and/or width.

Signal bounce Phenomenon that occurs when a transmitted signal is reflected by some feature of the environment causing directional distortion of the signal.

Time/depth recorders Transmitter option that consists of the combination of a pressure sensor to record depth, a timer, and a memory chip. Used on aquatic animals to obtain and store data on diving activities and behavior.

Tip switch A small glass tube containing a bead of liquid mercury, wire leads, and a microprocessor. The bead of mercury moves back and forth within the tube, making contact with the wire leads, thus sending information to the microprocessor on movement of the animal. The electronics are useful in studies designed to monitor animal activity or mortality.

Transmitter An instrument package that consists of electronic components that produce a radio signal, an antenna that radiates the signal, and a power source.

Triangulation A technique for locating animals where the investigator estimates the location of an animal by using two or more compass bearings obtained from receiving locations remote from the transmitter.

Yagi An antenna that is directional and consists of a horizontal boom and cross elements. There can be from 2 to 16 elements, with increasing numbers of elements providing increased directionality and higher gain.

Bibliography

Harris, R. B., Fancy, S. G., Douglas, D. C., Garner, G. W., Amstrup, S. C., McCabe, T. R., and Park,

L. F. (1990). "Tracking Wildlife by Satellite: Current Systems and Performance." Washington, D.C.: U.S. Dept. Int. Fish and Wildl. Tech. Report 30.

Kenward, R. (1987). "Wildlife Radio Tagging: Equipment, Field Techniques and Data Analysis." New York: Academic Press.

Mech, L. D. (1983). "Handbook of Animal Radio-Tracking." Minneapolis, MN: Univ. of Minnesota Press.

White, G. C., and Garrott, R. A. (1990). "Analysis of Wildlife Radio-Tracking Data." New York: Academic Press.

Zoological Parks

Cheryl S. Asa
St. Louis Zoological Park

I. Conservation
II. Animal Care
III. Research
IV. Education
V. Entertainment or Recreation

The role of the modern zoo has changed dramatically. From early beginnings more than a 100 years ago as menageries in which exotic animals were presented in small cages to a curious public, through the mid-20th century when zoogoers were entertained by the antics of costumed chimpanzees, zoos have emerged as centers of conservation education and research and as bastions for endangered species. In fact, zoos today might aptly be called modern arks.

I. CONSERVATION

Captive breeding is not the answer for all endangered species. In fact, there is a fear that reliance on captive breeding may shift attention from the more immediate need to preserve and maintain habitat. When programs are successful, the public may mistakenly believe that the problems are solved. Yet captive breeding programs also can focus attention on the plight of animals in general, raising public awareness and encouraging support of *in situ* projects and habitat preservation. There is no question that captive breeding can be a useful tool in conserving endangered species. That zoos cannot save all animals does not mean that they should not contribute to saving the few that they

can. [*See* CAPTIVE BREEDING AND MANAGEMENT OF ENDANGERED SPECIES.]

A. AAZPA

The American Association of Zoological Parks and Aquariums (AAZPA), founded in 1924, shortened its logo in January 1994 to AZA, the American Association of Zoos and Aquariums (Table I). This organization accredits North American zoos, sponsors conferences, and coordinates programs for its members. It represents 162 institutions and more than 6500 individual members in the United States and Canada. From its recently established conservation center, conservation biologists and specialists in government affairs, development, and education help manage and coordinate AZA's various programs. In 1980, its board of directors voted unanimously to set wildlife conservation as the highest priority.

AZA publishes an Annual Report on Conservation and Science, which contains progress reports for its Species Survival Plans (SSPs) and advisory groups. In addition, it contains a listing of both *in situ* and *ex situ* conservation projects sponsored by member institutions, as well as scientific publications. The 1993 edition contained more than 1100 conservation projects and nearly 500 publications.

TABLE I
Zoo Conservation Organization Acronyms

AAZPA: American Association of Zoological Parks and Aquariums

AZA: American Association of Zoos and Aquariums (formerly AAZPA)

AAZK: American Association of Zoo Keepers

AZV: Association of Zoo Veterinarians

ALAZA: Association of Latin American Zoos and Aquariums

ARAZPA: Australasian Regional Association of Zoological Parks and Aquariums

ARKS: Animal Record-Keeping System

CAMP: Conservation Assessment and Management Plans

CAUZ: Consortium of Aquariums, Universities, and Zoos

CAZPA: Canadian Association of Zoological Parks and Aquariums

CBSG: Captive Breeding Specialist Group

CEF: Conservation Endowment Fund

EEP: European Endangered Species Programme

FIG: Fauna Interest Group

GCAP: Global Captive Management Plan

ISIS: International Species Inventory System

IUDZG: International Union of Directors of Zoological Gardens

MEDARKS: Medical Animal Record-Keeping System

PAAZAB: Pan African Association of Zoological Gardens, Aquaria, and Botanical Gardens

PHVA: Population and Habitat Viability Analysis

SAG: Scientific Advisory Group

SPARKS: Single Population Analysis and Record-Keeping System

SPMAG: Small Population Management Advisory Group

SSP: Species Survival Plan

TAG: Taxon Advisory Group

WCMC: Wildlife Conservation and Management Committee

ZCOG: Zoo Conservation Outreach Group

To assist in financing cooperative conservation and related scientific and educational activities of its members, the AZA has established the Conservation Endowment Fund. Other organizations, such as the American Association of Zoo Keepers and the Association of Avian Veterinarians, as well as several zoos, also sponsor grant competition for conservation-related programs.

B. IUCN

To more effectively establish goals for its conservation programs, the AZA works closely with the International Union for the Conservation of Nature and Natural Resources (also called The World Conservation Union), the world's largest conservation organization. The IUCN, which publishes the "Red Data Book" of endangered, threatened, and vulnerable species, monitors the status of populations in the wild through its Species Survival Commission (SSC) specialist groups. [*See* CONSERVATION AGREEMENTS, INTERNATIONAL.]

In its policy statement on captive breeding, the IUCN has declared that "habitat protection alone is not sufficient if the expressed goal of the World Conservation Strategy, the maintenance of biotic diversity, is to be achieved. Establishment of self-sustaining captive populations and other supportive intervention will be needed to avoid the loss of many species, especially those at high risk in greatly reduced, highly fragmented, and disturbed habitats. Captive breeding programs need to be established before species are reduced to critically low numbers, and thereafter need to be coordinated internationally according to sound biological principles, with a view to the maintaining or reestablishment of viable populations in the wild." [*See* DUTIES TO ENDANGERED SPECIES.]

C. CBSG

The coordination of programs between AZA and IUCN is directed by the IUCN's Captive Breeding Specialist Group (CBSG), which promotes management of the wild and captive populations as one, especially as pertains to issues of population genetics and demographics. The CBSG, in addition to fostering international cooperation, performs population and habitat viability analyses (PHVAs) to assess the probabilities of extinction of small populations of endangered animals. It also sponsors conservation assessment and management plans (CAMPs), which attempt to integrate *in situ* and *ex situ* (captive) programs involving input from representatives from the taxon-based specialist groups of the IUCN, the International Council for Bird Preservation, and AZA's taxon advisory groups. The resulting broad-based recommendations for management and research are then published for distribution as a global captive action recommendations (GCAR).

D. IUDZG

Zoos also coordinate globally through the International Union of Directors of Zoological Gardens, also called the World Zoo Organization. IUDZG, along with IUCN and CBSG, recently published "The World Zoo Conservation Strategy," which outlines the role of zoos and aquaria of the world in global conservation. Their primary stated aim is "to support the conservation of species, natural habitats, and ecosystems." However, when captive breeding programs are recommended, they should complement, not substitute for, other conservation activities. The publication also advocates integration of zoo conservation programs with those of other conservation organizations.

E. WCMC

The AZA's Wildlife Conservation and Management Committee oversees the various conservation and scientific programs, in consultation with the director of conservation and science, makes recommendations concerning policies and procedures, and helps guide the development of programs.

F. SSP

The foundation program for captive species conservation is the AZA's Species Survival Plan. The SSP, conceived by Dr. Tom Foose in the late 1970s and established in 1981, addresses genetic and demographic issues associated with the maintenance of small populations of a species in captivity for extended periods of time. Through the SSP, individual animals can be managed collectively as one population through the cooperation of individual zoos (Table II).

Each SSP is directed by a species coordinator under the advice of a propagation group, whose members represent institutions that have signed memoranda of cooperation. Advisors in genetics, veterinary medicine, nutrition, reproduction, or behavior may also participate. The group develops

TABLE II
SSP Species

Mammals	
Addax	Lion-tailed macaque
African elephant	Maned wolf
African wild dog	Mexican wolf
Arabian oryx	Okapi
Asian elephant	Orangutan
Asian lion	Red panda
Asian small-clawed otter	Red wolf
Asian wild horse	Rodriguez fruit bat
Barasingha	Ruffed lemur
Black-footed ferret	Scimitar-horned oryx
Black lemur	Siberian tiger
Black rhinoceros	Snow leopard
Bonobo	Spectacled bear
Chacoan peccary	Sumatran rhinoceros
Cheetah	Sumatran tiger
Chimpanzee	Tree kangaroo
Chinese alligator	White rhinoceros
Clouded leopard	
Cotton-topped tamarin	Reptiles
Drill	Aruba Island
Gaur	rattlesnake
Gibbons	Chinese alligator
Goeldi's monkey	Dumeril's ground boa
Golden lion tamarin	Puerto Rican crested
Gorilla	toad
Greater one-horned rhinoceros	Radiated tortoise
Grevy's zebra	Virgin Islands boa
	Invertebrates
	Partula snail

a master plan, a regional breeding strategy, which results from extensive genetic and demographic analyses based on studbook data. From the master plan, annual breeding recommendations are made to minimize the deleterious genetic effects which could result from the small population size. Recommendations are updated yearly to accommodate the changing population. In addition to the studbook, a husbandry manual is prepared for each species to address current techniques and problems and to indicate areas in need of research.

Currently there are more than 70 SSPs, with a goal of 100 by the end of the century, and nearly 200 species have studbooks. Yet these figures do not approach the number of species expected to meet extinction in the coming decades. However, zoos can ensure the survival of at least some of the more imperiled and significant of these species by maintaining viable captive populations. Indeed, the ultimate goal of SSPs is the reintroduction of animals into former ranges.

G. TAG

On a slightly broader scale, the regional taxon advisory groups are charged with making conservation recommendations for groups of related species, often at the level of the family. For example, the equid advisory group considers all species in the family of horse-like animals, i.e., wild horses, zebras, and asses. Even the membership of the TAG is broader, including representatives from other conservation organizations and field biologists, as well as zoo professionals. There currently are more than 40 TAGs.

Each TAG, on the recommendation of its IUCN/SSC counterpart, assesses the need for captive propagation of particular species, evaluates the potentially available captive space, and recommends new species for studbooks and SSPs. The outcome of these exercises is the development of a regional collection plan for the taxon under its umbrella. As suggested in the AZA Conservation Resource Guide, the following criteria are often used:

1. Current and anticipated captive space available
2. Current captive population size and composition
3. Ability to breed successfully in captivity
4. Status in the wild
5. Sufficient number of founders available
6. Usefulness of the taxon to save habitat and other syntropic taxa as a "flagship" or "keystone" species around which a viable conservation program might be built
7. Uniqueness of the taxon in terms of phylogeny, adaptive strategy, interactions and coevolution with other taxa, ecological approach to survival, cultural appeal, or scientific significance
8. Ability to survive in human-altered ecosystems that are now ubiquitous
9. Probability of successful reintroduction

Especially at the TAG level, an effort is made to communicate with the corresponding IUCN/SSC specialist group which evaluates the status of the same taxon in the wild.

H. FIG

Recognizing that each species can exist only as an integral part of its ecosystems and that habitat destruction is the primary force behind species endangerment, AZA recently established fauna interest groups, which help coordinate conservation activities in specific geographic locations. FIGs foster cooperation and communication with foreign zoos, as well as governmental and nongovernmental organizations, to support local nature preserves, zoos, and aquariums, to conduct field research and public education, and to assist in the transfer of technology and equipment. FIGs currently exist for Indonesia/Malaysia, Madagascar, Brazil, the Caribbean, Meso-America, Paraguay, and Zaire.

I. SAG

Other scientific advisory groups have been formed to consider issues of behavior and husbandry, veterinary sciences, genome banking, contraception, reintroduction, and small population management (genetics and demography). These groups promote collaborations among zoos and between zoos and university biologists, advise AZA committees, and communicate with scientific societies and with IUCN/SSC specialist groups.

J. Reintroduction

Although reintroduction into the wild may be the goal of zoo biologists, at present it is only a dream for most species. Even where adequate captive populations exist, areas with a suitable, stable habitat are nearly impossible to find. There have been some notable successes, starting with the American bison which faced extinction early in this century. Bison herds roam the western United States today thanks to the efforts of the New York Zoological Society. More recently, the Arabian oryx, Przewalski horse, red wolf, and golden-lion tamarin have been reintroduced into native habitat from zoo stock.

II. ANIMAL CARE

Although conservation programs are given the highest priority by AZA, the primary function of most zoo professionals is caring for animals. At the most basic level, this care consists of providing food and clean housing for the animals in their charge. Keepers contribute most of the day-to-day care, under the guidance of curators. Support staff typically include a veterinarian and perhaps a nutritionist.

Historically, keepers and curators learned their trade through experience, with instruction provided by superiors. More recently, the possibility for exchange of information among zoos has increased by attendance at conferences and via publications such as the "International Zoo Yearbook" and the journal *Zoo Biology*. Surveys may also be employed to assess aspects of husbandry, then analyzed, and the results distributed to participants. The *Journal of Zoo and Wildlife Medicine,* the official publication of the Association of Zoo Veterinarians (AZV), provides a forum for presentation of medical cases and research.

More specific guidelines for animal care are contained in husbandry manuals, which are produced for species covered by SSPs with an eye to preserving traits that are adaptive to the natural environment in which the species evolved (Table III).

TABLE III

AZA Conservation Resource Guide: Suggested Husbandry Manual Content

Housing and enclosure requirements	Management
Containment barriers	Identification methods
Shelter requirements	Capture and restraint
Substrate/topography	Transport method
Water source	Pest control
Special furnishings	
Temperature/humidity	Nutrition
Enclosure size	Feeding schedule
Handling facilities	Feeding location
Utilities (e.g., lights)	Nutrient content
Solitary or group housing	Food items
Utilities	Nutritional needs
Reproduction	Behavior and social organization
Gestation/incubation period	Optimal social group
Breeding/birth season	Age at dispersal
Age-specific fecundity	Seasonal changes
Birth sex ratio	Mating system
Birth weights	Behavioral ontogeny
Growth characteristics	Parental care
Neonatal mortality	Interspecific tolerance
Age at weaning/fledging/ sexual maturity	Stress assessment
Estrous cycle	Health
Environmental effects	Inoculations
Semen collection methods	Neonatal exams
Semen cryopreservation	Parasites
Artificial insemination	Major diseases
Estrous cycle control	Common injuries
Embryo transfer	Physiological norms
Embryo cryopreservation	Immobilization
Contraception	Life span
	Necropsy protocol

A. Behavioral Enrichment

Especially for primates, but increasingly for other taxonomic groups as well, natural behavioral patterns are being considered in setting management routines. The goal is to allow animals to engage in species-typical activities as much as possible. This strategy depends first on knowledge gained from field studies and then on tests in captivity. Because most species spend the better part of each day hunting or foraging for food, it is easy to understand why they might be bored if presented with their entire day's ration at one time and are able to consume it in a few minutes. Perhaps the simplest method of extending the feeding period is to broadcast food throughout the enclosure along with straw or other bedding material so that the animals are required to search for each food item. This can be especially effective if food is presented more than once per day. More elaborate methods include boards with holes into which raisins can be placed, to be dug out by the animals, and artificial honey combs for dipping into with sticks or fingers. Research in this area is very active, as evidenced by the breadth of studies presented at a symposium on behavioral enrichment held in 1993.

B. Genetic Management

Advances in the fields of molecular and population genetics are being applied to the management of captive populations. Recognition among zoo professionals that inbreeding allowed the expression of deleterious alleles provided the initial impetus for genetic management. More recently, consideration is being given to maximizing the heterozygosity of each species as a population, including equalization of the genetic contribution of each founder. These goals are accomplished through recommendations from SSP propagation groups, based on data from studbooks and the International Species Inventory System.

C. Contraception

The responsible management of the vast majority of captive animals includes contraception at some time during their reproductive lives. In many cases, zoos have become the victims of their own successes. Progress in areas of nutrition, husbandry, and veterinary care has meant higher reproductive rates and increased longevity. Without physical separation or contraception, the consequence is the production of surplus animals that tax already limited resources.

Although physical separation is effective, it is an undesirable alternative for several reasons: it is stressful to the animals, requires more space, and presents an unnatural image of the animals' social organization to the public. A large part of the mission of the modern zoo is education and instilling in the public an appreciation for the animals they come to see. There is a much higher educational value of a species in as natural a social setting as possible.

Contraception allows the maintenance of natural social groups without unwanted pregnancies. It also allows for spaced pregnancies for the health of both the mother and her offspring, and for sound genetic management to reduce inbreeding and to equalize founder representation.

Contraception choices include most of those available for humans such as steroid implants, birth control pills, and injections. In addition, the zoo community supports research into alternative techniques such as vas plugs, contraceptive vaccines, and male birth control pills. The contraceptive advisory group makes recommendations based on research and on information maintained in its international wildlife contraception database.

D. Assisted Reproductive Techniques

Although not practical in all situations, techniques such as artificial insemination and embryo transfer can be important for species that breed poorly in captivity and for more effective genetic management. The long-term benefit to the genetic health, and perhaps to the survival, of some species may depend on the ability to freeze genetic material. Frozen sperm and embryos can be much more efficiently shipped among zoos and between the captive and wild populations, effecting outbreeding without transporting animals. Banks of frozen germ plasm also can serve as genetic reservoirs for future use, preserving today's gene pool for tomorrow.

Unfortunately, early optimism about the simple transfer of technology tested and in practice with humans and many domestic species has waned. Species differences demand that methodologies be tailored for each new application. Considerable research effort is needed before routine application of these techniques is possible for conservation programs.

III. RESEARCH

Most television viewers have been introduced to field research. There are millions of feet of film of animals in the wild, often told through the perspective of the biologist studying them. The validity of field research is obvious: studying an animal under the conditions in which it evolved and to which it has adapted. In contrast, the validity of captive research is often questioned because of the unnatural or artificial conditions in which the animals live.

However, there are disadvantages inherent to field studies that do not apply to captive research, that might not at first be apparent. First, variables, such as climate and photoperiod, cannot be controlled in the wild in order to study their effect on biological parameters. Dietary intake and genetic relationships cannot be easily determined. A further advantage to captive as compared to field studies is the accessibility of the subjects and their habituation to humans. In reality, a combination of captive and field studies can provide the most complete picture of a species. Scientists are now recognizing the benefits of collaborations between the field and captive investigations.

Disciplines that are typically part of captive research programs include behavior, physiology, anatomy, genetics, nutrition, and veterinary medicine. The emphasis is frequently on applied research or problem-solving, often in collaboration with medical centers or with commercial enterprises in the development of veterinary and nutritional products. However, basic research is more likely to be carried out in concert with university scientists.

IV. EDUCATION

At least 600,000,000 people, approximately 10% of the current world population, annually visit the more than 1000 zoos worldwide. The yearly zoo, aquarium, and wildlife park attendance in the United States and Canada is more than 112,000,000, greater than the number at football, baseball, and hockey games combined.

Although these visitors go to zoos primarily for recreation, their presence gives zoos the opportunity to increase public awareness about endangered species and conservation. Books and movies cannot substitute for a living animal. Experiencing animals first hand produces much deeper and more long-lasting impressions, and makes the visitor more receptive to information about the animals' plight in the wild. Because the zoo-going audience is very diverse, which presents a challenge to zoo educators, it offers the opportunity to reach people with a very wide range of attitudes.

The traditional form that zoo education took was merely informing visitors about basic biological characteristics of the animals through signage. Today, although basic biology is still covered, it is combined with a conservation theme. The most effective programs include information about how the public can help.

In addition to its function of teaching the public, the education of zoo professionals is also necessary. Although each zoo assumes the responsibility of training its own personnel, the AZA sponsors programs for conservation education, such as the St.

Louis Zoo's Conservation Academy, which includes courses for SSP coordinators, studbook keepers, and animal husbandry. Workshops and symposia also are held regularly on topics such as bioethics, environmental enrichment, and creating master plans.

V. ENTERTAINMENT OR RECREATION

These are the primary reasons people visit zoos. It is likely that they will continue to attend only if they are pleased with their experience, which can present a problem to zoo professionals. Conservation efforts may be in conflict with entertainment. For example, not all endangered species are popular with the public. Typical parents want their children to see the standard lion, tiger, elephant, and giraffe, but these are not the most endangered species.

Furthermore, successful captive breeding often depends on maintaining more than one pair, but space taken up by additional pairs is then not available for other species, limiting the variety that can be displayed. Another disappointment to many zoo-goers is the trend away from "animal shows." However, modern zoo philosophy views most shows as demeaning to the animals and as not fostering an appreciation of the animals' true nature and worth.

But, ultimately, the existence of zoos depends on the goodwill and interest of the public, forcing some balance or compromise between public expectation and successful conservation. The most satisfactory answer would be the increasing appreciation by the public of the need for animal conservation, fostered by insights gained at the zoo.

Glossary

Assisted reproductive techniques Procedures such as artificial insemination and embryo transfer used to enable particular animals to reproduce or to enhance their reproductive rate.

Captive breeding Maintaining and reproducing species in zoos or other facilities as an adjunct to field conservation programs.

Founders Original animals in a population that supply the foundation for that population's gene pool.

Inbreeding Reproduction between closely related individuals, which allows expression of recessive alleles that might be deleterious.

Reintroduction Releasing animals back into the wild after a period in captivity.

Surplus animals Individuals in a population that are well-represented genetically and so are no longer needed for reproduction.

Bibliography

DeBlieu, J. (1991). "Meant to be Wild: The Struggle to Save Endangered Species." Golden, CO: Fulcrum Publishing.

Gipps, J. H. W., ed. (1991). "Beyond Captive Breeding: Reintroducing Endangered Mammals to the Wild." Symposium of the Zoological Society of London 62. Oxford: Clarendon Press.

Gold, D. (1988). "Zoo: A Behind the Scenes Look at the Animals and the People Who Care for Them." Chicago: Contemporary Books.

Page, J. (1990). "Zoo: The Modern Ark." New York: Facts on File.

Tudge, C. (1992). "Last Animals at the Zoo: How Mass Extinction Can Be Stopped." Washington, DC: Island Press.

Contributors

John F. Addicott
Ecology of Mutualism
  Department of Biological Sciences
  University of Alberta
  Edmonton, Alberta
  Canada T6G 2E9

Walter H. Adey
Controlled Ecologies
  Marine Systems Laboratory
  National Museum of Natural History
  Smithsonian Institution
  Washington, DC 20560

James K. Agee
Park and Wilderness Management
  University of Washington
  Seattle, Washington 98195

Lewis M. Alexander
Management of Large Marine Ecosystems
  University of Rhode Island
  Kingston, Rhode Island 02881

Miguel A. Altieri
Agroecology
  Laboratory of Biological Control
  Department of Environment Science, Policy,
    and Management
  University of California at Berkeley
  Berkeley, California 94720

W. Scott Armbruster
Plant–Animal Interactions
  Department of Biology and Wildlife
  and Institute of Arctic Biology
  University of Alaska
  Fairbanks, Alaska 94775

Cheryl S. Asa
Zoological Parks
  St. Louis Zoological Park
  St. Louis, Missouri 63110

Peter M. Attiwill
Nutrient Cycling in Forests
  School of Botany
  University of Melbourne
  Parkville, Victoria 3052
  Australia

Gwendolyn Bachman
Ecological Energetics of Terrestrial Vertebrates
  Department of Biology
  University of California at Los Angeles
  Los Angeles, California 90024

Michael J. Benton
Mass Extinction, Biotic and Abiotic
  Department of Geology
  University of Bristol
  Bristol BS8 1RJ
  United Kingdom

Fikret Berkes
*Community-Based Management of Common
Property Resources*
  Natural Resources Institute
  The University of Manitoba
  Winnipeg, Manitoba
  Canada R3T 2N2

Alan A. Berryman
Biological Control
  Department of Entomology
  Washington State University
  Pullman, Washington 99164

L. C. Bliss
*Industrial Development, Arctic
Northern Polar Ecosystems*
  Department of Botany
  University of Washington
  Seattle, Washington 98195

Nigel J. Bunce
Aspects of the Environmental Chemistry of Elements
  Department of Chemistry and Biochemistry
  University of Guelph
  Guelph, Ontario
  Canada N1G 2W1

I. C. Burke
Great Plains, Climate Variability
  Department of Forest Sciences
  Colorado State University
  Fort Collins, Colorado 80523

Harold E. Burkhart
Modeling Forest Growth
  Department of Forestry
  Virginia Polytechnic Institute and State
    University
  Blacksburg, Virginia 24061

John Cairns, Jr.
Restoration Ecology
  University Center for Environmental and
    Hazardous Materials Studies
  Virginia Polytechnic Institute and State
    University
  Blacksburg, Virginia 24061

W. A. Calder
*Ecology of Size, Shape, and Age Structure in
Animals*
  University of Arizona
  Tucson, Arizona 85721

Gerard M. Capriulo
Marine Microbial Ecology
  Environmental Science Department
  State University of New York at Purchase
  Purchase, New York 10577

James R. Carey
Insect Demography
  Department of Entomology
  University of California at Davis
  Davis, California 95616

Sandra J. Carlson
Systematics
  Department of Geology
  University of California at Davis
  Davis, California 95616

Ghassan Chebbo
Pollution of Urban Wet Weather Discharges
  CERGRENE
  Ecole Nationale des Ponts et Chaussées
  93167 Noisy le Grand
  France

Norman L. Christensen
Fire Ecology
  School of the Environment
  Duke University
  Durham, North Carolina 27708

George W. Cox
Galápagos Islands
  Department of Biology
  San Diego State University
  San Diego, California 92182

Clinton J. Dawes
Intertidal Ecology
  Department of Biology
  University of South Florida
  Tampa, Florida 33620

Donald L. DeAngelis
*Equilibrium and Nonequilibrium Concepts in
Ecological Models*
  Department of Mathematics
  The University of Tennessee
  Knoxville, Tennessee 3799

David R. DeWalle
Forest Hydrology
  Department of Forest Hydrology
    School of Forest Resources
  and Environmental Resources Research
    Institute

The Pennsylvania State University
University Park, Pennsylvania 16802

Durell C. Dobbins
Biodegradation of Pollutants
BioTrol, Inc.
Eden Prairie, Minnesota 55344

M. I. Dyer
Biosphere Reserves
Institute of Ecology
University of Georgia
Athens, Georgia 30602

W. T. Edmondson
Eutrophication
Department of Zoology
University of Washington
Seattle, Washington 98195

Scott A. Elias
Packrat Middens, Archives of Desert Biotic History
Institute for Arctic and Alpine Research
University of Colorado
Boulder, Colorado 80309

R. W. Enser
Plant Conservation
Rhode Island Natural Heritage Program
Providence, Rhode Island 02903

Douglas H. Erwin
Diversity Crises in the Geologic Past
Department of Paleobiology
National Museum of Natural History
Smithsonian Institution
Washington, DC 20560

Douglas J. Futuyma
Speciation
Department of Ecology and Evolution
State University of New York at Stony Brook
Stony Brook, New York 11794

W. S. Fyfe
Global Anthropogenic Influences
Department of Earth Sciences
University of Western Ontario
London, Ontario
Canada N6A 5B7

Madhav Gadgil
Traditional Conservation Practices
Centre for Ecological Sciences
Indian Institute of Science
Bangalore 560012
India

Robert A. Garrott
Wildlife Radiotelemetry
Department of Wildlife Ecology
University of Wisconsin—Madison
Madison, Wisconsin 53706

Thomas F. Gesell
Environmental Radioactivity and Radiation
Physics Department
Idaho State University
Pocatello, Idaho 83209

James P. Gibbs
Hydrologic Needs of Wetland Animals
School of Forestry and Environmental Studies
Yale University
New Haven, Connecticut 06511

Edward P. Glenn
Terrestrial Halophytes
Environmental Research Laboratory
University of Arizona
Tucson, Arizona 85706

H. Charles Godfray
Sex Ratio
Imperial College at Silwood Park
Ascot, Berks
United Kingdom

N. F. Gray
Wastewater Treatment Biology
Environmental Sciences Unit
Trinity College
University of Dublin
Dublin 2
Ireland

Lance H. Gunderson
Everglades, Human Transformations of a Dynamic Ecosystem
Department of Zoology
University of Florida
Gainesville, Florida 32611

Raymond P. Guries
Forest Genetics
Department of Forestry
University of Wisconsin—Madison
Madison, Wisconsin 53706

Yitzchak Gutterman
Seed Dispersal, Germination, and Flowering Strategies of Desert Plants
The Jacob Blaustein Institute for Desert Research

and Department of Life Sciences
Ben-Gurion University of the Negev
Sede Boker Campus 84990
Israel

Donald A. Hammer
Water Quality Improvement Functions of Wetlands
Hammer Resources, Inc.
Norris, Tennessee 37828

Garrett Hardin
Tragedy of the Commons
Department of Biological Sciences
University of California at Santa Barbara
Santa Barbara, California 93106

A. W. Hawkes
Ecotoxicology
Department of Environmental Toxicology
Institute of Wildlife and Environmental
Toxicology
Clemson University
Pendleton, South Carolina 29670

R. J. Haynes
Nitrous Oxide Budget
New Zealand Institute for Crop and
Food Research
Christchurch
New Zealand

Joel T. Heinen
*Conservation Agreements, International
Nature Preserves*
Department of Environmental Studies
College of Arts and Sciences
Florida International University
Miami, Florida 33199

George Hidy
Acid Rain
Electric Power Research Institute
Palo Alto, California 94303

R. J. Hobbs
Landscape Ecology
Commonwealth Scientific and Industrial
Research Organization
Division of Wildlife and Ecology
Midland, WA 6056
Australia

A. A. Hoffmann
Environmental Stress and Evolution
Department of Genetics and Human Variation

La Trobe University
Bundoora, Victoria 3083
Australia

Osmund Holm-Hansen
*Atmospheric Ozone and the Biological Impact of
Solar Ultraviolet Radiation*
Scripps Institution of Oceanography
University of California at San Diego
La Jolla, California 92093

Kent E. Holsinger
Conservation Programs for Endangered Plant Species
Department of Ecology and Evolutionary
Biology
University of Connecticut
Storrs, Connecticut 06269

Kurt Hostettmann
Plant Sources of Natural Drugs and Compounds
School of Pharmacy
Institute of Pharmacognosy and
Phytochemistry
University of Lausanne
CH-1015 Lausanne
Switzerland

R. A. Houghton
Deforestation
The Woods Hole Research Center
Woods Hole, Massachusetts 02543

G. L. Hutchinson
*Nitrogen Cycle Interactions with Global Change
Processes*
U.S. Department of Agriculture
Agricultural Research Service
Fort Collins, Colorado 80522

J. Ishizaka
Continental Shelf Ecosystems
National Institute for Resources and
Environment
Ibaraki 305
Japan

John Jaenike
Parasitism, Ecology
Department of Biology
University of Rochester
Rochester, New York 14627

R. L. Jefferies
Tunda, Arctic and Sub-Arctic
Department of Botany

University of Toronto
Toronto, Ontario
Canada M5S 3B2
Arthur T. Johnson
Biological Engineering for Sustainable Biomass
Production
  Biological Resources Engineering Program
  Department of Agricultural Engineering
  University of Maryland
  College Park, Maryland 20742
Carl F. Jordan
Nutrient Cycling in Tropical Forests
  School of Ecology
  University of Georgia
  Athens, Georgia 30602
Patrick Kangas
Biological Engineering for Sustainable Biomass
Production
  Natural Resources Management Program
  Department of Agricultural Engineering
  University of Maryland
  College Park, Maryland 20742
Ludger Kappen
Plant Ecophysiology
  Botanical Institute
  University of Kiel
  D-24098 Kiel
  Germany
R. J. Kendall
Ecotoxicology
  Department of Environmental Toxicology
  Institute of Wildlife and Environmental
    Toxicology
  Clemson University
  Pendleton, South Carolina 29670
M. A. K. Khalil
Greenhouse Gases in the Earth's Atmosphere
  Department of Environmental Science and
    Engineering
  Oregon Graduate Institute
  Portland, Oregon 97291
K. Killham
Rhizosphere Ecophysiology
  Department of Plant and Soil Science
  University of Aberdeen
  Aberdeen AB9 2UE
  Scotland, United Kingdom

J. C. Koella
Ecology and Genetics of Virulence and Resistance
  ETH Zürich
  Experimental Ecology
  ETH-Zentrum NW
  CH-80952 Zürich
  Switzerland
T. T. Kozlowski
Physiological Ecology of Forest Stands
  Department of Environmental Science, Policy,
    and Management
  University of California at Berkeley
  Berkeley, California 94720
Charles J. Krebs
Population Regulation
  Department of Zoology
  University of British Columbia
  Vancouver, British Columbia
  Canada V6T 1Z4
N. Kulkarni
Forest Insect Control
  Entomology Division
  Tropical Forest Research Institute
  Jabalpur, 482001
  India
Russell Lande
Population Viability Analysis
  Department of Biology
  University of Oregon
  Eugene, Oregon 97403
J. J. Landsberg
Forest Canopies
  Commonwealth Scientific and Industrial
    Research Organization
  Centre for Environmental Mechanics
  Canberra, ACT 2601
  Australia
W. K. Lauenroth
Great Plains, Climate Variability
  Department of Rangeland Ecosystem Science
  Colorado State University
  Fort Collins, Colorado 80523
Mary Allessio Leck
Seed Banks
  Department of Biology
  Rider University
  Lawrenceville, New Jersey 08648

Li-Yin Lin
Wastewater Treatment for Inorganics
City of Niagara Falls
Niagara Falls, New York 14302

J. W. Lloyd
Sahara
School of Earth Sciences
University of Birmingham
Edgbaston, Birmingham
United Kingdom B15 2TT

John B. Loomis
Biodiversity, Economic Appraisal
Department of Agriculture and Resource
Economics
Colorado State University
Fort Collins, Colorado 80523

Dan Lubin
*Atmospheric Ozone and the Biological Impact of
Solar Ultraviolet Radiation*
Scripps Institution of Oceanography
University of California at San Diego
La Jolla, California 92093

Fred T. Mackenzie
Biogeochemistry
Department of Oceanography
School of Ocean and Earth Science and
Technology
University of Hawaii
Honolulu, Hawaii 96822

Mark R. Macnair
Speciation and Adaptive Radiation
Department of Biological Sciences
University of Exeter
Exeter EX4 4PS
United Kingdom

J. G. McColl
Forest Clear-Cutting, Soil Response
Environmental Science, Policy, and
Management Department
Ecosystem Sciences Division
University of California at Berkeley
Berkeley, California 94720

M. Meili
Biogeochemical Cycles
Institute of Earth Sciences
University of Uppsala
S-75236 Uppsala
Sweden

P. B. Meshram
Forest Insect Control
Entomology Division
Tropical Forest Research Institute
Jabalpur, 482001
India

L. Scott Mills
Keystone Species
Department of Fisheries and
Wildlife
University of Idaho
Moscow, Idaho 83844

M. K. Misra
Ecological Energetics of Ecosystems
Department of Botany
Ecology and Floristics Laboratory
Berhampur University
Berhampur, 760007 Orissa
India

Jeffry B. Mitton
Insect Interactions with Trees
Department of Biology
University of Colorado
Boulder, Colorado 80309

Coleen L. Moloney
Antarctic Marine Food Webs
Department of Zoology
University of Cape Town
Rondebosch 7700
South Africa

Berrien Moore III
Global Carbon Cycle
Institute for the Study of Earth, Oceans, and
Space
University of New Hampshire
Durham, New Hampshire 03824

Phillip A. Morin
*Captive Breeding and Management of Endangered
Species*
Department of Anthropology
University of California at Davis
Davis, California 95616

K. J. Murphy
Aquatic Weeds
The Centre for Research in Environmental
Science and Technology
Department of Botany
University of Glasgow

Glasgow G12 8QQ
Scotland

Eviatar Nevo
Evolution and Extinction
Institute of Evolution
University of Haifa
Mount Carmel, Haifa 31905
Israel

Sharon E. Nicholson
Sahel, West Africa
Department of Meteorology
Florida State University
Tallahassee, Florida 32306

Stephen Nicol
Krill, Antarctic
Australian Antarctic Division
Kingston, Tasmania 7053
Australia

S. K. Nisanka
Ecological Energetics of Ecosystems
Department of Botany
Ecology and Floristics Laboratory
Berhampur University
Berhampur, 760007 Orissa
India

Richard B. Norgaard
Biodiversity, Processes of Loss
Energy & Resources Group
University of California at Berkeley
Berkeley, California 94720

M. P. North
*Virgin Forests and Endangered Species, Northern
Spotted Owl and Mt. Graham Red Squirrel*
College of Forest Resources
University of Washington
Seattle, Washington 98195

P. E. O'Sullivan
Paleolimnology
Department of Environmental Sciences
University of Plymouth
Drake Circus
Plymouth PL4 8AA
United Kingdom

T. Ohta
Molecular Evolution
Department of Population Genetics
National Institute of Genetics
Mishima 411
Japan

Margery L. Oldfield
Biodiversity, Values and Uses
The Sealuck Foundation
Islip, New York 11751

Gordon H. Orians
Ecology, Aggregate Variables
Department of Zoology
University of Washington
Seattle, Washington 98195

Mark Pagel
Species Diversity
Department of Zoology
University of Oxford
Oxford OX1 3PS
United Kingdom

P. A. Parsons
Environmental Stress and Evolution
University of Adelaide
Australia

T. R. Parsons
Marine Biology, Human Impacts
Pollution Impacts on Marine Biology
Department of Oceanography
University of British Columbia
Vancouver, British Columbia
Canada V6T 1Z4

E. Paterson
Rhizosphere Ecophysiology
Department of Plant and Soil Science
University of Aberdeen
Aberdeen AB9 2UE
Scotland, United Kingdom

S. C. Pathak
Forest Insect Control
Department of Biological Sciences
Rani Durgavati University
Jabalpur 482001
India

Daniel Pauly
Ocean Ecology
International Center for Living Aquatic
  Resources Management
Manila
Philippines
and Fisheries Centre
University of British Columbia
Vancouver, British Columbia
Canada V6T 1Z4

Craig M. Pease
Population Viability Analysis
  Department of Zoology
  University of Texas
  Austin, Texas 78712
R. Perissinotto
Marine Productivity
  Department of Zoology and Entomology
  Southern Ocean Group
  Rhodes University
  Grahamstown 6140
  South Africa
Glenn Perrigo
*Mammalian Reproductive Strategies, Biology of
Infanticide*
  College of Arts and Science
  Division of Biological Sciences
  University of Missouri—Columbia
  Columbia, Missouri 65211
David A. Perry
Forests, Competition and Succession
  Department of Forest Science
  Oregon State University
  Corvallis, Oregon 97331
Michel Pichon
Coral Reef Ecosystems
  Ecole Pratique des Hautes Etudes
  URA CNRS 1453
  Université de Perpignan
  66860 Perpignan
  France
Roger A. Pielke
Atmosphere–Terrestrial Ecosystem Modeling
  Department of Atmospheric Science
  Colorado State University
  Fort Collins, Colorado 80523
Olivier Potterat
Plant Sources of Natural Drugs and Compounds
  School of Pharmacy
  Institute of Pharmacognosy and
    Phytochemistry
  University of Lausanne
  CH-1015 Lausanne
  Switzerland
G. T. Prance
Biodiversity
  Royal Botanical Gardens

  Kew, Richmond
  Surrey TW9 3AB
  United Kingdom
Elaine Mae Prins
Biomass Burning
  Cooperative Institute for Meteorological
    Satellite Studies
  University of Wisconsin—Madison
  Madison, Wisconsin 53706
T. R. Rainwater
Ecotoxicology
  Department of Environmental Toxicology
  Institute of Wildlife and Environmental
    Toxicology
  Clemson University
  Pendleton, South Carolina 29670
P. S. Ramakrishnan
Shifting Cultivation
  School of Environmental Sciences
  Jawaharlal Nehru University
  New Dehli 110067
  India
E. A. S. Rattray
Rhizosphere Ecophysiology
  Department of Plant and Soil Science
  University of Aberdeen
  Aberdeen AB9 2UE
  Scotland, United Kingdom
Alan F. Raybould
Wild Crops
  Furzebrook Research Station
  Institute of Terrestrial Ecology
  Wareham, Dorset BH20 5AS
  United Kingdom
William A. Reiners
Land–Ocean Interactions in the Coastal Zone
  Department of Botany
  University of Wyoming
  Laramie, Wyoming 82071
Curtis J. Richardson
Wetlands Ecology
  Duke Wetland Center
  School of the Environment
  Duke University
  Durham, North Carolina 27708
Holmes Rolston III
Duties to Endangered Species

Department of Philosophy
Colorado State University
Fort Collins, Colorado 80523

Sheila M. Ross[1]
Ecological Restoration
Department of Geography
University of Bristol
Bristol BS8 1S5
United Kingdom

Jelte Rozema
Plants and High CO_2
Department of Ecology and Ecotoxicology
Faculty of Biology
Vrije University
1081 HV Amsterdam
The Netherlands

James R. Runkle
Deciduous Forests
Department of Biological Sciences
Wright State University
Dayton, Ohio 45435

Peter G. Ryan
Antarctic Marine Food Webs
Percy Fitzpatrick Institute of African
Ornithology
University of Cape Town
Rondebosch 7700
South Africa

Agnes Saget
Pollution of Urban Wet Weather Discharges
CERGRENE
Ecole Nationale des Ponts et Chaussées
93167 Noisy le Grand
France

I. R. Sanders
Grassland Ecology
Botanisches Institut
Universität Basel
CH-4056 Basel
Switzerland

S. D. Schemnitz
Wildlife Management

Department of Fishery and Wildlife Sciences
New Mexico State University
Las Cruces, New Mexico 88003

Don W. Schloesser
Introduced Species, Zebra Mussels in North America
National Biological Service
Great Lakes Science Center
Ann Arbor, Michigan 48105

P. Schmid-Hempel
Ecology and Genetics of Virulence and Resistance
ETH Zürich
Experimental Ecology
ETH-Zentrum NW
CH-80952 Zürich
Switzerland

Robert A. Schmidt
Forest Pathology
Department of Forest Pathology
School of Forest Resources and Conservation
University of Florida
Gainesville, Florida 32611

William Seaman, Jr.
Artificial Habitats for Fish
Department of Fisheries and Aquatic Sciences
and Florida Sea Grant College Program
University of Florida
Gainesville, Florida 32611

Charles R. C. Sheppard
Biological Communities of Tropical Oceans
Department of Biological Sciences
University of Warwick
Conventry CV4 7AL
United Kingdom

Kenneth Sherman
Management of Large Marine Ecosystems
NOAA/NMFS/NEFSC
Narragansett Laboratory
Narragansett, Rhode Island 02882

Philip W. Signor
Evolutionary History of Biodiversity
Department of Geology
University of California at Davis
Davis, California 95616

Daniel Simberloff
Introduced Species
Department of Biological Science
Florida State University
Tallahassee, Florida 32306

[1] Present Address: SGS Environmental
Yorkshire House
Liverpool, Merseyside L3 9AG
United Kingdom

David M. Smith
Forest Stand Regeneration, Natural and Artificial
School of Forestry and Environmental Studies
Yale University
New Haven, Connecticut 06520

George D. Smith
Soil Management in the Tropics
Queensland Department of Primary Industries
Toowoomba, Queensland 4350
Australia

William H. Smith
Air Pollution and Forests
School of Forestry and Environmental Studies
Yale University
New Haven, Connecticut 06511

Stephen A. Spongberg
Arboreta
The Arnold Arboretum of Harvard University
Jamaica Plain, Massachusetts 02130

V. R. Squires
Farming, Dryland
Department of Agricultural and Natural
Resource Sciences
University of Adelaide
Roseworthy, South Australia 5371
Australia

Mark Stafford Smith
Deserts, Australian
Commonwealth Scientific and Industrial
Research Organization
Division of Wildlife and Ecology
Alice Springs, NT 0870
Australia

Judy Stamps
Territoriality
School of Evolution and Ecology
University of California at Davis
Davis, California 95616

David W. Steadman
Ecosystem Integrity, Impact of Traditional Peoples
New York State Museum
Albany, New York 12230

D. W. Stephens
Foraging Strategies
School of Biological Sciences
University of Nebraska at Lincoln
Lincoln, Nebraska 68588

David S. G. Thomas
Desertification, Causes and Processes
Sheffield Centre for International Drylands
Research
Department of Geography
University of Sheffield
Sheffield S10 2TN
United Kingdom

James H. Thorp
Invertebrates, Freshwater
Department of Biology
University of Louisville
Louisville, Kentucky 40292

R. W. Tiner
Wetland Restoration and Creation
Department of Plant and Soil Sciences
University of Massachusetts
Amherst, Massachusetts 01003

Rosie Trevelyan
Species Diversity
Department of Zoology
University of Oxford
Oxford OX1 3PS
United Kingdom

John F. Vallentine
Foraging by Ungulate Herbivores
Range Ecology and Grazing
Department of Botany & Range Science
Brigham Young University
Provo, Utah 84602

John L. Vankat
Island Biogeography, Theory and Applications
Department of Botany
Miami University
Oxford, Ohio 45056

Sandra Vehrencamp
Ecological Energetics of Terrestrial Vertebrates
University of California at San Diego
La Jolla, California 92093

H. A. Verhoef
Soil Ecosystems
Department of Ecology and Ecotoxicology
Vrije Universiteit
1081 HV Amsterdam
The Netherlands

Raimo Virkkala
Bird Communities

National Board of Waters and the
  Environment
Nature Conservation Research Unit
FIN-00101 Helsinki
Finland

J. V. Ward
River Ecology
  Department of Biology
  Colorado State University
  Fort Collins, Colorado 80523

John H. Werren
Sex Ratio
  Department of Biology
  University of Rochester
  Rochester, New York 14627

Neil E. West
Deserts
  Department of Rangeland Resources and the
    Ecology Center
  Utah State University
  Logan, Utah 84322

Robert G. Wetzel
Limnology, Inland Aquatic Ecosystems
  Department of Biological Sciences
  University of Alabama
  Tuscaloosa, Alabama 35487

G. E. Wickens
Agriculture and Grazing on Arid Lands
  Royal Botanic Gardens
  Kew, Richmond
  Surrey TW9 3AB
  United Kingdom

Henry M. Wilbur
Life Histories
  University of Virginia
  Charlottesville, Virginia 22903

Garrison Wilkes
Gene Banks
Germplasm Conservation

Department of Biology
University of Massachusetts at Boston
Boston, Massachusetts 02125

Jack E. Williams
Fish Conservation
  Bureau of Land Management
  Intermountain Research Station
  Boise, Idaho 83702

Patricia A. Williams
Evolutionary Taxonomy versus Cladism
  Department of History and International
    Studies
  Virginia State University
  Petersburg, Virginia 23806

K. O. Winemiller
Fish Ecology
  Department of Wildlife and Fisheries Science
  College of Agriculture and Life Sciences
  Texas A&M University
  College Station, Texas 77843

P. C. Withers
Ecophysiology
  Department of Zoology
  The University of Western Australia
  Nedlands, WA 6009
  Australia

Jianguo Wu
Island Biogeography, Theory and Applications
  Biological Sciences Center
  Desert Research Institute
  University of Nevada System
  Reno, Nevada 89506

J. B. Yavitt
Bog Ecology
  Department of Natural Resources
  New York State College of Agriculture and
    Life Sciences
  Cornell University
  Ithaca, New York 14853

Subject Index

E

Evolution, molecular (*continued*)
  genetic instructions, **1:**719, **1:**721
  protein instructions, **1:**721
Evolutionary aggregates
  adaptive, **1:**584–585
  taxonomic, **1:**584
Evolutionary gardens, **2:**195
Evolutionary significant units (ESUs), defined, **2:**41, **2:**47
Evolutionary species concept, cladist, **1:**760
Evolutionary systematics
  classification in, **3:**398
  goals, **3:**398
Evolutionary taxonomy
  vs. cladism, **1:**764
  classification criteria, **1:**764–765
  divergences in, **1:**765–766
  founder principle, **1:**766
  monophyly, **1:**764
Evolutionary time, habitat diversity, **3:**389
Evolutionary values, biodiversity, **1:**222–224
Exclusive Economic Zones, **2:**460, **3:**15
Existence value, defined, **1:**200
Exogenous forces, defined, **1:**292
Exoskeletons
  defined, **3:**35
  from packrat middens, **3:**20–21
Exotic germplasm, **2:**185–186
Exotic species
  defined, **1:**209
  introductions, **1:**202, **3:**106
  management attempts, **3:**572–573
Experimental ecosystems, *see* Controlled ecologies
Exploitation competition, defined, **1:**338
Extinctions, *see also* Endangered plants; Endangered
    species; Mass extinctions; Population viability
    analysis
  anthropogenic, **1:**524–525, **1:**527
  background, **1:**513–**1:**515, **2:**524
  and biological evolution, **1:**522–523
  birds, **1:**343, **1:**513
  birds on oceanic islands, **1:**642–646
  cascade, **1:**224
  causes, **1:**363, **1:**390–391
  Cenozoic, **1:**732
  and climate variations, **2:**52
  climatic causes, **1:**738–740
  current, **1:**733, **1:**737–738, **1:**740
  defined, **2:**47
  endangerment indicators, **1:**387–388, **1:**390–391
  end-Devonian, **1:**731
  and environmental stress, **1:**682–684
  and equilibrium theory, **1:**373
  and evolution, **1:**512, **1:**514
  Galápagos Islands, **2:**176–177
  and genetic erosion, **2:**183

grassland species, **2:**234
Holocene, **1:**733–**1:**734
insects, **1:**190, **1:**511
irreversible decisions concerning, **1:**198
island-resident species, **3:**386
K-T (end Cretaceous), **1:**732, **1:**737–738
late Pleistocene, **1:**637–639, **1:**733–**1:**734
late Precambrian, **1:**724
legal protections, **1:**519–520
mammals, **1:**637–639
marine animals, **2:**475–477
Mesozoic, **1:**731–732
Miocene, **1:**732–733
natural, **1:**524–525, **1:**527, **1:**641
Paleozoic, **1:**729–731
and parasitism, **3:**69–70
patterns, **1:**728–733, **1:**735–737
Pliocene, **1:**732–733
population fluctuation, **1:**690
population viability analysis, **1:**389–390, **1:**400
Precambrian, **1:**729
predisposing characteristics, **1:**508, **1:**514
preventing, *see* Conservation
principle, table of, **1:**728
rates, **1:**183, **1:**190–191, **1:**201, **1:**357–**1:**358, **1:**386
and speciation, **1:**522–523, **3:**367
theories of, **1:**735–737
Extirpation, defined, **2:**47
Extraterrestrial impacts
  and end-Cretaceous mass extinction, **1:**511–512
  and mass extinctions, **1:**507, **1:**511–**1:**514
Extrinsic value, biodiversity, **1:**212–230
Eyes, UV radiation effects, **1:**161

F

Fàahia archaeological site (Huahine island), **1:**642
FACE technique, *see* Free air carbon dioxide enrichment
    technique
Facilitation
  and community structure, **2:**137
  of succession, **2:**137, **2:**144–146
Facultative agents, as population restraints, **3:**186–187
Facultative long-day plants, flowering strategies,
    3:309–310
Facultative short-day plants, flowering strategies,
    3:311–312
Families, losses during mass extinctions, **2:**524–526
Farmers, participation in soil management, **3:**350–351
Farming, *see* Agricultural systems; Agriculture;
    Agroecology; Crop management; Cultivation;
    Grazing systems

G

I

N

O

P

S

X

Y

Index of Related Titles

Landscape Ecology

Biological Control; Deforestation; Ecological Enegetics of Ecosystems; Equilibrium and Nonequilibrium Concepts in Ecological Models; Introduced Species; Nutrient Cycling in Forests; Plant Ecophysiology; Soil Ecosystems

Life Histories

Sex Ratio

Limnology, Inland Aquatic Ecosystems

Aquatic Weeds; Biogeochemical Cycles; River Ecology; Wetlands Ecology

M

Mammalian Reproductive Strategies, Biology of Infanticide

Evolution and Extinction; Sex Ratio; Territoriality

Management of Large Marine Ecosystems

Biodiversity; Conservation Agreements, International; Continental Shelf Ecosystems; Coral Reef Ecosystems; Marine Biology, Human Impacts

Marine Biology, Human Impacts

Anarctic Marine Food Webs; Coral Reef Ecosystems; Evolution and Extinction; Fish Conservation; Krill, Antarctic; Marine Productivity; Ocean Ecology; Pollution Impacts on Marine Biology

Marine Microbial Ecology

Marine Productivity; Ocean Ecology

Marine Productivity

Global Carbon Cycle; Marine Biology, Human Impacts; Marine Microbial Ecology; Ocean Ecology

Mass Extinction, Biotic and Abiotic

Duties to Endangered Species; Evolution and Extinction

Modeling Forest Growth

Forests, Competition and Succession

Molecular Evolution

Evolution and Extinction

N

Nature Preserves

Biosphere Reserves; Conservation Agreements, International; Park and Wilderness Management

Nitrogen Cycle Interactions with Global Change Processes

Biogeochemical Cycles; Greenhouse Gases in the Earth's Atmosphere; Nitrous Oxide Budget; Rhizosphere Ecophysiology; Soil Ecosystems

Nitrous Oxide Budget

Atmospheric Ozone and the Biological Impact of Solar Ultraviolet Radiation; Global Carbon Cycle; Greenhouse Gases in the Earth's Atmosphere; Nitrogen Cycle Interactions with Global Change Processes

Northern Polar Ecosystems

Bird Communities; Marine Productivity; Ocean Ecology; Soil Ecosystems; Tundra, Arctic and Sub-Arctic

Nutrient Cycling in Forests

Nitrogen Cycle Interactions with Global Change Processes; Rhizosphere Ecophysiology; Soil Ecosystems

Nutrient Cycling in Tropical Forests

Biomass Burning; Forest Clear-Cutting, Soil Response; Global Carbon Cycle; Nutrient Cycling in Forests; Rhizosphere Ecophysiology; Soil Ecosystems

O

Ocean Ecology

Community-Based Management of Common Property Resources; Coral Reef Ecosystems; Fish Conservation; Fish Ecology

P

Packrat Middens, Archives of Desert Biotic History

Ecophysiology

Paleolimnology

Eutrophication; Limnology, Inland Aquatic Ecosystems

Parasitism, Ecology

Equilibrium and Nonequilibrium Concepts in Ecological Models

Park and Wilderness Management

Conservation Agreements, International; Fire Ecology; Fish Conservation; Forest Stand Regeneration, Natural and Artificial; Global Anthropogenic Influences; Nature Preserves; Wildlife Management

Physiological Ecology of Forest Stands

Forest Canopies; Forests, Competition and Succession; Forest Stand Regeneration, Natural and Artificial; Insect Interactions with Trees; Seed Banks

Plant–Animal Interactions

Conservation Programs for Endangered Plant Species; Global Carbon Cycle; Plant Ecophysiology

ISBN 0-12-226733-8

9 780122 267338

90018